Foundations of Engineering Mechanics

Series Editors: V.I. Babitsky, J. Wittenburg

Foundations of Engineering Mechanics

Series Editors: Vladimir I. Babitsky, Loughborough University, UK
Jens Wittenburg, Karlsruhe University, Germany

Further volumes of this series can be found on our homepage: springer.com

Slivker, V.I.
Mechanics of Structural Elements, 2007
ISBN 978-3-540-44718-4

Elsoufiev, S.A.
Strength Analysis in Geomechanics, 2007
ISBN 978-3-540-37052-9

Awrejcewicz, J., Krysko, V.A.,
Krysko, A.V.
Thermo-Dynamics of Plates and Shells,
2007
ISBN 978-3-540-37261-8

Wittbrodt, E., Adamiec-Wojcik, I.,
Wojciech, S.
Dynamics of Flexible Multibody Systems, 2006
ISBN 3-540-32351-1

Aleynikov, S.M.
Spatial Contact Problems in Geotechnics,
2006
ISBN 3-540-25138-3

Skubov, D.Y., Khodzhaev, K.S.
Non-Linear Electromechanics, 2006
ISBN 3-540-25139-1

Feodosiev, V.I.,
Advanced Stress and Stability Analysis Worked Examples, 2005
ISBN 3-540-23935-9

Lurie, A.I.
Theory of Elasticity, 2005
ISBN 3-540-24556-1

Sosnovskiy, L.A.,
TRIBO-FATIGUE · Wear-Fatigue Damage and its Prediction, 2005
ISBN 3-540-23153-6

Andrianov, I.V., Awrejcewicz, J.,
Manevitch, L.I. (Eds.)
Asymptotical Mechanics of Thin-Walled Structures, 2004
ISBN 3-540-40876-2

Ginevsky, A.S., Vlasov, Y.V.,
Karavosov, R.K.
Acoustic Control of Turbulent Jets, 2004
ISBN 3-540-20143-2,

Kolpakov, A.G.
Stressed Composite Structures, 2004
ISBN 3-540-40790-1

Shorr, B.F.
The Wave Finite Element Method, 2004
ISBN 3-540-41638-2

Svetlitsky, V.A.
Engineering Vibration Analysis - Worked Problems 1, 2004
ISBN 3-540-20658-2

Babitsky, V.I., Shipilov, A.
Resonant Robotic Systems, 2003
ISBN 3-540-00334-7

Le xuan Anh,
Dynamics of Mechanical Systems with Coulomb Friction, 2003
ISBN 3-540-00654-0

Nagaev, R.F.
Dynamics of Synchronising Systems,
2003
ISBN 3-540-44195-6

Neimark, J.I.
Mathematical Models in Natural Science and Engineering, 2003
ISBN 3-540-43680-4

Perelmuter, A.V., Slivker, V.I.
Numerical Structural Analysis, 2003
ISBN 3-540-00628-1

Lurie, A.I.,
Analytical Mechanics, 2002
ISBN 3-540-42982-4

Manevitch, L.I., Andrianov, I.V.,
Oshmyan, V.G.
Mechanics of Periodically Heterogeneous Structures, 2002
ISBN 3-540-41630-7

Babitsky, V.I., Krupenin, V.L.
Vibration of Strongly Nonlinear Discontinuous Systems, 2001
ISBN 3-540-41447-9

Landa, P.S.
Regular and Chaotic Oscillations, 2001
ISBN 3-540-41001-5

(Continued after index)

V. Slivker

Mechanics of Structural Elements

Theory and Applications

with 93 Figures and 28 Tables

Springer

Series Editors:

V.I. Babitsky
Department of Mechanical Engineering
Loughborough University
Loughborough LE11 3TU, Leicestershire
United Kingdom

J. Wittenburg
Institut für Technische Mechanik
Universität Karlsruhe (TH)
Kaiserstraße 12
76128 Karlsruhe
Germany

Author:

Vladimir Slivker
ISC Giprostroymost
ul.Yablochkova 7
197198 St.
Petersburg, Russia
e-mail: slivker@gpsm.ru

ISSN print edition: 1612-1384

ISBN-10: 3-540-44718-0
ISBN-13: 978-3-540-44718-4 **Springer Berlin Heidelberg New York**

Library of Congress Control Number: 2006932401

Springer is a part of Springer Science+Business Media

springer.com

Typesetting: Data conversion by the author and SPi
Cover-Design: deblik, Berlin
Printed on acid-free paper SPIN: 11671046 62/3100/SPi - 5 4 3 2 1 0

PREFACE

The presentment should be as simple as possible, but not a bit simpler.

Albert Einstein

Introduction

The power of the variational approach in mechanics of solids and structures follows from its versatility: the approach is used both as a universal tool for describing physical relationships and as a basis for qualitative methods of analysis [1]. And there is yet another important advantage inherent in the variational approach – the latter is a crystal clear, pure and unsophisticated source of ideas that help build and establish numerical techniques for mechanics. This circumstance was realized thoroughly and became especially important after the advanced numerical techniques of structural mechanics, first of all the finite element method, had become a helpful tool of the modern engineer. Certainly, it took some time after pioneering works by Turner, Clough and Melos until the finite element method was understood as a numerical technique for solving mathematical physics problems; nowadays no one would attempt to question an eminent role played by the variational approach in the process of this understanding. It is a combination of intuitive engineer thinking and a thoroughly developed mathematical theory of variational calculus which gave the finite element method an impulse so strong that its influence can still be felt.

It would be too rash to say that there are few publications or books on the subject matter discussed in this book. It suffices to list such names of prominent mathematicians and mechanicians as Leibenzon [2], Mikhlin [3], Washizu [4], Rectoris [5], Rozin [6] … – the ellipsis shows that this list could be continued. So, a person can be thought of as overmuch confident (even arrogant) to follow the listed authors and other recognized personalities, who furrowed up their way through the ocean of variational principles in mechanics long ago, and to make the venture of writing another book on the same subject. The words said by English physicist H.Bondy come into mind in this regard [7]:

"A book is a wonderful thing, but, honestly, there are too many books; so the readers have a hard time, and the authors maybe harder".

However, every book written is worth its readers' audience. Some of the books (by Mikhlin or Rectoris) are intentionally oriented at mathematical aspects of variational solutions, while others (by Leibenzon, Washizu, Rozin) have a clear and pure mechanical accent.

Obviously, when an author is in process of writing a book like this one, there is a difficult issue that constantly crosses the way: who are the potential readers of the book and how to keep to their interests. K. Rektorys [5] is totally right by stating that it is quite a fancy matter how to make a book useful for both the mathematician and the engineer because:

"...the said reader categories often have opposite opinions about a book like this, so they advance totally different requirements to it, which cannot be satisfied at the same time. For example, one can hardly accommodate oneself to the wish of the mathematician and provide a book written very concisely where the theory would be evolved at a quick pace".

This is a matter of choice, and the choice in this book is unambiguous: The book is oriented at people who took (or intend to take) their engineering degree and also have a certain awareness of mathematics — generally, within the curriculum of the present mathematical education given to students of engineering at universities.

Here follows a short list of skills and knowledge that the reader of the book should possess. The reader is believed to have acquaintance with a standard set of solid mechanics subjects included in the curriculum on engineering at any university — strength of materials, structural mechanics, basics of elasticity theory — and to know something about basic notions of the calculus of variations. The concepts like a functional, Euler equations for one, principal and natural boundary conditions, the Lagrangian multiplier rule for a functional's point of stationarity when additional conditions are present, some others are assumed to be known to the reader and understood by him. The reader is also supposed to have mastered the basics of linear algebra; as for the calculi, the Gauss–Ostrogradski formula is used everywhere in several variations without additional explanation. Also, the author believes the reader will not have any difficulties with the differentiation of a function with respect to its vector argument; this operation can be met in the book a few times.

The author wanted to restrict the requirements to the mathematical skill of the reader, therefore the book does not use basics of tensor analysis even in cases when the tensors would be totally relevant. All that the reader should know about the subject is how to sum over repeated indices.

The author keeps to the needs of the engineers and tries to avoid where possible the lure of discussing delicate mathematical issues — for example, the very important notion of space completeness. However, the reader is assumed to know simple things about the Hilbert spaces. It is possible that mathematical purists might find this style of presentment inadmissible... well, then we refer to the following opinion by Bertran Russell:

"A book must be either strict or simple. These two requirements are not compatible".

Speaking briefly, this book is addressed to the engineers rather than the mathematicians; however, to the engineers who have a taste for mathematical formulations and methods of engineering analysis based thereupon, even though the methods are not presented in their pure mathematical form.

Speaking about the potential reader, the author already mentioned the engineers and researchers (first of all) and wishes to add senior students of engineering who intend to make their career in close connection with engineering analysis. Post-graduates of specialities related to mechanical strength are welcome, too. I do hope the professors of the same specialities will be able to find the book useful in some way for their lectures or topical seminars.

The discussion of the book's contents by chapters is omitted; a look at the table of contents is enough to have a clear idea of the subject. Also, the reader should noticc that thc book pays cqual attcntion to gcncral formulations of variational problems and to the variational treatment of particular classes of mechanical problems. Therefore the book can be both (1) a guide to deeper study of variational principles and methods in mechanics of solids and structures and (2) a practical manual for the engineer.

The variational principles of structural mechanics can be presented in a variety of ways. One of the approaches suggests that particular variations of the basic principles can be derived one from another by formal mathematical transformations such as Legendre transform, Friedrichs transform, Lagrange transform. This approach is used systematically in [8], for example. But the same variational principles can be derived independently, too, so that the connections between the respective functionals are established later, maybe using the same mathematical transforms. For methodical reasons, one of which is the orientation of the book at the reader educated in engineering, the book uses the second approach.

Obviously, it is not necessary to consider all thinkable variational formulations in one book (nor it is possible because the volume of the book is limited). The scientific journals never cease publishing more and more papers on the subject, which is an evidence that the topic is far from being exhausted. This book presents only some most important and popular formulations; the author has chosen those as useful for both the general theory of structural mechanics and the construction of numerical algorithms that solve application problems.

It should be specified that all structural mechanics formulations in this book are strictly linear[1]. These are the considerations why this limitation has been adopted:

- First, the variational formulations and methods of solutions in the linear analysis are self-contained. The author thinks it is a good methodical approach to treat most important features of the variational methods in the linear formulation without making things too complicated by introducing nonlinear effects.
- Second, one should keep in mind that the solution of nonlinear problems is based in most cases on a reduction to a sequence of linear solutions.
- Third, and most essential, the nonlinear analysis is both practically important and very specific. Therefore the respective problems deserve a separate detailed treatment in a separate publication.

The above said is an actual promise, given by the author to his reader audience, to prepare a book as soon as possible which will be dedicated particularly to formulations and methods of solution in nonlinear structural analysis.

The author wishes to give one excuse for terminology used in the book. The book makes extensive use of a number of abbreviations such as: SSS for 'stress and strain state', PSS for 'plane stress state', FEM for 'finite element method'. The author is aware that a number of experts in mechanics of solids and structures (MSS) feel bad about the abbreviations like these. But even the last abbreviation is used in the title of a respectable academic journal, so is it not an evidence that abbreviations are recognized by the mechanicians and can be used in publications? As for the sense of proportion, it is the reader who will judge.

---

[1] That is why we do not distinguish between the strain energy of a system and its complementary energy; this difference becomes essential in the nonlinear analysis.

Remarks on references to publications

The fact that no list of literature references can by any means claim its completeness is a very traditional excuse made by authors; I don't even feel myself obliged to make that excuse again.

The only thing worth mentioning here is the purpose of the references made in the book. Actually, there can be more than one purpose. However, if the reader seeks to find the author's reasoning on historical priorities in the book, there will be a disappointment. This is not because the author underestimates the historical component in the development of the scientific thinking. On the contrary, the author feels so deep a respect to the science he is engaged in that he cannot declare himself the historian of that science even to a slightest degree[2]. Generally, the problem of priorities is both complicated and very delicate, and sometimes it just cannot be resolved so that no one has bad feelings about the historical unfairness of the solution. Historians of science belonging to different scientific schools are often devoted to strictly opposite opinions[3]. It is better here to step aside from the priority problem and the related issue — how to name particular scientific achievements based on their historical precedence. I just note that the references to publications are given chiefly for the reader to be able to find more information on a particular topic covered in the book. Another purpose of listing the references is to give the reader an idea what sources were used by the author in order to present particular topics of the subject.

How to read this book

Strictly speaking, the reader is not required to follow the recommendations given below. The method of reading depends on the qualification of the reader and on the goals he has in mind when he is going to spend his time for studying the suggested material.

For the beginning, the reader is asked to read the first three chapters of the book. Chapters 4 through 8 present formulations of particular classes of problems based on the general variational principles. If the reader feels sufficiently knowledgeable about those formulations, or if he has no

---

[2] However, the author feels he has a right sometimes to express his point of view on the priority issue, too, especially when that point of view is quite well grounded.

[3] Just for example, recall arguments between the adherents of the priorities of Newton and Leibnitz in the invention of differential calculus.

interest in those for some reason, the reader can skip the chapters entirely or partially with little effect on the further understanding.

Chapter 9 contains an introduction to the Ritz method, intended for engineers and researchers in mechanics. A well-prepared reader can skip the chapter or just take a look at it. However, Section **9**.3 of it contains some new information not represented in monographs until now.

Chapters 10 and 11 are intended for engineers or researchers interested in the frequency spectrum analysis and the stability of equilibrium of structures.

The appendices give some general mathematics which, though sometimes relate indirectly to the main presentment, can be of help for the reader who does not feel like following literary references simultaneously with perusing this book. For example, Appendix F presents a brief but complete description of the theory of curvilinear coordinates. We recommend that even a prepared reader familiarize himself with this appendix in order to master the system of designations which is used in many places of the book.

The appendices include also sections which present something different from general mathematics. Those sections discuss certain specific details or particularize issues of a theory; they are intentionally removed from the main presentment in order not to overload it.

Before studying the plate bending theories in Chapter 5, one is recommended to look through Section **4**.7 dedicated to planar curvilinear bars. It will help to understand better at least an important section on the static-geometric analogy in the plate theory, especially in connection with the formulation of so-called boundary conditions for deformations.

The book does not abound in examples, so we recommend not to miss ones that the book does have. Generally, the examples presented in the book are not intended to coach a student for solving typical problems like piece of cake. The examples are there to provide an explanatory material that helps look at a problem at a different angle. This special role played by examples in the cognitive process in mathematics and engineering is well-known and traditional in scientific papers. The role was emphasized many times in works by a great expert in teaching mathematics and mechanics, A.N. Krylov. In his well-known book [9], A. Krylov refers to words by I. Newton: "*in the study of sciences, examples are no less educational than rules*".

In most of the cases, all statements of theoretical nature are provided in the book along with a detailed background. If there are any violations of the rule, they are intentional — it is the reader who is invited to complete the demonstration. This is not to save space; this is to ensure a better education. Exercises of this kind help master the theory much better and grow one's creative potential.

Acknowledgements

The book now lying on your table would not be written and published:

- If the author had not felt the support (and sometimes a push from behind) of his friends and colleagues. Without giving a long list of names, I would like to say that the name at the head of it would be that of my old friend and co-author in many publications, Anatoly V. Perelmuter.
- If the "GiproStroyMost Sankt-Peterburg" closed corporation in the person of its Director General, Yuri P. Lipkin, had not shared the belief in the practical value of the book with the author and had not supported the spirit of this venture.
- And the last (but not the least), if the author had not been fortunate to meet Leonid A. Rozin in his time, who had become both a teacher and a friend. The influence of works and the personality of Mr. Rosin on the scientific interests of the author (obvious to anybody who knows both the publications by L.A. Rozin and this book) and even on the author's style of thinking in the field of mechanics has been a decisive factor.

The author is much obliged to D.V. Dereviankin and D.A. Maslov who helped much in the preparation of the book's text. The former, being just a student, made some important numerical calculations at the request of the author, which then were included in the book. The same person created all the figures, after having mastered the advanced tools of modern computer graphics. The latter has developed a software entitled GeomyX which calculates a full set of geometric properties of thin-walled bar cross-sections[4].

I.D. Evzerov did so much as read the book in manuscript. His remarks were so constructive that the author totally followed them to change (and thus improve) the original version. V.S. Karpilovsky read through some chapters of the book and made useful comments on those.

The author did not fail to use a repeated occasion of discussing many particularities of the thin-walled bar theory with a recognized expert in this area of structural mechanics, E.A. Beilin. Prof. Beilin felt a kind interest to

[4] The GeomyX software is licensed by "GiproStroyMost Saint-Petersburg" closed corporation. Further information about the software can be found at http://geomyx.gpsm.ru.

the work by the author which he demonstrated in those discussions, and the author readily admits how much help it was in the preparation of the respective sections of the book.

Finally, the SCAD Group company, that the author has the bonds of creative cooperation with, was (and is) always a great initiator of the author's thinking in the development of numerical algorithms for structural analysis. Results of that thinking are included in the book to some extent.

Expressing my warmest gratitude to the above mentioned people and institutions is both my pleasant obligation and privilege.

Some general rules for designations used in this book

The following numbering of formulas and references to those is used. Within one section, the formulas are numbered by two numbers separated with periods. First number is No. of a section in a particular chapter and second one is No. of the formula in the subsection. When there is a reference to a formula from a different chapter, the number of that chapter is indicated in addition; for a better visual recognition, No. of the chapter is bold-faced. For example, (2.3) means a reference to Formula (2.3) from Section 2 of the current chapter, while (**1**.2.3) is Formula (2.3) from Section 2 of Chapter 1.

The author is deeply convinced by the whole course of his student's and professor's experience that a well-thought system of designations is one of important educational components in the presentment of any physical theory which uses mathematics. A good system of designations will help both master the theory and remember results presented in formulas. On the contrary, a babelized, disorganized system of designations will only repel the student. The sensible designations help the students concentrate on the ideas of the subject without distracting their minds to recalling each time the meaning of symbols used in new formulas. Therefore the author worked hard to introduce such designations which would be mnemonic and systematic rather than chaotic, without deviating much from ones commonly used in the science. The reader will judge how well the author did his work.

Vectors, matrices, and tensors are printed in bold face.

The matrix and vector transposition is denoted by the superscript ^{T}. The same mark is used to denote differential operators conjugate in the Lagrangian sense.

Both the identity operator and the identity matrix are denoted as $\boldsymbol{I}$.

An overstrike is used consistently to denote given values, such as given volumetric forces $\bar{X}$. However, this designation rule is not strict over the whole book. If there are any deviations from the rule, they will be specified.

An upper tilde, $\tilde{}$, over a certain quantity shows that a perturbed value of the quantity is under consideration.

An asterisk $_*$ used as a subscript shows that the subscripted quantity is exact.

An example:

$\tilde{\mathsf{E}}_*$ is an exact value of the strain energy in a system which is different from a given one by some perturbation.

It has been already mentioned that the index transformations make use of a common convention of summing over repeated, so-called mute, indices[5]. In cases when an index is repeated in a formula but is not a mute one, the summing will not be done, and the index will be indicated after the respective formula in parentheses, accompanied by an exclamation mark. For example, in the following formula

$$c^{ij} = a^{ij} b_j \qquad (j\ !)$$

there is no summing over the index j, which is quite obvious because the mute index cannot participate in the formation of an aggregate in the left part.

Also, in many cases we do not even comment on the carrying of indices from upside down and vice versa because the tensor analysis says it is quite admissible in orthogonal Cartesian coordinates where covariant and contravariant components of the tensors are indistinguishable.

It is worth mentioning here also that the figures will use a common rule of depicting internal forces (stresses) in elements of a system by double arrows. The external forces are represented by single arrows. A moment vector is denoted, as a rule, by a right-hand corkscrew.

Form of representation of equations used in the book

It is a tradition in the solid mechanics, to use an index form for tensor relations. On the other hand, a number of authors (such as A.I. Lurie [10] and his school, Truesdell [11] etc.) are oriented at a so-called index-free tensor representation. In Russian-language papers on structural mechanics of bar systems, basic equations were written in a matrix form since the

[5] Some authors prefer “deaf indices” to “mute indices”. In order to make peace, maybe we should call them the “deaf-mute” indices?

publication of a book by A.F. Smirnov [12]; this commonly used form has become a prevailing one by now. Undoubtedly, each one of the forms should be allowed to exist because every one of them brings along both its advantages and its shortcomings. The argument over these advantages and shortcomings can be both very long and totally fruitless.

Certainly, it is the author of a publication who should choose what form of equations to use in a particular publication; this choice is influenced by tradition, by the way the author thinks, and not in the least by the opinion of the majority. However, the main thing in this choice is still the adequacy of the mathematical theory for particular problems treated by the publication.

The basic form of representation used in this book is derived from the general operator form of governing equations of structural mechanics (not necessarily of bar systems) where matrices and vectors are widely employed. This way seems concise, visually convenient, and universal in the variational formulations of the problems; also, engineering-educated people find this form quite apprehensible. The systems of designations closest to that used in this book include one employed in works by L.A. Rosin [6], [13], one used by T. Belytschko, Wing Kam Liu, B. Moran [14], and one found in a well-know three-volume encyclopedic edition by Zienkiewicz & Taylor [15] on the finite element method. However, when the author deemed it reasonable to switch to a different form, there was no hesitation.

And one more terminology note. We will call a column matrix a vector. However, one should keep in mind that a mathematical object represented by the respective column matrix is not necessarily an actual vector, i.e. a physical object with appropriate transformations of its components between different coordinate systems. The actual object can be a tensor or even a scalar.

List of key designations

Designations of functionals

The Arial regular font is used to designate the functionals.

A – a virtual work of external forces;
B – a virtual work of internal forces;
– Bolotin functional in the equilibrium stability analysis;
– Brian–Treftz functional in the linearized equilibrium stability analysis;
E – a potential energy of strain;
G – Gurtin functional;

H – Herrmann functional in the plate bending analysis;
K – Castigliano functional;
L – Lagrange functional (a full potential energy of a system);
N – Novozhilov functional in the equilibrium stability analysis;
P – a functional of physical relations;
R – Reissner (Hellinger–Reissner) functional;
S – a general designation for functionals in the equilibrium stability analysis;
T – Timoshenko functional in Saint-Venant problem of a prism torsion;
– a kinetic energy of a mechanical system;
W – Washizu (Hu–Washizu) functional;

Γ – a functional of boundary conditions;
Φ – a generalized (parameterized) mixed functional;

Π_s – a force potential (a potential of static actions);
Π_k – a kinematical potential (a potential of kinematical actions);
Π_{s0} – a potential of initial strains, a force one;
Π_{k0} – a potential of initial strains, a kinematical one;

r – Raleigh functional (Raleigh ratio) in the spectral problem.

Designations of sets

The italic ArtScript font is used to designate the sets.

P – a set of physically admissible SSS fields;

U_k – a set of kinematically admissible SSS fields;
U_{ko} – a set of uniformly kinematically admissible SSS fields;
$U_{k/2}$ – a set of kinematically semi-admissible SSS fields;
$U_{s/2}$ – a set of statically semi-admissible SSS fields;
$P_{k/2}$ – a set of physically and kinematically semi-admissible SSS fields;
P_k – a set of physically and kinematically admissible SSS fields;
P_{ko} – a set of physically and uniformly kinematically admissible SSS fields;

$P_{s/2}$ – a set of physically and statically semi-admissible SSS fields;
P_s – a set of physically and statically semi-admissible SSS fields;
P_{so} – a set of physically and uniformly statically admissible SSS fields;

R – a set of rigid displacements of a mechanical system;

R_A – a set of rigid displacements of an elastic body;
R_K – a set of rigid displacements of an elastic medium;
R_o – a set of uniformly kinematically admissible rigid displacements of a mechanical system;
R_{Ao} – a set of uniformly kinematically admissible rigid displacements of an elastic body

$$R \subseteq R_A, \quad R_{Ao} \subseteq R_A, \qquad R_o = R \cap R_{Ao};$$

K – Castigliano energy space;
L – Lagrange energy space;
F – a parameterized energy space;
$\square_n$ – Euclid space of dimension n.

Designations of fields and operators

F – an arbitrary SSS field with stresses $\boldsymbol{\sigma}$, strains $\boldsymbol{\varepsilon}$, displacements $\mathbf{u}$,

$$F = \{\boldsymbol{\sigma}, \boldsymbol{\varepsilon}, \mathbf{u}\};$$

V – a field of external actions with external forces $\bar{\boldsymbol{X}}$ distributed over the area Ω, with edge forces $\bar{\boldsymbol{p}}$ and given edge displacements $\bar{\boldsymbol{u}}$ specified on the boundary Γ,

$$V = \{\bar{\boldsymbol{X}}, \bar{\boldsymbol{p}}, \bar{\boldsymbol{u}}\};$$

$\boldsymbol{O}$ – a general designation of a zero operator (annihilator);
$\boldsymbol{I}$ – a general designation of an identity operator;

$\boldsymbol{A}$ – a differential operator of geometry,

$$\boldsymbol{A}\mathbf{u} = \boldsymbol{\varepsilon};$$

$\boldsymbol{B}$ – a differential operator of equilibrium, commonly $\boldsymbol{B} = \boldsymbol{A}^{\mathrm{T}}$

$$\boldsymbol{A}^{\mathrm{T}}\boldsymbol{\sigma} + \boldsymbol{K}\mathbf{u} = \bar{\boldsymbol{X}};$$

$\boldsymbol{K}$ – an algebraic operator of stiffness of a medium in which an elastic body is placed;
$\boldsymbol{C}$ – an algebraic operator of a physical law of linear elasticity,

$$\boldsymbol{\sigma} = \boldsymbol{C}\boldsymbol{\varepsilon}, \qquad \boldsymbol{\varepsilon} = \boldsymbol{C}^{-1}\boldsymbol{\sigma};$$

$\boldsymbol{L}$ – a differential operator of equilibrium in displacements (Lame operator)

$$\boldsymbol{L} = \boldsymbol{A}^{\mathrm{T}}\boldsymbol{C}\boldsymbol{A} + \boldsymbol{K}, \qquad \boldsymbol{L}\mathbf{u} = \bar{\boldsymbol{X}};$$

$\boldsymbol{S}$ – a differential operator of strain compatibility (Saint-Venant operator),

$$\boldsymbol{S}\boldsymbol{\varepsilon} = \mathbf{0}, \qquad \boldsymbol{S}\boldsymbol{A} = \boldsymbol{O};$$

– a differential operator of stability of equilibrium;

$\boldsymbol{\Phi}$ – a vector of stress functions

$$\boldsymbol{\sigma} = \boldsymbol{S}^{\mathsf{T}}\boldsymbol{\Phi}, \qquad \boldsymbol{A}^{\mathsf{T}}\boldsymbol{S}^{\mathsf{T}} = \boldsymbol{O};$$

$\boldsymbol{M}$ – a differential operator of compatibility in the stress functions

$$\boldsymbol{M} = \boldsymbol{S}\boldsymbol{C}^{-1}\boldsymbol{S}^{\mathsf{T}}, \qquad \boldsymbol{M}\boldsymbol{\Phi} = \boldsymbol{O};$$

$\boldsymbol{\Omega}$ – a matrix differential operator of rotation;

$\boldsymbol{E}_p$ – an operator of extracting static edge conditions;

$\boldsymbol{E}_u$ – an operator of extracting kinematical edge conditions,

$$\boldsymbol{E}_p + \boldsymbol{E}_u = \boldsymbol{I}, \quad \boldsymbol{E}_p\boldsymbol{E}_u = \boldsymbol{E}_u\boldsymbol{E}_p = \boldsymbol{O};$$

$\boldsymbol{H}_\sigma$ – an operator of transforming internal forces $\boldsymbol{\sigma}$ into edge forces $\boldsymbol{p}$,

$$\boldsymbol{p} = \boldsymbol{H}_\sigma\boldsymbol{\sigma};$$

$\boldsymbol{H}_u$ – an operator of transforming internal displacements $\mathbf{u}$ into edge displacements $\boldsymbol{u}$,

$$\boldsymbol{u} = \boldsymbol{H}_u\mathbf{u}.$$

Some of other designations

$\boldsymbol{\omega} = \left\|\left[\omega_x, \omega_y, \omega_z\right]\right\|^{\mathsf{T}}$ – a vector of rotations;

$\boldsymbol{n} = \|[n_x, n_y, n_z]\|^{\mathsf{T}}$ – a vector of an exterior normal to the boundary Γ of the area Ω.

References

1. Abovski NP, Andreyev NP, Deruga AP (1978) Variational principles in elasticity and shell theory (in Russian). Nauka, Moscow
2. Belytschko T, Wing Kam Liu, Moran B (2001) Nonlinear Finite Elements for Continua and Structures. John Wiley & Sons, Chichester New York Weinheim Brisbane Singapore Toronto
3. Berdichevsky VL (1983) Variational principles in mechanics of continua (in Russian). Nauka, Moscow

4. Bondy H (1967) Assumption and myths in physical theory. Cambridge at the University Press, Cambridge
5. Krylov AN (1950) About some differential equations of mathematical physics applicable in engineering (in Russian). GosTekhTeorIzdat, Moscow Leningrad
6. Leibenzon LS (1951) Collected papers. Vol. 1. Theory of elasticity (in Russian). USSR Acad. Sci. Publ., Moscow
7. Lurie AI (1980) Nonlinear theory of elasticity (English translation). North-Holland, New York
8. Mikhlin SG (1970) Variational methods in mathematical physics (in Russian). Nauka, Moscow
9. Rektorys K (1980) Variational Methods in Mathematics, Science and Engineering. – Riedel Publishing Company, Dordreht Holland/Boston USA/London- England
10. Rozin LA (1978) Variational formulations of problems for elastic systems (in Russian). Len. State Univ. Publ., Leningrad
11. Rozin LA (1998) Problems of elasticity and numerical methods for their solution (in Russian). Publ. House of St.-Pet. State Tech. Univ., St.-Petersburg
12. Smirnov AF (1958) Stability and oscillations of structures (in Russian). TransZhelDorIzdat, Moscow
13. Truesdell C (1972) A first course of rational continuum mechanics. The Jons Hopkins University, Baltimore, Maryland
14. Washizu K (1982) Variational methods in elasticity and plasticity. Pergamon Press, Oxford New York Toronto Sydney Paris Frankfurt
15. Zienkiewicz OC, Taylor RL (1989) The Finite Element Method. Vol.1. McGraw-Hill Book Company, London

CONTENTS

1 BASIC VARIATIONAL PRINCIPLES OF STATICS AND GEOMETRY IN STRUCTURAL MECHANICS

We have the right as well as are obliged to subject all our definitions to critical analysis from the standpoint of their application and revise them (fundamentally, if need be) if they do not work for us.

Young L (1969) Lectures on the calculus of variations and optimal control theory. W.B. Saunders company, Philadelphia London Toronto

1.1 Preliminaries

Let an area Ω with a piecewise smooth boundary Γ be defined in a k-dimensional space. Structural mechanics deals with one, two, and three-dimensional problems only, therefore we take the case of $k \leq 3$.

Consider a linear set $\mathcal{M}$ with elements $\boldsymbol{a}$, $\boldsymbol{b}$, $\boldsymbol{c}$... which are functions of points $x \in \Omega$. The elements of the set $\mathcal{M}$ will be assumed to exist as scalar, vector or tensor functions. We assume also that for every couple of elements, $\boldsymbol{a}$ and $\boldsymbol{b}$, from the set $\mathcal{M}$, two bilinear functionals, $(\boldsymbol{a}, \boldsymbol{b})$ and $(\boldsymbol{a}, \boldsymbol{b})_\Gamma$, can be defined which acquire finite values. The functionals are specified by these formulas:

$$(\boldsymbol{a}, \boldsymbol{b}) = \int_\Omega \boldsymbol{a} \cdot \boldsymbol{b} \, d\Omega, \tag{1.1}$$

$$(\boldsymbol{a}, \boldsymbol{b})_\Gamma = \int_\Gamma \boldsymbol{a} \cdot \boldsymbol{b} \, d\Gamma, \tag{1.2}$$

where $\boldsymbol{a} \cdot \boldsymbol{b}$ means a scalar expression calculated according to the rule

$$\boldsymbol{a} \cdot \boldsymbol{b} = \begin{cases} ab & \text{– for scalar quantities,} \\ a^i b_i & \text{– for vector quantities,} \\ a^{ij} b_{ij} & \text{– for tensor quantities.} \end{cases} \tag{1.3}$$

Here and further we use a common rule: the same indices on different levels are used for summation.

The one-dimensional version of the area Ω is an interval, $[x_1,x_2]$, over which an independent variable, x, can vary, so that

$$(\boldsymbol{a}, \boldsymbol{b}) = \int_{x_1}^{x_2} \boldsymbol{a} \cdot \boldsymbol{b} \; dx \,, \qquad (\boldsymbol{a}, \boldsymbol{b})_\Gamma = (\boldsymbol{a} \cdot \boldsymbol{b})_{x=x_2} + (\boldsymbol{a} \cdot \boldsymbol{b})_{x=x_1} \,. \tag{1.4}$$

The latter relationship can be rewritten also as

$$(\boldsymbol{a}, \boldsymbol{b})_\Gamma = (+1\boldsymbol{a} \cdot \boldsymbol{b})_{x=x_2} - (-1\boldsymbol{a} \cdot \boldsymbol{b})_{x=x_1} = [n\boldsymbol{a} \cdot \boldsymbol{b}]\Big|_{x_1}^{x_2} \,. \tag{1.5}$$

Here n is the cosine of the angle between the x axis and the external normal to the interval $[x_1,x_2]$ in its end points, so that

$$n = \begin{cases} 1 & \text{at } x = x_2 \,, \\ -1 & \text{at } x = x_1 \,. \end{cases}$$

Note that the bilinear functional $(\boldsymbol{a}, \boldsymbol{b})$ can be treated as a scalar product on the linear set $\mathcal{M}$, while the functional $(\boldsymbol{a}, \boldsymbol{b})_\Gamma$ generates a scalar product on the linear set $\mathcal{M}_\Gamma$ which consists of functions defined on the boundary Γ. Indeed, both functionals are linear with respect to either argument and symmetric (insensitive to the order of their arguments). Also, the result is nonnegative when the arguments are equal:

$$(\boldsymbol{a}, \boldsymbol{a}) \geq 0, \;\; (\boldsymbol{a}, \boldsymbol{a})_\Gamma \geq 0.$$

Moreover, the condition $(\boldsymbol{a}, \boldsymbol{a}) = 0$ implies the equality $\boldsymbol{a} = \boldsymbol{0}$ where $\boldsymbol{0}$ is a zero element of the $\mathcal{M}$ set, that is, all components of $\boldsymbol{a}$ are zero functions on Ω. Similarly, $(\boldsymbol{a}, \boldsymbol{a})_\Gamma = 0$ implies $\boldsymbol{a} = \boldsymbol{0}$ where $\boldsymbol{0}$ is a zero element of the set $\mathcal{M}_\Gamma$, i.e. all components of $\boldsymbol{a}$ are functions with zero values on Γ.

1.1.1 Formally conjugate differential operators

Now we consider two linear sets, $\mathcal{N}$ and $\mathcal{M}$, such that for any element $\boldsymbol{a} \in \mathcal{N}$ a differential operation $\boldsymbol{A}$ is defined with its range of values in $\mathcal{M}$, and for any element $\boldsymbol{b} \in \mathcal{M}$ a differential operation $\boldsymbol{B}$ is defined with its range of values in $\mathcal{N}$, in other words,

$$\boldsymbol{a} \in \mathcal{N} \Rightarrow \boldsymbol{A}\boldsymbol{a} \in \mathcal{M} \qquad \text{and} \qquad \boldsymbol{b} \in \mathcal{M} \Rightarrow \boldsymbol{B}\boldsymbol{b} \in \mathcal{N} \,. \tag{1.6}$$

Further on we will follow the terminology common in mechanics and use the words *differential operators* with objects like $\boldsymbol{A}$ and $\boldsymbol{B}$ thus treating them as symbolic differentiation operations, although in mathematics [3] an operator is a bigger notion than a simple differentiation expression.

The differential operators $\boldsymbol{A}$ and $\boldsymbol{B}$ are called *formally conjugate* (sometimes Lagrange-conjugate) if they satisfy (1.6) and the following relationship holds:

$$(\boldsymbol{Aa}, \boldsymbol{b}) = (\boldsymbol{a}, \boldsymbol{Bb}) + (\boldsymbol{A}_\gamma \boldsymbol{a}, \boldsymbol{B}_\gamma \boldsymbol{b})_\Gamma \tag{1.7}$$

where $\boldsymbol{A}_\gamma$ and $\boldsymbol{B}_\gamma$ are certain linear differential operations such that the bilinear functional $(\boldsymbol{A}_\gamma \boldsymbol{a}, \boldsymbol{B}_\gamma \boldsymbol{b})_\Gamma$ makes sense for all elements $\boldsymbol{a}$ from $\mathcal{N}$ and all elements $\boldsymbol{b}$ from $\mathcal{M}$.

Further we will use the word *conjugate* for this relationship between the operators, always assuming the Lagrange-type conjugation. Also, an operator, $\boldsymbol{B}$, conjugate to the operator $\boldsymbol{A}$ will be denoted as $\boldsymbol{A}^\mathrm{T}$, that is, $\boldsymbol{B} = \boldsymbol{A}^\mathrm{T}$, if the relationship (1.7) holds[1].

Let's give an example of a matrix differential operation which conforms to a certain operator $\boldsymbol{A}$ and its conjugate operator $\boldsymbol{A}^\mathrm{T}$

$$\boldsymbol{A} = \left\|\begin{bmatrix} \dfrac{d(\ )}{dx} & \dfrac{(\ )}{\rho} \\ \dfrac{d}{dx}\left(\dfrac{(\ )}{\rho}\right) & -\dfrac{d^2(\ )}{dx^2} \end{bmatrix}\right\|, \quad \boldsymbol{A}^\mathrm{T} = \left\|\begin{bmatrix} -\dfrac{d(\ )}{dx} & -\dfrac{1}{\rho}\dfrac{d(\ )}{dx} \\ \dfrac{(\ )}{\rho} & -\dfrac{d^2(\ )}{dx^2} \end{bmatrix}\right\| \tag{1.8}$$

where $\rho = \rho(x)$ is a sufficiently smooth function of the independent variable[2].

The linear sets $\mathcal{N}$ and $\mathcal{M}$ both will be a set of vectors of the type $\boldsymbol{a} = \|[a_1(x), a_2(x)]\|^\mathrm{T}$ with their components being sufficiently smooth functions defined on the interval $[0, l]$. The following holds for two vectors $\boldsymbol{a}$ and $\boldsymbol{b}$:

[1] We do not use a common mathematical notation for conjugate operators using $*$; instead, we choose $^\mathrm{T}$, not just because the asterisk $*$ is reserved in our book for marking quantities related to an exact solution of a problem. The matter is that the conjugation operation is a generalization of a matrix transposition which is usually denoted by the symbol $^\mathrm{T}$. There is no confusion; just remember that this symbol applied to a differential operator means something bigger than the mere transposition of a matrix.

[2] As we will see, this example gives the operators of geometry and equilibrium in the problem of bending of a planar curvilinear beam with its curvature radius, $\rho(x)$, variable along the arc coordinate x – see Section **4**.7.

$$\boldsymbol{Aa} = \begin{bmatrix} a_1' + \dfrac{a_2}{\rho} \\ \left(\dfrac{a_1}{\rho}\right)' - a_2'' \end{bmatrix}, \qquad \boldsymbol{A}^{\mathsf{T}}\boldsymbol{b} = \begin{bmatrix} -b_1' - \dfrac{b_2'}{\rho} \\ \dfrac{b_1}{\rho} - b_2'' \end{bmatrix} \tag{1.9}$$

where the stroke means the differentiation with respect to x. The scalar products of our interest can be represented now as

$$(\boldsymbol{Aa}, \boldsymbol{b}) = \int_0^l \left\{ \left(a_1' + \frac{a_2}{\rho} \right) b_1 + \left[\left(\frac{a_1}{\rho} \right)' - a'' \right] b_2 \right\} dx \,,$$

$$(\boldsymbol{a}, \boldsymbol{A}^{\mathsf{T}}\boldsymbol{b}) = \int_0^l \left\{ \left(-b_1' - \frac{b_2'}{\rho} \right) a_1 + \left(\frac{b_1}{\rho} - b_2'' \right) a_2 \right\} dx \,.$$

Integrating these products by parts will transform them into the following:

$$(\boldsymbol{Aa}, \boldsymbol{b}) = \int_0^l \left(-a_1 b_1' + \frac{a_2 b_1}{\rho} - \frac{a_1 b_2'}{\rho} + a_2' b_2' \right) dx + \left[a_1 b_1 + \frac{a_1 b_2}{\rho} - a_2' b_2 \right]_0^l ,$$

$$(\boldsymbol{a}, \boldsymbol{A}^{\mathsf{T}}\boldsymbol{b}) = \int_0^l \left(-a_1 b_1' + \frac{a_2 b_1}{\rho} - \frac{a_1 b_2'}{\rho} + a_2' b_2' \right) dx + \left[-a_2 b_2' \right]_0^l$$

which makes it clear that $(\boldsymbol{Aa}, \boldsymbol{b})$ and $(\boldsymbol{a}, \boldsymbol{A}^{\mathsf{T}}\boldsymbol{b})$ are different only in non-integral terms; this is where the mutual conjugation of the $\boldsymbol{A}$ and $\boldsymbol{A}^{\mathsf{T}}$ operators can be seen.

Confining ourselves to this only example [3], now we give general rules how to construct a conjugate operator, $\boldsymbol{A}^{\mathsf{T}}$, for a given original operator $\boldsymbol{A}$. The rules are simple [4] and consist of the following two operations:

- the matrix $\boldsymbol{A}$ of symbolic differentiations is transposed;
- every term of the type $F_\alpha(x)D^\alpha(\,)$ in the matrix obtained by the transposition is replaced by its conjugate term in the form $(-1)^\alpha D^\alpha\,[F_\alpha(x)(\,)]$ where D^α represents a symbolic form of the α-order differentiation, that is,

---

[3] L. Young [13] gives a convincing reference which proves that one example is enough for every rule. This is an experience of "*…a nine-year old girl, a lady indeed, who solved only one of summation exercises from her homework and wrote in her writing-book that the others can be solved similarly*".

$$D^{\alpha}(\ )=\frac{\partial^{|\alpha|}(\ )}{\partial x_1^{\alpha_1}\cdots\partial x_k^{\alpha_k}},\quad |\alpha|=\alpha_1+\ldots+\alpha_k\ .$$

With one dimension ($k = 1$) we can use the integration by parts, while with two or three dimensions we can use the Gauss–Ostrogradsky formula to check that these rules really produce the conjugate operator.

To complete this section, we note two simple but important properties of the conjugation operation. First, the definition (1.7) implies directly that the conjugation is mutual, that is, the conjugate operator of a conjugate operator is identical to the original operator:

$$(\boldsymbol{A}^{\mathsf{T}})^{\mathsf{T}}=\boldsymbol{A}\ . \tag{1.10}$$

Second, the conjugate of the product of two operators is equal to the product of the operators conjugate of the original cofactors, placed in the reverse order, that is,

$$(\boldsymbol{AB})^{\mathsf{T}}=\boldsymbol{B}^{\mathsf{T}}\boldsymbol{A}^{\mathsf{T}}\ . \tag{1.11}$$

Both properties are quite similar to the properties of the usual matrix transposition.

1.2 Basic integral identity

Let $\boldsymbol{\sigma}$ be a somehow ordered set of functions which determine the stress state of a mechanical system (a full set of stresses or internal forces).

For example, in three-dimensional elasticity this $\boldsymbol{\sigma}$ will be a stress tensor with its components σ^{ij} referred to a particular system of axes, (x_1,x_2,x_3), which will be treated as a rectangular right-oriented Cartesian coordinate system if not stated otherwise. For further applications it will be convenient to use $\boldsymbol{\sigma}$ as a six-dimensional column vector, $\boldsymbol{\sigma}=\|[\,\sigma^{11},\sigma^{22},\sigma^{33},\sigma^{12},\sigma^{23},\sigma^{31}]\|^{\mathsf{T}}$, remembering its symmetry and its tensor nature. Actually, the tensor nature of the stresses is relevant only when doing a transformation of the coordinate system.

Another example is a Timoshenko beam in bending. Here $\boldsymbol{\sigma}$ will be a two-component column vector, $\boldsymbol{\sigma}=\|[Q,M]\|^{\mathsf{T}}$, Q being the shear force and M the bending moment in a cross-section of the beam. When discussing and posing any particular problem of structural mechanics, we will assume that the ordering rules for the components of the internal stresses/forces in $\boldsymbol{\sigma}$ are defined and known.

Let $\boldsymbol{\varepsilon}$ be a set of components of a strain tensor (vector) which is in energy reciprocity with $\boldsymbol{\sigma}$. The energy reciprocity means that the usual scalar product $\boldsymbol{\sigma}\cdot\boldsymbol{\varepsilon}$ in the sense of (1.3) will yield an expression of the elastic deformation energy.

For example, in three-dimensional elasticity $\boldsymbol{\varepsilon}$ means the strain tensor ε_{ij} or the column vector $\boldsymbol{\varepsilon} = \|[\varepsilon_{11},\varepsilon_{22},\varepsilon_{33},\gamma_{12},\gamma_{23},\gamma_{31}]\|^{\mathsf{T}}$ where $\gamma_{ij} = 2\varepsilon_{ij}$ at $i \neq j$. It is easy to notice that the vector notation for the set of the components of the stress and strain tensors will produce the following representation of the scalar product:

$$\boldsymbol{\sigma}\cdot\boldsymbol{\varepsilon} = \boldsymbol{\sigma}^{\mathsf{T}}\boldsymbol{\varepsilon} = \boldsymbol{\varepsilon}^{\mathsf{T}}\boldsymbol{\sigma}\,. \tag{2.1}$$

In the Timoshenko beam bending, $\boldsymbol{\varepsilon} = \|[\gamma, \chi]\|^{\mathsf{T}}$ where γ is a shear strain, χ is a bending strain (the derivative of the slope of the beam's cross-sections).

A set of governing equations for the analysis of a mechanical system under static loading will be presented in the following form:

$$\boldsymbol{B\sigma} + \boldsymbol{K}\mathbf{u} = \bar{\boldsymbol{X}} \qquad \text{equilibrium equations} \qquad (2.2\text{-}a)$$

$$\boldsymbol{A}\mathbf{u} = \boldsymbol{\varepsilon} \qquad \text{geometric equations} \qquad (2.2\text{-}b)$$

$$\boldsymbol{\sigma} = \boldsymbol{C\varepsilon} \ \text{ or } \ \boldsymbol{\varepsilon} = \boldsymbol{C}^{-1}\boldsymbol{\sigma} \qquad \text{physical equations} \qquad (2.2\text{-}c)$$

Here

- $\mathbf{u} = \|[u_i]\|$ is a displacement vector;
- $\bar{\boldsymbol{X}} = \|[\bar{X}^i]\|$ is a vector of given external forces per unit of volume of an elastic body;
- $\boldsymbol{K} = \{k_{ij}\}$ is a tensor of elasticity coefficients of a medium in which the deformable solid in question is put;
- $\boldsymbol{C} = \{C^{ijkl}\}$ and $\boldsymbol{C}^{-1} = \{D_{ijkl}\}$ are algebraic, mutually inverse operators which represent the respective tensors of coefficients of elasticity and compliance for the material of the deformable system.

The operators $\boldsymbol{C}, \boldsymbol{C}^{-1}$ and $\boldsymbol{K}$ are symmetric, and $\boldsymbol{C}$ and $\boldsymbol{C}^{-1}$ are also positive definite, while operator $\boldsymbol{K}$ is nonnegative (which is sometimes referred to as 'positive semi-definite') in every point of the area Ω. The conditions thus formulated can be reduced to the following requirements:

$$C^{ijkl} = C^{jikl} = C^{klij}\,, \qquad M_c a^{ij} a_{ij} \geq C^{ijkl} a_{ij} a_{kl} \geq m_c a^{ij} a_{ij} \quad (m_c > 0),$$

$$D_{ijkl} = D_{klij} = D_{jikl}\,, \qquad M_d a^{ij} a_{ij} \geq D_{ijkl} a^{ij} a^{kl} \geq m_d a^{ij} a_{ij} \quad (m_d > 0),$$

$$k^{ij} = k^{ji}\,, \qquad k^{ij} b_i b_j \geq m_k b_i b^i \quad (m_k \geq 0), \tag{2.3}$$

where

- $\boldsymbol{a} = \{a^{ij}\}$ is an arbitrary symmetric tensor of second rank;
- $\boldsymbol{b} = \{b^i\}$ is a vector;
- $M_c = 1/m_d$, $M_d = 1/m_c$, m_k are scalars independent of coordinates, tensor $\boldsymbol{a}$, and vector $\boldsymbol{b}$.

Further we will refer to the operator $\boldsymbol{B}$ as an *operator of equilibrium* and to the operator $\boldsymbol{A}$, which is purely geometric and relates the strains with the displacements, as an *operator of geometry.*

The system of governing equations (2.2) should be supplemented with boundary conditions; in the operator-based form they can be written as

$$\boldsymbol{E}_p(\boldsymbol{H}_\sigma\boldsymbol{\sigma} - \bar{\boldsymbol{p}}) = \boldsymbol{0} \qquad \text{(static boundary conditions)} \qquad (2.4\text{-}a)$$

$$\boldsymbol{E}_u(\boldsymbol{H}_u\mathbf{u} - \bar{\boldsymbol{u}}) = \boldsymbol{0} \qquad \text{(kinematic boundary conditions)} \qquad (2.4\text{-}b)$$

and should be specified on the boundary Γ of the area Ω.

The formulas (2.4) use the notation:

- $\bar{\boldsymbol{p}}$ for a vector of given external boundary forces;
- $\bar{\boldsymbol{u}}$ for a vector of given external boundary displacements.

We will assume that the components of these two vectors are represented simultaneously either in a coordinate system global for the whole structure, (x_1, x_2, x_3) or as decompositions by axes of a local basis in each point of the boundary Γ. The only local coordinate basis that we will use will be a right-oriented orthonormalized triple of vectors $(\boldsymbol{n}, \boldsymbol{t}, \boldsymbol{b})$ where $\boldsymbol{n}$ is a unit vector of the exterior (with respect to the area Ω) normal to the boundary Γ; $\boldsymbol{t}$ is a unit vector tangential to Γ. The condition of orthogonality and the right orientation of the triple $(\boldsymbol{n}, \boldsymbol{t}, \boldsymbol{b})$ define the third unit vector $\boldsymbol{b}$ as a vector product, $\boldsymbol{b} = \boldsymbol{n}\times\boldsymbol{t}$. In two-dimensional elasticity, the global coordinate system consists of the couple of axes, (x_1, x_2), and the local basis is made up by the unit vectors $(\boldsymbol{n}, \boldsymbol{t})$ where the positive direction of the vector $\boldsymbol{t}$ tangential to the boundary Γ is defined in such way that the condition $\boldsymbol{n}\times\boldsymbol{t} = \boldsymbol{i}_1\times\boldsymbol{i}_2$ holds, where $\boldsymbol{i}_1$ and $\boldsymbol{i}_2$ are unit vectors of the respective axes x_1 and x_2.

Algebraic operators $\boldsymbol{E}_p$ and $\boldsymbol{E}_u$ take the boundary conditions in each point of the boundary Γ and extract those of them which are actually specified in a particular problem. The operators are symmetric; they make up a decomposition of the identity operator $\boldsymbol{I}$ into an orthogonal sum, that is,

$$\boldsymbol{E}_p + \boldsymbol{E}_u = \boldsymbol{I}, \qquad \boldsymbol{E}_p\boldsymbol{E}_u = \boldsymbol{E}_u\boldsymbol{E}_p = \boldsymbol{O}\,, \qquad (2.5)$$

where $\boldsymbol{O}$ is a zero (annihilating) operator.

The relation (2.5) implies the *idempotency* of the boundary condition extraction operators:

$$\boldsymbol{E}_p\boldsymbol{E}_p = \boldsymbol{E}_p\,, \qquad \boldsymbol{E}_u\boldsymbol{E}_u = \boldsymbol{E}_u\,. \tag{2.6}$$

It should be obvious that the operators are defined on the whole boundary Γ, but they may have different values on different pieces of the boundary.

If we introduce the designation $\boldsymbol{p}$ for the boundary force vector and $\boldsymbol{u}$ for the boundary displacement vector, i.e. if we assume

$$\boldsymbol{p} = \boldsymbol{H}_\sigma \boldsymbol{\sigma}\,, \qquad \boldsymbol{u} = \boldsymbol{H}_u \mathbf{u} \in \Gamma \tag{2.7}$$

then the conditions (2.5) show that one and only one boundary condition can be specified for each component of the two vectors: a static one (for the $\boldsymbol{p}$ vector components) or a kinematic one (for the $\boldsymbol{u}$ vector components).

It should be clear also that the couple of the predefined vectors, $\overline{\boldsymbol{p}}$ and $\overline{\boldsymbol{u}}$, and the calculated couple of vectors, $\boldsymbol{p}$ and $\boldsymbol{u}$, are represented by their decompositions by the coordinate axes of the same basis, either global or local, and only in this case the respective components of the vectors can be compared according to (2.4).

Let's explain this by an example of planar elasticity. Suppose that a part of the boundary Γ is subject to external forces $\overline{p}^t$ tangential to the contour Γ and displacements $\overline{u}_n$ in the direction of the external normal to Γ (Fig. 1.1).

This means not all components of the vectors are specified in the local coordinates. In this particular case the decomposition of the vectors $\overline{\boldsymbol{p}}$ and $\overline{\boldsymbol{u}}$ by the axes of the local basis will give

$$\overline{\boldsymbol{p}} = \left\|\begin{bmatrix} \otimes \\ \overline{p}^t \end{bmatrix}\right\|, \qquad \overline{\boldsymbol{u}} = \left\|\begin{bmatrix} \overline{u}_n \\ \otimes \end{bmatrix}\right\| \tag{2.8}$$

where the symbol $\otimes$ designates values of components not known beforehand. Obviously, on this piece of the boundary Γ the algebraic operators $\boldsymbol{E}_p$ and $\boldsymbol{E}_u$ are described by the matrices of the following kind:

$$\boldsymbol{E}_p = \left\|\begin{bmatrix} 0 & 0 \\ 0 & 1 \end{bmatrix}\right\|, \qquad \boldsymbol{E}_u = \left\|\begin{bmatrix} 1 & 0 \\ 0 & 0 \end{bmatrix}\right\|, \tag{2.9}$$

to exclude the undefined quantities $\otimes$ from the boundary conditions (2.4).

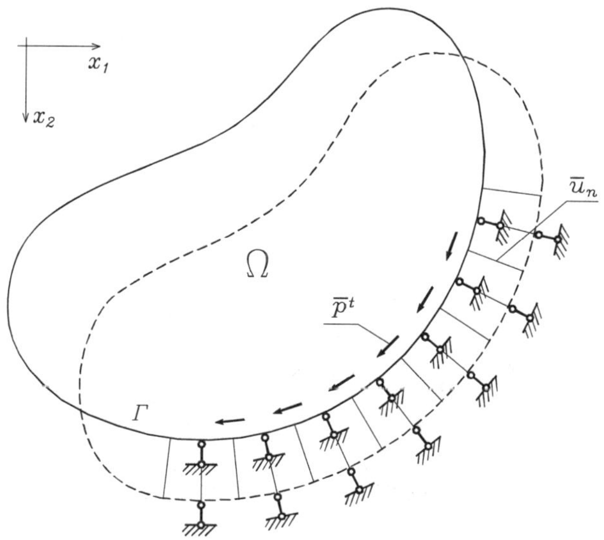

Fig. 1.1. Mixed boundary conditions on a piece of the boundary Γ of the area Ω

If $\overline{\boldsymbol{p}}$ and $\overline{\boldsymbol{u}}$ are specified in a local coordinate system, and if the components of the boundary force vector, $\boldsymbol{p}$, and of the boundary displacement vector, $\boldsymbol{u}$, are transformed by formal calculation according to (2.7) into quantities expressed in the global coordinate system,

$$\boldsymbol{p} = \|[p^1, p^2]\|^{\mathsf{T}}, \qquad \boldsymbol{u} = \|[u_1, u_2]\|^{\mathsf{T}},$$

then, before using the vectors $\boldsymbol{p}$ and $\boldsymbol{u}$ in the boundary conditions

$$\boldsymbol{E}_p(\boldsymbol{p} - \overline{\boldsymbol{p}}) = \boldsymbol{0}, \qquad \boldsymbol{E}_u(\boldsymbol{u} - \overline{\boldsymbol{u}}) = \boldsymbol{0} \quad \in \Gamma,$$

they should be converted to the local coordinate system following common rules. The reason for this is the sensitivity of the boundary condition extraction operators, $\boldsymbol{E}_p$ and $\boldsymbol{E}_u$, to the coordinate system; they are constructed in such way that they are not invariant to the coordinates, instead they keep track of a coordinate system in which the predefined boundary forces and displacements are specified.

If there are no mixed boundary conditions anywhere on the contour Γ, then the latter can be divided into two parts, Γ_p and Γ_u, so that

$$\Gamma = \Gamma_p \cup \Gamma_u, \quad \Gamma_p \cap \Gamma_u = \varnothing \tag{2.10}$$

and

$$\boldsymbol{E}_p = \boldsymbol{I}\,,\; \boldsymbol{E}_u = \boldsymbol{O} \in \Gamma_p\,, \qquad \boldsymbol{E}_p = \boldsymbol{O}\,,\; \boldsymbol{E}_u = \boldsymbol{I} \in \Gamma_u\,. \tag{2.11}$$

When (2.10) and (2.11) hold, the boundary conditions (2.4) can be written as

$$\boldsymbol{H}_\sigma \boldsymbol{\sigma} - \bar{\boldsymbol{p}} = \boldsymbol{0} \quad \in \Gamma_p \qquad \text{(static boundary conditions)} \tag{2.12-a}$$

$$\boldsymbol{H}_u \mathbf{u} - \bar{\boldsymbol{u}} = \boldsymbol{0} \quad \in \Gamma_u \qquad \text{(kinematic boundary conditions)} \tag{2.12-b}$$

In problems where the equilibrium operator $\boldsymbol{B}$ and the geometry operator $\boldsymbol{A}$ contain derivatives of orders higher than first, the vector of the boundary displacements, $\boldsymbol{u}$, exceeds the vector of internal displacements, $\mathbf{u}$, by the number of its components: the kinematic boundary conditions on the boundary Γ may contain both values of the vector function $\mathbf{u}$ itself and some derivatives of the components. The matrix differential operator $\boldsymbol{H}_u$ transforms the vector of internal displacements $\mathbf{u}$, to a space of vectors of the boundary displacements $\boldsymbol{u}$. When needed, this operator can perform a transformation of the boundary displacement vector to a local coordinate system.

The matrix differential operator $\boldsymbol{H}_\sigma$ works in a similar way. It transforms an internal force vector, $\boldsymbol{\sigma}$, to the space of boundary force vectors $\boldsymbol{p}$. In a particular case when the equilibrium and geometry operators contain differential operations of at most first order and the local coordinate system coincides with the global one, $\boldsymbol{H}_u$ is an identity operator, that is,

$$\boldsymbol{H}_u = \boldsymbol{I}. \tag{2.13}$$

In the same situation, the algebraic operator $\boldsymbol{H}_\sigma$ contains components of the unit vector $\boldsymbol{n}$ of an exterior normal to the boundary Γ with respect to the global coordinate system, i.e. the cosines of angles between the vector $\boldsymbol{n}$ and the direction of the respective axis x_i, in other words, the quantities

$$n_i = \cos(\boldsymbol{n}, x_i) = (\boldsymbol{i}_i, \boldsymbol{n}) \tag{2.14}$$

where $\boldsymbol{i}_i$ is the unit vector of the x_i axis.

As it will be seen from formulations of various types of structural mechanics problems, in all cases the operators of equilibrium and geometry at the given boundary conditions (2.4) satisfy a so-called *basic integral identity*

$$(\boldsymbol{A}\mathbf{u}, \boldsymbol{\sigma}) = (\mathbf{u}, \boldsymbol{B}\boldsymbol{\sigma}) + (\boldsymbol{H}_\sigma \boldsymbol{\sigma}, \boldsymbol{H}_u \mathbf{u})_\Gamma\,. \tag{2.15}$$

This equation can be validated directly by constructing the appropriate equilibrium and geometry operators. However, as we will see further, the basic integral identity (2.15) is actually true and fundamental for all linear problems of structural mechanics. It implies all basic theorems given further below. In particular, comparing (2.15) and (1.7) will give an immediate result that the equilibrium operator and the geometry operator are mutually conjugate:

$$\boldsymbol{B} = \boldsymbol{A}^{\mathrm{T}} . \tag{2.16}$$

The latter relationship allows us to designate the equilibrium operator further as $\boldsymbol{A}^{\mathrm{T}}$ without noticing this fact additionally[4].

Taking into account the designations from (2.7) will convert the basic integral identity into the following:

$$(A\mathbf{u}, \boldsymbol{\sigma}) = (\mathbf{u}, A^{\mathrm{T}}\boldsymbol{\sigma}) + (\boldsymbol{p}, \boldsymbol{u})_{\Gamma} . \tag{2.17}$$

Now, we can take the properties (2.5) and (2.6) of the boundary condition extraction operators into account and use a chain of obvious transformations

$$(\boldsymbol{p}, \boldsymbol{u})_{\Gamma} = ((\boldsymbol{E}_p + \boldsymbol{E}_u)\boldsymbol{p}, (\boldsymbol{E}_p + \boldsymbol{E}_u)\boldsymbol{u})_{\Gamma} = (\boldsymbol{E}_p\boldsymbol{p}, \boldsymbol{E}_p\boldsymbol{u})_{\Gamma} + (\boldsymbol{E}_u\boldsymbol{p}, \boldsymbol{E}_u\boldsymbol{u})_{\Gamma} \tag{2.18}$$

to represent the same integral identity as

$$(A\mathbf{u}, \boldsymbol{\sigma}) = (\mathbf{u}, A^{\mathrm{T}}\boldsymbol{\sigma}) + (\boldsymbol{E}_p\boldsymbol{p}, \boldsymbol{E}_p\boldsymbol{u})_{\Gamma} + (\boldsymbol{E}_u\boldsymbol{p}, \boldsymbol{E}_u\boldsymbol{u})_{\Gamma} . \tag{2.19}$$

This is the form of the basic integral identity which we will use most often throughout the book.

1.3 Various types of stress and strain fields

We will say that a whole set of stresses $\boldsymbol{\sigma}$, strains $\boldsymbol{\varepsilon}$, and displacements $\mathbf{u}$ determine a stress-and-strain state (SSS) of a mechanical system. As the stresses, strains, and displacements of the mechanical system may vary from point to point, another proper term will be a field or a distribution of the stresses, the strains, or the displacements. All three fields together will be called a *stress-and-strain field.* Further below, whenever the nature of a field is not specified, we will understand that the field is actually a stress-and-strain field. An arbitrary field of this kind will be denoted as F, and an expression like $F = \{\boldsymbol{\sigma}, \boldsymbol{\varepsilon}, \mathbf{u}\}$ should be read as a *field F consisting of stresses* $\boldsymbol{\sigma}$, *strains* $\boldsymbol{\varepsilon}$, *and displacements* $\mathbf{u}$. In their turn, the stresses $\boldsymbol{\sigma}$, the strains $\boldsymbol{\varepsilon}$, and the displacements $\mathbf{u}$ will be referred to as *elements of the*

[4] Nearly everywhere. Rare exceptions will be specified.

field F. It should be emphasized that the elements of a field are not supposed to relate to one another anyhow in the most general case.

Now, let's introduce more notions and definitions which will be useful for further presentment.

We will say that external forces $\boldsymbol{X}$, distributed over the volume of a body Ω, and contour forces $\boldsymbol{p}$ are in *static conformance* to a field $F = \{\boldsymbol{\sigma}, \boldsymbol{\varepsilon}, \mathbf{u}\}$ if the following holds:

$$\boldsymbol{X} = \boldsymbol{A}^{\mathsf{T}}\boldsymbol{\sigma} + \boldsymbol{K}\mathbf{u} \in \Omega, \qquad \boldsymbol{E}_p\,\boldsymbol{p} = \boldsymbol{E}_p\boldsymbol{H}_\sigma\boldsymbol{\sigma} \in \Gamma, \tag{3.1}$$

or, in another words, if the forces $\boldsymbol{X}$ and $\boldsymbol{p}$ obey the equations of equilibrium inside the body Ω and the static boundary conditions on the surface Γ.

A field $F_s = \{\boldsymbol{\sigma}_s, \boldsymbol{\varepsilon}_s, \mathbf{u}_s\}$ will be called *statically admissible* if the given external forces $\bar{\boldsymbol{X}}$ and $\bar{\boldsymbol{p}}$ are in static conformance to the stresses $\boldsymbol{\sigma}_s$ and displacements $\mathbf{u}_s$, or, in other words,

$$\bar{\boldsymbol{X}} = \boldsymbol{A}^{\mathsf{T}}\boldsymbol{\sigma}_s + \boldsymbol{K}\mathbf{u}_s \in \Omega, \qquad \boldsymbol{E}_p\,\bar{\boldsymbol{p}} = \boldsymbol{E}_p\boldsymbol{H}_\sigma\boldsymbol{\sigma}_s \in \Gamma. \tag{3.2}$$

The "s" subscript emphasizes the static admissibility of the field and its elements. This definition implies that the property of static admissibility of a field does not depend on the strains $\boldsymbol{\varepsilon}_s$.

A field $F_{so} = \{\boldsymbol{\sigma}_{so}, \boldsymbol{\varepsilon}_{so}, \mathbf{u}_{so}\}$ will be called *homogeneously statically admissible* if internal forces in Ω created by this field are self-balanced while on the boundary Γ homogeneous static boundary conditions hold in such locations and in such directions where the original problem has the respective static boundary conditions. To put it otherwise, the elements of the field F_{so} satisfy the relationships

$$\boldsymbol{A}^{\mathsf{T}}\boldsymbol{\sigma}_{so} + \boldsymbol{K}\mathbf{u}_{so} = \mathbf{0} \in \Omega, \qquad \boldsymbol{E}_p\boldsymbol{H}_\sigma\boldsymbol{\sigma}_{so} = \mathbf{0} \in \Gamma. \tag{3.3}$$

The additional subscript "o" emphasizes the homogeneous static admissibility of the field and its elements.

A field $F_k = \{\boldsymbol{\sigma}_k, \boldsymbol{\varepsilon}_k, \mathbf{u}_k\}$ will be called *kinematically admissible* if the displacements $\mathbf{u}_k$ and the strains $\boldsymbol{\varepsilon}_k$ inside the area Ω satisfy the geometric equations and the kinematic boundary conditions on the boundary Γ. In other words, the elements of the field F_k satisfy the following relationships:

$$\boldsymbol{\varepsilon}_k = \boldsymbol{A}\mathbf{u}_k \in \Omega, \qquad \boldsymbol{E}_u(\boldsymbol{H}_u\mathbf{u}_k - \bar{\boldsymbol{u}}) = \mathbf{0} \in \Gamma. \tag{3.4}$$

The subscript "k" emphasizes the kinematic admissibility of the field and its elements.

A field $F_{\text{ko}} = \{\boldsymbol{\sigma}_{\text{ko}}, \boldsymbol{\varepsilon}_{\text{ko}}, \mathbf{u}_{\text{ko}}\}$ will be called *homogeneously kinematically admissible* if the displacements $\mathbf{u}_{\text{ko}}$ and the strains $\boldsymbol{\varepsilon}_{\text{ko}}$ satisfy the geometric equations inside the area Ω and the homogeneous kinematic boundary conditions on the boundary Γ. In other words, the elements of the field F_{ko} meet the conditions

$$\boldsymbol{\varepsilon}_{\text{ko}} = \boldsymbol{A}\mathbf{u}_{\text{ko}} \in \Omega, \qquad \boldsymbol{E}_u \boldsymbol{H}_u \mathbf{u}_{\text{ko}} = \mathbf{0} \quad \in \Gamma. \tag{3.5}$$

A field $F_{\text{s/2}} = \{\boldsymbol{\sigma}_{\text{s/2}}, \boldsymbol{\varepsilon}_{\text{s/2}}, \mathbf{u}_{\text{s/2}}\}$ will be called *statically semi-admissible* if the elements of the field satisfy the static boundary conditions

$$\boldsymbol{E}_p \overline{\boldsymbol{p}} = \boldsymbol{E}_p \boldsymbol{H}_\sigma \boldsymbol{\sigma}_{\text{s/2}} \quad \in \Gamma. \tag{3.6}$$

The equations of equilibrium in the volume of the body, Ω, are not required to hold with such a field.

A field $F_{\text{k/2}} = \{\boldsymbol{\sigma}_{\text{k/2}}, \boldsymbol{\varepsilon}_{\text{k/2}}, \mathbf{u}_{\text{k/2}}\}$ will be called *kinematically semi-admissible* if the displacements $\mathbf{u}_{\text{k/2}}$ and the strains $\boldsymbol{\varepsilon}_{\text{k/2}}$ satisfy the geometric equations inside the area Ω,

$$\boldsymbol{\varepsilon}_{\text{k/2}} = \boldsymbol{A}\mathbf{u}_{\text{k/2}} \in \Omega. \tag{3.7}$$

There are no requirements to the values of the displacements $\mathbf{u}_{\text{k/2}}$ or their derivatives on the boundary Γ.

It will be useful to introduce, together with the definitions of stress-and-strain fields, the notion of a *field of external actions*, $V = \{\overline{\boldsymbol{X}}, \overline{\boldsymbol{p}}, \overline{\boldsymbol{u}}\}$, which will mean a set of given external forces, $\overline{\boldsymbol{X}}$, distributed over the area Ω, contour forces $\overline{\boldsymbol{p}}$ specified on the boundary Γ, and given boundary displacements $\overline{\boldsymbol{u}}$.

1.4 The general principle of statics and geometry

Consider two states of a mechanical system; one of those will be named the state 1 and the other the state 2. Let the state 1 be defined by the field $F_1 = \{\boldsymbol{\sigma}_1, \boldsymbol{\varepsilon}_1, \mathbf{u}_1\}$ and the state 2 by the field $F_2 = \{\boldsymbol{\sigma}_2, \boldsymbol{\varepsilon}_2, \mathbf{u}_2\}$. For now, let's think both fields are absolutely independent.

Assuming $\boldsymbol{\sigma} = \boldsymbol{\sigma}_1$ and $\mathbf{u} = \mathbf{u}_2$ in the basic integral identity (2.19) and taking into account the conjugation between the geometry and equilibrium operators will give

$$(\boldsymbol{\sigma}_1, \boldsymbol{A}\mathbf{u}_2) = (\boldsymbol{A}^{\mathrm{T}}\boldsymbol{\sigma}_1, \mathbf{u}_2) + (\boldsymbol{E}_p \boldsymbol{p}_1, \boldsymbol{E}_p \boldsymbol{u}_2)_\Gamma + (\boldsymbol{E}_u \boldsymbol{p}_1, \boldsymbol{E}_u \boldsymbol{u}_2)_\Gamma \tag{4.1}$$

where the contour stresses, $\boldsymbol{p}_1 = \boldsymbol{H}_\sigma \boldsymbol{\sigma}_1$, are taken from the state 1 and the contour displacements, $\boldsymbol{u}_2 = \boldsymbol{H}_u \mathbf{u}_2$, from the state 2.

By adding external volumetric forces, $\boldsymbol{X}_1 = \boldsymbol{A}^{\mathsf{T}} \boldsymbol{\sigma}_1 + \boldsymbol{K}\mathbf{u}_1$, which are in static conformance to the field F_1, we make the relationship (4.1) look like

$$(\boldsymbol{X}_1, \mathbf{u}_2) + (\boldsymbol{E}_p \boldsymbol{p}_1, \boldsymbol{E}_p \boldsymbol{u}_2)_\Gamma + (\boldsymbol{E}_u \boldsymbol{p}_1, \boldsymbol{E}_u \boldsymbol{u}_2)_\Gamma = (\boldsymbol{\sigma}_1, \boldsymbol{A}\mathbf{u}_2) + (\boldsymbol{K}\mathbf{u}_1, \mathbf{u}_{2\text{-}}) \,. \tag{4.2}$$

Actually, (4.2) is an equivalent of the basic integral identity (2.19). Now, by giving the states 1 and 2 a certain sense we can derive various versions and mechanical treatments of the basic integral identity in the form (4.2).

First of all, note that the left part in (4.2) is a *virtual work* A_{12} *of the external forces* in static conformance with the state 1 on the displacements of the state 2 corresponding to those forces,

$$\mathsf{A}_{12} = (\boldsymbol{X}_1, \mathbf{u}_2) + (\boldsymbol{E}_p \boldsymbol{p}_1, \boldsymbol{E}_p \boldsymbol{u}_2)_\Gamma + (\boldsymbol{E}_u \boldsymbol{p}_1, \boldsymbol{E}_u \boldsymbol{u}_2)_\Gamma \,. \tag{4.3}$$

We should add some explanation to the definition of the virtual work of external forces according to (4.3). The matter is that the external contour forces, $\boldsymbol{p}_1$, include the full set of the contour forces in the state 1, together with the contour forces $\boldsymbol{E}_u \boldsymbol{p}_1$ in such locations and in such directions where the kinematic boundary conditions are specified. In other words, the expression of A_{12} from (4.3) includes the work of the contour reactive forces of the state 1 in the form $(\boldsymbol{E}_u \boldsymbol{p}_1, \boldsymbol{E}_u \boldsymbol{u}_2)_\Gamma$. This part of the work can be different from zero because the field of displacements, $\mathbf{u}_2$, is not required to be homogeneously kinematically admissible, thus it does not have to satisfy homogeneous kinematic boundary conditions.

Finally, it should be noted that the expression (4.3) is not an actual work of the external forces of the state 1 but a *virtual work* which would be done by the forces hypothetically if these forces could be somehow maintained in the body when the displacements $\mathbf{u}_2$ appeared in our deformable body.

Next, let's define the right part of (4.2) as a *virtual work* B_{12} *of the internal forces (stresses)* of the state 1 on the respective displacements of the state 2, taken with the opposite sign.

$$\mathsf{B}_{12} = -(\boldsymbol{\sigma}_1, \boldsymbol{A}\mathbf{u}_2) - (\boldsymbol{K}\mathbf{u}_1, \mathbf{u}_2) \,. \tag{4.4}$$

Note that the internal forces in the mechanical system in question are of a dual nature. On one hand, we can treat the stresses $\boldsymbol{\sigma}$ as internal forces which appear in the material of an elastic body. Another source of the internal forces is related to the forces

$$\boldsymbol{r} = \boldsymbol{K}\mathbf{u} \tag{4.5}$$

created by an elastic medium in which the deformable solid is placed. The forces of this kind are *reactions* of the elastic medium sometimes called a *response of an elastic foundation.*

As one can see from (4.4), the expression of the virtual work of the internal forces, B_{12}, consists of two components where the first term is interpreted as a virtual work of the stresses $\boldsymbol{\sigma}_1$ and the second term is treated as a virtual work of the internal forces of the elastic medium. Note also that the assumed independence of the three elements of the field F_2 makes it impossible, in the general case, to identify the virtual work of the stresses $\boldsymbol{\sigma}_1$ in the form $-(\boldsymbol{\sigma}_1, \boldsymbol{A}\mathbf{u}_2)$ with the scalar product, $-(\boldsymbol{\sigma}_1, \boldsymbol{\varepsilon}_2)$, of the stresses of the field F_1 and the strains of the field F_2. These two can be identified with each other only when the field F_2 is at least semi-admissible kinematically.

Now the relationships (4.3) and (4.4) turn (4.2) into

$$\mathsf{A}_{12} + \mathsf{B}_{12} = 0, \tag{4.6}$$

and this form of it is called the *general principle of statics and geometry* [8]. This entitlement for (4.6) is totally justified because, first, the relationship can be proved without any physical state equations and it does not have anything in common with any thermodynamic considerations being of pure static and geometric nature; second, it is formulated on the basis of two arbitrarily chosen states of the mechanical system.

An inverse proposition can be formulated just as easily: if (4.6) holds for a certain state 1, then this relationship implies the equilibrium equations for the state both in the volume of the body Ω and on its boundary Γ. To see this, note that (4.6) is a brief form of the relationship (4.2) which we assume to hold for a certain set of external forces $\boldsymbol{X}_1$ and $\boldsymbol{p}_1$ not subject to any additional conditions so far. Transforming (4.2) with help of the basic integral identity in the form (2.15) and regrouping its terms will give

$$(\boldsymbol{A}^{\mathsf{T}}\boldsymbol{\sigma}_1 + \boldsymbol{K}\mathbf{u}_1 - \boldsymbol{X}_1, \mathbf{u}_2) + (\boldsymbol{H}_\sigma\boldsymbol{\sigma}_1 - \boldsymbol{p}_1, \boldsymbol{u}_2)_\Gamma = 0\,. \tag{4.7}$$

Assuming that the relationship (4.7) holds for <u>any</u> fields of displacements $\mathbf{u}_2$, we conclude that the scalar products from (4.7) should be equal to zero separately as well. The properties of the scalar products imply that the field $F_1 = \{\boldsymbol{\sigma}_1, \boldsymbol{\varepsilon}_1, \mathbf{u}_1\}$ is balanced by the volumetric forces $\boldsymbol{X}_1$ and the contour forces $\boldsymbol{p}_1$.

Thus, in terms of mechanics the general principle of statics and geometry (GPSG) can be read as follows:

> *It is necessary and sufficient for a linearly deformable mechanical system to remain in equilibrium in a certain state that the overall virtual work of all external and internal forces of this state of the system be equal to zero on any displacements.*

It should be emphasized once more that we speak about the *virtual* work, and the two states of the system which participate in the GPSG formulation are not related to each other in any way.

This formulation of the general principle of the statics and geometry operates the following three categories:

- an expression of the work of internal forces (WIF);
- equilibrium equations (EE);
- geometric equations (GE).

The fact that the general principle of statics and geometry contains implicitly the geometric equations becomes obvious after we represent the virtual work of the stresses $\boldsymbol{\sigma}_1$ as $-(\boldsymbol{\sigma}_1, \boldsymbol{\varepsilon}_2)$; this can be done only if the geometric equations $\boldsymbol{\varepsilon}_2 = \boldsymbol{A}\mathbf{u}_2$ hold.

There is a tradition in courses of structural mechanics and strength of materials: the definition of the virtual work of internal forces is not introduced in axiomatic manner on the basis of an expression like (4.4); instead, certain direct mechanical considerations are used. Let's connect the considerations of that kind with our general case of the geometrically linear structural analysis and do the following reasoning.

Divide mentally a deformable solid body into a set of elementary bodies of small volume, Ω_i, so that $\Sigma\Omega_i = \Omega$ where the sum is taken over the whole set of the elementary bodies. Let Γ_i be a boundary surface of i-th elementary body. Next, replace the action of adjacent bodies on a certain elementary body by a vector of forces, $\boldsymbol{p}_{1i}$, resulted from the stresses $\boldsymbol{\sigma}_1$,

$$\boldsymbol{p}_{1i} = \boldsymbol{H}_\sigma \boldsymbol{\sigma}_{1i} \tag{4.8}$$

and acting in every point of the boundary surface Γ_i. Now, for the i-th elementary body the set of the forces $\boldsymbol{p}_{1i}$ becomes a set of external forces (with respect to this i-th body only) which perform their work on the displacements $\boldsymbol{u}_{2i} = \boldsymbol{H}_u \mathbf{u}_{2i}$. It is assumed implicitly that the internal stresses are self-balanced inside each of the elementary bodies in the sense that $\boldsymbol{A}^\mathsf{T}\boldsymbol{\sigma}_{1i} + \boldsymbol{K}\mathbf{u}_{1i} = \mathbf{0} \;\; \in \Omega_i$. Then the virtual work $d\tilde{\mathsf{B}}_{12}$ of the forces $\boldsymbol{p}_{1i}$ on the displacements of the body, $\mathbf{u}_{2i}$, is defined by the integral

$$d\tilde{\mathsf{B}}_{12} = \int_{\Gamma_i} \boldsymbol{p}_{1i} \cdot \boldsymbol{u}_{2i} \, d\Gamma \,. \tag{4.9}$$

Now we use the basic integral identity in the form (2.17) in application to the elementary body; let's transform the integral into the form

$$d\tilde{\mathsf{B}}_{12} = \int_{\Omega_i} \boldsymbol{A}\mathbf{u}_{2i} \cdot \boldsymbol{\sigma}_{1i} d\Omega - \int_{\Omega_i} \mathbf{u}_{2i} \cdot \boldsymbol{A}^{\mathsf{T}} \boldsymbol{\sigma}_{1i} d\Omega = \int_{\Omega_i} \boldsymbol{A}\mathbf{u}_{2i} \cdot \boldsymbol{\sigma}_{1i} d\Omega + \int_{\Omega_i} \boldsymbol{K}\mathbf{u}_{1i} \cdot \mathbf{u}_{2i} d\Omega .$$

This allows for the above assumption $\boldsymbol{A}^{\mathsf{T}}\boldsymbol{\sigma}_{1i} = -\boldsymbol{K}\mathbf{u}_{1i} \quad \in \Omega_i$.

Now we sum the expression of the work thus obtained over all elementary bodies and derive this:

$$\tilde{\mathsf{B}}_{12} = (\boldsymbol{\sigma}_1, \boldsymbol{A}\mathbf{u}_2) + (\boldsymbol{K}\mathbf{u}_1, \mathbf{u}_2) . \qquad (4.10)$$

The quantity $\tilde{\mathsf{B}}_{12}$ is defined here as a virtual work of internal forces, but it is calculated with the assumption that the internal forces have been formally converted to external ones. If we compare (4.10) and (4.4), we will have to admit that B_{12} and $\tilde{\mathsf{B}}_{12}$ can be matched only by assuming $\mathsf{B}_{12} = -\tilde{\mathsf{B}}_{12}$. This logical transformation is usually based on the statement that the internal forces resist to deformations in the body and thus perform a negative work[5]. This reasoning is logically vulnerable, which is obvious to anybody who ever tried to explain to an inquisitive student why it is necessary (!?) to alter the sign in (4.10) when deriving the expression of the virtual work of the internal forces.

Actually, the four categories introduced above are related to one another in such way that any three of them can be postulated, and then the fourth one will result from the three. This circumstance was noticed by L.A. Rozin; he composed the following table which we borrow from [8]:

GPSG	EE	WIF	GE
Implied	Postulated	Postulated	Postulated
Postulated	Implied	Postulated	Postulated
Postulated	Postulated	Implied	Postulated
Postulated	Postulated	Postulated	Implied

First two rows of the table conform to the verbal formulation of the general principle of statics and geometry given above; it is assumed that the virtual work of the internal forces is expressed by the relationship (4.4) while the geometric equations are as in (2.2-*b*). The statements contained

[5] Here's a characteristic quote on this subject [2]: "*The work done by internal forces is negative because the internal forces tend to resist to displacements caused by the external forces; the internal forces are directed oppositely to the displacements of their application points*". The essential point is that the subject being discussed is actually a virtual work of the internal forces.

in the third and fourth rows of the table can be proved to be true, too. We will not dwell on this; instead, we recommend [8] to the reader who wants details.

The notion of the virtual work can belong to the category of constructible concepts; one of the confirmations of this fact is that some publications (see [12], or [9], or [10], for example) treat the principle of virtual displacements in the statics of the deformable solid as an equality between the virtual work of the external forces and the virtual work of the internal forces, rather than an equality of the sum of the works to zero. Of course, this requires redefining the notion of the virtual work of the internal forces by introducing the expression (4.4) with the opposite sign. As it could be expected, no further logical contradiction comes from this redefinition of the virtual work of internal forces. Therefore it should be clear that one cannot really justify the necessity of minus in the right part of (4.4) by any logical reasoning.

1.4.1 The principle of virtual displacements as an implication of the general principle of statics and geometry

It is obvious that while the relationship (4.2) holds for arbitrary displacements $\mathbf{u}_2$ as long as the forces $\boldsymbol{X}_1$ and $\boldsymbol{p}_1$ balance the internal stresses in the system in its state 1, it is going to hold also for a particular case when the state 2 of the deformable system is assumed to be a homogeneously kinematically admissible state, so that $\mathbf{u}_2 = \mathbf{u}_{\text{ko}}$, which gives us immediately

$$(\boldsymbol{X}_1,\, \mathbf{u}_{\text{ko}}) + (\boldsymbol{E}_p \boldsymbol{p}_1,\, \boldsymbol{E}_p \boldsymbol{u}_{\text{ko}})_\Gamma = (\boldsymbol{\sigma}_1,\, \boldsymbol{\varepsilon}_{\text{ko}}) + (\boldsymbol{K}\mathbf{u}_1,\, \mathbf{u}_{\text{ko}})\,. \tag{4.11}$$

This relationship takes into account that a homogeneously kinematically admissible field meets the condition $\boldsymbol{E}_u \boldsymbol{u}_{\text{ko}} = \mathbf{0} \in \Gamma$.

Now, let's show that the equality (4.11) being held is enough to keep the equilibrium in the volume of the body and on a part of its surface where static boundary conditions are specified. To see this, substitute (4.1) and regroup the terms:

$$(\boldsymbol{A}^{\mathsf{T}}\boldsymbol{\sigma}_1 + \boldsymbol{K}\mathbf{u}_1 - \boldsymbol{X}_1,\, \mathbf{u}_{\text{ko}}) + (\boldsymbol{E}_p \boldsymbol{H}_\sigma \boldsymbol{\sigma}_1 - \boldsymbol{E}_p \boldsymbol{p}_1\,,\, \boldsymbol{E}_p \boldsymbol{u}_{\text{ko}})_\Gamma = 0\,. \tag{4.12}$$

We use the same reasoning as before and come to the principle of virtual displacements:

> *It is necessary and sufficient for a linearly deformable mechanical system to remain in equilibrium that the overall virtual work of all external and internal forces on any homogeneously kinematically admissible displacements be equal to zero.*

Some courses of structural mechanics advance another requirement when formulating the principle of virtual displacements: the displacements on which the virtual work is calculated must be infinitesimal. This is relevant when we deal with a geometric nonlinearity (second-order analysis). The requirement that homogeneously kinematically admissible displacements should be infinitesimal is not essential in first-order analysis and can be omitted. Even though this is obvious, some authors were confused and suggested various justifications. However, in the linear analysis this principle is totally independent from any particular cases of the nonlinear analysis, therefore no justification is required.

It should be said that the direct application of the principle of virtual displacements sometimes produces a quicker and nicer solution that the use of equilibrium equations. Here is an example to illustrate this statement Consider a problem shown in Fig. 1.2.

The bars in intermediate nodes of the system are connected as in scissors; two couples of bars come to one node, so that two bars that form a couple and belong to one straight line are attached stiffly while the couples are joined by hinges.

We are required to find the stress in the bar that connects nodes 0 and 1 in this simple, statically determinate system. We could use the equilibrium equations in nodes and move from the system's right end to its left end to find the stress (which we denote as S_{01}); however, the direct application of the principle of virtual displacements will bring us to our goal much faster.

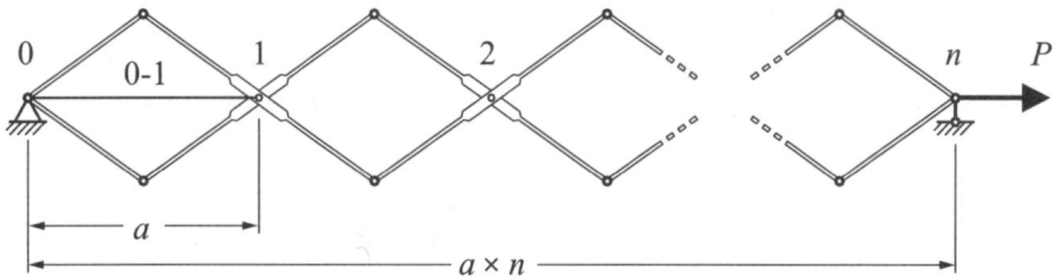

Fig. 1.2. An example of application of the principle of virtual displacements

The system is statically determinate, therefore its internal forces (stresses) in its elements do not depend on physical properties of the elements. That's why we can assume, without limiting the generality, that all bars except for 0-1 are perfectly rigid, that is, their strains are zero – as is the virtual work of the internal forces which appear in the bars under external forces. As a result, the virtual work of all internal forces in the system can be expressed as $-S_{01}\Delta$ where Δ is a virtual (homogeneously kinematically admissible) displacement of the node 1 along the x-axis. It is easy to see that this displacement of the node 1 causes all cells of the system to distort their size equally, so the difference between the displacements of any two adjacent nodes will be equal to Δ, and the last node, n, will acquire the displacement $n\Delta$. Thus, the virtual work of the external forces will be equal to $Pn\Delta$. The principle of virtual displacements displacements gives

$$Pn\Delta - S_{01}\Delta = 0$$

and, accordingly, $S_{01} = nP$.

1.4.2 The principle of virtual stress increments as an implication of the general principle of statics and geometry

A principle reciprocal with the principle of virtual displacements will be derived again from the general principle of statics and geometry, but this time the state 1 will be a homogeneously statically admissible state; this assumption allows us to rewrite (4.2) in the form

$$(\boldsymbol{E}_u \boldsymbol{p}_{\text{so}}, \boldsymbol{E}_u \boldsymbol{u}_2)_\Gamma = (\boldsymbol{\sigma}_{\text{so}}, \boldsymbol{A}\mathbf{u}_2) + (\boldsymbol{K}\mathbf{u}_{\text{so}}, \mathbf{u}_2)\,. \tag{4.13}$$

We choose only the class of kinematically admissible states from all possible kinds of the state 2 of the system. Then, according to (3.4), the relationship (4.13) will become

$$(\boldsymbol{E}_u \boldsymbol{p}_{\text{so}}, \boldsymbol{E}_u \overline{\boldsymbol{u}})_\Gamma = (\boldsymbol{\sigma}_{\text{so}}, \boldsymbol{\varepsilon}_2) + (\boldsymbol{K}\mathbf{u}_{\text{so}}, \mathbf{u}_2)\,. \tag{4.14}$$

The scalar product in the left part of (4.14) can be interpreted as a virtual work, A_{12}, of all external forces of the state 1 on real displacements of the system (there are no active external forces in the state 1, hence the zero virtual work done by them). If we postulate the expression for the virtual work of the internal forces of the state 1 on the displacements of the state 2 directly via the strains,

$$\mathsf{B}_{12} = -(\boldsymbol{\sigma}_{\text{so}}, \boldsymbol{\varepsilon}_2) - (\boldsymbol{K}\mathbf{u}_{\text{so}}, \mathbf{u}_2)\,, \tag{4.15}$$

we will come to a formulation of the principle of virtual stress increments:

It is necessary and sufficient for a certain state 2 of a linearly deformable mechanical system to be kinematically admissible that the sum of the virtual work, A_{12} *, of all external forces of any homogeneously statically admissible state 1 on the real displacements of the system and the virtual work,* B_{12}*, of the internal forces of the same state on the displacements of the state 2 be equal to zero.*

The necessity is obvious because this is just a verbal presentation of the proved equality (4.14), therefore we start with proving the sufficiency.

So, let (4.14) hold for a certain, arbitrary so far, state 2 with any field $F_{so} = \{\boldsymbol{\sigma}_{so}, \boldsymbol{\varepsilon}_{so}, \mathbf{u}_{so}\}$ which satisfies the conditions (3.3). Let's prove that the state 2 is kinematically admissible. To do it, we take a scalar product of the first condition in (3.3) and an arbitrary vector $\boldsymbol{\lambda}$ and then subtract the obtained expression, equal to zero, from the right part of (4.14). The result is

$$(\boldsymbol{\sigma}_{so}, \boldsymbol{\varepsilon}_2) + (\boldsymbol{K}\mathbf{u}_{so}, \mathbf{u}_2) - (\boldsymbol{\lambda}, \boldsymbol{A}^{\mathsf{T}}\boldsymbol{\sigma}_{so}) - (\boldsymbol{\lambda}, \boldsymbol{K}\mathbf{u}_{so}) - (\boldsymbol{E}_u \boldsymbol{p}_{so}, \boldsymbol{E}_u\, \overline{\boldsymbol{u}}\,)_\Gamma = 0. \quad (4.16)$$

With this identity we can already assume that the field $\{\boldsymbol{\sigma}_{so}, \boldsymbol{\varepsilon}_{so}, \mathbf{u}_{so}\}$ does not satisfy the first condition in (3.3) because the condition itself is a corollary from (4.16) as the vector $\boldsymbol{\lambda}$ is an arbitrary one. In particular, we can assume $\boldsymbol{\lambda} = \mathbf{u}_2$ and use the basic integral formula (2.19) to derive

$$(\boldsymbol{\lambda}, \boldsymbol{A}^{\mathsf{T}}\boldsymbol{\sigma}_{so}) = (\mathbf{u}_2, \boldsymbol{A}^{\mathsf{T}}\boldsymbol{\sigma}_{so}) = (\boldsymbol{\sigma}_{so}, \boldsymbol{A}\mathbf{u}_2) - (\boldsymbol{p}_{so}, \boldsymbol{H}_u\mathbf{u}_2)_\Gamma\,,$$

which will yield the following after substituting to (4.16):

$$(\boldsymbol{\sigma}_{so}, \boldsymbol{\varepsilon}_2) - (\boldsymbol{\sigma}_{so}, \boldsymbol{A}\mathbf{u}_2) + (\boldsymbol{p}_{so}, \boldsymbol{H}_u\mathbf{u}_2)_\Gamma - (\boldsymbol{E}_u \boldsymbol{p}_{so}, \boldsymbol{E}_u\, \overline{\boldsymbol{u}}\,)_\Gamma = 0.$$

Now we use the second condition in (3.3) and the fact that $\boldsymbol{E}_p + \boldsymbol{E}_u = \boldsymbol{I}$ to transform the latter identity into the following:

$$(\boldsymbol{\sigma}_{so}, \boldsymbol{\varepsilon}_2 - \boldsymbol{A}\mathbf{u}_2) - (\boldsymbol{E}_u \boldsymbol{p}_{so}, \boldsymbol{E}_u(\overline{\boldsymbol{u}} - \boldsymbol{u}_2))_\Gamma = 0. \quad (4.17)$$

As (4.17) should hold for <u>any</u> homogeneously statically admissible field $\{\boldsymbol{\sigma}_{so}, \boldsymbol{\varepsilon}_{so}, \mathbf{u}_{so}\}$, we can conclude that either scalar product from the left part of (4.17) should be equal to zero, thus

$$\boldsymbol{\varepsilon}_2 - \boldsymbol{A}\mathbf{u}_2 = \mathbf{0} \quad \in \Omega\,, \qquad \boldsymbol{E}_u(\overline{\boldsymbol{u}} - \boldsymbol{u}_2) = \mathbf{0} \;\in \Gamma, \quad (4.18)$$

which are the same requirements (3.4) of the kinematic admissibility of the state 2.

1.4.3 Theorem of field orthogonality

We can derive the following statement from the formula (4.13) as a particular case:

> *The virtual work of internal stresses of a homogeneously statically admissible state (field) on the displacements of a homogeneously kinematically admissible state (field) is equal to zero.*

To put it another way,

$$(\boldsymbol{\sigma}_{\text{so}}, \boldsymbol{\varepsilon}_{\text{ko}}) + (\boldsymbol{K}\mathbf{u}_{\text{so}}, \mathbf{u}_{\text{ko}}) = 0\,. \tag{4.19}$$

To see this, we use the formula (4.13) and take a homogeneously kinematically admissible state as the state 2. Seeing that $\boldsymbol{E}_u \boldsymbol{u}_{\text{ko}} = \boldsymbol{0} \in \Gamma$, we have (4.19).

The formulation of the field orthogonality theorem can be expressed in another way:

> *If the virtual work of the internal stresses of a certain state 1 of a system on the displacements of any homogeneously kinematically admissible state of the same system is equal to zero, then the state 1 is homogeneously statically admissible.*

In other words, the equality

$$(\boldsymbol{\sigma}_1, \boldsymbol{\varepsilon}_{\text{ko}}) + (\boldsymbol{K}\mathbf{u}_1, \mathbf{u}_{\text{ko}}) = 0, \tag{4.20}$$

which is true for an arbitrary homogeneously kinematically admissible field $F_{\text{ko}} = \{\boldsymbol{\sigma}_{\text{ko}}, \boldsymbol{\varepsilon}_{\text{ko}}, \mathbf{u}_{\text{ko}}\}$, implies that the state 1 is homogeneously statically admissible. It is clear that this formulation of the orthogonality theorem follows from the principle of virtual displacements.

Finally,

> *If the virtual work of the internal stresses of any homogeneously statically admissible state on the displacements of a certain state 2 of the system is equal to zero, then the state 2 is homogeneously kinematically admissible.*

In other words, the equality

$$(\boldsymbol{\sigma}_{\text{so}}, \boldsymbol{\varepsilon}_2) + (\boldsymbol{K}\mathbf{u}_{\text{so}}, \mathbf{u}_2) = 0, \tag{4.21}$$

which is true for an arbitrary homogeneously statically admissible state $F_{\text{so}} = \{\boldsymbol{\sigma}_{\text{so}}, \boldsymbol{\varepsilon}_{\text{so}}, \mathbf{u}_{\text{so}}\}$, implies that the state 2 is homogeneously kinematically admissible. It is clear also that the latter formulation is a corollary from the principle of virtual stress increments.

All three framed statements above, taken together, will be called a *theorem of field orthogonality*. Further on, after we introduce the notion of an energy metrics, the fundamental relationship (4.19) will be treated also as a condition of orthogonality between a homogeneously statically admissible field, F_{so}, and a homogeneously kinematically admissible field, F_{ko}, in this metrics. This is why we have the right to name the theorem in the way we do in the title of this subsection.

Let's give a tiny illustration of the field orthogonality theorem using a three-bar truss as an example (Fig. 1.3).

In this case, the "stress" vector, $\boldsymbol{\sigma} = |[N_1,\ N_2,\ N_3]|^{\mathrm{T}}$, consists of longitudinal forces/stresses in the bars of the truss which are assumed positive when in tension. The "strain" vector corresponding to the $\boldsymbol{\sigma}$ vector, $\boldsymbol{\varepsilon} = |[\Delta_1,\ \Delta_2,\ \Delta_3]|^{\mathrm{T}}$, consists of the elongations of the same bars. If we denote as U and V the displacements of the only free node of the truss in the respective directions x and y, we find that the homogeneously kinematically admissible strains, $\boldsymbol{\varepsilon}_{ko}$, can be calculated as

$$\boldsymbol{\varepsilon}_{ko} = |[\mathrm{U}\sin\alpha + \mathrm{V}\cos\alpha,\ \mathrm{V},\ -\mathrm{U}\sin\alpha + \mathrm{V}\cos\alpha]|^{\mathrm{T}}.$$

The condition of equilibrium of the free node in its projections onto the axes x and y in the absence of external loads will give the homogeneously statically admissible stresses in the bars:

$$\boldsymbol{\sigma}_{so} = |[\mathrm{A},\ -2\mathrm{A}\cos\alpha,\ \mathrm{A}]|^{\mathrm{T}}$$

where A is an arbitrary constant.

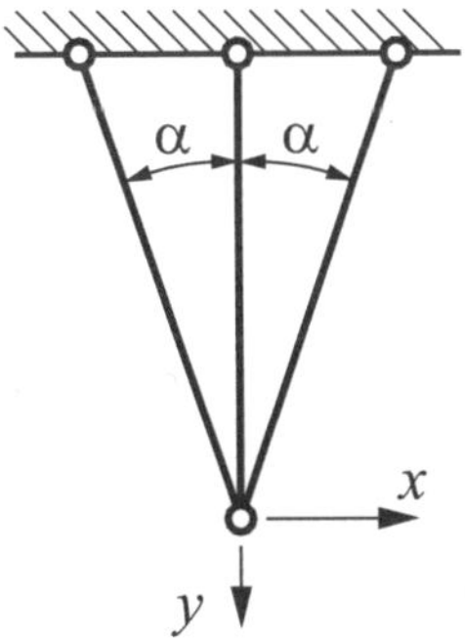

Fig. 1.3. An illustration to the theorem of field orthogonality

Direct substitution helps verify that the equality $(\boldsymbol{\sigma}_{so},\ \boldsymbol{\varepsilon}_{ko}) = \boldsymbol{\sigma}^{\mathrm{T}}_{so}\,\boldsymbol{\varepsilon}_{ko} = 0$ holds; this corresponds to (4.19) because $\boldsymbol{K} = \boldsymbol{O}$ in this case.

1.4.4 Integral identity by Papkovich

P.F. Papkovich [6] derived an integral identity in the theory of elasticity, which he called a *general expression of the work of external forces*. Papkovich used this identity to derive various relationships and theorems applicable to the variational formulations in elasticity. Now we will show how this identity looks for a general elastic body all equations of which are written in the operator form that we use in this chapter.

Consider the following four independent states of a mechanical system. Let us assume that in the state 1 we know all external forces (active and reactive), such as: a vector of external volumetric forces, $\boldsymbol{X}_1$, and a vector of surface forces, $\boldsymbol{p}_1$. The forces are supposed to be known on the whole boundary surface of the body, including places where the kinematic conditions are specified in the original problem. Only a displacement field, $\mathbf{u}_2$, is supposed to be known in the state 2. The state 3 of the system is defined by a stress field, $\boldsymbol{\sigma}_3$, only. Finally, we know only a strain field, $\boldsymbol{\varepsilon}_4$, in the state 4. The four listed states of the system are in no relation with one another, that is, they are arbitrary.

Following the expression (4.3), we define a virtual work, A_{12}, of the external forces of the state 1 on the respective displacements of the state 2 as

$$\mathsf{A}_{12} = (\boldsymbol{X}_1, \mathbf{u}_2) + (\boldsymbol{E}_p \boldsymbol{p}_1, \boldsymbol{E}_p \boldsymbol{u}_2)_\Gamma + (\boldsymbol{E}_u \boldsymbol{p}_1, \boldsymbol{E}_u \boldsymbol{u}_2)_\Gamma \ . \tag{4.22}$$

Note once more that the term $\boldsymbol{E}_u \boldsymbol{p}_1$ means the reactive forces in the state 1 which appear in the exact locations of the contour Γ and in the exact directions where the kinematic conditions are defined.

According to Papkovich, the expression of A_{12} can be represented also in the form

$$\mathsf{A}_{12} = (\boldsymbol{X}_1 - \boldsymbol{A}^{\mathsf{T}}\boldsymbol{\sigma}_3, \mathbf{u}_2) + (\boldsymbol{\sigma}_3, \boldsymbol{A}\mathbf{u}_2 - \boldsymbol{\varepsilon}_4) - (\boldsymbol{E}_p \boldsymbol{H}_\sigma \boldsymbol{\sigma}_3 - \boldsymbol{E}_p \boldsymbol{p}_1, \boldsymbol{E}_p \boldsymbol{u}_2)_\Gamma -$$

$$- (\boldsymbol{E}_u \boldsymbol{H}_\sigma \boldsymbol{\sigma}_3 - \boldsymbol{E}_u \boldsymbol{p}_1 \, , \boldsymbol{E}_u \boldsymbol{u}_2)_\Gamma + (\boldsymbol{\sigma}_3, \boldsymbol{\varepsilon}_4). \tag{4.23}$$

The equivalence of the expressions (4.22) and (4.23) can be checked directly by using the basic integral identity (2.19) with the second and third states of the system, that is, in the form

$$(\boldsymbol{A}^{\mathsf{T}}\boldsymbol{\sigma}_3, \mathbf{u}_2) = (\boldsymbol{\sigma}_3, \boldsymbol{A}\mathbf{u}_2) - (\boldsymbol{E}_p \boldsymbol{H}_\sigma \boldsymbol{\sigma}_3, \boldsymbol{E}_p \boldsymbol{u}_2)_\Gamma - (\boldsymbol{E}_u \boldsymbol{H}_\sigma \boldsymbol{\sigma}_3, \boldsymbol{E}_u \boldsymbol{u}_2)_\Gamma \, . \tag{4.24}$$

As can be seen, substituting (4.24) to the right part of (4.23) will annihilate all terms related to the third and fourth states, and the rest will coincide with (4.22).

Further, taking mechanical states with particular properties as each of the four independent states of the system can turn (4.23) into various relationships and theorems related to the variational formulations of structural mechanics [6].

1.5 Final comments to Chapter 1

First of all, we have to explicate our usage of the term *principle.* The literature on the structural mechanics uses the words *principle* and *law* sometimes as mutual equivalents. One can occasionally run into an expression like "law of virtual displacements". In principle[6], this should not make any confusion if the context makes it clear. Nevertheless, in order to avoid ambiguity in terms we had better adhere to the linguistic basis once established. We prefer the terminology advocated by L.A. Rozin. A quote from [8] helps explain what we mean:

"*Variational formulations of problems can be represented either as integral identities which mechanics treats as variational principles and refers to as variational equations, or as requirements that certain functionals should be stationary, which mechanics treats as variational laws*".

As the definitions of the static and kinematic semi-admissibility of stress and strain fields are new for the reader[7], we should emphasize the differences between the notions once again. The criterion for the static semi-admissibility is that the static boundary conditions should be met while the equilibrium equations in the volume of a body are neglected. On the contrary, the kinematic semi-admissibility uses the criterion of the geometric equations satisfied inside the body while the boundary conditions of the geometric type may be violated.

Further, as we use the different notions of stress and strain fields, it would be appropriate for us to name the two basic principles of statics and geometry as the *principle of homogeneously kinematically admissible displacements and the principle of homogeneously statically admissible stresses*, to emphasize the reciprocity of the principles by their very names. However, there is a long-living historical tradition of naming these principles otherwise (the principle of virtual displacements and the principle of virtual increments of the stress state) which is widely used in literature, so any deviations from this tradition are undesirable. The *virtual* displacements in mechanics are displacements which do not conflict with

[6] Another use of the term *principle*, and of a totally different meaning, too.

[7] Strictly speaking, one of the notions, the static semi-admissibility of a stress field, was introduced by us earlier in the paper [11].

constraints of a particular mechanical system. In our situation they are equivalent to displacements from a homogeneously kinematically admissible field.

It should be said that the notion of the homogeneously kinematically admissible displacements (as well as just kinematically admissible displacements) actually needs a more accurate definition. Formal mathematics advances two requirements relevant to this notion:

- the displacements should satisfy the respective kinematic boundary conditions;
- the displacements should be sufficiently smooth functions.

Different definitions of a problem's solution (a classic, strong, weak, semi-weak solution [1]) require a different smoothness of the displacements, therefore the very set of the kinematically admissible displacements will change together with the definition of the solution. In the context that we use, we require the smoothness corresponding to the semi-weak solution definition. This admits the $\boldsymbol{A}\mathbf{u}$ operation, at the least.

The theorem of field orthogonality was obtained here as a corollary of the principle of virtual displacements and the principle of virtual variations of the stress state. The reader can make sure (we do not dwell on it here) that the inverse relationship is also true: the theorem of field orthogonality implies the two basic versions of the variational principles. In that sense the theorem of field orthogonality can be treated as a fundamental static-geometric statement.

And the final note. As we have said before, the general principle of statics and geometry written as an integral identity (4.2) is a good starting point for obtaining various partial but in a certain way important versions of the basic variational principles. All of those are derived from (4.2) by giving particular meanings to the states 1 and 2 of a mechanical system, which states were quite arbitrary from the beginning. This is the way they derive a whole bunch of partial variational principles in [8] where the interested reader can find the details. We confine ourselves intentionally to only two: the principle of virtual displacements and the principle of virtual variations of the stress state. These two principles can be called 'basic' due to the following reasons. First, these two were the first to appear, and many known works on elasticity and/or structural mechanics discuss nothing more in their presentment of variational principles of statics and geometry – refer to [5] for an example. Second, as we will show in the next chapter, these two principles are in tight connection with two classic variational laws: the Lagrange variational principle and the Castigliano variational principle, which we also call basic ones and which have an important property of extremality. Third, all the other variational principles and laws (which we call *supplementary*) can be derived from the basic ones by means of various known transformations such as: the Lagrange transformation, the Legendre transformation, the Friedrichs transformation. More details on the use of these transformations can be found in the book by L.A. Rozin [7].

Finally, we should mention a so-called Vorobiov identity [5] formulated for a general nonlinear analysis. In the particular case of linear elasticity this identity becomes

$$(\boldsymbol{A}\mathbf{u}_{\mathrm{k}}, \boldsymbol{\sigma}_{\mathrm{s}}) = (\mathbf{u}_{\mathrm{k}}, \bar{\boldsymbol{X}}) + (\boldsymbol{E}_p \bar{\boldsymbol{p}}, \boldsymbol{E}_p \boldsymbol{u}_{\mathrm{k}})_{\Gamma} + (\boldsymbol{E}_u \boldsymbol{H}_{\sigma}\boldsymbol{\sigma}_{\mathrm{s}}, \boldsymbol{E}_u \bar{\boldsymbol{u}})_{\Gamma} .$$

Taking into account the static admissibility of the stresses $\boldsymbol{\sigma}_{\mathrm{s}}$ and the kinematic admissibility of the displacements $\mathbf{u}_{\mathrm{k}}$ will transform this identity into the following:

$$(\boldsymbol{A}\mathbf{u}_{\mathrm{k}}, \boldsymbol{\sigma}_{\mathrm{s}}) = (\mathbf{u}_{\mathrm{k}}, \boldsymbol{A}^{\mathrm{T}}\boldsymbol{\sigma}_{\mathrm{s}}) + (\boldsymbol{E}_p \boldsymbol{H}_{\sigma}\boldsymbol{\sigma}_{\mathrm{s}}, \boldsymbol{E}_p \boldsymbol{u}_{\mathrm{k}})_{\Gamma} + (\boldsymbol{E}_u \boldsymbol{H}_{\sigma}\boldsymbol{\sigma}_{\mathrm{s}}, \boldsymbol{E}_u \boldsymbol{u}_{\mathrm{k}})_{\Gamma} .$$

Now it is easy to see that the Vorobiov identity in linear elasticity is just another form of the same basic integral identity (2.19) in application to statically admissible stresses and kinematically admissible displacements.

References

1. Descloux J (1973) Methode des elements finis. Departement de mathematiques. Lausanne, Suisse
2. Filin AP (1978) Applied mechanics of deformable solids, Vol. 2 (in Russian). Nauka, Moscow
3. Mikhlin SG (1970) Variational methods in mathematical physics (in Russian) Nauka, Moscow
4. Mikhlin SG (1977) Linear equations in partial derivatives (in Russian). Vysshaya shkola, Moscow
5. Novozhilov VV (1958) Theory of elasticity (in Russian). Sudpromgiz, Leningrad
6. Papkovich PF (1939) Theory of elasticity (in Russian). Oborongiz, Leningrad
7. Rozin LA (1978) Variational formulations of problems for elastic systems (in Russian). Leningrad State Univ., Leningrad
8. Rozin LA (1998) Problems of elasticity and numerical methods for their solution (in Russian). St. Petersburg Technical University, St. Petersburg
9. Rzhanitsyn AR (1982) Structural mechanics (in Russian). Vysshaya shkola, Moscow
10. Shames IH, Dym CL (1991) Energy and Finite Element Methods in Structural Mechanics. Taylor & Francis

11. Slivker VI (1982) Ritz method in problems of elasticity, based on sequentially minimizing two functionals (in Russian). Trans. Acad. Sci. USSR, Mech. of Solids 2: 57–64
12. Washizu K (1982) Variational methods in elasticity and plasticity. Pergamon Press, Oxford New York Toronto Paris Frankfurt
13. Young L (1969) Lectures on the calculus of variations and optimal control theory. W.B. Saunders company, Philadelphia London Toronto

2 BASIC VARIATIONAL PRINCIPLES OF STRUCTURAL MECHANICS

The history of mechanics and physics is a history of attempts to explain things that happen around us in the world, using a small number of laws and universal principles. The most successful and fruitful attempts are based on the idea that phenomena we observe have certain extreme properties and these general principles sought for are of variational nature...
V.L. Berdichevsky. Variational principles in mechanics of continua. Moscow, "Nauka", 1983, p.16

2.1 Energy space

2.1.1 Physically admissible fields

Up to this point, we dealt solely with static and geometric properties of a deformable body (assuming that the problems are posed as geometrically linear, first-order). All theorems given in Chapter 1 are invariant with respect to the properties of the deformable solid's material. From now on, we will assume that the material of the body obeys a linear law of relation between the stresses (internal forces) and strains,

$$\boldsymbol{\sigma} = \boldsymbol{C}\boldsymbol{\varepsilon} \quad \text{or} \quad \boldsymbol{\varepsilon} = \boldsymbol{C}^{-1}\boldsymbol{\sigma}\,. \tag{1.1}$$

Now, let's extract a separate class from the set of all thinkable stress-and-strain fields: a set $\mathscr{P}$ of all <u>*physically admissible*</u> fields, $F_{\mathrm{p}} = \{\boldsymbol{\sigma}_{\mathrm{p}}, \boldsymbol{\varepsilon}_{\mathrm{p}}, \mathbf{u}_{\mathrm{p}}\}$, which have a linear relation between their stresses and their strains in the form (1.1). All fields dealt with in this chapter will be physically admissible.

An important role in further proceedings will be played by the set $\mathscr{R}_{\mathrm{k}/2}$. Let's say that an arbitrary field, $F = \{\boldsymbol{\sigma}, \boldsymbol{\varepsilon}, \mathbf{u}\}$, belongs to the set $\mathscr{R}_{\mathrm{k}/2}$ if the said field is physically admissible and kinematically semi-admissible at the same time. Thus, our definition permits us to represent any field $F \in \mathscr{P}_{k/2}$ as

$$F = \{\boldsymbol{CA}\mathbf{u}, \boldsymbol{A}\mathbf{u}, \mathbf{u}\}, \tag{1.2}$$

so the field is defined thoroughly by a given vector of displacements, $\mathbf{u}$. As can be easily seen, the $\mathscr{P}_{k/2}$ set is a linear set.

Two subsets will be introduced on the set $\mathscr{P}_{k/2}$: $\mathscr{P}_{k} \subseteq \mathscr{P}_{k/2}$ and $\mathscr{P}_{ko} \subseteq \mathscr{P}_{k/2}$. The set $\mathscr{P}_{k}$ consists of kinematically admissible fields, and the set $\mathscr{P}_{ko}$ consists of homogeneusly kinematically admissible fields. Note that $\mathscr{P}_{ko}$ is a linear set while the set $\mathscr{P}_{k}$ is not. Thus, according to its definition, the set $\mathscr{P}_{ko}$ consists of all fields of the kind

$$F = \{\boldsymbol{CA}\mathbf{u}, \boldsymbol{A}\mathbf{u}, \mathbf{u}\} \quad \text{where} \quad \boldsymbol{E}_u \boldsymbol{H}_u \mathbf{u} = \mathbf{0} \in \Gamma. \tag{1.3}$$

Consequently, this linear set is fully defined by a specific displacement vector $\mathbf{u}$.

Next, we take all physically admissible fields and extract a set, $\mathscr{P}_{s/2}$, of statically semi-admissible fields. We introduce a subset $\mathscr{P}_{s} \subseteq \mathscr{P}_{s/2}$ on the set $\mathscr{P}_{s/2}$, which will consist of statically admissible fields. Finally, the set $\mathscr{P}_{so} \subseteq \mathscr{P}$ will consist of homogeneously statically admissible fields. Note that $\mathscr{P}_{so}$ is a linear set while the sets $\mathscr{P}_{s}$ and $\mathscr{P}_{s/2}$ are not. So, according to the definition, $F \in \mathscr{P}_{so}$ if it is of the type

$$F = \{\boldsymbol{\sigma}, \boldsymbol{C}^{-1}\boldsymbol{\sigma}, \mathbf{u}\} \quad \text{where} \quad \boldsymbol{A}^{\mathsf{T}}\boldsymbol{\sigma} + \boldsymbol{K}\mathbf{u} = \mathbf{0} \;\in \Omega \;\text{ and }\; \boldsymbol{E}_p \boldsymbol{H}_\sigma \boldsymbol{\sigma} = \mathbf{0} \;\in \Gamma. \tag{1.4}$$

Thus, the linear set $\mathscr{P}_{so}$ is completely defined by two known vectors, the displacement vector $\mathbf{u}$ and the stress vector $\boldsymbol{\sigma}$.

A field, $F_* = \{\boldsymbol{\sigma}_*, \boldsymbol{\varepsilon}_*, \mathbf{u}_*\}$, will be called a *true field* (or a *real field*) if it is statically admissible, kinematically admissible, and physically admissible, all at the same time. This is the case when all equations (**1**.2.2) and boundary conditions (**1**.2.4) hold. So, any field from the intersection of the sets $\mathscr{P}_{s}$ and $\mathscr{P}_{k}$ is true by definition:

$$F_* \in \mathscr{P}_{s} \cap \mathscr{P}_{k}. \tag{1.5}$$

2.1.2 Betty theorem

Let us consider two arbitrary but physically admissible fields, F_1 and F_2, and write an expression of the virtual work, B_{12}, of the internal forces of the state 1 of a system on the displacements of the state 2, and an expression of the virtual work, B_{21}, of the internal forces of the state 2 of the same system on the displacements of the state 1. We have

$$\mathsf{B}_{12} = -(\boldsymbol{\sigma}_1, \boldsymbol{\varepsilon}_2) - (\boldsymbol{K}\mathbf{u}_1, \mathbf{u}_2)\,, \quad \mathsf{B}_{21} = -(\boldsymbol{\sigma}_2, \boldsymbol{\varepsilon}_1) - (\boldsymbol{K}\mathbf{u}_2, \mathbf{u}_1)\,. \tag{1.6}$$

Now we use (1.1) and take into account the symmetry of the algebraic operators $\boldsymbol{C}$ and $\boldsymbol{K}$ to make a conclusion that

$$\mathsf{B}_{12} = \mathsf{B}_{21}\,. \tag{1.7}$$

In other words, the virtual work B_{12} of internal forces of the state 1 of a mechanical system on the respective displacements of the state 2 of the same system is equal to the virtual work B_{21} of internal forces of the state 2 of the system on the respective displacements of the state 1.

The relationship (1.7) is called a *theorem of work reciprocity* (Betty theorem). We emphasize once again that the Betty theorem can be proven true for physically admissible fields only.

Now, let both states of the system belong to the linear set $\mathscr{P}_{k/2}$. Then the expression of the virtual work of the internal forces can be represented in the form (**1**.4.4). If each of the two states of the system is assumed to correspond to a set of statically conforming external forces, then the general principle of static and geometry in the form (**1**.4.6) yeilds

$$\mathsf{A}_{12} = \mathsf{A}_{21}\,. \tag{1.8}$$

In other words, the virtual work of all external (active and reactive) forces of the state 1 of a system on the respective displacements of the state 2 of the same system is equal to the virtual work of all external (active and reactive) forces of the state 2 of the system on the respective displacements of the state 1. Thus, for the Betty theorem in the form (1.8) to be true, the following three conditions must be met:

- each of the two states taken into account must be physically admissible;
- the two states must be kinematically semi-admissible;
- each of the two must be statically admissible (each one under its own forces).

However, the most frequent formulation (and application) of the Betty theorem is a partial case when the states 1 and 2 of a system are fully kinematically admissible. But then the respective fields, F_1 and F_2, will contain true components of the stresses, strains, and displacements which conform to the external forces of the state 1 and state 2, respectively. The result is the following formulation of the Betty theorem:

The virtual work, A_{12}, of the external forces of state 1 of a system on actual displacements of state 2 of the same system is equal to the virtual work, A_{21}, of the external forces of state 2 on the actual displacements of state 1.

Let's give a brief illustration of the Betty theorem's application which will remind us that the calculation of the virtual work of the external forces must take into account the work of the reactive contour forces. Consider a flexible bar of the length l clamped at its left end and simply supported at its right end (Fig. 2.1).

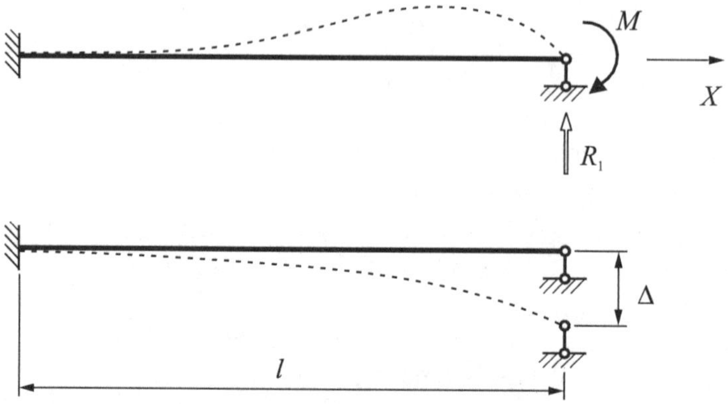

Fig. 2.1. An illustration to the Betty theorem

Let the bar be loaded by an end moment, M, in its state 1 and let a kinematic external action be specified for the state 2: a sag of the right support by a length of Δ. It is easy to find that the deflection v_2 of the beam in its second state is

$$v_2 = \frac{3\Delta}{2l^2}\left(x^2 - \frac{x^3}{3l}\right), \qquad \text{so} \qquad \theta_2 = \frac{dv_2}{dx}(l) = \frac{3\Delta}{2l},$$

where θ_2 is a slope of the right end of the bar in its second state.

We denote the reaction of the right support in the first state of the system as R_1. The virtual work A_{12} of all external (active and reactive) forces of the 1st state of the system on the respective displacements of the 2nd state is $\mathsf{A}_{12} = M\theta_2 - R_1\Delta$. On the other hand, it is easy to see that the reactive forces of the 2nd state do not perform any work on the displacements of the 1st state, i.e. $\mathsf{A}_{21} = 0$. Thus, the Betty theorem gives $M\theta_2 - R_1\Delta = 0$, so

$$R_1 = \frac{3M}{2l}.$$

2.1.3 Energy of strain. Clapeyron theorem

Consider a true field, $F_1 = \{\boldsymbol{\sigma}, \boldsymbol{\varepsilon}, \mathbf{u}\}$, which conforms to given external actions, $V = \{\bar{\boldsymbol{X}}, \bar{\boldsymbol{p}}, \bar{\boldsymbol{u}}\}$, applied to an elastic mechanical system. In the vicinity of the true field F_1, we consider a perturbed state (state 2) with its true field $F_2 = \{\boldsymbol{\sigma} + \delta\boldsymbol{\sigma}, \boldsymbol{\varepsilon} + \delta\boldsymbol{\varepsilon}, \mathbf{u} + \delta\mathbf{u}\}$ that conforms to new external actions already perturbed, $V + \delta V = \{\bar{\boldsymbol{X}} + \delta\bar{\boldsymbol{X}}, \bar{\boldsymbol{p}} + \delta\bar{\boldsymbol{p}}, \bar{\boldsymbol{u}} + \delta\bar{\boldsymbol{u}}\}$. It is easy to notice that $F_2 = F_1 + \delta F$ and that the field of variations $\delta F = \{\delta\boldsymbol{\sigma}, \delta\boldsymbol{\varepsilon}, \delta\mathbf{u}\}$ belongs to the linear set $\mathscr{R}_{k/2}$.

As the system transits to its perturbed state, the external forces perform a certain amount of work, δA. The energy consumption for doing this work will increase the internal energy of the elastic body caused by the body's deformation. When calculating the work δA, one does not have to allow for the work of the external force increments on the displacement increments because this contribution is a second-order infinitesimal comparing to the work of the external forces of the state 1 on the displacement increments. Thus,

$$\delta \mathsf{A} = (\bar{\boldsymbol{X}}, \delta\mathbf{u}) + (\boldsymbol{E}_p\bar{\boldsymbol{p}}, \boldsymbol{E}_p\delta\boldsymbol{u})_\Gamma + (\boldsymbol{E}_u\boldsymbol{p}, \boldsymbol{E}_u\delta\bar{\boldsymbol{u}})_\Gamma . \tag{1.9}$$

The expression that stands in the right part of (1.9) can be treated also as a virtual work of all external forces of the state 1 on variations of the displacements. The general principle of statics and geometry makes it possible to replace the said work by the virtual work of the internal forces of the state 1 on the same displacements. To put it another way,

$$\delta \mathsf{A} = (\boldsymbol{\sigma}, \delta\boldsymbol{\varepsilon}) + (\boldsymbol{K}\mathbf{u}, \delta\mathbf{u}) , \tag{1.10}$$

which takes into account the kinematic semi-admissibility of the variations $\delta\boldsymbol{\varepsilon}$ and $\delta\mathbf{u}$.

The laws of thermodynamics (for example, see [1, 2]) imply that adiabatic or isothermic reversible elastic deformation transforms the whole change in the internal energy of an elastic mechanical system into its accumulated potential energy E. This lets us identify the increment of the work, δA, with the increment of the strain energy of an elastic mechanical system, δE

$$\delta \mathsf{E} = (\boldsymbol{\sigma}, \delta\boldsymbol{\varepsilon}) + (\boldsymbol{K}\mathbf{u}, \delta\mathbf{u}) . \tag{1.11}$$

The total energy of strain, E, accumulated in the mechanical system as it transits from its initial unstrained and unstressed state to the state 1, is determined in the general case by an integral of (1.11) along a given path of deformation. If this integral does not depend on the path of integration so that it depends only on the final state of the system (recall that the initial state is zero), then the quantities in the right part of (1.11) should be total differentials. Calculus of functions of multiple real variables says that the condition of existence of a total differential for the linear physical law (1.1) is essentially the symmetry of the algebraic operators $\boldsymbol{C}$ and $\boldsymbol{K}$, which is the same condition we stipulated earlier in Section **1**.2. If the strain energy happens not to depend on how the system transits to its final state, then we can choose a simplest, so-called proportional mode of loading in which all external (both static and kinematic) actions vary proportionally to a single time-like parameter, t, which changes through the interval from 0 to 1. But then, because of the linearity of the problem in any arbitrary "moment of time" t, the actual state of the system will be described by the field $F_t = \{t\boldsymbol{\sigma}, t\boldsymbol{\varepsilon}, t\mathbf{u}\}$, so with this loading mode the variations of the kinematic parameters are proportional to the increment of "time" dt, i.e. $\delta\boldsymbol{\varepsilon} = \boldsymbol{\varepsilon}dt$, $\delta\mathbf{u} = \mathbf{u}dt$. Thus the increment of the strain energy, B, will become

$$\delta\mathsf{E} = [(\boldsymbol{C}\boldsymbol{\varepsilon}, \boldsymbol{\varepsilon}) + (\boldsymbol{K}\mathbf{u}, \mathbf{u})]\, t dt\,. \tag{1.12}$$

As the scalar quantity in the brackets above does not depend on t, integrating (1.12) with respect to t from zero to one will give the following expression for the *strain energy*, E:

$$\mathsf{E} = \tfrac{1}{2}\,(\boldsymbol{C}\boldsymbol{\varepsilon}, \boldsymbol{\varepsilon}) + \tfrac{1}{2}\,(\boldsymbol{K}\mathbf{u}, \mathbf{u})\,. \tag{1.13}$$

As one can see from (1.13), the strain energy accumulated in a deformable system consists of two terms. The first term in (1.13) is an energy accumulated in the deformable body itself, while the second term is a part of the energy stored in an elastic medium that surrounds the body.

Now we return to the general principle of statics and geometry, to find that the virtual work A_{11} of the external forces of the state 1 of the system on the actual displacements of the same state is equal to the doubled potential energy of strain,

$$\mathsf{A}_{11} = 2\mathsf{E}\,. \tag{1.14}$$

This relation is known as the Clapeyron theorem.

When using the Clapeyron theorem, one should keep in mind that the work of the external forces, A_{11}, must inlcude also the work of reactive contour forces, i.e. the work of reactions of imposed constraints in those places and in those directions where the given external actions are displacements of the constraints.

Now, let us give another derivation of the Clapeyron theorem, also to demonstrate an application of the Papkovich identity. To do this, assume that all four states of a system are the same in (**1**.4.23), which gives us the right to identify all of the states with the subscript 1, that is,

$$\mathsf{A}_{11} = (\boldsymbol{X}_1 - \boldsymbol{A}^{\mathsf{T}}\boldsymbol{\sigma}_1, \mathbf{u}_1) + (\boldsymbol{\sigma}_1, \boldsymbol{A}\mathbf{u}_1 - \boldsymbol{\varepsilon}_1) - (\boldsymbol{E}_p\boldsymbol{H}_\sigma\boldsymbol{\sigma}_1 - \boldsymbol{E}_p\boldsymbol{p}_1, \boldsymbol{E}_p\boldsymbol{u}_1)_\Gamma -$$
$$- (\boldsymbol{E}_u\boldsymbol{H}_\sigma\boldsymbol{\sigma}_1 - \boldsymbol{E}_u\boldsymbol{p}_1, \boldsymbol{E}_u\boldsymbol{u}_1)_\Gamma + (\boldsymbol{\sigma}_1, \boldsymbol{\varepsilon}_1).$$

If we assume also that the state 1 is a true state of the system, then all external actions will be the same as given ones, i.e.

$$\boldsymbol{X}_1 = \overline{\boldsymbol{X}} \in \Omega, \qquad \boldsymbol{E}_p(\boldsymbol{p}_1 - \overline{\boldsymbol{p}}) = \boldsymbol{0} \quad \in \Gamma, \qquad \boldsymbol{E}_u(\boldsymbol{u}_1 - \overline{\boldsymbol{u}}) = \boldsymbol{0} \in \Gamma,$$

$$\text{where} \qquad \boldsymbol{u}_1 = \boldsymbol{H}_u\mathbf{u}_1 \,.$$

Also, the second, third, and fourth terms in the right part of the expression of A_{11} will disappear when a true state of the system is used, which gives

$$\mathsf{A}_{11} = (\boldsymbol{X}_1 - \boldsymbol{A}^{\mathsf{T}}\boldsymbol{\sigma}_1, \mathbf{u}_1) + (\boldsymbol{\sigma}_1, \boldsymbol{\varepsilon}_1).$$

Recalling that for a true state of the system $\boldsymbol{X}_1 - \boldsymbol{A}^{\mathsf{T}}\boldsymbol{\sigma}_1 = \boldsymbol{K}\mathbf{u}_1$ and $\boldsymbol{\sigma}_1 = \boldsymbol{C}\boldsymbol{\varepsilon}_1$, we take (1.13) into account and arrive at the Clapeyron theorem (1.14) again.

2.1.4 Rigid displacements

If displacements $\mathbf{u}$ do not cause any strain, that is, if $\boldsymbol{A}\mathbf{u} = \mathbf{0}$, then the displacements of this kind are usually called collectively a *rigid displacement*, though a *rigid displacement for the elastic body* would be more exact. With a rigid displacement of an elastic body, the part of the energy E_A,

$$\mathsf{E}_A = \tfrac{1}{2}\,(\boldsymbol{C}\boldsymbol{\varepsilon}, \boldsymbol{\varepsilon})\,, \tag{1.15}$$

stored by the body itself will be equal to zero[1]. However, the presence of an elastic surrounding medium can make the mechanical system accumulate the energy in this medium due to the presence of the term E_K,

[1] Here and further we use the subscript $_A$ to emphasize the connection of a quantity with the geometry operator $\boldsymbol{A}$, which means that the quantity relates not to the whole mechanical system but only to the elastic body mentally deprived of the surrounding elastic medium. Everything related to the elastic medium is subscripted by $_K$ which shows that the quantity in question depends on the operator $\boldsymbol{K}$.

$$\mathsf{E}_K = \tfrac{1}{2}\,(\boldsymbol{K}\mathbf{u}, \mathbf{u})\,, \tag{1.16}$$

in the energy expression (1.13), so that $\mathsf{E} = \mathsf{E}_A + \mathsf{E}_K$.

We use energy considerations to demand that both terms in the energy expression (1.13) could not be negative in any state of the mechanical system belonging to the linear set $\mathscr{R}_{k/2}$, i.e. $\mathsf{E}_A \geq 0$ and $\mathsf{E}_K \geq 0$. Moreover, for all displacements **u** which cause a nonzero strain **ε**, the energy E_A must be strictly positive, and thus the quadratic form $(\boldsymbol{C}\boldsymbol{\varepsilon}, \boldsymbol{\varepsilon})$ must be positive definite. Physically, the strict positivity of E_A postulates a stable behavior of the material of a linearly elastic body [9]. This is where the requirement of positive definiteness of the algebraic operator $\boldsymbol{C}$ formulated in Section **1**.2 comes from.

We will call displacements **u** which do not change the potential energy of the surrounding medium a *rigid displacement for the elastic medium.* In other words, $(\boldsymbol{K}\mathbf{u}, \mathbf{u}) = 0$ on a rigid displacement for the medium. Displacements of this kind may exist if the algebraic operator $\boldsymbol{K}$ is positive semi-definite but is not positive definite, which is quite admissible. This can be represented from the mechanical point of view as a medium which provides an elastic resistance to the displacements of a body's points along one of the coordinate axes only and does not resist to the displacements in the directions orthogonal to this coordinate. In a particular case, the ealstic medium can be totally absent; this conforms to a zero operator $\boldsymbol{K}$.

The following statement is true: for a displacement **u** to be a rigid displacement for the elastic medium, it is necessary and sufficient that the vector $\boldsymbol{K}\mathbf{u}$ be a zero vector, i.e.,

$$\boldsymbol{K}\mathbf{u} = \mathbf{0}\,. \tag{1.17}$$

The sufficiency is obvious because when (1.17) is true for a certain vector **u,** it immediately implies $(\boldsymbol{K}\mathbf{u}, \mathbf{u}) = 0$.

Now, let $\mathbf{u}_o$ be a vector of rigid displacements for the elastic medium, that is, let $(\boldsymbol{K}\mathbf{u}_o, \mathbf{u}_o) = 0$. We can assume, without limiting the generality, this vector to be normalized to the unit value, in the sense that $(\mathbf{u}_o, \mathbf{u}_o) = 1$. Let's prove this implies $\boldsymbol{K}\mathbf{u}_o = \mathbf{0}$.

Thus, we are considering a set of values of the quadratic functional, $\mathrm{f}(\mathbf{u}) = (\boldsymbol{K}\mathbf{u}, \mathbf{u})$, on all **u** vectors which draw a sphere of a unit radius, that is, which satisfy the condition

$$(\mathbf{u}, \mathbf{u}) = 1. \tag{1.18}$$

The limitations (**1**.2.3) related to the algebraic operator $\boldsymbol{K}$ give

$$\mathrm{f}(\mathbf{u}) \geq 0 \tag{1.19}$$

for any vectors $\mathbf{u}$ including those meeting the condition (1.18). Therefore it should be clear that the least possible value of $f(\mathbf{u})$ is zero, and this minimum is achieved, in particular, on the vector $\mathbf{u}_o$ because $f(\mathbf{u}_o) = (\boldsymbol{K}\mathbf{u}_o, \mathbf{u}_o) = 0$ by definition. Now let's find necessary conditions of the extremum of the functional $f(\mathbf{u})$ under the limitations (1.18).

We use standard Lagrange's method of undetermined multipliers λ to reduce this conditional extremum problem to the problem of finding a point of stationarity of an auxiliary functional, $F(\mathbf{u},\lambda)$,

$$F(\mathbf{u},\lambda) = (\boldsymbol{K}\mathbf{u}, \mathbf{u}) - \lambda[(\mathbf{u}, \mathbf{u}) - 1] \tag{1.20}$$

where the Lagrangian multiplier λ and the vector $\mathbf{u}$ will be varied independently. With this functional, we don't have to impose the limitations (1.18) on the vector $\mathbf{u}$ beforehand because the limitations follow immediately from the conditions of stationarity of $F(\mathbf{u},\lambda)$. To see this, we vary the functional $F(\mathbf{u},\lambda)$ with respect to $\mathbf{u}$ and to λ and then equal its first variation to zero thus obtaining

$$2(\boldsymbol{K}\mathbf{u}, \delta\mathbf{u}) - 2\lambda(\mathbf{u}, \delta\mathbf{u}) - [(\mathbf{u}, \mathbf{u}) - 1]\delta\lambda = 0\,, \tag{1.21}$$

wherefrom we can see that the necessary conditions for the stationarity of $F(\mathbf{u},\lambda)$ are

$$(\boldsymbol{K}\mathbf{u} - \lambda\mathbf{u}, \delta\mathbf{u}) = 0\,, \qquad (\mathbf{u}, \mathbf{u}) - 1 = 0\,. \tag{1.22}$$

As the variations of $\delta\mathbf{u}$ are assumed arbitrary here, the first of the stationary conditions in (1.22) is equivalent to the requirement

$$\boldsymbol{K}\mathbf{u} = \lambda\mathbf{u} \;\in\Omega\,, \tag{1.23}$$

that is, $\mathbf{u}$ is one of the eigenvectors and λ is its respective eigenvalue of the operator $\boldsymbol{K}$. But according to the given condition, $\mathbf{u}_o$ is a stationary point of the functional $f(\mathbf{u})$, consequently,

$$\boldsymbol{K}\mathbf{u}_o = \lambda\mathbf{u}_o \in\Omega\,, \tag{1.24}$$

and this minimum is equal to zero, hence $f(\mathbf{u}_o) = (\boldsymbol{K}\mathbf{u}_o, \mathbf{u}_o) = \lambda = 0$. Now the desirable relationship, $\boldsymbol{K}\mathbf{u}_o = \mathbf{0}$, follows directly from (1.24).

Note that the physical meaning of the requirement that $\boldsymbol{K}\mathbf{u} = \mathbf{0}$ on a rigid displacement for the elastic medium is quite clear because this is the only case when the displacements $\mathbf{u}$ do not cause any reactions, $\boldsymbol{r} = \boldsymbol{K}\mathbf{u}$, in the elastic medium.

A *rigid displacement of a mechanical system* will refer to a displacement, $\mathbf{u}$, which is a rigid displacement both for the elastic body and the surrounding medium. Thus, the set of rigid displacements of a mechanical system is an intersection of the set of rigid displacements for the elastic body and that for the elastic medium. This set can be empty in a particular case. Obviously, by definition $\mathsf{E} = 0$ for a rigid displacement of a mechanical system. Also, each of the sets of rigid displacements is obviously linear.

To give an example, let's consider rigid displacements for a three-dimensional elastic body. Geometric considerations immediately produce the fact that the general expression of the rigid displacements of an elastic body in the vector form is

$$\mathbf{u} = \mathbf{u}_o + \boldsymbol{\theta} \times \mathbf{r} \tag{1.25}$$

where $\mathbf{u}_o$ is a vector of translational displacements of a point selected as the coordinate origin; $\mathbf{r}$ is a position vector of an arbitrary point of the body in which the displacement $\mathbf{u}$ is determined; $\boldsymbol{\theta}$ is a rotation vector of the body. By introducing a Cartesian coordinate system, (x, y, z), and decomposing the vector product into the respective components, we can represent (1.25) in a matrix form,

$$\mathbf{u} = \mathrm{u}_{ox}\left\|\begin{bmatrix}1\\0\\0\end{bmatrix}\right\| + \mathrm{u}_{oy}\left\|\begin{bmatrix}0\\1\\0\end{bmatrix}\right\| + \mathrm{u}_{oz}\left\|\begin{bmatrix}0\\0\\1\end{bmatrix}\right\| + \theta_x\left\|\begin{bmatrix}0\\-z\\y\end{bmatrix}\right\| + \theta_y\left\|\begin{bmatrix}z\\0\\-x\end{bmatrix}\right\| + \theta_z\left\|\begin{bmatrix}-y\\x\\0\end{bmatrix}\right\|, \tag{1.26}$$

where $\mathrm{u}_{ox}, \mathrm{u}_{oy}, \mathrm{u}_{oz}$ are components of the vector $\mathbf{u}_o$; $\theta_x, \theta_y, \theta_z$ are components of the vector $\boldsymbol{\theta}$; x, y, z are components of the vector $\mathbf{r}$ in the selected coordinate system.

Seeing that the six quantities $\mathrm{u}_{ox}, \mathrm{u}_{oy}, \mathrm{u}_{oz}, \theta_x, \theta_y, \theta_z$ can have arbitrary independent values, one can see from (1.26) that the number of dimensions in the linear set of rigid displacements for a three-dimensional elastic body is six.

We will postulate an important property for the operator of geometry, $\boldsymbol{A}$:

> *the set of linearly independent solutions of the homogeneous equation of geometry,* $\boldsymbol{A}\mathbf{u} = \mathbf{0}$, *must be finite-dimensional.*

This is equivalent to adopting the requirement that the set of rigid displacements for an elastic body, hence for the whole mechanical system, should be finite-dimensional. The Laplace operator is a simple example of an operator that does not satisfy this requirement. Indeed, any harmonic function on Ω is a solution of the homogeneous Laplace equation, and the set of harmonic functions is known to be infinite-dimensional. Consequently, the Laplace operator cannot be a geometry operator according to the postulated statement. If we assumed the possibility for the homogeneous equation with the geometry operator to have an infinite number of linearly independent solutions, this would mean mechanically that the system was *internally unstable* so that the instability could not be eliminated by a finite number of externally imposed constraints. This kind of (linear) mechanical systems will be further excluded from consideration.

So, let $\mathbf{u}_i$ $(i = 1, \ldots, n_A)$ be a full set of linearly independent solutions of the homogeneous equation of geometry, $\boldsymbol{A}\mathbf{u} = \mathbf{0}$, which is called a *fundamental system* for the operator $\boldsymbol{A}$ in mathematics; the linear span of the fundamental system is called a *kernel* of the operator $\boldsymbol{A}$. The mechanical interpretation of the kernel of the operator $\boldsymbol{A}$ is a linear set of the rigid displacements for the elastic body which we will denote as $\mathscr{R}_A$.

Note that the field $F \in \mathscr{F}_{k/2}$ with the displacement component $\mathbf{u} \in \mathscr{R}_A$ has the form $F = \{\mathbf{0}, \mathbf{0}, \mathbf{u}\}$. To see this, note that its kinematical semi-admissibility implies $\boldsymbol{\varepsilon} = \boldsymbol{A}\mathbf{u}$, knowing that $\mathbf{u} \in \mathscr{R}_A$ gives $\boldsymbol{\varepsilon} = \mathbf{0}$, and the physical admissibility entails also $\boldsymbol{\sigma} = \boldsymbol{C}\boldsymbol{\varepsilon} = \mathbf{0}$.

If we clear the fundamental system $\mathbf{u}_i$ $(i = 1, \ldots, n_A)$ of all displacements $\mathbf{u}_j$ which generate nonzero energy in the elastic surrounding medium, that is, displacements which meet the condition $(\boldsymbol{K}\mathbf{u}_j, \mathbf{u}_j) > 0$, then the displacements not removed by this operation from the fundamental system, in the amount of n, will form a linear set, $\mathscr{R}$, of rigid displacements for the mechanical system. The dimensionality, n, of this set is within $0 \le n \le n_A$.

Now let's truncate both linear sets, $\mathscr{R}_A$ and $\mathscr{R}$, futher by keeping only homogeneously kinematically admissible displacements. All these operations will produce linear sets of homogeneously kinematically admissible rigid displacements for the elastic body and for the mechanical system, which we denote as $\mathscr{R}_{A\text{o}}$ и $\mathscr{R}_\text{o}$, respectively, with the respective dimensionalities $n_{A\text{o}}$ and n_o. Obviously, the method of construction implies the following embedding relations of these linear sets:

$$\mathscr{R} \subseteq \mathscr{R}_A\,, \qquad \mathscr{R}_{A\text{o}} \subseteq \mathscr{R}_A\,, \qquad \mathscr{R}_\text{o} = \mathscr{R} \cap \mathscr{R}_{A\text{o}}\,,$$

$$0 \le n \le n_A, \qquad 0 \le n_{A\text{o}} \le n_A, \qquad 0 \le n_\text{o} \le n, \qquad 0 \le n_\text{o} \le n_{A\text{o}}\,.$$

If the linear set $\mathscr{R}_\text{o}$ is an empty set, $\mathscr{R}_\text{o} = \varnothing$, or, equivalently, if $n_\text{o} = 0$, then the system is called *kinematically unstable* (also *substatic*), otherwise *kinematically stable*, and the dimensionality n_o of the linear set $\mathscr{R}_\text{o}$ is called a *degree of kinematic instability* of the system.

Thus, the degree of kinematic instability of a system is equal to the number, n_o, of linearly independent displacements which do not conflict with the external constraints and which make the strain energy of the system equal to zero.

Let us illustrate how the linear sets of rigid displacements introduced above appear in practice by a simple example of bending of a three-span beam with hinged connection between its bars, as shown in Fig. 2.2. The same figure shows how the beam is attached to its supports. In its middle span, the beam is supported by an elastic compliant Winkler-type foundation with a nonzero coefficient of subgrade reaction.

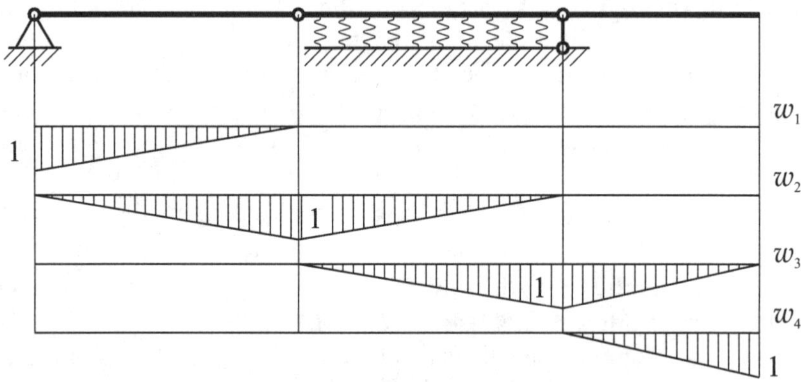

Fig. 2.2. Rigid displacements for a three-span beam

The vector of displacements, **u**, in this problem is a function of lateral displacements of the beam's axis, $w(x)$. The strain **ε** will be the curvature of its axis, χ, in the deformed state. It is easy to see that the linear set of rigid displacements of the elastic body, $\mathscr{R}_A$, is made up by four piecewise-linear functions w_i represented by their plots in Fig. 2.2.

Now let us show what is included in each of the linear sets of rigid displacements introduced above and what dimensionality they have. We have:

$$\mathscr{R}_A = \{ w_1, w_2, w_3, w_4\}, \quad n_A = 4; \qquad \mathscr{R}_{Ao} = \{ w_2, w_4\}, \quad n_{Ao} = 2;$$

$$\mathscr{R} = \{ w_1, w_4\}, \quad n = 2; \qquad \mathscr{R}_o = \{ w_4\}, \quad n_o = 1.$$

As $n_o \neq 0$, the mechanical system is kinematically unstable, and the degree of this instability is 1. This is quite obvious from mechanical viewpoint because the right span of the beam can rotate freely (without any resistance) around the hinge on its left end.

If we introduce the designation $\mathscr{R}_K$ for the set of rigid displacements for the elastic medium, then the definition will give

$$\mathscr{R} = \mathscr{R}_A \cap \mathscr{R}_K .$$

Further we will use the designations $\mathscr{R}_o$ and $\mathscr{R}_K$ also for representing a set of stress-and-strain fields of the type $F = \{\mathbf{0}, \mathbf{0}, \mathbf{u}\}$ with the respective displacement vector **u**. More exactly, $\mathscr{R}_0$ will designate a set of stress-and-strain fields of the type $F = \{\mathbf{0}, \mathbf{0}, \mathbf{u}\}$ with the displacement vector **u** which is a homogeneously kinematically admissible rigid displacement of the system. Similarly, $\mathscr{R}_K$ will designate a set of stress-and strain fields of the type $F = \{\mathbf{0}, \mathbf{0}, \mathbf{u}\}$ with the displacement vector **u** which is a rigid displacement for the elastic medium. Both sets are linear[2]. The definitions imply the following embedding relations between the sets:

$$\mathscr{R}_0 \subseteq \mathscr{K}_{\text{ko}}\,, \quad \mathscr{R}_K \subseteq \mathscr{S}_{\text{so}}\,, \quad \mathscr{R}_0 \subseteq \mathscr{R}_K\,. \tag{1.27}$$

Thus, all fields from the linear set $\mathscr{R}_0$ are at the same time homogeneously kinematically admissible and homogeneously statically admissible fields.

2.1.5 Strain compatibility conditions in the form of an integral identity

Theory of elasticity says [9] that a field of strains **ε** must satisfy continuity conditions the mathematical expression of which consists of the well-known strain compatibility equations by St.-Venant. The compatibility equations' being true is necessary and sufficient for a displacement field **u** to be restorable from a known strain field **ε**. On the other hand, the restoration of the displacements **u** from the strains **ε** can be treated as searching for a partial solution of the inhomogeneous system of equations $\boldsymbol{A}\mathbf{u} = \boldsymbol{\varepsilon}$.

Theory of differential equations says that the general solution of the inhomogeneous equation $\boldsymbol{A}\mathbf{u} = \boldsymbol{\varepsilon}$ can be represented as a sum of the general solution of the respective homogeneous equation and a partial solution of the homogeneous equation which we denote (in case it exists!) as $\mathbf{u}_0$. In other words, the general solution is of the type

$$\mathbf{u} = \mathbf{u}_0 + \sum_{i=1}^{n_A} C_i \mathbf{u}_i \tag{1.28}$$

with arbitrary constants C_i $(i = 1,\ldots, n_A)$.

[2] The dual meaning of the designations $\mathscr{R}_0$ and $\mathscr{R}_K$ does not cause any ambiguity. The context always makes it clear what exactly we are discussing: a set of displacement vectors or a set of stress-and-strain fields of the type $F = \{\mathbf{0}, \mathbf{0}, \mathbf{u}\}$.

As (1.28) is a general form of the geometry equation's solution, one can see that any two solutions of the geometry equation can differ from each other only by a rigid displacement for the elastic body[3].

However, right now we are more interested in the possibility to construct a partial solution of the geometry equation because it is the existence of the partial solution that ensures the restorability of the displacements from the strains, hence it is the strain compatibility condition.

The formulation of these conditions is facilitated by the main theorem of linear algebra [19]:

> *a system of linear algebraic equations, $\mathbf{Ax} = \mathbf{b}$, has a solution when and only when its right part, $\mathbf{b}$, is orthogonal to all solutions $\mathbf{y}$ of the homogeneous system of equations with the conjugate operator, $\boldsymbol{A}^{\mathsf{T}}\mathbf{y} = \mathbf{0}$, that is, under the condition $(\boldsymbol{y}, \boldsymbol{b}) = \boldsymbol{y}^{\mathsf{T}}\boldsymbol{b} = 0$.*

In our case the homogeneous equation with the conjugate operator is $\boldsymbol{A}^{\mathsf{T}}\boldsymbol{\sigma}_{\mathrm{o}} = \mathbf{0}$, that is, $\boldsymbol{\sigma}_{\mathrm{o}}$ is a field of self-balanced stresses on Ω. Note that the stresses $\boldsymbol{\sigma}_{\mathrm{o}}$ are assumed self-balanced without taking into account the forces in the elastic medium. Now it seems that the desirable strain compatibility condition can be represented in the form $(\boldsymbol{\sigma}_{\mathrm{o}}, \boldsymbol{\varepsilon}) = 0 \in \Omega$. However, borrowing the main theorem of algebra and using it in this situation is illegal. Just remember that the conjugation between the geometry operator and the equilibrium operator is understood formally, in the sense of the definition that includes also integrals over the boundary Γ of the area Ω. Therefore the correct formulation of the strain compatibility conditions as an integral identity is:

strains $\boldsymbol{\varepsilon}$ are compatible when and only when

$$(\boldsymbol{\sigma}_{\mathrm{o}}, \boldsymbol{\varepsilon}) = 0 \quad \in \Omega \tag{1.29}$$

for any stress field $\boldsymbol{\sigma}_{\mathrm{o}}$ self-balanced in the area Ω and on the boundary Γ, that is, for all $\boldsymbol{\sigma}_{\mathrm{o}}$ meeting the conditions

$$\boldsymbol{A}^{\mathsf{T}}\boldsymbol{\sigma}_{\mathrm{o}} = \mathbf{0} \ \in \Omega, \qquad \boldsymbol{H}_{\sigma}\boldsymbol{\sigma}_{\mathrm{o}} = \mathbf{0} \ \in \Gamma. \tag{1.30}$$

Let us prove the necessity first. If there exist displacements $\mathbf{u}$ such that $\boldsymbol{\varepsilon} = \boldsymbol{A}\mathbf{u}$ then using the basic integral identity (**1**.2.15) will yield

[3] We do not touch details of the conditions of unambiguity for a displacement field restored from known strains. A discussion on this problem, the solution of which depends on the degree of connectivity of the domain Ω, can be found in any detailed course of theory of elasticity.

$$(\boldsymbol{\sigma}_o, \boldsymbol{\varepsilon}) = (\boldsymbol{\sigma}_o, \boldsymbol{A}\mathbf{u}) = (\boldsymbol{A}^\mathsf{T}\boldsymbol{\sigma}_o, \mathbf{u}) + (\boldsymbol{H}_\sigma\boldsymbol{\sigma}_o, \boldsymbol{H}_u\mathbf{u})_\Gamma . \tag{1.31}$$

Applying (1.30) will annihilate both scalar products in the right part of (1.31).

Now let (1.29) be true at any $\boldsymbol{\sigma}_o$ satisfying the conditions (1.30). Multiplying the first of the equations in (1.30) scalarly by an arbitrary vector **u** and subtracting the expression thus obtained and equal to zero from (1.29) will give an equivalent identity:

$$(\boldsymbol{\sigma}_o, \boldsymbol{\varepsilon}) - (\boldsymbol{A}^\mathsf{T}\boldsymbol{\sigma}_o, \mathbf{u}) = 0 \quad \in\Omega . \tag{1.32}$$

We can assume for this identity that $\boldsymbol{\sigma}_o$ is not subject beforehand to any requirements in the area Ω because the equality $\boldsymbol{A}^\mathsf{T}\boldsymbol{\sigma}_o = \mathbf{0} \in\Omega$ itself is a corollary of the identity (1.32) as the vector **u** is arbitrary.

But $(\boldsymbol{A}^\mathsf{T}\boldsymbol{\sigma}_o, \mathbf{u}) = (\boldsymbol{\sigma}_o, \boldsymbol{A}\mathbf{u}) - (\boldsymbol{H}_\sigma\boldsymbol{\sigma}_o, \boldsymbol{H}_u\mathbf{u})_\Gamma$, and when this is substituted to (1.32) taking into account the second of the equalities in (1.30), we will have

$$(\boldsymbol{\sigma}_o, \boldsymbol{\varepsilon} - \boldsymbol{A}\mathbf{u}) = 0 , \tag{1.33}$$

wherefrom follows $\boldsymbol{\varepsilon} = \boldsymbol{A}\mathbf{u}$. Note that the proof of sufficiency uses actually the technique of Lagrangian multipliers well known in calculus of variations. This method lets us formally avoid additional conditions imposed on functions which are subject to variation.

The integral identity (1.29) is in close relation to the principle of virtual stress increments. However, these two have much in difference, too. First, and principal, while the principle of virtual stress increments defines conditions under which the displacements and the strains belong to the fully kinematically admissible class (one that takes into account the kinematic boundary conditions), the integral identity (1.29) being true ensures nothing more than the strain compatibility. That's why the integral identity (1.29) can be treated as a *general variational principle of strain compatibility*. Second, the principle of virtual stress increments contains the stress $\boldsymbol{\sigma}_{so}$ which belongs to a homogeneously statically admissible field while the formulation of the general variational principle of the strain compatibility contains the stress $\boldsymbol{\sigma}_o$ self-balanced in the sense of (1.30).

The strain compatibility conditions in the form of an integral identity (in terms that we use, we can talk also about the conditions under which the strains belong to a kinematically semi-admissible stress-and-strain field) have certain advantages over their differential analog, the St.-Venant equations. The matter is that the St.-Venant equations for elasticity contain twice differentiable strains which is too strict a requirement for the smoothness of the functions. At the same time, the integral form of the strain compatibility conditions does not require them to be differentiable even once.

At this point the following note should be made. One of substantial differences (though not the only one!) of one-dimensional problems from two- and three-dimensional ones (we mean the dimensionality k of the area Ω) is that, as we will show later, there is a wide class of one-dimensional problems where stresses $\boldsymbol{\sigma}_o$ self-balanced in the sense of (1.30) are always zero. Consequently, the general variational principle of strain compatibility in the form of the integral identity (1.29) is ensured for such problems. Thus, strains $\boldsymbol{\varepsilon}$ in problems of that kind are always compatible in the sense that there is always a vector of displacements, $\mathbf{u}$, which satisfies the geometric equation $\boldsymbol{\varepsilon} = \boldsymbol{A}\mathbf{u}$. This note corresponds thoroughly also to the continuity conditions in the form of differential equations which take place only at $k = 2$ and $k = 3$ but do not exist for one-dimensional problems.

2.1.6 Necessary conditions for an equilibrium state of a system to exist

Let $\mathbf{u}_i$ $(i = 1,\ldots, n_o)$ be a set of linearly independent, homogeneously kinematically admissible rigid displacements of a mechanical system. We use the principle of virtual displacements of statics and geometry in the form (**1**.4.11) and assume the state 1 to be the true state of the system under a given load and the state 2 to be any of fields $F_{2i} = \{\boldsymbol{0}, \boldsymbol{0}, \mathbf{u}_i\}$ $(i = 1,\ldots, n_o)$. The fields F_{2i} make the right part in (**1**.4.11) zero which gives the necessary condition for a balanced state of the system to exist, in the form

$$(\overline{\boldsymbol{X}}, \mathbf{u}_i) + (\boldsymbol{E}_p \overline{\boldsymbol{p}}, \boldsymbol{E}_p \boldsymbol{H}_u \mathbf{u}_i)_\Gamma = 0 . \qquad (1.34)$$

So, for an equilibrium state to exist under given external forces $\overline{\boldsymbol{X}}$ and $\overline{\boldsymbol{p}}$, the virtual work of the forces on all displacements $\mathbf{u}_i \in \mathscr{R}_o$ must be zero.

Let's take an example: a so-called second (Neumann) boundary-value problem for three-dimensional elasticity [6] in which the whole boundary of a body is subject to static boundary conditions (**1**.2.12-*a*) and there is no elastic medium, that is, $\boldsymbol{K} = \boldsymbol{O}$. Thus, in this case $\Gamma = \Gamma_p$, $\boldsymbol{E}_p = \boldsymbol{I}$, $\boldsymbol{H}_u = \boldsymbol{I}$. It is easy to notice that the linear set $\mathscr{R}_o$ for this problem has the number of dimensions $n_o = 6$, and the general form of any vector $\mathbf{u}$ from this set is given by the formula (1.25). The condition (1.34) can be reduced to the following requirement:

$$\int_{\Omega} \overline{X} \cdot (\mathbf{u}_o + \boldsymbol{\theta} \times \mathbf{r})\, d\Omega + \oint_{\Gamma} \overline{p} \cdot (\mathbf{u}_o + \boldsymbol{\theta} \times \mathbf{r})\, d\Gamma = 0\,. \tag{1.35}$$

Now using the known property of mixed product of three vectors, $\boldsymbol{a}$, $\boldsymbol{b}$ and $\boldsymbol{c}$ [3], in the form $\boldsymbol{a} \cdot (\boldsymbol{b}\times\boldsymbol{c}) = -\,\boldsymbol{b} \cdot (\boldsymbol{a}\times\boldsymbol{c})$, and seeing that the components of the vectors $\mathbf{u}_o$ and $\boldsymbol{\theta}$ are constant throughout the area Ω together with its boundary Γ lets us rewrite (1.35) as

$$\mathbf{u}_o \cdot \Big(\int_{\Omega} \overline{X} d\Omega + \oint_{\Gamma} \overline{p}\, d\Gamma \Big) - \boldsymbol{\theta} \cdot \Big(\int_{\Omega} \overline{X} \times \mathbf{r} d\Omega + \oint_{\Gamma} \overline{p} \times \mathbf{r}\, d\Gamma \Big) = 0$$

wherefrom the arbitrariness of the vectors $\mathbf{u}_o$ and $\boldsymbol{\theta}$ gives the known necessary conditions for the problem to be solvable:

$$\int_{\Omega} \overline{X} d\Omega + \oint_{\Gamma} \overline{p}\, d\Gamma = \mathbf{0}, \qquad \int_{\Omega} \overline{X} \times \mathbf{r} d\Omega + \oint_{\Gamma} \overline{p} \times \mathbf{r}\, d\Gamma = \mathbf{0}\,. \tag{1.36}$$

The physical meaning of the two requirements is obvious: for an equilibrium state of the system to exist, the resultant force and the resultant moment of all active external forces applied to the body must be equal to zero.

2.1.7 Theorem of a general form of an arbitrary physically admissible field

An arbitrary field $F \in \mathscr{P}$ can be decomposed into a sum of a homogeneously statically admissible field F_{so} and a homogeneously kinematically admissible field F_{ko},

$$F = F_{\mathrm{so}} + F_{\mathrm{ko}}\,, \quad F_{\mathrm{so}} \in \mathscr{P}_{\mathrm{so}}\,, \quad F_{\mathrm{ko}} \in \mathscr{P}_{\mathrm{ko}}\,. \tag{1.37}$$

This form means summation by elements, that is,

$$\boldsymbol{\sigma} = \boldsymbol{\sigma}_{\mathrm{so}} + \boldsymbol{\sigma}_{\mathrm{ko}}\,, \quad \boldsymbol{\varepsilon} = \boldsymbol{\varepsilon}_{\mathrm{so}} + \boldsymbol{\varepsilon}_{\mathrm{ko}}\,, \quad \mathbf{u} = \mathbf{u}_{\mathrm{so}} + \mathbf{u}_{\mathrm{ko}}\,. \tag{1.38}$$

The proof is based on the following consideration by S.G. Mikhlin [7].

Let some volumetric forces, $\overline{X}$, and contour forces, $\overline{p}$, be in static conformance to the field F, and let contour displacements $\overline{u}$ be caused by displacements $\mathbf{u} \in F$, that is,

$$\overline{X} = A^{\mathsf{T}}\boldsymbol{\sigma} + \boldsymbol{K}\mathbf{u} \;\in \Omega, \quad \boldsymbol{E}_p \overline{p} = \boldsymbol{E}_p \boldsymbol{H}_{\sigma}\boldsymbol{\sigma} \;\in \Gamma, \quad \boldsymbol{E}_u \overline{u} = \boldsymbol{E}_u \boldsymbol{H}_u \mathbf{u} \;\in \Gamma. \tag{1.39}$$

The elements of the field F_{ko} will be taken from the solution of the following auxiliary problem:

$$\boldsymbol{A}^{\mathsf{T}}\boldsymbol{\sigma}_{\mathrm{ko}} + \boldsymbol{K}\mathbf{u}_{\mathrm{ko}} = \bar{\boldsymbol{X}} \qquad \text{equilibrium equations,} \tag{1.40-a}$$

$$\boldsymbol{A}\mathbf{u}_{\mathrm{ko}} = \boldsymbol{\varepsilon}_{\mathrm{ko}} \qquad \text{geometric equations,} \tag{1.40-b}$$

$$\boldsymbol{\sigma}_{\mathrm{ko}} = \boldsymbol{C}\boldsymbol{\varepsilon}_{\mathrm{ko}} \qquad \text{physical equations,} \tag{1.40-c}$$

$$\boldsymbol{E}_p(\boldsymbol{H}_\sigma\boldsymbol{\sigma}_{\mathrm{ko}} - \bar{\boldsymbol{p}}) = \mathbf{0} \qquad \text{static boundary conditions,} \tag{1.40-d}$$

$$\boldsymbol{E}_u\boldsymbol{H}_u\mathbf{u}_{\mathrm{ko}} = \mathbf{0} \qquad \text{kinematic boundary conditions.} \tag{1.40-e}$$

The necessary conditions (1.34) for the solution of this problem to exist hold because the virtual work of the active external forces, $\bar{\boldsymbol{X}}$ and $\bar{\boldsymbol{p}}$, on homogeneously kinematically admissible rigid displacements of the system must be zero – otherwise the first two inequalities (1.39) could not hold. Now, all we have to prove is to check that the elements $\boldsymbol{\sigma}_{\mathrm{so}} = \boldsymbol{\sigma} - \boldsymbol{\sigma}_{\mathrm{ko}}$, $\boldsymbol{\varepsilon}_{\mathrm{so}} = \boldsymbol{\varepsilon} - \boldsymbol{\varepsilon}_{\mathrm{ko}}$, $\mathbf{u}_{\mathrm{so}} = \mathbf{u} - \mathbf{u}_{\mathrm{ko}}$ belong to a homogeneously statically admissible field, i.e. ensure the conditions (**1**.3.3). And indeed, we have

$$\boldsymbol{A}^{\mathsf{T}}\boldsymbol{\sigma}_{\mathrm{so}} + \boldsymbol{K}\mathbf{u}_{\mathrm{so}} = \boldsymbol{A}^{\mathsf{T}}(\boldsymbol{\sigma} - \boldsymbol{\sigma}_{\mathrm{ko}}) + \boldsymbol{K}(\mathbf{u} - \mathbf{u}_{\mathrm{ko}}) =$$
$$= (\boldsymbol{A}^{\mathsf{T}}\boldsymbol{\sigma} + \boldsymbol{K}\mathbf{u}) - (\boldsymbol{A}^{\mathsf{T}}\boldsymbol{\sigma}_{\mathrm{ko}} + \boldsymbol{K}\mathbf{u}_{\mathrm{ko}}) = \bar{\boldsymbol{X}} - \bar{\boldsymbol{X}} = \mathbf{0} \qquad \in\Omega,$$
$$\boldsymbol{E}_p\boldsymbol{H}_\sigma\boldsymbol{\sigma}_{\mathrm{so}} = \boldsymbol{E}_p\boldsymbol{H}_\sigma(\boldsymbol{\sigma} - \boldsymbol{\sigma}_{\mathrm{ko}}) = \boldsymbol{E}_p\,\bar{\boldsymbol{p}} - \boldsymbol{E}_p\,\bar{\boldsymbol{p}} = \mathbf{0} \qquad \in\Gamma,$$

which proves the theorem.

Note the circumstance that this general formulation makes the decomposition (1.37) sometimes ambiguous. To see this, let's consider the field $F_{\mathscr{R}\mathrm{o}} = \{\mathbf{0}, \mathbf{0}, \mathbf{u}_{\mathrm{o}}\} \in \mathscr{R}_{\mathrm{o}}$. As we mentioned above, the embedding relations (1.27) make the field $F_{\mathscr{R}\mathrm{o}}$ both a homogeneously statically admissible field and a homogeneously kinematically admissible field at the same time. Therefore, together with (1.37), the decomposition $F = F_{\mathrm{so1}} + F_{\mathrm{ko1}}$ is also true where the fields

$$F_{\mathrm{ko1}} = F_{\mathrm{ko}} + F_{\mathscr{R}\mathrm{o}}, \quad F_{\mathrm{so1}} = F_{\mathrm{so}} - F_{\mathscr{R}\mathrm{o}}$$

differ from the components of the decomposition (1.37) by an arbitrary field $F_{\mathscr{R}\mathrm{o}} \in \mathscr{R}_{\mathrm{o}}$.

Let's finally notice an obvious fact: if the original field F belongs to the linear set $\mathscr{F}_{\mathrm{k/2}}$, then $F_{\mathrm{so}} \in \mathscr{F}_{\mathrm{k/2}}$ also in the decomposition (1.37).

2.1.8 Lagrangian energy space

Now we are going to construct an important linear set for further consideration, $\mathscr{L}$, by removing all fields of $\mathscr{R}_o$ from the linear set $\mathscr{P}_{ko}$. This can be done by using the scalar product introduced earlier on the set $\mathscr{P}_{ko}$ with the formula (**1**.1.1). The metric produced by this scalar product will be called the *main metric*. Let's decompose the space thus obtained into a direct sum using the main metric:

$$\mathscr{P}_{ko} = \mathscr{L} \oplus \mathscr{R}_o \tag{1.41}$$

The latter means that for any field $F_{ko} = \{\boldsymbol{CA}\mathbf{u}_{ko}, \boldsymbol{A}\mathbf{u}_{ko}, \mathbf{u}_{ko}\}$ from $\mathscr{P}_{ko}$ with the displacement component $\mathbf{u}_{ko}$ the following representation holds: $F_{ko} = F_{\mathscr{L}} + F_{\mathscr{R}o}$, $F_{\mathscr{L}} = \{\boldsymbol{CA}\mathbf{u}_{\mathscr{L}}, \boldsymbol{A}\mathbf{u}_{\mathscr{L}}, \mathbf{u}_{\mathscr{L}}\} \in \mathscr{L}$, $F_{\mathscr{R}o} = \{\mathbf{0}, \mathbf{0}, \mathbf{u}_{\mathscr{R}o}\} \in \mathscr{R}_o$, $\mathbf{u}_{ko} = \mathbf{u}_{\mathscr{L}} + \mathbf{u}_{\mathscr{R}o}$, $\boldsymbol{A}\mathbf{u}_{\mathscr{R}o} = \mathbf{0}$, so that

$$(\mathbf{u}_{\mathscr{L}}, \mathbf{u}_{\mathscr{R}o}) = 0. \tag{1.42}$$

It is easy to understand that the linear set $\mathscr{L}$ is made up by all physically admissible and homogeneously kinematically admissible fields which ensure a positive strain energy. And indeed, if $\mathsf{E}(\mathbf{u}_{ko}) = 0$, then $\mathbf{u}_{ko}$ is a vector of homogeneously kinematically admissible rigid displacements of the system, hence the respective field $F_{ko} = \{\mathbf{0}, \mathbf{0}, \mathbf{u}_{ko}\}$ does not belong to the set $\mathscr{L}$.

Now we turn again to the example of second (Neumann) boundary-value problem of elasticity for three-dimensional bodies. We see that the orthogonality condition (1.42) in this case is written as:

$$\int_{\Omega} \mathbf{u}_{ko} \cdot (\mathbf{u}_o + \boldsymbol{\theta} \times \mathbf{r}) d\Omega = 0 \tag{1.43}$$

or, as $\mathbf{u}_o$ and $\boldsymbol{\theta}$ are arbitrary

$$\int_{\Omega} \mathbf{u}_{ko} d\Omega = \mathbf{0}, \qquad \int_{\Omega} \mathbf{u}_{ko} \times \mathbf{r} d\Omega = \mathbf{0}\,. \tag{1.44}$$

As there are no kinematical boundary conditions in this problem, the set $\mathscr{L}$ consists here of all possible (sufficiently smooth) vectors of displacements which satisfy the integral conditions (1.44). Obviously, the first condition in (1.44) restrains a rigid translational movement of the body and the second condition restrains its rigid rotation.

Let's introduce a symmetric bilinear functional on the set $\mathscr{P}$, we will denote it by broken brackets $<\,,\,>$:

$$<F_1, F_2> = \tfrac{1}{2}\,(\boldsymbol{C}\boldsymbol{\varepsilon}_1, \boldsymbol{\varepsilon}_2) + \tfrac{1}{2}\,(\boldsymbol{K}\mathbf{u}_1, \mathbf{u}_2) =$$

$$= \tfrac{1}{2}\,(\boldsymbol{C}^{-1}\boldsymbol{\sigma}_1, \boldsymbol{\sigma}_2) + \tfrac{1}{2}\,(\boldsymbol{K}\mathbf{u}_1, \mathbf{u}_2)\,. \tag{1.45}$$

Comparing (1.13) and (1.45) gives $<F, F> = \mathsf{E}(F)$ wherefrom the positivity of the energy E on $\mathscr{L}$ yields the scalar product on the set $\mathscr{L}$ produced by this bilinear functional. This scalar product is called an *energy (Lagrangian) scalar product.*

Introducing a scalar product on the linear set $\mathscr{L}$ allows us to treat this linear set as a Hilbert space which we will call a *Lagrangian energy space* and designate by the same letter, $\mathscr{L}$[4]. As it is usually done with the Hilbert spaces, we will demand additionally that the scalar product $<F, F> = \frac{1}{2}\,(\boldsymbol{CA}\mathbf{u}, \boldsymbol{A}\mathbf{u}) + \frac{1}{2}\,(\boldsymbol{K}\mathbf{u}, \mathbf{u})$ has a finite value for any field F from $\mathscr{L}$. Thus, the Lagrangian energy space $\mathscr{L} \subseteq \mathscr{P}_{\text{ko}} \subseteq \mathscr{P}$ consists of a set of fields of the type $F = \{\boldsymbol{CA}\mathbf{u}, \boldsymbol{A}\mathbf{u}, \mathbf{u}\}$ submitted to the following requirements:

- the displacements **u** must satisfy the homogeneous kinematic boundary conditions;
- the displacements **u** must be orthogonal, with respect to the main metric, to all homogeneously kinematically admissible rigid displacements of the system;
- the scalar product $<F, F> = \frac{1}{2}\,(\boldsymbol{CA}\mathbf{u}, \boldsymbol{A}\mathbf{u}) + \frac{1}{2}\,(\boldsymbol{K}\mathbf{u}, \mathbf{u})$ must have a finite value for any field F from $\mathscr{L}$.

Now let's return to the theorem of a general form of an arbitrary physically admissible field from the previous subsection. If the set of all homogeneously kinematically admissible fields is limited to the space $\mathscr{L}$, then the decomposition (1.37) can be proved to be the only possible one. To see this, we can assume the opposite situation and introduce two decompositions of a field F according to (1.37), with subscripts 1 and 2 for the component fields in these decompositions. Thus,

$$F_{\text{so1}} + F_{\text{ko1}} = F_{\text{so2}} + F_{\text{ko2}}, \qquad \text{so that } F_{\text{ko1}} \in \mathscr{L}, \quad F_{\text{ko2}} \in \mathscr{L}. \tag{1.46}$$

[4] The usual mathematical term for a linear set with a scalar product is a *pre-Hilbert* or *unitary space* [14]. A genuine Hilbert space is additionally required to be complete with respect to the metrics generated by the scalar product. However, as we already told in the preface to this book, we do not touch mathematical issues related to the space completeness. We imply that the conditions for all Hilbert spaces considered in this book to be complete, are fulfilled. Therefore here and further our reader (were he or she be more fastidious than one educated in engineering rather than mathematics could be expected) should not be embarrassed with the equivalence that we declare implicitly between the Hilbert and pre-Hilbert spaces.

Multiplying both parts of the equality (1.46) scalarly (in the energy metric) by an arbitrary field $F_{ko} \in \mathscr{L}$ and taking into account the theorem of field orthogonality (**1**.4.21) will give

$$< F_{ko}, F_{ko1} - F_{ko2} > = 0 . \tag{1.47}$$

As the selected field F_{ko} is arbitrary, this gives $F_{ko1} = F_{ko2}$, hence $F_{so1} = F_{so2}$.

By taking into account all that was said above, we can modify the decomposition (1.37) into

$$F = F_{so} + F_{\mathscr{L}} + F_{\mathscr{R}o}, \quad F_{\mathscr{L}} \in \mathscr{L}, \quad F_{\mathscr{R}o} \in \mathscr{R}_o, \tag{1.48}$$

the field F_{so} being orthogonal by energy to the two other fields, and the fields $F_{\mathscr{L}}$ and $F_{\mathscr{R}o}$ being orthogonal to each other in the main metric. By the way, the fields $F_{\mathscr{L}}$ and $F_{\mathscr{R}o}$ are orthogonal to each other not only in the main metric but also by energy. Indeed, $<F_{\mathscr{L}}, F_{\mathscr{R}o}> = ½\,(\boldsymbol{CA}\mathbf{u}_{\mathscr{L}}, \boldsymbol{A}\mathbf{u}_{\mathscr{R}o}) + ½\,(\boldsymbol{K}\mathbf{u}_{\mathscr{L}}, \mathbf{u}_{\mathscr{R}o})$. But $\boldsymbol{A}\mathbf{u}_{\mathscr{R}o} = 0$ and $\boldsymbol{K}\mathbf{u}_{\mathscr{R}o} = \mathbf{0}$ from the definition of a homogeneously kinematically admissible set of rigid displacements $\mathscr{R}_o$, hence $<F_{\mathscr{L}}, F_{\mathscr{R}o}> = 0$.

2.1.9 Prager–Synge identity

Let F_s and F_k be respective statically admissible and kinematically admissible fields from the linear set $\mathscr{P}$, and let F_* be a true stress-and-strain field. Then a so-called Prager–Synge identity [11] takes place:

$$\mathsf{E}(F_* - F_s) + \mathsf{E}(F_* - F_k) = \mathsf{E}(F_s - F_k) \tag{1.49}$$

which is useful to build error estimates for approximate solutions.

To see this, we perform a chain of identical transformations:

$$\begin{aligned} \mathsf{E}(F_* - F_s) + \mathsf{E}(F_* - F_k) &= \mathsf{E}(F_*) - 2<F_s, F_*> + \mathsf{E}(F_s) + \mathsf{E}(F_*) - \\ - 2<F_k, F_*> + \mathsf{E}(F_k) &= [\mathsf{E}(F_s) + \mathsf{E}(F_k) - 2<F_s, F_k>] + 2<F_s, F_k> - \\ &-2<F_s, F_*> - 2<F_k, F_*> + 2<F_*, F_*> = \\ &= \mathsf{E}(F_s - F_k) + 2<F_* - F_s, F_* - F_k>. \end{aligned} \tag{1.50}$$

Noting that $F_* - F_s = F_{so}$ and $F_* - F_k = F_{ko}$ are respective homogeneously statically admissible and homogeneously kinematically admissible fields and taking into account the field orthogonality theorem (**1**.4.21), we conclude that the second term in the right part of (1.50) is annihilated, hence (1.49).

2.2 Lagrange variational principle

2.2.1 Conservative external forces

Up to this point, we did not discuss the nature of external forces $\bar{\boldsymbol{X}}$ distributed over the volume of a body and contour forces $\bar{\boldsymbol{p}}$ which can depend (both in their value and in their direction), generally, on the displacements of the system. From here onward, we will assume that all external forces are *conservative* (or *potential*). The latter means that the work done by the forces during the process of deformation does not depend on the path of deformation; this is equivalent to the existence of two functions,

$$\Pi_\Omega = \Pi_\Omega(\mathbf{u}) \in \Omega\,, \qquad \Pi_\Gamma = \Pi_\Gamma(\boldsymbol{u}) \in \Gamma\,, \tag{2.1}$$

such that each of the force components of $\bar{\boldsymbol{X}}^i$ and $\bar{\boldsymbol{p}}^i$ permits the representation

$$\bar{\boldsymbol{X}}^i = \frac{\partial \Pi_\Omega}{\partial \mathbf{u}_i}\,, \qquad \bar{\boldsymbol{p}}^i = \frac{\partial \Pi_\Gamma}{\partial \boldsymbol{u}_i}\,. \tag{2.2}$$

The functions $\Pi_\Omega(\mathbf{u})$ and $\Pi_\Gamma(\boldsymbol{u})$ are called *densities of potentials of volumetric and surface forces*, respectively. The potential of all external (active) forces, Π_s, is defined as a sum of integrals of the above functions:

$$\Pi_s = \int_\Omega \Pi_\Omega(\mathbf{u})d\Omega + \int_\Gamma \Pi_\Gamma(\mathbf{u})d\Gamma. \tag{2.3}$$

Further we will refer to the potential Π_s as a *potential of static external actions*, to emphasize that it is different from a *potential of external actions*, Π_k, which will be introduced later and related to reactive forces that appear on the contour Γ where inhomogeneous kinematic boundary conditions are specified.

The simplest case of the conservative external forces $\bar{\boldsymbol{X}}$ and $\bar{\boldsymbol{p}}$ (the case we will confine ourselves to) includes so-called "dead" forces or loads[5]. External forces are called *dead* if their points of application in the Lagrangian (material) coordinate system do not change their locations in the process of the system's deformation, while the relations between the projections of those forces with respect to the Eulerian (spatial) coordinate

[5] Some examples of conservative external forces which are not dead loads are discussed in Appendix E.

system are also maintained constant. The densities of potentials of dead loads are especially simple to write:

$$\Pi_\Omega(\mathbf{u}) = \bar{\boldsymbol{X}} \cdot \mathbf{u}\,, \qquad \Pi_\Gamma(\boldsymbol{u}) = \boldsymbol{E}_p\, \bar{\boldsymbol{p}} \cdot \boldsymbol{u}\,. \tag{2.4}$$

As a result, the potential of static external actions, Π_s for dead external forces can be represented as

$$\Pi_s = \int_\Omega \bar{\boldsymbol{X}} \cdot \mathbf{u} d\Omega + \int_\Gamma \boldsymbol{E}_p \bar{\boldsymbol{p}} \cdot \boldsymbol{E}_p \boldsymbol{H}_u \mathbf{u} d\Gamma = (\bar{\boldsymbol{X}}, \mathbf{u}) + (\boldsymbol{E}_p \bar{\boldsymbol{p}}, \boldsymbol{E}_p \boldsymbol{H}_u \mathbf{u})_\Gamma\,. \tag{2.5}$$

2.2.2 Lagrange functional

First of all, we would like to note that any field F from the set $\mathscr{P}_{k/2}$ is unambiguously determined by the vector of displacements, $\mathbf{u}$, because $F = \{\boldsymbol{CA}\mathbf{u}, \boldsymbol{A}\mathbf{u}, \mathbf{u}\}$. Therefore any functional defined on that set or on a subset of it can be treated, without limiting the generality, as a functional which depends on the displacement vector $\mathbf{u}$ only.

Let us introduce a Lagrange functional, $\mathsf{L} = \mathsf{L}(F)$, on the set $\mathscr{P}_k \subseteq \mathscr{P}_{k/2}$, also called a *Lagrangian* or a *full potential energy of the system*

$$\mathsf{L}(\mathbf{u}) = \mathsf{E}(\mathbf{u}) - \Pi_s(\mathbf{u}) \tag{2.6}$$

where energy E is treated as a quadratic functional of $\mathbf{u}$, which gives the following according to (1.13):

$$\mathsf{E}(\mathbf{u}) = \tfrac{1}{2}\,(\boldsymbol{CA}\mathbf{u}, \boldsymbol{A}\mathbf{u}) + \tfrac{1}{2}\,(\boldsymbol{K}\mathbf{u}, \mathbf{u})\,.$$

Now let's show that the Lagrange functional L takes a minimum value on true fields F_*,

$$\mathsf{L}(F_*) \le \mathsf{L}(F), \tag{2.7}$$

for all $F \in \mathscr{P}_k$. By assuming $F = F_* + \delta F$, we can rewrite the desirable inequality (2.7) as

$$\mathsf{L}(F_* + \delta F) - \mathsf{L}(F_*) =$$

$$= \{\mathsf{E}(\mathbf{u}_* + \delta\mathbf{u}) - \mathsf{E}(\mathbf{u}_*)\} - \{\Pi_s(\mathbf{u}_* + \delta\mathbf{u}) - \Pi_s(\mathbf{u}_*)\} \ge 0\,. \tag{2.8}$$

Now let's evaluate the differences contained in (2.8) separately. We have

$$\mathsf{E}(\mathbf{u}_* + \delta\mathbf{u}) - \mathsf{E}(\mathbf{u}_*) = \mathsf{E}(\mathbf{u}_*) + 2\langle F_*, \delta F\rangle + \mathsf{E}(\delta\mathbf{u}) - \mathsf{E}(\mathbf{u}_*) =$$

$$= 2\langle F_*, \delta F\rangle + \mathsf{E}(\delta\mathbf{u})\,,$$

$$\Pi_s(\mathbf{u}_* + \delta\mathbf{u}) - \Pi_s(\mathbf{u}_*) = \Pi_s(\delta\mathbf{u}) .$$

Substituting all this in (2.8) will yield

$$\mathsf{L}(\mathbf{u}_* + \delta\mathbf{u}) - \mathsf{L}(\mathbf{u}_*) = \{2\langle F_*, \delta F\rangle - \Pi_s(\delta\mathbf{u})\} + \mathsf{E}(\delta\mathbf{u}) . \tag{2.9}$$

But the braced expression from (2.9) can be expanded as follows:

$$2\langle F_*, \delta F\rangle - \Pi_s(\delta\mathbf{u}) =$$

$$= (\boldsymbol{CA}\mathbf{u}_*, \boldsymbol{A}\delta\mathbf{u}) + (\boldsymbol{K}\mathbf{u}_*, \delta\mathbf{u}) - (\overline{\boldsymbol{X}}, \delta\mathbf{u}) - (\boldsymbol{E}_p\,\overline{\boldsymbol{p}}, \boldsymbol{E}_p\boldsymbol{H}_u\delta\mathbf{u})_\Gamma . \tag{2.10}$$

As the variation $\delta\mathbf{u}$ is a difference between two kinematically admissible displacements $\mathbf{u}$ and $\mathbf{u}_*$, $\delta\mathbf{u}$ will satisfy the homogeneous kinematic boundary conditions, that is, $\delta F \in \mathscr{P}_{\text{ko}}$. This gives us the right to refer to the principle of virtual displacements which makes the right part in (2.10) equal to zero. As a result, (2.9) becomes

$$\mathsf{L}(F_* + \delta F) - \mathsf{L}(F_*) = \mathsf{E}(\delta\mathbf{u}) . \tag{2.11}$$

But $\mathsf{E}(\delta\mathbf{u}) \ge 0$ at any vector $\delta F \in \mathscr{P}_{\text{ko}}$, and thus the inequality (2.7) is proved.

The inverse statement is also true: a field F_1 on which the Lagrange functional $\mathsf{L}(F)$ is minimum,

$$\mathsf{L}(F_1) \le \mathsf{L}(F), \tag{2.12}$$

is a true fields among all $F \in \mathscr{P}_{\text{k}}$, that is, $F_1 = F_*$.

To validate this statement, it suffices to prove that a stress-and-strain field F_1 from $\mathscr{P}_{\text{k}}$ with the displacement component $\mathbf{u}_1$, which we will call the state 1 of the system, satisfies both the equilibrium equations and the static boundary conditions. By introducing δF as a difference between the fields, $\delta F = F - F_1$, and doing the same as before, we can come to an evaluation of the difference of the Lagrange functional's values

$$\mathsf{L}(F) - \mathsf{L}(F_1) =$$

$$= (\boldsymbol{CA}\mathbf{u}_1, \boldsymbol{A}\delta\mathbf{u}) + (\boldsymbol{K}\mathbf{u}_1, \delta\mathbf{u}) - (\overline{\boldsymbol{X}}, \delta\mathbf{u}) - (\boldsymbol{E}_p\,\overline{\boldsymbol{p}}, \boldsymbol{E}_p\boldsymbol{H}_u\delta\mathbf{u})_\Gamma + \mathsf{E}(\delta\mathbf{u}). \tag{2.13}$$

Now it is convenient to represent the variation δF as $\delta F = \alpha F_{\text{ko}}$ where F_{ko} is a field from $\mathscr{P}_{\text{ko}}$ and α is an arbitrary numerical parameter, which makes it possible to rewrite (2.13) as

$$\mathsf{L}(F) - \mathsf{L}(F_1) = \alpha\{(\boldsymbol{CA}\mathbf{u}_1, \boldsymbol{A}\mathbf{u}_{\text{ko}}) + (\boldsymbol{K}\mathbf{u}_1, \mathbf{u}_{\text{ko}}) - (\overline{\boldsymbol{X}}, \mathbf{u}_{\text{ko}}) -$$

$$- (\boldsymbol{E}_p\,\overline{\boldsymbol{p}}, \boldsymbol{E}_p\boldsymbol{H}_u\mathbf{u}_{\text{ko}})_\Gamma\} + \alpha^2\mathsf{E}(\mathbf{u}_{\text{ko}}). \tag{2.14}$$

According to the condition (2.12), the right part in (2.14) must be nonnegative at any, even infinitesimal, value of α. But this can be true only if the factor at α raised to power one is zero:

$$(\boldsymbol{CA}\mathbf{u}_1, \boldsymbol{A}\mathbf{u}_{\text{ko}}) + (\boldsymbol{K}\mathbf{u}_1, \mathbf{u}_{\text{ko}}) - (\bar{\boldsymbol{X}}, \mathbf{u}_{\text{ko}}) - (\boldsymbol{E}_p\bar{\boldsymbol{p}}, \boldsymbol{E}_p\boldsymbol{H}_u\mathbf{u}_{\text{ko}})_\Gamma = 0\,.$$

If we consider the equality above from the standpoint of the principle of virtual displacements, we will conclude that the state 1 satisfies all the equilibrium equations. So, it yields the following correct formulation of the Lagrange principle:

A stress-and-strain field in a linear deformable mechanical system is a true field if and only if this field makes the Lagrange functional, L, *of the system take a least possible value among all physically and kinematically admissible stress-and-strain fields.*

Note that the principle of virtual displacements is invariant with respect to the physical equations while the Lagrange variational principle is essentially based on the physical admissibility of stress-and-strain fields which we compare.

Note the circumstance that the set $\mathscr{P}_\text{k}$ which we search for the minimum of the Lagrange functional is not linear under nonzero external kinematical actions; this can be inconvenient. The natural measure of deviation of an approximation F_h from the exact solution F_* can be the norm of the difference $\|F_* - F_h\|$ in the energy metric where

$$\|F_* - F_h\| = \sqrt{<(F_* - F_h),(F_* - F_h)>} = \sqrt{\mathsf{E}(F_* - F_h)}\,.$$

In this regard, we would like to give another equivalent formulation of the Lagrange variational principle. To do this, first we will transform the functional to conditions of its variations on the Lagrangian energy space $\mathscr{L}$. Let's fix an arbitrary field $F_\text{k} \in \mathscr{P}_\text{k}$ and then represent any field from $F \in \mathscr{P}_\text{k}$ in the form

$$F = F_\text{k} + F_\text{ko} = F_\text{k} + F_\mathscr{L} + F_{\mathscr{R}\text{o}}\,, \qquad F_\mathscr{L} \in \mathscr{L}, \qquad F_{\mathscr{R}\text{o}} \in \mathscr{R}_\text{o}\,. \tag{2.15}$$

Now, substituting (2.15) to (2.6) and seeing that $\mathsf{E}(F_{\mathscr{R}\text{o}}) = 0$ and $<F_\mathscr{L}, F_{\mathscr{R}\text{o}}> = 0$ will give

$$\mathsf{L}(F) = \mathsf{E}(\mathbf{u}_\mathscr{L}) - \Pi_\text{s}(\mathbf{u}_\mathscr{L}) + 2<F_\text{k}, F_\mathscr{L}> + C \tag{2.16}$$

where the constant C independent of the vector $\mathbf{u}_\mathscr{L}$ is

$$C = \mathsf{E}(\mathbf{u}_\text{k}) - \Pi_\text{s}(\mathbf{u}_\text{k}) - \Pi_\text{s}(\mathbf{u}_\mathscr{R})\,.$$

Here we take into account the fact that $<F_\text{k}, F_{\mathscr{R}\text{o}}> = 0$.

By varying the Lagrange functional in the form (2.16) on the Lagrangian energy space $\mathscr{L}$ and using the same reasoning as before, we can find its minimum "point" $F_{*\mathscr{L}}$ from $\mathscr{L}$. As a result, we have $\mathbf{u}_* = \mathbf{u}_k + \mathbf{u}_{*\mathscr{B}}$ up to an inessential homogeneously kinematically admissible rigid displacement $\mathbf{u}_{\mathscr{R}0}$ where $\mathbf{u}_{*\mathscr{L}}$ gives the functional (2.16) its least value on the Lagrangian energy space $\mathscr{L}$.

Pay attention to the fact that the inequality (2.7) is unstrict and allows the equality, too, which follows from the inequality $\mathsf{E}(\delta F) \geq 0$ at any field $\delta F \in \mathscr{R}_{k0}$. However, if we limit the variability of the displacement vector by requiring that the fields to be compared should belong to the energy space $\mathscr{L}$ and thus assuming $\delta F \in \mathscr{L}$, then $\mathsf{E}(\delta F) > 0$ and the inequality (2.7) will become a strict one,

$$\mathsf{L}(F_*) < \mathsf{L}(F)\,. \tag{2.17}$$

This is where we immediately obtain the theorem of uniqueness of the solution known in elasticity as the *Kirchhoff theorem.* Indeed, if there existed two points of minimum of the Lagrange functional, say, $F_{*1} \in \mathscr{L}$ and $F_{*2} \in \mathscr{L}$, then (2.17) would produce two mutually contradictory inequalities, $\mathsf{L}(F_{*1}) < \mathsf{L}(F_{*2})$ and $\mathsf{L}(F_{*2}) < \mathsf{L}(F_{*1})$.

So, the solution of the problem (**1**.2.2), (**1**.2.4) is unique up to homogeneously kinematically admissible rigid displacements.

2.3 Castigliano variational principle

2.3.1 Castigliano functional

While the Lagrange variational principle is related to the principle of virtual displacements, there is also a reciprocal variational principle for it, one based on the principle of virtual stress increments and called a *Castigliano principle*. The Castigliano functional (or *Castiglianian*, as one likes), $\mathsf{K} = \mathsf{K}(F)$, can be represented in the form

$$\mathsf{K}(\boldsymbol{\sigma},\mathbf{u}) = \tfrac{1}{2}\,(\boldsymbol{C}^{-1}\boldsymbol{\sigma}, \boldsymbol{\sigma}) + \tfrac{1}{2}\,(\boldsymbol{K}\mathbf{u}, \mathbf{u}) - (\boldsymbol{E}_u\boldsymbol{H}_\sigma\boldsymbol{\sigma}, \boldsymbol{E}_u\bar{\boldsymbol{u}})_\Gamma\;. \tag{3.1}$$

The linear part of the Castigliano functional, that depends only on the stresses $\boldsymbol{\sigma}$, can be interpreted mechanically as a *potential of the kinematic external actions*, $\Pi_k(\boldsymbol{\sigma})$, while its quadratic part can be treated as the strain energy of the system, $\mathsf{E}(\boldsymbol{\sigma},\mathbf{u})$, in which the energy of the elastic body itself is expressed via the stresses $\boldsymbol{\sigma}$,

$$\mathsf{E}(\boldsymbol{\sigma},\mathbf{u}) = \tfrac{1}{2}\,(\boldsymbol{C}^{-1}\boldsymbol{\sigma}, \boldsymbol{\sigma}) + \tfrac{1}{2}\,(\boldsymbol{K}\mathbf{u}, \mathbf{u})\,, \quad \Pi_{\mathrm{k}}(\boldsymbol{\sigma}) = (\boldsymbol{E}_u \boldsymbol{H}_\sigma \boldsymbol{\sigma}, \boldsymbol{E}_u\, \overline{\boldsymbol{u}}\,)_\Gamma \tag{3.2}$$

where

$$\mathsf{K}(\boldsymbol{\sigma},\mathbf{u}) = \mathsf{E}(\boldsymbol{\sigma},\mathbf{u}) - \Pi_{\mathrm{k}}(\boldsymbol{\sigma})\,. \tag{3.3}$$

Let $F_* = \{\boldsymbol{\sigma}_*, \boldsymbol{\varepsilon}_*, \mathbf{u}_*\}$ be a true stress-and-strain field. The following statement holds:

$$\mathsf{K}_* = \mathsf{K}(F_*) \le \mathsf{K}(F_{\mathrm{s}}) \tag{3.4}$$

for any field $F_{\mathrm{s}} = \{\boldsymbol{\sigma}_{\mathrm{s}}, \boldsymbol{C}^{-1}\boldsymbol{\sigma}_{\mathrm{s}}, \mathbf{u}_{\mathrm{s}}\} \in \mathscr{P}_{\mathrm{s}}$.

So, we have to compare values of the Castigliano functional on two fields, F_* and F_{s}.

Assume $\delta F = F_{\mathrm{s}} - F_* = \{\boldsymbol{\sigma}_{\mathrm{s}} - \boldsymbol{\sigma}_*, \boldsymbol{C}^{-1}(\boldsymbol{\sigma}_{\mathrm{s}} - \boldsymbol{\sigma}_*), \mathbf{u}_{\mathrm{s}} - \mathbf{u}_*\}$. Obviously, the field of variations, δF, is a homogeneously statically admissible field, and this permits us to write the components of the field as $\delta F = \alpha\{\boldsymbol{\sigma}_{\mathrm{so}}, \boldsymbol{C}^{-1}\boldsymbol{\sigma}_{\mathrm{so}}, \mathbf{u}_{\mathrm{so}}\}$ with an arbitrary factor α. Now we have

$$\begin{aligned}\mathsf{K}(F_{\mathrm{s}}) - \mathsf{K}(F_*) &= \mathsf{K}(F_* + \alpha F_{\mathrm{so}}) - \mathsf{K}(F_*) = \\ = \alpha\{(\boldsymbol{C}^{-1}\boldsymbol{\sigma}_*, \boldsymbol{\sigma}_{\mathrm{so}}) &+ (\boldsymbol{K}\mathbf{u}_*, \mathbf{u}_{\mathrm{so}}) - (\boldsymbol{E}_u \boldsymbol{H}_\sigma \boldsymbol{\sigma}_{\mathrm{so}}, \boldsymbol{E}_u\, \overline{\boldsymbol{u}}\,)_\Gamma\} + \\ &+ \alpha^2\, \tfrac{1}{2}\,\{(\boldsymbol{C}^{-1}\boldsymbol{\sigma}_{\mathrm{so}}, \boldsymbol{\sigma}_{\mathrm{so}}) + (\boldsymbol{K}\mathbf{u}_{\mathrm{so}}, \mathbf{u}_{\mathrm{so}})\}.\end{aligned} \tag{3.5}$$

Due to the principle of virtual stress increments, the factor at α to the power of one becomes zero while the factor at α^2 is nonnegative.

Now let's prove an inverse statement: a field $F_1 = \{\boldsymbol{\sigma}_1, \boldsymbol{C}^{-1}\boldsymbol{\sigma}_1, \mathbf{u}_1\} \in \mathscr{P}_{\mathrm{s}}$ on which the Castigliano functional $\mathsf{K}(F_{\mathrm{s}})$ takes a minimum value among all fields $F \in \mathscr{P}_{\mathrm{s}}$,

$$\mathsf{K}(F_1) \le \mathsf{K}(F), \tag{3.6}$$

is a true field, that is, $F_1 = F_*$.

It suffices to prove that the field $F_1 \in \mathscr{P}_{\mathrm{s}}$ which gives the minimum to the Castigliano functional is a kinematically admissible field because any field from the intersection $\mathscr{P}_{\mathrm{s}} \cap \mathscr{P}_{\mathrm{k}}$ is a true field by definition. With this goal in mind, we introduce a field of variations, δF, as a difference between fields $\delta F = F - F_1$ and evaluate the difference between the values of the Castigliano functional:

$$\mathsf{K}(F) - \mathsf{K}(F_1) = (\boldsymbol{C}^{-1}\boldsymbol{\sigma}_1, \delta\boldsymbol{\sigma}) + (\boldsymbol{K}\mathbf{u}_1, \delta\mathbf{u}) - (\boldsymbol{E}_u \boldsymbol{H}_\sigma \delta\boldsymbol{\sigma}, \boldsymbol{E}_u\, \overline{\boldsymbol{u}}\,)_\Gamma + \mathsf{E}(\delta F). \tag{3.7}$$

The field of variations δF is a homogeneously statically admissible field, therefore this field can be represented as $\delta F = \alpha F_{\text{so}}$ with a certain field $F_{\text{so}} \in \mathscr{P}_{\text{so}}$ and an arbitrary numerical parameter α. The result is that (3.7) will become

$$\mathsf{K}(F) - \mathsf{K}(F_1) =$$

$$= \alpha\{(\boldsymbol{C}^{-1}\boldsymbol{\sigma}_1, \boldsymbol{\sigma}_{\text{so}}) + (\boldsymbol{K}\mathbf{u}_1, \mathbf{u}_{\text{so}}) - (\boldsymbol{E}_u\boldsymbol{H}_\sigma\boldsymbol{\sigma}_{\text{so}}, \boldsymbol{E}_u\overline{\boldsymbol{u}})_\Gamma\} + \alpha^2\mathsf{E}(F_{\text{so}}) . \qquad (3.8)$$

The condition of nonnegativity of the right part in (3.8) implies that at any α

$$(\boldsymbol{C}^{-1}\boldsymbol{\sigma}_1, \boldsymbol{\sigma}_{\text{so}}) + (\boldsymbol{K}\mathbf{u}_1, \mathbf{u}_{\text{so}}) - (\boldsymbol{E}_u\boldsymbol{H}_\sigma\boldsymbol{\sigma}_{\text{so}}, \boldsymbol{E}_u\overline{\boldsymbol{u}})_\Gamma = 0 . \qquad (3.9)$$

But the equality (3.9) is nothing but a mathematical formulation of the principle of virtual stress increments, wherefrom we have that the field F_1 is kinematically admissible.

So we finally compe up with the following formulation of the Castigliano variational principle:

A stress-and-strain field in a linear deformable mechanical system is a true field if and only if this field makes the Castigliano functional, K, *of the system take the least value among all stress-and-strain fields, both physically and statically admissible.*

Note that the principle of virtual stress increments is invariant with respect to the physical equations while the Castigliano variational principle is essentially based on the physical admissibility of the stress-and-strain fields being compared.

2.3.2 Castigliano energy space

We indicated above that the energy $\mathsf{E}(\boldsymbol{\sigma},\mathbf{u}) \geq 0$ on the linear set $\mathscr{P}_{\text{so}}$. Now let's ask ourselves a question what conditions are needed for the energy $\mathsf{E}(\boldsymbol{\sigma},\mathbf{u})$ to be strictly positive on any field $F \in \mathscr{P}_{\text{so}}$ not identical to zero. The expression of the energy in (3.2) shows directly that if the stresses $\boldsymbol{\sigma}$ are nonzero then the respective energy $\mathsf{E}(\boldsymbol{\sigma},\mathbf{u})$ is strictly positive, therefore the equality $\mathsf{E}(\boldsymbol{\sigma},\mathbf{u}) = 0$ can hold only on fields of the type $F = \{\mathbf{0}, \mathbf{0}, \mathbf{u}\}$ among all fields from $\mathscr{P}_{\text{so}}$. However, as this field must be homogeneously statically admissible, $\boldsymbol{A}^{\mathsf{T}}\mathbf{0} + \boldsymbol{K}\mathbf{u} = \mathbf{0} \in \Omega$, we have immediately that $\boldsymbol{K}\mathbf{u} = \mathbf{0}$, and $\mathsf{E}(\boldsymbol{\sigma},\mathbf{u}) = 0$ on all fields of this kind. The set of all fields of the type $F = \{\mathbf{0}, \mathbf{0}, \mathbf{u}\}$ under the restriction $\boldsymbol{K}\mathbf{u} = \mathbf{0}$ was introduced earlier as a linear set denoted as $\mathscr{R}_K$. It is also obvious that $\mathscr{R}_K \subset \mathscr{P}_{\text{so}}$.

Now we can introduce the linear set $\mathscr{K}$ as a truncation of the set $\mathscr{P}_{so}$ at the expense of all fields from $\mathscr{R}_K$ and represent $\mathscr{P}_{so}$ as a direct sum

$$\mathscr{P}_{so} = \mathscr{K} \oplus \mathscr{R}_K \tag{3.10}$$

where the orthogonality is understood in the sense of the main metric. In other words, for any field $F_{so} \in \mathscr{P}_{so}$ with the displacement component $\mathbf{u}_{so}$ the following representation holds:

$$F_{so} = F_{\mathscr{K}} + F_{\mathscr{R}K},\ F_{\mathscr{K}} \in \mathscr{K},\ F_{\mathscr{R}K} \in \mathscr{R}_K, \quad \text{и при этом} \quad (\mathbf{u}_{\mathscr{K}}, \mathbf{u}_{\mathscr{R}K}) = 0\,. \tag{3.11}$$

As a result, the bilinear functional < , >

$$<F_1, F_2> = ½\,(\boldsymbol{C}^{-1}\boldsymbol{\sigma}_1, \boldsymbol{\sigma}_2) + ½\,(\boldsymbol{K}\mathbf{u}_1, \mathbf{u}_2)\,, \tag{3.12}$$

which is defined on the whole set $\mathscr{P}_{so}$, can be treated as an *energy scalar product* (*in the Castigliano sense*) on the set $\mathscr{K}$, and the linear set itself will become a Hilbert space which we will call a *Castigliano energy space*.

Thus, the Castigliano energy space $\mathscr{K} \subseteq \mathscr{P}_{so}$ consists of a set of fields, $F = \{\boldsymbol{\sigma}, \boldsymbol{C}^{-1}\boldsymbol{\sigma}, \mathbf{u}\}$, which meet the following requirements:

- the displacements $\mathbf{u}$ and stresses $\boldsymbol{\sigma}$ satisfy the homogeneous equilibrium equations in the area Ω and homogeneous static boundary conditions;
- the displacements $\mathbf{u}$ are orthogonal, in the main metric, to all rigid displacements for the elastic medium;
- the scalar product $<F, F> = ½(\boldsymbol{C}^{-1}\boldsymbol{\sigma}, \boldsymbol{\sigma}) + ½(\boldsymbol{K}\mathbf{u}, \mathbf{u})$ takes a finite value for any field $F \in \mathscr{K}$.

Notice a particular but important case when there is no elastic medium whatsoever, that is, $\boldsymbol{K} = \boldsymbol{O}$. In this situation the set $\mathscr{R}_K$ consists of all possible fields of the type $\{\mathbf{0}, \mathbf{0}, \mathbf{u}\}$. But then the Castigliano energy space $\mathscr{K}$ is a set of fields of the type

$$\{\boldsymbol{\sigma}, \boldsymbol{C}^{-1}\boldsymbol{\sigma}, \mathbf{0}\},$$

with quadratically summable[6] stresses $\boldsymbol{\sigma}$ which satisfy the conditions $\boldsymbol{A}^{\mathsf{T}}\boldsymbol{\sigma} = \mathbf{0} \in \Omega$, $\boldsymbol{E}_p\boldsymbol{H}_\sigma\boldsymbol{\sigma} = \mathbf{0} \in \Gamma$.

Now, in the same way as we did with the Lagrange functional, we can formulate the Castigliano variational principle in another equivalent manner, by transforming the functional to the conditions of its variation on the Castigliano energy space, $\mathscr{K}$. To achieve this goal, we represent an arbitrary field F from $\mathscr{P}_s$ as

[6] The requirement of quadratic summability of the stresses is equivalent to the requirement of finiteness of the scalar product, $(\boldsymbol{C}^{-1}\boldsymbol{\sigma}, \boldsymbol{\sigma})$.

$$F = F_s + F_{\mathcal{K}} + F_{\mathcal{R}K}\,, \tag{3.13}$$

where F_s is a certain fixed field from $\mathcal{P}_s$, $F_{\mathcal{K}} \in \mathcal{K}$, $F_{\mathcal{R}K} \in \mathcal{R}_K$. Substituting (3.13) to (3.3) and taking into account the equalities $\mathsf{E}(F_{\mathcal{R}K}) = 0$ and $<F_{\mathcal{K}}, F_{\mathcal{R}K}> = 0$ will give

$$\mathsf{K}(F) = \mathsf{E}(F_{\mathcal{K}}) - \Pi_k(F_{\mathcal{K}}) + 2<F_s, F_{\mathcal{K}}> + C \tag{3.14}$$

where the C constant independent of the field F_K is defined as

$$C = \mathsf{E}(F_s) - \Pi_k(F_s) - \Pi_k(F_{\mathcal{R}K})\,,$$

where we use the relation $< F_s, F_{\mathcal{R}K} > = 0$.

In the Castigliano functional represented in the form (3.14), the comparison involves fields belonging to the energy space $\mathcal{K}$.

It is interesting to note that the Castigliano variational principle follows immediately from the Prager–Synge identity in the particular case when the problem has homogeneous kinematic boundary conditions. And indeed, in this case the field F_k in (1.49) can be assumed zero and then the Prager–Synge identity will become

$$\mathsf{E}(F_* - F_s) + \mathsf{E}(F_*) = \mathsf{E}(F_s)\,. \tag{3.15}$$

But at $\boldsymbol{E}_u \overline{\boldsymbol{u}} = \boldsymbol{0} \in \Gamma$ the Castigliano functional becomes as simple as $\mathsf{K}(F) = \mathsf{E}(F)$, and (3.15) entails the inequality $\mathsf{E}(F_*) \le \mathsf{E}(F_s)$ which is equivalent to the inequality $\mathsf{K}(F_*) \le \mathsf{K}(F_s)$. This is the method of validation that the book [7] adopts for the Castigliano principle.

We could follow a similar method to validate the Lagrange variational principle in an opposite particular case: when the given external actions upon the system are only the displacements on the fixed part of the boundary Γ and there are no external static (force) actions. In that case the Prager–Synge identity can make use of the zero field F_s which gives

$$\mathsf{E}(F_* - F_k) + \mathsf{E}(F_*) = \mathsf{E}(F_k)\,. \tag{3.16}$$

If there are no active external forces, the Lagrange functional becomes simpler, $\mathsf{L}(F) = \mathsf{E}(F)$, and it follows from (3.16) that $\mathsf{L}(F_*) \le \mathsf{L}(F_k)$.

Now, a final note. As it was demonstrated above, the condition of minimum of the Castgliano functional implies an integral identity which is actually the principle of virtual stress increments from the mechanical standpoint. In its turn, the latter in an integral form of the strain compatibility equations and the kinematical boundary conditions. Consequently, one may expect that the St.-Venant compatibility equations are actually Euler equations for the Castigliano functional from the standpoint of classic calculus of variations, while the kinematic boundary conditions are *natural boundary conditions* for the same functional[7]. The direct proof of this fact, which is not based on the principle of virtual stress increments as an integral equivalent of the compatiblity conditions, can be found in such works as a book by L.S. Leibenzon [5].

2.4 Sensitivity of the strain energy to modifications of a system

2.4.1 First theorem of the strain energy minimum

First of all, let's consider values of the Lagrange and Castigliano functionals on the solutions of a problem. To shorten the notation, we will use asterisks with exact values of all involved functionals, for example, $\mathsf{L}_* = \mathsf{L}(\mathbf{u}_*)$. Now, turning to the values of the potentials Π_s according to (2.5) and Π_k according to (3.2) at $\mathbf{u} = \mathbf{u}_*$ and $\boldsymbol{\sigma} = \boldsymbol{\sigma}_*$,

$$\Pi_{s*} = (\overline{\boldsymbol{X}}, \mathbf{u}_*) + (\boldsymbol{E}_p \overline{\boldsymbol{p}}, \boldsymbol{E}_p \boldsymbol{H}_u \mathbf{u}_*)_\Gamma , \qquad \Pi_{k*} = (\boldsymbol{E}_u \boldsymbol{H}_\sigma \boldsymbol{\sigma}_*, \boldsymbol{E}_u \overline{\boldsymbol{u}})_\Gamma ,$$

we note that the sum of the potentials $\Pi_{s*} + \Pi_{k*}$ can be understood as a virtual work of all external (active and reactive) forces on actual displacements of the system. But then we will have the following from the Clapeyron theorem (1.14):

$$2\mathsf{E}_* = \Pi_{s*} + \Pi_{k*} . \tag{4.1}$$

Using the formulas (2.6) and (3.3) makes it possible to come up with the following expressions for the values of the Lagrange and Castigliano functionals on the solution of a problem:

$$\mathsf{L}_* = \mathsf{E}_* - \Pi_{s*} = -\mathsf{E}_* + \Pi_{k*} , \qquad \mathsf{K}_* = \mathsf{E}_* - \Pi_{k*} = -\mathsf{E}_* + \Pi_{s*} , \tag{4.2}$$

hence an important relationships:

$$\boxed{\mathsf{L}_* + \mathsf{K}_* = 0} . \tag{4.3}$$

Thus, the values of the Lagrange and Castigliano functionals on the solution of a problem are always equal in their absolute values and opposite by their signs. This is exactly why approximate solutions obtained

[7] As it was said in the preface to this book, we assume the reader's familiarity with basics of calculus of variations, so expressions like "*Euler equations*" or "*natural boundary conditions*" need not be explained here.

with the Ritz method where either the Lagrange functional or the Castigliano functional is used, respectively, enables one to approach the exact solution of a problem from both sides.

Let's sum up the results obtained here into Table 2.1 which shows exact values of the functionals we use in the general case and in some particular cases important for further presentment.

Table 2.1

	External actions applied to the system		
	Force and kinematic	Force only $\Pi_k = 0$	Kinematic only $\Pi_s = 0$
L_*	$E_* - \Pi_{s*} = - E_* + \Pi_{k*}$	$- E_*$	E_*
K_*	$E_* - \Pi_{k*} = - E_* + \Pi_{s*}$	E_*	$- E_*$
E_*	$\frac{1}{2}(\Pi_{s*} + \Pi_{k*})$	$\frac{1}{2}\Pi_{s*}$	$\frac{1}{2}\Pi_{k*}$

A known book by Southwell on the theory of elasticity [18] proves the following statement, a formulation of the so-called "first theorem of energy minimum":

> *If there are displacements specified in a body that observes Hooke's law, then the deformation which corresponds to the equilibrium configuration has a lower elastic energy than the deformation which corresponds to any other configuration (one that does not satisfy the equilibrium equations) in which the displacements have the same given values* [8].

It is easy to see that the above statement or theorem is just a corollary of the Lagrange variational principle applied to the case of purely kinematic external actions because under such conditions the following holds:

$$E_* = L_* \le L(\mathbf{u}) = E(\mathbf{u}) \quad \forall\, \mathbf{u} \in \mathscr{P}_k.$$

[8] Here we present a literal formulation of first theorem of the energy minimum given by Southwell. Of course, this formulation assumes an implicit condition that the comparison involves not all possible states (configurations) of a system but only those which are kinematically admissible.

2.4.2 Remarks on the effect of additional constraints (kinematical and force)

Kinematical constraints

What is usually called a constraint in structural mechanics is a limitation imposed upon allowed displacements of the system. If those limitations are formulated mathematically as linear homogeneous equations for the displacements of the system's points, then we call them linear retaining constraints. Here we will confine ourselves to this type of constraints, calling them *kinematic(al) constraints*. The linear retaining kinematical constraints do not violate the superposition principle according to which the sum of results of various independent loads on a system is equal to the overal result of their combined application[9]. In an extended definition, the kinematical constraints are any limitations of the allowed displacements of the system.

Now let's consider an original mechanical system described mathematically by the equations (**1**.2.2) and the boundary conditions (**1**.2.4). Together with the system thus specified, we will consider also a perturbed mechanical system (under the same external load) which will differ from the original one only by the presence of additional kinematical constraints imposed upon the allowable displacements of the system. Practical sources of such constraints can be things like additional external supports or fixations; however, this is not the only source of the kinematical constraints. Internal constraints are possible, too, which are implemented, say, by making particular elements or parts of a system so stiff that those elements can be treated as perfectly rigid, i.e. nondeformable, in the design model. An example of imposing of an internal kinematical constraint on a system is a three-dimensional elasticity problem for an incompressible material; if the material is isotropic, this is formally equivalent to making the Poisson ratio equal to ½. Anyway, an additional kinematic constraint can be treated mathematically as a truncation of the linear set of homogeneously kinematically admissible displacements, $\mathscr{P}_{\text{ko}}$. Making this truncation is equivalent to requiring that the homogeneously kinematically admissible displacements should be orthogonal in the main metric to a certain vector of displacements, $\mathbf{v} \in \mathscr{P}_{\text{ko}}$, which depends on a particular constraint, that is,

---

[9] Here 'the result' means a stress-and-strain field and/or any linear functional of its elements. Obviously, no linear superposition can take place for nonlinear functional such as the strain energy of a system.

$$(\mathbf{v}, \mathbf{u}_{\text{ko}}) = 0\,, \qquad \forall\, \mathbf{u}_{\text{ko}} \in \mathscr{P}_{\text{ko}}\,. \tag{4.4}$$

This definition of a "constraint plane" (4.4) includes a vector, **v**, which we call a *constraint vector*.

As a result, the linear set $\tilde{\mathscr{P}}_{\text{ko}}$ of homogeneously kinematically admissible displacements for the modified system (the system with the constraint) becomes a reduction of the set $\mathscr{P}_{\text{ko}}$ for the original system, that is, $\tilde{\mathscr{P}}_{\text{ko}} \subseteq \mathscr{P}_{\text{ko}}$, and so the Lagrange energy space $\mathscr{L}$ will narrow to $\tilde{\mathscr{L}}$. If we turn to the Lagrange variational principle and take the form (2.16) of it into account, we will understand that

$$\mathsf{L}_* \le \tilde{\mathsf{L}}_* \tag{4.5}$$

where $\tilde{\mathsf{L}}_*$ is an exact value of the Lagrangian functional on the solution of the perturbed problem.

The inequality (4.5) follows obviously from the consideration that

$$\mathsf{L}_* = \min_{\{\mathbf{u} \in \mathscr{L}\}} \mathsf{L}(\mathbf{u}), \qquad \tilde{\mathsf{L}}_* = \min_{\{\mathbf{u} \in \tilde{\mathscr{L}} \subseteq \mathscr{L}\}} \mathsf{L}(\mathbf{u}),$$

and the functional's minimum cannot become lower on a narrower set comparing to the original wider set where this minimum takes place.

To find out the qualitative effect of an additional kinematic constraint on the behavior of the Castigliano functional, we will use the identity (4.3) obtained above which is true for any mechanical system and thus for the system perturbed by adding a new constraint,

$$\tilde{\mathsf{L}}_* + \tilde{\mathsf{K}}_* = 0\,. \tag{4.6}$$

Now we derive from (4.6) and (4.3) that $(\tilde{\mathsf{L}}_* - \mathsf{L}_*) + (\tilde{\mathsf{K}}_* - \mathsf{K}_*) = 0$ which produces the following if we take the inequality (4.5) into account:

$$\mathsf{K}_* \ge \tilde{\mathsf{K}}_*\,. \tag{4.7}$$

Thus, additional kinematic constraints affect the behavior of the Lagrange and Castigliano functionals in opposite ways.

Note that the equality in (4.5) and (4.7) can be achieved only if the true displacement vector, $\mathbf{u}_*$, for the original problem does not have the respective component belonging to the plane of the additional constraint, that is, under the condition that $\mathbf{u}_*$ and the constraint vector **v** are orthogonal:

$$(\mathbf{v}, \mathbf{u}_*) = 0\,. \tag{4.8}$$

To illustrate this with an example, we will present a solution of the *"Oxford problem"* formulated as follows[10].

Two series of two consecutive experiments are dedicated to investigation of an initially unstressed elastic body. In the first series:

- Experiment "*a*": given forces $P_1, P_2, \ldots, P_k$ are applied at points $a_1, a_2, \ldots, a_k$;
- Experiments "*b*": points $b_1, b_2, \ldots, b_n$ are subjected to given displacements $u_1, u_2, \ldots, u_n$.

In the second series, the body is subjected to some constraints which prevent its points from moving. We have to prove that introducing the constraints makes the elastic energy E accumulated in the boyd under the conditions of Experiment "*a*" either remains unchanged or lowers while the elastic energy accumulated in the boyd under the conditions of Experiment "*b*" either remains unchanged or rises.

It becomes obvious immediately why this statement is true if we consider the inequalities (4.5), (4.7) and Table 2.1 above for particular values of the Lagrange and Castigliano functionals where

in Experiment "*a*": $\mathsf{E}_* = \mathsf{K}_* \geq \tilde{\mathsf{K}}_* = \tilde{\mathsf{E}}_*$;

in Experiment "*b*": $\mathsf{E}_* = \mathsf{L}_* \leq \tilde{\mathsf{L}}_* = \tilde{\mathsf{E}}_*$.

Force constraints

It is useful to introduce a separate class of *force* (or *static*) constraints opposite to the kinematic constraints: the force constraints are limitations imposed on forces/stresses in elements of a system. Just as with the kinematic constraints, we will deal with linear force constraints only; the said limitations will be expressed as linear homogeneous equations[11]. It is

---

[10] Southwell [18], who provides us with the formulation below, refers to final exams in Oxford in 1939 where this problem comes from. The solution given there, in a statement with some additional limitations of constraint types, occupies nearly two pages, while using the variational principles in their modern understanding makes the solution fit in one phrase of four lines of text which immediately follows the statement of the problem.

[11] The case of linear inhomogeneous equations of force constraints does not introduce anything new; the only difference is that the nonzero right part of such an equation would be interpreted as a given external action.

easy to see that the force constraints do not violate the superposition principle either.

It is often necessary to allow for force constraints in the structural design practice. For example, one may have to analyze a structure with pre-stresses when the stresses in some of the elements of the structure must acquire certain predefined values. A peculiar example of the force constraints appears in the analysis of structures supported by a system of hydraulic jacks [10]. All jacks of the common hydraulic system should have the same pressure of liquid, and thus we have linear homogeneous equations between the reactions applied to the structure by the jacks. The force constraint technique is useful also to construct some technical theories based on certain static hypotheses. Characteristic examples are momentless or semi-momentless theories of shells. Seeing all that was said, it seems quite useful to separate the force constraints into a particular class where they will exist apart from kinematical constraints. It is even more useful to do so within the scope of issues we discuss here because, as we will see later, the responses of a deformable system to additional kinematic or additional static constraints are exactly opposite.

Now let's do some analysis how additional force constraints affect the behavior of our basic functionals. It is more convenient to start with the Castigliano functional rather than the Lagrange one. First of all, note that adding one more force constraint to the system can be treated as a truncation of the linear set of homogeneously statically admissble stress-andstrain fields $\mathscr{P}_{so}$ to its linear subset $\tilde{\mathscr{P}}_{so} \subset \mathscr{P}_{so}$ the elements of which satisfy the equation of the added force constraint. The Castigliano energy space $\mathscr{K}$ will be respectively truncated to its subspace $\tilde{\mathscr{K}}$. Repeating the earlier reasoning gives us estimates of the functionals for a problem perturbed by adding a force constraint:

$$\mathsf{K}_* \le \tilde{\mathsf{K}}_*, \quad \mathsf{L}_* \ge \tilde{\mathsf{L}}_* . \tag{4.9}$$

Comparing the inequality (4.9) with the inequalities (4.5) and (4.7) shows us that the effect of force constraints on changes which happen in the values of the basic functionals is exactly opposite to the effect of kinematical constraints.

Now let us collect all obtained results together and put them conveniently in Table 2.2 where some useful cases of external loads are selected. We remind that the values of the functionals for a configuration perturbed by a kinematic or force constraint are marked with tilde above "~", while the same functionals for the original configuration are not marked in any way.

Table 2.2

Type of perturbation	External actions applied to a system: Force and kinematic	Force only $\Pi_k = 0$	Kinematic only $\Pi_s = 0$
Kinematical constraints	$L_* \le \tilde{L}_*$	$E_* \ge \tilde{E}_*$	$E_* \le \tilde{E}_*$
	$K_* \ge \tilde{K}_*$		
Force constraints	$L_* \ge \tilde{L}_*$	$E_* \le \tilde{E}_*$	$E_* \ge \tilde{E}_*$
	$K_* \le \tilde{K}_*$		

We give a simple example below to illustrate the effect of force constraints.

Fig. 2.3-*a* shows a two-span continuous beam under a concentrated force P applied to the middle of the right span. The static analysis of this simple statically indeterminate system with redundance 1 gives the following support reactions:

$$R_1 = -\frac{3}{32}P, \quad R_2 = \frac{22}{32}P, \quad R_3 = \frac{13}{32}P.$$

The reactions are positive if they look upward. The bending moment $M(x)$ in the beam is shown as a distribution diagram in Fig. 2.3-*b*. As we know from any course of structural mechanics of bar systems [4], [12], the energy of strain accumulated in the beam that experiences bending is given by the integral

$$E_* = \int_0^{2l} \frac{M^2}{2EI} dx$$

where EI is the bending stiffness of the beam.

Simple calculations yield (here we assume the stiffness EI to remain constant over the beam's length):

$$E_* = \frac{23}{32 \times 48} \frac{P^2 l^3}{EI} \approx 0.015 \frac{P^2 l^3}{EI}.$$

Now let's modify the statement of the problem by adding a force constraint into the system. We will assume that the reactions of supports 2 and 3 should meet the condition of equality, that is,

$$R_2 - R_3 = 0.$$

The mechanical implementation of this force constraint can be imagined as putting points 2 and 3 of the beam on two ends of an equal-sided nondeformable swing or a sway-beam — see Fig. 2.3-*c*.

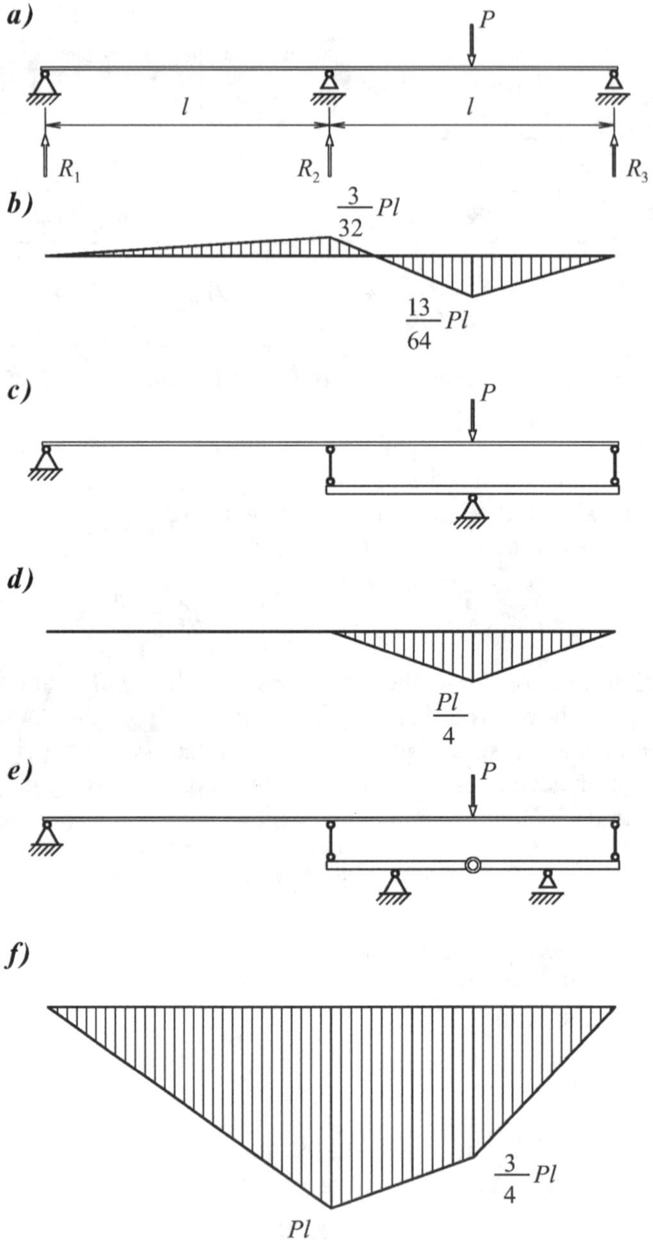

Fig. 2.3. An example of imposed force constraints

Introducing the force constraint makes the system statically determinate immediately because now the equilibrium equations are enough to determine the reactions. The calculation gives

$$R_1 = 0, \; R_2 = \frac{1}{2}P, \qquad R_3 = \frac{1}{2}P.$$

The moment distribution that conforms to the modified system is shown in Fig. 2.3-*d*. Now, calculating the strain energy of the modified system yields

$$\tilde{\mathsf{E}}_* = \frac{1}{48}\frac{P^2 l^3}{EI} \approx 0.021\frac{P^2 l^3}{EI},$$

wherefrom one can see that the inequality $\mathsf{E}_* \leq \tilde{\mathsf{E}}_*$ holds.

Another version of a force constraint can be implemented by making the reactions of the supports 2 and 3 exactly opposite, or

$$R_2 + R_3 = 0.$$

One of possible mechanical models that implement this force constraint is a system shown in Fig. 2.3-*e* where two sway-beams (equal-sided swings) connected with a hinge are used. The reactions of the supports are easy to calculate:

$$R_1 = P, \; R_2 = -\frac{3}{2}P, \qquad R_3 = \frac{3}{2}P.$$

The distribution of the bending moments in the beam modified with the above force constraint is shown in Fig. 2.3-*f*.

The energy of strain for this system is

$$\tilde{\mathsf{E}}_* = \frac{39}{48}\frac{P^2 l^3}{EI} \approx 0.813\frac{P^2 l^3}{EI}$$

which is obviously greater than E_*.

In the problem above, the system shown in Fig. 2.3-*a* is derived from the modified system as in Fig. 2.3-*c* by adding one kinematical constraint to the latter – by prohibiting the rotation of the sway-beam around its support. Similarly, the original system in Fig. 2.3-*a* is derived from the system in Fig. 2.3-*e* by restraining one of the sway-beams in the latter – the other sway-beam loses its freedom to rotate automatically.

Mechanically, when we impose a force constraint on a system, therewith we always remove one kinematical constraint. And vice versa, introducing one kinematical constraint into a system is equivalent to removing one force constraint.

It may seem this reciprocity of kinematical and force constraints makes the separate consideration of force constraints unnecessary because force constraints can always be transformed into kinematical ones. This is actually true, and force constraints are seldom introduced in courses of mechanics. However, sometimes a direct analysis of the effect of force constraints on the behavior of a mechanical system may be more convenient, that's why we separated these constraints into a new class.

2.4.3 Build-up of a system

In this section we will deal with a qualitative sensitivity of our basic functionals to changes in the area Ω itself in statements of elasticity problems. In short, we will compare the values of the Lagrange and Castigliano functionals for the original body's configuration and for one where the area Ω is built up to $\Omega + \delta\Omega$.

The build-up of the system can be studied directly; however, it is simpler to reduce the problem to the already established facts concerning the effect of additional constraints. We cut the area Ω by a certain surface into two sub-areas so that $\Omega = \Omega_1 + \Omega_2$ as shown in Fig. 2.4. The surface which divides the Ω area into two sub-areas will be denoted as Γ_{12}.

We begin our consideration with a system consisting of two disconnected sub-areas, Ω_1 and Ω_2. Under given external loads the Lagrange functional on this system will consist of a sum of the functionals $\mathsf{L}_{1*} + \mathsf{L}_{2*}$ calculated separately for each of the said sub-areas.

Fig. 2.4. An area divided into two sub-areas

It is easy to see that the Lagrange functional for the whole area Ω, that is, for the system shown in Fig. 2.4 on the left, is a sum of functionals, $\mathsf{L} = \mathsf{L}_1 + \mathsf{L}_2$, and the minimum of this sum should be searched for under the additional condition that the displacements of each of the separate systems must be equal on the separation surface, Γ_{12}. These conditions written as

$$\boldsymbol{u}_1 = \boldsymbol{u}_2 \quad \in \Gamma_{12} \tag{4.10}$$

are nothing but one of possible forms of kinematic constraints[12] .

As the sum $L_{1*} + L_{2*}$ (just as $K_{1*} + K_{2*}$) is the value of the respective functional on the system consisting of separate disconnected areas, Ω_1 and Ω_2, we come up with the following estimates immediately:

$$L_{1*} + L_{2*} \leq L_* \ , \quad K_{1*} + K_{2*} \geq K_* \ . \tag{4.11}$$

Assuming the area Ω_1 to be the original area Ω and the area Ω_2 to be the perturbation $\delta\Omega$, and supposing that $\delta\Omega$ is not subjected to any external loads, we can use $L_{2*} = 0$ and $K_{2*} = 0$ in (4.11). Changing the designations to those we used for the original and perturbed configurations of problems will give

$$L_* \leq \tilde{L}_* , \quad K_* \geq \tilde{K}_* \ . \tag{4.12}$$

Thus, building up a system cannot decrease the Lagrange functional or increase the Castigliano functional. Table 2.3 shows conveniently how building up a system affects the behavior of its basic functionals.

Table 2.3

	External actions applied to the system		
Type of perturbation	Force and kinematical	Force only $\Pi_k = 0$	Kinematical only $\Pi_s = 0$
Building up a system	$L_* \leq \tilde{L}_*$ $K_* \geq \tilde{K}_*$	$E_* \geq \tilde{E}_*$	$E_* \leq \tilde{E}_*$

Taking into account the functional dependence of the basic functionals on the area Ω, we can rewrite the same results in another form, as the following inequalities:

$$\frac{dL_*}{d\Omega} \geq 0, \qquad \frac{dK_*}{d\Omega} \leq 0 \tag{4.13}$$

which are sometimes more convenient.

To conclude this section, we note that the inequalities considered here are strict, as a rule. The equality will take place if and only if there is no stressed state caused by attaching the area $\delta\Omega$ to the original area. One

---

[12] As we talk about finding a point of minimum of the Lagrange functional, we do not have to require that the stresses match on the common boundary, Γ_{12}, of the areas. Stress matching conditions for the areas are implied by the Lagrange functional's stationarity conditions on a set of functions that satisfy the relations (4.10).

should keep in mind that first we “glue” the unstressed areas and only then apply external loads to the area Ω as a part of the merged area $\Omega + \delta\Omega$.

We leave the proof of this fact to the reader for an exercise.

2.4.4 Modification of stiffness properties of a system

Now let’s see how values of the functionals are affected by changes in the tensors $\boldsymbol{C}$ and $\boldsymbol{K}$ which participate in the governing equations (**1**.2.2). We consider an original elastic problem and a perturbed problem different from the original one in that the perturbations of these algebraic operators,

$$\delta\boldsymbol{C} = \tilde{\boldsymbol{C}} - \boldsymbol{C}, \qquad \delta\boldsymbol{K} = \tilde{\boldsymbol{K}} - \boldsymbol{K}, \tag{4.14}$$

are positive semi-definite in each point of the area Ω, that is, the following conditions hold:

$$\boldsymbol{a}^{\mathsf{T}}\delta\boldsymbol{C}\boldsymbol{a} \geq 0, \qquad \boldsymbol{b}^{\mathsf{T}}\delta\boldsymbol{K}\boldsymbol{b} \geq 0 \tag{4.15}$$

for any vectors $\boldsymbol{a}$ and $\boldsymbol{b}$ of the appropriate dimensionality.

Let’s write out explicit expressions of the Lagrange functionals for the original and perturbed problems. Based on (2.6), (2.5), and (1.13), we have

$$\mathsf{L}(\mathbf{u}) = \tfrac{1}{2}(\boldsymbol{C}\boldsymbol{A}\mathbf{u}, \boldsymbol{A}\mathbf{u}) + \tfrac{1}{2}(\boldsymbol{K}\mathbf{u}, \mathbf{u}) - (\overline{\boldsymbol{X}}, \mathbf{u}) + (\boldsymbol{E}_p\overline{\boldsymbol{p}}, \boldsymbol{E}_p\boldsymbol{H}_u\mathbf{u})_\Gamma ,$$

$$\tilde{\mathsf{L}}(\mathbf{u}) = \tfrac{1}{2}(\tilde{\boldsymbol{C}}\boldsymbol{A}\mathbf{u}, \boldsymbol{A}\mathbf{u}) + \tfrac{1}{2}(\tilde{\boldsymbol{K}}\mathbf{u}, \mathbf{u}) - (\overline{\boldsymbol{X}}, \mathbf{u}) + (\boldsymbol{E}_p\overline{\boldsymbol{p}}, \boldsymbol{E}_p\boldsymbol{H}_u\mathbf{u})_\Gamma$$

thus, taking into account (4.15), we have the estimate:

$$\tilde{\mathsf{L}}(\mathbf{u}) - \mathsf{L}(\mathbf{u}) = \tfrac{1}{2}(\delta\boldsymbol{C}\boldsymbol{A}\mathbf{u}, \boldsymbol{A}\mathbf{u}) + \tfrac{1}{2}(\delta\boldsymbol{K}\mathbf{u}, \mathbf{u}) \geq 0 . \tag{4.16}$$

Also, (4.16) gives the inequality

$$\tilde{\mathsf{L}}(\mathbf{u}) \geq \mathsf{L}(\mathbf{u}) \geq min\, \mathsf{L}(\mathbf{u}) = \mathsf{L}_* .$$

But the latter is true for any vector $\mathbf{u}$ which means

$$\mathsf{L}_* \leq \tilde{\mathsf{L}}_*, \qquad \mathsf{K}_* \geq \tilde{\mathsf{K}}_* . \tag{4.17}$$

Obviously, the second inequality in (4.17) follows from the first and from the identity (4.3).

Thus, when we make an elastic body and/or its surrounding elastic medium stiffer, the Lagrange functional cannot become bigger while the Castigliano functional cannot become smaller. If we consider force and kinematic actions upon the system separately, we can formulate this conclusion in a more obvious way from the physical standpoint:

> *When the stiffness properties of an elastic body and/or its surrounding elastic medium are increased, the strain energy of the system, E_*, cannot rise under a purely force load and cannot lower under a purely kinematic action.*

The conditions of modification of a mechanical system in the form of inequalities (4.15) are guaranteed when, for example, one increases the material's modulus of elasticity and/or stiffness coefficients of the elastic medium in the diagonal matrix $\boldsymbol{K}$. It is especially curious, however, to track the effect of changing such property as the Poisson ratio on the behavior of the system from the integral viewpoint (energy).

In a three-dimensional elasticity problem [2] written in the matrix/vector form, the stiffness matrix of an isotropic material, $\boldsymbol{C}$, and its inverse compliance matrix, $\boldsymbol{C}^{-1}$, can be represented as

$$\boldsymbol{C} = \left\| \begin{bmatrix} \lambda+2\mu & \lambda & \lambda & 0 & 0 & 0 \\ \lambda & \lambda+2\mu & \lambda & 0 & 0 & 0 \\ \lambda & \lambda & \lambda+2\mu & 0 & 0 & 0 \\ 0 & 0 & 0 & \mu & 0 & 0 \\ 0 & 0 & 0 & 0 & \mu & 0 \\ 0 & 0 & 0 & 0 & 0 & \mu \end{bmatrix} \right\|,$$

$$\boldsymbol{C}^{-1} = \frac{1}{E} \left\| \begin{bmatrix} 1 & -\nu & -\nu & 0 & 0 & 0 \\ -\nu & 1 & -\nu & 0 & 0 & 0 \\ -\nu & -\nu & 1 & 0 & 0 & 0 \\ 0 & 0 & 0 & 2(1+\nu) & 0 & 0 \\ 0 & 0 & 0 & 0 & 2(1+\nu) & 0 \\ 0 & 0 & 0 & 0 & 0 & 2(1+\nu) \end{bmatrix} \right\|,$$

where the elastic Lame constants λ and μ are related to the engineering characteristics, the elasticity modulus E and the Poisson ratio ν, as follows:

$$E = \mu\frac{3\lambda+2\mu}{\lambda+\mu}, \qquad \nu = \frac{\lambda}{2(\lambda+\mu)}, \tag{4.18}$$

or, inversely,

$$\lambda = \frac{E\nu}{(1-2\nu)(1+\nu)}, \qquad \mu = \frac{E}{2(1+\nu)}. \tag{4.19}$$

Linear algebra [19] says the condition (4.15) of positive semi-definiteness of the matrix $\delta\boldsymbol{C}$ is equivalent to the requirement that all eigenvalues of it be nonnegative. We use direct calculation to expand the characteristic determinant of the matrix $\boldsymbol{C}$ and find out that all its eigenvalues ρ_i ($i = 1,\ldots, 6$) are roots of the polynomial

$$(\mu - \rho)^3 (2\mu - \rho)^2 (3\lambda + 2\mu - \rho) = 0 . \qquad (4.20)$$

One can see that increasing the Lame constants, λ and μ, will also increase these eigenvalues, so this change in the elastic properties will make the matrix $\delta\boldsymbol{C}$ positive semi-definite. However, when we vary the Poisson ratio by increasing its value from ν to $\nu + d\nu$, the increments in the first five of six (multiplicity taken into account) eigenvalues ρ_i are negative while the increment in the sixth eigenvalue is of the opposite sign. And indeed, the signs of those increments are determined by the signs of the derivatives $d\rho_i/d\nu$, which gives

$$\frac{d\mu}{d\nu} = -\frac{E}{2(1+\nu)^2} < 0, \qquad \frac{d(3\lambda+2\mu)}{d\nu} = \frac{2E}{(1-2\nu)^2} > 0. \qquad (4.21)$$

Thus, when we vary the Poisson ratio, the signs of the changes in the Lagrange and Castigliano functionals are not defined in the general case.

2.4.5 Perturbation of external actions

In this section we consider the behavior of our basic functionals when external force and/or kinematic actions undergo certain changes.

First, let's take the simplest case when all external loads vary proportionally. Suppose we know an exact solution, the field $F = \{\boldsymbol{\sigma}, \boldsymbol{\varepsilon}, \mathbf{u}\}$, under a given external load $V = \{ \overline{\boldsymbol{X}}, \overline{\boldsymbol{p}}, \overline{\boldsymbol{u}} \}$. Let's consider a perturbed state of the system different from the original one in that all three components of the external load are multiplied by a common factor t and thus can be written as $tV = \{t\overline{\boldsymbol{X}}, t\overline{\boldsymbol{p}}, t\overline{\boldsymbol{u}}\}$. The problem is linear, therefore the solution $\tilde{F}$ for the perturbed state is described by the field $\tilde{F} = tF$. It is easy to calculate that the values of all functionals on the solution of the perturbed problem are related quadratically (with respect to t) to the values of the same functionals on the solution of the original problem:

$$\tilde{\mathsf{E}}_* = t^2\mathsf{E}_*, \quad \tilde{\Pi}_{s*} = t^2\Pi_{s*}, \quad \tilde{\Pi}_{k*} = t^2\Pi_{k*}, \quad \tilde{\mathsf{L}}_* = t^2\mathsf{L}_*, \quad \tilde{\mathsf{K}}_* = t^2\mathsf{K}_*. \qquad (4.22)$$

To see this, consider that (4.22) follows from the definitions of the respective functionals and the chains of transformations

$$\tilde{\mathsf{E}}_* = <\tilde{F}, \tilde{F}> = <tF, tF> = t^2 <F, F> = t^2 \mathsf{E}_*,$$

$$\tilde{\Pi}_{s*} = (t\bar{X}, \tilde{\mathbf{u}}) + (\boldsymbol{E}_p t \bar{\boldsymbol{p}}, \boldsymbol{E}_p \tilde{\mathbf{u}})_\Gamma = (t\bar{X}, t\mathbf{u}) + (\boldsymbol{E}_p t \bar{\boldsymbol{p}}, \boldsymbol{E}_p t\boldsymbol{u})_\Gamma = t^2 \Pi_{s*},$$

$$\tilde{\Pi}_{k*} = (\boldsymbol{E}_u \boldsymbol{H}_\sigma \tilde{\boldsymbol{\sigma}}, \boldsymbol{E}_u t \bar{\boldsymbol{u}})_\Gamma = (\boldsymbol{E}_u t\boldsymbol{p}, \boldsymbol{E}_u t \bar{\boldsymbol{u}})_\Gamma = t^2 \Pi_{k*}.$$

Now let's analyze the effect of summing two independent loads. Suppose we know two states of the system with their loads, their resulting fields, and their respective values of the functionals marked by indices 1 and 2.

When the combined load is applied to the system, the resulting field, $F = F_1 + F_2$, will be the sum of the fields of state 1 and state 2 because of the linearity. Let's calculate the strain energy, E_*, accumulated by the system under the combined load:

$$\mathsf{E}_* = <F_1 + F_2, F_1 + F_2> = \mathsf{E}_{1*} + \mathsf{E}_{2*} + 2<F_1, F_2>.$$

The expression $2<F_1, F_2>$ is a virtual work (with the minus sign),

$$\mathsf{B}_{12} = -\mathsf{B}_{21},$$

of all internal forces of one of the two states on the displacements of the other state. Based on the general principle of statics and geometry, this same expression $2<F_1, F_2>$ can be interpreted also as a virtual work, $\mathsf{A}_{12} = \mathsf{A}_{21}$, of all external (active and reactive) forces of one of the two states on the displacements of the other state. So finally we have

$$\mathsf{E}_* = <F_1 + F_2, F_1 + F_2> = \mathsf{E}_{1*} + \mathsf{E}_{2*} + \mathsf{A}_{12}. \tag{4.23}$$

Similarly, on the basis of the formulas (2.5) and (3.2) we can prove that the force and kinematic potentials, Π_{s*} and Π_{k*}, permit the following representation under the combined load on the system:

$$\Pi_{s*} = \Pi_{s1*} + \Pi_{s2*} + \mathsf{A}_{s,12} + \mathsf{A}_{s,21}, \quad \Pi_{k*} = \Pi_{k1*} + \Pi_{k2*} + \mathsf{A}_{k,12} + \mathsf{A}_{k,21} \tag{4.24}$$

where $\mathsf{A}_{s,ij} = (\bar{X}_i, \mathbf{u}_j) + (\boldsymbol{E}_p \bar{\boldsymbol{p}}_i, \boldsymbol{E}_p \boldsymbol{u}_j)_\Gamma$, $\mathsf{A}_{k,ij} = (\boldsymbol{E}_u \boldsymbol{p}_i, \boldsymbol{E}_u \bar{\boldsymbol{u}}_j)_\Gamma$ are the virtual works of the respective active and reactive forces of state i on the full displacements of state j.

There is no cause to assume that $\mathsf{A}_{s,ij} = \mathsf{A}_{s,ji}$ or $\mathsf{A}_{k,ij} = \mathsf{A}_{k,ji}$ because the work reciprocity theorem is true only for the sum of the respective quantities,

$$\mathsf{A}_{ij} = \mathsf{A}_{s,ij} + \mathsf{A}_{k,ij} = \mathsf{A}_{s,ji} + \mathsf{A}_{k,ji} = \mathsf{A}_{ji}. \tag{4.25}$$

2.4.6 Second theorem of the strain energy minimum

Consider two independent actions on the system, one of which being purely kinematic (no active external forces) and the other purely force (no given nonzero displacements). The state of the system under the kinematic action will be the state 0 and that under the force action will be the state 1.

Now let's imagine a procedure of loading when the system is first subjected to the purely kinematic external load and then the purely force action is added. The state 0 created by the kinematic action can be treated as an initial state of the system from which to count further states caused by further loading. Based on the formula (4.23), we will derive an expression of the total strain energy accumulated in the system after the second phase of loading,

$$\mathsf{E}_* = \mathsf{E}_{0*} + \mathsf{E}_{1*} + \mathsf{A}_{01} \,.$$

Due to the specific nature of the loads, $\mathsf{A}_{01} = 0$. To see this, consider that the state 0 has only the reactions as external forces while the state 1 has the respective displacements equal to zero. Thus the latter relation becomes

$$\mathsf{E}_* = \mathsf{E}_{0*} + \mathsf{E}_{1*} \,, \tag{4.26}$$

which gives the following because the strain energy is positive:

$$\mathsf{E}_* \geq \mathsf{E}_{1*} \,. \tag{4.27}$$

Note that the inequality (4.27) has ben obtained with the assumption that the initial state of the system, state 0, was caused by a purely kinematic action. However, it is not hard to understand how this limitation can be avoided so that (4.27) could be true with any initial state of the system as long as that state is self-balanced (i.e. such that has no active external forces). It is not important how a particular initial self-balanced state was created (by given dislocations or temperature) as long as the equality $\mathsf{A}_{01} = 0$ holds and implies (4.27).

In this interpretation the inequality (4.27) expresses a so-called "*second theorem of energy minimum*" which we give here as it was formulated by Southwell [18]:

> *The value of energy* E_* *(total elastic energy accumulated in a body subjected to external forces) in a body with initial stresses is always greater than one in the same body without the initial stresses. The energy* E_* *has its minimum at* $\mathsf{E}_{0*} = 0$, *that is, if there are no initial stresses.*

This formulation of the second theorem of energy minimum is taken nearly literally from Southwell[13].

2.4.7 St.-Venant principle and its energy-based background

The earlier conclusion which we made about the effect of the structure's build-up on its strain energy gives much in the sense that it may be the basis for energy treatment of the St.-Venant principle well known in theory of elasticity. We present these considerations here below; the basic idea is borrowed from a paper by G. Janelidze and Y. Panovko [2] who, in their turn, refer to original results by Soutwell [17] and Zanaboni [20]. The reasoning by Zanaboni is repeated also in a book by V. Novatsky [8].

There are several known (equivalent) formulations of the St.-Venant principle. The one most convenient for us follows.

Let a deformable structure be subjected to a set of external forces in a vicinity of some point of it, and let this system of forces be statically equivalent to zero (a self-balanced system of forces). Then, as we move away from the said vicinity, we will see the stress state in the structure decay.

The meaning of the St.-Venant principle is the very fact of the stresses and strains decaying when getting farther from the location of applied loads. The behavior of this decay, i.e. its rate, cannot be established in a statement so general. Moreover, the decay rate of the stresses and strains depends much on what particular class of problems we are dealing with, but the very fact of the decay which is postulated by the St.-Venant principle can be validated on the basis of energy considerations.

Let an elastic deformable body, which occupies an area Ω with a boundary Γ, be loaded by external forces $\boldsymbol{X}$ different from zero and self-balanced in a certain limited area, Ω_p. Let us surround the loaded area Ω_p by a boundary Γ_1 and by another boundary Γ_2 wholly belonging to the area Ω as shown in Fig. 2.5.

It is essential here that the loaded area Ω_p is wholly contained in the area Ω_1 with the external boundary Γ_1, which is contained in the area with the external boundary Γ_2 in its turn, and the latter area is again contained in the original area Ω.

[13] Southwell calls the quantity E_* a *full energy*. However, there may be a confusion of terms because in the modern practice the full energy is not the strain energy but the Lagrange functional as it was said in Section **2**.2. Therefore we refer to it as a total energy rather than a full energy.

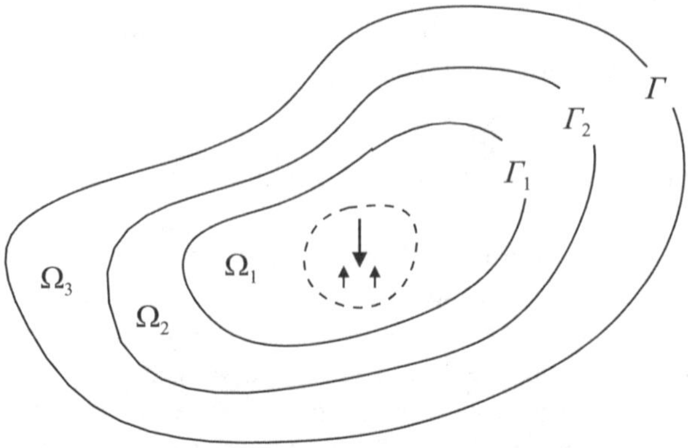

Fig. 2.5. To the validation of the St.-Venant principle

Let us use the designation Ω_2 for an area between the boundaries Γ_1 and Γ_2, and Ω_3 for an area between the boundaries Γ_2 and Γ. Next, we can denote interaction stresses on Γ_1 between the areas Ω_1 and Ω_2 as $\boldsymbol{p}_1$, and interaction stresses appearing on Γ_2 between the areas Ω_2 and Ω_3 as $\boldsymbol{p}_2$, and then the St.-Venant principle becomes just a statement that $\boldsymbol{p}_2$ is less than $\boldsymbol{p}_1$ in a certain sense.

The measure that we will use to compare these two stresses will not be based on their pointwise values; instead, we will use an integral characteristic of the stresses — an energy measure.

Now let's introduce the following designations:

- $\mathsf{E}(\boldsymbol{X})$ is a potential strain energy accumulated in the area Ω under given forces $\boldsymbol{X}$;
- $\mathsf{E}_1(\boldsymbol{X})$ is a potential strain energy accumulated in the area Ω_1, which is separated from the other part of the system, under the forces $\boldsymbol{X}$;
- $\mathsf{E}_i(\boldsymbol{p}_j)$ is a potential strain energy accumulated in the area Ω_i $(i = 1,2,3)$, which is separated from the other part of the system, under the forces $\boldsymbol{p}_j$ $(j = 1,2)$.

Let's show here that the total energy created by the surface forces $\boldsymbol{p}_2$, which are located farther from the loaded area, is less than the energy created by work of the surface forces $\boldsymbol{p}_1$ which are applied to the boundary Γ_1 nearer to the loaded area, i.e.

$$\mathsf{E}_2(\boldsymbol{p}_2) + \mathsf{E}_3(\boldsymbol{p}_2) < \mathsf{E}_1(\boldsymbol{p}_1) + \mathsf{E}_2(\boldsymbol{p}_1)\,. \tag{4.28}$$

This is what we will mean when saying 'stresses $\boldsymbol{p}_2$ are less than forces $\boldsymbol{p}_1$', and this is the same sense in which the stresses 'decay' farther away from the loaded area.

To prove the inequality (4.28), first we note that the strain energy $\mathsf{E}(\boldsymbol{X})$ accumulated in the original system can be represented as

$$\mathsf{E}(\boldsymbol{X}) = \{\mathsf{E}_1(\boldsymbol{X}) + \mathsf{E}_1(\boldsymbol{p}_1) + \mathsf{A}_1(\boldsymbol{p}_1,\boldsymbol{X})\} + \\ + \{\mathsf{E}_2(\boldsymbol{p}_1) + \mathsf{E}_2(\boldsymbol{p}_2) + \mathsf{A}_2(\boldsymbol{p}_1,\boldsymbol{p}_2)\} + \mathsf{E}_3(\boldsymbol{p}_2) \qquad (4.29)$$

where we add the following designations:

- $\mathsf{A}_1(\boldsymbol{p}_1, \boldsymbol{X})$ is a work done in the area Ω_1 by the forces $\boldsymbol{p}_1$ on displacements caused by the forces $\boldsymbol{X}$;
- $\mathsf{A}_2(\boldsymbol{p}_1, \boldsymbol{p}_2)$ is a work done in the area Ω_2 by the forces $\boldsymbol{p}_1$ on displacements caused by the forces $\boldsymbol{p}_2$.

The expression in the first brace in (4.29) is a total strain energy accumulated in Ω_1; the next brace contains the expression of an energy accumulated in Ω_2; the last term, $\mathsf{E}_3(\boldsymbol{p}_2)$, is an energy accumulated in Ω_3.

Now let's ask ourselves the question how the energy $\mathsf{E}(\boldsymbol{X})$ is going to change if we replace the actual boundary stresses $\boldsymbol{p}_1$ and $\boldsymbol{p}_2$ by stresses $\alpha\boldsymbol{p}_1$ and $\beta\boldsymbol{p}_2$ where α and β are numerical factors. It is easy to see that the total energy under this load will be

$$\mathsf{E}(\boldsymbol{X},\alpha,\beta) = \{\mathsf{E}_1(\boldsymbol{X}) + \alpha^2\mathsf{E}_1(\boldsymbol{p}_1) + \alpha\mathsf{A}_1(\boldsymbol{p}_1,\boldsymbol{X})\} + \\ +\{\alpha^2\mathsf{E}_2(\boldsymbol{p}_1) + \beta^2\mathsf{E}_2(\boldsymbol{p}_2) + \alpha\beta\mathsf{A}_2(\boldsymbol{p}_1, \boldsymbol{p}_2)\} + \beta^2\mathsf{E}_3(\boldsymbol{p}_2)\,,$$

so now we can treat $\mathsf{E}(\boldsymbol{X},\alpha,\beta)$ as a function of two arguments, α and β.

Note that the stresses $\boldsymbol{p}_1$ taken as a whole over the boundary Γ_1 are self-balanced, and the same is true about the stresses $\boldsymbol{p}_2$. Due to this fact we can use the Castigliano variational principle according to which the function $\mathsf{E}(\boldsymbol{X},\alpha,\beta)$ must have its minimum at $\alpha = 1$ and $\beta = 1$ because these values of the parameters conform to the true distribution of the stresses in the system.

Hence the following conditions must hold:

$$\left.\frac{\partial\mathsf{E}(\boldsymbol{X},\alpha,\beta)}{\partial\alpha}\right|_{\alpha=1,\,\beta=1} = 0, \qquad \left.\frac{\partial\mathsf{E}(\boldsymbol{X},\alpha,\beta)}{\partial\beta}\right|_{\alpha=1,\,\beta=1} = 0,$$

wherefrom two equations follow:

$$2\mathsf{E}_1(\boldsymbol{p}_1) + \mathsf{A}_1(\boldsymbol{p}_1, \boldsymbol{X}) + 2\mathsf{E}_2(\boldsymbol{p}_1) + \mathsf{A}_2(\boldsymbol{p}_1, \boldsymbol{p}_2) = 0,$$

$$2\mathsf{E}_2(\boldsymbol{p}_2) + \mathsf{A}_2(\boldsymbol{p}_1, \boldsymbol{p}_2) + 2\mathsf{E}_3(\boldsymbol{p}_2) = 0\,.$$

Resolving these equations with respect to the works $\mathsf{A}_1(\boldsymbol{p}_1, \boldsymbol{X})$ and $\mathsf{A}_2(\boldsymbol{p}_1, \boldsymbol{p}_2)$ yields

$$\mathsf{A}_2(\boldsymbol{p}_1, \boldsymbol{p}_2) = -2\mathsf{E}_2(\boldsymbol{p}_2) - 2\mathsf{E}_3(\boldsymbol{p}_2),$$

$$\mathsf{A}_1(\boldsymbol{p}_1, \boldsymbol{X}) = -2\mathsf{E}_1(\boldsymbol{p}_1) - 2\mathsf{E}_2(\boldsymbol{p}_1) + 2\mathsf{E}_2(\boldsymbol{p}_2) + 2\mathsf{E}_3(\boldsymbol{p}_2)\,.$$

Substituting these expressions to (4.29) gives the equality

$$\mathsf{E}(\boldsymbol{X}) - \mathsf{E}_1(\boldsymbol{X}) = -\,\mathsf{E}_1(\boldsymbol{p}_1) - \mathsf{E}_2(\boldsymbol{p}_1) + \mathsf{E}_2(\boldsymbol{p}_2) + \mathsf{E}_3(\boldsymbol{p}_2)\,. \tag{4.30}$$

But the energy $\mathsf{E}(\boldsymbol{X})$, comparing to $\mathsf{E}_1(\boldsymbol{X})$, is an energy in the system with the built-up area, so the data of Table 2.3 give $\mathsf{E}_1(\boldsymbol{X}) \geq \mathsf{E}(\boldsymbol{X})$. Moreover, in the case of our interest we can assume the strict inequality $\mathsf{E}_1(\boldsymbol{X}) > \mathsf{E}(\boldsymbol{X})$ to hold, because the strict equality conforms to the total absence of stresses beyond the area Ω_1 and the St.-Venant principle holds for sure. The relation (4.30) and the said inequality imply the inequality (4.28), which was to be proved.

Note that the paper [2] to which we referred above proves a bit different inequality,

$$\mathsf{E}_{1+2}(\boldsymbol{p}_2) + \mathsf{E}_3(\boldsymbol{p}_2) < \mathsf{E}_1(\boldsymbol{p}_1) + \mathsf{E}_{2+3}(\boldsymbol{p}_1)\,, \tag{4.31}$$

where $\mathsf{E}_{1+2}(\boldsymbol{p}_2)$ is an energy accumulated in the overall area $\Omega_1+\Omega_2$ under the action of $\boldsymbol{p}_2$, while $\mathsf{E}_{2+3}(\boldsymbol{p}_1)$ is an energy accumulated in the overall area $\Omega_2+\Omega_3$ under the action of $\boldsymbol{p}_1$.

Obviously, the difference between the inequalities (4.28) and (4.31) is not important from the standpoint of the St.-Venant principle's validation; the difference is just in a special construction and interpretation of the energy measure for the stresses $\boldsymbol{p}_1$ and $\boldsymbol{p}_2$, to be able to compare them adequately.

To conclude this section, we would like once again to draw the reader's attention to the paper [2] where the issue of applicability of the St.-Venant principle is discussed in great detail; the discussion includes an analysis of the stress attenuation rate for various classes of problems. We recommend that everybody deeply interested in the topic take a closer look at the paper [2] which, though of a considerable age, is still not at all obsolete.

2.5 Generalized forces and generalized displacements

2.5.1 Force actions. Castigliano theorem

Now let's consider a situation when a deformable system is subjected to m independent actions of purely force nature, $V_i = \{P_i\overline{\boldsymbol{X}}_i, P_i\overline{\boldsymbol{p}}_i,\ \boldsymbol{0}\}$ ($i = 1,\ldots, m$), each one being a set of loads distributed over the body's volume and over its surface. The class of force actions taken into consideration will be limited so far by one requirement only: we will assume that each of m actions V_i satisfies the conditions of existence of an equilibrium state; these conditions are that the virtual work of external forces $\overline{\boldsymbol{X}}_i$ and $\overline{\boldsymbol{p}}_i$ on any homogeneously kinematically admissible rigid displacements of the system must be zero — see Section **2**.1.6. Any set of actions of the type $V_i = \{P_i\overline{\boldsymbol{X}}_i, P_i\overline{\boldsymbol{p}}_i, \boldsymbol{0}\} = P_i\{\overline{\boldsymbol{X}}_i, \overline{\boldsymbol{p}}_i, \boldsymbol{0}\}$ from the admissible class can be treated as an external generalized force and the scalar factor P_i as a value of this generalized force. Then the increment of the generalized displacement, dq_i, that conforms to this generalized force is the coefficient in the expression of the work of this force, $P_i\, dq_i$.

Consider m *unit* states of the system, its i-th unit state with the stress-and-strain field F_i being the true state of the system that conforms to the action of only one, i-th, generalized unit force $\{\overline{\boldsymbol{X}}_i, \overline{\boldsymbol{p}}_i, \boldsymbol{0}\}$, that is, at $P_i = 1$. Obviously, all unit fields F_i belong to the class of homogeneously kinematically admissible fields. Moreover, we will assume that all of them belong to the Lagrangian energy space, $\mathscr{L}$, thus the energy scalar product $<F_i, F_j>$ makes sense.

Under these conditions the resulting expression of the strain energy of the system, E_*, under the combined action of all m external forces can be conveniently represented in the matrix form,

$$\mathsf{E}_* = <P_1F_1+\ldots+P_mF_m, P_1F_1+\ldots+P_mF_m> =$$

$$= \frac{1}{2}\left\|[P_1,\ldots,P_m]\right\| \left\|\begin{bmatrix} \mathsf{A}_{11} & \cdots & \mathsf{A}_{1m} \\ \vdots & \ddots & \vdots \\ \mathsf{A}_{m1} & \cdots & \mathsf{A}_{mm} \end{bmatrix}\right\| \left\|\begin{bmatrix} P_1 \\ \vdots \\ P_m \end{bmatrix}\right\| = \frac{1}{2}\boldsymbol{P}^{\mathsf{T}}\mathbf{A}\boldsymbol{P}, \tag{5.1}$$

where $\mathsf{A}_{ij} = 2<F_i, F_j>$ is the virtual work of i-th generalized unit force on displacements of the system caused by the action of j-th generalized unit force. Thus, the strain energy of the system is a homogeneous quadratic form that depends on the generalized forces through the matrix $\mathbf{A}$; the matrix can be represented by its components, energy scalar products (in the Lagrangian energy space), $\mathbf{A} = \|[2<F_i, F_j>]\|$.

Let the vector $\boldsymbol{q} = \|[q_1, q_2,\ldots, q_m]\|^{\mathsf{T}}$ be a vector of generalized displacements which conform to the vector of the generalized forces, $\boldsymbol{P} = \|[P_1,P_2,\ldots, P_m]\|^{\mathsf{T}}$. This means the strain energy, E_*, can be represented as

$$\mathsf{E}_* = \tfrac{1}{2}\,\boldsymbol{P}^{\mathsf{T}}\boldsymbol{q}\,. \tag{5.2}$$

Comparing (5.1) and (5.2) gives

$$\boldsymbol{q} = \mathbf{A}\boldsymbol{P}\,, \tag{5.3}$$

so the matrix **A** can be treated as a *compliance matrix of the system* which conforms to the vector of generalized forces, ***P***. Further we will use the ***D*** letter for the compliance matrix, so under the action of a pure force

$$\mathbf{A} = \boldsymbol{D} = \left\| \begin{bmatrix} \delta_{11} & \cdots & \delta_{1m} \\ \vdots & \ddots & \vdots \\ \delta_{m1} & \cdots & \delta_{mm} \end{bmatrix} \right\| = \left\| \begin{bmatrix} 2 < F_1, F_1 > & \cdots & 2 < F_1, F_m > \\ \vdots & \ddots & \vdots \\ 2 < F_m, F_1 > & \cdots & 2 < F_m, F_m > \end{bmatrix} \right\| \tag{5.4}$$

and

$$\mathsf{E}_* = \tfrac{1}{2}\, \boldsymbol{P}^{\mathsf{T}} \boldsymbol{D} \boldsymbol{P}. \tag{5.5}$$

The relationships (5.3) and (5.1) produce a formula for determining the vector of generalized displacements, ***q***

$$\boldsymbol{q} = \frac{\partial \mathsf{E}_*}{\partial \boldsymbol{P}}, \tag{5.6}$$

or, expanded into components,

$$q_i = \frac{\partial \mathsf{E}_*}{\partial P_i}. \tag{5.7}$$

The formula (5.7) is the essense of a theorem known in structural mechanics as a *Castigliano theorem*; it reads as follows:

> *A generalized displacement, q_i, which conforms to a generalized force, P_i, is equal to a partial derivative of the potential energy of strain in the system, E_*, with respect to that force.*

Obviously, the formulation of the Castigliano theorem treats the strain energy of the system, E_*, as a function (more exactly, a quadratic form) of the set of generalized forces, $P_1, P_2, \ldots, P_m$, applied to the body.

As can be seen, any arbitrary element $\delta_{ij} = 2<F_i, F_j>$ of the compliance matrix ***D*** can be treated as a generalized displacement corresponding to i-th generalized force P_i but caused by j-th generalized unit force, $P_j = 1$. The symmetry of scalar products implies that the coefficients δ_{ij} are insensitive to the swapping of their indices, $\delta_{ij} = \delta_{ji}$, and this fact is known in structural mechanics as a *Maxwell theorem* of displacement reciprocity.

2.5.2 Kinematic actions. Lagrange theorem

Now let a deformable system be subject to n independent, purely kinematic actions $V_i = \{\mathbf{0}, \mathbf{0}, q_i \bar{\boldsymbol{u}}_i\}$ $(i = 1,\ldots,n)$. Any set of external actions of the type $\{\mathbf{0}, \mathbf{0}, q_i \bar{\boldsymbol{u}}_i\} = q_i\{\bar{\boldsymbol{u}}_i\}$ can be treated as a given generalized displacement and the scalar factor q_i as a value of that displacement. There are no external active forces in either state, only reactive forces. A generalized (reactive) force, P_i, which corresponds to a generalized displacement, q_i, is a coefficient in the expression of the increment of the work, $P_i dq_i$, on the increment of that displacement, dq_i.

Consider n *unit* states of the system, such that i-th unit state with the stress-and-strain field F_i corresponds to a kinematic action upon the system in the form of a given i-th generalized unit displacement $\{\mathbf{0}, \mathbf{0}, \bar{\boldsymbol{u}}_i\}$, that is, at $q_i = 1$ and the other generalized displacements equal to zero, $q_j = 0$ $(j \neq i)$. Obviously, all unit fields F_i belong to the class of homogeneously statically admissible fields. Moreover, we suppose that all of them belong to the Castigliano energy space $\mathscr{K}$, so the energy scalar product is defined, $<F_i, F_j>$.

The resulting expression of the strain energy in the system, E_*, under the combined action of all external kinematic loads can be represented conveniently in the matrix form

$$
\begin{aligned}
\mathsf{E}_* &= <(q_1F_1+\ldots+q_nF_n),\ (q_1F_1+\ldots+q_nF_n)> = \\
&= \frac{1}{2}\,\|[q_1,\ldots,q_n]\| \left\| \begin{bmatrix} \mathsf{A}_{11} & \cdots & \mathsf{A}_{1n} \\ \vdots & \ddots & \vdots \\ \mathsf{A}_{n1} & \cdots & \mathsf{A}_{nn} \end{bmatrix} \right\| \left\| \begin{bmatrix} q_1 \\ \vdots \\ q_n \end{bmatrix} \right\| = \frac{1}{2}\boldsymbol{q}^{\mathsf{T}}\mathbf{A}\boldsymbol{q}
\end{aligned}
\tag{5.8}
$$

where $\mathsf{A}_{ij} = 2<F_i, F_j>$ is the virtual work of reactive generalized forces of i-th unit state of the system on the displacements of j-th unit state. Thus, the strain energy of the system is a homogeneous quadratic form depending on the generalized displacements through the matrix $\mathbf{A}$, which can be represented by its components, energy scalar products (in the Castigliano energy space), as $\mathbf{A} = \|[2<F_i, F_j>]\|$.

Let the vector $\boldsymbol{P} = \|[P_1, P_2,\ldots, P_m]\|^{\mathsf{T}}$ be a vector of reactive generalized forces which conforms to the vector of given generalized displacements, $\boldsymbol{q} = \|[q_1, q_2,\ldots, q_m]\|^{\mathsf{T}}$. This means that the strain energy E_* can be represented as

$$
\mathsf{E}_* = \tfrac{1}{2}\boldsymbol{P}^{\mathsf{T}}\boldsymbol{q}\,. \tag{5.9}
$$

Comparing (5.8) and (5.9) gives

$$\boldsymbol{P} = \mathbf{A}\boldsymbol{q}, \tag{5.10}$$

so under a purely kinematic action the $\mathbf{A}$ matrix can be treated as a *stiffness matrix of the system* which corresponds to the vector of generalized displacements $\boldsymbol{q}$. Further we will denote the stiffness matrix of the system as $\boldsymbol{R}$, therefore under a purely kinematic external action

$$\mathbf{A} = \boldsymbol{R} = \left\| \begin{bmatrix} r_{11} & \cdots & r_{1n} \\ \vdots & \ddots & \vdots \\ r_{n1} & \cdots & r_{nn} \end{bmatrix} \right\| = \left\| \begin{bmatrix} 2 < F_1, F_1 > & \cdots & 2 < F_1, F_n > \\ \vdots & \ddots & \vdots \\ 2 < F_n, F_1 > & \cdots & 2 < F_n, F_n > \end{bmatrix} \right\|, \tag{5.11}$$

and

$$\mathsf{E}_* = \tfrac{1}{2}\, \boldsymbol{q}^{\mathsf{T}} \boldsymbol{R}\, \boldsymbol{q}\,. \tag{5.12}$$

The relationships (5.8) and (5.10) produce a formula for determining the vector of reactive generalized forces $\boldsymbol{P}$,

$$\boldsymbol{P} = \frac{\partial \mathsf{E}_*}{\partial \boldsymbol{q}}, \tag{5.13}$$

or in the component form,

$$P_i = \frac{\partial \mathsf{E}_*}{\partial q_i}. \tag{5.14}$$

The formula (5.14) reciprocal with (5.7) is known as a *Lagrange theorem*:

> *A generaliled force, P_i, which conforms to a generalized displacement, q_i, is equal to the partial derivative of the system's potential energy of strain, E_*, with respect to this displacement.*

Keep in mind that in the formulation of the Lagrange theorem the strain energy of the system, E_*, is treated as a function (a quadratic form, to be exact) of a set of given generalized displacements, $q_1, q_2, \ldots, q_n$, treated as external actions applied to the system.

Obviously, any arbitrary element $r_{ij} = 2\langle F_i, F_j \rangle$ of the stiffness matrix $\boldsymbol{R}$ can be treated as a reaction in a constraint, which corresponds to i-th generalized displacement q_i but is caused by j-th generalized unit displacement $q_j = 1$. The symmetry of scalar product implies the insensitivity of the coefficients r_{ij} to the swapping of their indices, $r_{ij} = r_{ji}$, which fact is known in structural mechanics as a *Rayleigh theorem* of reaction reciprocity.

2.5.3 Inversion of stiffness and compliance matrices

Now let's return to purely static (force) actions applied to a system and assume additionally that all m external force actions (loads) are linearly independent. This means none of the loads $\{P_i\bar{\boldsymbol{X}}_i, P_i\bar{\boldsymbol{p}}_i\}$ $(i = 1,\dots, m)$ can be represented as a linear combination of the other $m - 1$ loads. It is quite clear this condition is necessary for the respective unit states defined by the stress-and-strain fields F_i $(i = 1,\dots, m)$ to be linearly independent as elements of the Lagrange energy space. This condition might seem sufficient also for linear independence between the fields F_i $(i = 1,\dots, m)$. However, this is not the case[14]. Therefore we will postulate the linear independence of all unit states of the system as a separate condition; this can be easily validated in practical problems. But then the compliance matrix $\boldsymbol{D}$ of the quadratic form (5.5), written as

$$\boldsymbol{D} = \left\| \begin{bmatrix} \delta_{11} & \cdots & \delta_{1m} \\ \vdots & \ddots & \vdots \\ \delta_{m1} & \cdots & \delta_{mm} \end{bmatrix} \right\| = \left\| \begin{bmatrix} 2 < F_1, F_1 > & \cdots & 2 < F_1, F_m > \\ \vdots & \ddots & \vdots \\ 2 < F_m, F_1 > & \cdots & 2 < F_m, F_m > \end{bmatrix} \right\|, \tag{5.15}$$

will be a *Gram matrix* for linearly independent elements [19], hence a nondegenerated and even positive definite matrix.

As we have established just above, the compliance matrix $\boldsymbol{D}$ is not degenerate, therefore the linear relationship (5.3) can be inversed to produce

$$\boldsymbol{P} = \boldsymbol{D}^{-1}\boldsymbol{q}\,, \tag{5.16}$$

so we can derive the following from (5.2):

$$\mathsf{E}_* = \tfrac{1}{2}\,\boldsymbol{q}^{\mathsf{T}}\boldsymbol{D}^{-1}\boldsymbol{q}\,. \tag{5.17}$$

The matrix $\boldsymbol{D}^{-1}$ (if it exists!) is also referred to as a *stiffness matrix of the system* that corresponds to the vector of generalized displacements, $\boldsymbol{q}$. However, the matrix $\boldsymbol{D}^{-1}$ does not describe the stiffness of the original system; rather, it describes the stiffness of a system derived from the given one by imposing m kinematic constraints which prevent the generalized displacements q_i $(i = 1,\dots, m)$.

Now let's repeat this kind of reasoning for the case of purely kinematic actions. Let all n unit states of the system be linearly independent. The necessary (not sufficient!) condition of it is that all n kinematic actions must be linearly independent. If all n unit states of the system are

---

[14] We leave it to the reader — to build an appropriate contrary example.

independent, then the stiffness matrix $\boldsymbol{R}$ of the quadratic form (5.12), written as

$$\boldsymbol{R} = \begin{Vmatrix} r_{11} & \cdots & r_{1n} \\ \vdots & \ddots & \vdots \\ r_{n1} & \cdots & r_{nn} \end{Vmatrix} = \begin{Vmatrix} 2<F_1,F_1> & \cdots & 2<F_1,F_n> \\ \vdots & \ddots & \vdots \\ 2<F_n,F_1> & \cdots & 2<F_n,F_n> \end{Vmatrix}, \qquad (5.18)$$

is a Gram matrix for linearly independent elements and thus is a nondegenerated and, moreover, positive definite matrix.

If the stiffness matrix $\boldsymbol{R}$ is non-degenerated, then the linear relation (5.10) can be inversed to derive

$$\boldsymbol{q} = \boldsymbol{R}^{-1}\boldsymbol{P}, \qquad (5.19)$$

thus (5.12) produces

$$\mathsf{E}_* = \tfrac{1}{2}\,\boldsymbol{P}^{\mathsf{T}}\boldsymbol{R}^{-1}\boldsymbol{P}. \qquad (5.20)$$

The $\boldsymbol{R}^{-1}$ matrix (if it exists!) is also referred to as a *compliance matrix of the system* that corresponds to the vector of generalized reactive forces $\boldsymbol{P}$. However, the matrix $\boldsymbol{R}^{-1}$ does not describe, actually, the compliance of the original system; instead, it describes that of a system derived from the original one by removing n kinematic consstraints which prevented the generalized displacements q_i $(i = 1,\ldots, n)$ earlier.

2.5.4 Lemma of constraints

Consider two mechanical systems, S_A and S_B, ordered in the sense that the *"higher"* system S_B is made of the *"lower"* system S_A by adding a few additional constraints to S_A[15], say, in the amount of n. Let F_A and F_B be the respective true stress-and-strain fields for those systems, which conform to the same external load V. for this couple of fields (F_A and F_B), we build a stress-and-strain field $F_{BA} = F_B - F_A$ which will be called a *difference field.*

Let $\mathscr{P}_{Ako}$ and $\mathscr{P}_{Aso}$ be the respective sets of homogeneously kinematically admissible and homogeneously statically admissible fields for the system S_A, and $\mathscr{P}_{Bko}$ and $\mathscr{P}_{Bso}$ be the similar sets for the system S_B. Here we will be interested only whether the difference field belongs to the respective linear set.

The following statement, which we call a *lemma of constraints*, is true:

[15] When talking about the constraints in this subsection, we will refer to kinematic constraints only.

Under any arbitrary action V, the difference field F_{BA} is a homogeneously kinematically admissible field for the lower system S_{A} and a homogeneously statically admissible field for the higher system S_{B}, or

$$F_{\mathrm{BA}} \in \mathscr{P}_{\mathrm{Ako}} \quad \text{and} \quad F_{\mathrm{BA}} \in \mathscr{P}_{\mathrm{Bso}} .$$

The system S_{B} is made of the system S_{A} by imposing n constraints on the latter, so we can assume the linear set $\mathscr{P}_{\mathrm{Bko}}$ to consist of elements of the set $\mathscr{P}_{\mathrm{Ako}}$ which satisfy the conditions of orthogonality,

$$(\mathbf{v}_i, \mathbf{u}_{\mathrm{ko}}) = 0 \quad (i = 1,\dots, n) \qquad \forall\, \mathbf{u}_{\mathrm{ko}} \in \mathscr{P}_{\mathrm{Bko}} , \tag{5.21}$$

where $\mathbf{v}_i$ is, as usual, the vector of reactions of i-th constraint. Hence the direct relation of embedding for the linear sets:

$$\mathscr{P}_{\mathrm{Ako}} \supset \mathscr{P}_{\mathrm{Bko}} . \tag{5.22}$$

The set of all physically admissible fields $\mathscr{P}$ is the same for the systems S_{A} and S_{B}. Therefore (5.22) and the theorem of a general form of any physically admissible field make also the opposite direction of embedding for the homogeneously statically admissible linear sets true:

$$\mathscr{P}_{\mathrm{Aso}} \subset \mathscr{P}_{\mathrm{Bso}} . \tag{5.23}$$

Note also that under a pure force action the resulting field for any system will be homogeneously kinematically admissible, while under a purely kinematic action it will be homogeneously statically admissible.

Further we associate the designation $\boldsymbol{X} = \|[X_1,\dots, X_n]\|^{\mathrm{T}}$ with a vector of reactions of the imposed constraints that appear in the system S_{B} and the designation $\boldsymbol{Z} = \|[Z_1,\dots, Z_n]\|^{\mathrm{T}}$ with a vector of the respective generalized displacements in the directions of the same constraints which appear in the system S_{A}. Based on the *principle of independent force action*, we can write[16]

$$F_{\mathrm{A}}(V) + F_{\mathrm{A}}(\boldsymbol{X}) = F_{\mathrm{B}}(V), \qquad F_{\mathrm{B}}(V) + F_{\mathrm{B}}(\boldsymbol{Z}) = F_{\mathrm{A}}(V). \tag{5.24}$$

The argument in the parentheses is an external action which creates the final field. By the way, (5.24) can produce a useful identity,

$$F_{\mathrm{A}}(\boldsymbol{X}) + F_{\mathrm{B}}(\boldsymbol{Z}) = 0 . \tag{5.25}$$

But the field $F_{\mathrm{A}}(\boldsymbol{X})$ is a resulting field of the system S_{A} under a purely static load, hence $F_{\mathrm{A}}(\boldsymbol{X}) \in \mathscr{P}_{\mathrm{Ako}}$. Similarly, $F_{\mathrm{B}}(\boldsymbol{Z})$ is a resulting field for the

[16] The reader will notice, undoubtedly, that the first equation in (5.24) is essentially the equation of the work (force) method for the system S_{B}. Similarly, the second equation in (5.24) can be understood as the equation of the displacement method for the system S_{A} .

system S_B under a purely kinematic action, hence $F_B(\boldsymbol{Z}) \in \mathscr{P}_{Bso}$. That's why we derive this directly from (5.24):

$$F_{BA} = F_B(V) - F_A(V) = - F_A(\boldsymbol{X}) \in \mathscr{P}_{Ako} ,$$

$$F_{BA} = F_B(V) - F_A(V) = - F_B(\boldsymbol{Z}) \in \mathscr{P}_{Bso} ,$$

and the statement is proved.

2.5.5 Mohr formula and its reciprocal

In many cases it may be interesting to find the value of a generalized displacement q_{ij} which conforms to i-th generalized force P_i but is caused by j-th generalized force P_j. This displacement can be represented in the form

$$q_{ij} = \delta_{ij} P_j \tag{5.26}$$

where δ_{ij} is the same displacement but caused by a unit generalized force $P_j = 1$. It is obvious from the previous presentment that $\delta_{ij} = \mathsf{A}_{ij} = 2\langle F_i, F_j\rangle$. Applying the formula (1.45) will give finally

$$\delta_{ij} = (\boldsymbol{C}\boldsymbol{\varepsilon}_i, \boldsymbol{\varepsilon}_j) + (\boldsymbol{K}\mathbf{u}_i, \mathbf{u}_j) = (\boldsymbol{C}^{-1}\boldsymbol{\sigma}_i, \boldsymbol{\sigma}_j) + (\boldsymbol{K}\mathbf{u}_i, \mathbf{u}_j) . \tag{5.27}$$

Here $\boldsymbol{\sigma}_i$, $\boldsymbol{\varepsilon}_i$, $\mathbf{u}_i$ are elements of the field F_i created in the system by the unit force $P_i = 1$, and $\boldsymbol{\sigma}_j$, $\boldsymbol{\varepsilon}_j$, $\mathbf{u}_j$ are elements of the field F_j created by the unit force $P_j = 1$.

The formula (5.27) is called a *Mohr formula* and is widely applicable in structural mechanics. As one can see, in its general form the formula works both for structural mechanics of bar systems and for elasticity, plates and shells, other applied theories of deformable solid mechanics.

Recall that both fields F_i and F_j are homogeneously kinematically admissible because both in the state i and in the state j the external actions are of purely force nature. But then, because of the theorem of field orthogonality, we can subtract an arbitrary homogeneously statically admissible field F_{so} in the Mohr formula from one of the two fields (say, from the field F_i) without violating the equality in (5.27):

$$\delta_{ij} = (\boldsymbol{C}^{-1}\boldsymbol{\sigma}_{iko}, \boldsymbol{\sigma}_j - \boldsymbol{\sigma}_{so}) + (\boldsymbol{K}\mathbf{u}_{iko}, \mathbf{u}_j - \mathbf{u}_{so}) \tag{5.28}$$

where $\boldsymbol{\sigma}_{so}$ amd $\mathbf{u}_{so}$ are elements of a homogeneously statically admissible field. In (5.28) the elements of the field F_i are subscripted additionally by $_{ko}$ in order to emphasize that the formula is true under the condition that the field F_i is homogeneously kinematically admissible.

This remarkable *invariance* of the Mohr formula with respect to a shift by an arbitrary homogeneously statically admissible field can be proved in another way [16, 17]; however, referring to the field orthogonality theorem is, in our opinion, the most general and direct way to validate this invariance[17].

To turn to the mechanical treatment of the Mohr's formula invariance, first we notice that the invariance with respect to a shift by F_{so} makes it possible to construct the shifted field (in our case, F_j) not for the given system (let it be called S_B here) but for a system S_A which is derived from the given one by removing some of the kinematic constraints. To see this, note that the lemma of constraints implies that the field $(F_j - F_{so})$ can be a true one but built for the system S_A rather than the given system S_B. To understand this, we should treat the shift field F_{so} as a difference field F_{BA}. The only limitation of the removed constraints is that a solution must exist for the system S_A under the action of the given generalized force P_j .

In particular, the analysis of bar structures is much facilitated by choosing this auxiliary system S_A appropriately from the class of statically determinate and kinematically stable systems derived from the given system by removing extra constraints.

This statement, very nice from mechanical viewpoint, helps sometimes obtain qualitative evaluations of the displacements without making bulky calculations. Let's give a simple example[18]. Fig. 2.6-*a* shows a planar frame and a load applied to the girder of the frame. Question: which side the girder is going to move to — right or left?

It is easy to depict a qualitative distribution of the bending moments M_g in the bars of the frame (Fig. 2.6-*b*).

---

[17] We cannot say for sure who was really the first to discover the invariance with respect to shift in the Mohr formula. The memoirs of a godfather of the structural mechanics in our country, I.M. Rabinovitch [13], refer to a paper by A.K. Vereschagin (1924) written by him when he was a student at Moscow Institute of Railway Engineers. This paper anticipates some statements of the modern structural mechanics formulated and understood thoroughly only decades later after the publication by Mr. Vereschagin. He suggested an idea (and a very advanced one for that time) that different principal systems can be used to analyze bar structures with the force method. It is easy to see that this is just another formulation of the Mohr formula's invariance with respect to shift.

[18] Professors of structural mechanics at engineering departments of universities know how fascinating the impression is that their students take from examples of such kind.

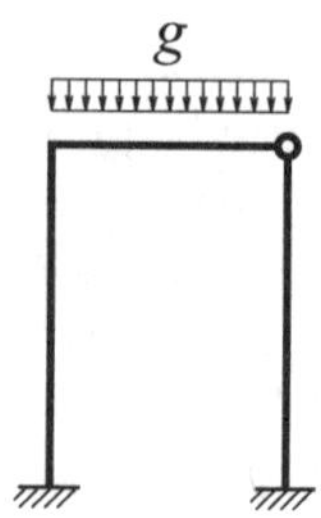

Fig. 2.6-*a*

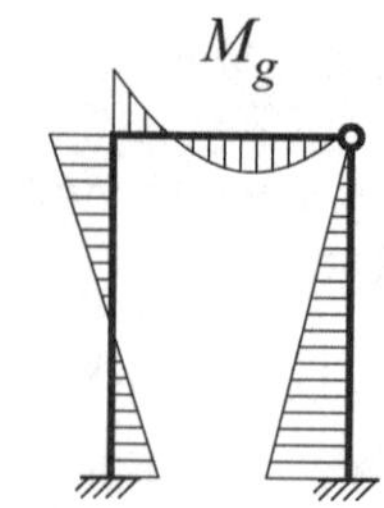

Fig. 2.6-*b*

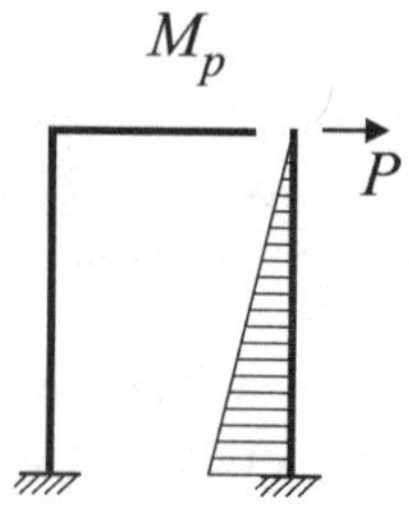

Fig. 2.6-*c*

The Mohr formula implies that the displacement we are interested with can be calculated as an integral of the product of moments M_g and M_P where M_P are the bending moments in the frame caused by a unit force, P, which is applied to the frame in the place and in the direction of our interest[19]. However, just as we have established above, the distribution of the moments M_P does not have to be taken from the original statically indeterminate frame; it can be taken from the frame with some of its constraints removed. In particular, it is convenient to build the M_P moment diagram for a statically determinate system, after cutting the frame apart in the location of the hinge as shown in Fig. 2.6-*c*. Now it is obvious that $(M_g\,,\, M_P) > 0$. This inequality shows that the girder of the frame moves in the same direction in which the P force is applied, that is, from left to right.

Now let $\mathbf{u}_1, \mathbf{u}_2, \ldots, \mathbf{u}_m$ be a set of given linearly independent fields of displacements of a system, and let $F_i = \{\boldsymbol{CA}\mathbf{u}_i, \boldsymbol{A}\mathbf{u}_i, \mathbf{u}_i\}$ $(i = 1,\ldots, m)$ be stress-and-strain fields created by these displacements. Each separate field of displacements, $\mathbf{u}_i$, can be treated as a unit kinematic action on the system, and the factor q_i at this field can be treated as a value of the respective generalized displacement. Let's assign the designation r_{ij} to a generalized force which corresponds to i-th generalized displacement but is caused by j-th generalized unit displacement q_j. Repeating the previous reasoning gives

$$r_{ij} = (\boldsymbol{C}\boldsymbol{\varepsilon}_i, \boldsymbol{\varepsilon}_j) + (\boldsymbol{K}\mathbf{u}_i, \mathbf{u}_j) = (\boldsymbol{C}^{-1}\boldsymbol{\sigma}_i, \boldsymbol{\sigma}_j) + (\boldsymbol{K}\mathbf{u}_i, \mathbf{u}_j)\,. \tag{5.29}$$

L.A. Rozin states that the formula (5.29) is reciprocal to the Mohr formula [16] because it enables us to restore "generalized loads" which cause a respective stress-and-strain field type.

[19] We assume that the effect of the longitudinal deformation in the bars can be ignored and that the bending stiffness, EI, is the same for all bars of the frame. Therefore, we can assume EI=1 for our qualitative evaluation.

These two formulas are different in their essense even though they have the same form. While in (5.27) the arguments $\boldsymbol{\sigma}_k$, $\boldsymbol{\varepsilon}_k$, $\mathbf{u}_k$ belong to stress-and-strain fields created by generalized unit forces, $P_k = 1$ $(k = i,j)$, in the formula (5.29) the same designations relate to fields created by generalized unit displacements, $q_k = 1$, at zero values of the other generalized displacements. Actually, the formula (5.29) refers to a system subjected to m external kinematic constraints rather than to the original system.

All external actions here are purely kinematic, so all fields F_i $(i = 1,\dots, m)$ are homogeneously statically admissible. Hence the invariance of the formula (5.29) with respect to a shift by any homogeneously kinematically admissible field F_{ko}. Thus, the formula (5.29) can be generalized into

$$r_{ij} = (\boldsymbol{C}^{-1}\boldsymbol{\sigma}_{iso}, \boldsymbol{\sigma}_j - \boldsymbol{\sigma}_{\text{ko}}) + (\boldsymbol{K}\mathbf{u}_{iso}, \mathbf{u}_j - \mathbf{u}_{\text{ko}}) \tag{5.30}$$

where $\boldsymbol{\sigma}_{\text{ko}}$, $\mathbf{u}_{\text{ko}}$ are elements of an arbitrary homogeneously kinematically admissible field.

2.6 Basic variational principles in problems with initial strains

Up to this point we assumed that when there are no external actions, the system is in a natural, unstressed and unstrained, state. This state was a reference point to calculate all components of the stress-and-strain field from — the stresses, the strains, and the displacements. However, this is not always convenient, and there are situations when this assumption is burdensome[20].

Let us assume that a system is subjected to both external actions *and internal kinematic actions*. The latter actions are defined as given initial strains $\boldsymbol{\varepsilon}_0$ which appear in practice because of inaccuracy of a structure's manufacturing or mounting, temperature deformation, shrinkage of an elastic body's material, creep and similar phenomena.

Here we believe that a certain hypothetic field of initial strains, $\boldsymbol{\varepsilon}_0$, is known, which would appear in elements of a structure if all internal constraints that provided the strain compatibility were removed. Therefore we would be more exact speaking about *free initial strains* rather than

---

[20] Some books on elasticity (for example, see [1]) treat this statement as a *hypothesis of a body's natural state* where no stress in the body is assumed to appear as long as there is no strain. We think, though, that calling it a 'postulate' rather than a 'hypothesis' would be more correct from the standpoint of terminology.

simple *initial strains*. For example, if we heat a three-dimensional elastic body from its natural state by an additional temperature T, the field of the free initial strains $\boldsymbol{\varepsilon}_0$ will be defined by the following well-known relationships

$$\boldsymbol{\varepsilon}_0 = \|[\varepsilon_{011}, \varepsilon_{022}, \varepsilon_{033}, \gamma_{012}, \gamma_{023}, \gamma_{031}]\|^{\mathsf{T}} = \|[\alpha \mathrm{T}, \alpha \mathrm{T}, \alpha \mathrm{T}, 0, 0, 0]\|^{\mathsf{T}} \qquad (6.1)$$

where α is a coefficient of heat expansion of the body's material.

Free initial strains, which do not satisfy the compatibility conditions by themselves, bring about elastic strains $\boldsymbol{\varepsilon}$ which ensure the continuity of the body together with $\boldsymbol{\varepsilon}_0$. If the displacements $\mathbf{u}$ are counted off from the original unstrained state of the system (its state before the initial strains have appeared), then the displacements will bring about the total strains $(\boldsymbol{\varepsilon}_0 + \boldsymbol{\varepsilon})$.

As a result, the mathematical formulation of the problem will change from the equations (**1**.2.2) into the equations (6.2),

$$\boldsymbol{A}^{\mathsf{T}}\boldsymbol{\sigma} + \boldsymbol{K}\mathbf{u} = \overline{\boldsymbol{X}} \qquad \text{equations of equilibrium,} \qquad (6.2\text{-}a)$$

$$\boldsymbol{A}\mathbf{u} = \boldsymbol{\varepsilon} + \boldsymbol{\varepsilon}_0 \qquad \text{geometric equations,} \qquad (6.2\text{-}b)$$

$$\boldsymbol{\sigma} = \boldsymbol{C}\boldsymbol{\varepsilon} \ \text{ or } \ \boldsymbol{\varepsilon} = \boldsymbol{C}^{-1}\boldsymbol{\sigma} \qquad \text{physical equations.} \qquad (6.2\text{-}c)$$

As for the boundary conditions (**1**.2.4), they do not depend on the presence or absence of the initial strain $\boldsymbol{\varepsilon}_0$ in the system, so they remain unchanged.

Now we will use the formulation of the problem in displacements. To do so, we exclude the stresses from the equilibrium equations, (6.2-*a*), using the equations (6.2-*c*) and then exclude the strains using (6.2-*b*). This will give

$$\boldsymbol{A}^{\mathsf{T}}\boldsymbol{C}\boldsymbol{A}\mathbf{u} + \boldsymbol{K}\mathbf{u} = \overline{\boldsymbol{X}} + \boldsymbol{A}^{\mathsf{T}}\boldsymbol{C}\boldsymbol{\varepsilon}_0 \,. \qquad (6.3)$$

By excluding the stresses from the boundary conditions in a similar way, we transform (**1**.2.4) into

$$\boldsymbol{E}_p(\boldsymbol{H}_\sigma \boldsymbol{C}\boldsymbol{A}\mathbf{u} - \overline{\boldsymbol{p}} - \boldsymbol{H}_\sigma \boldsymbol{C}\boldsymbol{\varepsilon}_0) = \mathbf{0} \qquad \text{static boundary conditions,} \qquad (6.4\text{-}a)$$

$$\boldsymbol{E}_u(\boldsymbol{H}_u\mathbf{u} - \overline{\boldsymbol{u}}) = \mathbf{0} \qquad \text{kinematic boundary conditions.} \qquad (6.4\text{-}b)$$

Equation (6.3) together with the boundary conditions, (6.4), will define a differential formulation of the problem in displacements if there are initial strains. Obviously, this problem is formally equivalent to finding a displacement field under modified external forces. The volumetric forces will be $(\overline{\boldsymbol{X}} + \boldsymbol{A}^{\mathsf{T}}\boldsymbol{C}\boldsymbol{\varepsilon}_0)$, and the surface forces on the boundary Γ will be

($\bar{p} + H_\sigma C\varepsilon_0$). Using this fact permits us to employ the Lagrange variational principle and modify the expression (2.6) of the Lagrangian functional into

$$\mathsf{L}(\mathbf{u}) = \mathsf{E}(\mathbf{u}) - \Pi_s(\mathbf{u}) - \Pi_{s0}(\mathbf{u}) \tag{6.5}$$

where

$$\Pi_{s0}(\mathbf{u}) = \int_\Omega A^\mathsf{T} C\,\varepsilon_0 \cdot \mathbf{u}\,d\Omega + \int_\Gamma E_p H_\sigma C\,\varepsilon_0 \cdot E_p H_u \mathbf{u}\,d\Gamma =$$

$$= (A^\mathsf{T} C\varepsilon_0, \mathbf{u}) + (E_p H_\sigma C\varepsilon_0, E_p H_u \mathbf{u})_\Gamma \;. \tag{6.6}$$

The quantity $\Pi_{s0}(\mathbf{u})$ can be naturally entitled a *force potential of initial strains*. Under the action of temperature, the free initial strains ε_0 are treated as initial temperature strains. In a particular case when an elastic three-dimensional body is heated by temperature T, this vector is defined by the expression (6.1).

Now let's see how the Castigliano variational principle will change in the case there are initial strains specified. To see this, we introduce a conditional stress, $\boldsymbol{\tau}$, equal to the desirable stress $\boldsymbol{\sigma}$ shifted by the vector $C\varepsilon_0$, and a conditional (total) strain, e, by assuming

$$\boldsymbol{\tau} = \boldsymbol{\sigma} + C\varepsilon_0, \qquad e = \varepsilon + \varepsilon_0. \tag{6.7}$$

The set of governing equations (6.2) in new variables (the components $\{\boldsymbol{\tau}, e, \mathbf{u}\}$ of the stress-and-strain field) will become

$A^\mathsf{T}\boldsymbol{\tau} + K\mathbf{u} = \bar{X} + A^\mathsf{T} C\varepsilon_0$ equations of equlibrium, (6.8-*a*)

$A\mathbf{u} = e$ geometric equations, (6.8-*b*)

$\boldsymbol{\tau} = Ce$ or $e = C^{-1}\boldsymbol{\tau}$ physical equations, (6.8-*c*)

while the boundary conditions (**1**.2.4) will turn into

$E_p(H_\sigma\boldsymbol{\tau} - \bar{p} - H_\sigma C\varepsilon_0) = \mathbf{0}$ static boundary conditions, (6.9-*a*)

$E_u(H_u\mathbf{u} - \bar{u}) = \mathbf{0}$ kinematic boundary conditions. (6.9-*b*)

The problem (6.8), (6.9) with respect to the stress-and-strain field $\{\boldsymbol{\tau}, e, \mathbf{u}\}$ is in no way different from the usual formulation (**1**.2.2), (**1**.2.4) if we assume the external volumetric forces to be equal to ($\bar{X} + A^\mathsf{T} C\varepsilon_0$) and the external surface forces defined on the boundary Γ to be equal to ($\bar{p} + H_\sigma C\varepsilon_0$).

Now we will derive the expression of the Castigliano functional for problems with initial strains from (3.3). We have

$$\mathsf{K}(\boldsymbol{\tau},\mathbf{u}) = \mathsf{E}(\boldsymbol{\tau},\mathbf{u}) - \Pi_k(\boldsymbol{\tau}) \tag{6.10}$$

where we use (3.2) and (6.7) and take into account the symmetry of the mutually inverse algebraic operators $\boldsymbol{D}$ and $\boldsymbol{C}$

$$\mathsf{E}(\boldsymbol{\tau},\mathbf{u}) = \tfrac{1}{2}(\boldsymbol{C}^{-1}\boldsymbol{\tau},\boldsymbol{\tau}) + \tfrac{1}{2}(\boldsymbol{K}\mathbf{u},\mathbf{u}) =$$

$$= \tfrac{1}{2}(\boldsymbol{C}^{-1}\boldsymbol{\sigma},\boldsymbol{\sigma}) + \tfrac{1}{2}(\boldsymbol{K}\mathbf{u},\mathbf{u}) + (\boldsymbol{\sigma},\, \boldsymbol{\varepsilon}_0) + \tfrac{1}{2}(\boldsymbol{C}\boldsymbol{\varepsilon}_0,\boldsymbol{\varepsilon}_0) = \mathsf{E}(\boldsymbol{\sigma},\mathbf{u}) + (\boldsymbol{\sigma},\boldsymbol{\varepsilon}_0) + C_1\,,$$

$$\Pi_k(\boldsymbol{\tau}) = (\boldsymbol{E}_u\boldsymbol{H}_\sigma\boldsymbol{\tau},\, \boldsymbol{E}_u\bar{\boldsymbol{u}})_\Gamma =$$

$$= (\boldsymbol{E}_u\boldsymbol{H}_\sigma\boldsymbol{\sigma},\, \boldsymbol{E}_u\bar{\boldsymbol{u}})_\Gamma + (\boldsymbol{E}_u\boldsymbol{H}_\sigma\boldsymbol{C}\boldsymbol{\varepsilon}_0,\, \boldsymbol{E}_u\bar{\boldsymbol{u}})_\Gamma = \Pi_k(\boldsymbol{\sigma}) + C_2\,, \tag{6.11}$$

the constants C_1 and C_2 (which do not depend on variables subject to variation) being equal to

$$C_1 = \tfrac{1}{2}\,(\boldsymbol{C}\boldsymbol{\varepsilon}_0,\, \boldsymbol{\varepsilon}_0), \qquad C_2 = (\boldsymbol{E}_u\boldsymbol{H}_\sigma\boldsymbol{C}\boldsymbol{\varepsilon}_0,\, \boldsymbol{E}_u\bar{\boldsymbol{u}})_\Gamma\,. \tag{6.12}$$

Now we return to the original components of the stress-and-strain fields $\{\boldsymbol{\sigma},\, \boldsymbol{\varepsilon},\, \mathbf{u}\}$ and rewrite the Castigliano functional (6.10) using (6.11), which gives

$$\boxed{\mathsf{K}(\boldsymbol{\sigma},\mathbf{u}) = \mathsf{E}(\boldsymbol{\sigma},\mathbf{u}) - \Pi_k(\boldsymbol{\sigma}) + \Pi_{k0}(\boldsymbol{\sigma})} \tag{6.13}$$

where

$$\Pi_{k0}(\boldsymbol{\sigma}) = \int_\Omega \boldsymbol{\sigma}\cdot\boldsymbol{\varepsilon}_0\, d\Omega = (\boldsymbol{\sigma},\boldsymbol{\varepsilon}_0)\,, \tag{6.14}$$

with the constants C_1 and C_2 omitted from the expression (6.13) because they are not essential when searching for points of stationarity of the functional $\mathsf{K}(\boldsymbol{\sigma},\mathbf{u})$.

The quantity $\Pi_{k0}(\boldsymbol{\sigma})$ can be conveniently entitled a *kinematic potential of initial strains*.

Notice that the set $\mathscr{R}_s$ of all physically and statically admissible stress-and-strain fields, which we search for a "point" of stationarity of the Castigliano functional, now consists of fields which meet the conditions

$$\boldsymbol{A}^{\mathsf{T}}\boldsymbol{\sigma} + \boldsymbol{K}\mathbf{u} = \bar{\boldsymbol{X}}\,, \quad \boldsymbol{\sigma} = \boldsymbol{C}(\boldsymbol{A}\mathbf{u} - \boldsymbol{\varepsilon}_0) \in \Omega\,, \quad \boldsymbol{E}_p(\boldsymbol{H}_\sigma\boldsymbol{\sigma} - \bar{\boldsymbol{p}}) = \mathbf{0} \in \Gamma\,. \tag{6.15}$$

Now let's discuss briefly how initial stresses affect the formulation of a problem. Let a body contain an initial stress field before any external forces are applied to it; this field will be denoted as $\boldsymbol{\sigma}_0$ and assumed to be self-balanced in the sense that it satisfies the relationships (1.30). If the kinematic parameters (the displacements $\mathbf{u}$ and strains $\boldsymbol{\varepsilon}$) are counted off from the initial stress state, and if we denote the stress increments as $\boldsymbol{\sigma}$ so that the full stress be equal to $(\boldsymbol{\sigma} + \boldsymbol{\sigma}_0)$, then the mathematical formulation of the problem with respect to the stress-and-strain fields of the type $\{\boldsymbol{\sigma}, \boldsymbol{\varepsilon}, \mathbf{u}\}$ will be reduced to the already known equations (**1**.2.2) with the boundary conditions (**1**.2.4). Further details are not discussed here.

2.7 Statically determinate and statically indeterminate systems

Earlier (see Section **2**.1.4 "Rigid displacements") we formulated a requirement that the set of solutions of a homogeneous equations with the geometry operator $\boldsymbol{A}$ should be finite-dimensional. However, it is important to notice that the requirement does not work with the conjugated operator $\boldsymbol{A}^{\mathsf{T}}$. The respective homogeneous equation for the equilibrium operator can have either finite or infinite number of linearly independent solutions.

First, let's consider the case when there is a finite number of linearly independent solutions of the homogeneous equation

$$\boldsymbol{A}^{\mathsf{T}}\boldsymbol{\sigma} = \mathbf{0}\,. \tag{7.1}$$

Let them be stresses $\boldsymbol{\sigma}_j$ $(j = 1,\ldots, m_A)$ where m_A is the number of linearly independent solutions of the homogeneous equation (7.1), i.e. a fundamental system for the equilibrium operator $\boldsymbol{A}^{\mathsf{T}}$, and let $\boldsymbol{\tau}$ be a partial solution of the inhomogeneous equilibrium equation

$$\boldsymbol{A}^{\mathsf{T}}\boldsymbol{\sigma} = \boldsymbol{X} \tag{7.2}$$

at given external forces $\boldsymbol{X}$. Of course, the equation (7.2) should be treated not as a full equation of equilibrium of the whole mechanical system but as a condition of equilibrium of the elastic body without the elastic medium that surrounds it.

The general solution of the inhomogeneous equation of equilibrium (7.2) can be represented in the form

$$\boldsymbol{\sigma} = \boldsymbol{\tau} + \sum_{j=1}^{m_A} \beta_j\, \boldsymbol{\sigma}_j\,. \tag{7.3}$$

Suppose there is no elastic medium, which is equivalent to the condition $\boldsymbol{K} = \boldsymbol{O}$. In that case, if the fundamental system $\boldsymbol{\sigma}_j$ $(j = 1,\ldots, m_A)$ is known and so is the partial solution $\boldsymbol{\tau}$, then what is left of the whole problem is to determine the coefficients β_j in the expansion (7.3)[21]. Some of the unknown coefficients (let there be r of them, $r \le m_A$) will be determined from the static boundary conditions, while the coefficients still unknown (in the number of m, $0 \le m = m_A - r$) will be obtained from the strain compatibility conditions and the kinematic boundary conditions. Obviously, if $m = 0$ then the true stress state of the system will be found from the equilibrium equations and the static boundary conditions only; in that case the system is called *statically determinate.* Otherwise the system is called *statically indeterminate* (or *redundant*), and the number $m > 0$ is called a *redundancy* of the system.

Mechanically, the coefficients β_j in the expansion (7.3) can be treated as values of generalized internal forces, and the respective generalized displacements will determine the degrees of freedom of the mechanical system. This is why a finite dimensionality of the fundamental system of the equilibrium operator $\boldsymbol{A}^{\mathsf{T}}$ always reduces the analysis of a structure to the analysis of a system having a finite number of degrees of freedom.

If we take unidimensional problems (in the sense that the area Ω is unidimensional), such as structural analysis of bar systems, the operator $\boldsymbol{A}^{\mathsf{T}}$ in this case will have the required properties. The critical role is played by the circumstance that all unknowns depend on one coordinate only, so the operator $\boldsymbol{A}^{\mathsf{T}}$ includes only ordinary (not partial) derivatives.

Section **2**.1.5 establishes that an arbitrary strain field, $\boldsymbol{\varepsilon}$, is compatible when and only when this field is orthogonal to any stress field, $\boldsymbol{\sigma}_o$, self-balanced in the sense that the following conditions hold for it:

$$\boldsymbol{A}^{\mathsf{T}}\boldsymbol{\sigma}_o = \mathbf{0} \ \in \Omega, \tag{7.4-a}$$

$$\boldsymbol{H}_\sigma \boldsymbol{\sigma}_o = \mathbf{0} \ \in \Gamma. \tag{7.4-b}$$

Due to the condition (7.4-*a*), the field of stresses $\boldsymbol{\sigma}_o$ belongs to the kernel of the equlibrium operator $\boldsymbol{A}^{\mathsf{T}}$ and thus is a linear combination of the fundamental system $\boldsymbol{\sigma}_j$ ($j = 1,\ldots, m_A$) of the equlibrium operator, that is,

[21] As it should be clear from the previous presentment, it is nearly always a trivial matter to construct the general solution of a homogeneous equation for the geometry operator $\boldsymbol{A}$ (the construction consists of searching for rigid displacements), while the complicated part is how to find a partial solution of the geometry equation, $\boldsymbol{A}\mathbf{u} = \boldsymbol{\varepsilon}$. With the equilibrium operator, $\boldsymbol{A}^{\mathsf{T}}$, the situation is exactly opposite. A partial solution of an inhomogeneous equation is quite easy to find, which is not true about the general solution of the homogeneous equation of equilibrium.

$$\boldsymbol{\sigma}_o = \sum_{j=1}^{m_A} \beta_j \, \boldsymbol{\sigma}_j \, . \tag{7.5}$$

But the condition (7.4-*b*) demands that $\boldsymbol{\sigma}_o$ satisfy the homogeneous static boundary conditions all over the boundary Γ. Mechanically, it ensures that the stresses in all *external kinematical constraints* imposed on the system are equal to zero.

If all boundary conditions of the type (7.4-*b*) contain exactly m_A of linearly independent ones, then all coefficients β_j in the expansion (7.5) are equal to zero. All one-dimensional problems of this, pretty important, class will be called *statically determinate problems*. In other words, a system is declared statically determinate if it is not possible to indicate a self-balanced and nonzero state of it.

Thus, the stresses $\boldsymbol{\sigma}_o$ self-balanced in the sense of (7.4) are always equal to zero for one-dimensional statically determinate problems.

A totally different situation takes place in two- and three-dimensional elasticity. In that case the kernel of the equilibrium operator $\boldsymbol{A}^{\mathrm{T}}$ is not a finite-dimensional linear set. A couple of important mechanical conclusions should be made in this regard. First, such problems are always statically indeterminate. Second, the two- and three-dimensional problems cannot be reduced to finite degrees of freedom in principle. Even if a problem *is* reduced (such as they always are in numerical analysis), this reduction can be treated only as an approximation. The infinite-dimensional kernel of the operator $\boldsymbol{A}^{\mathrm{T}}$ is replaced by its finite-dimensional approximate counterpart. Third and final, the strain compatibility conditions are an absolutely necessary component in the statements of two- and three-dimensional problems.

2.8 Final comments to Chapter 2

Let us return to the definition of the Castigliano functional according to (3.2) and (3.3). The traditional definition of the Castigliano functional [6, 20] does not contain the displacements as its functional argument, and $\mathsf{K} = \mathsf{K}(\boldsymbol{\sigma})$. In our case the dependence of the Castigliano functional on two functional arguments, $\mathsf{K} = \mathsf{K}(\boldsymbol{\sigma},\mathbf{u})$, seems unusual. However, this dependence is necessary to expand the definition so that it could cover the elastic foundation, too.

It follows from (3.3) and (3.2) that in an ultimate situation and in the absense of an elastic foundation ($\boldsymbol{K} = \boldsymbol{O}$) the displacements $\mathbf{u}$ are excluded from the list of functional arguments of the Castigliano functional. On the other hand, the functional $\mathsf{K}(\boldsymbol{\sigma},\mathbf{u})$ defined according to (3.3) keeps the critically important identity (4.3) true; this identity can be appropriately called a *dual relation* between the Lagrange and Castigliano functionals.

Of course, one could define the Castigliano functional formally as one depending on force-type variables only — by introducing the response of the elastic foundation, $\boldsymbol{r}$, according to (**1**.4.5) and thus having $\mathsf{K} = \mathsf{K}(\boldsymbol{\sigma}, \boldsymbol{r})$. However, that would mean seeking for the Castigliano functional's minimum on fields which contain an additional component of the responses $\boldsymbol{r}$ not included in the general definition of the stress-and-strain field. In other words, this interpretation would define the general stress-and-strain field, F, as a four-component field, $F = \{\boldsymbol{\sigma}, \boldsymbol{\varepsilon}, \mathbf{u}, \boldsymbol{r}\}$. A stress-and-strain field acceptable for consideration would have to satisfy the additional condition,

$$\boldsymbol{r} = \boldsymbol{K}\mathbf{u}\ .$$

The second difference of the definition of the Castigliano functional used here from one in common use [15] is that the functional here has the opposite sign; this makes the respective variational principle the principle of minimum instead of the traditional principle of maximum. The similar change is made to the formulation of the dual relation mentioned above which is expressed by the identity (4.3) in our case while traditionally it is an equality between the Castigliano functional and the Lagrange functional on the solution of a problem. One should understand that this change of sign is not critical; it is introduced only for convenience of further reasoning.

The lemma of constraints, in a formulation slightly different from one used here, is given in a textbook by L.A. Rozin [16] in connection with invariance of the Mohr formula which determines the displacements in the bar systems with respect to shear.

References

1. Filonenko-Boroditch MM (1965) Theory of elasticity. Dover, New York
2. Janelidze GY, Panovko YG (1950) St.-Venant principle and its applications to the analysis of plates and shells (in Russian). In: Spatial structural analysis, Moscow, Ministry of Construction Facilities of Engineering Industry. Vol 1: pp. 329–342
3. Kochin NE (1965) Vector calculus and basics of tensor calculus (in Russian). Nauka, Moscow

4. Lazarian VA (1972) Energy of strain and displacements of linear systems (in Russian). Naukova Dumka, Kiev
5. Leibenzon LS (1951) Variational methods for solving problems of elasticity (in Russian). In book: Academician L.S. Leibenson, collected works, vol. 1: Theory of elasticity. Academy of Sciences of USSR, Moscow
6. Mikhlin SG (1952) Problem of minimum of a quadratic functional (in Russian). Gostekhteorizdat, Moscow–Leningrad
7. Mikhlin SG (1964) Variational Methods in Mathematical Physics. Macmillan, New York
8. Novatsky V (1975) Theory of elasticity (in Russian). Mir, Moscow
9. Novozhilov VV (1953) Foundations of nonlinear theory of elasticity. Graylock Press, Rochester, N.Y.
10. Perelmuter AV, Slivker VI (2003) Numerical Structural Analysis. Springer-Verlag, Berlin Heidelberg New York
11. Prager W, Synge JL (1947) Approximations in elasticity based on the concept of function space. Quartely of Applied Mathematics, 5(3): 241–269.
12. Rabinovitch IM (1954) A course of structural mechanics of bar systems, part 2 (in Russian). Stroyizdat, Moscow
13. Rabinovitch IM (1984) Recollections (in Russian). Nauka, Moscow
14. Rektorys K (1985) Variational methods in mathematical physics and engineering (in Russian). Mir, Moscow
15. Rozin LA (1986) Theorems and methods of statics of deformable systems (in Russian). Leningrad University, Leningrad
16. Rozin LA (1990) Calculation of displacements in deformable systems (Maxwell–Mohr formulas) (in Russian). Leningrad Polytechnic Institute, Leningrad
17. Southwell RV (1923) On Castigliano's Theorem of Least Work and the Principle of St.-Venant. Phylosophical Magazine, 45: 193–212
18. Southwell RV (1941) An introduction to the theory of elasticity.
19. Strang G (1976) Linear algebra and its applications. Academic press, New York San Francisco London
20. Zanaboni (1937) Demonstrazione generale del Principio del de St.-Venant. Atti Acad. Lincei Roma, 25: 118–121

3 ADDITIONAL VARIATIONAL PRINCIPLES OF STRUCTURAL MECHANICS

Mechanics is no herbarium of dried leaves; it is a living and blooming garden!
Panovko YG (1985) Mechanics of deformable solids. Modern concepts, mistakes and paradoxes (in Russian). Nauka, Moscow

3.1 Reissner mixed variational principle

The previous chapter discussed two basic variational principles of structural mechanics: the Lagrange principle and the Castigliano principle. Both principles are similar in a most important way: certain functionals in their respective principles take minimum values on the solution of a problem, that is, both principles are of extreme nature. At the same time, these two basic variational principles are essentially different.

The Lagrange functional takes a minimum value on the solution of the problem among all physically and kinematically admissible fields, and the equations of equilibrium in the volume and on the surface of the body are corollaries (necessary conditions) of this minimality. The language of calculus of variations refers to this as Euler equations and natural boundary conditions.

The Castigliano functional is directly opposite. The Castigliano functional takes a minimum value on the solution of the problem among all physically and kinematically admissible fields, while the strain compatibility conditions and the kinematic boundary conditions are Euler equations and natural boundary conditions for the Castigliano functional.

So, these two basic variational principles are complementary and require that only a certain limited class of fields be considered — those that satisfy certain predefined conditions. In this regard, we would naturally like to construct such a functional (and its respective variational principle) that would take at least a stationary value if not an extreme one on the solution of the problem, but among stress-and-strain fields not submitted to any predefined conditions.

Also, the Lagrangian energy space, $\mathscr{L}$, is defined completely by the vector of displacements, **u**. If we assume the elastic medium to be absent, which corresponds to $\boldsymbol{K} = \boldsymbol{O}$, then the Castiglianian energy space, $\mathscr{K}$, will be completely defined by the stresses, $\boldsymbol{\sigma}$. Therefore in structural mechanics the problem of searching for a point of minimum of the Lagrangian functional defines the variational formulation of the original problem in displacements, while the problem of searching for a point of minimum of the Castiglianian functional defines the variational formulation of the problem in stresses. In opposition to this, variational principles where stress fields and displacement fields are treated as independent even if there is no elastic medium are usually referred to as *mixed variational principles.*

The first of functionals of this kind and its respective variational principle were built by E. Reissner [8], [9]. However, as even Reissner himself states [9], a similar principle in mathematics (classic calculus of variations) was known earlier to Hellinger [5] as a canonic form of the original problem. Therefore the Reissner principle is often referred to as the Hellinger–Reissner principle. Fair as this double entitlement could be from historical viewpoint, it was the work by Reissner that made this variational principle a common tool for variational formulations of problems in structural mechanics.

3.1.1 Reissner functional

First of all, let us introduce a linear set, $\mathscr{U}_{k/2}$, consisting of stress-and-strain fields which are kinematically semi-admissible but not necessarily physically admissible. Thus, according to the definition, an arbitrary field $F \in \mathscr{U}_{k/2}$ can be represented as $F = \{\boldsymbol{\sigma}, \boldsymbol{A}\mathbf{u}, \mathbf{u}\}$. Obviously, the linear set $\mathscr{P}_{k/2}$ introduced earlier is a subset of the linear set $\mathscr{U}_{k/2}$ or, exactly speaking, the intersection $\mathscr{P}_{k/2} = \mathscr{P} \cap \mathscr{U}_{k/2}$.

The following expression[1] on the linear set $\mathscr{U}_{k/2}$ will be adopted as a Reissner functional, or a Reissnerian as it can be titled,

$$\mathsf{R} = \tfrac{1}{2}(\boldsymbol{C}^{-1}\boldsymbol{\sigma}, \boldsymbol{\sigma}) - \tfrac{1}{2}(\boldsymbol{K}\mathbf{u}, \mathbf{u}) - (\boldsymbol{A}\mathbf{u}, \boldsymbol{\sigma}) + (\overline{\boldsymbol{X}}, \mathbf{u}) +$$
$$+ (\boldsymbol{E}_p\overline{\boldsymbol{p}}, \boldsymbol{E}_p\boldsymbol{u})_\Gamma + (\boldsymbol{E}_u\boldsymbol{p}, \boldsymbol{E}_u(\boldsymbol{u} - \overline{\boldsymbol{u}}))_\Gamma \ , \qquad (1.1)$$

[1] Based on certain, not very essential, considerations of convenience, we deviate from Reissner's definition of the functional R used in his papers by altering its sign.

where $\boldsymbol{p}$ и $\boldsymbol{u}$ designate, as before, the following vector functions defined on the boundary Γ:

$$\boldsymbol{p} = \boldsymbol{H}_\sigma \boldsymbol{\sigma}, \quad \boldsymbol{u} = \boldsymbol{H}_u \mathbf{u} \qquad \in \Gamma. \tag{1.2}$$

Now recalling the expressions of the force potential Π_s and kinematic potential Π_k of the external actions,

$$\Pi_s(\mathbf{u}) = (\bar{\boldsymbol{X}}, \mathbf{u}) + (\boldsymbol{E}_p \bar{\boldsymbol{p}}, \boldsymbol{E}_p \boldsymbol{u})_\Gamma, \qquad \Pi_k(\boldsymbol{\sigma}) = (\boldsymbol{E}_u \boldsymbol{p}, \boldsymbol{E}_u \bar{\boldsymbol{u}})_\Gamma, \tag{1.3}$$

we can rewrite the expression (1.1) as

$$\mathsf{R}_1(\boldsymbol{\sigma}, \mathbf{u}) = \tfrac{1}{2}(\boldsymbol{C}^{-1}\boldsymbol{\sigma}, \boldsymbol{\sigma}) - \tfrac{1}{2}(\boldsymbol{K}\mathbf{u}, \mathbf{u}) - (\boldsymbol{A}\mathbf{u}, \boldsymbol{\sigma}) + (\boldsymbol{E}_u \boldsymbol{p}, \boldsymbol{E}_u \boldsymbol{u})_\Gamma + \\ + \Pi_s - \Pi_k. \tag{1.4}$$

Expression (1.4) will be called, according to L.A. Rozin [11], *first form of the Reissner functional*, to differ it from its *second form* derived from (1.4) by using the basic integral formula (**1**.2.19). As a result, the Reissner functional can be rebuilt to look like

$$\mathsf{R}_2(\boldsymbol{\sigma}, \mathbf{u}) = \tfrac{1}{2}(\boldsymbol{C}^{-1}\boldsymbol{\sigma}, \boldsymbol{\sigma}) - \tfrac{1}{2}(\boldsymbol{K}\mathbf{u}, \mathbf{u}) - (\boldsymbol{A}^\mathsf{T}\boldsymbol{\sigma}, \mathbf{u}) - (\boldsymbol{E}_p \boldsymbol{p}, \boldsymbol{E}_p \boldsymbol{u})_\Gamma + \\ + \Pi_s - \Pi_k. \tag{1.5}$$

In (1.4) and (1.5), the subscript at the functional R indicates No. of the form of expression for the functional. Note that the first form of the Reissner functional contains the geometry operator $\boldsymbol{A}$, while the second form includes the equilibrium operator $\boldsymbol{A}^\mathsf{T}$.

Of course, both forms are equivalent in the sense that the values of $\mathsf{R}_1(\boldsymbol{\sigma}, \mathbf{u})$ and $\mathsf{R}_2(\boldsymbol{\sigma}, \mathbf{u})$ coincide on the same functions $\boldsymbol{\sigma}$ and $\mathbf{u}$. All the difference (which, however, can be important in the construction of numerical procedures!) is that the domains of the functionals overlap but do not coincide. The $\mathsf{R}_1(\boldsymbol{\sigma},\mathbf{u})$ functional is defined on sufficiently smooth displacements, $\mathbf{u}$, which at least allow the differential operation $\boldsymbol{A}$. However, for the $\mathsf{R}_2(\boldsymbol{\sigma}, \mathbf{u})$ functional to exist, the smoothness requirements to the displacements $\mathbf{u}$ are weakened while those to the stresses $\boldsymbol{\sigma}$ are strengthened, and the latter must allow the differential operation $\boldsymbol{A}^\mathsf{T}$.

In some of the formulations of problems in applied theories (bars, plates, shells), when the $\boldsymbol{A}$ operator contains differential operations of higher than first order (even), the third term, $(\boldsymbol{A}^\mathsf{T}\boldsymbol{\sigma}, \mathbf{u})$, included in the second form of the Reissner functional can be transformed by integration by parts. This involves removing half of the derivatives from the stresses $\boldsymbol{\sigma}$

and placing them on the displacements **u**[2]. This formal transformation in the problems of that kind will derive a third form of the Reissner functional. As the third form of the Reissner functional contains differential operations of lower orders than its first and second form, making use of it in numerical techniques (such as finite element method) can give some advantages because the elements of fields from $\mathscr{U}_{k/2}$ are required to satisfy only minimal conditions of smoothness. This means every vector function appearing in the definition of the functional R must be differentiable a sufficient number of times and must guarantee the existence of finite values of the integrals participating in the R functional. Accordingly, weaker requirements to smoothness are imposed on coordinate functions for the Ritz method; this will be discussed later in more detail.

Now let us show that the Reissner functional takes a stationary value on the solution of problem (**1**.2.2), (**1**.2.4). To do it, consider a perturbed field, $F_* + \delta F$, with a field of variations, δF, from the linear set $\mathscr{U}_{k/2}$ in the vicinity of a true field, $F_* = \{\boldsymbol{\sigma}_*, \boldsymbol{\varepsilon}_*, \mathbf{u}_*\}$. The latter requirement forces the perturbed field $F_* + \delta F$ to belong to the same set.

As a result, the value of the functional R in the vicinity of the true field can be represented as

$$\mathsf{R}(F_* + \delta F) = \mathsf{R}(F_*) + \delta\mathsf{R} + \tfrac{1}{2}\,\delta^2\mathsf{R} \tag{1.6}$$

where $\delta\mathsf{R}$ and $\delta^2\mathsf{R}$ are the respective first and second variations of the Reissner functional.

Based on the second form of the Reissner functional, we will have

$$\begin{aligned}\delta\mathsf{R} = (\boldsymbol{C}^{-1}\boldsymbol{\sigma}, \delta\boldsymbol{\sigma}) - (\boldsymbol{K}\mathbf{u}, \delta\mathbf{u}) - (\boldsymbol{A}^{\mathsf{T}}\boldsymbol{\sigma}, \delta\mathbf{u}) - \\ - (\boldsymbol{A}^{\mathsf{T}}\delta\boldsymbol{\sigma}, \mathbf{u}) - (\boldsymbol{E}_p\boldsymbol{p}, \boldsymbol{E}_p\delta\boldsymbol{u})_\Gamma - (\boldsymbol{E}_p\delta\boldsymbol{p}, \boldsymbol{E}_p\boldsymbol{u})_\Gamma + (\bar{\boldsymbol{X}}, \delta\mathbf{u}) + \\ + (\boldsymbol{E}_p\bar{\boldsymbol{p}}, \boldsymbol{E}_p\delta\boldsymbol{u})_\Gamma - (\boldsymbol{E}_u\delta\boldsymbol{p}, \boldsymbol{E}_u\bar{\boldsymbol{u}})_\Gamma\,,\end{aligned} \tag{1.7}$$

$$\delta^2\mathsf{R} = (\boldsymbol{C}^{-1}\delta\boldsymbol{\sigma}, \delta\boldsymbol{\sigma}) - (\boldsymbol{K}\delta\mathbf{u}, \delta\mathbf{u}) - 2(\boldsymbol{A}^{\mathsf{T}}\delta\boldsymbol{\sigma}, \delta\mathbf{u}) - 2(\boldsymbol{E}_p\delta\boldsymbol{p}, \boldsymbol{E}_p\delta\boldsymbol{u})_\Gamma\,. \tag{1.8}$$

Let us transform the expression of first variation $\delta\mathsf{R}$ by moving the differential operation $\boldsymbol{A}$ in the fourth term of $(\boldsymbol{A}^{\mathsf{T}}\delta\boldsymbol{\sigma}, \mathbf{u})$ to the displacement function **u** using the basic integral identity and then regrouping the terms. The result will be

[2] Exactly speaking, one can use integration by parts in one-dimensional problems. In two- or three-dimensional problems the 'integration by parts' operation is actually based on the Gauss-Ostrogradsky formula.

$$\delta\mathsf{R} = (\boldsymbol{C}^{-1}\boldsymbol{\sigma} - \boldsymbol{A}\mathbf{u}, \delta\boldsymbol{\sigma}) - (\boldsymbol{A}^{\mathsf{T}}\boldsymbol{\sigma} + \boldsymbol{K}\mathbf{u} - \overline{\boldsymbol{X}}, \delta\mathbf{u}) -$$
$$- (\boldsymbol{E}_p(\boldsymbol{p} - \overline{\boldsymbol{p}}), \boldsymbol{E}_p\delta\boldsymbol{u})_{\Gamma} + (\boldsymbol{E}_u\delta\boldsymbol{p}, \boldsymbol{E}_u(\boldsymbol{u} - \overline{\boldsymbol{u}}))_{\Gamma} . \tag{1.9}$$

As variation $\delta\mathsf{R}$ is taken in the vicinity of the true field F_*, according to (1.6) we should assume $\boldsymbol{\sigma} = \boldsymbol{\sigma}_*$, $\mathbf{u} = \mathbf{u}_* \in \Omega$ and $\boldsymbol{E}_p\boldsymbol{p} = \boldsymbol{E}_p\overline{\boldsymbol{p}}$, $\boldsymbol{E}_u\boldsymbol{u} = \boldsymbol{E}_u\overline{\boldsymbol{u}} \in \Gamma$ in (1.9). Now it becomes obvious that all terms in the right part of (1.9) are zeros. When the first variation of a functional is equal to zero at any $\delta\boldsymbol{\sigma}$ and $\delta\mathbf{u}$, it means the true field F_* is a "point" of stationarity of the Reissner functional.

The opposite statement is also true: the point of stationarity of the Reissner functional coincides with a true stress-and-strain field of the structure. To see this, we equal first variation $\delta\mathsf{R}$ as in (1.9) to zero and take into account the independence of variations $\delta\boldsymbol{\sigma}$ and $\delta\mathbf{u}$ to conclude that each of the four terms in the right part of (1.9) is equal to zero by itself. Consequently,

$$\boldsymbol{C}^{-1}\boldsymbol{\sigma} - \boldsymbol{A}\mathbf{u} = \mathbf{0}, \qquad \boldsymbol{A}^{\mathsf{T}}\boldsymbol{\sigma} + \boldsymbol{K}\mathbf{u} - \overline{\boldsymbol{X}} = \mathbf{0} \quad \in\Omega , \tag{1.10}$$

$$\boldsymbol{E}_p\boldsymbol{p} = \boldsymbol{E}_p\overline{\boldsymbol{p}} , \qquad \boldsymbol{E}_u\boldsymbol{u} = \boldsymbol{E}_u\overline{\boldsymbol{u}} \qquad \in\Gamma. \tag{1.11}$$

Thus, at the point of stationarity the stress-and-strain field is physically admissible, statically admissible, and kinematically admissible — that is, it is a true field. Equations (1.10) are Euler equations for the Reissner functional, and all boundary conditions (1.11) are natural for this functional.

Now let's turn to the analysis of extreme properties of the Reissner functional. First of all, we should recall that the Reissner functional is considered on the linear set $\mathscr{U}_{k/2}$, thus it is defined completely by only two elements of the stress-and-strain field — by the stresses and the displacements; this fact can be represented as $\mathsf{R} = \mathsf{R}(\boldsymbol{\sigma},\mathbf{u})$.

We fixate the displacements in the Reissner functional by assuming $\mathbf{u} = \mathbf{u}_*$ and, respectively, $\boldsymbol{E}_u\boldsymbol{u} = \boldsymbol{E}_u\overline{\boldsymbol{u}}$. By varying the stresses $\boldsymbol{\sigma}$ only (we should think the variations of displacements are zero both in the Ω area and on the Γ boundary in this special case) we derive from (1.8) that the second variation will be $\delta^2\mathsf{R} = (\boldsymbol{C}^{-1}\delta\boldsymbol{\sigma}, \delta\boldsymbol{\sigma}) > 0$. Consequently, this special case of variation will turn equation (1.6), considering that $\delta\mathsf{R} = 0$, into

$$\mathsf{R}(F_* + \delta F) - \mathsf{R}(F_*) > 0 , \tag{1.12}$$

so with this type of variation of the stress-and-strain fields the R functional will take a minimum value, that is,

$$\mathsf{R}(\boldsymbol{\sigma}_*,\mathbf{u}_*) < \mathsf{R}(\boldsymbol{\sigma},\mathbf{u}_*)\,. \tag{1.13}$$

On the contrary, if the stresses $\boldsymbol{\sigma} = \boldsymbol{\sigma}_*$ are fixated and so are $\boldsymbol{E}_p\boldsymbol{p} = \boldsymbol{E}_p\,\overline{\boldsymbol{p}}$, respectively, then the variation involves only the displacements, and (1.8) produces $\delta^2\mathsf{R} = -(\boldsymbol{K}\delta\mathbf{u},\,\delta\mathbf{u}) \le 0$. Similar reasoning leads to the inequality

$$\mathsf{R}(\boldsymbol{\sigma}_*,\mathbf{u}) \le \mathsf{R}(\boldsymbol{\sigma}_*,\mathbf{u}_*)\,. \tag{1.14}$$

Combining inequalities (1.13) and (1.14) gives a two-side estimate

$$\mathsf{R}(\boldsymbol{\sigma}_*,\mathbf{u}) \le \mathsf{R}(\boldsymbol{\sigma}_*,\mathbf{u}_*) < \mathsf{R}(\boldsymbol{\sigma},\mathbf{u}_*)\,. \tag{1.15}$$

If we use terminology of a branch of mathematics called "convex analysis" [10], [1], we can say that the point of stationarity of the Reissner functional is a *saddle point*. Note that if there is no elastic medium then the $\boldsymbol{K}$ operator is identical to zero, so varying the functional with the fixated stresses will give $\delta^2\mathsf{R} = 0$. In that particular case, we will have the following instead of (1.15)

$$\mathsf{R}(\boldsymbol{\sigma}_*,\mathbf{u}) = \mathsf{R}(\boldsymbol{\sigma}_*,\mathbf{u}_*) < \mathsf{R}(\boldsymbol{\sigma},\mathbf{u}_*)\,, \tag{1.16}$$

which is a definition of a *degenerate saddle point* of the functional.

The circumstance that $\mathsf{R}(\boldsymbol{\sigma}_*,\mathbf{u}) = \mathsf{R}(\boldsymbol{\sigma}_*,\mathbf{u}_*)$, and that consequently the Reissner functional's value remains the same whatever changes of the displacements may be, seems strange at first glance. However, all will become quite clear if we assume $\boldsymbol{\sigma} = \boldsymbol{\sigma}_*$ in the second form of the Reissner functional according to (1.5), not forgetting that the current case is $\boldsymbol{K} = \boldsymbol{O}$. As a result,

$$\mathsf{R}(\boldsymbol{\sigma}_*,\mathbf{u}) = \tfrac{1}{2}(\boldsymbol{C}^{-1}\boldsymbol{\sigma}_*,\,\boldsymbol{\sigma}_*) \;-(\boldsymbol{A}^{\mathsf{T}}\boldsymbol{\sigma}_*,\,\mathbf{u})\; +(\overline{\boldsymbol{X}}\,,\,\mathbf{u}) - (\boldsymbol{E}_u\boldsymbol{p}_*,\,\boldsymbol{E}_u\overline{\boldsymbol{u}}\,)_\Gamma\,,$$

and because $\boldsymbol{\sigma}_*$ satisfies the equilibrium equations by definition, the second and third terms will be mutually canceled. Thus,

$$\mathsf{R}(\boldsymbol{\sigma}_*,\mathbf{u}) = \tfrac{1}{2}(\boldsymbol{C}^{-1}\boldsymbol{\sigma}_*,\boldsymbol{\sigma}_*) - (\boldsymbol{E}_u\boldsymbol{p}_*,\,\boldsymbol{E}_u\overline{\boldsymbol{u}}\,)_\Gamma\,,$$

wherefrom it is immediately obvious that the $\mathsf{R}(\boldsymbol{\sigma}_*,\mathbf{u})$ functional does not depend on the displacements, so the second functional argument can be omitted, and we have $\mathsf{R}(\boldsymbol{\sigma}_*,\mathbf{u}) = \mathsf{R}_*$.

3.1.2 Principle of minimum for stresses

To characterize the stationarity point of the Reissner functional, it is useful to present another kind of reasoning the idea of which is borrowed from the original work by E. Reissner [9].

Suppose all states of the elastic system permitted for comparison satisfy the equilibrium equations both inside the body and on its boundary:

$$\boldsymbol{A}^{\mathsf{T}}\boldsymbol{\sigma} + \boldsymbol{K}\mathbf{u} - \bar{\boldsymbol{X}} = \mathbf{0} \quad \in \Omega, \qquad \boldsymbol{E}_p\boldsymbol{p} = \boldsymbol{E}_p\,\bar{\boldsymbol{p}} \quad \in \Gamma. \tag{1.17}$$

It means the variations of the displacements and stresses satisfy the respective homogeneous equations

$$\boldsymbol{A}^{\mathsf{T}}\delta\boldsymbol{\sigma} + \boldsymbol{K}\delta\mathbf{u} = \mathbf{0} \quad \in \Omega, \qquad \boldsymbol{E}_p\delta\boldsymbol{p} = \mathbf{0} \quad \in \Gamma.$$

By making a scalar product of the first of the equations with $\delta\mathbf{u}$ we derive $(\boldsymbol{A}^{\mathsf{T}}\delta\boldsymbol{\sigma}, \delta\mathbf{u}) = -(\boldsymbol{K}\delta\mathbf{u}, \delta\mathbf{u})$, which is then substituted to (1.8) to produce the following expression for the second variation:

$$\delta^2\mathsf{R} = (\boldsymbol{C}^{-1}\delta\boldsymbol{\sigma}, \delta\boldsymbol{\sigma}) + (\boldsymbol{K}\delta\mathbf{u}, \delta\mathbf{u}) \geq 0\,. \tag{1.18}$$

Actually, with this special kind of varying, we have a new functional which we designate R_σ; it is derived from R by using conditions (1.17). If we multiply scalarly the first of relationships (1.17) by $\mathbf{u}$ and the second by $\boldsymbol{u}$, then substituting the equalities thus produced to (1.5) will give

$$\mathsf{R}_\sigma = \tfrac{1}{2}(\boldsymbol{C}^{-1}\boldsymbol{\sigma}, \boldsymbol{\sigma}) + \tfrac{1}{2}(\boldsymbol{K}\mathbf{u}, \mathbf{u}) - (\boldsymbol{E}_u\boldsymbol{p}, \boldsymbol{E}_u\bar{\boldsymbol{u}})_\Gamma\,, \tag{1.19}$$

and this expression of $\mathsf{R}_\sigma(\boldsymbol{\sigma},\mathbf{u})$ is an exact copy of the Castigliano functional's, $\mathsf{K}(\boldsymbol{\sigma},\mathbf{u})$, according to (**2**.3.1). However, it would be a mistake to identify these two functionals with each other.

The matter is that the functionals $\mathsf{R}_\sigma(\boldsymbol{\sigma},\mathbf{u})$ and $\mathsf{K}(\boldsymbol{\sigma},\mathbf{u})$ are defined in different domains. The Castigliano functional is defined on the set $\mathscr{P}_S$ of physically and statically admissible stress-and-strain fields, that is, on fields of the kind $F = \{\boldsymbol{\sigma}, \boldsymbol{C}^{-1}\boldsymbol{\sigma}, \mathbf{u}\}$, which satisfy (1.17). The functional $\mathsf{R}_\sigma(\boldsymbol{\sigma},\mathbf{u})$ is defined on the set of statically admissible and kinematically semi-admissible stress-and-strain fields, that is, on fields of the type $F = \{\boldsymbol{\sigma}, \boldsymbol{A}\mathbf{u}, \mathbf{u}\}$ which satisfy the same equations (1.17).

As the second variation of R_σ in the vicinity of the true field according to (1.18) is nonnegative, we obtain the following estimate:

$$\mathsf{R}(\boldsymbol{\sigma}_*, \mathbf{u}_*) = \mathsf{R}_\sigma(\boldsymbol{\sigma}_*, \mathbf{u}_*) \leq \mathsf{R}_\sigma(\boldsymbol{\sigma}, \mathbf{u})\,. \tag{1.20}$$

In a particular case of zero operator $\boldsymbol{K}$, which was the object of consideration by Reissner, the functional R_σ does not depend on the displacements as it can be seen in (1.19), so the absence of an elastic medium turns the estimate (1.20) into

$$\mathsf{R}(\boldsymbol{\sigma}_*, \mathbf{u}_*) = \mathsf{R}_\sigma(\boldsymbol{\sigma}_*) \leq \mathsf{R}_\sigma(\boldsymbol{\sigma})\,, \tag{1.21}$$

and Reissner treats it in this form as a *variational principle of minimum for stresses*[3].

3.1.3 Principle of maximum for displacements

Now suppose that all states of the mechanical system allowed for comparison satisfy the physical equations and the kinematic boundary conditions

$$\boldsymbol{C}^{-1}\boldsymbol{\sigma} - \boldsymbol{A}\mathbf{u} = \mathbf{0} \in \Omega\,, \qquad \boldsymbol{E}_u\boldsymbol{u} = \boldsymbol{E}_u\bar{\boldsymbol{u}} \in \Gamma\,. \tag{1.22}$$

It will be natural to designate the functional $\mathsf{R}(\boldsymbol{\sigma},\mathbf{u})$ with elements of the stress-and-strain field which satisfy additional conditions (1.22) as $\mathsf{R}_u(\boldsymbol{\sigma},\mathbf{u})$. The first of equations (1.22) will give $\boldsymbol{\sigma} = \boldsymbol{C}\boldsymbol{A}\mathbf{u}$, and the second will yield $(\boldsymbol{E}_u(\boldsymbol{u}-\bar{\boldsymbol{u}}), \boldsymbol{E}_u\boldsymbol{p})_\Gamma = 0$. Using these equalities, we transform (1.5) to obtain

$$\mathsf{R}_u(\boldsymbol{\sigma},\mathbf{u}) = -\tfrac{1}{2}(\boldsymbol{C}\boldsymbol{A}\mathbf{u}, \boldsymbol{A}\mathbf{u}) - \tfrac{1}{2}(\boldsymbol{K}\mathbf{u}, \mathbf{u}) + (\bar{\boldsymbol{X}}, \mathbf{u}) + (\boldsymbol{E}_p\bar{\boldsymbol{p}}, \boldsymbol{E}_p\boldsymbol{u})_\Gamma\,. \tag{1.23}$$

Note first of all that the right part of (1.23) does not contain any stresses, therefore the first functional argument (stresses $\boldsymbol{\sigma}$) in R_u can be omitted. Further note that the right part of (1.23) contains an expression for the Lagrange functional $\mathsf{L}(\mathbf{u})$ taken with the minus sign. The sets of stress-and-strain fields on which the functionals $\mathsf{R}_u(\mathbf{u})$ and $\mathsf{L}(\mathbf{u})$ are defined coincide: in both cases we deal with physically and kinematically admissible fields. Therefore these two functionals can be identified with each other (up to their sign).

Using the ability of the Lagrange functional to take a minimum value on the solution of the problem as established in Chapter 2, we come up with the estimate

$$\mathsf{R}(\boldsymbol{\sigma}_*,\mathbf{u}_*) = \mathsf{R}_u(\mathbf{u}_*) = -\,\mathsf{L}(\mathbf{u}_*) \geq -\,\mathsf{L}(\mathbf{u}) = \mathsf{R}_u(\mathbf{u})\,. \tag{1.24}$$

The relationship thus obtained will be called, following E. Reissner, a *variational principle of maximum for displacements.*

Combining estimates (1.20) and (1.24) gives a useful two-sided estimate of the Reissner functional on the solution of a problem,

$$\mathsf{R}_u(\mathbf{u}) \leq \mathsf{R}(\boldsymbol{\sigma}_*,\mathbf{u}_*) \leq \mathsf{R}_\sigma(\boldsymbol{\sigma},\mathbf{u})\,, \tag{1.25}$$

---

[3] Actually, Reissner himself makes it the other way around: the principle of maximum for stresses and that of minimum for displacements. This difference takes place because we have altered the sign of the R functional comparing to that introduced by Reissner himself.

which can be written in a simpler way when $\boldsymbol{K} = \boldsymbol{O}$:

$$\mathsf{R}_u(\mathbf{u}) \le \mathsf{R}(\boldsymbol{\sigma}_*, \mathbf{u}_*) \le \mathsf{R}_\sigma(\boldsymbol{\sigma}) . \qquad (1.26)$$

Next, (1.24) shows that the value of the Reissner functional R_* on the solution of the problem is equal in magnitude and opposite in sign to the value of the Lagrange functional, so from identity (2.4.3) we immediately arrive at a conclusion that the values of the Reissner and Castigliano functionals on the solution of the problem are identical:

$$\mathsf{R}_* = \mathsf{K}_* . \qquad (1.27)$$

Identity (1.27) permits to extend the qualitative estimates, obtained in Section 2.4 for the effect of various kinds of perturbations in the formulation of the problem, to the value of the Reissner functional.

3.2 Principle of stationarity of the boundary conditions functional

Following E. Reissner [9], now we will build a functional the stationarity conditions for which will yield boundary conditions, both static and kinematic. We assume that the set of stress-and-strain fields, on which we consider the Reissner functional to be defined, is limited to the fields that are physically admissible and satisfy the equilibrium equations in the volume of the body (but not on its surface!)

$$\boldsymbol{C}^{-1}\boldsymbol{\sigma} - \boldsymbol{A}\mathbf{u} = \mathbf{0} , \qquad \boldsymbol{A}^{\mathsf{T}}\boldsymbol{\sigma} + \boldsymbol{K}\mathbf{u} - \overline{\boldsymbol{X}} = \mathbf{0} \quad \in \Omega . \qquad (2.1)$$

A functional derived from the Reissner functional under these conditions will be called a *functional of boundary conditions* and denoted by Γ. First of all, using (2.1) and the basic integral formula will give

$$(\boldsymbol{C}^{-1}\boldsymbol{\sigma}, \boldsymbol{\sigma}) = (\boldsymbol{A}\mathbf{u}, \boldsymbol{\sigma}) = (\boldsymbol{A}^{\mathsf{T}}\boldsymbol{\sigma}, \mathbf{u}) + (\boldsymbol{E}_p\boldsymbol{p}, \boldsymbol{E}_p\boldsymbol{u})_\Gamma + (\boldsymbol{E}_u\boldsymbol{p}, \boldsymbol{E}_u\boldsymbol{u})_\Gamma ,$$

$$(\boldsymbol{A}^{\mathsf{T}}\boldsymbol{\sigma}, \mathbf{u}) + (\boldsymbol{K}\mathbf{u}, \mathbf{u}) = (\overline{\boldsymbol{X}}, \mathbf{u}) .$$

Substituting these expressions to the second form of the Reissner functional according to (1.5) will give the desirable functional Γ:

$$\Gamma(\boldsymbol{\sigma}, \mathbf{u}) = -\tfrac{1}{2}(\overline{\boldsymbol{X}}, \mathbf{u}) - \tfrac{1}{2}(\boldsymbol{E}_p\boldsymbol{p}, \boldsymbol{E}_p\boldsymbol{u})_\Gamma + \tfrac{1}{2}(\boldsymbol{E}_u\boldsymbol{p}, \boldsymbol{E}_u\boldsymbol{u})_\Gamma + \\ + \Pi_s - \Pi_k . \qquad (2.2)$$

Involving expressions (2.2.5) and (2.3.2) for the static and kinematic potential of external actions, we turn functional $\Gamma = \Gamma(\boldsymbol{\sigma},\mathbf{u})$ into another equivalent form

$$\Gamma = \tfrac{1}{2}(\bar{\boldsymbol{X}}, \mathbf{u}) - \tfrac{1}{2}(\boldsymbol{E}_p \boldsymbol{p}, \boldsymbol{E}_p \boldsymbol{u})_\Gamma + \tfrac{1}{2}(\boldsymbol{E}_u \boldsymbol{p}, \boldsymbol{E}_u \boldsymbol{u})_\Gamma + \\ + (\boldsymbol{E}_p \bar{\boldsymbol{p}}, \boldsymbol{E}_p \boldsymbol{u})_\Gamma - (\boldsymbol{E}_u \boldsymbol{p}, \boldsymbol{E}_u \bar{\boldsymbol{u}})_\Gamma . \tag{2.3}$$

Now let us prove that the true stress-and-strain field of the system is a point of stationarity of the boundary conditions functional Γ. To see this, we present Γ in the vicinity of the true field F_* as

$$\Gamma(F_* + \delta F) = \Gamma(F_*) + \delta\Gamma + \tfrac{1}{2}\delta^2\Gamma . \tag{2.4}$$

Because of limitations (2.1) the variations of the stresses, $\delta\boldsymbol{\sigma}$, and the variations of the displacements, $\delta\mathbf{u}$, are not independent but are related as

$$\delta\boldsymbol{\sigma} = \boldsymbol{C}\boldsymbol{A}\delta\mathbf{u} , \qquad \boldsymbol{A}^{\mathrm{T}}\delta\boldsymbol{\sigma} = -\boldsymbol{K}\delta\mathbf{u} \quad \in\Omega . \tag{2.5}$$

Nor are independent the variations of the boundary conditions, $\delta\boldsymbol{p}$, and of the edge displacements, $\delta\boldsymbol{u}$. We will show that they are related through the condition

$$(\bar{\boldsymbol{X}}, \delta\mathbf{u}) = (\delta\boldsymbol{p}, \boldsymbol{u})_\Gamma - (\boldsymbol{p}, \delta\boldsymbol{u})_\Gamma . \tag{2.6}$$

And indeed, the basic integral formula helps represent the scalar products $(\delta\boldsymbol{p}, \boldsymbol{u})_\Gamma$ and $(\boldsymbol{p}, \delta\boldsymbol{u})_\Gamma$ using (2.1) and (2.5) as

$$(\delta\boldsymbol{p}, \boldsymbol{u})_\Gamma = (\boldsymbol{A}\mathbf{u}, \delta\boldsymbol{\sigma}) - (\boldsymbol{A}^{\mathrm{T}}\delta\boldsymbol{\sigma}, \mathbf{u}) = (\boldsymbol{A}\mathbf{u}, \boldsymbol{C}\boldsymbol{A}\delta\mathbf{u}) + (\boldsymbol{K}\mathbf{u}, \delta\mathbf{u}),$$

$$(\boldsymbol{p}, \delta\boldsymbol{u})_\Gamma = (\boldsymbol{A}\delta\mathbf{u}, \boldsymbol{\sigma}) - (\boldsymbol{A}^{\mathrm{T}}\boldsymbol{\sigma}, \delta\mathbf{u}) = (\boldsymbol{A}\delta\mathbf{u}, \boldsymbol{C}\boldsymbol{A}\mathbf{u}) - (\bar{\boldsymbol{X}} - \boldsymbol{K}\mathbf{u}, \delta\mathbf{u}).$$

Subtracting the second relationship from the first gives (2.6).

Now, taking (2.3) and allowing for (2.5) and (2.6), we go through a chain of transformations to find the first variation of the functional Γ,

$$\delta\Gamma = \tfrac{1}{2}(\bar{\boldsymbol{X}}, \delta\mathbf{u}) - \tfrac{1}{2}\delta(\boldsymbol{E}_p \boldsymbol{p}, \boldsymbol{E}_p \boldsymbol{u})_\Gamma + \tfrac{1}{2}\delta(\boldsymbol{E}_u \boldsymbol{p}, \boldsymbol{E}_u \boldsymbol{u})_\Gamma + (\boldsymbol{E}_p \bar{\boldsymbol{p}}, \boldsymbol{E}_p \delta\boldsymbol{u})_\Gamma - \\ - (\boldsymbol{E}_u \delta\boldsymbol{p}, \boldsymbol{E}_u \bar{\boldsymbol{u}})_\Gamma = \tfrac{1}{2}(\delta\boldsymbol{p}, \boldsymbol{u})_\Gamma - \tfrac{1}{2}(\boldsymbol{p}, \delta\boldsymbol{u})_\Gamma - \tfrac{1}{2}\delta(\boldsymbol{E}_p \boldsymbol{p}, \boldsymbol{E}_p \boldsymbol{u})_\Gamma + \\ + \tfrac{1}{2}\delta(\boldsymbol{E}_u \boldsymbol{p}, \boldsymbol{E}_u \boldsymbol{u})_\Gamma + (\boldsymbol{E}_p \bar{\boldsymbol{p}}, \boldsymbol{E}_p \delta\boldsymbol{u})_\Gamma - (\boldsymbol{E}_u \delta\boldsymbol{p}, \boldsymbol{E}_u \bar{\boldsymbol{u}})_\Gamma = \\ -(\boldsymbol{E}_u \boldsymbol{p}, \boldsymbol{E}_u \delta\boldsymbol{u})_\Gamma + (\boldsymbol{E}_u \delta\boldsymbol{p}, \boldsymbol{E}_u \boldsymbol{u})_\Gamma + (\boldsymbol{E}_p \bar{\boldsymbol{p}}, \boldsymbol{E}_p \delta\boldsymbol{u})_\Gamma - (\boldsymbol{E}_u \delta\boldsymbol{p}, \boldsymbol{E}_u \bar{\boldsymbol{u}})_\Gamma ,$$

hence

$$\delta\Gamma = (\boldsymbol{E}_u \delta\boldsymbol{p}, \boldsymbol{E}_u (\boldsymbol{u} - \bar{\boldsymbol{u}}))_\Gamma + (\boldsymbol{E}_p(\bar{\boldsymbol{p}} - \boldsymbol{p}), \boldsymbol{E}_p \delta\boldsymbol{u})_\Gamma . \tag{2.7}$$

As we are interested with the value of variation δΓ in the vicinity of the system's true state, we should assume $\boldsymbol{E}_u\boldsymbol{u} = \boldsymbol{E}_u\overline{\boldsymbol{u}}$ and $\boldsymbol{E}_p\boldsymbol{p} = \boldsymbol{E}_p\overline{\boldsymbol{p}} \in \Gamma$ in (2.7). As we can see from (2.7), δΓ is zero, which is an evidence that functional Γ has the point of stationarity.

It is easy to make sure that the inverse statement is true, too: a point of stationarity of the boundary conditions functional, Γ, coincides with the true stress-and-strain field of the system. And indeed, by equaling first variation δΓ from (2.7) to zero we arrive at the conclusion that boundary conditions (1.11) hold at the point of stationarity of functional Γ. All equations hold in the Ω area according to the definition of functional Γ, therefore the stress-and-strain field of the system coincides with the solution of the original problem at the point of stationarity of functional Γ.

Functional Γ is defined on a fairly limited class of fields — actually, so much limited that there are no Euler equations for this functional. The set of all necessary conditions of stationarity for Γ contains only natural boundary conditions as in (1.11).

Using formula (2.3), we find the value Γ_* of functional Γ on the problem's solution:

$$\Gamma_* = ½\Pi_{s*} - ½\Pi_{k*}, \tag{2.8}$$

or, referring to Table 2.1,

$$\Gamma_* = ½K_* - ½L_* = K_*. \tag{2.9}$$

The latter relationship helps extend the conclusions of Section **2**.4 to functional Γ.

3.3 A variational principle for physical relationships

Now, let's build a functional and its respective variational principle for which only physical equations will provide the conditions of stationarity. Again we use the second form of the Reissner functional according to (1.5), but now we define the functional on stress-and-strain fields which satisfy additional conditions

$$\boldsymbol{A}^{\mathrm{T}}\boldsymbol{\sigma} + \boldsymbol{K}\mathbf{u} - \overline{\boldsymbol{X}} = \mathbf{0} \quad \in\Omega, \tag{3.1}$$

$$\boldsymbol{E}_p\boldsymbol{p} = \boldsymbol{E}_p\overline{\boldsymbol{p}}, \quad \boldsymbol{E}_u\boldsymbol{u} = \boldsymbol{E}_u\overline{\boldsymbol{u}} \;\in \Gamma. \tag{3.2}$$

The functional obtained from this definition will be called a *functional of physical relationships* and denoted by P. Substituting (3.1) and (3.2) to (1.5) will give

$$\mathsf{P}(\boldsymbol{\sigma}, \mathbf{u}) = \tfrac{1}{2}(\boldsymbol{C}^{-1}\boldsymbol{\sigma}, \boldsymbol{\sigma}) + \tfrac{1}{2}(\boldsymbol{K}\mathbf{u}, \mathbf{u}) - (\boldsymbol{E}_u\boldsymbol{p}, \boldsymbol{E}_u\bar{\boldsymbol{u}})_\Gamma . \tag{3.3}$$

Its first variation $\delta\mathsf{P}$ is

$$\delta\mathsf{P} = (\boldsymbol{C}^{-1}\boldsymbol{\sigma}, \delta\boldsymbol{\sigma}) + (\boldsymbol{K}\mathbf{u}, \delta\mathbf{u}) - (\boldsymbol{E}_u\delta\boldsymbol{p}, \boldsymbol{E}_u\bar{\boldsymbol{u}})_\Gamma , \tag{3.4}$$

however, variations $\delta\boldsymbol{\sigma}$ and $\delta\mathbf{u}$ are not independent here because the displacements and stresses are related through equilibrium equation (3.1), hence

$$\boldsymbol{K}\delta\mathbf{u} = -\boldsymbol{A}^{\mathsf{T}}\delta\boldsymbol{\sigma} . \tag{3.5}$$

Considering the symmetry of algebraic operator $\boldsymbol{K}$, formula (3.5), and the basic integral identity, we can write a chain of equalities:

$$(\boldsymbol{K}\mathbf{u}, \delta\mathbf{u}) = (\mathbf{u}, \boldsymbol{K}\delta\mathbf{u}) = -(\mathbf{u}, \boldsymbol{A}^{\mathsf{T}}\delta\boldsymbol{\sigma}) = -(\boldsymbol{A}\mathbf{u}, \delta\boldsymbol{\sigma}) + (\delta\boldsymbol{p}, \boldsymbol{u})_\Gamma .$$

Further, $(\delta\boldsymbol{p}, \boldsymbol{u})_\Gamma = (\boldsymbol{E}_u\delta\boldsymbol{p}, \boldsymbol{E}_u\boldsymbol{u})_\Gamma + (\boldsymbol{E}_p\delta\boldsymbol{p}, \boldsymbol{E}_p\boldsymbol{u})_\Gamma$. But boundary conditions (3.2) yield $\boldsymbol{E}_p\delta\boldsymbol{p} = \mathbf{0} \ \in\Gamma$, while $\boldsymbol{E}_u\boldsymbol{u} = \boldsymbol{E}_u\bar{\boldsymbol{u}} \in\Gamma$. Therefore

$$(\boldsymbol{K}\mathbf{u}, \delta\mathbf{u}) = -(\boldsymbol{A}\mathbf{u}, \delta\boldsymbol{\sigma}) + (\boldsymbol{E}_u\delta\boldsymbol{p}, \boldsymbol{E}_u\bar{\boldsymbol{u}})_\Gamma , \tag{3.6}$$

and the expression for $\delta\mathsf{P}$ will acquire its final form

$$\delta\mathsf{P} = (\boldsymbol{C}^{-1}\boldsymbol{\sigma} - \boldsymbol{A}\mathbf{u}, \delta\boldsymbol{\sigma}). \tag{3.7}$$

Now it is obvious that the point of stationarity of the functional of physical relationships coincides with the true stress-and-strain field of the system.

Substituting the true stress and strain fields to (3.3) as functional arguments will yield

$$\mathsf{P}_* = \mathsf{E}_* - \Pi_{k*} = \mathsf{K}_* . \tag{3.8}$$

3.4 Hu–Washizu mixed variational principle

A generalization of the mixed Reissner principle is a variational principle known in literature as the Hu–Washizu principle [16]. This one is widely used in a well-known book by Japanese scientist K. Washizu [17] where it is called a *generalized variational principle*. Further we will refer to the respective functional as a Washizu functional and designate it by W.

The basic idea behind the introduction of W (which can be also called a *Washizuan* functional) is quite simple; it is to construct a functional the

Euler equations and the natural boundary conditions for which would cover the full set of all governing equations (**1**.2.2) and all boundary conditions (**1**.2.4). Obviously, this functional should depend on three functional arguments: stresses, strains, and displacements.

The Washizu functional, W, that depends on all three elements of the stress-and-strain field, will be based on the following expression:

$$\mathsf{W}(\boldsymbol{\sigma},\boldsymbol{\varepsilon},\mathbf{u}) = \tfrac{1}{2}(\boldsymbol{C\varepsilon},\boldsymbol{\varepsilon}) + \tfrac{1}{2}(\boldsymbol{K}\mathbf{u},\mathbf{u}) + (\boldsymbol{A}\mathbf{u},\boldsymbol{\sigma}) - (\boldsymbol{\sigma},\boldsymbol{\varepsilon}) - \\ - \Pi_s + \Pi_k - (\boldsymbol{E}_u\boldsymbol{p}, \boldsymbol{E}_u\boldsymbol{u})_\Gamma. \tag{4.1}$$

To find the necessary conditions of stationarity of the Washizu functional, we establish its first variation

$$\delta\mathsf{W} = \\ = (\boldsymbol{C\varepsilon},\delta\boldsymbol{\varepsilon}) + (\boldsymbol{K}\mathbf{u},\delta\mathbf{u}) + \delta(\boldsymbol{A}\mathbf{u},\boldsymbol{\sigma}) - \delta(\boldsymbol{\sigma},\boldsymbol{\varepsilon}) - \delta\Pi_s + \delta\Pi_k - \delta(\boldsymbol{E}_u\boldsymbol{p},\boldsymbol{E}_u\boldsymbol{u})_\Gamma = \\ = (\boldsymbol{A}\mathbf{u}-\boldsymbol{\varepsilon},\delta\boldsymbol{\sigma}) + (\boldsymbol{C\varepsilon}-\boldsymbol{\sigma},\delta\boldsymbol{\varepsilon}) + (\boldsymbol{A}^{\mathrm{T}}\boldsymbol{\sigma}+\boldsymbol{K}\mathbf{u}-\overline{\boldsymbol{X}},\delta\mathbf{u}) + (\boldsymbol{p},\delta\boldsymbol{u})_\Gamma - \\ - (\boldsymbol{E}_p\overline{\boldsymbol{p}},\boldsymbol{E}_p\delta\boldsymbol{u})_\Gamma + (\boldsymbol{E}_u\delta\boldsymbol{p},\boldsymbol{E}_u\overline{\boldsymbol{u}})_\Gamma - (\boldsymbol{E}_u\delta\boldsymbol{p},\boldsymbol{E}_u\boldsymbol{u})_\Gamma - (\boldsymbol{E}_u\boldsymbol{p},\boldsymbol{E}_u\delta\boldsymbol{u})_\Gamma.$$

Gathering all terms that contain variations of the same quantities will give

$$\delta\mathsf{W} = (\boldsymbol{A}\mathbf{u}-\boldsymbol{\varepsilon},\delta\boldsymbol{\sigma}) + (\boldsymbol{C\varepsilon}-\boldsymbol{\sigma},\delta\boldsymbol{\varepsilon}) + (\boldsymbol{A}^{\mathrm{T}}\boldsymbol{\sigma}+\boldsymbol{K}\mathbf{u}-\overline{\boldsymbol{X}},\delta\mathbf{u}) + \\ + (\boldsymbol{E}_p(\boldsymbol{p}-\overline{\boldsymbol{p}}),\boldsymbol{E}_p\delta\boldsymbol{u})_\Gamma - (\boldsymbol{E}_u\delta\boldsymbol{p},\boldsymbol{E}_u(\boldsymbol{u}-\overline{\boldsymbol{u}}))_\Gamma. \tag{4.2}$$

By equaling the first variation, δW, to zero we conclude that all equations (**1**.2.2) are Euler equations for the Washizu functional and all boundary conditions are natural boundary conditions for the same functional.

To characterize the stationarity point of the Washizu functional, we compose its second variation's expression, that is, the variation of variation $\delta^2\mathsf{W} = \delta(\delta\mathsf{W})$. Thus,

$$\delta^2\mathsf{W} = (\boldsymbol{A}\delta\mathbf{u},\delta\boldsymbol{\sigma}) - 2(\delta\boldsymbol{\varepsilon},\delta\boldsymbol{\sigma}) + (\boldsymbol{C}\delta\boldsymbol{\varepsilon},\delta\boldsymbol{\varepsilon}) + (\boldsymbol{A}^{\mathrm{T}}\delta\boldsymbol{\sigma},\delta\mathbf{u}) + \\ + (\boldsymbol{K}\delta\mathbf{u},\delta\mathbf{u}) + (\boldsymbol{E}_p\delta\boldsymbol{p},\boldsymbol{E}_p\delta\boldsymbol{u})_\Gamma - (\boldsymbol{E}_u\delta\boldsymbol{p},\boldsymbol{E}_u\delta\boldsymbol{u})_\Gamma. \tag{4.3}$$

If we fixate the stresses by assuming $\boldsymbol{\sigma} = \boldsymbol{\sigma}_*$ and, respectively, $\delta\boldsymbol{\sigma} = \mathbf{0}$, $\delta\boldsymbol{p} = \mathbf{0}$, we will have $\delta^2\mathsf{W} = (\boldsymbol{C}\delta\boldsymbol{\varepsilon},\delta\boldsymbol{\varepsilon}) + (\boldsymbol{K}\delta\mathbf{u},\delta\mathbf{u})$. As both terms are not negative, we have the estimates

$$\mathsf{W}(\boldsymbol{\sigma}_*,\boldsymbol{\varepsilon}_*,\mathbf{u}_*) \le \mathsf{W}(\boldsymbol{\sigma}_*,\boldsymbol{\varepsilon}_*,\mathbf{u}) \le \mathsf{W}(\boldsymbol{\sigma}_*,\boldsymbol{\varepsilon},\mathbf{u}),$$

$$\mathsf{W}(\boldsymbol{\sigma}_*,\boldsymbol{\varepsilon}_*,\mathbf{u}_*) \le \mathsf{W}(\boldsymbol{\sigma}_*,\boldsymbol{\varepsilon},\mathbf{u}_*) \le \mathsf{W}(\boldsymbol{\sigma}_*,\boldsymbol{\varepsilon},\mathbf{u}). \qquad (4.4)$$

If we fixate the strains and displacements, thus assuming $\delta\boldsymbol{\varepsilon} = \mathbf{0}$, $\delta\boldsymbol{u} = \mathbf{0}$, then we will have $\delta^2\mathsf{W} = 0$. As a result, we arrive at the following estimates by taking (4.4) into account:

$$\mathsf{W}(\boldsymbol{\sigma},\boldsymbol{\varepsilon}_*,\mathbf{u}_*) = \mathsf{W}(\boldsymbol{\sigma}_*,\boldsymbol{\varepsilon}_*,\mathbf{u}_*) \le \mathsf{W}(\boldsymbol{\sigma}_*,\boldsymbol{\varepsilon}_*,\mathbf{u}) \le \mathsf{W}(\boldsymbol{\sigma}_*,\boldsymbol{\varepsilon},\mathbf{u}). \qquad (4.5)$$

So, point $F_* = \{\boldsymbol{\sigma}_*,\boldsymbol{\varepsilon}_*,\mathbf{u}_*\}$ of stationarity of the Washizu functional W is a degenerate saddle point.

The value of the Washizu functional, W_*, on the solution of the problem can be derived from (4.1) by replacing the functional arguments with their values from the true stress-and-strain field. As a result, we will have

$$\mathsf{W}_* = \mathsf{E}_* - \Pi_{s*} = \mathsf{L}_*. \qquad (4.6)$$

3.5 A generalized mixed variational principle

A lot of attention has been paid in recent years to mixed variational formulations of problems and to their respective software implementations, especially to those related to a mixed form of the finite element method. Solutions developed for most various problems evidence high efficiency of mixed approaches; there are numerous confirmations from numerical experiments. However, wide practical implementations of the mixed approaches based mainly on Reissner-type functionals are restrained by a vexatious circumstance: the Reissner functional does not have an extremum in its stationarity point. This is a source of difficulties in two ways:

- First, it is difficult to offer a theoretical basis for numerical algorithms because the Reissner functional is not a positive definite one, thus it does not generate a metric[4];
- Second, there are computational difficulties, too, because the governing Ritz-type simultaneous equations do not have the property of positive definiteness.

The said difficulties were the reason why a generalized mixed functional [13], [14] different from one by Reissner was constructed. This functional,

[4] Form the standpoint and in terms of convex analysis, one could say that the Reissner functional is not a convex one.

or, more exactly, a parametric family of functionals, has a minimum in its point of stationarity under certain (pretty loose) conditions. Particular cases from the respective variational principle are the Lagrange and Reissner principles, therefore we have the right to call this principle a generalized one[5].

It is interesting to note that the Euler equations for that functional are linear combinations of the equilibrium equations and the physical relationships after being differentiated. Similar linear combinations participate also in the static-type natural boundary conditions. Seeing that the said variational principle has not been sufficiently addressed to in monographic literature (except for a book by L.A. Rozin [12]), we should give more background for it and use a stricter mathematical approach.

3.5.1 A generalized solution of a problem

Let $\mathscr{U}_k$ be a set of all possible kinematically admissible stress-and-strain fields, and let $\mathscr{U}_{ko}$ be a set of homogeneously kinematically admissible fields.

Definition

A generalized solution of problem (**1**.2.2), (**1**.2.4) will be called a field, $F_* = \{\boldsymbol{\sigma}_*, \boldsymbol{A}\mathbf{u}_*, \mathbf{u}_*\}$, such that for every field $F_o = \{\boldsymbol{\sigma}_{ko}, \boldsymbol{A}\mathbf{u}_{ko}, \mathbf{u}_{ko}\}$ from $\mathscr{U}_{ko}$ the following integral identity holds:

$$\kappa(\boldsymbol{CA}\mathbf{u}_*, \boldsymbol{A}\mathbf{u}_{ko}) + (1-\kappa)(\boldsymbol{C}^{-1}\boldsymbol{\sigma}_*, \boldsymbol{\sigma}_{ko}) - (1-\kappa)(\boldsymbol{A}\mathbf{u}_*, \boldsymbol{\sigma}_{ko}) +$$
$$(2\kappa-1)(\boldsymbol{K}\mathbf{u}_*, \mathbf{u}_{ko}) - (2\kappa-1)\Pi_s(\mathbf{u}_{ko}) = 0 \qquad (5.1)$$

where κ is an arbitrary numerical parameter, $\kappa \in [0,1]$.

Actually, identity (5.1) defines a family of generalized solutions which depend on parameter κ. Note that integral identity (5.1) contains the principle of virtual displacements as a particular case; it suffices to assume $\kappa = 1$ to see this. On the other hand, at $\kappa = 0$ integral identity (5.1) becomes a Reissner variational equation written for the elements of set $\mathscr{U}_k$.

Now let's show that field F_*, which is the solution of problem (**1**.2.2), (**1**.2.4) in the usual sense, is at the same time a generalized solution, that is,

[5] To be exact, we should note that the variational principle to be considered below implies a particular version of the Reissner variational principle rather than its general case. In the said particular case the functional is defined on kinematically admissible displacement fields.

satisfies the integral identity (5.1). And indeed, according to the principle of virtual displacements we have

$$(\boldsymbol{CA}\mathbf{u}_*, \boldsymbol{A}\mathbf{u}_{ko}) + (\boldsymbol{K}\mathbf{u}_*, \mathbf{u}_{ko}) - \Pi_s(\mathbf{u}_{ko}) = 0\,. \tag{5.2}$$

Further, from (**1**.2.2) $\boldsymbol{CA}\mathbf{u}_* = \boldsymbol{\sigma}_*$ or $\boldsymbol{C}^{-1}\boldsymbol{\sigma}_* = \boldsymbol{A}\mathbf{u}_*$, so, the following identity holds:

$$(\boldsymbol{CA}\mathbf{u}_*, \boldsymbol{A}\mathbf{u}_{ko}) + (\boldsymbol{C}^{-1}\boldsymbol{\sigma}_*, \boldsymbol{\sigma}_{ko}) - (\boldsymbol{\sigma}_*, \boldsymbol{A}\mathbf{u}_{ko}) - (\boldsymbol{A}\mathbf{u}_*, \boldsymbol{\sigma}_{ko}) = 0\,. \tag{5.3}$$

By multiplying (5.2) by $(2\kappa - 1)$ and (5.3) by $(1 - \kappa)$ and summing the results we arrive at identity (5.1).

Now let (5.1) hold. We show that if the generalized solution in the sense of (5.1) is smooth enough, then it is the solution of problem (**1**.2.2), (**1**.2.4) in the usual sense.

To see this, we recall that $\boldsymbol{\sigma}_{ko}$ from (5.1) is arbitrary, and this implies $\boldsymbol{C}^{-1}\boldsymbol{\sigma}_* = \boldsymbol{A}\mathbf{u}_*$, hence identity (5.3) holds. Multiplying (5.3) by $(1 - \kappa)$ and subtracting the result from (5.1), we obtain identity (5.2) after canceling the common factor $(2\kappa - 1)$, and this is nothing but a mathematical notation for the principle of virtual displacements[6]. The principle of virtual displacements implies the equilibrium equations in displacements, provided the $\mathbf{u}_*$ vector is smooth enough [7].

3.5.2 A generalized mixed functional

Let us introduce a family of functional on the $\mathscr{U}_k$ set, such that depend on parameter κ

$$\Phi(\boldsymbol{\sigma},\mathbf{u}) = \frac{\kappa}{2}(\boldsymbol{CA}\mathbf{u}, \boldsymbol{A}\mathbf{u}) + \frac{1-\kappa}{2}(\boldsymbol{C}^{-1}\boldsymbol{\sigma}, \boldsymbol{\sigma}) - (1-\kappa)(\boldsymbol{A}\mathbf{u}, \boldsymbol{\sigma}) +$$

$$+\frac{2\kappa-1}{2}(\boldsymbol{K}\mathbf{u}, \mathbf{u}) - (2\kappa - 1)\Pi_s(\mathbf{u})\,. \tag{5.4}$$

By varying the Φ functional on the $\mathscr{U}_k$ set, we can obtain its stationarity conditions in the form

$$\delta\Phi = \kappa(\boldsymbol{CA}\mathbf{u}, \boldsymbol{A}\delta\mathbf{u}) + (1-\kappa)(\boldsymbol{C}^{-1}\boldsymbol{\sigma}, \delta\boldsymbol{\sigma}) - (1-\kappa)(\boldsymbol{A}\mathbf{u}, \delta\boldsymbol{\sigma}) -$$

$$-(1-\kappa)(\boldsymbol{A}\delta\mathbf{u}, \boldsymbol{\sigma}) + (2\kappa-1)(\boldsymbol{K}\mathbf{u}, \delta\mathbf{u}) - (2\kappa-1)\delta\Pi_s(\mathbf{u}) = 0\,. \tag{5.5}$$

Variations $\delta\boldsymbol{\sigma}$ and $\delta\mathbf{u}$ must not let the varied states leave the $\mathscr{U}_k$ set, so they must belong to the linear set $\mathscr{U}_{ko}$. Thus, we can identify these

[6] We suppose $\kappa \neq ½$.

variations with arbitrary elements from the $\mathscr{U}_{ko}$ set, that is, with $\boldsymbol{\sigma}_{ko}$ and $\mathbf{u}_{ko}$, respectively. After substituting $\delta\boldsymbol{\sigma} = \boldsymbol{\sigma}_{ko}$ and $\delta\mathbf{u} = \mathbf{u}_{ko}$ to (5.5) we will have the integral identity (5.1).

Thus, the problems of searching for a generalized solution that would satisfy the integral identity (5.1) and for a stationarity point of the Φ functional on the $\mathscr{U}_k$ set are equivalent.

Now we are going to prove the following theorem:

If there exists a generalized solution, F_*, in the sense of (5.1), then, provided

$$\tfrac{1}{2} < \kappa < 1, \tag{5.6}$$

this solution gives a minimum to functional Φ.

To see this, we produce the following for any field $F_{ko} \in \mathscr{U}_{ko}$ by taking (5.1) into account:

$$\Phi(F_* + F_{ko}) - \Phi(F_*) =$$

$$= \frac{\kappa}{2}(\boldsymbol{CA}\mathbf{u}_{ko}, \boldsymbol{A}\mathbf{u}_{ko}) + \frac{1-\kappa}{2}(\boldsymbol{C}^{-1}\boldsymbol{\sigma}_{ko}, \boldsymbol{\sigma}_{ko}) - (1-\kappa)(\boldsymbol{A}\mathbf{u}_{ko}, \boldsymbol{\sigma}_{ko}) +$$

$$+ \frac{2\kappa-1}{2}(\boldsymbol{K}\mathbf{u}_{ko}, \mathbf{u}_{ko}), \tag{5.7}$$

and we should prove only that the right part in (5.7) is not negative.

It can be verified directly that the right part in (5.7) can be rewritten in the equivalent form

$$\frac{1-\kappa}{2}(\boldsymbol{C}(\boldsymbol{C}^{-1}\boldsymbol{\sigma}_{ko} - \boldsymbol{A}\mathbf{u}_{ko}), (\boldsymbol{C}^{-1}\boldsymbol{\sigma}_{ko} - \boldsymbol{A}\mathbf{u}_{ko})) + \frac{2\kappa-1}{2}(\boldsymbol{CA}\mathbf{u}_{ko}, \boldsymbol{A}\mathbf{u}_{ko}) +$$

$$+ \frac{2\kappa-1}{2}(\boldsymbol{K}\mathbf{u}_{ko}, \mathbf{u}_{ko}) =$$

$$= \frac{1-\kappa}{2}(\boldsymbol{C}(\boldsymbol{C}^{-1}\boldsymbol{\sigma}_{ko} - \boldsymbol{A}\mathbf{u}_{ko}), (\boldsymbol{C}^{-1}\boldsymbol{\sigma}_{ko} - \boldsymbol{A}\mathbf{u}_{ko})) + (2\kappa-1)\mathsf{E}(\mathbf{u}_{ko}), \tag{5.8}$$

where each term is not negative if (5.6) holds. Immediately we have the estimate

$$\Phi(F_* + F_{ko}) \geq \Phi(F_*), \tag{5.9}$$

and this completes the proof.

If the energy of deformation, $\mathsf{E}(\mathbf{u}_{ko})$, is a strictly positive value for any nonzero vector $\mathbf{u}_{ko} \in \mathscr{U}_{ko}$, then (5.9) will become a strict inequality[7]

[7] We remind that the strict inequality $\mathsf{E}(\mathbf{u}_{ko}) > 0$ takes place in the case when the set of rigid displacements for the system is empty.

$$\Phi(F_* + F_{\text{ko}}) > \Phi(F_*) . \tag{5.10}$$

Under these conditions, and also when inequality (5.6) holds, the generalized solution in the sense of (5.1) is unique. Supposing the opposite, that is, assuming that there are two different fields F_{*1} and F_{*2} which minimize functional Φ, we use (5.10) to derive two mutually contradicting inequalities $\Phi(F_{*1}) < \Phi(F_{*2})$ and $\Phi(F_{*2}) < \Phi(F_{*1})$.

3.5.3 A connection between the Lagrange & Reissner functionals and the generalized mixed functional

It is fairly easy to establish a connection between three functionals: the Lagrange one L, the Reissner one R, and the mixed one Φ:

$$\Phi = \kappa\mathsf{L} + (1 - \kappa)\mathsf{R} . \tag{5.11}$$

Formula (5.11) can be validated by direct substitution[8]. This formula proves to be very useful when establishing the expressions of the mixed functional Φ in various particular cases. By the way, here we can see that the mixed functional Φ can be written in any of three forms (first, second, third) depending on which form is taken by the Reissner functional participating in (5.11).

The form of the mixed functional Φ, $\Phi = \mathsf{L}+2\mathsf{R}$, suggested originally in the paper [13], is a particular case of formula (5.11) derived later in [14]. This form can be derived from (5.11) by assuming $\kappa = {}^2/_3$ and disregarding the common factor ${}^1/_3$ inessential for the stationarity point.

Based on formula (5.11) and taking into account (1.27) and (**2**.4.3), we can find the value of the Φ functional on the solution of the problem:

$$\Phi_* = \kappa\mathsf{L}_* + (1 - \kappa)\mathsf{K}_* = (2\kappa - 1)\mathsf{L}_* .$$

It is convenient to redefine functional Φ, by assuming the following instead of (5.11):

$$\boxed{\Phi = \frac{1}{2\kappa - 1}\left[\kappa\mathsf{L} + (1-\kappa)\mathsf{R}\right].} \tag{5.12}$$

This is a form of functional Φ we will use further. In that case the value of functional Φ on the solution of the problem will coincide with the exact value of the Lagrange functional, that is,

[8] It is the convenient structure of this formula that made us alter the sign of the Reissner functional.

$$\Phi_* = L_*. \tag{5.13}$$

L.A. Rozin drew our attention to another connection between the three functionals. If we introduce a new parameter α by assuming

$$\alpha = \frac{1-\kappa}{2\kappa-1}, \qquad \kappa = \frac{1+\alpha}{2\alpha+1}, \tag{5.14}$$

then formula (5.12) with the new parameter will become

$$\Phi = L + \alpha(L + R). \tag{5.15}$$

Condition (5.6) for the new parameter α turns into

$$0 < \alpha < \infty . \tag{5.16}$$

If we sum the expressions of the Lagrange and Reissner functionals and consider the kinematic boundary conditions, we will have the functional

$$\begin{aligned} Q = L + R &= \tfrac{1}{2}(\boldsymbol{C}^{-1}\boldsymbol{\sigma}, \boldsymbol{\sigma}) + \tfrac{1}{2}(\boldsymbol{CA}\mathbf{u}, \boldsymbol{A}\mathbf{u}) - (\boldsymbol{A}\mathbf{u}, \boldsymbol{\sigma}) = \\ &= \tfrac{1}{2}(\boldsymbol{C}(\boldsymbol{C}^{-1}\boldsymbol{\sigma} - \boldsymbol{A}\mathbf{u}), (\boldsymbol{C}^{-1}\boldsymbol{\sigma} - \boldsymbol{A}\mathbf{u})), \end{aligned} \tag{5.17}$$

and formula (5.15) can be rewritten as

$$\Phi = L + \alpha Q . \tag{5.18}$$

Functional Q can be used as a functional of least squares built for physical equations.

Generally speaking, the idea of supplementing known functionals with additional terms which have the meaning of least-squares functionals helps produce more versions of mixed functionals with similar properties. For example, the same functional Q could be added to the Castigliano functional. More details on this kind of possibilities and other application aspects of various least-squares functionals can be found in the book [12].

The generalized mixed functional Φ, formally introduced above, contains as a particular case both the Reissner functional (at $\kappa = 0$) and the Lagrange functional (at $\kappa = 1$). Note also that the end points of our interval of interest, $\tfrac{1}{2} < \kappa < 1$, are singular points for functional Φ. And indeed, at $\kappa = 0.5$ functional Φ defined by the original formula (5.11) acquires the form of the Q functional because we can derive the following from (5.11):

$$\Phi_{\kappa=0.5} = \tfrac{1}{2} L + \tfrac{1}{2} R = \tfrac{1}{2} Q . \tag{5.19}$$

Functional Q implies only the physical relationships as Euler equations. On the other hand, at $\kappa = 1$ functional Φ degenerates because it becomes a Lagrangian functional and does not contain stresses.

3.5.4 Parametrized energy space

There is a standard technique with which we can reformulate the variational problem for functional Φ on the set $\mathscr{U}_k$ as an equivalent problem of minimizing Φ on the linear set $\mathscr{U}_{ko}$. Let's fixate a certain field, $F_k = \{\mathbf{0}, A\mathbf{v}, \mathbf{v}\}$, from set $\mathscr{U}_k$. Then an arbitrary field, $F \in \mathscr{U}_k$, can be represented as $F = F_k + F_{ko}$, $F_{ko} = \{\boldsymbol{\sigma}, A\mathbf{u}, \mathbf{u}\} \in \mathscr{U}_{ko}$. Using formula (5.12), we can specify the Φ functional as one subject to variation on the $\mathscr{U}_{ko}$ linear set:

$$\Phi = \frac{\kappa}{2(2\kappa-1)}(\boldsymbol{C}A\mathbf{u}, A\mathbf{u}) + \frac{1-\kappa}{2(2\kappa-1)}(\boldsymbol{C}^{-1}\boldsymbol{\sigma}, \boldsymbol{\sigma}) - \frac{1-\kappa}{2\kappa-1} +$$

$$+ (A\mathbf{u}, \boldsymbol{\sigma}) + \frac{1}{2}(\boldsymbol{K}\mathbf{u}, \mathbf{u}) - \Pi_s(\mathbf{u}) + \frac{\kappa}{2\kappa-1}(\boldsymbol{C}A\mathbf{u}, A\mathbf{v}) -$$

$$- \frac{1-\kappa}{2\kappa-1}(A\mathbf{v}, \boldsymbol{\sigma}) + (\boldsymbol{K}\mathbf{u}, \mathbf{v}) + C \tag{5.20}$$

where C is a constant that depends on the fixated vector $\mathbf{v}$. We introduce a bilinear form, $b(F_1,F_2)$, on the linear set $\mathscr{U}_{ko}$, that depends on two fields $F_1 \in \mathscr{U}_{ko}$ and $F_2 \in \mathscr{U}_{ko}$, by assuming

$$b(F_1,F_2) = \frac{\kappa}{2(2\kappa-1)}(\boldsymbol{C}A\mathbf{u}_1, A\mathbf{u}_2) + \frac{1-\kappa}{2(2\kappa-1)}(\boldsymbol{C}^{-1}\boldsymbol{\sigma}_1, \boldsymbol{\sigma}_2) -$$

$$- \frac{1-\kappa}{2(2\kappa-1)}(A\mathbf{u}_1, \boldsymbol{\sigma}_2) - \frac{1-\kappa}{2(2\kappa-1)}(A\mathbf{u}_2, \boldsymbol{\sigma}_1) + \frac{1}{2}(\boldsymbol{K}\mathbf{u}_1, \mathbf{u}_2) .$$

This form, under the conditions

$$\mathsf{E}(\mathbf{u}_{ko}) > 0 \quad \forall\, \mathbf{u}_{ko} \in \mathscr{U}_{ko},\ \mathbf{u}_{ko} \neq \mathbf{0} \text{ и } ½ < \kappa < 1, \tag{5.21}$$

satisfies all axioms of scalar product and thus generates a metric in $\mathscr{U}_{ko}$ because

$$b(F_1,F_1) = \frac{1-\kappa}{2(2\kappa-1)}(\boldsymbol{C}(\boldsymbol{C}^{-1}\boldsymbol{\sigma}_1 - A\mathbf{u}_1), (\boldsymbol{C}^{-1}\boldsymbol{\sigma}_1 - A\mathbf{u}_1)) + \mathsf{E}(\mathbf{u}_1) > 0$$

if $\mathbf{u}_1 \neq \mathbf{0}$.

For condition (5.21) to hold, it suffices to demand that all displacements $\mathbf{u}_{ko}$ allowed for comparison be orthogonal in the main metric to all homogeneously kinematically admissible rigid displacements $\mathbf{u}_{\mathscr{R}o}$

$$(\mathbf{u}_{ko}, \mathbf{u}_{\mathscr{R}o}) = 0. \tag{5.22}$$

A set of all stress-and-strain fields of the type $F = \{\boldsymbol{\sigma}, \boldsymbol{A}\mathbf{u}_{\text{ko}}, \mathbf{u}_{\text{ko}}\}$ with homogeneously kinematically admissible displacements $\mathbf{u}_{\text{ko}}$ that satisfy conditions (5.22) makes up a space with the scalar product generated by the bilinear functional $b(F_1,F_1)$. Completing this space produces a Hilbert space, $\mathscr{F}$, which we will call a *parametrized energy space* conforming to functional Φ.

Obviously, the parametrization means that the κ parameter is introduced. This energy space permits to estimate the error of approximate solutions based on minimization of functional Φ by the respective parametrized energy norm.

Condition (5.22) can be treated now as a decomposition of $\mathscr{U}_{\text{ko}}$ into a direct sum

$$\mathscr{U}_{\text{ko}} = \mathscr{F} \oplus \mathscr{R}_{\text{o}}\,, \tag{5.23}$$

where $\mathscr{R}_{\text{o}}$ means a set of stress-and-strain fields of the type $F = \{\mathbf{0}, \mathbf{0}, \mathbf{u}_{\mathscr{R}\text{o}}\}$ under the condition $\boldsymbol{A}\mathbf{u}_{\mathscr{R}\text{o}} = \mathbf{0}$.

Thus, parametrized energy space $\mathscr{F} \subseteq \mathscr{U}_{\text{ko}}$ consists of a set of fields of the type $F = \{\boldsymbol{\sigma}, \boldsymbol{A}\mathbf{u}, \mathbf{u}\}$ which obey the following requirements:

- displacements $\mathbf{u}$ satisfy homogeneous kinematic boundary conditions;
- displacements $\mathbf{u}$ are orthogonal in the main metric to all homogeneously kinematically admissible rigid displacements of the system;
- scalar product

$$\langle F, F\rangle =$$

$$= \frac{\kappa}{2(2\kappa-1)}(\boldsymbol{C}\boldsymbol{A}\mathbf{u}, \boldsymbol{A}\mathbf{u}) + \frac{1-\kappa}{2(2\kappa-1)}(\boldsymbol{C}^{-1}\boldsymbol{\sigma}, \boldsymbol{\sigma}) - \frac{1-\kappa}{2\kappa-1}(\boldsymbol{A}\mathbf{u}, \boldsymbol{\sigma}) + \frac{1}{2}(\boldsymbol{K}\mathbf{u}_1, \mathbf{u}_2)$$

takes a finite value on any field F from $\mathscr{F}$.

Now let's find Euler equations and natural boundary conditions for functional Φ. To do it, we turn to formula (5.5) which states a condition for the Φ functional's first variation to be equal to zero. Using the basic integral formula (**1**.2.19), we transform formula (5.5) into the following variational equation:

$$\delta\Phi = \kappa(\boldsymbol{A}^{\mathsf{T}}\boldsymbol{C}\boldsymbol{A}\mathbf{u}, \delta\mathbf{u}) - (1-\kappa)(\boldsymbol{A}^{\mathsf{T}}\boldsymbol{\sigma}, \delta\mathbf{u}) + (2\kappa-1)(\boldsymbol{K}\mathbf{u}, \delta\mathbf{u}) -$$

$$-(2\kappa-1)(\bar{\boldsymbol{X}}, \delta\mathbf{u}) + (1-\kappa)(\boldsymbol{C}^{-1}\boldsymbol{\sigma}, \delta\boldsymbol{\sigma}) - (1-\kappa)(\boldsymbol{A}\mathbf{u}, \delta\boldsymbol{\sigma}) -$$

$$-(2\kappa-1)(\boldsymbol{E}_p\bar{\boldsymbol{p}}, \boldsymbol{E}_p\delta\boldsymbol{u})_\Gamma + \kappa(\boldsymbol{E}_p\boldsymbol{H}_\sigma\boldsymbol{C}\boldsymbol{A}\mathbf{u}, \boldsymbol{E}_p\delta\boldsymbol{u})_\Gamma - (1-\kappa)(\boldsymbol{E}_p\boldsymbol{p}, \boldsymbol{E}_p\delta\boldsymbol{u})_\Gamma = 0\,.$$

As the $\delta\mathbf{u}$ and $\delta\boldsymbol{\sigma}$ variations are arbitrary, we have the Euler equations and the natural boundary conditions for functional Φ,

$$\kappa \boldsymbol{A}^{\mathrm{T}}\boldsymbol{C}\boldsymbol{A}\mathbf{u} - (1-\kappa)\boldsymbol{A}^{\mathrm{T}}\boldsymbol{\sigma} + (2\kappa - 1)\boldsymbol{K}\mathbf{u} - (2\kappa - 1)\bar{\boldsymbol{X}} = \mathbf{0} \quad \in \Omega, \tag{5.24}$$

$$\boldsymbol{C}^{-1}\boldsymbol{\sigma} - \boldsymbol{A}\mathbf{u} = \mathbf{0} \quad \in \Omega, \tag{5.25}$$

$$\boldsymbol{E}_p[-(2\kappa - 1)\bar{\boldsymbol{p}} + \kappa \boldsymbol{H}_\sigma \boldsymbol{C}\boldsymbol{A}\mathbf{u} - (1-\kappa)\boldsymbol{p}] = \mathbf{0} \quad \in \Gamma, \tag{5.26}$$

equivalent to the original equations of the problem, (**1**.2.2), (**1**.2.4); the geometric equation (**1**.2.2-*b*) and the kinematic boundary condition (**1**.2.4-*b*) hold because $\mathbf{u} \in \mathcal{U}_k$ (the main boundary condition for the Φ functional). The said equivalence of the equations can be easily established by noticing that (5.24) is a linear combination of the equilibrium equations and physical equations (5.25) with the $\boldsymbol{A}^{\mathrm{T}}\boldsymbol{C}$ operator applied to them. A similar linear combination of the static boundary conditions and the physical equations participates in the natural boundary condition (5.26).

3.6 Gurtin's variational principle

The practice of structural design knows cases when an engineer wants to know only some of the components of the stress-and-strain state in a structure rather than all of them. In particular, in many problems the distribution of stresses is the main thing to know while the strains/displacements are of less interest or of no importance at all for that particular purpose.

On the other hand, traditional approaches of numerical structural analysis based on variational formulations, such as the variation-difference method, finite element method, or Ritz method with smooth approximants, in their best developed forms require either that both stress and kinematic fields be found together using mixed-type functionals or that the strains/displacements be found first and then the stresses be derived from the calculated results.

There is a natural wish to find the stress distribution directly by minimizing the Castigliano functional, but more often than not this approach does not work because of the difficulty to build statically admissible stress fields[9].

The said difficulties can be overcome by making the variational formulation of the problem "rougher" — demanding that the equilibrium

[9] We mean the case when there is no elastic medium in the problem's statement. Then the Castigliano functional depends only on the stresses in the system and does not depend on the displacements.

equations hold in a weak (integral) sense rather than exactly; this will yield a functional with a penalty derived from M. Gurtin's functional for a dynamic problem in convolutions [4]. Later it was found that the Gurtin-type functionals could be extended to the problems of statics, too, if there was an elastic medium with a non-degenerate operator $\boldsymbol{K}$. This kind of problems is most often encountered in beam- and slab-like constructions lying on elastic Winkler-type foundations.

The extension of the approach to the problems with no elastic medium gave birth to the idea of formal introduction of an elastic medium with a non-degenerate operator $\boldsymbol{K}$ to the mechanical model. This artificial medium plays the part of a penalty in the problem's statement, which distorts the solution a little bit but in return provides a convenient transition to the formulation of the variational problem in stresses only [3].

3.6.1 Gurtin's functional

As it was said before, M. Gurtin has constructed a governing functional for problems of dynamics in convolutions, which depends only on the stress vector, $\boldsymbol{\sigma}$, and (which is important!) this Gurtin functional is defined on stress fields not subject beforehand to any equations or boundary conditions. In order not to deviate from the main topic, we will not discuss the statement of a dynamical problem in convolutions, including its variational statement. Instead we will give the respective functional for the static problems with the non-degenerate operator $\boldsymbol{K}$ and then give a validation for the variational equation that corresponds to that functional[10].

However, before doing all that, it is useful to transform the formulation of the original problem (**1**.2.2), (**1**.2.4) into such form that does not include any displacements and stresses. It can be done because we assume the operator $\boldsymbol{K}$ to be non-degenerate. And indeed, the equation of equilibrium (**1**.2.2-*a*) gives

$$\mathbf{u} = \boldsymbol{K}^{-1}(\bar{\boldsymbol{X}} - \boldsymbol{A}^{\mathrm{T}}\boldsymbol{\sigma}) \,. \tag{6.1}$$

On the other hand, by excluding the strains from geometric equations (**1**.2.2-*b*) using physical equations (**1**.2.2-*c*) we obtain

$$\boldsymbol{A}\mathbf{u} = \boldsymbol{C}^{-1}\boldsymbol{\sigma} \,, \tag{6.2}$$

[10] People interested with the statement and validation of the dynamic problem in convolutions can find all necessary information in any of the works [11], [12], [4].

so replacing displacements **u** by their expressions from (6.1) will give a single equation containing the stresses only:

$$\boldsymbol{A}\boldsymbol{K}^{-1}(\overline{\boldsymbol{X}} - \boldsymbol{A}^{\mathsf{T}}\boldsymbol{\sigma}) - \boldsymbol{C}^{-1}\boldsymbol{\sigma} = \boldsymbol{0}\,. \qquad (6.3)$$

Now boundary conditions (1.2.4) can be rewritten in stresses using (6.2) and (6.3):

$$\boldsymbol{E}_p(\boldsymbol{p} - \overline{\boldsymbol{p}}) = \boldsymbol{0}, \qquad \in\Gamma \quad \text{– static boundary conditions,} \qquad (6.4\text{-}a)$$

$$\boldsymbol{E}_u[\boldsymbol{H}_u\boldsymbol{K}^{-1}(\overline{\boldsymbol{X}} - \boldsymbol{A}^{\mathsf{T}}\boldsymbol{\sigma}) - \overline{\boldsymbol{u}}\,] = \boldsymbol{0}, \in\Gamma \quad \text{– kinematic boundary conditions.} \qquad (6.4\text{-}b)$$

Now let's turn to the variational formulation of problem (6.3), (6.4), and consider the following functional G_1 – a Gurtin functional[11]

$$\mathsf{G}_1(\boldsymbol{\sigma}) = \tfrac{1}{2}(\boldsymbol{C}^{-1}\boldsymbol{\sigma}, \boldsymbol{\sigma}) + \tfrac{1}{2}(\boldsymbol{K}^{-1}\boldsymbol{A}^{\mathsf{T}}\boldsymbol{\sigma}, \boldsymbol{A}^{\mathsf{T}}\boldsymbol{\sigma}) - (\boldsymbol{A}\boldsymbol{K}^{-1}\overline{\boldsymbol{X}}, \boldsymbol{\sigma}) \;+$$
$$+ (\boldsymbol{E}_p(\boldsymbol{p} - \overline{\boldsymbol{p}}), \boldsymbol{E}_p\boldsymbol{H}_u\boldsymbol{K}^{-1}\boldsymbol{A}^{\mathsf{T}}\boldsymbol{\sigma})_\Gamma + (\boldsymbol{E}_u\boldsymbol{p}, \boldsymbol{E}_u(\boldsymbol{H}_u\boldsymbol{K}^{-1}\overline{\boldsymbol{X}} - \overline{\boldsymbol{u}}\,))_\Gamma\,. \qquad (6.5)$$

By varying the G_1 functional with respect to stresses **σ** we derive

$$\delta\mathsf{G}_1 = (\boldsymbol{C}^{-1}\boldsymbol{\sigma}, \delta\boldsymbol{\sigma}) + (\boldsymbol{K}^{-1}\boldsymbol{A}^{\mathsf{T}}\boldsymbol{\sigma}, \boldsymbol{A}^{\mathsf{T}}\delta\boldsymbol{\sigma}) - (\boldsymbol{A}\boldsymbol{K}^{-1}\overline{\boldsymbol{X}}, \delta\boldsymbol{\sigma}) \;+$$
$$+ (\boldsymbol{E}_p\delta\boldsymbol{p}, \boldsymbol{E}_p\boldsymbol{H}_u\boldsymbol{K}^{-1}\boldsymbol{A}^{\mathsf{T}}\boldsymbol{\sigma})_\Gamma \;+$$
$$+ (\boldsymbol{E}_p(\boldsymbol{p} - \overline{\boldsymbol{p}}), \boldsymbol{E}_p\boldsymbol{H}_u\boldsymbol{K}^{-1}\boldsymbol{A}^{\mathsf{T}}\delta\boldsymbol{\sigma})_\Gamma - (\boldsymbol{E}_u\delta\boldsymbol{p}, \boldsymbol{E}_u(\boldsymbol{H}_u\boldsymbol{K}^{-1}\overline{\boldsymbol{X}} - \overline{\boldsymbol{u}}\,))_\Gamma\,.$$

Now we transform this expression using the main integral formula:

$$\delta\mathsf{G}_1 = (\boldsymbol{C}^{-1}\boldsymbol{\sigma} + \boldsymbol{A}\boldsymbol{K}^{-1}\boldsymbol{A}^{\mathsf{T}}\boldsymbol{\sigma} - \boldsymbol{A}\boldsymbol{K}^{-1}\overline{\boldsymbol{X}}, \delta\boldsymbol{\sigma}) \;+$$
$$+ (\boldsymbol{E}_u\delta\boldsymbol{p}, \boldsymbol{E}_u\boldsymbol{H}_u\boldsymbol{K}^{-1}(\overline{\boldsymbol{X}} - \boldsymbol{A}^{\mathsf{T}}\boldsymbol{\sigma}))_\Gamma \;-$$
$$- (\boldsymbol{E}_u\delta\boldsymbol{p}, \boldsymbol{E}_u\overline{\boldsymbol{u}}\,)_\Gamma + (\boldsymbol{E}_p(\boldsymbol{p} - \overline{\boldsymbol{p}}), \boldsymbol{E}_p\boldsymbol{H}_u\boldsymbol{K}^{-1}\boldsymbol{A}^{\mathsf{T}}\delta\boldsymbol{\sigma})_\Gamma. \qquad (6.6)$$

Equaling the $\delta\mathsf{G}_1$ variation to zero gives the Euler equation and the natural boundary conditions for the Gurtin functional which repeat (6.3) and (6.4) exactly.

The variational formulation of the problem (6.3), (6.4), based on the Gurtin functional as in (6.5), has a shortcoming: it has overly strict requirements to the smoothness of the vector of external actions, $\overline{\boldsymbol{X}}$, and to the tensor of compliance of the elastic medium, $\boldsymbol{K}^{-1}$, because their components in (6.5) are differentiated (subject to the action of the matrix

---

[11] The $_1$ subscript emphasizes that we mean here the first form of the Gurtin functional.

differential operator of geometry, $\boldsymbol{A}$). This shortcoming can be removed by applying the basic integral formula to the $(\boldsymbol{A}\boldsymbol{K}^{-1}\bar{\boldsymbol{X}}, \boldsymbol{\sigma})$ term. As a result, we obtain the second form of the Gurtin functional

$$\mathsf{G}_2 = \tfrac{1}{2}(\boldsymbol{C}^{-1}\boldsymbol{\sigma}, \boldsymbol{\sigma}) + \tfrac{1}{2}(\boldsymbol{K}^{-1}\boldsymbol{A}^{\mathsf{T}}\boldsymbol{\sigma}, \boldsymbol{A}^{\mathsf{T}}\boldsymbol{\sigma}) - (\boldsymbol{K}^{-1}\bar{\boldsymbol{X}}, \boldsymbol{A}^{\mathsf{T}}\boldsymbol{\sigma}) +$$

$$+ (\boldsymbol{E}_p(\boldsymbol{p} - \bar{\boldsymbol{p}}), \boldsymbol{E}_p\boldsymbol{H}_u\boldsymbol{K}^{-1}\boldsymbol{A}^{\mathsf{T}}\boldsymbol{\sigma})_\Gamma - (\boldsymbol{E}_p\boldsymbol{p}, \boldsymbol{E}_p\boldsymbol{H}_u\boldsymbol{K}^{-1}\bar{\boldsymbol{X}})_\Gamma -$$

$$- (\boldsymbol{E}_u\boldsymbol{p}, \boldsymbol{E}_u\bar{\boldsymbol{u}})_\Gamma . \tag{6.7}$$

In the general case, the Gurtin functional does not have an extremum in the stationarity point. To see this, let $\boldsymbol{\sigma}_*$ be the exact solution of problem (6.3), (6.4). The increment of the Gurtin functional, G, from point $\boldsymbol{\sigma}_*$ to point $\boldsymbol{\sigma}_* + \delta\boldsymbol{\sigma}$ will be equal to $\tfrac{1}{2}\delta^2\mathsf{G}$, so the sign of this increment is defined by that of the second variation, $\delta^2\mathsf{G}$

$$\delta^2\mathsf{G} = (\boldsymbol{C}^{-1}\delta\boldsymbol{\sigma}, \delta\boldsymbol{\sigma}) + (\boldsymbol{K}^{-1}\boldsymbol{A}^{\mathsf{T}}\delta\boldsymbol{\sigma}, \boldsymbol{A}^{\mathsf{T}}\delta\boldsymbol{\sigma}) + 2(\boldsymbol{E}_p\delta\boldsymbol{p}, \boldsymbol{E}_p\boldsymbol{H}_u\boldsymbol{K}^{-1}\boldsymbol{A}^{\mathsf{T}}\delta\boldsymbol{\sigma})_\Gamma . \tag{6.8}$$

As the $\delta\boldsymbol{\sigma}$ variations are arbitrary, they can be chosen in such way that $\delta\boldsymbol{p} = \boldsymbol{0} \in \Gamma$, and then $\delta^2\mathsf{G} > 0$.

On the other hand, the $\delta\boldsymbol{\sigma}$ tensor can be selected in such way that Euler equation (6.3) hold for it at $\bar{\boldsymbol{X}} = \boldsymbol{0}$

$$\boldsymbol{A}\boldsymbol{K}^{-1}\boldsymbol{A}^{\mathsf{T}}\delta\boldsymbol{\sigma} + \boldsymbol{C}^{-1}\delta\boldsymbol{\sigma} = \boldsymbol{0} . \tag{6.9}$$

By making a scalar product of this equality with $\delta\boldsymbol{\sigma}$ in the Ω area and using the basic integral formula, we derive

$$(\boldsymbol{K}^{-1}\boldsymbol{A}^{\mathsf{T}}\delta\boldsymbol{\sigma}, \boldsymbol{A}^{\mathsf{T}}\delta\boldsymbol{\sigma}) + (\boldsymbol{C}^{-1}\delta\boldsymbol{\sigma}, \delta\boldsymbol{\sigma}) + (\delta\boldsymbol{p}, \boldsymbol{H}_u\boldsymbol{K}^{-1}\boldsymbol{A}^{\mathsf{T}}\delta\boldsymbol{\sigma})_\Gamma = 0 .$$

If we further limit the stress variations taken into consideration by requiring that $\boldsymbol{E}_u\delta\boldsymbol{p} = \boldsymbol{0} \in \Gamma$, this equality can be rewritten as

$$(\boldsymbol{E}_p\delta\boldsymbol{p}, \boldsymbol{E}_p\boldsymbol{H}_u\boldsymbol{K}^{-1}\boldsymbol{A}^{\mathsf{T}}\delta\boldsymbol{\sigma})_\Gamma = - (\boldsymbol{K}^{-1}\boldsymbol{A}^{\mathsf{T}}\delta\boldsymbol{\sigma}, \boldsymbol{A}^{\mathsf{T}}\delta\boldsymbol{\sigma}) - (\boldsymbol{C}^{-1}\delta\boldsymbol{\sigma}, \delta\boldsymbol{\sigma}),$$

and the expression for $\delta^2\mathsf{G}$ for this special case of variation will become

$$\delta^2\mathsf{G} = - (\boldsymbol{K}^{-1}\boldsymbol{A}^{\mathsf{T}}\delta\boldsymbol{\sigma}, \boldsymbol{A}^{\mathsf{T}}\delta\boldsymbol{\sigma}) - (\boldsymbol{C}^{-1}\delta\boldsymbol{\sigma}, \delta\boldsymbol{\sigma}) < 0 . \tag{6.10}$$

Thus, different methods of variation of the stresses makes the second variation, $\delta^2\mathsf{G}$, have opposite signs, therefore there is no extremum in the point of stationarity of the G functional.

Further simplifications in the Gurtin functional can be done by narrowing the set of stress fields allowed for comparison — limiting it to the statically semi-admissible fields, that is, stress fields which satisfy the static boundary conditions. Then the G functional can be rewritten as

$$\mathsf{G}_3 = ½(\boldsymbol{C}^{-1}\boldsymbol{\sigma}, \boldsymbol{\sigma}) + ½(\boldsymbol{K}^{-1}\boldsymbol{A}^{\mathsf{T}}\boldsymbol{\sigma}, \boldsymbol{A}^{\mathsf{T}}\boldsymbol{\sigma}) - (\boldsymbol{K}^{-1}\overline{\boldsymbol{X}}, \boldsymbol{A}^{\mathsf{T}}\boldsymbol{\sigma}) - \\ - (\boldsymbol{E}_u\boldsymbol{p}, \boldsymbol{E}_u\overline{\boldsymbol{u}})_\Gamma + C \qquad (6.11)$$

where $C = -(\boldsymbol{E}_p\overline{\boldsymbol{p}}, \boldsymbol{E}_p\boldsymbol{H}_u\boldsymbol{K}^{-1}\overline{\boldsymbol{X}})_\Gamma$ is an additive constant that does not affect the search for a stationarity point of the functional. Expression (6.11) will be called the third form of the Gurtin functional.

The last term in the right part of (6.8) is absent in the third form of the Gurtin functional, consequently, the variational principle of minimum holds for this functional defined on statically semi-admissible stress fields.

Let's introduce two more functionals:

$$\mathsf{K}_\sigma = ½(\boldsymbol{C}^{-1}\boldsymbol{\sigma}, \boldsymbol{\sigma}) - (\boldsymbol{E}_u\boldsymbol{p}, \boldsymbol{E}_u\overline{\boldsymbol{u}})_\Gamma, \quad \mathsf{Q}_\sigma = (\boldsymbol{K}^{-1}(\boldsymbol{A}^{\mathsf{T}}\boldsymbol{\sigma} - 2\overline{\boldsymbol{X}}), \boldsymbol{A}^{\mathsf{T}}\boldsymbol{\sigma}). \qquad (6.12)$$

The K_σ functional is formally a Castigliano functional for the elastic body, while Q_σ can be treated as a functional of quadratic residual in the equations of equilibrium of the elastic body in the absence of an elastic medium, where the $\boldsymbol{K}^{-1}$ matrix plays the role of a weight function. This treatment of the Q_σ functional becomes clear when we represent the second formula in (6.12) as

$$\mathsf{Q}_\sigma = (\boldsymbol{K}^{-1}(\boldsymbol{A}^{\mathsf{T}}\boldsymbol{\sigma} - \overline{\boldsymbol{X}}), (\boldsymbol{A}^{\mathsf{T}}\boldsymbol{\sigma} - \overline{\boldsymbol{X}})) - (\boldsymbol{K}^{-1}\overline{\boldsymbol{X}}, \overline{\boldsymbol{X}})$$

and omit the constant $(\boldsymbol{K}^{-1}\overline{\boldsymbol{X}}, \overline{\boldsymbol{X}})$ that appears in this transformation of Q_σ because the constant is not essential for the stationarity.

Comparing (6.11) and (6.12) yields a simple formula for the Gurtin functional defined on statically semi-admissible stress fields:

$$\mathsf{G}_3 = \mathsf{K}_\sigma + ½\mathsf{Q}_\sigma, \qquad \mathsf{Q}_\sigma = (\boldsymbol{K}^{-1}(\boldsymbol{A}^{\mathsf{T}}\boldsymbol{\sigma} - \overline{\boldsymbol{X}}), (\boldsymbol{A}^{\mathsf{T}}\boldsymbol{\sigma} - \overline{\boldsymbol{X}})). \qquad (6.13)$$

It is useful for determining expressions of the Gurtin functional's third form in particular problems.

This treatment of the Gurtin functional's third problem has one more meaning: it produces an idea that the functional can be used in problems where there is no elastic medium in the original statement ($\boldsymbol{K} = \boldsymbol{O}$). If we used the variational principle by Castigliano with the functional K_σ as a basic tool in the problems of this kind, we would have to demand that stresses admitted to the comparison satisfy the equilibrium conditions in the volume of the body and static conditions on the boundary, that is,

$$\boldsymbol{A}^{\mathsf{T}}\boldsymbol{\sigma} - \overline{\boldsymbol{X}} = \boldsymbol{0} \ \in \Omega, \qquad \boldsymbol{E}_p\boldsymbol{p} = \boldsymbol{E}_p\overline{\boldsymbol{p}} \ \in \Gamma. \qquad (6.14)$$

Assuming $\boldsymbol{K} = k\boldsymbol{I}$, we reduce the problem of a conditional minimum of the K_σ functional with limitations (6.14) to the problem of an unconditional minimum of a functional with a penalty,

$$\mathsf{G}_3 = \mathsf{K}_\sigma + \frac{1}{2k}\left((A^{\mathsf{T}}\boldsymbol{\sigma} - \bar{X}), (A^{\mathsf{T}}\boldsymbol{\sigma} - \bar{X}) \right), \tag{6.15}$$

where the first limitation in (6.14) is taken into account integrally by introducing a penalty term to the functional itself, and the second limitation is taken into account explicitly by narrowing the set of stresses admitted to the comparison.

At $k \to 0$ ($1/k$ is a penalty factor) the solution of the problem of minimization of (6.15) tends to the solution of the original problem [15].

Note that one of useful applications of the penalty method in structural analysis is to use the penalty functional for formal reduction of the differentiation order under the integral, to weaken the requirements to smoothness of coordinate functions in the Ritz method [3].

3.7 Geometric interpretation of functionals used in structural mechanic

It is useful to give a geometric interpretation to some of the functionals employed in structural mechanics. Obviously, an analysis of this kind can be done in a simplest way on a system that has one degree of freedom. As in this case we are not going to involve a space of more than three dimensions, the analysis will be also easier to understand and imagine.

3.7.1 Generalized mixed functional

So, let's consider a spring of stiffness c and compliance $d = 1/c$. Let a compressive force, P, be applied to the spring. We denote by u the shortening of the spring and by N the stress that develops in it. Then

$$\mathsf{L} = \tfrac{1}{2}\, cu^2 - Pu\,, \qquad \mathsf{R} = \tfrac{1}{2}\, N^2 d - Nu + Pu\,, \tag{7.1}$$

and according to formula (5.11) the generalized mixed functional, Φ, for this problem will be

$$\Phi = \frac{\kappa}{2}\, cu^2 + \frac{1-\kappa}{2} N^2 d - (1-\kappa)\, Nu - (2\kappa - 1)\, Pu\,. \tag{7.2}$$

This simple example can be used to track visually how the Φ functional changes its behavior when the κ parameter varies.

Figs. 3.1–3.5 show axonometric images of the $\Phi(N,u)$ surface at $\kappa = 0$, $\tfrac{1}{4}$, $\tfrac{1}{2}$, $\tfrac{3}{4}$, 1, respectively, in the case of $c = 1$, $P = \tfrac{1}{2}$.

The images show clearly a "saddle" (a negative Gaussian curvature of the surface) in the stationarity point at $0 < \kappa < ½$. On the other hand, at $½ < \kappa < 1$ the stationarity point realizes a minimum (the surface has a positive Gaussian curvature).

$\kappa = 0$ Reissnerian

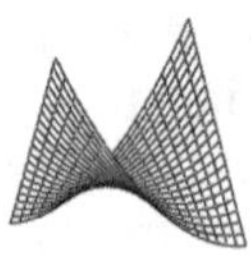

Fig. 3.1.

$\kappa = ¼$

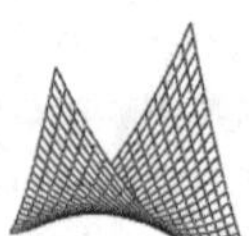

Fig. 3.2.

$\kappa = ½$

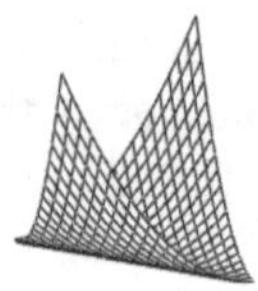

Fig. 3.3.

$\kappa = ¾$

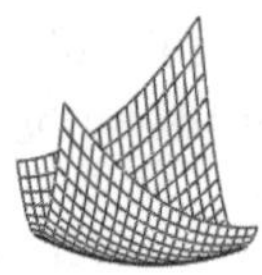

Fig. 3.4.

$\kappa = 1$ Lagrangian

Fig. 3.5.

Shapes of the surface graphs of the $\Phi(N,u)$ functional at different values of κ

Based on the expression of the Φ functional from (7.2), we can find conditions for its minimum in the following form:

$$\frac{\partial \Phi}{\partial u} = \kappa\, c\, u - (1-\kappa)\, N - (2\kappa - 1)P = 0, \tag{7.3}$$

$$\frac{\partial \Phi}{\partial N} = -\,(1-\kappa)\, u + (1-\kappa)\, dN = 0\,. \tag{7.4}$$

We can always choose a dimensionality of the participating quantities in such way that $c = 1$, $d = 1$. In this case the matrix of the simultaneous equations (7.3), (7.4) is

$$\left\| \begin{bmatrix} \kappa & \kappa-1 \\ \kappa-1 & 1-\kappa \end{bmatrix} \right\|, \tag{7.5}$$

and its eigenvalues are

$$\lambda_{1,2} = \frac{1 \pm \sqrt{1-4(1-\kappa)(2\kappa-1)}}{2}\,. \tag{7.6}$$

Fig. 3.6 shows a dependence of the maximum, λ_2, and the minimum, λ_1, eigenvalues of the matrix from (7.6) on the κ parameter, where it can be seen also that, as it could be expected, matrix (7.5) is positive definite at $½ < \kappa < 1$.

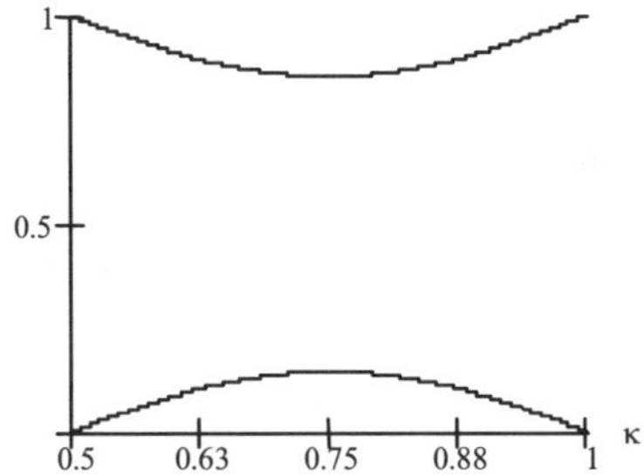

Fig. 3.6. Dependence of λ_1 and λ_2 on the κ parameter.

The arbitrariness in choosing parameter κ should be used to improve somehow the quality of the problem's solution. For example, one may require that the system of equations be stablest possible.

As it is known [4], an adequate measure of stability of a system of equations can be a so-called conditionality number, H, of that system's matrix:

$$H = \frac{|\lambda_{max}|}{|\lambda_{min}|}, \tag{7.7}$$

and the stablest system possible is one where the conditionality number is minimum.

As it can be seen in Fig. 3.6, the conditionality number H is minimum at $\kappa = 3/4$, consequently, on the interval $1/2 < \kappa < 1$ the system of equations (7.3), (7.4) is stablest at the central point of the admissible interval, that is, it is least sensible to the error of computation at $\kappa = 3/4$. By the way, the optimum parameter in this sense, $\kappa = 3/4$, corresponds to the value of α equal to $\alpha = 1/2$ according to (5.17).

Thus, we have

$$\Phi = \tfrac{1}{2}\,(3\mathsf{L} + \mathsf{R})\,. \tag{7.8}$$

The ½ factor, itself inessential for the stationarity point of the Φ functional, is used here because it makes the exact value of the functional, Φ_*, identified with L_*

$$\Phi_* = \mathsf{L}_*\,, \tag{7.9}$$

which can be useful in comparisons of energy norms of different approximate solutions of the problem.

An alluring hypothesis appears at this point: what if this value of the parameter, $\kappa = 3/4$, always conforms to a stablest system of Ritz equations? However, the analysis shown below makes this hypothesis totally invalid.

So, let us consider a mechanical system with n degrees of freedom. Let $\boldsymbol{C}$ be a stiffness matrix of the system, $\boldsymbol{C}^{-1}$ a compliance matrix, $\mathbf{u}$ a vector of displacements, $\boldsymbol{N}$ a vector of internal generalized forces, $\boldsymbol{P}$ a vector of external generalized forces which conform to the selected degrees of freedom. Let's write the expressions of the Lagrange functional, L, and the Reissner functional, R, in the matrix form. We have

$$\mathsf{L} = \tfrac{1}{2}\,\mathbf{u}^{\mathsf{T}}\boldsymbol{C}\mathbf{u} - \boldsymbol{P}^{\mathsf{T}}\mathbf{u}\,, \quad \mathsf{R} = \tfrac{1}{2}\,\boldsymbol{N}^{\mathsf{T}}\boldsymbol{C}^{-1}\boldsymbol{N} - \boldsymbol{N}^{\mathsf{T}}\mathbf{u} + \boldsymbol{P}^{\mathsf{T}}\mathbf{u}\,. \tag{7.10}$$

From the general formula (5.11) we derive this expression of the functional:

$$\Phi = \frac{\kappa}{2}\,\mathbf{u}^{\mathsf{T}}\boldsymbol{C}\mathbf{u} + \frac{1-\kappa}{2}\boldsymbol{N}^{\mathsf{T}}\boldsymbol{C}^{-1}\boldsymbol{N} - (1-\kappa)\boldsymbol{N}^{\mathsf{T}}\mathbf{u} - (2\kappa-1)\boldsymbol{P}^{\mathsf{T}}\mathbf{u}\,. \tag{7.11}$$

Conditions of stationarity of the Φ functional help obtain the following system of equations

$$\left\|\begin{bmatrix} \kappa\boldsymbol{C} & (\kappa-1)\boldsymbol{I} \\ (\kappa-1)\boldsymbol{I} & (1-\kappa)\boldsymbol{C}^{-1} \end{bmatrix}\right\| \left\|\begin{bmatrix} \mathbf{u} \\ \boldsymbol{N} \end{bmatrix}\right\| = \left\|\begin{bmatrix} (2\kappa-1)\boldsymbol{P} \\ \mathbf{0} \end{bmatrix}\right\|, \tag{7.12}$$

and our task here is to find the conditionality number, H, of the matrix of equation system (7.12); this matrix will be designated by $\boldsymbol{G}$:

$$\boldsymbol{G} = \left\|\begin{bmatrix} \kappa\boldsymbol{C} & (\kappa-1)\boldsymbol{I} \\ (\kappa-1)\boldsymbol{I} & (1-\kappa)\boldsymbol{C}^{-1} \end{bmatrix}\right\|. \tag{7.13}$$

To determine the eigenvalues, λ, of the $\boldsymbol{G}$ matrix, we equal to zero the respective characteristic determinant, that is, we assume

$$\begin{vmatrix} \kappa\boldsymbol{C} - \lambda\boldsymbol{I} & (\kappa-1)\boldsymbol{I} \\ (\kappa-1)\boldsymbol{I} & (1-\kappa)\boldsymbol{C}^{-1} - \lambda\boldsymbol{I} \end{vmatrix} = 0\,.$$

Using the Schur formulas [2], we transform this characteristic equation into

$$\begin{aligned} &|(\kappa\boldsymbol{C} - \lambda\boldsymbol{I})[(1-\kappa)\boldsymbol{C}^{-1} - \lambda\boldsymbol{I})] - (\kappa-1)^2\boldsymbol{I}| = \\ &= |-\lambda[(1-\kappa)\boldsymbol{C}^{-1} + \kappa\boldsymbol{C}] + [(1-\kappa)(2\kappa-1)+\lambda^2]\boldsymbol{I}| = 0. \end{aligned} \tag{7.14}$$

After introducing the designation

$$\chi = \frac{(1-\kappa)(2\kappa-1)+\lambda^2}{\lambda} \tag{7.15}$$

we can rewrite equation (7.14) as

$$|[(1-\kappa)\boldsymbol{C}^{-1}+\kappa\boldsymbol{C}]-\chi\boldsymbol{I}|=0,$$

consequently, χ is an eigenvalue of the matrix $(1-\kappa)\boldsymbol{C}^{-1}+\kappa\boldsymbol{C}$.

Further, matrix $\boldsymbol{C}$ is symmetric and positive definite, therefore all its eigenvalues, μ_i ($i=1,\ldots,n$), are strictly positive, so

$$0<\mu_1\leq\mu_2\leq\ldots\leq\mu_n\,.$$

It is easy to understand that eigenvalues χ_i of the matrix $(1-\kappa)\boldsymbol{C}^{-1}+\kappa\boldsymbol{C}$ and eigenvalues μ_i of matrix $\boldsymbol{C}$ are related as[12]

$$\chi_i=\frac{1-\kappa}{\mu_i}+\kappa\mu_i\,. \tag{7.16}$$

Replacing the χ_i parameter with its expression from (7.15), we obtain a relationship between λ_i and μ_i

$$\frac{(1-\kappa)(2\kappa-1)+\lambda_i^2}{\lambda_i}=\frac{1-\kappa}{\mu_i}+\kappa\mu_i\,. \tag{7.17}$$

For each index $i=1,\ldots,n$, formula (7.17) gives two values of λ_i, which is quite appropriate because the order of matrix $\boldsymbol{G}$ is $2n$. By solving equation (7.17), which is quadratic with respect to λ_i we derive

$$\lambda_i=\frac{1-\kappa+\kappa\mu_i^2\pm\sqrt{(1-\kappa)^2+\kappa^2\mu_i^4-2(1-\kappa)(3\kappa-2)\mu_i^2}}{2\mu_i}\,. \tag{7.18}$$

Without limiting the generality, we can assume that the major eigenvalue of the $\boldsymbol{C}$ matrix is equal to one, that is, $\mu_n=1$. Obviously, it can always be done by scaling the $\boldsymbol{C}$ matrix appropriately.

If we denote by λ_i^+ the expression in (7.18) with the plus sign before the root and by λ_i^- the same expression with the minus sign, we can represent the desirable conditionality number, H, of the $\boldsymbol{G}$ matrix as a function of parameter κ in the following form:

---

[12] This follows from the consideration that both matrices, $\boldsymbol{C}$ and $\boldsymbol{C}^{-1}$, can be reduced to their diagonal form by the same similarity transformation, $\boldsymbol{T}^{-1}\boldsymbol{C}\boldsymbol{T}=\boldsymbol{\Lambda}$ and $\boldsymbol{T}^{-1}\boldsymbol{C}^{-1}\boldsymbol{T}=\boldsymbol{\Lambda}^{-1}$. Here $\boldsymbol{\Lambda}$ is a diagonal matrix with elements $\mu_1,\ldots,\mu_n$. Consequently, $\boldsymbol{T}^{-1}[(1-\kappa)\boldsymbol{C}^{-1}+\kappa\boldsymbol{C}]\boldsymbol{T}=(1-\kappa)\boldsymbol{\Lambda}^{-1}+\kappa\boldsymbol{\Lambda}$, hence (7.16).

$$H(\kappa) = \frac{\max_i \lambda_i^+}{\min_j \lambda_j^-} =$$

$$= \max_i \max_j \frac{\mu_j[1-\kappa+\kappa\mu_i^2+\sqrt{(1-\kappa)^2+\kappa^2\mu_i^4-2(1-\kappa)(3\kappa-2)\mu_i^2}\,]}{\mu_i[1-\kappa+\kappa\mu_j^2-\sqrt{(1-\kappa)^2+\kappa^2\mu_j^4-2(1-\kappa)(3\kappa-2)\mu_j^2}\,]}. \quad (7.19)$$

It is not possible to find the general expression of the κ parameter on the allowable interval $\frac{1}{2} < \kappa < 1$, at which the $H(\kappa)$ function has its minimum. Therefore we involve a numerical experiment.

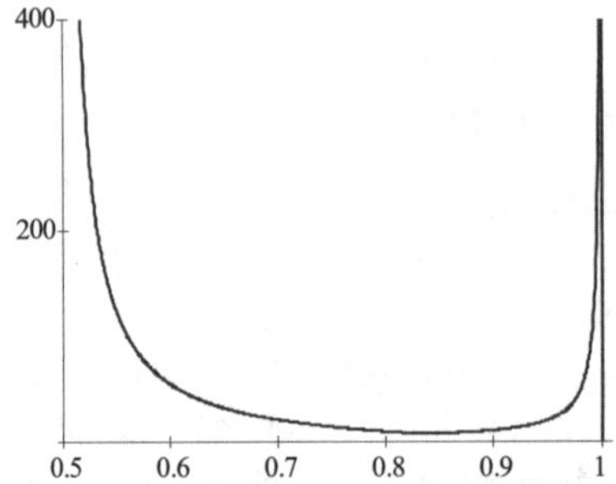

Fig. 3.7. An example of the $H = H(\kappa)$ dependence

For simplicity, we assume $n = 2$, and let $\mu_1 = 0.2$ and $\mu_2 = 1$. We can use the well-known ***Mathcad*** software to build a graph of the dependence $H = H(\kappa)$ on the basis of formula (7.19) at the values of the parameters indicated above. The result is shown in Fig. 3.7.

The graph in Fig. 3.7 shows that the minimum of function $H(\kappa)$ is achieved in this case at κ slightly greater than $\kappa = \frac{3}{4}$ (approximately, at $\kappa = 0{,}84$).

3.7.2 A remark on the Gurtin functional

Now let's use an equally simple system to track the behavior of the Gurtin functional, G. Let a package of two springs connected in parallel, one of which being a 'master' with stiffness c and the other being a 'slave' with stiffness k, be subjected to an external force, P (Fig. 3.8).

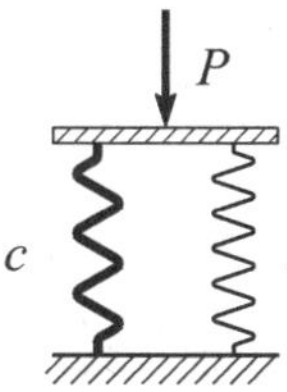

Fig. 3.8. A package of two parallel springs

Let u be a shortening of the package of springs, as before, and let N be a force that develops in the main spring. Obviously,

$$u = \frac{P}{c+k}, \qquad N = \frac{Pc}{c+k}. \tag{7.20}$$

In this case the master spring is an elastic body while the slave spring is an elastic medium, and we have

$$\mathsf{K}_\sigma = \frac{1}{2}N^2 d, \quad \mathsf{Q}_\sigma = \frac{1}{k}(N-P)^2, \quad d = 1/c\,, \tag{7.21}$$

wherefrom, using formula (6.13), we derive

$$\mathsf{G} = \frac{1}{2}N^2 d + \frac{1}{2k}(N-P)^2. \tag{7.22}$$

Here G is a quadratic parabola with its point of minimum found from the condition

$$\frac{d\mathsf{G}}{dN} = 0 \qquad \text{or} \qquad Nd + \frac{N-P}{k} = 0\,,$$

wherefrom (7.20) follows. By the way, it is easy to calculate in this case that

$$\mathsf{G}_* = \frac{P^2}{2(c+k)}. \tag{7.23}$$

3.8 Final comments to Chapter 3

Gurtin himself was interested, as it was said before, in a variational formulation of the dynamics problem in convolutions rather than the static variational problem; this is how he obtained his functional G. A similar functional was obtained later, after Gurtin, by using the penalty method.

The same functional was built independently by M.D. Nikolsky [6] and used by him effectively in numerical procedures for static structural analysis of constructions supported by elastic Winkler-type foundations. In this regard, titling the G functional in problems of statics as the Gurtin functional is not quite unquestionable, but is totally acceptable in consideration of priority.

References

1. Ecklund I, Temam R (1979) Convex analysis and variational problems (in Russian). Mir, Moscow
2. Gantmakher FR (1967) Theory of matrices (in Russian). Nauka, Moscow
3. Gordon LA, Shoikhet BA (1972) On a method of numerical analysis of thin plates and shells (in Russian). Bull. VNIIG, Vol.98: 34-42
4. Gurtin ME (1964) Variational Principles for Linear Elastodynamics. Arch. Rat. Mech. Anal., N16: 34-50
5. Hellinger E (1914) Die allgemeine Ansätze der Mechanik der Kontinua. Enzycl. Math. Wiss., **IV**, 4: 654-655
6. Nikolsky MD (1980) To the formulation of the principle of minimum of complementary work in linear elasticity (in Russian). In collected papers: Mechanics of materials and transport constructions. Leningrad Inst. Railway Engng., pp. 108-117
7. Pobedrya BE (1979) Some general theorems of mechanics of deformable solids (in Russian). Appl. Math and Mech., Vol.43, 3: 531–541
8. Reissner E (1950) On a variational theorem in elasticity. J. Math. And Phys., **29**, N2: 90-95
9. Reissner E (1961) On some variational theorems of elasticity (in Russian). In collected papers: Problems of mechanics of continua. In honor of 70th anniversary of Academician N.I. Muskhelishvili. Academy of Science of USSR, Moscow, pp. 328–337
10. Rockafellar R (1973) Convex analysis (in Russian). Mir, Moscow
11. Rozin LA (1978) Variational formulations of problems for elastic systems (in Russian). Len. State Univ. Publ., Leningrad
12. Rozin LA (1998) Problems of elasticity and numerical methods for their solution (in Russian). St.-Petersburg: Publish. House of St.-Pet. State Tech. Univ.
13. Slivker VI (1982) On a mixed variational formulation of problems for elastic systems (in Russian). Bull. Acad. Sci. USSR, Mechanics of deformable solids, N4: 88-97
14. Slivker VI (1983) A mixed variational principle of structural mechanics, generated by a convex functional (in Russian). Researches on theoretical

principles of structural analysis. Collected papers of universities. Leningrad Institute of Civil Engineering, pp. 114-121

15. Strang G, Fix GJ (1973) An analysis of the finite element method. Prentice-Hall Inc., Englewood Cliffs, N.Y.
16. Tonti E (1969) Variational principles in elasticity (in Russian). Collected translations "Mechanics", Moscow, 5: 124-138
17. Washizu K (1968) Variational methods in elasticity and plasticity. Pergamon Press, Oxford
18. Wilkinson JH (1965) The algebraic eigenvalue problem. Clarendon Press, Oxford

4 PARTICULAR CLASSES OF PROBLEMS IN STRUCTURAL MECHANICS – part 1

Construction of a model of any physical process begins with laws of conservation. If we may say so, the laws of conservation are a foundation, a basis of any model or simulation.

Moiseyev NN (1979) Mathematics makes an experiment (in Russian). Nauka, Moscow

4.1 Variations of the operator formulations in structural mechanics

This subsection will present a brief discussion on some variations of the problem statements in their differential forms.

The contents of this book concerns mainly the variational formulations of mechanical problems, and the respective differential forms are of secondary importance, therefore we do not dwell on advantages and disadvantages of those formulations; we only give a minimum amount of information which we will need in further presentment.

4.1.1 Statement of a problem in displacements

First of all, let us write out the original problem statement (**1**.2.2)-(**1**.2.4), to have it right here instead of referring to previous chapters:

$$\boldsymbol{A}^{\mathrm{T}}\boldsymbol{\sigma} + \boldsymbol{K}\mathbf{u} = \overline{\boldsymbol{X}} \qquad \text{equations of equilibrium,} \qquad (1.1\text{-}a)$$

$$\boldsymbol{A}\mathbf{u} = \boldsymbol{\varepsilon} \qquad \text{geometric equations,} \qquad (1.1\text{-}b)$$

$$\boldsymbol{\sigma} = \boldsymbol{C}\boldsymbol{\varepsilon} \ \text{ or } \ \boldsymbol{\varepsilon} = \boldsymbol{C}^{-1}\boldsymbol{\sigma} \qquad \text{physical equations.} \qquad (1.1\text{-}c)$$

The simultaneous governing equations of the problem (1.1) should be supplemented with boundary conditions which look as follows in their operator form:

$\boldsymbol{E}_p(\boldsymbol{H}_\sigma\boldsymbol{\sigma} - \bar{\boldsymbol{p}}) = \mathbf{0}$ static boundary conditions, (1.2-*a*)

$\boldsymbol{E}_u(\boldsymbol{H}_u\mathbf{u} - \bar{\boldsymbol{u}}) = \mathbf{0}$ kinematic boundary conditions. (1.2-*b*)

If we exclude strains $\boldsymbol{\varepsilon}$ from geometric equations (1.1-*b*) using physical equations (1.1-*c*), then stresses $\boldsymbol{\sigma}$ can be expressed directly via the displacements,

$$\boldsymbol{\sigma} = \boldsymbol{C}\boldsymbol{A}\mathbf{u}\,, \tag{1.3}$$

which, substituted to (1.1-*a*), produces equations of equilibrium in displacements:

$$\boldsymbol{A}^{\mathsf{T}}\boldsymbol{C}\boldsymbol{A}\mathbf{u} + \boldsymbol{K}\mathbf{u} = \bar{\boldsymbol{X}}\,. \tag{1.4}$$

Accordingly, boundary conditions (1.2) can be expressed via the displacements only, using (1.3):

$\boldsymbol{E}_p(\boldsymbol{H}_\sigma\boldsymbol{C}\boldsymbol{A}\mathbf{u} - \bar{\boldsymbol{p}}) = \mathbf{0}$ static boundary conditions, (1.5-*a*)

$\boldsymbol{E}_u(\boldsymbol{H}_u\mathbf{u} - \bar{\boldsymbol{u}}) = \mathbf{0}$ kinematic boundary conditions (1.5-*b*)

The statement of an elastic problem in displacements is usually associated with the name of Lame, therefore further we will call the formal differential operator

$$\boldsymbol{L} = \boldsymbol{A}^{\mathsf{T}}\boldsymbol{C}\boldsymbol{A} + \boldsymbol{K} \tag{1.6}$$

a *Lame operator*. Actually, the Lame operator equation is an equation of equilibrium of a mechanical system written in displacements,

$$\boldsymbol{L}\mathbf{u} = \bar{\boldsymbol{X}}. \tag{1.7}$$

Being supplemented by boundary conditions (1.5), operator equation (1.7) will be a full operator statement of the original problem in displacements. After the problem is solved, that is, after a field of displacements is calculated, the strains will be found from geometric equations (1.1-*b*), and then the strains will be used to find the stresses from physical equations (1.1-*c*).

4.1.2 Statement of a problem in stresses

Let $\boldsymbol{S}$ be a Saint-Venant matrix differential operator — a strain compatibility operator. This means that strains $\boldsymbol{\varepsilon}$ ensure the continuity of the body by satisfying the homogeneous equations with the $\boldsymbol{S}$ operator. In the mathematical language we could say that the strains are compatible if and only if they belong to the kernel of the Saint-Venant operator,

$$\boldsymbol{S}\boldsymbol{\varepsilon} = \mathbf{0}. \tag{1.8}$$

If we replace the strains in Saint-Venant equations (1.8) by their expressions via the displacements according to (1.1-*b*), we will have the identity,

$$\boldsymbol{S}\boldsymbol{A}\mathbf{u} = \mathbf{0}, \tag{1.9}$$

because strains of the type $\boldsymbol{\varepsilon} = \boldsymbol{A}\mathbf{u}$ do not violate the continuity of a deformable body according to the very definition of the strain compatibility. So, operator $\boldsymbol{SA}$ is an annulling operator for any vector $\mathbf{u}$, in other words,

$$\boldsymbol{S}\boldsymbol{A} = \boldsymbol{O}. \tag{1.10}$$

Now, using the conjugation operation on both parts of (1.10), we will have also

$$\boldsymbol{A}^{\mathsf{T}}\boldsymbol{S}^{\mathsf{T}} = \boldsymbol{O}. \tag{1.11}$$

To return to the discussion on statements of problems in stresses, we consider two separate variations of the problems. The first variation is where there is no elastic foundation, that is, where we can assume $\boldsymbol{K} = \boldsymbol{O}$ in (1.1-*a*). If we replace the strains in (1.8) by the stresses using Hooke's law, then we will have a set of equations in the Ω area where only the stresses participate as unknowns:

$$\boldsymbol{A}^{\mathsf{T}}\boldsymbol{\sigma} = \bar{\boldsymbol{X}} \quad \in \Omega\,, \tag{1.12-a}$$

$$\boldsymbol{S}\boldsymbol{C}^{-1}\boldsymbol{\sigma} = \mathbf{0} \quad \in \Omega\,. \tag{1.12-b}$$

If the boundary conditions along the whole boundary, Γ, are of a static type, then they define a differential statement of the problem in stresses together with equations (1.12). If there are kinematic boundary conditions, the differential statement in displacements is difficult.

Now let's turn to second option. The second variation has the operator of elastic foundation, $\boldsymbol{K}$, positive definite in the whole area Ω. But then the operator is reversible, and we can derive the following from the equation of equilibrium (1.1-*a*):

$$\mathbf{u} = \boldsymbol{K}^{-1}(\bar{\boldsymbol{X}} - \boldsymbol{A}^{\mathsf{T}}\boldsymbol{\sigma})\,. \tag{1.13}$$

Substituting this to the rest of equations (1.1) and making some elementary transformations, we have

$$(\boldsymbol{A}\boldsymbol{K}^{-1}\boldsymbol{A}^{\mathsf{T}} + \boldsymbol{C}^{-1})\boldsymbol{\sigma} = \boldsymbol{A}\boldsymbol{K}^{-1}\,\overline{\boldsymbol{X}}\,. \tag{1.14}$$

All boundary conditions can be easily written in stresses in this case, too

$\boldsymbol{E}_p(\boldsymbol{H}_\sigma\boldsymbol{\sigma} - \overline{\boldsymbol{p}}) = \boldsymbol{0}$ static boundary conditions, (1.15-*a*)

$\boldsymbol{E}_u[\boldsymbol{H}_u\boldsymbol{K}^{-1}(\overline{\boldsymbol{X}} - \boldsymbol{A}^{\mathsf{T}}\boldsymbol{\sigma}) - \overline{\boldsymbol{u}}\,] = \boldsymbol{0}$ kinematic boundary conditions. (1.15-*b*)

4.2 Spatial elasticity

We suppose that our reader is familiar with basics of theory of elasticity, therefore we present principal relationships of linear spatial elasticity here without their derivation. By using a designation, which is common in tensor analysis, of a derivative with respect to coordinate x_i as subscript i ($i = 1,2,3$) of the variable of differentiation after a comma,

$$\frac{\partial()}{\partial x_i} = ()_{,i}\,,$$

we can rewrite those relationships as follows:

$-\sigma^{ij}_{,j} + k^{ij}u_j = \overline{X}^i$ equations of equilibrium, (2.1-*a*)

$\varepsilon_{ij} = \frac{1}{2}\,(u_{i,j} + u_{j,i})$ geometric equations, (2.1-*b*)

$\sigma^{ij} = C^{ijkl}\,\varepsilon_{kl}$ or $\varepsilon_{ij} = D_{ijkl}\,\sigma^{kl}$ physical equations. (2.1-*c*)

We represent the stress tensor, $\boldsymbol{\sigma} = \{\sigma^{ij}\}$, the strain tensor, $\boldsymbol{\varepsilon} = \{\varepsilon_{ij}\}$, and the displacement vector, **u**, in a matrix form where we use a symmetry of the $\boldsymbol{\sigma}$ and $\boldsymbol{\varepsilon}$ tensors:

$$\boldsymbol{\sigma} = |[\,\sigma^{11}, \sigma^{22}, \sigma^{33}, \sigma^{12}, \sigma^{23}, \sigma^{31}]|^{\mathsf{T}}\,,$$

$$\boldsymbol{\varepsilon} = |[\varepsilon_{11}, \varepsilon_{22}, \varepsilon_{33}, \gamma_{12}, \gamma_{23}, \gamma_{31}]|^{\mathsf{T}} \quad \text{where } \gamma_{ij} = 2\varepsilon_{ij} \text{ for } i \neq j,$$

$$\mathbf{u} = |[u_1, u_2, u_3]|^{\mathsf{T}}\,. \tag{2.2}$$

Now all three relationships in (2.1) can have the matrix form as in (1.1), where the matrix differential operator of geometry, $\boldsymbol{A}$, is as follows in its component representation:

$$
\boldsymbol{A} = \begin{bmatrix} \partial/\partial x_1 & 0 & 0 \\ 0 & \partial/\partial x_2 & 0 \\ 0 & 0 & \partial/\partial x_3 \\ \partial/\partial x_2 & \partial/\partial x_1 & 0 \\ 0 & \partial/\partial x_3 & \partial/\partial x_2 \\ \partial/\partial x_3 & 0 & \partial/\partial x_1 \end{bmatrix} = \begin{bmatrix} ()_{,1} & 0 & 0 \\ 0 & ()_{,2} & 0 \\ 0 & 0 & ()_{,3} \\ ()_{,2} & ()_{,1} & 0 \\ 0 & ()_{,3} & ()_{,2} \\ ()_{,3} & 0 & ()_{,1} \end{bmatrix}. \qquad (2.3)
$$

But then the equilibrium operator conjugate in the Lagrangian sense, $\boldsymbol{A}^{\mathsf{T}}$, will become as follows, according to common rules of operator conjugation (see §**1**.2):

$$
\boldsymbol{A}^{\mathsf{T}} = \begin{bmatrix} -\partial/\partial x_1 & 0 & 0 & -\partial/\partial x_2 & 0 & -\partial/\partial x_3 \\ 0 & -\partial/\partial x_2 & 0 & -\partial/\partial x_1 & -\partial/\partial x_3 & 0 \\ 0 & 0 & -\partial/\partial x_3 & 0 & -\partial/\partial x_2 & -\partial/\partial x_1 \end{bmatrix} \qquad (2.4)
$$

where it really corresponds to equilibrium equations (2.1-*a*).

Let's pay more attention to the strain compatibility equations which are referred to by theory of elasticity as Saint-Venant equations. After being expanded into separate components, they will become [12]

$$S_{ij} = \in_{ikm} \in_{jln} \varepsilon_{kl,mn} = 0.$$

Here S_{ij} are components of a so-called *incompatibility tensor*, $\in_{ikm}$ is a Levi–Chivita symbol. The Levi–Chivita symbol, $\in_{ikm}$, is defined by unit vectors $\boldsymbol{i}_1$, $\boldsymbol{i}_2$, $\boldsymbol{i}_3$ of Cartesian rectangular coordinate system (x_1, x_2, x_3) using well-known formulas

$$\in_{ikm} = \boldsymbol{i}_i \bullet (\boldsymbol{i}_k \times \boldsymbol{i}_m) .$$

The Saint-Venant matrix differential operator, $\boldsymbol{S}$, of strain compatibility can be represented as

$$\boldsymbol{S}=\begin{bmatrix} 0 & \frac{\partial^2}{\partial x_3^2} & \frac{\partial^2}{\partial x_2^2} & 0 & -\frac{\partial^2}{\partial x_2 \partial x_3} & 0 \\ \frac{\partial^2}{\partial x_3^2} & 0 & \frac{\partial^2}{\partial x_1^2} & 0 & 0 & -\frac{\partial^2}{\partial x_3 \partial x_1} \\ \frac{\partial^2}{\partial x_2^2} & \frac{\partial^2}{\partial x_1^2} & 0 & -\frac{\partial^2}{\partial x_1 \partial x_2} & 0 & 0 \\ 0 & 0 & -\frac{\partial^2}{\partial x_1 \partial x_2} & -\frac{\partial^2}{2\partial x_3^2} & \frac{\partial^2}{2\partial x_3 \partial x_1} & \frac{\partial^2}{2\partial x_2 \partial x_3} \\ -\frac{\partial^2}{\partial x_2 \partial x_3} & 0 & 0 & \frac{\partial^2}{2\partial x_3 \partial x_1} & -\frac{\partial^2}{2\partial x_1^2} & \frac{\partial^2}{2\partial x_1 \partial x_2} \\ 0 & -\frac{\partial^2}{\partial x_3 \partial x_1} & 0 & \frac{\partial^2}{2\partial x_2 \partial x_3} & \frac{\partial^2}{2\partial x_1 \partial x_2} & -\frac{\partial^2}{2\partial x_2^2} \end{bmatrix}.$$

In the matrix form of a general anisotropic case, matrix $\boldsymbol{C}$ of the material's stiffness coefficients will include 21 independent coefficients, seeing that it is symmetric. Its inverse matrix, $\boldsymbol{C}^{-1}$, of compliance coefficients of the elastic body's material has the same kind of representation.

$$\boldsymbol{C}=\begin{bmatrix} c^{11} & c^{12} & c^{13} & c^{14} & c^{15} & c^{16} \\ & c^{22} & c^{23} & c^{24} & c^{25} & c^{26} \\ & & c^{33} & c^{34} & c^{35} & c^{36} \\ & & & c^{44} & c^{45} & c^{46} \\ & & & & c^{55} & c^{56} \\ & & & & & c^{66} \end{bmatrix}, \quad \boldsymbol{C}^{-1}=\begin{bmatrix} d_{11} & d_{12} & d_{13} & d_{14} & d_{15} & d_{16} \\ & d_{22} & d_{23} & d_{24} & d_{25} & d_{26} \\ & & d_{33} & d_{34} & d_{35} & d_{36} \\ & & & d_{44} & d_{45} & d_{46} \\ & & & & d_{55} & d_{56} \\ & & & & & d_{66} \end{bmatrix}.$$

The elements of the $\boldsymbol{C}$ and $\boldsymbol{C}^{-1}$ matrices below the main diagonal are not shown in this matrix form because they can be easily restored from the condition of symmetry.

In a particular but practically important case of an isotropic elastic material, the components of these tensors, $\boldsymbol{C}=\{C^{ijkl}\}$ and $\boldsymbol{C}^{-1}=\{D_{ijkl}\}$, are expressed via two independent constants [13]:

$$C^{ijkl}=\frac{E}{1+\nu}\left[\frac{\nu}{1-2\nu}\delta^{ij}\delta^{kl}+\frac{1}{2}(\delta^{ik}\delta^{jl}+\delta^{jk}\delta^{il})\right],$$

$$D_{ijkl}=\frac{1}{E}\left[-\nu\delta_{ij}\delta_{kl}+\frac{1+\nu}{2}(\delta_{ik}\delta_{jl}+\delta_{jk}\delta_{il})\right],$$

where $\delta_{ij} = \delta^{ij}$ is Kronecher's delta.

This is how it looks in the matrix representation:

$$\boldsymbol{C} = \begin{Vmatrix} \lambda+2\mu & \lambda & \lambda & 0 & 0 & 0 \\ \lambda & \lambda+2\mu & \lambda & 0 & 0 & 0 \\ \lambda & \lambda & \lambda+2\mu & 0 & 0 & 0 \\ 0 & 0 & 0 & \mu & 0 & 0 \\ 0 & 0 & 0 & 0 & \mu & 0 \\ 0 & 0 & 0 & 0 & 0 & \mu \end{Vmatrix},$$

$$\boldsymbol{C}^{-1} = \frac{1}{E}\begin{Vmatrix} 1 & -\nu & -\nu & 0 & 0 & 0 \\ -\nu & 1 & -\nu & 0 & 0 & 0 \\ -\nu & -\nu & 1 & 0 & 0 & 0 \\ 0 & 0 & 0 & 2(1+\nu) & 0 & 0 \\ 0 & 0 & 0 & 0 & 2(1+\nu) & 0 \\ 0 & 0 & 0 & 0 & 0 & 2(1+\nu) \end{Vmatrix},$$

where Lame elastic constants λ and μ are related to technical characteristics of the material, elasticity modulus E and Poisson ratio ν, through the following dependencies:

$$E - \mu\frac{3\lambda+2\mu}{\lambda+\mu}, \qquad \nu = \frac{\lambda}{2(\lambda+\mu)}, \tag{2.5}$$

or, inversely,

$$\lambda = \frac{E\nu}{(1-2\nu)(1+\nu)}, \qquad \mu = \frac{E}{2(1+\nu)}. \tag{2.6}$$

Now let us discuss boundary conditions. Let $\boldsymbol{n} = \|[n_1,\ n_2,\ n_3]\|^{\mathrm{T}}$ be a vector of an external normal to surface Γ that bounds area Ω occupied by an elastic body. In each point of boundary Γ, three boundary conditions are posed: either static or kinematic, or both. Let $(\boldsymbol{n}, \boldsymbol{t}, \boldsymbol{b})$ be a right-hand Cartesian triple of axes of a local coordinate system placed in each point of the boundary surface, Γ, in the way described in Section **1**.2. Let us assume that the vectors of given boundary forces, $\overline{\boldsymbol{p}}$, and displacements, $\overline{\boldsymbol{u}}$, can be represented by either their projections on the global coordinate axes or their projections onto the local axes. In the first case we have

$$\overline{\boldsymbol{p}} = \|[\,\overline{p}^1, \overline{p}^2, \overline{p}^3\,]\|^{\mathrm{T}}, \qquad \overline{\boldsymbol{u}} = \|[\,\overline{u}_1, \overline{u}_2, \overline{u}_3\,]\|^{\mathrm{T}}, \tag{2.7}$$

while in the second case

$$\overline{\boldsymbol{p}} = \|[\,\overline{p}^n, \overline{p}^t, \overline{p}^b\,]\|^{\mathrm{T}}, \qquad \overline{\boldsymbol{u}} = \|[\,\overline{u}_n, \overline{u}_t, \overline{u}_b\,]\|^{\mathrm{T}}. \tag{2.8}$$

In each point of the boundary surface, each coordinate component has one of its two respective quantities specified: either a component of the force vector, $\overline{\boldsymbol{p}}$, or a component of the displacement vector, $\overline{\boldsymbol{u}}$. The set of specified boundary conditions is defined by the contents of operators $\boldsymbol{E}_p$ and $\boldsymbol{E}_u$ which were called earlier *boundary condition extraction operators*. For the spatial elasticity, these algebraic operators are

$$\boldsymbol{E}_p = \left\|\begin{bmatrix} e_{p1} & 0 & 0 \\ 0 & e_{p2} & 0 \\ 0 & 0 & e_{p3} \end{bmatrix}\right\|, \qquad \boldsymbol{E}_u = \left\|\begin{bmatrix} e_{u1} & 0 & 0 \\ 0 & e_{u2} & 0 \\ 0 & 0 & e_{u3} \end{bmatrix}\right\|, \tag{2.9}$$

or

$$\boldsymbol{E}_p = \left\|\begin{bmatrix} e_{pn} & 0 & 0 \\ 0 & e_{pt} & 0 \\ 0 & 0 & e_{pb} \end{bmatrix}\right\|, \qquad \boldsymbol{E}_u = \left\|\begin{bmatrix} e_{un} & 0 & 0 \\ 0 & e_{ut} & 0 \\ 0 & 0 & e_{ub} \end{bmatrix}\right\|, \tag{2.10}$$

where representation (2.9) corresponds to the global coordinate system and (2.10) to the local one.

Each of the diagonal components of matrix $\boldsymbol{E}_p$ is equal to either one or zero, and the same requirement relates also to the diagonal components of matrix $\boldsymbol{E}_u$, where, according to conditions (**1**.2.5),

$$e_{p\alpha} + e_{u\alpha} = 1 \qquad (\alpha = 1,2,3;\ \text{or}\ \alpha = n,t,b)\,. \tag{2.11}$$

As for matrix $\boldsymbol{H}_\sigma$, it can be written in the following form, where the boundary conditions are specified in the global coordinates:

$$\boldsymbol{H}_\sigma = \left\|\begin{bmatrix} n_1 & 0 & 0 & n_2 & 0 & n_3 \\ 0 & n_2 & 0 & n_1 & n_3 & 0 \\ 0 & 0 & n_3 & 0 & n_2 & n_1 \end{bmatrix}\right\|, \tag{2.12}$$

where n_i is a component of the $\boldsymbol{n}$ vector of the external normal to boundary Γ with respect to the global axis x_i. Interestingly enough, the structures of matrices $\boldsymbol{A}^{\mathrm{T}}$ and $\boldsymbol{H}_\sigma$ are identical; matrix $\boldsymbol{H}_\sigma$ can be derived from matrix $\boldsymbol{A}^{\mathsf{T}}$ by formally replacing the differentiation by the respective component of vector $\boldsymbol{n}$.

And indeed, the matrix product $\boldsymbol{H}_\sigma\boldsymbol{\sigma}$ produces a vector, $\|[\sigma^{1j}n_j, \sigma^{2j}n_j, \sigma^{3j}n_j]\|^{\mathsf{T}}$, of boundary forces acting on a spot with the normal $\boldsymbol{n}$, and the coordinates of that vector refer to the global coordinates. As a result, boundary conditions (1.2) in their component form will be

$$e_{pi}(\sigma^{ij}n_j - \overline{p}^i) = 0 \qquad (i = 1,2,3), \tag{2.13-a}$$

$$e_{ui}(u_i - \overline{u}_i) = 0 \qquad (i = 1,2,3)\,. \tag{2.13-b}$$

The same boundary conditions in the equivalent matrix form are

$$\boldsymbol{E}_p(\boldsymbol{H}_\sigma\boldsymbol{\sigma} - \overline{\boldsymbol{p}}) = \boldsymbol{0}\ , \tag{2.14-a}$$

$$\boldsymbol{E}_u(\boldsymbol{H}_u\mathbf{u} - \overline{\boldsymbol{u}}) = \boldsymbol{0}\,. \tag{2.14-b}$$

It follows from (2.13-*b*) that the vector of boundary displacements, $\boldsymbol{u}$, coincides with the vector of displacements, $\mathbf{u}$, on boundary Γ, hence a conclusion — matrix $\boldsymbol{H}_u$ is an identity matrix of third order. Of course, this conclusion is true if the boundary conditions are formulated in the global coordinate system. Note that the geometry and equilibrium operators for spatial elasticity, $\boldsymbol{A}$ and $\boldsymbol{A}^{\mathsf{T}}$, contain differential operators of first order at the most, and this is why operator $\boldsymbol{H}_u$ is an identity operator in this case.

If the boundary conditions are specified by their components along the axes of a local basis, then a matrix of coordinate transformation, $\boldsymbol{\Lambda}$, can be introduced for convenience:

$$\boldsymbol{\Lambda} = \left\| \begin{bmatrix} \lambda_1^n & \lambda_2^n & \lambda_3^n \\ \lambda_1^t & \lambda_2^t & \lambda_3^t \\ \lambda_1^b & \lambda_2^b & \lambda_3^b \end{bmatrix} \right\| = \left\| \begin{bmatrix} n_1 & n_2 & n_3 \\ t_1 & t_2 & t_3 \\ b_1 & b_2 & b_3 \end{bmatrix} \right\| \tag{2.15}$$

its components being projections of the local basis' unit vectors onto the axes of the global coordinate system. Then, by expanding the vector of boundary forces and the vector of boundary displacements over the local basis, we will have the following instead of (2.13):[1]

$$e_{p\alpha}(p^\alpha - \overline{p}^\alpha) = 0\,, \qquad p^\alpha = \lambda_i^\alpha\,\sigma^{ij}n_j \qquad (\alpha = n, t, b)\,, \tag{2.16-a}$$

[1] Note that the $e_{pi},\ldots, e_{u\alpha}$ quantities are not components of a tensor, therefore the rule of summation over repeating indexes does not apply to them.

$$e_{u\alpha}(u_\alpha - \overline{u}_\alpha) = 0\,, \qquad u_\alpha = \lambda^i_\alpha\, u_i \qquad (\alpha = n, t, b)\,. \qquad (2.16\text{-}b)$$

So, in the formulation of the boundary conditions with respect to the local coordinate system, the vector of boundary forces, $\boldsymbol{p}$, and the vector of boundary displacements, $\boldsymbol{u}$, must be converted beforehand to the local coordinate system. This conversion can be done with formal matrix transformations by left-multiplying matrices $\boldsymbol{H}_\sigma$ and $\boldsymbol{H}_u$ which participate in (2.14) by the coordinate transformation matrix, $\boldsymbol{\Lambda}$ — that is, by assuming

$$\boldsymbol{H}_\sigma = \boldsymbol{\Lambda}\left\|\begin{bmatrix} n_1 & 0 & 0 & n_2 & 0 & n_3 \\ 0 & n_2 & 0 & n_1 & n_3 & 0 \\ 0 & 0 & n_3 & 0 & n_2 & n_1 \end{bmatrix}\right\|, \qquad \boldsymbol{H}_u = \boldsymbol{\Lambda}. \qquad (2.17)$$

The basic integral identity (**1**.2.17) for spatial elasticity can be represented in components as

$$\frac{1}{2}\int_\Omega \sigma^{ij}(u_{i,j} + u_{j,i})d\Omega = -\int_\Omega \sigma^{ij}_{\ ,j} u_i d\Omega + \oint_\Gamma \sigma^{ij} n_j u_i d\Gamma\,. \qquad (2.18)$$

Now let us write out explicit expressions of some functionals of structural mechanics in application to spatial (three-dimensional) elasticity.

4.2.1 Lagrange functional

$$\mathsf{L}(\mathbf{u}) = \iint_\Omega \left[\frac{1}{8}C^{ijkl}(u_{i,j} + u_{j,i})(u_{k,l} + u_{l,k}) + \frac{1}{2}k^{ij}u_i u_j - u_i \overline{X}^i\right] d\Omega\, -$$

$$- \oint_\Gamma e_{pi} u_i \overline{p}^i d\Gamma\,. \qquad (2.19)$$

This functional is defined on a set of vector functions $\mathbf{u}$ which satisfy kinematic boundary conditions (2.13-*b*) or (2.16-*b*) and assure a finite value of the energy integral, E,

$$\mathsf{E}(\mathbf{u}) = \frac{1}{2}\iint_\Omega \left[\frac{1}{4}C^{ijkl}(u_{i,j} + u_{j,i})(u_{k,l} + u_{l,k}) + k^{ij}u_i u_j\right] d\Omega\,. \qquad (2.20)$$

Euler equations for the L functional are the Lame equations which are especially simple in the case of an isotropic material [11]:

$$-G\left(\Delta u_i + \frac{1}{1-2\nu}\frac{\partial\theta}{\partial x_i}\right) + k^{ij}u_j = \overline{X}^i \tag{2.21}$$

where G is a shear modulus and θ is a dilatation,

$$G = \frac{E}{2(1+\nu)}, \qquad \theta = \frac{\partial u_1}{\partial x_1} + \frac{\partial u_2}{\partial x_2} + \frac{\partial u_3}{\partial x_3}. \tag{2.22}$$

Static boundary conditions are natural boundary conditions for the L functional.

The variational formulation of the problem based on minimizing the Lagrange functional conforms to the operator statement of the problem in displacements.

4.2.2 Reissner functional

According to (**3**.1.4) and (**3**.1.5), we present expressions for the first and second forms of the Reissner functional in an expanded representation for three-dimensional elasticity:

$$\mathsf{R}_1(\boldsymbol{\sigma},\mathbf{u}) = \int_\Omega \left[\frac{1}{2}D_{ijkl}\sigma^{ij}\sigma^{kl} - \frac{1}{2}k^{ij}u_iu_j - \frac{1}{2}(u_{i,j}+u_{j,i})\sigma^{ij} + u_i\overline{X}^i\right]d\Omega +$$
$$+\oint_\Gamma e_{pi}u_i\overline{p}^i d\Gamma - \oint_\Gamma e_{ui}(\overline{u}_i - u_i)\sigma^{ij}n_j d\Gamma, \tag{2.23}$$

$$\mathsf{R}_2(\boldsymbol{\sigma},\mathbf{u}) = \int_\Omega \left(\frac{1}{2}D_{ijkl}\sigma^{ij}\sigma^{kl} - \frac{1}{2}k^{ij}u_iu_j + u_i\sigma^{ij}_{,j} + u_i\overline{X}^i\right)d\Omega +$$
$$+\oint_\Gamma e_{pi}u_i(\overline{p}^i - \sigma^{ij}n_j)d\Gamma - \oint_\Gamma e_{ui}\overline{u}_i\sigma^{ij}n_j d\Gamma. \tag{2.24}$$

The Reissner functional is defined on a set of stress tensors and displacement vectors not subject to any predefined boundary conditions. The requirements to smoothness of the integrand functions in the Reissner functionals are based on the existence of finite values of the integrals in the respective expressions of R_1 and R_2.

Euler equations for functional R are: equilibrium equations (2.1-*a*) and physical equations (2.1-*c*). All boundary conditions play the part of natural boundary conditions for functional R .

4.2.3 Castigliano functional

$$\mathsf{K}(\boldsymbol{\sigma},\mathbf{u}) = \frac{1}{2}\int_{\Omega}(D_{ijkl}\sigma^{ij}\sigma^{kl} + k^{ij}u_i u_j)d\Omega - \oint_{\Gamma} e_{ui}\bar{u}_i\sigma^{ij}n_j d\Gamma . \quad (2.25)$$

The Castigliano functional is defined on a set of stress tensors and displacement vectors which satisfy equilibrium equations (2.1-*a*) and static boundary conditions (2.13-*a*) or (2.16-*a*).

Euler equations for functional K are equations of strain compatibility in stresses (1.12-*b*), and its natural boundary conditions are kinematic-type boundary conditions (2.13-*b*) or (2.16-*b*).

4.2.4 Gurtin functional (third form)

$$\mathsf{G}_3(\boldsymbol{\sigma}) = \int_{\Omega}\left(\frac{1}{2}D_{ijkl}\sigma^{ij}\sigma^{kl} + \frac{1}{2}b_{ij}\sigma^{ik}_{,k}\sigma^{jl}_{,l} + b_{ij}\sigma^{jk}_{,k}\bar{X}^i\right)d\Omega -$$

$$-\oint_{\Gamma} e_{ui}\bar{u}_i\sigma^{ij}n_j d\Gamma , \quad (2.26)$$

where $\|[b_{ij}]\| = \boldsymbol{K}^{-1}$.

Gurtin functional in its third form is defined on a set of stress tensors which satisfy static boundary conditions (2.13-*a*) or (2.16-*a*). Finite values are supposed to exist for all integrals included in the expression of G_3. Obviously, the Gurtin functional is defined only for problems for which the $\boldsymbol{K}$ operator is not degenerate.

4.3 Plane elasticity

There are two versions of two-dimensional, or plane, elasticity: *plane stress* and *plain deformation.* Area Ω in plane elasticity is a two-dimensional region which we assume to lie in the (*x*, *y*)-plane, just to be definite. In both plane elasticity versions all desirable functions depend on two coordinates, which are *x* and *y* according to our convention.

A plane stress (state) is a kind of a deformed state in a body where the stresses on all planes parallel to the (*x*, *y*)-plane are assumed to be zero, and the stressed state at all points on the perpendicular to that plane is the same. We can write the condition for the plane stress to take place as follows, by denoting the normal stresses as σ^i and the tangential stresses as τ^{ij}:

$$\sigma^z = 0, \quad \tau^{yz} = \tau^{zx} = 0, \quad \sigma^x = \sigma^x(x,y), \quad \sigma^y = \sigma^y(x,y), \quad \tau^{xy} = \tau^{xy}(x,y).$$

The plane stress takes place, for example, in a thin plate subject to loads constant over its thickness and parallel to the plate's plane.

A plane deformation (plane strain state) is a kind of a deformed state in a body where the displacements of all points of this body are parallel to the same plane — to the (x, y)-plane under our convention. Introducing the z-axis as an axis orthogonal to the (x, y)-plane and denoting the respective components of the three-dimensional displacement vector by u, v, w, we can write the conditions for a plane deformation to take place as follows:

$$u = u(x, y)\,, \quad v = v(x, y)\,, \quad w = 0\,.$$

The desirable stresses, strains, and displacements, together with external forces distributed over a two-dimensional area Ω, can be represented in the matrix form as follows:

$$\boldsymbol{\sigma} = \|[\sigma^x, \sigma^y, \tau^{xy}]\|^{\mathsf{T}}, \quad \boldsymbol{\varepsilon} = \|[\varepsilon_x, \varepsilon_y, \gamma_{xy}]\|^{\mathsf{T}}, \quad \mathbf{u} = \|[u, v]\|^{\mathsf{T}},$$
$$\bar{\boldsymbol{X}} = \|[\bar{X}, \bar{Y}]\|^{\mathsf{T}}. \tag{3.1}$$

Now let us write out the basic equations of plane elasticity in coordinates [5], confining ourselves to the case of no elastic medium ($\boldsymbol{K} = \boldsymbol{O}$). Differential equations of equilibrium are:

$$-\frac{\partial \sigma^x}{\partial x} - \frac{\partial \tau^{xy}}{\partial y} = \bar{X}\,, \qquad -\frac{\partial \tau^{yx}}{\partial x} - \frac{\partial \sigma^y}{\partial y} = \bar{Y}\,. \tag{3.2}$$

Geometric equations (Cauchy relationships), which establish a connection between the strains and the displacements, are:

$$\varepsilon_x = \frac{\partial u}{\partial x}\,, \qquad \varepsilon_y = \frac{\partial v}{\partial y}\,, \qquad \gamma_{xy} = \frac{\partial v}{\partial x} + \frac{\partial u}{\partial y}\,. \tag{3.3}$$

Physical equations, which we write here for an isotropic material in plane stress, are:

$$\sigma^x = \frac{E}{1-\nu^2}(\varepsilon_x + \nu\,\varepsilon_y), \quad \sigma^y = \frac{E}{1-\nu^2}(\varepsilon_y + \nu\,\varepsilon_x), \quad \tau^{xy} = \frac{E}{2(1+\nu)}\gamma_{xy}\,, \tag{3.4}$$

or

$$\varepsilon_x = \frac{1}{E}(\sigma^x - \nu\sigma^y), \quad \varepsilon_y = \frac{1}{E}(\sigma^y - \nu\sigma^x), \quad \gamma_{xy} = \frac{2(1+\nu)}{E}\tau^{xy}\,. \tag{3.5}$$

In the plane deformation problem, the structure of the physical equations is maintained, and all one should do to be able to apply relationships (3.4) and (3.5) to the plane deformation is to replace the material constants, E with E_o and ν with ν_o, as follows:

$$E_o = \frac{E}{1-\nu^2}, \qquad \nu_o = \frac{\nu}{1-\nu}. \tag{3.6}$$

Now let us formulate boundary conditions, confining ourselves to the global coordinate system.

static boundary conditions	kinematic boundary conditions
$e_{px}(\sigma^x n_x + \tau^{xy} n_y - \bar{p}^x) = 0,$	$e_{ux}(u - \bar{u}) = 0,$
$e_{py}(\tau^{xy} n_x + \sigma^y n_y - \bar{p}^y) = 0,$	$e_{uy}(v - \bar{v}) = 0.$ (3.7)

The set of governing equations and boundary conditions of the plane elasticity can be represented in the matrix form if we assume the following in addition to (3.1):

$$\bar{\boldsymbol{p}} = \|[\bar{p}^x, \bar{p}^y]\|^{\mathsf{T}}, \qquad \boldsymbol{H}_u = \boldsymbol{I},$$

$$\boldsymbol{A} = \left\|\begin{bmatrix} \partial/\partial x & 0 \\ 0 & \partial/\partial y \\ \partial/\partial y & \partial/\partial x \end{bmatrix}\right\|, \qquad \boldsymbol{A}^{\mathsf{T}} = \left\|\begin{bmatrix} -\partial/\partial x & 0 & -\partial/\partial y \\ 0 & -\partial/\partial y & -\partial/\partial x \end{bmatrix}\right\|,$$

$$\boldsymbol{H}_\sigma = \left\|\begin{bmatrix} n_x & 0 & n_y \\ 0 & n_y & n_x \end{bmatrix}\right\|,$$

$$\boldsymbol{C} = \frac{E}{1-\nu^2}\left\|\begin{bmatrix} 1 & \nu & 0 \\ \nu & 1 & 0 \\ 0 & 0 & \frac{1-\nu}{2} \end{bmatrix}\right\|, \qquad \boldsymbol{C}^{-1} = \frac{1}{E}\left\|\begin{bmatrix} 1 & -\nu & 0 \\ -\nu & 1 & 0 \\ 0 & 0 & 2(1+\nu) \end{bmatrix}\right\|.$$

By performing matrix operations according to formula (1.6), we can find the Lame operator for plane elasticity:

$$\boldsymbol{L} = \boldsymbol{A}^{\mathrm{T}}\boldsymbol{C}\boldsymbol{A} = \frac{E}{1-\nu^2}\left\|\begin{matrix} -\nabla^2 + \dfrac{1+\nu}{2}\dfrac{\partial^2}{\partial y^2} & -\dfrac{1+\nu}{2}\dfrac{\partial^2}{\partial x \partial y} \\ -\dfrac{1+\nu}{2}\dfrac{\partial^2}{\partial x \partial y} & -\nabla^2 + \dfrac{1+\nu}{2}\dfrac{\partial^2}{\partial x^2} \end{matrix}\right\|. \tag{3.8}$$

The strain compatibility condition for plane elasticity consists of a single equation:

$$\frac{\partial^2 \varepsilon_x}{\partial y^2} + \frac{\partial^2 \varepsilon_y}{\partial x^2} - \frac{\partial^2 \gamma_{xy}}{\partial x \partial y} = 0,$$

so the Saint-Venant matrix differential operator, $\boldsymbol{S}$, can be written in the form

$$\boldsymbol{S} = \left\| \partial^2/\partial y^2 \quad \partial^2/\partial x^2 \quad -\partial^2/\partial x \partial y \right\|.$$

Basic integral identity (**1**.2.17) for plane elasticity can be represented in components as

$$\int_{\Omega}(\sigma^x \varepsilon_x + \sigma^y \varepsilon_y + \tau^{xy}\gamma_{xy})d\Omega = -\int_{\Omega}[(\sigma^x_{,x} + \tau^{xy}_{,y})u + (\sigma^y_{,y} + \tau^{xy}_{,x})v]d\Omega +$$

$$+\oint_{\Gamma}[(\sigma^x n_x + \tau^{xy} n_y)u + (\sigma^y n_y + \tau^{xy} n_x)v]d\Gamma\,. \tag{3.9}$$

4.3.1 Lagrange functional

Let us write out an expression of the Lagrange functional in application to an isotropic material in plane stress:

$$\mathsf{L}(\mathbf{u}) = \frac{E}{2(1-\nu^2)}\int_{\Omega}\left[u^2_{,x} + v^2_{,y} + 2\nu u_{,x}v_{,y} + \frac{1-\nu}{2}(u_{,y} + v_{,x})^2\right]d\Omega -$$

$$-\int_{\Omega}(u\overline{X} + v\overline{Y})d\Omega - \oint_{\Gamma}(e_{px}u\overline{p}^x + e_{py}v\overline{p}^y)d\Gamma\,. \tag{3.10}$$

This functional is defined on a set of vector functions $\mathbf{u}$ which meet the kinematic boundary conditions and assure a finite value of the energy integral, E,

$$\mathsf{E}(\mathbf{u}) = \frac{E}{2(1-\nu^2)} \iint_\Omega \left[u_{,x}^2 + v_{,y}^2 + 2\nu u_{,x} v_{,y} + \frac{1-\nu}{2}(u_{,y} + v_{,x})^2 \right] d\Omega . \qquad (3.11)$$

Euler equations for functional L are the equilibrium equations in displacements,

$$\frac{E}{1-\nu^2}\left(-\nabla^2 u + \frac{1+\nu}{2}\frac{\partial^2 u}{\partial y^2} - \frac{1+\nu}{2}\frac{\partial^2 v}{\partial x \partial y} \right) = \bar{X} ,$$

$$\frac{E}{1-\nu^2}\left(-\frac{1+\nu}{2}\frac{\partial^2 u}{\partial x \partial y} - \nabla^2 v + \frac{1+\nu}{2}\frac{\partial^2 v}{\partial x^2} \right) = \bar{Y} . \qquad (3.12)$$

The static boundary conditions are natural ones for functional L .

4.3.2 Castigliano functional

$$\mathsf{K}(\boldsymbol{\sigma}) = \frac{1}{2E} \iint_\Omega [(\sigma^x)^2 - 2\nu\sigma^x\sigma^y + (\sigma^y)^2 + 2(1+\nu)(\tau^{xy})^2] d\Omega \; -$$

$$- \oint_\Gamma [e_u(\sigma^x n_x + \tau^{xy} n_y)\bar{u} + e_v(\sigma^y n_y + \tau^{xy} n_x)\bar{v}] d\Gamma . \qquad (3.13)$$

Functional $\mathsf{K}(\boldsymbol{\sigma})$ is defined on a set of stress fields $\boldsymbol{\sigma}$ that satisfy equilibrium equations (3.2) and static boundary conditions from (3.7).

Euler equations for the Castigliano functional are equations of strain compatibility in stresses, which can be written in the following form for plane elasticity after some transformations using the equilibrium equations [5]:

$$\frac{\partial^2 \sigma^x}{\partial x^2} - 2\frac{\partial^2 \tau^{xy}}{\partial x \partial y} + \frac{\partial^2 \sigma^y}{\partial y^2} - \nu\Delta(\sigma^x + \sigma^y) = 0 .$$

4.3.3 Reissner functional

Based on formulas (**3**.1.4) and (**3**.1.5), here we present expressions of the Reissner functional in its first and second form for plane elasticity. The expanded representation is

$$R_1(\boldsymbol{\sigma},\mathbf{u}) = \frac{1}{2E}\int_\Omega [(\sigma^x)^2 - 2\nu\sigma^x\sigma^y + (\sigma^y)^2 + 2(1+\nu)(\tau^{xy})^2]d\Omega -$$

$$-\int_\Omega [\sigma^x u_{,x} + \sigma^y v_{,y} + \tau^{xy}(u_{,y} + v_{,x})]d\Omega + \int_\Omega [u\bar{X} + v\bar{Y}]d\Omega +$$

$$+\oint_\Gamma (e_{px}\bar{p}^x u + e_{py}\bar{p}^y v)d\Gamma +$$

$$+\oint_\Gamma [e_u(u-\bar{u})(\sigma^x n_x + \tau^{xy}n_y) + e_v(v-\bar{v})(\sigma^y n_y + \tau^{xy}n_x)]d\Gamma\,, \qquad (3.14)$$

$$R_2(\boldsymbol{\sigma},\mathbf{u}) = \frac{1}{2E}\int_\Omega [(\sigma^x)^2 - 2\nu\sigma^x\sigma^y + (\sigma^y)^2 + 2(1+\nu)(\tau^{xy})^2]d\Omega +$$

$$+\int_\Omega [u(\sigma^x_{,x} + \tau^{xy}_{,y}) + v(\sigma^y_{,y} + \tau^{xy}_{,x})]d\Omega + \int_\Omega [u\bar{X} + v\bar{Y}]d\Omega +$$

$$+\oint_\Gamma [e_{px}u(\bar{p}^x - \sigma^x n_x - \tau^{xy}n_y) + e_{py}v(\bar{p}^y - \sigma^y n_y - \tau^{xy}n_x)]d\Gamma -$$

$$-\oint_\Gamma [e_u\bar{u}(\sigma^x n_x + \tau^{xy}n_y) + e_v\bar{v}(\sigma^y n_y + \tau^{xy}n_x)]d\Gamma\,. \qquad (3.15)$$

Euler equations for the R equations are: equilibrium equations (2.1-*a*) and physical equations (2.1-*c*). All boundary conditions are natural for functional R.

4.4 Lengthwise deformation of a straight bar

We consider a problem of tension/compression of a straight bar. Let x be a lengthwise axis of the bar, and let $q(x)$ be an intensity of a load distributed along the axis of the bar and acting in the direction of the x-axis.

If we assume that the beginning of the coordinate system coincides with one of the bar's ends and the whole length of the bar is l, then concentrated forces applied to the respective ends of the bar can be denoted by $\bar{P}_0$ and $\bar{P}_l$ (Fig. 4.1). Let $u(x)$ be a function of lengthwise displacements of the bar's sections during its deformation, and let $N(x)$ be a longitudinal force assumed to be positive when the bar is in tension. We denote by $A = A(x)$ a function that shows a change in the area of a cross-section of the bar. In addition, we assume that the lengthwise deformation of the bar is restrained by an elastic bed with its reaction coefficient $k = k(x)$.

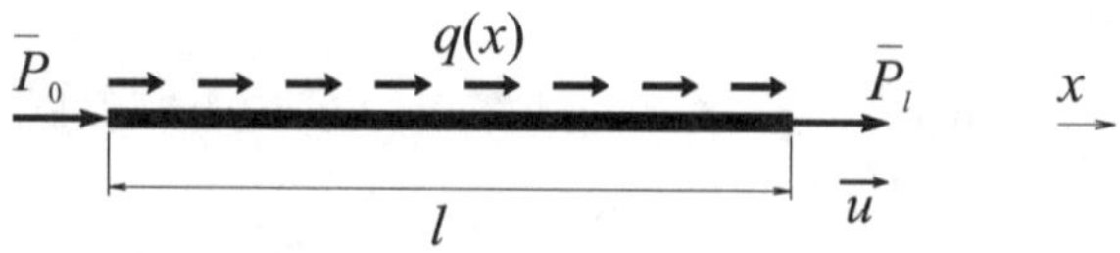

Fig. 4.1. Tension or compression of a straight bar

The vectors of stresses **σ**, strains **ε**, and displacements **u**, which define the stress-and-strain distribution in the bar, consist in this case of one component each:

$$\boldsymbol{\sigma} = \|[N]\|, \qquad \boldsymbol{\varepsilon} = \|[\varepsilon]\|, \qquad \mathbf{u} = \|[u]\|, \tag{4.1}$$

where $\varepsilon = \dfrac{du}{dx}$ is a relative lengthwise strain, positive when in tension. The set of governing equations is

$$-\frac{dN}{dx} + ku = q \qquad \text{equation of equilibrium,}$$

$$\frac{du}{dx} = \varepsilon \qquad \text{geometric equation,}$$

$$N = EA\varepsilon \qquad \text{physical equation.}$$

These equations acquire their usual matrix form (1.1) if we assume this in addition to (4.1):

$$\boldsymbol{A} = \left\|\left[\frac{d}{dx}\right]\right\|, \qquad \boldsymbol{A}^{\mathrm{T}} = \left\|\left[-\frac{d}{dx}\right]\right\|, \qquad \boldsymbol{C} = \|[EA]\|, \qquad \boldsymbol{C}^{-1} = \left\|\left[\frac{1}{EA}\right]\right\|. \tag{4.2}$$

In this (one-dimensional) problem, area Ω occupied by the elastic body is a straight segment $[0,l]$, and boundary Γ of area Ω consists of two points, $x = 0$ and $x = l$, so that the direction cosines, n, of external normal $\boldsymbol{n}$ to boundary Γ at these points are equal, respectively, to:

$$n(0) = -1, \qquad n(l) = 1. \tag{4.3}$$

Boundary conditions in points $x = 0$ and $x = l$ are formulated as

static boundary conditions	kinematic boundary conditions	
$e_p\,[N(0) + \bar{P}_0\,] = 0,$	$e_u\,[u(0) - \bar{u}\,(0)] = 0,$	
$e_p\,[-N(l) + \bar{P}_l\,] = 0,$	$e_u\,[u(l) - \bar{u}\,(l)] = 0.$	(4.4)

Parameters of boundary condition extraction, e_p and e_u, are specified on each end of the bar and are either 0 or 1 so that $e_p + e_u = 1$. It is also obvious that matrix operators $\boldsymbol{E}_p$ and $\boldsymbol{E}_u$ are first-order matrices, that is,

$$\boldsymbol{E}_p = \|[e_p]\| , \qquad \boldsymbol{E}_u = \|[e_u]\| . \tag{4.5}$$

The basic integral identity for the bar tension/compression problem is

$$\int_0^l \frac{du}{dx} N dx = -\int_0^l \frac{dN}{dx} u dx + [e_p N u]|_0^l + [e_u N u]|_0^l , \tag{4.6}$$

By assuming $\boldsymbol{H}_\sigma = \|[n]\|$ and $\boldsymbol{H}_u = \boldsymbol{I} = \|[1]\|$ and using formulas (**1**.1.4) and (**1**.1.5), we can write the non-integral terms in the right part of (4.6) in the operator form:

$$(\boldsymbol{H}_\sigma \boldsymbol{\sigma}, \boldsymbol{u})_\Gamma = (nNu)_{x=0} + (nNu)_{x=l} = (Nu)_{x=l} - (Nu)_{x=0} = [Nu]|_0^l .$$

Also, expressions of the force and kinematic potentials can be represented as

$$\Pi_s(\mathbf{u}) = (\overline{\boldsymbol{X}}, \mathbf{u}) + (\boldsymbol{E}_p \overline{\boldsymbol{p}}, \boldsymbol{E}_p \boldsymbol{H}_u \mathbf{u})_\Gamma =$$

$$= \int_0^l q u dx + (e_p \overline{P} u)_{x=0} + (e_p \overline{P} u)_{x=l} = \int_0^l q u dx + [e_p n \overline{P} u]|_0^l ,$$

$$\Pi_k(\boldsymbol{\sigma}) = (\boldsymbol{E}_u \boldsymbol{H}_\sigma \boldsymbol{\sigma}, \boldsymbol{E}_u \overline{\boldsymbol{u}})_\Gamma = (e_u \overline{u}\, nN)_{x=0} + (e_u \overline{u}\, nN)_{x=l} = [e_u N \overline{u}]|_0^l , \tag{4.7}$$

where we allow for the fact that $n^2 = 1$ according to (4.3).

As a result, the Lagrange functional in components is

$$\mathsf{L}(u) = \int_0^l \left(\frac{1}{2} EAu_{,x}^2 + \frac{1}{2} ku^2 - qu\right) dx - [e_p n \overline{P} u]|_0^l . \tag{4.8}$$

Euler equations of the functional consist of one equilibrium equation in displacements,

$$-\frac{d}{dx}\left(EA \frac{du}{dx}\right) + ku = q , \tag{4.9}$$

and the natural boundary conditions consist of the static boundary conditions.

The Castigliano functional in components is

$$\mathsf{K}(N,u) = \frac{1}{2} \int_0^l \left(\frac{N^2}{EA} + ku^2 \right) dx - [e_u N \overline{u}]|_0^l . \tag{4.10}$$

We will also present expressions of first and second forms of the Reissner functional. Based on general expressions for those functionals according to (**3**.1.4) and (**3**.1.5), and taking (4.7) into account, we have

$$\mathsf{R}_1(N,u) =$$

$$= \int_0^l \left(\frac{N^2}{2EA} - \frac{1}{2} k u^2 - u_{,x} N + qu \right) dx - [e_u N(\bar{u} - u)]|_0^l + [e_p n \bar{P} u]|_0^l ,$$

$$\mathsf{R}_2(N,u) =$$

$$= \int_0^l \left(\frac{N^2}{2EA} - \frac{1}{2} k u^2 + u N_{,x} + qu \right) dx - [e_u N \bar{u}]|_0^l + [e_p n(\bar{P} - nN)u]|_0^l . \quad (4.11)$$

Finally, we present an expression of the Gurtin functional in its third form

$$\mathsf{G}_3(N) = \int_0^l \left(\frac{N^2}{2EA} + \frac{1}{2k} N_{,x}^2 + \frac{1}{2k} q N_{,x} \right) dx - [e_u N \bar{u}]|_0^l , \quad (4.12)$$

where functions $N(x)$ allowed for comparison must satisfy the main boundary conditions:

$$e_p (\bar{P} + N)_{x=0} = 0 , \qquad e_p (\bar{P} - N)_{x=l} = 0 .$$

4.5 Bernoulli-type beam on elastic foundation

Let us consider a problem of bending of a straight bar (beam). We introduce a Cartesian coordinate system, (x, y, z), and match its x-axis with the longitudinal axis of the beam being bent. We assume the beam to bend in the (x, y)-axis, that is, in the plane of drawing that coincides with the plane of loading (a plane in which external loads are applied). If we direct third axis, z, of the right-hand Cartesian triple (x, y, z) from the observer to the plane of drawing, then the arrangement of the x and y axes in the drawing plane will be such as shown in Fig. 4.2. Here we assume that the origin of the coordinate system is matched with one of the ends (the left one) of the bar, and the whole length of the bar is l.

Let $q(x)$ be an intensity of a lateral load distributed along the bar and acting in the y-axis direction, and let $m(x)$ be an intensity of an external moment-type load distributed over the beam's length. External force and moment actions specified on the respective ends of the bar will be denoted by $\bar{Q}_0$ and $\bar{Q}_l$, $\bar{M}_0$ and $\bar{M}_l$, respectively (Fig. 4.2).

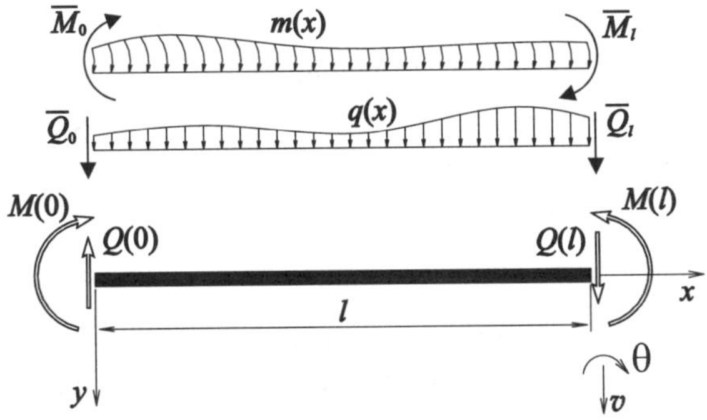

Fig. 4.2. External actions, internal forces, and displacements in a bar in bending

Let $v(x)$ be lateral displacements (deflections) of the bar, $\theta(x)$ its cross-sections' slopes, $Q(x)$ its lateral shear force, $M(x)$ its bending moment in the cross-section. Fig. 4.2 shows assumed positive directions of the internal forces, displacements, and external load actions.

As commonly known, the classic theory of bending of a straight bar (Bernoulli–Euler theory) is based on the following three assumptions (sometimes they say, hypotheses):

- a hypothesis of *planar sections*, according to which all sections normal to the beam's axis in its undeformed state will remain planar (undistorted) and perpendicular to the beam's axis during its bending;
- a hypothesis of *no pressure*, according to which the longitudinal fibers of the beam do not interact with one another in their normal directions, in other words, $\sigma^y = \sigma^z = 0$ is assumed;
- a hypothesis of *no shear*, according to which the shear strains are so small that the work of the tangential stresses, τ, in the expression of the strain energy, E, can be neglected comparing to the work of the normal stresses, σ^x [2].

[2] Strictly speaking, the no-shear hypothesis is not a self-contained assumption. It is a corollary from the planar sections hypothesis. And indeed, it is not hard to understand that a section can remain planar (undistorted) only in the case the shear strain, γ, is constant over the whole height of the section. But that constant should be equal to zero as long as the shear at the level of at least one fiber is zero; the latter *is* zero in any of points of the section's contour, for example, in the uppermost fiber. Hence the work of the tangential stresses, τ, is zero, that is, $\tau\gamma = 0$. Nevertheless, here we make a separate hypothesis out of this assumption, only because further it will be the only hypothesis to be thrown away in order to construct a theory of beam bending by Timoshenko on the basis of the variational approach; the second hypothesis will be kept intact and the first will be weakened.

The planar sections hypothesis is purely kinematic; it permits to use geometric considerations to immediately obtain an expression of the slopes, θ, of the beam's cross-sections, and at the same time to establish a relationship between $\theta(x)$ and a change in the curvature of the beam's axis, χ,

$$\theta = v', \quad \chi = -\theta' = -v'', \tag{5.1}$$

where the longitudinal strain, ε_x, is a uniform linear function of the y coordinate of a fiber of the section where strain ε_x is calculated,

$$\varepsilon_x = \chi y. \tag{5.2}$$

Curvature χ is assumed to be positive when the beam's axis bends in such way that its convexity looks towards the increasing y coordinate (towards bigger deflections v).

Relation (5.2) is essentially a mathematical form of the planar sections hypothesis. To see this, recall that any section of the bar should move as a perfectly rigid body according to this hypothesis, therefore the function of longitudinal displacements, $u(x,y)$, of the section's points satisfy a plane relationship, that is,

$$u(x,y) = u(x,0) - \theta(x)y,$$

where $u(x,0)$ are displacements along the x-axis of points on the beam's axis. Now we take (5.1) and find the following from the Cauchy relationships for the general three-dimensional elasticity, which establish a relation between the displacements and the strains $\varepsilon_x = \partial u/\partial x$,

$$\varepsilon_x = \varepsilon_0 + \chi y, \tag{5.3}$$

where $\varepsilon_0 = \partial u(x,0)/\partial x$ is a relative lengthwise strain at the points of the section with the coordinate $y = 0$. Further, by assuming the physical law for normal stresses σ^x in the form[3]

$$\sigma^x = E\varepsilon_x, \tag{5.4}$$

where E is an elasticity modulus of the beam's material, we can find the overall longitudinal force N that develops in an arbitrary section of the beam:

$$N = \int_A \sigma^x dA = E(\varepsilon_0 A + \chi S_z), \qquad S_z = \int_A y dA, \tag{5.5}$$

[3] Quite obviously, relationship (5.4) follows immediately from the general form of Hooke's law for a three-dimensional isotropic elastic body and from the hypothesis of no pressure between the beam's longitudinal fibers.

where A is an area of the beam's cross-section.

If we choose the coordinate system's origin in such way that the longitudinal axis, x, passes through the centers of gravity of the beam's cross-sections, then the static moment, S_z, of any section with respect to the z-axis will be zero. However, there are no external longitudinal forces among those acting on the flexural beam, therefore the condition of equilibrium of the beam in the projection onto the x-axis gives $N = 0$. Substituting this condition to (5.5) gives $\varepsilon_0 = 0$ and thus (5.3) turns into (5.2), which was to be proved.

We neglect the work of tangential stresses τ according to the no-shear hypothesis and calculate the strain energy of the beam, E_A, in bending:

$$\mathsf{E}_A = \int_0^l \int_A \frac{\sigma^x \varepsilon_x}{2} dA dx = \frac{1}{2} \int_0^l \int_A E\varepsilon_x^2 dA dx = \frac{1}{2} \int_0^l v''^2 E \int_A y^2 dA dx =$$

$$= \frac{1}{2} \int_0^l EIv''^2 dx. \tag{5.6}$$

Here $A = A(x)$ is an area of the beam's cross-section, which is variable over the beam's length. As usual, here we denote by $I = I(x)$ a moment of inertia of the beam's cross-section, generally variable along the beam:

$$I(x) = \int_A y^2 dA.$$

Supposing that the beam lies on a Winkler-type elastic bed with the subgrade reaction coefficient $k(x)$, we calculate also an energy accumulated in this elastic bed:

$$\mathsf{E}_K = \frac{1}{2} \int_0^l kv^2 dx.$$

Let the parameters of boundary condition extraction, e_Q, e_M, e_v and e_θ, be specified on each of the beam's ends, and let them take the values of either 0 or 1 so that $e_Q + e_v = 1$ and $e_M + e_\theta = 1$. The meaning of these parameters is defined by its respective subscript. For example, $e_Q = 1$ if the beam has an external concentrated force, $\bar{Q}$, specified on the respective end, while displacement v is not known at the same place.

Similarly, $e_\theta = 1$ if the beam has the cross-section's slope, $\bar{\theta}$, specified on its respective end, while the bending moment, M, at the same point is unknown and will be obtained from the solution of the problem.

The potential of external static actions, $\Pi_s(v)$, is a virtual work of all external forces on displacements v and slopes $\theta = v'$, hence it can be written in its general form as

$$\Pi_s(v) = \int_0^l qv dx + \int_0^l mv' dx + e_Q(0)\,\bar{Q}_0\, v(0) +$$

$$+ e_Q(l)\,\bar{Q}_l\, v(l) + e_M(0)\,\bar{M}_0\, v'(0) + e_M(l)\,\bar{M}_l\, v'(l) =$$

$$= \int_0^l qv dx + \int_0^l mv' dx + [e_Q n\bar{Q}v]\big|_0^l + [e_M n\bar{M}v']\big|_0^l . \tag{5.7}$$

As a result, the Lagrange functional (the full potential energy of a system) for a flexural beam which lies on an elastic bed will be

$$\mathsf{L}(v) = \frac{1}{2}\int_0^l [EIv''^2 + kv^2]dx -$$

$$-\int_0^l qv dx - \int_0^l mv' dx - [e_Q n\bar{Q}v]\big|_0^l - [e_M n\bar{M}v']\big|_0^l . \tag{5.8}$$

By varying functional L and using the integration by parts, we derive

$$\delta\mathsf{L} = \int_0^l [(EIv'')'' + kv - q + m']\delta v dx +$$

$$+\left[(EIv'' - e_M n\bar{M})\delta v'\right]\big|_0^l - \left[[(EIv'')' + m + e_Q n\bar{Q}]\delta v\right]\big|_0^l .$$

As it was noted in Chapter 2, the Lagrange functional is defined on kinematically admissible fields of displacements, therefore its admissible variation, δv, is different from zero at the end of the beam if $e_Q = 1$ and $e_v = 0$. Similarly, variation $\delta v'$ is different from zero at the end of the beam if $e_M = 1$ and $e_\theta = 0$. Considering this, the variation of the Lagrange functional on kinematically admissible displacements will finally become

$$\delta\mathsf{L} = \int_0^l [(EIv'')'' + kv - q + m']\delta v dx +$$

$$+\left[e_M(EIv'' - n\bar{M})\delta v'\right]\Big|_0^l - \left[e_Q[(EIv'')' + m + n\bar{Q}]\delta v\right]\Big|_0^l .$$

Equaling first variation, $\delta\mathsf{L}$, to zero gives an Euler equation,

$$\boxed{(EIv'')'' + kv = q - m',} \tag{5.9}$$

which is a differential equation of the beam's equilibrium written in displacements. This also gives natural (static) boundary conditions

$$e_M(nEIv'' - \bar{M}) = 0, \qquad e_Q[n(EIv'')' + nm + \bar{Q}] = 0 \tag{5.10}$$

at ends points $x = 0$ and $x = l$.

The formulation of boundary conditions (5.10) which follow from condition $\delta\mathsf{L} = 0$ makes use of the obvious identity $n^2 = 1$. The physical meaning of the boundary conditions is quite clear. The first of them makes the bending moment, M, in the end section of the beam equal (allowing for a sign convention) to an external concentrated moment, $\bar{M}$, specified in the same section. The second of the boundary conditions concerns the shear force, Q, that develops in the end section of the beam and must be equal to a given external force $\bar{Q}$. To see this, we consider the equilibrium of an elementary piece of the beam having the length of dx (Fig. 4.3),

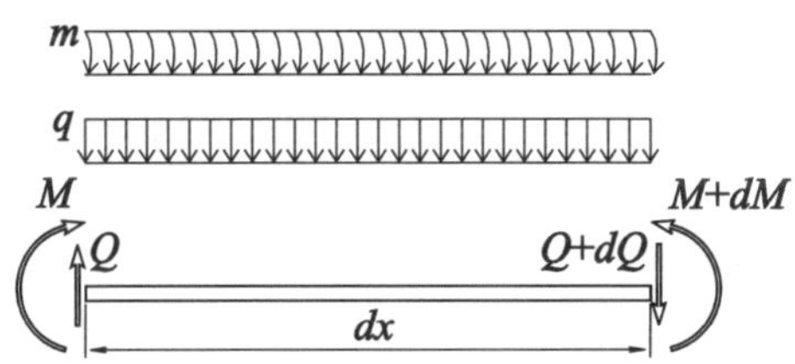

Fig. 4.3. An elementary piece of the beam

and arrive at the following two differential relationships for shear forces Q and bending moments M:

$$Q' = -q, \qquad M' = Q + m. \tag{5.11}$$

Further, bending moment M in an arbitrary section of the beam is (by definition) an integral characteristic of first order with respect to normal stresses σ^x, that is,

$$M = \int_A \sigma^x y dA = \int_A E\chi y^2 dA = \chi EI, \tag{5.12}$$

hence using (5.1) and (5.11) we have

$$M = -EIv'', \qquad Q = -(EIv'')' - m. \tag{5.13}$$

Now we take (5.13) and make static boundary conditions (5.10) look as follows:

$$e_M(nM+\bar{M})=0, \qquad e_Q(nQ-\bar{Q})=0. \tag{5.14}$$

We want to convert all relationships of the Bernoulli–Euler theory of beam bending to a common matrix form.

The vectors of stresses **σ**, strains **ε**, and displacements **u**, which determine the stress-and-strain state of a beam, consist each of one component in the present case:

$$\boldsymbol{\sigma} = \|[M]\|, \quad \boldsymbol{\varepsilon} = \|[\chi]\|, \quad \mathbf{u} = \|[v]\|, \tag{5.15}$$

the only "strain" component ε being a change in the curvature of the beam's axis during the bending, $\chi = -v''$, and the only "stress" component σ being the bending moment, M.

The set of the governing equations for the problem will consist of

$$-\frac{d^2M}{dx^2}+kv=q-m' \qquad \text{equilibrium equation,}$$

$$-\frac{d^2v}{dx^2}=\chi \qquad \text{geometric equation,}$$

$$M=EI\chi \qquad \text{physical equation.}$$

These equations will take our usual matrix form (1.1) if we assume the following in addition to (5.15):

$$\boldsymbol{A}=\left\|\left[-\frac{d^2}{dx^2}\right]\right\|, \quad \boldsymbol{A}^{\mathsf{T}}=\left\|\left[-\frac{d^2}{dx^2}\right]\right\|, \quad \boldsymbol{C}=\|[EI]\|, \quad \boldsymbol{C}^{-1}=\left\|\left[{}^{1}\!/\!_{EI}\right]\right\|,$$

$$\boldsymbol{K}=\|[k]\|, \quad \bar{\boldsymbol{X}}=\|[\,q-m'\,]\|. \tag{5.16}$$

Now let us re-formulate the boundary conditions in the matrix form. The vectors of edge displacements and forces, $\boldsymbol{u}$ and $\boldsymbol{p}$, and their respective vectors of given edge displacements, $\bar{\boldsymbol{u}}$, and forces, $\bar{\boldsymbol{p}}$, are written as second-order column matrices:

$$\boldsymbol{u}=\left\|\begin{bmatrix} v \\ v' \end{bmatrix}\right\|, \qquad \boldsymbol{p}=\left\|\begin{bmatrix} nM' \\ -nM \end{bmatrix}\right\|, \qquad \bar{\boldsymbol{u}}=\left\|\begin{bmatrix} \bar{v} \\ \bar{\theta} \end{bmatrix}\right\|, \qquad \bar{\boldsymbol{p}}=\left\|\begin{bmatrix} \bar{Q}+nm \\ \bar{M} \end{bmatrix}\right\|. \tag{5.17}$$

The matrix operators of boundary condition extraction, $\boldsymbol{E}_p$ and $\boldsymbol{E}_u$, are square matrices of second order:

$$\boldsymbol{E}_p = \left\| \begin{bmatrix} e_Q & 0 \\ 0 & e_M \end{bmatrix} \right\|, \qquad \boldsymbol{E}_u = \left\| \begin{bmatrix} e_v & 0 \\ 0 & e_\theta \end{bmatrix} \right\|. \tag{5.18}$$

Further we assume

$$\boldsymbol{H}_\sigma = \left\| \begin{bmatrix} n\dfrac{d}{dx} \\ -n \end{bmatrix} \right\|, \qquad \boldsymbol{H}_u = \left\| \begin{bmatrix} 1 \\ \dfrac{d}{dx} \end{bmatrix} \right\|, \tag{5.19}$$

which enables us to represent the boundary conditions in the matrix form, too — in the form of equations (1.2).

Now let us verify that the basic integral identity holds. First of all, recall (**1**.1.5) and write

$$(\boldsymbol{p},\boldsymbol{u})_\Gamma = (\boldsymbol{H}_\sigma\boldsymbol{\sigma}, \boldsymbol{H}_u\mathbf{u})_\Gamma = (nM'v - nMv')_{x=0} + (nM'v - nMv')_{x=l} =$$

$$= [M'v]\big|_0^l - [Mv']\big|_0^l .$$

Thus,

$$(\boldsymbol{A}\mathbf{u},\boldsymbol{\sigma}) = (-v'', M) =$$

$$= -\int_0^l v''M dx = -\int_0^l vM''dx - [Mv']\big|_0^l + [M'v]\big|_0^l = (\mathbf{u}, \boldsymbol{A}^{\mathsf{T}}\boldsymbol{\sigma}) + (\boldsymbol{p}, \boldsymbol{u})_\Gamma ,$$

which is the basic integral identity as in (**1**.2.17).

Now we want to derive expressions for the Castigliano and Reissner functionals based on a general operator representation. First of all, we determine the potential of external kinematic actions, $\Pi_k(\boldsymbol{\sigma})$, from (**2**.3.2). We have

$$\Pi_k(\boldsymbol{\sigma}) = (\boldsymbol{E}_u\boldsymbol{H}_\sigma\boldsymbol{\sigma}, \boldsymbol{E}_u\, \bar{\boldsymbol{u}})_\Gamma =$$

$$= \|[e_v nM', -e_\theta nM]\| \cdot \left\| \begin{bmatrix} \bar{v} \\ \bar{\theta} \end{bmatrix} \right\|_{x=0} + \|[e_v nM', -e_\theta nM]\| \cdot \left\| \begin{bmatrix} \bar{v} \\ \bar{\theta} \end{bmatrix} \right\|_{x=l} =$$

$$= [e_v M'\bar{v}]\big|_0^l - [e_\theta M\bar{\theta}]\big|_0^l . \tag{5.20}$$

Further, using formula (**2**.3.3) helps derive the Castigliano functional for the Bernoulli–Euler beam bending as

$$\mathsf{K}(M,v) = \frac{1}{2}\int_0^l \frac{M^2}{EI}dx + \frac{1}{2}\int_0^l kv^2 dx + [e_v M'\bar{v}]\big|_0^l + [e_\theta M\bar{\theta}]\big|_0^l . \tag{5.21}$$

In this case the set $\mathscr{R}_S$ of physically and statically admissible stress-and-strain fields consists of all possible couples of functions M and v which satisfy the following equations at the interval $[0,l]$:

$$-M''+kv=q-m', \qquad M=-EIv'', \tag{5.22}$$

and meet the static boundary conditions at each end of the interval:

$$e_M(nM+\bar{M})=0, \qquad e_Q(nM'-nm-\bar{Q})=0. \tag{5.23}$$

According to (**3**.1.4) and (**3**.1.5), we define the Reissner functional in its first and second forms as follows:

$$\mathrm{R}_1(M,v)=$$

$$=\frac{1}{2}\int_0^l\frac{M^2}{EI}dx-\frac{1}{2}\int_0^l kv^2dx+\int_0^l v''Mdx+\int_0^l qvdx+\int_0^l mv'dx+$$

$$+[e_vM'(v-\bar{v})]|_0^l-[e_\theta M(v'-\bar{\theta})]|_0^l+[e_Qn\bar{Q}v]|_0^l+[e_Mn\bar{M}v']|_0^l, \tag{5.24}$$

$$\mathrm{R}_2(M,v)=$$

$$=\frac{1}{2}\int_0^l\frac{M^2}{EI}dx-\frac{1}{2}\int_0^l kv^2dx+\int_0^l vM''dx+\int_0^l qvdx+\int_0^l mv'dx-$$

$$-[e_vM'\bar{v}]|_0^l+[e_\theta M\bar{\theta}]|_0^l-[e_Q(M'-n\bar{Q})v]|_0^l+[e_M(M+n\bar{M})v']|_0^l. \tag{5.25}$$

While the Reissner functional in its first form requires that displacements v be twice differentiable, its second form requires the differentiability of the same kind from moments M.

Requirements of smoothness to those two functions can be symmetrized, and thus we will arrive at the third form of Reissnerian with equal conditions of differentiability for both displacements and moments. In order to do this, we use integration by parts on the third integral in either $\mathrm{R}_1(M,v)$ or $\mathrm{R}_2(M,v)$, which gives

$$\mathrm{R}_3(M,v)=$$

$$=\frac{1}{2}\int_0^l\frac{M^2}{EI}dx-\frac{1}{2}\int_0^l kv^2dx-\int_0^l v'M'dx+\int_0^l qvdx+\int_0^l mv'dx+$$

$$+[e_vM'(v-\bar{v})]|_0^l+[e_\theta M\bar{\theta}]|_0^l+[e_Qn\bar{Q}v]|_0^l+[e_M(M+n\bar{M})v']|_0^l. \tag{5.26}$$

All boundary conditions (both static and kinematic) are natural boundary conditions for the Reissner functional in any form of it.

Now we will finally give an expression in components for the Gurtin functional in its third form, assuming the functional to be defined on statically semi-admissible stress fields. Using the general formula (**3**.6.11), we obtain the following for a Bernoulli–Euler beam:

$$\mathsf{G}_3(M) = \frac{1}{2}\int_0^l \frac{M^2}{EI}dx + \frac{1}{2}\int_0^l \frac{M''^2}{k}dx + \int_0^l \frac{M''(q-m')}{k}dx -$$

$$-[e_v M'\overline{v}]\Big|_0^l + [e_\theta M\overline{\theta}]\Big|_0^l . \qquad (5.27)$$

We emphasize again that functional G_3 takes its minimum on the solution of the problem among a set of functions $M(x)$ which satisfy static boundary conditions (5.14). The obvious thing is that the $M(x)$ functions allowed for comparison must be smooth enough — more exactly, they are required to have quadratically summable second derivatives.

4.6 Timoshenko-type beam on elastic foundation

We want to formulate a problem of analysis for a flexural beam where shear strains are taken into account. This is a so-called theory of beams by Timoshenko. This theory is an enhancement of the Bernoulli–Euler beam theory; it abandons the third hypothesis used in the construction of the classic beam theory. In other words, the governing equations will now contain a refinement that allows for a work of tangential stresses; that's why the Timoshenko beam theory can be treated as a beam bending theory that takes shear into account.

Expression (5.6) acquires an additional term of the strain energy in the flexural beam, so

$$\mathsf{E}_A = \int_0^l\int_A \frac{\sigma^x\varepsilon_x}{2}dA\,dx + \int_0^l\int_A \frac{\tau^{xy}\gamma_{xy}}{2}dA\,dx\,, \qquad (6.1)$$

where τ^{xy} are tangential stresses developing in the sections of the beam and γ_{xy} are their respective shear strains.

According to the planar sections hypothesis, any cross-section of the beam is treated as an undeformable body that does not allow any displacements other than rigid ones[4]. It immediately follows from here that shear strain γ_{xy} can be a constant only for all points of a particular section. To see this, consider that the u and v displacements along the x and y axes can be written as follows on the basis the condition of rigid displacements and the sign convention, respectively:

$$u(x,y) = u(x,0) - \theta(x)y, \qquad v(x,y) = v(x),$$

hence

$$\gamma_{xy}(x,y) = \frac{\partial u}{\partial y} + \frac{\partial v}{\partial x} = v'(x) - \theta(x), \tag{6.2}$$

wherefrom we can see that the shear strains do not depend on the y-coordinate.

On the other hand, the distribution of tangential stresses τ^{xy} over a cross-section is determined by well-known Zhuravsky's formula[5]

$$\tau^{xy} = \frac{QS_o}{Ib}, \tag{6.3}$$

where we introduce these designations in addition to the previous ones:

- b is a width of a cross-section (its size in the direction of the z-axis) at a point with a given coordinate y where the tangential stress, τ^{xy}, is to be determined;
- S_o is a static moment of a cutoff part of the section which is above (or below) the fibers in question with a fixed y-coordinate.

According to Hooke's law,

$$\gamma_{xy} = \frac{\tau^{xy}}{G}, \tag{6.4}$$

where G is a shear modulus. Obviously, (6.4) is in general contradiction with relationship (6.2) because stresses τ^{xy} vary with the height of the section (they depend on the y-coordinate). In order to eliminate this contradiction, we will think that formula (6.2) determines only a certain shear strain averaged over the section which we will denote by γ without any additional indexes, that is, we assume

---

[4] The Timoshenko theory of beams adopts the planar sections hypothesis in a weakened form: any cross-section of a beam is assumed to stay planar after the deformation, but the cross-section is not required to remain perpendicular to the deformed axis of the beam.

[5] Comments to Zhuravsky's formula can be found in Appendix B.

$$\gamma(x) = v'(x) - \theta(x)\,. \tag{6.5}$$

This section-average shear γ will be found from the condition of energy equivalence between tangential stresses, τ^{xy}, and their integral characteristic, Q. We want to calculate a part of the strain energy in (6.1) per unit of length of the beam, which conforms to the work of the tangential stresses. Now, using the theorem of average value from integral calculus, we have

$$\frac{1}{2}\int_A \tau^{xy}\gamma_{xy}\,dA = \frac{1}{2}\gamma\int_A \tau^{xy}\,dA = \frac{1}{2}Q\,\gamma \qquad \text{where} \qquad Q = \int_A \tau^{xy}\,dA\,. \tag{6.6}$$

The section-average shear strain, γ, should be in a natural linear relationship with an overall tangential stress in the section, that is, shear force Q,

$$\gamma = \frac{Q}{GF_y}\,, \tag{6.7}$$

where F_y is a certain quantity of the `area' dimensionality which we will call a *shear area*. Here subscript $_y$ emphasizes that we analyze the shear in the (x, y)-plane.

Substituting expressions (6.3) and (6.4) to the left part of (6.6) gives

$$\frac{1}{2}\int_A \tau^{xy}\gamma_{xy}\,dA = \frac{1}{2}\int_A \frac{(\tau^{xy})^2}{G}\,dA = \frac{1}{2}\,\frac{Q^2}{GI^2}\int_A \frac{S_o^2}{b^2}\,dA =$$

$$\frac{1}{2}Q = \frac{Q}{GA}\left(\frac{A}{I^2}\int_A \frac{S_o^2}{b^2}\,dA\right). \tag{6.8}$$

Comparing (6.8) with (6.6) and (6.7) yields shear area F_y, where μ is a so-called *section shape factor* :

$$F_y = \frac{1}{\mu}A, \qquad \mu = \frac{A}{I^2}\int_A \frac{S_o^2}{b^2}\,dA\,. \tag{6.9}$$

As a result, (6.1) produces an expression for the strain energy stored in the beam and represented as a functional of internal forces M and Q:

$$\mathsf{E}_A(M, Q) = \int_0^l\int_A \frac{(\sigma^x)^2}{2E}\,dA\,dx + \int_0^l\int_A \frac{(\tau^{xy})^2}{2G}\,dA\,dx =$$

$$= \int_0^l \frac{M^2}{2EI}\,dx + \int_0^l \frac{Q^2}{2GF_y}\,dx\,. \tag{6.10}$$

Further, internal forces M and Q are related through elasticity dependences

$$M = EI\chi, \qquad Q = GF_y\gamma \tag{6.11}$$

to kinematic parameters χ and γ which can be treated as components of a strain vector[6]

$$\chi = -\theta', \qquad \gamma = v' - \theta. \tag{6.12}$$

Substituting (6.11) to (6.10) will make the expression of E_A a functional of displacements,

$$\mathsf{E}_A(v,\theta) = \int_0^l \frac{EI\theta'^2}{2}dx + \int_0^l \frac{GF_y(v'-\theta)^2}{2}dx. \tag{6.13}$$

Now we are finally able to write out the Lagrange functional (the full potential energy of the system) for a Timoshenko flexural bar lying on an elastic bed:

$$\mathsf{L}(v,\theta) = \frac{1}{2}\int_0^l [EI\theta'^2 + GF_y(v'-\theta)^2 + kv^2]dx - \int_0^l qv dx - \int_0^l m\theta dx -$$

$$-[e_Q n\bar{Q}v]\big|_0^l - [e_M n\bar{M}\theta]\big|_0^l. \tag{6.14}$$

Taking (6.14) as a basis, we want to write an expression of first variation of the Lagrange functional on kinematically admissible displacements:

$$\delta\mathsf{L} = \int_0^l \{[-(EI\theta')' - GF_y(v'-\theta) - m]\delta\theta +$$

$$+[-(GF_y(v'-\theta))' + kv - q]\delta v\}dx +$$

$$+\left[e_M(EI\theta' - n\bar{M})\delta\theta\right]\Big|_0^l + \left[e_Q[GF_y(v'-\theta) - n\bar{Q}]\delta v\right]\Big|_0^l. \tag{6.15}$$

By equaling first variation $\delta\mathsf{L}$ to zero, we derive Euler equations

$$\boxed{\begin{aligned} &-(GF_y v')' + kv + (GF_y\theta)' = q, \\ &-GF_y v' + GF_y\theta - (EI\theta')' = m, \end{aligned}} \tag{6.16}$$

---

[6] We note that in the Timoshenko beam bending analysis the flexural strain parameter, χ, cannot have a simple geometric interpretation such as a curvature of the beam's axis in the deformed state. Relationships (6.12) imply that $1/\rho = -v'' = \chi - \gamma'$ where ρ is a radius of curvature of the beam's axis when deformed.

which are simultaneous differential equations of equilibrium for a Timoshenko beam, written in displacements. The natural boundary conditions for the L functional are

$$\left[e_Q[GF_y(v'-\theta)-n\bar{Q}]\right]\Big|_0^l = 0, \qquad \left[e_M(EI\theta'-n\bar{M})\right]\Big|_0^l = 0, \tag{6.17}$$

and are boundary conditions of static type.

Collecting together all relationships derived above, we will have a full set of governing equations for the bending of a Timoshenko beam,

$$-Q'+kv=q\,, \quad -Q+M'=m \qquad \text{equilibrium equations,} \tag{6.18-a}$$

$$\gamma=v'-\theta\,, \quad \chi=-\theta' \qquad \text{geometric equations,} \tag{6.18-b}$$

$$Q=GF_y\gamma\,, \quad M=EI\chi \qquad \text{physical equations.} \tag{6.18-c}$$

Boundary conditions at points $x=0$ and $x=l$ include

static boundary conditions	kinematic boundary conditions
$nQ-\bar{Q}=0 \ \in\Gamma_Q$,	$v-\bar{v}=0 \ \in\Gamma_v$,
$nM+\bar{M}=0 \ \in\Gamma_M$,	$\theta-\bar{\theta}=0 \ \in\Gamma_\theta$.

(6.19)

Now we want to unify all relationships of the Timoshenko beam bending theory into the same matrix form.

The vectors of stresses $\boldsymbol{\sigma}$, strains $\boldsymbol{\varepsilon}$, and displacements $\mathbf{u}$, which define a stress-and-strain state in a bar, contain two components each in this case:

$$\boldsymbol{\sigma}=\left\|\begin{bmatrix}Q\\M\end{bmatrix}\right\|, \quad \boldsymbol{\varepsilon}=\left\|\begin{bmatrix}\gamma\\\chi\end{bmatrix}\right\|, \quad \mathbf{u}=\left\|\begin{bmatrix}v\\\theta\end{bmatrix}\right\|. \tag{6.20}$$

All equations will have the usual matrix form (1.1) if, in addition to (6.20), we assume

$$\boldsymbol{A}=\left\|\begin{bmatrix}\dfrac{d}{dx} & 1\\ 0 & -\dfrac{d}{dx}\end{bmatrix}\right\|, \quad \boldsymbol{A}^{\mathsf{T}}=\left\|\begin{bmatrix}-\dfrac{d}{dx} & 0\\ -1 & \dfrac{d}{dx}\end{bmatrix}\right\|, \quad \boldsymbol{K}=\left\|\begin{bmatrix}k & 0\\0 & 0\end{bmatrix}\right\|,$$

$$\boldsymbol{C}=\left\|\begin{bmatrix}GF_y & 0\\ 0 & EI\end{bmatrix}\right\|, \quad \boldsymbol{C}^{-1}=\left\|\begin{bmatrix}1/GF_y & 0\\ 0 & 1/EI\end{bmatrix}\right\|, \quad \overline{\boldsymbol{X}}=\left\|\begin{bmatrix}q\\m\end{bmatrix}\right\|. \tag{6.21}$$

Now let us formulate the boundary conditions in the matrix form. The vectors of edge displacements and forces, $\boldsymbol{u}$ and $\boldsymbol{p}$, and their respective vectors of given edge displacements and stresses, $\bar{\boldsymbol{u}}$ and $\bar{\boldsymbol{p}}$, are column matrices of second order:

$$\boldsymbol{u} = \left\| \begin{bmatrix} v \\ \theta \end{bmatrix} \right\|, \qquad \boldsymbol{p} = \left\| \begin{bmatrix} nQ \\ -nM \end{bmatrix} \right\|, \qquad \bar{\boldsymbol{u}} = \left\| \begin{bmatrix} \bar{v} \\ \bar{\theta} \end{bmatrix} \right\|, \qquad \bar{\boldsymbol{p}} = \left\| \begin{bmatrix} \bar{Q} \\ \bar{M} \end{bmatrix} \right\|. \tag{6.22}$$

Exactly in the same way as in the Bernoulli–Euler beam bending analysis, the matrix operators of boundary conditions extraction, $\boldsymbol{E}_p$ and $\boldsymbol{E}_u$, are square matrices of second order:

$$\boldsymbol{E}_p = \left\| \begin{bmatrix} e_Q & 0 \\ 0 & e_M \end{bmatrix} \right\|, \qquad \boldsymbol{E}_u = \left\| \begin{bmatrix} e_v & 0 \\ 0 & e_\theta \end{bmatrix} \right\|. \tag{6.23}$$

Further we assume

$$\boldsymbol{H}_\sigma = \left\| \begin{bmatrix} n & 0 \\ 0 & -n \end{bmatrix} \right\|, \qquad \boldsymbol{H}_u = \boldsymbol{I}, \tag{6.24}$$

which makes it possible to represent the boundary conditions in the matrix form, too — as equations (1.2).

Let us check that the basic integral identity holds. First of all, (**1**.1.5) gives

$$(\boldsymbol{p}, \boldsymbol{u})_\Gamma = (\boldsymbol{H}_\sigma \boldsymbol{\sigma}, \boldsymbol{H}_u \mathbf{u})_\Gamma = (nQv - nM\theta)_{x=0} + (nQv - nM\theta)_{x=l} =$$
$$= [Qv]\big|_0^l - [M\theta]\big|_0^l .$$

Thus,

$$(\boldsymbol{A}\mathbf{u}, \boldsymbol{\sigma}) = (v' - \theta, Q) + (-\theta', M) = \int_0^l [(v' - \theta)Q - \theta' M]dx =$$

$$= \int_0^l [vQ' - \theta(Q - M')]dx + [Qv]\big|_0^l - [M\theta]\big|_0^l = (\mathbf{u}, \boldsymbol{A}^{\mathrm{T}}\boldsymbol{\sigma}) + (\boldsymbol{p}, \boldsymbol{u})_\Gamma , \tag{6.25}$$

so we can see that the basic integral identity as in (**1**.2.17) holds.

Based on the general operator form of representation, we want to obtain expressions of the Castigliano and Reissner functionals. First of all, we take (**2**.3.2) and determine the potential of kinematic external actions, $\Pi_k(\boldsymbol{\sigma})$. We have

$$\Pi_k(\boldsymbol{\sigma}) = (\boldsymbol{E}_u\boldsymbol{H}_\sigma\boldsymbol{\sigma}, \boldsymbol{E}_u\,\overline{\boldsymbol{u}})_\Gamma =$$

$$= \|[e_v nQ, -e_\theta nM]\|\cdot\left\|\begin{bmatrix}\overline{v}\\ \overline{\theta}\end{bmatrix}\right\|_{x=0} + \|[e_v nQ, -e_\theta nM]\|\cdot\left\|\begin{bmatrix}\overline{v}\\ \overline{\theta}\end{bmatrix}\right\|_{x=l} =$$

$$= [e_v Q\overline{v}]\big|_0^l - [e_\theta M\overline{\theta}]\big|_0^l\,. \tag{6.26}$$

Further, we use formula (**2**.3.3) to determine the Castigliano functional for the Timoshenko beam bending analysis as

$$\mathsf{K}(Q,M,v) = \frac{1}{2}\int_0^l\left(\frac{Q^2}{GF_y}+\frac{M^2}{EI}\right)dx + \frac{1}{2}\int_0^l kv^2dx -$$

$$- [e_v Q\overline{v}]\big|_0^l + [e_\theta M\overline{\theta}]\big|_0^l\,. \tag{6.27}$$

In this case the set, $\mathscr{R}_s$, of physically and statically admissible stress-and-strain fields consists of all possible sets of functions Q, M, and v which satisfy equations (6.18-*a*) on the interval $[0,l]$ and the static boundary conditions on each end of the interval.

According to (**3**.1.4) and (**3**.1.5), we define the Reissner functional in its first and second forms as

$$\mathsf{R}_1(Q,M,v) = \frac{1}{2}\int_0^l\left(\frac{Q^2}{GF_y}+\frac{M^2}{EI}\right)dx - \frac{1}{2}\int_0^l kv^2dx -$$

$$-\int_0^l[(v'-\theta)Q-\theta'M]dx + \int_0^l qvdx + \int_0^l m\theta dx +$$

$$+[e_v Q(v-\overline{v})]\big|_0^l - [e_\theta M(\theta-\overline{\theta})]\big|_0^l + [e_Q n\overline{Q}v]\big|_0^l + [e_M n\overline{M}\theta]\big|_0^l\,, \tag{6.28}$$

$$\mathsf{R}_2(Q,M,v) = \frac{1}{2}\int_0^l\left(\frac{Q^2}{GF_y}+\frac{M^2}{EI}\right)dx - \frac{1}{2}\int_0^l kv^2dx -$$

$$-\int_0^l[vQ'-\theta(Q-M')]dx + \int_0^l qvdx + \int_0^l m\theta dx -$$

$$-[e_v Q\overline{v}]\big|_0^l + [e_\theta M\overline{\theta}]\big|_0^l - [e_Q(Q-n\overline{Q})v]\big|_0^l + [e_M(M+n\overline{M})\theta]\big|_0^l\,. \tag{6.29}$$

Note that, unlike the Bernoulli beam case, there is no third form of the Reissner functional for a Timoshenko beam. This is because all functional arguments participate in the Reissner functional with their derivatives of first order at the most.

4.6.1 Another remark on kinematic boundary conditions for a Timoshenko beam

As can be seen from (6.19), the kinematic boundary conditions for a Timoshenko beam involve displacements v and slopes θ of the beam's cross-sections. Meanwhile, authors of various publications sometimes try to reanimate an erroneous suggestion that the boundary conditions should be formulated for first derivative of the displacement function, v', rather than for slope θ. However, boundary conditions which include v' (or the shear angle, γ) do not conform to a physically correct problem statement. The boundary conditions like these violate fundamental laws of mechanics so the reciprocity theorems do not work anymore[7].

It is simplest to demonstrate this by an example of a cantilever beam (Fig. 4.4) in its two states.

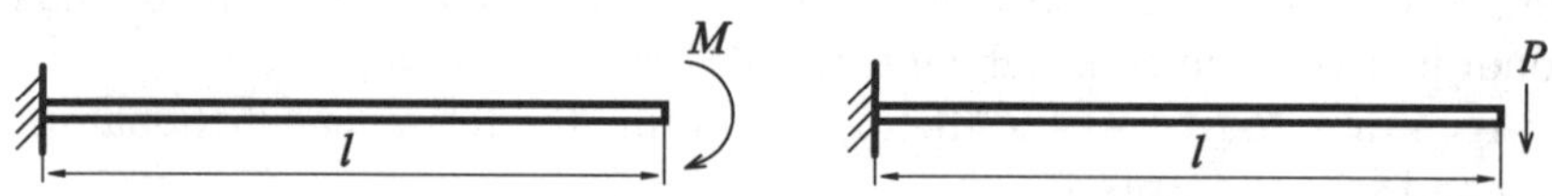

Fig. 4.4. Two states of a cantilever beam

In its state 1, the beam is loaded by an end moment M, and in its state 2 by a force P applied to its free end. First, we imagine that the boundary conditions $v(0) = 0$ and $v'(0) = 0$ are formulated for the clamped section $x = 0$. It is easy to make sure by direct substitution that the respective solutions for load cases 1 and 2 will be

$$v_1 = \frac{M}{2EI}x^2, \qquad \theta_1 = \frac{M}{EI}x;$$

$$v_2 = \frac{Pl^3}{2EI}\left(\frac{x^2}{l^2} - \frac{x^3}{3l^3}\right), \qquad \theta_2 = \frac{Pl^2}{EI}\left(-\frac{EI}{l^2 GF_y} + \frac{x}{l} - \frac{x^2}{2l^2}\right),$$

because these solutions satisfy both the differential equations of the problem and all boundary conditions of it.

If now we calculate the work of external forces in each state on the respective displacements from the other state, we will have

[7] In the mathematical language it sounds like these boundary conditions are not self-conjugated anymore.

$$A_{12} = \frac{MPl^2}{EI}\left(-\frac{EI}{l^2GF_y}+\frac{1}{2}\right), \quad A_{21} = \frac{PMl^2}{2EI},$$

and, obviously, these values are not equal, which contradicts to the work reciprocity theorem (Betty theorem). On the other hand, if we pose the boundary conditions correctly in the location where the beam is clamped, that is, if we assume $v(0) = 0$ and $\theta(0) = 0$, then the solution for the load in the form of an edge moment will remain the same whereas the state of the system under a concentrated force (state 2) will change to become

$$v_2 = \frac{Pl^3}{EI}\left(\frac{EI}{l^2GF_y}\frac{x}{l}-\frac{x^3}{6l^3}+\frac{x^2}{2l^2}\right), \qquad \theta_2 = \frac{Pl^2}{EI}\left(\frac{x}{l}-\frac{x^2}{2l^2}\right),$$

so

$$A_{12} = \frac{MPl^2}{2EI}, \qquad A_{21} = \frac{PMl^2}{2EI},$$

and the Betty theorem holds.

4.7 Planar curvilinear bar, shear ignored

Consider a planar curvilinear bar all cross-sections of which are orthogonal to a certain planar curve that belongs to the (x_1,x_2)-plane of a Cartesian coordinate system, (x_1,x_2,x_3). This planar curve will be called an *axis of the bar*, and the (x_1,x_2)-plane a *plane of the* (*curvilinear*) *bar*.

Let us assume that any cross-section of the bar orthogonal to the bar's axis is crossed by the axis in its center of gravity. One of principal central axes of inertia of the bar's section (z-axis, to be definite) will be deemed to belong to the plane of the bar. We will also assume that the plane of the bar is at the same time its *plane of loading*. The plane of loading (or the load plane) is a plane in which all external forces are applied[8].

The limitations formulated above permit us to deal with displacements and strains in the bar's plane only, without considering any behavior of the bar out of the plane.

Let the axis of the bar be a smooth curve defined by the following vector equation in the parametric representation:

---

[8] When a moment load is involved, its vector must be orthogonal to the load plane. This should be clear because a moment load is actually a force couple consisting of two forces in opposite directions, which belong to the same plane.

$$\mathbf{M} = \mathbf{M}(s) \tag{7.1}$$

where:

- **M** is a radius vector of the current point, M, of the bar's axis;
- s is a scalar parameter, being the length of arc counted from a certain initial point with the coordinate $s = 0$ in a direction assumed positive for the movement along the curve.

Each point of the bar's axis will conform to a local natural trihedron consisting of a triple of unit vectors, $(\boldsymbol{n}, \boldsymbol{t}, \boldsymbol{b})$. Unit vector $\boldsymbol{t}$ will be defined as a vector of length one, tangential to the bar's axis and directed towards increasing values of parameter s. From a well-known formula of vector analysis we have

$$\boldsymbol{t} = \frac{d\mathbf{M}}{ds}. \tag{7.2}$$

Unit vector $\boldsymbol{n}$, called a *normal*, belongs to the plane of the bar's axis, is orthogonal to the tangential unit vector $\boldsymbol{t}$, and is directed towards the convexity of the bar's axis, that is, away from the center of curvature of the axis. Unit vector $\boldsymbol{b}$, called a *binormal*, is orthogonal to the plane of the bar. As the $(\boldsymbol{n}, \boldsymbol{t}, \boldsymbol{b})$ trihedron has a right-hand orientation, each of the unit vectors can be represented via a vector product of the two others:

$$\boldsymbol{n} = \boldsymbol{t}\times\boldsymbol{b}, \quad \boldsymbol{t} = \boldsymbol{b}\times\boldsymbol{n}, \quad \boldsymbol{b} = \boldsymbol{n}\times\boldsymbol{t}. \tag{7.3}$$

The conventions for the positive directions of the tangential vector $\boldsymbol{t}$ and normal $\boldsymbol{n}$ are given above; relationship (7.3) enables us to define also an unambiguous positive direction for binormal $\boldsymbol{b}$.

And vice versa, if we demand from the very beginning that the positive direction of binormal $\boldsymbol{b}$ coincide with the direction of increasing for third coordinate x_3 of the fixed coordinate system, (x_1,x_2,x_3), in this way we will define the above said positive direction of the movement along the bar's axis because $\boldsymbol{t} = \boldsymbol{b}\times\boldsymbol{n}$. For example, for a piece of a curvilinear bar shown in Fig. 4.5, binormal $\boldsymbol{b}$ looks into the drawing away from the reader.

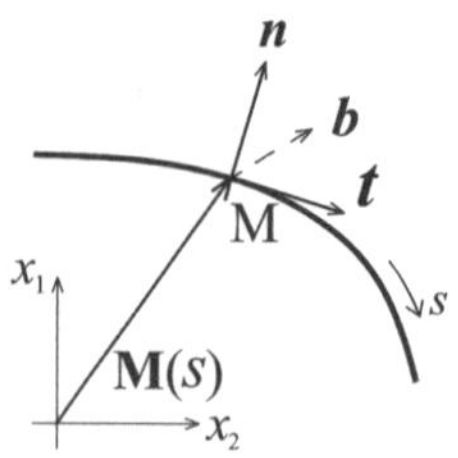

Fig. 4.5. A natural trihedron in the current point, M, of the bar's axis

A set of all points of the (x_1,x_2)-plane at the distance z from the bar's axis along its normal (taking into account the sign of z) will be called a *z-fiber* of the bar. In other words, z-fiber will be a curve defined by radius vector $\mathbf{N} = \mathbf{M} + z\boldsymbol{n}$ at a fixed value of z. The combination of parameters (z,s) can be conveniently treated as Lagrangian (material) coordinates in the two-dimensional analysis of deformation of a planar curvilinear bar, and further we will call z a *normal coordinate* and s an *arc coordinate.*

As we move along the bar's axis, unit vectors $\boldsymbol{n}$ and $\boldsymbol{t}$ will change (their direction), while binormal $\boldsymbol{b}$ will remain a constant vector[9].

Further we will need formulas for differentiating unit vectors $\boldsymbol{n}$ and $\boldsymbol{t}$ with respect to the arc coordinate. The formulas following below are used for unit vector differentiation — these are called Frenet formulas (see, for example, [6])[10]

$$\frac{d\boldsymbol{t}}{ds} = -k\boldsymbol{n}\,, \qquad \frac{d\boldsymbol{n}}{ds} = k\boldsymbol{t}\,, \tag{7.4}$$

where k is a curvature of the bar's axis — a quantity inverse to the radius of curvature, ρ, that is, $k = 1/\rho$.

---

[9] Strictly speaking, binormal $\boldsymbol{b}$ does not change its direction as long as the arc coordinate keeps the sign of the curvature of the bar's axis. When the curvature alters its sign in a point of zero curvature on the axis (a contraflexure), the positive direction of the binormal will become its opposite. The point of contraflexure itself is a singular point because there is no definite direction of normal $\boldsymbol{n}$ in it. Of course, one could build a theory of a curvilinear bar the axis of which contains contraflexure points, too. We will, however, omit those complicating circumstances and will assume for the sake of simplicity that the bar's axis does not contain any contraflexure points. Moreover, we will assume that the deformed state of the bar does not alter the sign (direction) of the curvature in any point of the bar's axis, nor it contains any points of zero curvature.

[10] In [6] the Frenet formulas contain the curvature with an opposite sign. The explanation for this is that the convention for the positive direction of normal $\boldsymbol{n}$ in [6] is opposite to the convention we use here. However, the formulas of differentiation (7.4) can be easily validated. The length of vector $\boldsymbol{n}$ is one by definition, therefore $(\boldsymbol{n}, \boldsymbol{n}) = 1$. We can differentiate this equality with respect to parameter s to find $(\boldsymbol{n}, d\boldsymbol{n}/ds) = 0$, hence the mutual orthogonality of vector $\boldsymbol{n}$ and vector $d\boldsymbol{n}/ds$. Consequently, this latter vector is collinear with vector $\boldsymbol{t}$, so it can be represented as $d\boldsymbol{n}/ds = a\boldsymbol{t}$ where a is a certain scalar. The value of this scalar, equal to the curvature, can be easily found out from simple geometric considerations. The second Frenet formula can be proved in the same way.

4.7.1 Geometric equations

The basis for an engineering theory of planar curvilinear bars is the same kinematic and static assumptions or hypotheses which were used to construct the Bernoulli–Euler theory of beams.

First of all, let's turn to kinematic relationships and express the strain of a z-fiber via displacements of the bar's axis. In order to do this, we extract an elementary piece of the bar bounded by the axis and by the z-fiber of the bar together with two cross-sections orthogonal to the axis at the points with the arc coordinates s and $s + ds$. A curvilinear quadrangle which we obtain in this way is shown in Fig. 4.6 with Lagrangian coordinates of its respective angle points as shown here: M(s, 0), N(s, z), P(s+ds, 0), Q(s+ds, z). The same figure depicts the selected element in its deformed state; the respective angle points of the quadrangle in the deformed state are denoted by the same letters but with the additional subscript 1.

According to the planar sections hypothesis, straight segments MN and PQ remain straight in their deformed state, keep their length, and remain orthogonal to the deformed axis of the bar. These straight segments in their deformed state are denoted as M_1N_1 and P_1Q_1, respectively, in Fig. 4.6.

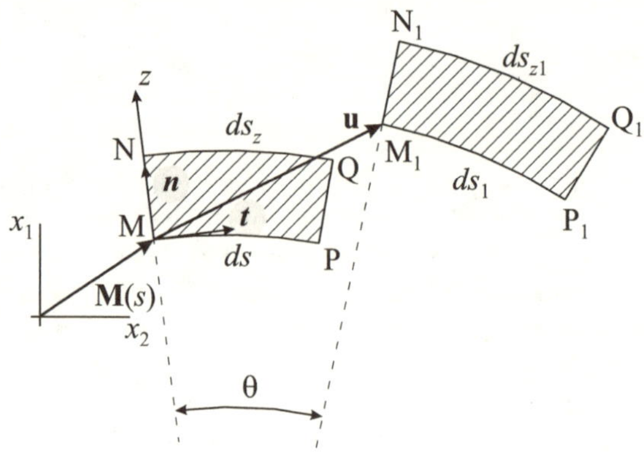

Fig. 4.6. A displacement of an elementary piece of the curvilinear bar

Let the current point, M, of the bar's axis have got displacements corresponding to vector $\mathbf{u} = w\boldsymbol{n} + v\boldsymbol{t}$. It is common to refer to w as a *normal displacement* and to v as a *tangential displacement*.

Now let us explain the designations of arc differentials shown in Fig. 4.6:

ds is the length of arc MP; ds_1 is the length of arc M_1P_1;
ds_z is the length of arc NQ; ds_{z1} is the length of arc N_1Q_1.

Obviously,

$$\frac{ds_z}{\rho + z} = \frac{ds}{\rho} \qquad \text{or} \qquad ds_z = (1 + kz)\, ds. \tag{7.5}$$

Let ε be a relative strain of the z-fiber and let ε_0 be a relative strain at the level of the bar's axis (that is, at $z = 0$). Obviously, the arc differentials introduced above are related to one another as

$$ds_1 = (1 + \varepsilon_0)\, ds, \qquad ds_{z1} = (1 + \varepsilon)\, ds_z = (1 + \varepsilon)(1 + kz)\, ds. \tag{7.6}$$

We will denote by k_1 a curvature of the bar's axis in its deformed state. We use formula (7.5) in application to the deformed state to express the length of arc N_1Q_1 via element ds in another way:

$$ds_{z1} = (1 + k_1 z)\, ds_1 = (1 + k_1 z)(1 + \varepsilon_0)\, ds\ , \tag{7.7}$$

Comparing two expressions for ds_{z1} from (7.6) and (7.7) gives the equality

$$\varepsilon(1 + kz) = \varepsilon_0(1 + k_1 z) + k_1 z - kz\ . \tag{7.8}$$

However, it is more convenient to work with a quantity denoted by χ — a variation of the curvature of the bar's axis — rather than the new curvature of the axis, k_1, after assuming

$$\chi = k_1 - k\ . \tag{7.9}$$

Substituting (7.9) to (7.8) and making some transformations, we have

$$\varepsilon = \varepsilon_0\left(1 + \frac{\chi z}{1 + kz}\right) + \frac{\chi z}{1 + kz}\ .$$

We want to estimate the value of $\chi z/(1+kz)$ in comparison to one. It is easy to notice that this aggregate characterizes a distortion of the z-fiber caused by the bending of the bar (without the longitudinal deformation of the bar's axis, that is, at $\varepsilon_0 = 0$). Consequently, this aggregate conforms to a strain by its order of magnitude, so in the small-strain theory it can be neglected when added to one. As a result, we have this formula:

$$\varepsilon = \varepsilon_0 + \frac{\chi z}{1 + kz}\ . \tag{7.10}$$

Formula (7.10) defines a law of variation of the relative strains, ε, over the z-coordinate in the general case of a curvilinear bar. Further simplifications can be made seeing that the bar in question has a small initial curvature.

When we build kinematic relationships for a curvilinear bar, for our convenience we want to distinguish between three principal versions of the theory depending on the order of smallness of a dimensionless parameter,

kh, where h is a height of the bar's section, that is, a maximum size of the section in the direction of axis $\boldsymbol{n}$.

Following this classification criterion, we adopt this separation of all curvilinear bars into:

- *small-curvature bars*, for which this estimate takes place: $kh << 1$; it shows that we can neglect the value of kh in comparison to one in various relationships;
- *medium-curvature bars*, for which parameter kh is less than one but not so much that we would be able to neglect it when added to one without losing a good deal of accuracy. We think, however, that the estimate $(kh)^2 << 1$ takes place, and it says we can neglect the value $(kh)^2$ comparing to one in various relationships;
- *big-curvature bars*, with which formula (7.10) must be used without any simplification.

We begin our discussion with curvilinear bars of medium curvature. Formulas related to small-curvature bars will be derived further using more simplifications.

We expand expression $1/(1+kz)$ into a series over the powers of z:

$$\frac{1}{1+kz} = 1 - kz + \ldots \tag{7.11}$$

If we follow V.Z. Vlasov [14] and keep only two first terms in expansion (7.11), which is consistent with the estimate $(kh)^2 << 1$, then expression (7.10) for the strain in the z-fiber will become as simple as

$$\varepsilon = \varepsilon_0 + \chi z(1 - kz)\,. \tag{7.12}$$

Parameters ε_0 and χ permit to restore the strains at any point of a planar curvilinear bar, so in this sense they can be called *components of strains in a curvilinear bar*. However, for reasons which will be discussed later, we deem it reasonable to use a more exact entitlement with these parameters: *components of strains in a curvilinear bar according to Vlasov*.

Our task is now to establish geometric relationships which would derive the bar's strain components, ε_0 and χ, from given components of the displacements of its axis, v and w.

We introduce a new trihedron, $(\boldsymbol{n}_1, \boldsymbol{t}_1, \boldsymbol{b}_1)$, together with the old one, $(\boldsymbol{n}, \boldsymbol{t}, \boldsymbol{b})$. The new trihedron will have the same meaning but relate to the deformed state of the bar. Note that $\boldsymbol{b}_1 = \boldsymbol{b}$ because of the conditions stated in footnote [9]. Further, exactly as formula (7.2) holds for the original position of the bar's axis, its deformed state is subject to the relationship

$$\boldsymbol{t}_1 = \frac{d\mathbf{M}_1}{ds_1} = \frac{d\mathbf{M}_1}{ds}\frac{ds}{ds_1} = \frac{d(\mathbf{M} + v\boldsymbol{t} + w\boldsymbol{n})}{ds}\frac{1}{(1+\varepsilon_0)} =$$

$$= \frac{1}{(1+\varepsilon_0)}(\boldsymbol{t} + v'\boldsymbol{t} + w'\boldsymbol{n} + v\frac{d\boldsymbol{t}}{ds} + w\frac{d\boldsymbol{n}}{ds}).$$

Here and further in this section we use a stroke to denote the differentiation with respect to parameter s.

Now we want to use the Frenet formulas to derive this:

$$\boldsymbol{t}_1 = \frac{1+a}{(1+\varepsilon_0)}\boldsymbol{t} + \frac{b}{(1+\varepsilon_0)}\boldsymbol{n}, \tag{7.13}$$

where we denote

$$a = v' + kw, \qquad b = w' - kv. \tag{7.14}$$

We can use (7.13) to obtain an expression of strain ε_0 at the level of the bar's axis. In order to do this, we recall that $\boldsymbol{t}_1$ is a unit vector.

Let's bother the ashes of the Samian Sage, Pythagoras:

$$(1 + \varepsilon_0)^2 = (1 + a)^2 + b^2,$$

or, after transformations,

$$\varepsilon_0(1 + \frac{\varepsilon_0}{2}) = a + \frac{1}{2}(a^2 + b^2).$$

To remain within the small-strain theory, we neglect the value of $\varepsilon_0/2$ comparing to one and have

$$\varepsilon_0 = a + ½(a^2 + b^2). \tag{7.15}$$

Formula (7.15) defines a relative strain at the level of the axis of a curvilinear bar and is based on geometrically nonlinear (second-order) considerations. In the case of first-order analysis (geometrically linear), which is our subject at the moment, we should omit quadratic displacement terms in the right part of (7.15), to have finally

$$\varepsilon_0 = v' + kw. \tag{7.16}$$

Returning to formula (7.13), we define unit vector $\boldsymbol{n}_1$ as a unit normal to the deformed axis of the bar. We have

$$\boldsymbol{n}_1 = \boldsymbol{t}_1 \times \boldsymbol{b} = \frac{1+a}{(1+\varepsilon_0)}\boldsymbol{n} - \frac{b}{(1+\varepsilon_0)}\boldsymbol{t}. \tag{7.17}$$

Now we use the Frenet formulas on the axis of the bar in its deformed state and derive these relationships:

$$\frac{d\boldsymbol{t}_1}{ds_1} = -\, k_1\boldsymbol{n}_1, \qquad \frac{d\boldsymbol{n}_1}{ds_1} = k_1\boldsymbol{t}_1. \tag{7.18}$$

The new curvature of the bar's axis, k_1, can be obtained by making a scalar product of the second equality in (7.18) and vector $\boldsymbol{t}_1$, to derive

$$k_1 = (\boldsymbol{t}_1, \frac{d\boldsymbol{n}_1}{ds_1}) = \frac{1}{(1+\varepsilon_0)}(\,\boldsymbol{t}_1, \frac{d\boldsymbol{n}_1}{ds}\,). \tag{7.19}$$

Formal differentiation of (7.17) and application of the Frenet formulas gives

$$\frac{d\boldsymbol{n}_1}{ds} = A\boldsymbol{t} + B\boldsymbol{n} \tag{7.20}$$

where

$$A = k\frac{1+a}{(1+\varepsilon_0)} - \frac{d}{ds}\frac{b}{(1+\varepsilon_0)} = \frac{k(1+a)-b'}{1+\varepsilon_0} + \frac{b\varepsilon_0'}{(1+\varepsilon_0)^2},$$

$$B = k\frac{b}{(1+\varepsilon_0)} + \frac{d}{ds}\frac{1+a}{(1+\varepsilon_0)} = \frac{kb+a'}{1+\varepsilon_0} - \frac{(1+a)\varepsilon_0'}{(1+\varepsilon_0)^2}. \tag{7.21}$$

Substituting (7.13) and (7.20) to formula (7.19) and taking (7.9) into account, we have an expression for the variation of the curvature of the bar's axis, χ,

$$\chi = \frac{1}{(1+\varepsilon_0)^2}[A(1+a) + Bb] - k\,. \tag{7.22}$$

Replacing the expressions of parameters A and B in (7.22) by those from (7.21), we make some transformations and arrive at the equality

$$\chi(1+\varepsilon_0)^3 = k[(1+a)^2 + b^2 - (1+\varepsilon_0)^3] - b'(1+a) + a'b\,.$$

But the expression in the brackets here is equal to

$$(1+a)^2 + b^2 - (1+\varepsilon_0)^3 = (1+\varepsilon_0)^2 - (1+\varepsilon_0)^3 = -\,\varepsilon_0(1+\varepsilon_0)^2 = -\,\varepsilon_0(1+2a+a^2+b^2),$$

consequently,

$$\chi = \frac{-k\varepsilon_0(1+2a+a^2+b^2) - b'(1+a) + a'b}{(1+\varepsilon_0)^3}. \tag{7.23}$$

Formula (7.23) is exact in the sense that no simplifications — omissions of small quantities — were used to derive it.

Adopting the small-strain theory, we can assume the denominator in formula (7.23) to be equal to one, and to take the value for ε_0 in the numerator from formula (7.15). Keeping terms of at most second order of the displacements and their derivatives in the expression thus obtained, we have

$$\chi = -ka - b' - \frac{k}{2}(5a^2 + b^2) - b'a + a'b \,. \tag{7.24}$$

Finally, in the geometrically linear (first-order) analysis, we omit second-order terms of displacements in (7.24). As a result, replacing parameters a and b with their expressions via the displacement components as in (7.14) gives a final equation:

$$\chi = -w'' - k^2 w + k v'. \tag{7.25}$$

Further we want to manipulate another convenient geometric parameter which has the meaning of a slope, θ, of a current cross-section in the bar. It is easy to notice that geometric considerations give $\sin\theta = -(\boldsymbol{n}, \boldsymbol{t}_1)$. We suppose slope θ to be small, therefore its sine can be replaced by the value of the slope itself, and after involving (7.13) we have

$$\theta = -b = -w' + kv. \tag{7.26}$$

Formulas (7.16), (7.25), and (7.26) make up a complete set of geometric equations which establish relationships between the displacements and the strains in the theory of planar curvilinear bars. They are a particular case of geometric relationships obtained by V.Z. Vlasov in his engineering theory of cylindrical shells [14], therefore we will call them *Vlasov's geometric relationships.*

There is another possible approach to geometric relationships for a planar curvilinear bar which has its roots in the theory of shells by Love [8] and adopted by most authoritative experts in the thin shell theory in our country [10], [4].

The second approach is different in two points. First, and most important, the initial curvature of the bar's axis is assumed to be so small that the estimate $(kh)^2 \ll 1$ should be replaced by a stronger estimate, $kh \ll 1$. The classification given above treats such case as curvilinear bars of small curvature. Second, the theory (which we will call a Kirchhoff–Klebsch theory) uses a different parameter of flexural strain, other than the variation of the bar's axis curvature $\chi = k_1 - k$ but close enough to it, namely

$$\kappa = \chi + k\varepsilon_0\,, \tag{7.27}$$

which gives the following after substituting formulas (7.16) and (7.25)

$$\kappa = -w'' + (kv)'. \tag{7.28}$$

The κ quantity will be called a Kirchhoff–Klebsch parameter of flexural strain[11]. Taking into account (7.27), we can rewrite formula (7.12) as

$$\varepsilon = \varepsilon_0 + \chi z(1 - kz) = \varepsilon_0(1 - kz + k^2z^2) + \kappa z(1 - kz),$$

so, in order to remain consistent in our omitting infinitesimal terms, we should replace both expressions in the parentheses by one and thus obtain a linear distribution of the strains along the z-coordinate,

$$\varepsilon = \varepsilon_0 + \kappa z\,. \tag{7.29}$$

Now let us collect all geometry-related results that we have obtained and place them together in Table 4.1 for convenience of browsing. Please note again that all formulas in Table 4.1 refer to the geometrically linear (first-order) analysis.

Table 4.1

Big-curvature bars	*Medium-curvature bars* $(kh)^2 << 1$	*Small-curvature bars* $kh << 1$
$\varepsilon = \varepsilon_0 + \dfrac{\chi z}{1+kz}$	$\varepsilon = \varepsilon_0 + \chi z - \chi k z^2$	$\varepsilon = \varepsilon_0 + \kappa z$
$\varepsilon_0 = v' + kw$ $\theta = -w' + kv$ $\chi = -k\varepsilon_0 + \theta'$	$\varepsilon_0 = v' + kw$ $\theta = -w' + kv$ $\chi = -k\varepsilon_0 + \theta'$	$\varepsilon_0 = v' + kw$ $\theta = -w' + kv$ $\kappa = \theta'$
$\chi = -w'' - k^2w + k'v$	$\chi = -w'' - k^2w + k'v$	$\kappa = -w'' + (kv)'$
$\varepsilon = v' + kw +$ $+(-w'' - k^2w + k'v)\dfrac{z}{1+kz}$	$\varepsilon = v' + kw +$ $+(-w'' - k^2w + k'v)z(1-kz)$	$\varepsilon = v' + kw +$ $+(-w'' + kv' + k'v)z$

[11] Strictly speaking, Love treats the flexural strain parameter, κ, in Section 259 of his known work [8] exactly as a variation of the bar's axis curvature rather than a quantity close enough to but different from χ, based on and referring to previous results by Klebsch and even earlier works by Kirchhoff. This treatment, though not quite accurate, is used by him also for building the general theory of shells in Section 326. See formulas (4.27) in Novozhilov's [10].

4.7.2 Equations of equilibrium

The set of internal forces (stresses), N, Q, M, is shown in Fig. 4.7. Let q_t(s) and q_n(s) be components of intensities of external loads distributed along the bar's axis and acting along the respective tangential and normal directions, and let m(s) be an intensity of a distributed moment load. External concentrated static actions specified on the ends of the bar will be denoted by $\bar{N}_0$ and $\bar{Q}_l$, $\bar{M}_0$ and $\bar{M}_l$, respectively (Fig. 4.7).

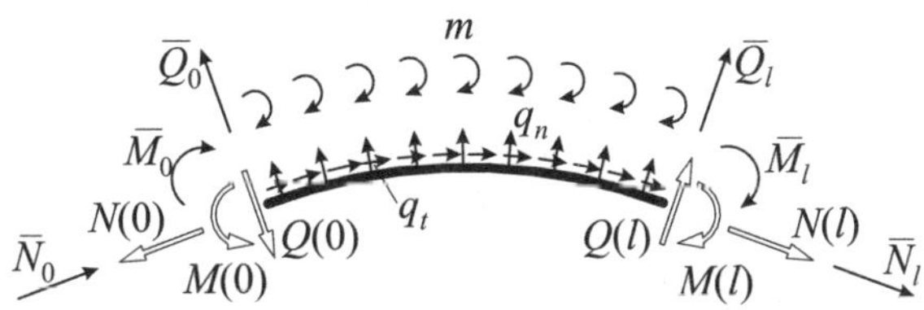

Fig. 4.7. Internal and external forces acting on a curvilinear bar

This figure shows all internal and external force actions in their positive directions.

We want to consider the equilibrium of an elementary piece of the curvilinear bar, ds long, in projections onto the tangential, ***t***, and the normal, ***n***, directions and in moments with respect to one of the bar's ends; in this way we obtain the following differential relationships between internal force quantities:

$$-N' - kQ = q_t, \qquad kN - Q' = q_n, \qquad Q - M' = m. \tag{7.30}$$

Equations (7.30) have the meaning of differential equations of equilibrium; they are called *Kirchhoff equations* [3].

If we use the third equation in (7.30) to exclude the shear force, Q, from first two Kirchhoff equations, then the equilibrium equations will contain two unknown functions, longitudinal force N and bending moment M,

$$-N' - kM' = q_t + km, \qquad kN - M'' = q_n + m', \tag{7.31}$$

and these will be also called Kirchhoff equations.

Based on the hypothesis of no pressure between the longitudinal fibers of the bar, we adopt the physical law of relation between longitudinal stress σ and strain ε in its simplest form, a linear relation:

$$\sigma = E\varepsilon. \tag{7.32}$$

Further, the internal forces – longitudinal force N and bending moment M are integral characteristics of the stresses; they are calculated by the following formulas:

- for medium-curvature bars (Vlasov's theory),

$$N=\int_A \sigma dA=\int_A E(\varepsilon_0+\chi z-\chi k z^2)dA=\varepsilon_0 EA-\chi kEI,$$

$$M=\int_A \sigma z dA=\int_A E(\varepsilon_0+\chi z-\chi k z^2)z dA=$$

$$=\chi EI-\chi kEI_2=\chi EI(1-k\frac{I_2}{I}), \tag{7.33}$$

- for small-curvature bars (Kirchhoff–Klebsch theory),

$$N=\int_A \sigma dA=\int_A E(\varepsilon_0+\kappa z)dA=\varepsilon_0 EA,$$

$$M=\int_A \sigma z dA=\int_A E(\varepsilon_0+\kappa z)z dA=\kappa EI, \tag{7.34}$$

- for big-curvature bars,

$$N=\int_A \sigma dA=\int_A E(\varepsilon_0+\frac{\chi z}{1+kz})dA=\varepsilon_0 EA+\chi ES_\rho,$$

$$M=\int_A \sigma z dA=\int_A E(\varepsilon_0+\frac{\chi z}{1+kz})z dA=\chi EI_\rho. \tag{7.35}$$

Formulas (7.33) involve a new geometric property of a cross-section in addition to standard ones (area of section A and moment of inertia I) – *a second-order moment of inertia,* I_2.

Generally, let I_n be called a *moment of inertia of n-th order* if

$$I_n=\int_A z^{n+1}dA \qquad \text{so that} \qquad I_2=\int_A z^3 dA. \tag{7.36}$$

The integration in (7.33) to (7.35) over the area takes into account the fact that the z-coordinate is counted from the section's center of gravity along the principal central axis, therefore the integral of z to first power (a static moment of the section's area) becomes zero. Note, by the way, that for cross-sections with two axes of symmetry the moment of inertia of second order (as well as of any even order) is zero. It is quite clear because $dA = b(z)dz$ where $b(z)$ is a size of the section in the direction perpendicular to the z-axis (width of the section). For a bisymmetric section, $b(z) = b(-z)$, hence $I_n = 0$ for an even n. However, if there is no bisymmetry, the kI_2/I ratio is small comparing to one. For example, it is easy to calculate that $kI_2/I = 2kh/15$ for a section in the shape of an equilateral triangle of height h. This is a reason why we can adopt the additional estimate $kI_2/I \ll 1$ in the theory of medium-curvature bars. But then the second of relationships (7.33) will be simplified, so finally we arrive at this in the Vlasov theory:

$$N = \varepsilon_0 EA - \chi kEI, \qquad M = \chi EI. \tag{7.37}$$

Further, (7.35) introduces two more geometric properties: a static moment, S_ρ, of a *reduced section* with respect to the level of the center of gravity of the original section, and a moment of inertia of the reduced section, I_ρ, with respect to the same axis. The reduced section of a curvilinear bar is a section derived from a given one by formally decreasing its width, b, $(1+kz)$ times at the level of the z-fiber. In other words,

$$S_\rho = \int_A \frac{z}{1+kz} dA, \qquad I_\rho = \int_A \frac{z^2}{1+kz} dA. \tag{7.38}$$

By the way, these two geometric properties can be easily expressed via the reduced section's area, A_ρ, because

$$A_\rho = \int_A \frac{1}{1+kz} dA, \qquad kS_\rho = \int_A \frac{1+kz-1}{1+kz} dA = A - A_\rho,$$

$$kI_\rho = \int_A \frac{z+kz^2-z}{1+kz} dA = -S_\rho.$$

Physical relationships (7.35) for big-curvature bars should be rewritten now in a final convenient form:

$$N = \varepsilon_0 EA - \chi kEI_\rho, \qquad M = \chi EI_\rho. \tag{7.39}$$

We present all physical relationships for three classes of curvilinear bars together in Table 4.2. The second column of the same table presents expressions of the stresses/forces via the displacements, obtained by substituting strain expressions from the respective column of Table 4.1 to the physical relationships.

Table 4.2

Big-curvature bars	
$N = \varepsilon_0 EA - \chi kEI_\rho$	$N = (v' + kw)EA + (w'' + k^2 w - k'v)kEI_\rho$
$M = \chi EI_\rho$	$M = (-w'' - k^2 w + k'v)EI_\rho$
Medium-curvature bars (according to Vlasov) $(kh)^2 \ll 1$	
$N = \varepsilon_0 EA - \chi kEI$	$N = (v' + kw)EA + (w'' + k^2 w - k'v)kEI$
$M = \chi EI$	$M = (-w'' - k^2 w + k'v)EI$

Small-curvature bars (according to Kirchhoff–Klebsch) $kh \ll 1$	
$N = \varepsilon_0 EA$	$N = (v' + kw)EA$
$M = \kappa EI$	$M = (-w'' + k'v + kv')EI$

Comparing formulas related to the bars of big and medium curvature, we notice that they are identical up to the replacement of the section's moment of inertia, I, by the reduced section's moment of inertia, I_ρ, and vice versa. As for the bars of small curvature, the differences from the first two theories are more substantial because the very structure of the formulas is subject to changes.

We can replace the internal forces in equilibrium equations (7.31) by their representations via the displacement components, v and w, according to Table 4.2 and collect terms related to particular displacements to arrive, after some transformations, at a governing system of two simultaneous differential equations with respect to the displacements:

$$\boxed{\begin{aligned} L_{11}v + L_{12}w &= q_t + km, \\ L_{21}v + L_{22}w &= q_n + m', \end{aligned}} \tag{7.40}$$

which has a meaning of simultaneous equations of equilibrium in displacements (a Lame-type system of equations, if it is more to the reader's liking) for the analysis of the behavior of a curvilinear bar in its plane of curvature. The differential operators, L_{11}, L_{12}, L_{21} and L_{22}, for different versions of the theory are given below – see Table 4.3. In particular, for a bar of a round shape with the radius R and constant cross-section these operators become simpler, as shown in Table 4.4.

Tables 4.3 and 4.4 do not contain explicit expressions of operators L_{ij} for big-curvature bars. But they are not needed, actually, because those operators coincide with the respective Vlasov operators after replacing I with I_ρ. Equations (7.40) for medium-curvature bars are a particular case of V.Z. Vlasov's equations derived by him for a circular cylindrical shell [14].

Table 4.3

Medium-curvature bars (according to Vlasov) $(kh)^2 \ll 1$	
$L_{11}v = -(EAv')' + (k')^2 EIv$	$L_{12}w = -(kEAw)' - k'EI(w'' + k^2 w)$
$L_{21}v = kEAv' - k^2 k'EIv - (k'EIv)''$	$L_{22}w = k^2 EAw + k^2 EI(w'' + k^2 w) + [EI(w'' + k^2 w)]''$

Small-curvature bars (according to Kirchhoff–Klebsch) $kh \ll 1$	
$L_{11}v = -(EAv')' - k[EI(kv')]'$	$L_{12}w = -(kEAw)' + k(EIw'')'$

Small-curvature bars (according to Kirchhoff–Klebsch) $kh \ll 1$	
$L_{21}v = kEAv' - [EI(kv)']''$	$L_{22}w = k^2EAw + (EIw'')''$

Table 4.4

Medium-curvature bars (according to Vlasov) $(kh)^2 \ll 1$	
$L_{11}v = -EAv''$	$L_{12}w = -kEAw'$
$L_{21}v = kEAv'$	$L_{22}w = k^2EAw + EI(w^{IV} + 2k^2w'' + k^4w)$

Small-curvature bars (according to Kirchhoff–Klebsch) $kh \ll 1$	
$L_{11}v = -(EA + k^2EI)v''$	$L_{12}w = kEIw''' - kEAw'$
$L_{21}v = -kEIv''' + kEAv'$	$L_{22}w = EIw^{IV} + k^2EAw$

Here we call upon V.Z. Vlasov to speak for himself:

"*In Table 11 we can see*[12] *that differential operators located in symmetric positions with respect to the main diagonal terms have the same expressions. This symmetry of shell equations, which was noted for the first time in a number of our papers, is in complete accordance with basic theorems of elasticity, such as the work reciprocity theorem by Betty.*"

And further, on page 214:

"*…we believe that the system of differential equations with a symmetric matrix*[13] *is in full conformance to basic laws of energy in statics of solid elastic bodies.*

Due to this symmetry that follows from the reciprocity law, the problem of equilibrium of an elastic shell can be represented in the form of purely integral or integral-differential equations with their kernels unexceptionally symmetric. In addition, the analysis of natural oscillations of a shell that makes use of a symmetric matrix of the governing differential equations can always be reduced to a secular equation which gives only real values for all the frequencies in the oscillation spectrum".

An analysis given below indeed gives a confirmation of the statement that the reciprocity laws are satisfied; however, we do not feel that the

---

[12] Table 11 on page 212 of the work [14] contains a system of differential equations of equilibrium for a circular cylindrical shell. As we noted above, a particular case of that system of equations developed by V.Z. Vlasov in application to a bar of a circular shape and a constant cross-section is the system of differential equations (7.40) with operators L_{ij}, shown in Table 4.4 for medium-curvature bars.

[13] Underlined by us.

grounds presented by Prof. Vlasov are convincing. His statement seems to be rather a conjecture based on a perfect intuition of a prominent mechanician.

What draws an immediate attention is that the symmetry of the differential equations for a curvilinear bar takes place only if the bar has a constant curvature and a constant cross-section; in addition, the sign should be altered in one of the equations of system (7.40). But this formal symmetry is unimportant — and the more so seeing that it vanishes for general equations defined in Table 4.3. What *is* important is the fact that the matrix differential operator for equations (7.40) is self-conjugated in the Lagrangian sense. This statement can be easily validated by recalling rules for construction of a conjugated operator indicated in Section **1**.1. But even the formal self-conjugation of the differential operator is only a necessary but not sufficient condition for the fundamental reciprocity theorems to hold. Formulations of the boundary conditions are equally important.

V.Z. Vlasov does not write out explicit formulas for the static boundary conditions, but he gives the following verbal explanation [14]:

"*If boundary conditions for a shell are specified in forces, then there must be four such conditions in each point of the contour line. Of those conditions, two will relate to the normal and shearing forces and will conform to two static conditions in plane elasticity...*"

Obviously, in the last sentence of the quotation V.Z. Vlasov means edge conditions for shear force Q and longitudinal force N in application to a curvilinear bar.

It is easy to see that the matrix differential operator for the Kirchhoff–Klebsch theory is also self-conjugated in the Lagrangian sense.

Before we begin a direct consideration of the edge conditions, we want to convert all relationships obtained above to a matrix form by introducing the vectors of displacements $\mathbf{u}$, strains $\boldsymbol{\varepsilon}$, "stresses" $\boldsymbol{\sigma}$, and given external forces $\overline{X}$. We will need to distinguish between two strain vectors which correspond to two theories: $\boldsymbol{\varepsilon}_V$ will be a strain vector of Vlasov's theory, and $\boldsymbol{\varepsilon}_K$ will be a strain vector of the Kirchhoff–Klebsch theory

$$\mathbf{u} = \|[v, w]\|^{\mathsf{T}}, \quad \boldsymbol{\varepsilon}_V = \|[\varepsilon_0, \chi]\|^{\mathsf{T}}, \quad \boldsymbol{\varepsilon}_K = \|[\varepsilon_0, \kappa]\|^{\mathsf{T}},$$

$$\boldsymbol{\sigma} = \|[N, M]\|^{\mathsf{T}}, \quad \overline{\boldsymbol{X}} = \|[q_t + km, q_n + m']\|^{\mathsf{T}}. \tag{7.41}$$

Based on the formulas derived above, further we present the matrix differential operators of geometry, $\boldsymbol{A}_V$, for the Vlasov theory and $\boldsymbol{A}_K$ for the Kirchhoff–Klebsch theory, the general (Kirchhoff's) operator of equilibrium, $\boldsymbol{B}$, and the respective operators of physical relationships, $\boldsymbol{C}_\mathrm{V}$ and $\boldsymbol{C}_\mathrm{K}$. Thus we have

$$\boldsymbol{A}_\mathrm{V} = \left\|\begin{bmatrix} \dfrac{d(\,)}{ds} & k(\,) \\ \dfrac{dk}{ds}(\,) & -\dfrac{d^2(\,)}{ds^2} - k^2(\,) \end{bmatrix}\right\|, \quad \boldsymbol{A}_\mathrm{K} = \left\|\begin{bmatrix} \dfrac{d(\,)}{ds} & k(\,) \\ \dfrac{d(k(\,))}{ds} & -\dfrac{d^2(\,)}{ds^2} \end{bmatrix}\right\|,$$

$$\boldsymbol{B} = \left\|\begin{bmatrix} -\dfrac{d(\,)}{ds} & -k\dfrac{d(\,)}{ds} \\ k(\,) & -\dfrac{d^2(\,)}{ds^2} \end{bmatrix}\right\|, \quad \boldsymbol{C}_\mathrm{V} = \left\|\begin{bmatrix} EA & -kEI \\ 0 & EI \end{bmatrix}\right\|,$$

$$\boldsymbol{C}_\mathrm{V}^{-1} = \left\|\begin{bmatrix} 1/EA & k/EA \\ 0 & 1/EI \end{bmatrix}\right\|, \quad \boldsymbol{C}_\mathrm{K} = \left\|\begin{bmatrix} EA & 0 \\ 0 & EI \end{bmatrix}\right\|, \quad \boldsymbol{C}_\mathrm{K}^{-1} = \left\|\begin{bmatrix} 1/EA & 0 \\ 0 & 1/EI \end{bmatrix}\right\|. \tag{7.42}$$

Here we have

$$\boldsymbol{\varepsilon}_\mathrm{V} = \boldsymbol{A}_\mathrm{V}\mathbf{u}, \quad \boldsymbol{\varepsilon}_\mathrm{K} = \boldsymbol{A}_\mathrm{K}\mathbf{u}, \qquad \boldsymbol{B}\mathbf{u} = \overline{\boldsymbol{X}}, \quad \boldsymbol{\sigma} = \boldsymbol{C}_\mathrm{V}\,\boldsymbol{\varepsilon}_\mathrm{V}, \quad \boldsymbol{\sigma} = \boldsymbol{C}_\mathrm{K}\,\boldsymbol{\varepsilon}_\mathrm{K}. \tag{7.43}$$

4.7.3 Stresses conjugate to the Vlasov vector of strains

First of all, we cannot help but notice that in Vlasov's selection of governing vectors the equilibrium operator, $\boldsymbol{B}$, is not conjugate to the geometry operator, $\boldsymbol{A}_\mathrm{V}$, which makes us return to the original designation of $\boldsymbol{B}$ for an equilibrium operator. Also, Vlasov's mutually inverse algebraic operators $\boldsymbol{C}_\mathrm{V}$ and $\boldsymbol{C}_\mathrm{V}^{-1}$ for physical relationships in (7.42) are not symmetric. This occurs because the vector of stresses, $\boldsymbol{\sigma}$, and the vector of strains, $\boldsymbol{\varepsilon}_\mathrm{V}$, have been chosen arbitrarily and do not match each other in the sense of energy. At the same time, for the Kirchhoff–Klebsch theory

$$\boldsymbol{B} = \boldsymbol{A}_\mathrm{K}^\mathrm{T} \qquad \text{and} \qquad \boldsymbol{C}_\mathrm{K} = \boldsymbol{C}_\mathrm{K}^\mathrm{T}. \tag{7.44}$$

It is easy to prove that if we introduce a generalized force denoted by T according to formula

$$T = N + kM, \tag{7.45}$$

then we can define a vector of stresses, $\boldsymbol{\sigma}_\mathrm{V} = \|[T, M]\|^\mathrm{T}$, *conjugated by energy* to Vlasov's vector of strains, $\boldsymbol{\varepsilon}_\mathrm{V} = \|[\varepsilon_0, \chi]\|^\mathrm{T}$.

The stress/force T defined by formula (7.45) can be conveniently named a *generalized longitudinal force*. It can be expressed via displacements v and w based on the formulas from Table 4.2 as follows:

$$T = (v' + kw)EA. \tag{7.46}$$

The conjugation by energy between strain and stress vectors will be understood in the sense that the following two conditions must be met:

- the operators of geometry and equilibrium must be mutually conjugated;
- the physical relationships must be symmetric and positive definite.

From the mechanical standpoint, the generalized forces and their respective strains are vectors conjugated by energy if their scalar product has the meaning of a virtual work of internal forces.

And indeed, linear transformation (7.45) can be represented as follows in the matrix form:

$$\boldsymbol{\sigma}_V = \boldsymbol{\Lambda\sigma}, \quad \boldsymbol{\Lambda} = \left\|\begin{bmatrix} 1 & k \\ 0 & 1 \end{bmatrix}\right\|, \quad \boldsymbol{\Lambda}^{-1} = \left\|\begin{bmatrix} 1 & -k \\ 0 & 1 \end{bmatrix}\right\|, \tag{7.47}$$

so in the new force variables, $\boldsymbol{\sigma}_V$, the equations of equilibrium $\boldsymbol{B\sigma} = \bar{\boldsymbol{X}}$ and the physical relationships $\boldsymbol{C}_V\boldsymbol{\varepsilon}_V = \boldsymbol{\sigma}$ will be written respectively as

$$\boldsymbol{B\Lambda}^{-1}\boldsymbol{\sigma}_V = \bar{\boldsymbol{X}}, \qquad \boldsymbol{\Lambda C}_V\, \boldsymbol{\varepsilon}_V = \boldsymbol{\sigma}_V. \tag{7.48}$$

But the $\boldsymbol{B\Lambda}^{-1}$ operator is equal to

$$\boldsymbol{B\Lambda}^{-1} = \left\|\begin{bmatrix} -\dfrac{d(\;)}{ds} & \dfrac{dk}{ds}(\;) \\ k(\;) & -\dfrac{d^2(\;)}{ds^2} - k^2(\;) \end{bmatrix}\right\|, \tag{7.49}$$

and now we find that

$$\boldsymbol{A}_V^{\mathrm{T}} = \boldsymbol{B\Lambda}^{-1}, \qquad \boldsymbol{C}_K = \boldsymbol{\Lambda C}_V, \tag{7.50}$$

where operator $\boldsymbol{A}_V^{\mathrm{T}}$ is conjugate (in the Lagrangian sense) to geometry operator $\boldsymbol{A}_V$, and operators $\boldsymbol{C}_K$ and $\boldsymbol{C}_K^{-1}$ are symmetric and positive definite as can be seen from (7.42).

So, the equations of equilibrium in stresses $\boldsymbol{\sigma}_V = \|[T, M]\|^{\mathrm{T}}$ conjugate to strains $\boldsymbol{\varepsilon}_V$ will be

$$-T' + k'M = q_t + km, \qquad kT - M'' - k^2M = q_n + m'. \tag{7.51}$$

Now we use general formula (1.6) to build Lame's matrix differential operator $\boldsymbol{L}_{\mathrm{V}}$ for the theory of bars of medium and big curvature as a consecutive product of the equilibrium operator, the operator of physical relationships, and the operator of geometry, that is, $\boldsymbol{L}_{\mathrm{V}}=\boldsymbol{B}\boldsymbol{C}_{\mathrm{V}}\boldsymbol{A}_{\mathrm{V}}$. By noting that $\boldsymbol{B}\boldsymbol{C}_{\mathrm{V}}=(\boldsymbol{B}\boldsymbol{\Lambda}^{-1})(\boldsymbol{\Lambda}\boldsymbol{C}_{\mathrm{V}}) = \boldsymbol{A}_{\mathrm{V}}^{\mathrm{T}}\boldsymbol{C}_{\mathrm{K}}$ and recalling properties of conjugate operators (**1**.1.10) and (**1**.1.11), we derive

$$\boldsymbol{L}_{\mathrm{V}}=\boldsymbol{B}\boldsymbol{C}_{\mathrm{V}}\boldsymbol{A}_{\mathrm{V}}= \boldsymbol{A}_{\mathrm{V}}^{\mathrm{T}}\boldsymbol{C}_{\mathrm{K}}\boldsymbol{A}_{\mathrm{V}}=(\boldsymbol{A}_{\mathrm{V}}^{\mathrm{T}}\boldsymbol{C}_{\mathrm{K}}\boldsymbol{A}_{\mathrm{V}})^{\mathrm{T}}=\boldsymbol{L}_{\mathrm{V}}^{\mathrm{T}}. \tag{7.52}$$

Now (7.52) makes it clear why operator $\boldsymbol{L}_{\mathrm{V}}$ for Vlasov's system of equation is formally self-conjugated.

Consider the basic integral identity in the form of (**1**.2.17) in application to stresses $\boldsymbol{\sigma}_{\mathrm{V}}$. We have

$$(\boldsymbol{A}_{\mathrm{V}}\mathbf{u},\,\boldsymbol{\sigma}_{\mathrm{V}})=\int_0^l[\varepsilon_0 T+\chi M]ds=$$

$$=\int_0^l[(v'+kw)T+(-w''-k^2w+k'v)M]ds=$$

$$=\int_0^l[v(-T'+k'M)+w(kT-M''-k^2M]ds+[vT]\big|_0^l+[wM']\big|_0^l-$$

$$-[w'M]\big|_0^l=(\mathbf{u},\,\boldsymbol{A}_{\mathrm{V}}^{\mathrm{T}}\boldsymbol{\sigma}_{\mathrm{V}})+(\boldsymbol{p},\boldsymbol{u})_{\Gamma}. \tag{7.53}$$

Obviously, the basic integral identity (7.53) holds if, for example, we take $\boldsymbol{p}=\boldsymbol{p}_a$ and $\boldsymbol{u}=\boldsymbol{u}_a$ for edge forces and edge displacements, where

$$\boldsymbol{p}_a=\|[\,nT,\,nM',\,nM]\|^{\mathrm{T}},\qquad \boldsymbol{u}_a=\|[v,\,w\,,\,\psi]\|^{\mathrm{T}},\qquad \psi=-w', \tag{7.54}$$

ψ being a kinematic parameter which we will call a *generalized slope.*

In (7.54) and further on we denote by n the cosine of an angle between the external normal to a current section and the direction of unit vector $\boldsymbol{t}$, so

$$n(l)=1,\qquad n(0)=-1.$$

At this point we have to note that the distribution of edge forces $\boldsymbol{p}_a$ and edge displacements $\boldsymbol{u}_a$ according to (7.54) is not the only possible one. Basic integral identity (7.53) will not be violated if instead (7.54) we take second version (let's call it "version b") of the edge forces and displacements by assuming $\boldsymbol{p}=\boldsymbol{p}_b$ and $\boldsymbol{u}=\boldsymbol{u}_b$ in this way:

$$\boldsymbol{p}_b=\|[\,nN,\,n\,M'\,,\,nM]\|^{\mathrm{T}},\qquad \boldsymbol{u}_b=\|[v,\,w\,,\,\theta]\|^{\mathrm{T}},\qquad \theta=-w'+kv\,. \tag{7.55}$$

These two versions of the boundary conditions are related as follows:

$$\boldsymbol{p}_a = \boldsymbol{\Lambda}_\Gamma \boldsymbol{p}_b, \qquad \boldsymbol{u}_a = (\boldsymbol{\Lambda}_\Gamma^{-1})^\mathsf{T} \boldsymbol{u}_b,$$

$$\boldsymbol{\Lambda}_\Gamma = \left\| \begin{bmatrix} 1 & 0 & k \\ 0 & 1 & 0 \\ 0 & 0 & 1 \end{bmatrix} \right\|, \qquad (\boldsymbol{\Lambda}_\Gamma^{-1})^\mathsf{T} = \left\| \begin{bmatrix} 1 & 0 & 0 \\ 0 & 1 & 0 \\ -k & 0 & 1 \end{bmatrix} \right\|, \tag{7.56}$$

wherefrom we have an important relationship

$$(\boldsymbol{p}_a, \boldsymbol{u}_a)_\Gamma = (\boldsymbol{\Lambda}_\Gamma \boldsymbol{p}_b, (\boldsymbol{\Lambda}_\Gamma^{-1})^\mathsf{T} \boldsymbol{u}_b)_\Gamma = (\boldsymbol{\Lambda}_\Gamma^{-1} \boldsymbol{\Lambda}_\Gamma \boldsymbol{p}_b, \boldsymbol{u}_b)_\Gamma = (\boldsymbol{p}_b, \boldsymbol{u}_b)_\Gamma . \tag{7.57}$$

From the mechanical standpoint we can interpret this as follows:

the virtual work of contour forces, $\boldsymbol{p}_a$, on their respective contour displacements, $\boldsymbol{u}_a$, is equal to the virtual work of contour forces $\boldsymbol{p}_b$ of their respective contour displacements $\boldsymbol{u}_b$.

Note that this fact is not a variation of the work reciprocity theorem by Betty, though the verbal formulations of the two statements are very similar. It is a separate independent proposition. In the Betty theorem, indexes "a" and "b" refer to two states of a system, while here the same indexes refer to two possible formulations of the edge conditions for the same state of the system.

Now let's turn to the basic integral identity for the $\boldsymbol{A}_\mathrm{K}$ operator. Instead of (7.53) we have

$$(\boldsymbol{A}_\mathrm{K}\mathbf{u}, \boldsymbol{\sigma}) = \int_0^l [\varepsilon_0 N + \kappa M] ds = \int_0^l [(v' + kw)N + (-w'' + (kv)')M] ds =$$

$$= \int_0^l [v(-N' - kM') + w(kN - M''] ds + [vN]\big|_0^l -$$

$$- [(w' - kv)M]\big|_0^l + [wM']\big|_0^l = (\mathbf{u}, \boldsymbol{A}_\mathrm{K}^\mathsf{T} \boldsymbol{\sigma}) + (\boldsymbol{p}, \boldsymbol{u})_\Gamma . \tag{7.58}$$

And again we can choose to assume either $\boldsymbol{p} = \boldsymbol{p}_b$ and $\boldsymbol{u} = \boldsymbol{u}_b$ according to (7.55) or $\boldsymbol{p} = \boldsymbol{p}_a$ and $\boldsymbol{u} = \boldsymbol{u}_a$ according to (7.54).

Thus, the boundary conditions can be formulated in a dual way both in Vlasov's theory and in one by Kirchhoff and Klebsch, and both versions of the boundary conditions never contradict to the laws of energy in statics. The most popular version is one we denoted by index "b".

In many cases both versions of the boundary condition formulations are equivalent. For example, if the end of a curvilinear bar is fully clamped, the conditions on that end, either $\boldsymbol{u}_b = \mathbf{0}$ or $\boldsymbol{u}_a = \mathbf{0}$, are implied by each other because of (7.56) and the nondegeneracy of matrix $\boldsymbol{\Lambda}_\Gamma$. However, this may not be the case in some other combinations of boundary conditions. For example, let us consider the case when the end of a bar is restrained from rotation, can move tangentially, and cannot move in the direction normal to the bar's axis (Fig. 4.8-*c*).

Clear and convenient mechanical interpretations of version *a* and version *b* of the boundary conditions are shown in Fig. 4.8-*a* and 4.8-*b*, respectively. The boundary condition as in Fig. 4.8-*a* should be treated in a limit sense: two constraints that prohibit normal displacements are installed at an infinitesimal distance. In Fig. 4.8-*b* it is a rotation of the bar's section which is prohibited.

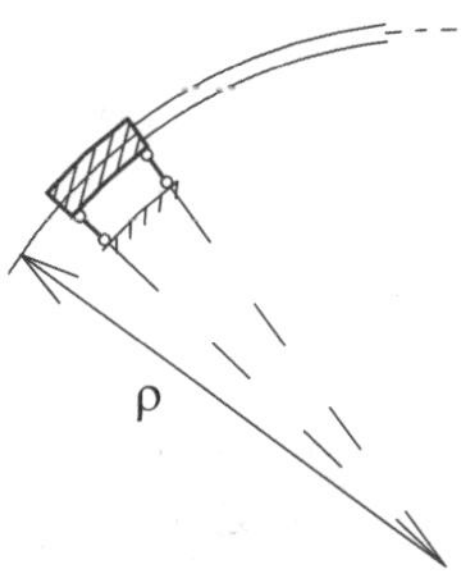

Fig. 4.8-*a*. Boundary conditions according to version *a*

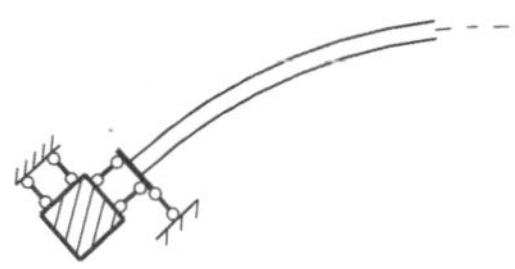

Fig. 4.8-*b*. Boundary conditions according to version *b*

Fig. 4.8-*c*. A slipping fixation at the bar's end

So, for the homogeneous boundary conditions we have

version *a* (Fig. 4.8-*a*): $w = 0, \quad \psi = 0, \quad T = 0,$

version *b* (Fig. 4.8-*b*): $w = 0, \quad \theta = 0, \quad N = 0,$

or, talking displacements,

- for Vlasov's theory:

version *a* (Fig. 4.8-*a*):

$$w = 0, \quad w' = 0, \quad v' = 0,$$

version *b* (Fig. 4.8-*b*):

$$w = 0\,, \quad -w' + kv = 0, \quad v'EA + (w'' - kv')kEI = 0.$$

- for Kirchhoff–Klebsch theory:

version *a* (Fig. 4.8-*a*):

$$w = 0, \quad w' = 0, \quad v'EA + (w'' - kv')kEI = 0,$$

version *b* (Fig. 4.8-*b*):

$$w = 0\,, \quad -w' + kv = 0, \quad v' = 0.$$

Both formulations of the boundary conditions are valid.

It should be said that we could go an opposite way in Vlasov's theory: first, take Kirchhoff's stress vector $\boldsymbol{\sigma} = \|[N, M]\|^{\mathrm{T}}$ into consideration and then deal with a vector of strains conjugated to it by energy, $\boldsymbol{\varepsilon}$. We do not present this approach here because it will give nothing new in the final expressions. Now we can write a matrix form of both versions of the boundary conditions and re-formulate the boundary conditions in a common form as in (**1**.2.4). Table 4.5 presents matrix operators of boundary condition extraction, $\boldsymbol{E}_p$ and $\boldsymbol{E}_u$, and matrix differential operators, $\boldsymbol{H}_u$ and $\boldsymbol{H}_\sigma$, for converting the displacement vector, $\mathbf{u} = \|[v,w]\|^{\mathrm{T}}$, and the stress vector, $\boldsymbol{\sigma} = \|[N, M]\|^{\mathrm{T}}$ or $\boldsymbol{\sigma}_{\mathrm{V}} = \|[T, M]\|^{\mathrm{T}}$, to the respective vector of edge displacements, $\boldsymbol{u}$, and the vector of edge forces, $\boldsymbol{p}$. Note that Table 4.5 presents two matrix operators $\boldsymbol{H}_\sigma$ for each of two possible versions of the boundary condition formulations; one of the operators transforms Vlasov's stress vector $\boldsymbol{\sigma}_{\mathrm{V}}$ and the other transforms the Kirchhoff–Klebsch stress vector $\boldsymbol{\sigma}$ into the edge force vector, $\boldsymbol{p}$.

Table 4.5

		Version "a" of boundary conditions	*Version "b" of boundary conditions*
$\boldsymbol{E}_u$		$\begin{bmatrix} e_v & 0 & 0 \\ 0 & e_w & 0 \\ 0 & 0 & e_\psi \end{bmatrix}$	$\begin{bmatrix} e_v & 0 & 0 \\ 0 & e_w & 0 \\ 0 & 0 & e_\theta \end{bmatrix}$
$\boldsymbol{E}_p$		$\begin{bmatrix} e_T & 0 & 0 \\ 0 & e_Q & 0 \\ 0 & 0 & e_M \end{bmatrix}$	$\begin{bmatrix} e_N & 0 & 0 \\ 0 & e_Q & 0 \\ 0 & 0 & e_M \end{bmatrix}$
$\boldsymbol{H}_u$		$\begin{bmatrix} 1 & 0 \\ 0 & 1 \\ 0 & -\partial/\partial s \end{bmatrix}$	$\begin{bmatrix} 1 & 0 \\ 0 & 1 \\ k & -\partial/\partial s \end{bmatrix}$
$\boldsymbol{H}_\sigma$	$\boldsymbol{\sigma}_{\mathrm{V}} = \begin{bmatrix} T \\ M \end{bmatrix}$	$\begin{bmatrix} n & 0 \\ 0 & n\dfrac{\partial}{\partial s} \\ 0 & n \end{bmatrix}$	$\begin{bmatrix} n & -kn \\ 0 & n\dfrac{\partial}{\partial s} \\ 0 & n \end{bmatrix}$
	$\boldsymbol{\sigma} = \begin{bmatrix} N \\ M \end{bmatrix}$	$\begin{bmatrix} n & kn \\ 0 & n\dfrac{\partial}{\partial s} \\ 0 & n \end{bmatrix}$	$\begin{bmatrix} n & 0 \\ 0 & n\dfrac{\partial}{\partial s} \\ 0 & n \end{bmatrix}$

A question may arise: why would Vlasov's theory (or any theory, for that matter) absolutely have to derive such types of stresses and strains that should be conjugated by energy? The answer is simple. There are formal techniques of building variational principles for conjugated force/stress and kinematical parameters, and in the next sections we will develop those principles for curvilinear bars.

4.7.4 Variational principles for curvilinear bars

First of all, we want to define an expression for the strain energy, E. In Vlasov's theory we have the following, marking the respective functionals by subscript V:

$$\mathsf{E}_V(\mathbf{u}) = \frac{1}{2}(\boldsymbol{\sigma}_V, \boldsymbol{\varepsilon}_V) = \frac{1}{2}(\boldsymbol{\Lambda C}_V\, \boldsymbol{\varepsilon}_V, \boldsymbol{\varepsilon}_V) = \frac{1}{2}(\boldsymbol{C}_K \boldsymbol{A}_V\mathbf{u}, \boldsymbol{A}_V\mathbf{u}) =$$

$$= \frac{1}{2}\int_0^l [EA(v' + kw)^2 + EI(-w'' - k^2 w + k'v)^2]ds\,, \tag{7.59}$$

$$\mathsf{E}_V(\boldsymbol{\sigma}) = \frac{1}{2}(\boldsymbol{\sigma}_V, \boldsymbol{\varepsilon}_V) = \frac{1}{2}(\boldsymbol{\sigma}_V, \boldsymbol{C}_K^{-1}\boldsymbol{\sigma}_V) = \frac{1}{2}\int_0^l \left[\frac{(N+kM)^2}{EA} + \frac{M^2}{EI}\right]ds\,. \tag{7.60}$$

We can follow another approach: take a direct definition of the strain energy and omit small quantities from it as necessary. And indeed, we can notice that an element of volume of a curvilinear bar is equal to $d\Omega = (1 + kz)dAds$. By summing the strain energy accumulated in all elementary volumes $d\Omega$ we obtain

$$\mathsf{E}_V(\mathbf{u}) = \frac{1}{2}\int_0^l\int_A E\varepsilon^2(1+kz)dAds = \frac{1}{2}\int_0^l\int_A E[\varepsilon_0 + \chi z(1-kz)]^2(1+kz)dAds =$$

$$= \frac{1}{2}\int_0^l\int_A E[\varepsilon_0^2(1+kz) + 2\varepsilon_0\chi z(1-k^2z^2) + \chi^2 z^2(1-k^2z^2)(1+kz)]dAds\,.$$

Here we can neglect the quantity k^2z^2 in comparison to one and then integrate over the cross-section area, A, to derive

$$\mathsf{E}_V(\mathbf{u}) = \frac{1}{2}\int_0^l [EA\varepsilon_0^2 + E(I + kI_2)\chi^2]ds\,.$$

But, as we stated before, for medium-curvature bars we adopt the additional estimate, $kI_2/I \ll 1$, so within that theory

$$\mathsf{E}_V(\mathbf{u}) = \frac{1}{2}\int_0^l (EA\varepsilon_0^2 + EI\chi^2)ds\,. \tag{7.61}$$

Quite expectably, formula (7.61) coincides with formula (7.59), which becomes obvious after replacing strain parameters ε_0 and χ in (7.61) by their expressions via the displacement vector's components according to the middle column in Table 4.1.

Now recall that all formulas for medium-curvature curvilinear bars (Vlasov's theory) are valid also for the big-curvature bars. We only need to replace the moment of inertia of the bar's cross-section, I, by that of the reduced section, I_ρ, defined by (7.38). Hence formulas (7.59), (7.60) work for the big-curvature bars, too, after the said replacement.

Similarly, we want to establish an expression of the strain energy, E_K, that conforms to the Kirchhoff–Klebsch theory:

$$\mathsf{E}_K(\mathbf{u}) = \frac{1}{2}(\boldsymbol{\sigma}, \boldsymbol{\varepsilon}_K) = \frac{1}{2}(\boldsymbol{C}_K\,\boldsymbol{\varepsilon}_K, \boldsymbol{\varepsilon}_K) = \frac{1}{2}(\boldsymbol{C}_K\,\boldsymbol{A}_K\mathbf{u}, \boldsymbol{A}_K\mathbf{u}) =$$

$$= \frac{1}{2}\int_0^l [EA(v' + kw)^2 + EI(-w'' + kv' + k'v)^2]ds\,, \tag{7.62}$$

$$\mathsf{E}_K(\boldsymbol{\sigma}) = \frac{1}{2}(\boldsymbol{\sigma}, \boldsymbol{\varepsilon}_K) = \frac{1}{2}(\boldsymbol{\sigma}, \boldsymbol{C}_K^{-1}\boldsymbol{\sigma}) = \frac{1}{2}\int_0^l \left(\frac{N^2}{EA} + \frac{M^2}{EI}\right)ds\,. \tag{7.63}$$

Let q_t(s) and q_n(s) be components of intensities of external loads distributed along the bar's axis, which act in the respective normal and tangential directions, and let m(s) be an intensity of a distributed moment load. We will denote external concentrated force/moment static actions on the ends of the bar as $\bar{Q}_0$ and $\bar{Q}_l$, $\bar{N}_0$ and $\bar{N}_l$, $\bar{M}_0$ and $\bar{M}_l$, respectively (Fig. 4.7).

A full list of parameters of the edge condition extraction, specified on each end of the curvilinear bar, consists of (e_T, e_Q, e_M, e_v, e_w, e_ψ) for version "*a*" and of (e_N, e_Q, e_M, e_v, e_w, e_θ) for version "*b*" of the boundary conditions. These parameters are components of the respective matrices $\boldsymbol{E}_p$ and $\boldsymbol{E}_u$ (Table 4.5) and satisfy the following conditions on each end:

- version *a*: $e_T + e_v = 1$, $e_Q + e_w = 1$, $e_M + e_\psi = 1$,
- version *b*: $e_N + e_v = 1$, $e_Q + e_w = 1$, $e_M + e_\theta = 1$.

The meanings of these parameters are obvious.

As usual, we define the potential of external static actions, $\Pi_s(v,w)$, as the virtual work of external forces on displacements v, w and slopes $\theta = -w' + kv$. We have

$$\Pi_s(v,w) = \int_0^l q_t v ds + \int_0^l q_n w ds + \int_0^l m(-w' + kv) ds +$$

$$+[e_N n\bar{N} v]|_0^l + [e_Q n\bar{Q} w]|_0^l + [e_M n\bar{M}(-w' + kv)]|_0^l . \qquad (7.64)$$

However, non-integral terms in (7.64) can be represented in another way after re-grouping, as

$$[e_T n\bar{T} v]|_0^l + [e_Q n\bar{Q} w]|_0^l + [e_M n\bar{M}(-w')]|_0^l . \qquad (7.65)$$

The method of representation of the work of external contour forces as in (7.65) corresponds to the first form (version *a*) of the boundary forces and displacements according to (7.54). On the contrary, the non-integral terms in (7.64) define the same work but correspond to the second form (version *b*) of the boundary forces and displacements according to (7.55).

In addition to the potential of external static actions, we will need also an expression for the potential of external kinematic actions, $\Pi_k(\boldsymbol{\sigma})$. Based on formula (**2**.3.2) and involving data from Table 4.5, we have

$$\Pi_k(\boldsymbol{\sigma}) = (\boldsymbol{E}_u \boldsymbol{p}, \boldsymbol{E}_u \bar{\boldsymbol{u}})_\Gamma = [e_v T\bar{v}]|_0^l + [e_w M'\bar{w}]|_0^l + [e_\psi M\bar{\psi}]|_0^l , \qquad (7.66\text{-}a)$$

or, in another form,

$$\Pi_k(\boldsymbol{\sigma}) = [e_v N\bar{v}]|_0^l + [e_w M'\bar{w}]|_0^l + [e_\theta M\bar{\theta}]|_0^l . \qquad (7.66\text{-}b)$$

Formula (7.66-*a*) deals with the first form (version *a*) of the edge conditions for the curvilinear bar, and formula (7.66-*b*) relates to the second form (version *b*). Note that both expressions of kinematic potential $\Pi_k(\boldsymbol{\sigma})$ are the same for both the Vlasov stress vector, $\boldsymbol{\sigma}_V$, and the Kirchhoff–Klebsch stress vector, $\boldsymbol{\sigma}$.

Lagrange functional

Now we want to build the Lagrange functional, $L(v,w) = E(v,w) - \Pi_s(v,w)$. We have, for Vlasov's theory,

$$\mathsf{L}_{\mathrm{V}}(v,w) = \frac{1}{2}\int_0^l [EA(v'+kw)^2 + EI(-w''-k^2w+k'v)^2]ds -$$

$$-\int_0^l q_t v ds - \int_0^l q_n w ds - \int_0^l m(-w'+kv)ds -$$

$$-[e_N n\bar{N}v]|_0^l - [e_Q n\bar{Q}w]|_0^l - [e_M n\bar{M}(-w'+kv)]|_0^l . \tag{7.67}$$

For the Kirchhoff–Klebsch theory, we have

$$\mathsf{L}_{\mathrm{K}}(v,w) = \frac{1}{2}\int_0^l [EA(v'+kw)^2 + EI(-w''+kv'+k'v)^2]ds -$$

$$-\int_0^l q_t v ds - \int_0^l q_n w\ ds - \int_0^l m(-w'+kv)ds -$$

$$-[e_N n\bar{N}v]|_0^l - [e_Q n\bar{Q}w]|_0^l - [e_M n\bar{M}(-w'+kv)]|_0^l . \tag{7.68}$$

To formulate the variational problem carefully, we need to define a set, $\mathscr{P}_{\mathrm{k}}$, of physically and kinematically admissible stress-and-strain fields which will be searched for the minimum of the Lagrange functional. This is where the choice of a particular form of boundary conditions is important. For example, the $\mathscr{P}_{\mathrm{k}a}$ set in case *a* will consist of functions v and w which satisfy the following conditions at end points 0 and l:

$$e_v v = \bar{v}, \quad e_w w = \bar{w}, \quad e_\psi(-w') = \bar{\psi} . \tag{7.69-a}$$

In case *b*, the $\mathscr{P}_{\mathrm{k}b}$ set of functions v and w allowed for comparison will consist of functions satisfying the following requirements at the same end points:

$$e_v v = \bar{v}, \quad e_w w = \bar{w}, \quad e_\theta(-w'+kv) = \bar{\theta} . \tag{7.69-b}$$

These sets are obviously different, that is, $\mathscr{P}_{\mathrm{k}a} \neq \mathscr{P}_{\mathrm{k}b}$.

By varying Lagrange functionals $\mathsf{L}_{\mathrm{V}}(v,w)$ and $\mathsf{L}_{\mathrm{K}}(v,w)$ one time on the $\mathscr{P}_{\mathrm{k}a}$ set and second time on the $\mathscr{P}_{\mathrm{k}b}$ set and making the first variation equal to zero, we find that the Euler equations for these functionals are equations (7.40) where differential operators L_{ij} are defined by the respective rows of Table 4.3.

As we can see, the Euler equations for the two functionals do not depend on which set is used, $\mathscr{P}_{\mathrm{k}a}$ or $\mathscr{P}_{\mathrm{k}b}$, but they are different for two different functionals $\mathsf{L}_{\mathrm{V}}(v,w)$ and $\mathsf{L}_{\mathrm{K}}(v,w)$. As for the natural boundary conditions, they do depend on which set is chosen, $\mathscr{P}_{ka}$ and $\mathscr{P}_{kb}$. See Table 4.6 for details of the natural boundary conditions for these functionals.

Table 4.6

	for functional $\mathsf{L}_V(v,w)$
$\mathscr{P}_{ka}$	$e_T[(v'+kw)EA - n\bar{T}]=0$ $e_Q\{[(-w''-k^2w+k'v)EI]'-n\bar{Q}+m\}=0$ $e_M[(-w''-k^2w+k'v)EI-n\bar{M}]=0$
$\mathscr{P}_{kb}$	$e_N[(v'+kw)EA+(w''+k^2w-k'v)kEI-n\bar{N}]=0$ $e_Q\{[(-w''-k^2w+k'v)EI]'-n\bar{Q}+m\}=0$ $e_M[(-w''-k^2w+k'v)EI-n\bar{M}]=0$

	for functional $\mathsf{L}_K(v,w)$
$\mathscr{P}_{ka}$	$e_T[(v'+kw)EA+(-w''+k'v+kv')kEI-n\bar{T}]=0$ $e_Q\{[(-w''+k'v+kv')EI]'-n\bar{Q}+m\}=0$ $e_M[(-w''+k'v+kv')EI-n\bar{M}]=0$
$\mathscr{P}_{kb}$	$e_N[(v'+kw)EA-n\bar{N}]=0$ $e_Q\{[(-w''+k'v+kv')EI]'-n\bar{Q}+m\}=0$ $e_M[(-w''+k'v+kv')EI-n\bar{M}]=0$

Castigliano functional

We want to derive an expression for the Castigliano functional, so we use general formula (**2**.3.3) for this purpose. The strain energy for a medium-curvature bar is defined by (7.60), and the potential of external kinematic actions is defined by (7.66). As a result, we have this for Vlasov's theory:

$$\mathsf{K}_V(T,M)=\frac{1}{2}\int_0^l\left(\frac{T^2}{EA}+\frac{M^2}{EI}\right)ds-[e_vT\bar{v}]\big|_0^l-[e_wM'\bar{w}]\big|_0^l-[e_\psi M\bar{\psi}]\big|_0^l, \quad (7.70)$$

or, to put it another way,

$$\mathsf{K}_V(N,M)=\frac{1}{2}\int_0^l\left[\frac{(N+kM)^2}{EA}+\frac{M^2}{EI}\right]ds-$$

$$-[e_vN\bar{v}]\big|_0^l-[e_wM'\bar{w}]\big|_0^l-[e_\theta M\bar{\theta}]\big|_0^l. \quad (7.71)$$

A particular form of the Castigliano functional should be chosen depending on what static boundary conditions we have.

In the first case, the $\mathscr{P}_{sa}$ set of physically and statically admissible fields of stresses consists of functions T and M which satisfy the following conditions at end points 0 and l

$$e_T(nT-\bar{T})=0, \qquad e_Q(nQ-\bar{Q}+nm)=0, \qquad e_M(nM-\bar{M})=0. \quad (7.72)$$

In the second case, the variation is applied to functions N and M from set $\mathscr{P}_{sb}$, which at end points 0 and l satisfy

$$e_N(nN-\bar{N})=0, \qquad e_Q(nQ-\bar{Q}+nm)=0, \qquad e_M(nM-\bar{M})=0. \quad (7.73)$$

Similarly to (7.70) and (7.71), we will write also expressions for the Castigliano functional which correspond to the curvilinear bar theory by Kirchhoff–Klebsch:

$$\mathsf{K}_K(T,M)=\frac{1}{2}\int_0^l\left[\frac{(T-kM)^2}{EA}+\frac{M^2}{EI}\right]ds-$$

$$-[e_v T\bar{v}]|_0^l-[e_w M'\bar{w}]|_0^l-[e_\psi M\bar{\psi}]|_0^l, \quad (7.74)$$

or, to put it another way,

$$\mathsf{K}_K(N,M)=\frac{1}{2}\int_0^l\left(\frac{N^2}{EA}+\frac{M^2}{EI}\right)ds-[e_v N\bar{v}]|_0^l-[e_w M'\bar{w}]|_0^l-[e_\theta M\bar{\theta}]|_0^l. \quad (7.75)$$

Reissner functional

We want to find an expression of the Reissner functional in its first form. To do it, we use general formula (**3**.1.4) in application to Vlasov's theory of curvilinear bars by assuming $\boldsymbol{K}=\boldsymbol{O}$, $\boldsymbol{\sigma}=\boldsymbol{\sigma}_V$, $\boldsymbol{A}=\boldsymbol{A}_V$, and also $\boldsymbol{C}=\boldsymbol{C}_K$. The latter substitution has been introduced because the elasticity relationship that connects vectors $\boldsymbol{\sigma}_V$ and $\boldsymbol{\varepsilon}_V$ is $\boldsymbol{C}_K\boldsymbol{\varepsilon}_V=\boldsymbol{\sigma}_V$ as it follows from (7.47) and (7.49). So,

$$\mathsf{R}_{V1}(\boldsymbol{\sigma}_V,\mathbf{u})=\frac{1}{2}(\boldsymbol{C}_K^{-1}\boldsymbol{\sigma}_V,\boldsymbol{\sigma}_V)-(\boldsymbol{A}_V\mathbf{u},\boldsymbol{\sigma}_V)+\Pi_s-\Pi_k+(\boldsymbol{E}_u\boldsymbol{p}_1,\boldsymbol{E}_u\boldsymbol{u}_1)_\Gamma. \quad (7.76)$$

An additional numeric subscript here marks No. of the Reissner functional's form. Therefore a designation like R_{V1} should be read as *first form of the Reissner functional for the Vlasov theory*.

Switching to coordinates in (7.76) gives

$$\mathrm{R}_{\mathrm{V1}}(T,M,v,w) =$$

$$= \frac{1}{2}\int_0^l \left(\frac{T^2}{EA} + \frac{M^2}{EI}\right) ds - \int_0^l [(v' + kw)T + (-w'' - k^2 w + k'v)M] ds +$$

$$+ \int_0^l q_t v\, ds + \int_0^l q_n w\, ds + \int_0^l m(-w' + kv)\, ds +$$

$$+ [e_N n\bar{N} v]\big|_0^l + [e_Q n\bar{Q} w]\big|_0^l + [e_M n\bar{M}(-w' + kv)]\big|_0^l +$$

$$+ [e_v N(v - \bar{v})]\big|_0^l + [e_w M'(w - \bar{w})]\big|_0^l + [e_\theta M(\theta - \bar{\theta})]\big|_0^l . \quad (7.77)$$

Having obtained first form of the Reissner functional, it is easy to derive its second and third form by moving the respective derivatives from the displacements to the stresses in the integrands. If need be, the reader can do this simple operation by himself.

In exactly the same way we can derive the Reissner functional for the theory of curvilinear bars by Kirchhoff–Klebsch. Here we present an expression of the second form of that functional:

$$\mathrm{R}_{\mathrm{K2}}(N,M,v,w) =$$

$$= \frac{1}{2}\int_0^l \left[\frac{N^2}{EA} + \frac{M^2}{EI} + v(N' + kM') - w(kN - M'') + q_t v + q_n w\right] ds +$$

$$+ [e_N(n\bar{N} - N)v]\big|_0^l + [e_Q(n\bar{Q} - M')w]\big|_0^l -$$

$$-[e_M(n\bar{M} + M)(-w' + kv)]\big|_0^l - [e_v N\bar{v}]\big|_0^l - [e_w M'\bar{w}]\big|_0^l + [e_\theta M\bar{\theta}]\big|_0^l . \quad (7.78)$$

4.7.5 Remark on comparison of solutions in various formulations of curvilinear-bar-related problems

So, we have three basic engineering theories for a curvilinear bar (theories of big-curvature, medium-curvature, and small-curvature bars) and two versions or forms of boundary conditions for each. It is a natural question how much solutions of problems differ in different theories, and another — how much a solution changes after switching from one boundary condition form to the other.

We want to introduce a criterion for comparing the solutions correctly and reasonably. This criterion will be an energy error of a solution which enables us to make comparisons integrally without minding local differences.

The very concept of the energy error of a theory was introduced to structural analysis in a systematic form by V.V. Bolotin [2], and he was the first to use it extensively. However, as Bolotin himself indicated, this concept was used implicitly by other authors earlier.

We introduce a dimensionless parameter, c, following Vlasov [14]:

$$c^2 = \frac{k^2 I}{A}, \tag{7.79}$$

which will be convenient for our further presentment[14]. One should keep in mind that, when dealing with big-curvature bars, formula (7.79) should contain reduced moment of inertia I_ρ according to (7.38) in place of moment of inertia I.

We suppose for simplicity that a curvilinear bar has statically determinate fixations on its ends, to be able to obtain internal forces/stresses in the bar from the static equations only; in this way the forces (longitudinal force N and bending moment M) will not depend on which version of theory is used. But then, using formulas (7.60) and (7.63), we can calculate the relative energy error, η, introduced to the solution by switching from one theory to another. So we have

$$\eta = \frac{\mathsf{E}_V - \mathsf{E}_K}{\mathsf{E}_K} = \frac{\int_0^l \frac{2kNM + k^2M^2}{EA} ds}{\int_0^l \left(\frac{N^2}{EA} + \frac{M^2}{EI} \right) ds} = \frac{\int_0^l \frac{T^2 - N^2}{EA} ds}{\int_0^l \left(\frac{N^2}{EA} + \frac{M^2}{EI} \right) ds}. \tag{7.80}$$

This shows immediately that the energy error of switching between the theories is zero for a moment-free stress state in the curvilinear bar. In the opposite particular case, that is, in a pure bending ($N = 0$), this error can be easily calculated as well. It is equal to something of order of magnitude c^2 with an averaged value of this parameter along the bar. Consequently, we can expect that switching to the refined Vlasov theory will give a fairly small refinement comparing to the Kirchhoff–Klebsch model in the sense of energy (averagely, integrally). However, this is not the case in the sense of a local behavior. It is these data (such as stresses in particular points) which are of special practical interest for an engineer. That's why

[14] When dealing with the theory of cylindrical shells, Vlasov assumes $c^2 = k^2h^2/12$ where h is a thickness of a shell. In application to a curvilinear bar this gives (7.79).

switching to the refined Vlasov theory is justified for medium-curvature bars.

It is not possible to obtain an explicit analytic solution for a general curvilinear bar. However, we can do that by confining ourselves to bars which have a circular arc shape of a radius R and by introducing an additional simplification of a constant cross-section along the whole bar.

We switch from the arc coordinate, s, to a dimensionless angular coordinate, $\varphi = s/R$, and use Table 4.3 for operators L_{ij} to rewrite equations (7.40) as:

- for big-curvature and medium-curvature bars,

$$
\begin{aligned}
-\frac{d^2 v}{d\varphi^2} - \frac{dw}{d\varphi} &= \frac{R^2}{EA}\left(q_t + \frac{m}{R}\right),\\
\frac{dv}{d\varphi} + c^2\frac{d^4 w}{d\varphi^4} + 2c^2\frac{d^2 w}{d\varphi^2} + (1+c^2)w &= \frac{R^2}{EA}\left(q_n + \frac{dm}{Rd\varphi}\right),
\end{aligned}
\tag{7.81}
$$

- for small-curvature bars,

$$
\begin{aligned}
-(1+c^2)\frac{d^2 v}{d\varphi^2} + c^2\frac{d^3 w}{d\varphi^3} - \frac{dw}{d\varphi} &= \frac{R^2}{EA}\left(q_t + \frac{m}{R}\right),\\
-c^2\frac{d^3 v}{d\varphi^3} + \frac{dv}{d\varphi} + c^2\frac{d^4 w}{d\varphi^4} + w &= \frac{R^2}{EA}\left(q_n + \frac{dm}{Rd\varphi}\right).
\end{aligned}
\tag{7.82}
$$

As we want to be consistent in removing negligible terms, we are tempted to omit parameter c^2 as small comparing to one because $c^2 < (kh)^2 \ll 1$. However, let us not hurry but see what consequences will follow from this kind of simplification of simultaneous equations (7.81) and (7.82). It is easy to notice that the self-conjugation of the Lame differential operator will be maintained in this simplification. But that is not enough; just recall words by N.N. Moiseyev given as an epigraph to this chapter.

We begin with system of equations (7.81) simplified in the said way. Let's consider a half-ring ($0 \le \varphi \le \pi$) and let a true field of displacements be described by functions

$$
v = \cos\varphi, \quad w = \sin\varphi. \tag{7.83}
$$

It is easy to see that both equations of the simplified system (7.81) will be satisfied by assuming

$$m = 0, \quad q_t = 0, \quad q_n = -c^2 \frac{EA}{R^2} \sin\varphi. \tag{7.84}$$

The edge displacements will be

$$v(0) = 1, \quad w(0) = 0, \quad \theta(0) = 0, \quad v(\pi) = -1, \quad w(\pi) = 0, \quad \theta(\pi) = 0,$$

and all internal forces in the bar are identical to zero.

Now we calculate the work, A, of all external forces on the respective displacements caused by the same forces to find that the reactive forces do not do any work in places where the bar is fixed. As a result,

$$\mathsf{A} = \int_0^\pi q_n w R d\varphi = -c^2 \frac{EA}{R} \int_0^\pi \sin^2 \varphi d\varphi .$$

Because $\mathsf{A} < 0$, we find that removing the term $c^2 w$ in the second equation of (7.81) violates a most fundamental law of mechanics, the energy conservation law.

It is exactly the same case with bars of small curvature. And indeed, if we find that the displacements vary according to (7.83), then the simplified equations (7.82) will be satisfied, too, but under the following load:

$$m = 0, \quad q_t = -c^2 \frac{EA}{R^2} \cos\varphi, \quad q_n = 0. \tag{7.85}$$

The same reasoning as before gives again the inequality $\mathsf{A} < 0$ which contradicts the energy conservation law.

The above considerations force us into using the governing equations for curvilinear bars exactly in the same form they are written in (7.81) and (7.82).

Let's give a general solution of the homogeneous simultaneous equations from (7.81) and (7.82) for reference. We have

- for medium-curvature and big-curvature bars

$$\begin{aligned} v = {} & C_1 \cos\varphi - C_2 \sin\varphi - C_3 (\sin\varphi - \varphi \cos\varphi) - \\ & - C_4 (\cos\varphi + \varphi \sin\varphi) - (1+c^2) C_5 \varphi + C_6, \\ w = {} & C_1 \sin\varphi + C_2 \cos\varphi + C_3 \varphi \sin\varphi + C_4 \varphi \cos\varphi + C_5 , \end{aligned} \tag{7.86}$$

- for small-curvature bars

$$v = C_1 \cos\varphi - C_2 \sin\varphi - C_3 \left(\frac{1-c^2}{1+c^2} \sin\varphi - \varphi \cos\varphi \right) -$$

$$- C_4 \left(\frac{1-c^2}{1+c^2} \cos\varphi + \varphi \sin\varphi \right) - C_5 \varphi + C_6,$$

$$w = C_1 \sin\varphi + C_2 \cos\varphi + C_3 \varphi \sin\varphi + C_4 \varphi \cos\varphi + C_5 . \qquad (7.87)$$

Representations (7.86) and (7.87) can be validated by direct substitution.

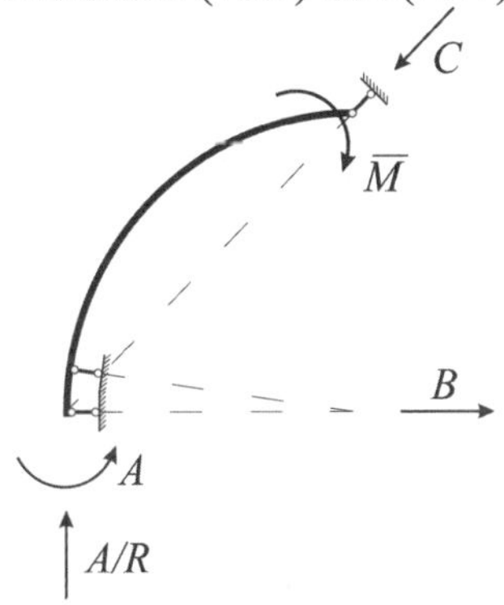

Fig. 4.9-*a*

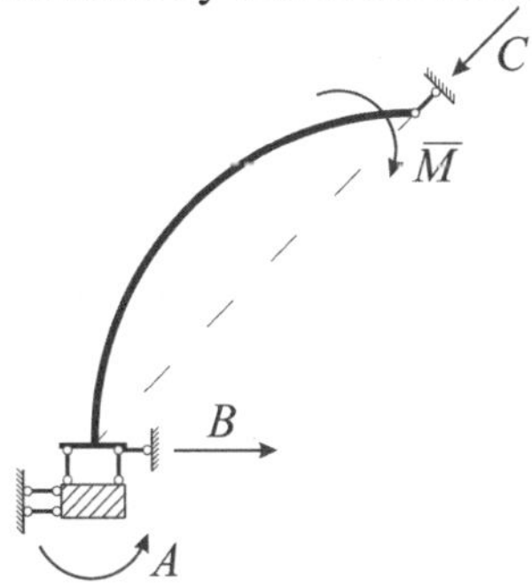

Fig. 4.9-*b*

Now let us analyze the effect of a method of specification of the boundary conditions on the solution of a curvilinear bar problem. We will use a particular example shown in Fig. 4.9-*a* and Fig. 4.9-*b*.

So, we deal with two states of a curvilinear bar (a quarter of a circle) under the same load (an end moment, $\overline{M}$), which we will call State *a* and State *b*. In State *a*, one of the bar's ends (at $s = 0$) has homogeneous boundary conditions according to version *a* as we defined earlier. In State *b* the same end of the bar has homogeneous boundary conditions according to version *b*.

Letters *A*, *B* and *C* denote in these figures the reactive forces which can be easily found from the equilibrium equations (both problems are statically determinate).

In State *a* we have

$$A_a = \overline{M}, \quad B_a = \frac{\overline{M}}{R}, \quad C_a = \frac{\overline{M}}{R}\sqrt{2}$$

thus the forces in the bar's sections are equal to the following in State *a*:

$$M_a = \bar{M}\,(\sin\varphi + \cos\varphi), \qquad N_a = -\frac{\bar{M}}{R}\,(\sin\varphi + \cos\varphi).$$

On the other hand, for the problem in Fig. 4.9-*b* we have

$$A_b = \bar{M}\,, \quad B_b = 0, \quad C_b = 0,$$

thus the forces in the bar's sections in State *b* are

$$M_b = \bar{M}\,, \quad N_b = 0.$$

There are obvious and noticeable differences in the solutions of these two problems.

4.8 Planar curvilinear bar, shear considered

4.8.1 Geometry of a curvilinear bar revisited

It is useful to return to the geometric relationships for a planar curvilinear bar, now using formal mathematical transformations. Of course, this way of deriving geometric relationships is less obvious because it does not involve direct geometric considerations. However, its advantage is that it is based on a strict mathematical formalism.

Having this purpose in mind, first of all we will write out basic geometric relationships for two-dimensional elasticity in an orthogonal curvilinear coordinate system, referring to [9] (see also Appendix F).

So, let (α_1, α_2) be an orthogonal coordinate system defined on a plane. Let H_1, H_2 be Lame parameters which correspond to those coordinates. This means aggregates $H_1 d\alpha_1$ and $H_2 d\alpha_2$ are increments of arc lengths which conform to the respective coordinate increments, $d\alpha_1$ and $d\alpha_2$.

In application to a planar curvilinear bar, we take the normal coordinate z as coordinate α_1, and coordinate α_2 will be the arc coordinate s. The respective Lame parameters will be

$$H_1 = 1, \qquad H_2 = 1 + kz.$$

General formulas for the strain components in the curvilinear coordinate system, (α_1, α_2), will be [9]

$$\varepsilon_1 = \frac{1}{H_1}\frac{\partial u_1}{\partial \alpha_1} + \frac{1}{H_1 H_2}\frac{\partial H_1}{\partial \alpha_2}u_2, \qquad \varepsilon_2 = \frac{1}{H_2}\frac{\partial u_2}{\partial \alpha_2} + \frac{1}{H_2 H_1}\frac{\partial H_2}{\partial \alpha_1}u_1,$$

$$\gamma_{12} = \frac{H_2}{H_1}\frac{\partial}{\partial \alpha_1}\left(\frac{u_2}{H_2}\right) + \frac{H_1}{H_2}\frac{\partial}{\partial \alpha_2}\left(\frac{u_1}{H_1}\right). \tag{8.1}$$

Here:

- ε_1 is a relative elongation in the direction of coordinate α_1;
- ε_2 is a relative elongation in the direction of coordinate α_2;
- γ_{12} is a shear angle between coordinate lines α_1 and α_2;
- u_1, u_2 are components of a displacement vector, $\mathbf{u}_z = u_1\boldsymbol{e}_1 + u_2\boldsymbol{e}_2$, with respect to unit vectors $\boldsymbol{e}_2$ and $\boldsymbol{e}_3$ tangential to coordinate lines α_1 and α_2.

In our case coordinate α_1 is the normal coordinate z and coordinate α_2 is the tangential coordinate s. This means that unit vectors $\boldsymbol{e}_1$ and $\boldsymbol{e}_2$ should be $\boldsymbol{e}_1 = \boldsymbol{n}$ and $\boldsymbol{e}_2 = \boldsymbol{t}$, and displacements u_1, u_2 should be $u_1 = w_z(z,s)$ and $u_2 = v_z(z,s)$, where v_z and w_z are the tangential and normal components of the displacement vector, $\mathbf{u}_z = w_z\boldsymbol{n} + v_z\boldsymbol{t}$, of an arbitrary point with coordinates (z,s).

The respective substitutions of dependent and independent variables will give

$$\varepsilon_s = \frac{1}{1+kz}\frac{\partial v_z}{\partial s} + \frac{1}{1+kz}\frac{\partial(1+kz)}{\partial z}w_z = \frac{1}{1+kz}\left(v_z' + kw_z\right),$$

$$\varepsilon_z = \frac{\partial w_z}{\partial z} + \frac{1}{1+kz}\frac{\partial(1)}{\partial s}v_z = \frac{\partial w_z}{\partial z},$$

$$\gamma_{sz} = \frac{1}{1+kz}\frac{\partial}{\partial s}\left(\frac{w_z}{1}\right) + \frac{1+kz}{1}\frac{\partial}{\partial z}\left(\frac{v_z}{1+kz}\right) = \frac{1}{1+kz}w_z' + \frac{\partial v_z}{\partial z} - \frac{k}{1+kz}v_z. \tag{8.2}$$

To simplify the formulas, we will use a stroke for differentiation with respect to tangential coordinate s.

The basic kinematic hypothesis used to build technical theories of bars (the planar sections hypothesis) can be represented by two formal mathematical requirements:

$$\varepsilon_z = 0, \quad \gamma_{sz} = 0. \tag{8.3}$$

And indeed, if the first requirement is met, then the normals to the bar's axis have invariable lengths. The second requirement ensures the unchanged orthogonality of the coordinate lines in a deformed state of a bar because the respective shear angle is assumed to remain zero.

As in Section **4**.7, we denote by $w(s)$ and $v(s)$ the normal and tangential components of the displacement vector, $\mathbf{u} = w\boldsymbol{n} + v\boldsymbol{t}$, in an arbitrary point of the bar's axis. In other words, we assume by definition

$$v(s) = v_z(0,s), \quad w(s) = w_z(0,s). \tag{8.4}$$

From the first requirement in (8.3) we find that function $w_z(z,s)$ does not depend on z, consequently

$$w = w_z\,. \tag{8.5}$$

Further, the condition of no shear, $\gamma_{sz} = 0$, leads to a differential equation for function v_z:

$$(1+kz)\frac{\partial v_z}{\partial z} - kv_z = -w'\,. \tag{8.6}$$

The right part of equation (8.6) contains a function depending only on coordinate s but not on coordinate z. Therefore we have actually an ordinary differential equation of first order, and its general solution can be found easily by introducing an integrating multiplier [7]. We omit quite elementary transformations and present a final solution of the equation:

$$v_z = -w'\frac{z-1}{1+k} + (1+kz)C(s)\,, \tag{8.7}$$

where $C(s)$ is an arbitrary function of s. Representation (8.7) can be validated directly by substituting to equation (8.6).

Assuming $z = 0$ in (8.7), we have

$$C(s) = v - \frac{w'}{1+k}\,,$$

which gives the following after being substituted to (8.7):

$$v_z = -w'z + v(1+kz)\,. \tag{8.8}$$

If we now substitute relationships (8.5) and (8.8) to the expression of longitudinal strains ε according to the first formula in (8.2) and take into account the estimates of smallness of dimensionless parameter kh adopted in the small-curvature and medium-curvature bar theories, we will arrive at the formulas obtained earlier and given in the last row of Table 4.1. This fact evidences that results obtained in different ways in Section **4**.7.1 and in this section are identical.

4.8.2 Allowing for shear deformations

Recalling expression (7.26) for slope θ of the bar's section, we rewrite formula (8.8) as

$$v_z = v + \theta z\,. \tag{8.9}$$

Keeping the kinematic relationships in the form (8.5) and (8.9), we will adopt the planar sections hypothesis in a weaker formulation. Starting from this point, we will assume further on that the normal to the bar's axis does not change its length and remains rectilinear; however, it is not obliged to stay perpendicular to the bar's axis after the deformation. It means that the slope of the normal, θ, is not defined by formula (7.26) anymore but is treated as an independent kinematical parameter which describes the deformed state of the bar. Under these conditions, substituting (8.5) and (8.9) to the expression of the shear strain in the third relationship from (8.2) gives

$$\gamma_{sz} = \frac{w' - kv + \theta}{1 + kz}\,. \tag{8.10}$$

Thus, based on the kinematical relationships, we know now that the shear angle varies along the height of the curvilinear bar's section according to a hyperbolic law. Earlier we found out for a rectilinear bar that its shear angle is constant over the height of its section, which also follows from (8.10) if we assume $k = 0$.

Now we want to calculate a strain energy stored in an elementary volume, $d\Omega = b(1{+}kz)dzds$, of the bar's material, which is caused by the work of tangential stresses. This energy is

$$\tfrac{1}{2}\,\tau^{sz}\,\gamma_{sz}\,b(1{+}kz)dzds = \tfrac{1}{2}\,\tau^{sz}(\,w' - kv + \theta\,)bdzds.$$

By integrating the above expression over the height of the bar's cross-section, we obtain energy $d\mathsf{E}_\tau$ of the tangential stresses per unit of length of the bar's axis. Obviously, it is

$$d\mathsf{E}_\tau = \tfrac{1}{2}\,(\,w' - kv + \theta\,)\int_{-h/2}^{h/2} \tau^{sz} b dz = \tfrac{1}{2}Q\gamma,$$

where Q is a shear force in the section of interest, and γ is a shear angle at the level of the bar's axis, that is, $\gamma = \gamma_{sz}(0,s) = w' - kv + \theta$. Shear angle γ is related to the shear force via a linear dependence; we assume

$$\gamma = \frac{Q}{GF}$$

where F is a certain quantity yet to be found; we will call it (exactly as for a Timoshenko beam) *a shear area*. But then

$$d\mathsf{E}_\tau = \frac{Q^2}{2GF}. \tag{8.11}$$

Turning to Hooke's law for shear stresses and strains,

$$\tau^{sz} = G\gamma_{sz},$$

we have

$$d\mathsf{E}_\tau = \frac{1}{2}\int_{-h/2}^{h/2} \frac{(\tau^{sz})^2}{G} b(1+kz)dz. \tag{8.12}$$

Appendix B presents derivations of formulas for tangential stresses which develop in curvilinear bars. Here we place those formulas in Table 4.7. Based on the same considerations as we used in the justification of the Timoshenko bar theory, we will keep only the main term – an analogue of Zhuravsky's formula for rectilinear bars.

Table 4.7

Big-curvature bars	*Medium-curvature bars*	*Small-curvature bars*
$\tau^{sz} = \dfrac{QS_o}{bI_\rho(1+kz)^2}$	$\tau^{sz} = \dfrac{QS_o}{bI}(1-2kz)$	$\tau^{sz} = \dfrac{QS_o}{bI}$
$\dfrac{A}{I_\rho^2}\displaystyle\int_{-h/2}^{h/2}\frac{S_o^2}{b(1+kz)}dz$	$\dfrac{A}{I^2}\displaystyle\int_{-h/2}^{h/2}\frac{S_o^2}{b}(1-2kz)(1+kz)dz$	$\dfrac{A}{I^2}\displaystyle\int_{-h/2}^{h/2}\frac{S_o^2}{b}(1+kz)dz.$

Comparing (8.11) and (8.12), after having substituted expressions of τ^{sz} from Table 4.7 to formula (8.12) of $d\mathsf{E}_\tau$, we have a general formula for calculating the shear area:

$$F = \frac{1}{\mu}A. \tag{8.13}$$

The section shape factor, μ, for this case is given in the second row of Table 4.7. Note that for a symmetric section, which is the only case here, the integral

$$\int_{-h/2}^{h/2} \frac{S_o^2}{b} z^n dz$$

is zero at any odd n. If we consider the previous estimate of smallness, $(kh)^2 \ll 1$, for medium-curvature bars, we will understand that the same formula (6.8) which worked for rectilinear bars is true for the section's shape factor, μ, of small-curvature and medium-curvature bars.

To give an example, we will show what the section's shape factor μ is for a big-curvature bar of a rectangular cross-section $b \times h$ the axis of which is a piece of a circle of a radius R. Fig. 4.10 presents a graph of the dependence $\mu = \mu(kh)$ built for this particular case after results of calculation by the formulas given above.

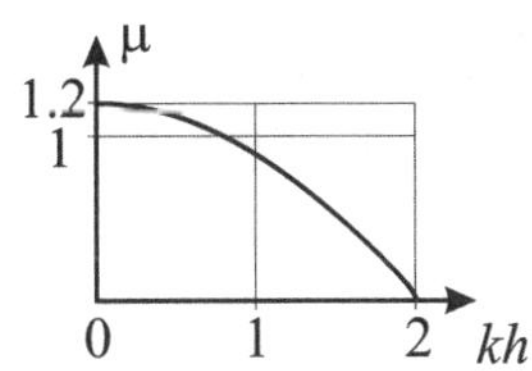

Fig. 4.10. Shape factor μ for the section of a curvilinear bar

As can be seen from this graph, at $kh = 0$ the section's shape factor is equal to the known value $\mu = 6/5$ and is close to it in a vicinity of the zero argument.

Now we can write out an expression of the strain energy accumulated in a curvilinear bar of the length l accounting for the work of shear strains. We use formulas (7.60) and (7.63) adding the respective term from (8.11) to the integrands to allow for the work of the shear strain.

Bars of medium and big curvature

All formulas presented in this section will be written for medium-curvature bars. However, they will be fitting for big-curvature bars, too, after replacing moment of inertia I with reduced moment of inertia I_ρ.

So we have

$$\mathsf{E}_V(\boldsymbol{\sigma}) = \frac{1}{2}\int_0^l \left[\frac{(N+kM)^2}{EA} + \frac{M^2}{EI} + \frac{Q^2}{GF}\right] ds \,. \tag{8.14}$$

Further, internal forces N, Q, M are related via the elasticity equations

$$N = \varepsilon_0 EA - \chi kEI, \qquad Q = GF\gamma, \qquad M = \chi EI \tag{8.15}$$

with kinematic parameters ε_o, γ, and χ which can be treated as components of a strain vector expressed via displacements v, w и θ by these formulas:

$$\varepsilon_0 = v' + kw, \qquad \gamma = w' - kv + \theta, \qquad \chi = -kv' - k^2 w + \theta'. \tag{8.16}$$

Substituting (8.15) and (8.16) to (8.14) lets us represent the expression of energy E_V as a functional of the displacements,

$$\mathsf{E}_V(v,w,\theta) = \frac{1}{2}\int_0^l (\varepsilon_0^2 EA + \chi^2 EI + \gamma^2 GF)ds =$$

$$= \frac{1}{2}\int_0^l [(v' + kw)^2 EA + (-kv' - k^2 w + \theta')^2 EI + (w' - kv + \theta)^2 GF]ds\,. \tag{8.17}$$

Now we can write a final Lagrange functional for a medium-curvature bar where the shear work is allowed for:

$$\mathsf{L}_V(v,w,\theta) =$$

$$= \frac{1}{2}\int_0^l [(v' + kw)^2 EA + (-kv' - k^2 w + \theta')^2 EI + (w' - kv + \theta)^2 GF]ds \; -$$

$$- \int_0^l q_t v ds - \int_0^l q_n w ds - \int_0^l m\theta ds -$$

$$-[e_N n\bar{N} v]\big|_0^l - [e_Q n\bar{Q} w]\big|_0^l - [e_M n\bar{M}\theta]\big|_0^l\,. \tag{8.18}$$

By varying functional $\mathsf{L}_V(v,w,\theta)$ on the set of kinematically admissible displacements and equaling its first variation to zero, we arrive at Euler equations and natural boundary conditions.

Euler equations for functional L_V according to (8.18), which are in essence differential equations of equilibrium in displacements for medium-curvature bars (Lame-type equations), are

$$\boxed{\begin{aligned} L_{11}v + L_{12}w + L_{13}\theta &= q_t\,, \\ L_{21}v + L_{22}w + L_{23}\theta &= q_n\,, \\ L_{31}v + L_{32}w + L_{33}\theta &= m\,. \end{aligned}} \tag{8.19}$$

Differential operators L_{ij} are listed in Table 4.8 . In this table we denote, as before,

$$c^2 = \frac{k^2 I}{A}\,.$$

Table 4.8

$L_{11}v =$ $= -[EA(1+c^2)v']' +$ $+k^2GFv$	$L_{12}w =$ $= -[kEA(1+c^2)w]' -$ $- kGFw'$	$L_{13}\theta =$ $= (kEI\theta')' - kGF\theta$
$L_{21}v =$ $= kEA(1+c^2)v' + (kGFv)'$	$L_{22}w =$ $= k^2EA(1+c^2)w -$ $-(GFw')'$	$L_{23}\theta =$ $= -k^2EI\theta' - (GF\theta)'$
$L_{31}v =$ $= (kEIv')' - kGFv$	$L_{32}w =$ $= (k^2EIw)' + GFw'$	$L_{33}\theta =$ $= -(EI\theta')' + GF\theta$

We are eager to omit the c^2 parameter (negligible comparing to one) in these equations for medium-curvature bars. However, wouldn't this operation contradict the basic theorems of energy conservation, as it was the case with the shear-free theory of curvilinear bars?

The matter is that, unlike the shear-free curvilinear bar theory, here we are entitled to do so without violating any energy laws, theorems, or principles. This conclusion could be justified by formal reasoning, but actually we can arrive at it using really simple considerations. As we can see in the formulas of Table 4.8, neglecting parameter c^2 in comparison to one is equivalent to the mere replacement of the bar's section area, A, by a fairly close value, $A/(1 + c^2)$, which is quite admissible.

What about natural boundary conditions that this functional generates? They are, as they should be, static-type boundary conditions, and their formulation depends on which set, $\mathscr{P}_{ka}$ or $\mathscr{P}_{kb}$, of kinematically admissible displacements we choose.

But let us discuss this issue a bit later. Now we want to convert all relationships of the medium-curvature bar theory where shear is allowed to a matrix form. The vectors of stresses $\boldsymbol{\sigma}$, strains $\boldsymbol{\varepsilon}$, and displacements $\mathbf{u}$, which define a stress-and-strain state of a curvilinear bar, contain three components each:

$$\boldsymbol{\sigma} = \left\|\begin{bmatrix} T = N + kM \\ Q \\ M \end{bmatrix}\right\|, \quad \boldsymbol{\varepsilon} = \left\|\begin{bmatrix} \varepsilon_o \\ \gamma \\ \chi \end{bmatrix}\right\|, \quad \mathbf{u} = \left\|\begin{bmatrix} v \\ w \\ \theta \end{bmatrix}\right\|. \tag{8.20}$$

All equations will acquire a common operator form if we assume the following in addition to (8.20):

$$A = \begin{Vmatrix} \dfrac{d}{ds} & k & 0 \\ -k & \dfrac{d}{ds} & 1 \\ -k\dfrac{d}{ds} & -k^2 & \dfrac{d}{ds} \end{Vmatrix}, \quad A^{\mathrm{T}} = \begin{Vmatrix} -\dfrac{d}{ds} & -k & k\dfrac{d}{ds}+\dfrac{dk}{ds} \\ k & -\dfrac{d}{ds} & -k^2 \\ 0 & 1 & -\dfrac{d}{ds} \end{Vmatrix},$$

$$C = \begin{Vmatrix} EA & 0 & 0 \\ 0 & GF & 0 \\ 0 & 0 & EI \end{Vmatrix}, \quad C^{-1} = \begin{Vmatrix} 1/EA & 0 & 0 \\ 0 & 1/GF & 0 \\ 0 & 0 & 1/EI \end{Vmatrix}, \quad \bar{X} = \begin{Vmatrix} q_t \\ q_n \\ m \end{Vmatrix}. \qquad (8.21)$$

It can be checked that general formula (1.6) for Lame matrix differential operator $L = A^{\mathrm{T}}CA$ works in this case, too[15]. To see this, we make formal mathematical transformations and find that all six components of operator L coincide with data of Table 4.8. Now we demand that the basic integral identity in the form of (**1**.2.17) hold. We have

$$(A\mathbf{u}, \boldsymbol{\sigma}) =$$

$$= (v'+kw, T) + (-kv + w' + \theta, Q) + (-kv' - k^2w + \theta', M) =$$

$$= \int_0^l [(v'+kw)T + (-kv' - k^2w + \theta')M + (-kv + w' + \theta)Q]ds =$$

$$= \int_0^l \{v[-T' - kQ + (kM)'] + w[kT - Q' - k^2M] + \theta[Q - M']\}ds +$$

$$+ [v(T - kM)]\Big|_0^l + [wQ]\Big|_0^l + [\theta M]\Big|_0^l = (\mathbf{u}, A^{\mathrm{T}}\boldsymbol{\sigma}) + (\boldsymbol{p}, \boldsymbol{u})_\Gamma. \qquad (8.22)$$

We can see from (8.22) that, in order for the basic integral identity to hold, we can adopt any of the representations of the work of contour forces:

- version *a*: $(\boldsymbol{p}, \boldsymbol{u})_\Gamma = [vT]\Big|_0^l + [wQ]\Big|_0^l + [\psi M]\Big|_0^l,$
- version *b*: $(\boldsymbol{p}, \boldsymbol{u})_\Gamma = [vN]\Big|_0^l + [wQ]\Big|_0^l + [\theta M]\Big|_0^l,$

the generalized slope, ψ, being defined as follows in curvilinear bars where shear is allowed:

[15] In theories of curvilinear bars considered here, we assume no elastic foundation or bed present, therefore the K operator has been excluded from the general formula (1.6) in application to the curvilinear bars.

$$\psi = \theta - kv\,, \tag{8.23}$$

which is different from the generalized slope expression, $\psi = -w'$, used in the theory of curvilinear bars with no shear.

Let's formulate boundary conditions in the matrix form. As should be clear now, the vector of edge displacements, $\boldsymbol{u}$, and the vector of edge forces, $\boldsymbol{p}$, for curvilinear bars where shear is allowed for can be also represented in either of two versions: a or b. Here we have

$$\boldsymbol{u}_a = \left\|\begin{bmatrix} v \\ w \\ \psi \end{bmatrix}\right\|, \quad \boldsymbol{p}_a = \left\|\begin{bmatrix} nT \\ nQ \\ nM \end{bmatrix}\right\|, \quad \overline{\boldsymbol{u}}_a = \left\|\begin{bmatrix} \overline{v} \\ \overline{w} \\ \overline{\psi} \end{bmatrix}\right\|, \quad \overline{\boldsymbol{p}}_a = \left\|\begin{bmatrix} \overline{T} \\ \overline{Q} \\ \overline{M} \end{bmatrix}\right\|,$$

$$\boldsymbol{u}_b = \left\|\begin{bmatrix} v \\ w \\ \theta \end{bmatrix}\right\|, \quad \boldsymbol{p}_b = \left\|\begin{bmatrix} nN \\ nQ \\ nM \end{bmatrix}\right\|, \quad \overline{\boldsymbol{u}}_b = \left\|\begin{bmatrix} \overline{v} \\ \overline{w} \\ \overline{\theta} \end{bmatrix}\right\|, \quad \overline{\boldsymbol{p}}_b = \left\|\begin{bmatrix} \overline{N} \\ \overline{Q} \\ \overline{M} \end{bmatrix}\right\|. \tag{8.24}$$

Note that the equality between the virtual works of edge forces holds in the following form, exactly as one for curvilinear bars where no shear is taken into account:

$$(\boldsymbol{p}_a\,, \boldsymbol{u}_a)_\Gamma = (\boldsymbol{p}_b\,, \boldsymbol{u}_b)_\Gamma\,.$$

Now we are able to write out a matrix form of both versions of the boundary conditions in a common representation defined by (**1**.2.4). Table 4.9 below lists the respective matrix operators of boundary condition extraction together with operators of transition to edge displacements and edge forces.

Now we have matrix expressions for all needed operators, so we can use the general formulas of the functionals of structural mechanics given in Chapters 2 and 3 to easily obtain these functionals in their component form. The reader can make those simple operations himself, if need be.

Table 4.9

		Version "a" of boundary conditions	*Version "b" of boundary conditions*
$\boldsymbol{E}_u$		$\begin{bmatrix} e_v & 0 & 0 \\ 0 & e_w & 0 \\ 0 & 0 & e_\psi \end{bmatrix}$	$\begin{bmatrix} e_v & 0 & 0 \\ 0 & e_w & 0 \\ 0 & 0 & e_\theta \end{bmatrix}$
$\boldsymbol{E}_p$		$\begin{bmatrix} e_T & 0 & 0 \\ 0 & e_Q & 0 \\ 0 & 0 & e_M \end{bmatrix}$	$\begin{bmatrix} e_N & 0 & 0 \\ 0 & e_Q & 0 \\ 0 & 0 & e_M \end{bmatrix}$
$\boldsymbol{H}_u$	$\mathbf{u} = \begin{bmatrix} v \\ w \\ \theta \end{bmatrix}$	$\begin{bmatrix} 1 & 0 & 0 \\ 0 & 1 & 0 \\ -k & 0 & 1 \end{bmatrix}$	$\begin{bmatrix} 1 & 0 & 0 \\ 0 & 1 & 0 \\ 0 & 0 & 1 \end{bmatrix}$
$\boldsymbol{H}_\sigma$	$\boldsymbol{\sigma} = \begin{bmatrix} T \\ Q \\ M \end{bmatrix}$	$\begin{bmatrix} n & 0 & 0 \\ 0 & n & 0 \\ 0 & 0 & n \end{bmatrix}$	$\begin{bmatrix} n & 0 & -kn \\ 0 & n & 0 \\ 0 & 0 & n \end{bmatrix}$

Bars of small curvature

Formula (7.63) is a starting point for the small-curvature bars. We add an elementary work of the tangential stresses to the integrand and have

$$\mathsf{E}_K(\boldsymbol{\sigma}) = \frac{1}{2}\int_0^l \left(\frac{N^2}{EA} + \frac{M^2}{EI} + \frac{Q^2}{GF} \right) ds \,. \tag{8.25}$$

But the physical law for the small-curvature bars is

$$N = \varepsilon_0 EA, \qquad Q = GF\gamma, \qquad M = \kappa EI. \tag{8.26}$$

Kinematic parameters ε_o, γ, and κ are components of a strain vector, expressed via displacements v, w, and θ using these formulas

$$\varepsilon_0 = v' + kw, \qquad \gamma = w' - kv + \theta, \qquad \kappa = \theta'. \tag{8.27}$$

Substituting (8.15) and (8.16) to (8.14) permits us to represent the expression of energy E_K as a functional of displacements,

$$\mathsf{E}_K(v,w,\theta) = \frac{1}{2}\int_0^l (\varepsilon_o^2 EA + \kappa^2 EI + \gamma^2 GF)ds =$$

$$= \frac{1}{2}\int_0^l [(v' + kw)^2 EA + \theta'^2 EI + (w' - kv + \theta)^2 GF]ds\,. \tag{8.28}$$

The final form of the Lagrange functional for a bar of small curvature, that takes into account the work of shearing stresses, is

$$\mathsf{L}_K(v,w,\theta) =$$

$$= \frac{1}{2}\int_0^l [(v' + kw)^2 EA + \theta'^2 EI + (w' - kv + \theta)^2 GF]ds \; -$$

$$- \int_0^l q_t v ds \; - \int_0^l q_n w ds \; - \int_0^l m\theta ds \; -$$

$$-[e_N n\bar{N}v]|_0^l - [e_Q n\bar{Q}w]|_0^l - [e_M n\bar{M}\theta]|_0^l\,. \tag{8.29}$$

Now we vary functional $\mathsf{L}_K(v,w,\theta)$ on a set of kinematically admissible displacements and equal its first variation to zero. In this way we derive Euler equations and natural boundary conditions. The Euler equations for this functional are written in the same operator form (8.19) which we used for the big-curvature and medium-curvature bars. However, differential operators L_{ij} for the small-curvature bars are different. The operators are given in Table 4.10

Table 4.10

$L_{11}v =$ $= -[EAv']' + k^2 GFv$	$L_{12}w =$ $= -[kEAw]' - kGFw'$	$L_{13}\theta =$ $= -kGF\theta$
$L_{21}v =$ $= kEAv' + (kGFv)'$	$L_{22}w =$ $= k^2 EAw - (GFw')'$	$L_{23}\theta =$ $= -(GF\theta)'$
$L_{31}v =$ $= -kGFv$	$L_{32}w =$ $= GFw'$	$L_{33}\theta =$ $= -(EI\theta')' + GF\theta$

We want to switch to a matrix form and represent the vectors of stresses **σ**, strains **ε**, and displacements **u**, which describe the stress-and-strain state of a small-curvature bar, as

$$\boldsymbol{\sigma} = \begin{Vmatrix} N \\ Q \\ M \end{Vmatrix}, \quad \boldsymbol{\varepsilon} = \begin{Vmatrix} \varepsilon_0 \\ \gamma \\ \kappa \end{Vmatrix}, \quad \mathbf{u} = \begin{Vmatrix} v \\ w \\ \theta \end{Vmatrix}. \tag{8.30}$$

All equations acquire a common operator form if in addition to (8.30) we assume

$$\boldsymbol{A} = \begin{Vmatrix} \frac{d}{ds} & k & 0 \\ -k & \frac{d}{ds} & 1 \\ 0 & 0 & \frac{d}{ds} \end{Vmatrix}, \quad \boldsymbol{A}^{\mathsf{T}} = \begin{Vmatrix} -\frac{d}{ds} & -k & 0 \\ k & -\frac{d}{ds} & 0 \\ 0 & 1 & -\frac{d}{ds} \end{Vmatrix},$$

$$\boldsymbol{C} = \begin{Vmatrix} EA & 0 & 0 \\ 0 & GF & 0 \\ 0 & 0 & EI \end{Vmatrix}, \quad \boldsymbol{C}^{-1} = \begin{Vmatrix} 1/EA & 0 & 0 \\ 0 & 1/GF & 0 \\ 0 & 0 & 1/EI \end{Vmatrix}, \quad \bar{\boldsymbol{X}} = \begin{Vmatrix} q_t \\ q_n \\ m \end{Vmatrix}. \tag{8.31}$$

It is easy to check that general formula (1.6) for the Lame matrix differential operator, $\boldsymbol{L} = \boldsymbol{A}^{\mathsf{T}}\boldsymbol{C}\boldsymbol{A}$, works for the small-curvature bars, too. And indeed, after making formal matrix transformations we find that all six components of the $\boldsymbol{L}$ operator are the same as in Table 4.10.

Now we demand that the basic integral identity in the form of (**1**.2.17) hold. We have

$$(\boldsymbol{A}\mathbf{u}, \boldsymbol{\sigma}) = (v'+kw, N) + (-kv + w' + \theta, Q) + (\theta', M) =$$

$$= \int_0^l \{v[-N - kQ] + w[kN - Q'] + \theta[Q - M']\} ds +$$

$$+ [vN]\big|_0^l + [wQ]\big|_0^l + [\theta M]\big|_0^l =$$

$$= (\mathbf{u}, \boldsymbol{A}^{\mathsf{T}}\boldsymbol{\sigma}) + (\boldsymbol{p}, \boldsymbol{u})_\Gamma. \tag{8.32}$$

And again we can see from (8.32) that, for the basic integral identity to hold, we can adopt any of the two forms of the work of contour forces –

- version *a*: $(\boldsymbol{p}, \boldsymbol{u})_\Gamma = [vT]\big|_0^l + [wQ]\big|_0^l + [\psi M]\big|_0^l$,
- version *b*: $(\boldsymbol{p}, \boldsymbol{u})_\Gamma = [vN]\big|_0^l + [wQ]\big|_0^l + [\theta M]\big|_0^l$,

where the generalized slope, ψ, is defined by the same formula (8.23).

The vector of edge displacements, $\boldsymbol{u}$, and the vector of edge forces, $\boldsymbol{p}$, for the small-curvature bars where shear is taken into account can be also represented by two different versions: *a* and *b* according to the previously derived formulas (8.24). It is easy to see that Table 4.9, with its boundary condition extraction operators and operators of transition to edge displacements and to edge forces, is applicable to the small-curvature bars as well. The only thing to change is to replace the last row of Table 4.9 by a rows shown below in Table 4.11.

Table 4.11

		Version "a" of boundary conditions	*Version "b" of boundary conditions*
$\boldsymbol{H}_\sigma$	$\boldsymbol{\sigma}=\begin{bmatrix} N \\ Q \\ M \end{bmatrix}$	$\begin{bmatrix} n & 0 & -kn \\ 0 & n & 0 \\ 0 & 0 & n \end{bmatrix}$	$\begin{bmatrix} n & 0 & 0 \\ 0 & n & 0 \\ 0 & 0 & n \end{bmatrix}$

4.8.3 Estimation of changes in the strain energy introduced by switching to the refined theory

Let us try to compare the orders of magnitude of refinements introduced to the strain energy of a curvilinear bar:

- by taking shear into account;
- by switching from the theory of small-curvature bars to the theory of medium-curvature bars.

For the convenience of comparison, here we give expressions of the respective energies again; those are represented as quadratic functionals of internal forces in Table 4.12

Table 4.12

	Shear ignored	Shear taken into account
Small-curvature bars	$\mathsf{E}_{11} = \frac{1}{2}\int_0^l \left(\frac{N^2}{EA}+\frac{M^2}{EI}\right)ds$	$\mathsf{E}_{12} = \frac{1}{2}\int_0^l \left(\frac{N^2}{EA}+\frac{M^2}{EI}+\frac{Q^2}{GF}\right)ds$
Medium-curvature bars	$\mathsf{E}_{21} = \frac{1}{2}\int_0^l \left(\frac{T^2}{EA}+\frac{M^2}{EI}\right)ds$	$\mathsf{E}_{12} = \frac{1}{2}\int_0^l \left(\frac{T^2}{EA}+\frac{M^2}{EI}+\frac{Q^2}{GF}\right)ds$

Recall that, according to (7.45),

$$T = N + kM.$$

We will suppose for simplicity that the curvilinear bar is subject to end forces and/or displacements only, which will let us treat the equations of equilibrium (7.30) as homogeneous – there is no external distributed load upon the bar. Thus,

$$-N' - kQ = 0, \qquad kN - Q' = 0, \qquad Q - M' = 0\,. \tag{8.33}$$

We assume

$$\mathsf{e}_{12} = \mathsf{E}_{12} - \mathsf{E}_{11}\,, \qquad \mathsf{e}_{21} = \mathsf{E}_{21} - \mathsf{E}_{11}\,, \qquad \mathsf{e}_{22} = \mathsf{E}_{22} - \mathsf{E}_{11}\,. \tag{8.34}$$

Taking the homogeneous equations of equilibrium (8.33) into account and using Table 4.12, we have

$$\mathsf{e}_{12} = \frac{1}{2}\int_0^l \frac{Q^2}{GF}ds = \frac{1}{2}\int_0^l \frac{M'^2}{GF}ds\,,$$

$$\mathsf{e}_{21} = \frac{1}{2}\int_0^l \frac{2kNM + k^2M^2}{EA}ds = \frac{1}{2}\int_0^l \frac{2M''M + k^2M^2}{EA}ds\,. \tag{8.35}$$

If we assume the bar to have a constant cross-section, then the integration by parts will turn the expression of e_{21} into

$$\mathsf{e}_{21} = \frac{1}{2}\int_0^l \frac{k^2M^2 - 2M'^2}{EA}ds + \left[\frac{M'M}{EA}\right]_0^l.$$

Now we want to take another assumption that the boundary conditions make the non-integral term in the above formula equal to zero; also, we adopt an approximate estimate, $2GF \approx EA$. Then we have

$$\mathsf{e}_{12} + \mathsf{e}_{21} = \frac{1}{2}\int_0^l \frac{k^2M^2}{EA}ds\,. \tag{8.36}$$

But the integral in the right part of formula (8.36) is a fairly small quantity comparing to the main expression of energy E_{11}. And indeed, this addition can be estimated as

$$\mathsf{E}_{11} + \frac{1}{2}\int_0^l \frac{k^2M^2}{EA}ds = \frac{1}{2}\int_0^l \left(\frac{N^2}{EA} + \frac{M^2}{EI} + \frac{k^2M^2}{EA}\right)ds =$$

$$= \frac{1}{2} \int_0^l \left(\frac{N^2}{EA} + \frac{(1+c^2)M^2}{EI} \right) ds$$

where the dimensionless parameter $c^2 < (kh)^2 << 1$ is defined by (7.79).

In this way we arrive at an important qualitative conclusion that the refinements in the energy, e_{12} and e_{21}, are approximately equal but opposite in sign, so that $\mathsf{e}_{12} \geq 0$ while $\mathsf{e}_{21} \leq 0$.

This conclusion can be interpreted from the mechanical standpoint as follows:

> *taking the shear deformations into account makes the stiffness of a structure integrally lower, while switching from the theory of small-curvature bars to the theory of medium-curvature bars makes the structure integrally stiffer.*

This conclusion is especially important in problems where the engineer is interested mainly with integral properties of a structure, such as the spectrum of natural oscillation frequencies.

4.9 Final comments to Chapter 4

This chapter, as well as the following four chapters, is dedicated to particular classes of problems or analyses encountered in structural mechanics, basically in their variational formulations.

As we can see, the variational approach lets us use the strictness and persuasiveness of mathematical formality, which would be hardly possible otherwise. Interestingly, it was the mathematically formal character of calculus of variations that helped reveal certain aberrations which somehow had crept into traditional foundations and particular theses of structural analysis, and this occured a lot of times in the history of this science. We will encounter examples of this kind many times in the forthcoming chapters. Here we can refer to Section **4**.6.1 which discusses one of popular mistakes in the formulation of kinematic boundary conditions for the Timoshenko beam.

Similarly, in the analysis of curvilinear bars the variational approach reveals a natural possibility for the formulation of boundary conditions in two versions, both being noncontradictory and self-consistent in the physical and mathematical sense.

Finally, a direct variational formulation of applied mechanical problems is needed for building effective numerical methods of structural design.

That is, it is necessary for solving problems which appear in the course of development and improvement of contemporary engineering software.

References

1. Bolotin VV (1965) On reducing three-dimensional elastic stability problems to one and two dimensions (in Russian). In: Problems of stability in structural mechanics. Stroyizdat, Moscow, pp. 166 – 180
2. Bolotin VV (1966) On estimation of errors in equations of applied theory of elasticity (in Russian). In: Structural mechanics. In honor of 80th anniversary of I.M. Rabinovich. Stroyizdat, Moscow, pp. 297 – 302
3. Dinnik AN (1955) Buckling. Torsion (in Russian). USSR Acad. Sci. Publishing House, Moscow
4. Goldenweiser AL (1953) Theory of thin elastic shells (in Russian). Gostekhteorizdat, Moscow
5. Kiseliov VA (1976) Plane elasticity (in Russian). Vysshaya Shkola, Moscow
6. Kochin NE (1965) Vector calculus and basics of tensor calculus (in Russian). Nauka, Moscow,
7. Korn GA, Korn TM (1968) Mathematical handbook for scientists and engineers. McGraw-Hill Book Company
8. Love AEH (1944) The mathematical theory of elasticity. Dover Publications, New York
9. Novozhilov VV (1958) Theory of elasticity (in Russian). Sudpromgiz, Leningrad
10. Novozhilov VV (1962) Theory of thin shells (in Russian). Sudpromgiz, Leningrad
11. Papkovich PF (1939) Theory of elasticity (in Russian). Oborongiz, Leningrad Moscow
12. Rabotnov YN (1979) Mechanics of deformable solids (in Russian). Physmathgiz, Moscow
13. Sneddon JN, Berry DS (1958) The classic theory of elasticity. Springer-Verlag, Berlin Gőttingen Heidelberg
14. Vlasov VZ (1962) Selected works. Volume 1 (in Russian). USSR Acad. Sci. Publishing House, Moscow

5 PARTICULAR CLASSES OF PROBLEMS IN STRUCTURAL MECHANICS – part 2

Theory is a wonderful thing, but you cannot do as much as feed a dog with it.

Grinzo L (1996) ZEN of Windows 95 programming. Coriolis Group Books

5.1 Thin plate bending — Kirchhoff–Love theory

Let us consider the bending of a thin plate, of a thickness h. We assume that a set of points dividing the thickness of the plate into two equal parts forms a plane. This plane is usually referred to as a *median plane* of a plate.

We introduce a right-hand Cartesian coordinate system, (X,Y,Z), in such manner that the plane of the (X,Y)-axes coincides with the plate's median plane.

In the theory of plates we usually call a *normal* a piece of a straight line, all points of which have the same x and y coordinates while the z coordinate varies from $-h/2$ to $h/2$. For the sake of certainty, we will call the face surface of the plate with the coordinate $z = -h/2$ a *top (upper) face* and the surface with the coordinate $z = h/2$ a *bottom (lower) face* of the plate. Thus we can say that the median surface is equidistant from the top and bottom faces of the plate.

A classic technical theory of bending of thin plates, traditionally associated with the names of Kirchhoff and Love, is based on the following three assumptions (hypotheses):

- a *straight-normals assumption*, according to which any normal to the median surface (plane) of an undeformed plate will remain rectilinear, will keep its length, and will remain orthogonal to the *median surface* that the median plane becomes after the plate is deformed;
- a *no-pressure assumption*, according to which all planes of a plate which are parallel to its median plane do not interact in their normal direction; in other words, we suppose $\sigma^{zz} = 0$;

- a *no-shear assumption*, according to which the shear strains, γ_{xz} and γ_{yz}, are so small that the work of the tangential stresses, τ^{xz} and τ^{yz}, in the expression of energy E can be neglected in comparison to the work of stresses σ^{xx}, σ^{yy}, and τ^{xy}.

The straight-normals assumption is purely kinematic. Its mechanical statement is that a normal to a plate behaves during the bending as if it were a perfectly solid body.

Thus, the mathematical equivalents of the above assumptions are the requirements that

$$\varepsilon_{zz} = 0, \quad \gamma_{xz} = 0, \quad \gamma_{yz} = 0, \quad \sigma^{zz} = 0.$$

As usual, we denote by u, v, w the components of displacements of any arbitrary point of a plate with respect to the X, Y, Z axes, accordingly. As we have $\varepsilon_{zz} = \partial w/\partial z$ from general three-dimensional geometric relationships, we now know that the w function of the lateral displacements of the plate's points is independent of the z coordinate. In other words,

$$w = w(x,y). \tag{1.1}$$

In the plate's deformed state, a normal to its median surface will acquire a translational displacement toward Z equal to w and slopes θ_x and θ_y with respect to axes X and Y, correspondingly. The positive directions of slopes θ_x and θ_y are defined by the right-hand screw rule. Hence, simple geometric considerations give the following representation of the displacements of an arbitrary point of the plate with the coordinates (x,y,z):

$$u = u_0 + z\theta_y, \qquad v = v_0 - z\theta_x \tag{1.2}$$

where $u_0 = u(x,y,0)$ and $v_0 = v(x,y,0)$ are displacements along X and Y, respectively, of points of the plate belonging to the median plane.

Now we are able to construct expressions for the other components of the strain tensor. As $\varepsilon_{xx} = \partial u/\partial x$, $\varepsilon_{yy} = \partial v/\partial y$, and $\gamma_{xy} = \partial u/\partial y + \partial v/\partial x$, we use this and (1.2) to produce

$$\varepsilon_{xx} = \varepsilon_{0xx} + z\theta_{y,x}, \qquad \varepsilon_{yy} = \varepsilon_{0yy} - z\theta_{x,y}, \quad \gamma_{xy} = \gamma_{0xy} + z(\theta_{y,y} - \theta_{x,x}), \tag{1.3}$$

where

$$\varepsilon_{0xx} = \partial u_0/\partial x, \qquad \varepsilon_{0yy} = \partial v_0/\partial y, \qquad \gamma_{0xy} = \partial u_0/\partial y + \partial v_0/\partial x .$$

Simple relationships can be established between the slopes of the normal, θ_x and θ_y, on one hand, and the function of the plate's lateral displacements, $w(x,y)$, on the other hand, by using the condition of no shear γ_{xz} and γ_{yz}. And indeed, we have

$$\gamma_{xz} = \frac{\partial u}{\partial z} + \frac{\partial w}{\partial x} = \theta_y + w_{,x} = 0\,, \quad \gamma_{yz} = \frac{\partial v}{\partial z} + \frac{\partial w}{\partial y} = -\theta_x + w_{,y} = 0\,.$$

Consequently,

$$\theta_x = w_{,y}\,, \quad -\theta_y = w_{,x}\,. \tag{1.4}$$

By the way, (1.4) immediately gives

$$\theta_{x,x} + \theta_{y,y} = 0\,. \tag{1.5}$$

The latter means we can introduce a normal's slope vector, $\boldsymbol{\theta} = \theta_x \boldsymbol{i}_x + \theta_y \boldsymbol{i}_y$, where $\boldsymbol{i}_x$ and $\boldsymbol{i}_y$ are unit vectors of the X and Y axes, and rewrite relationship (1.5) in a form invariant with respect to a coordinate system,

$$\operatorname{div} \boldsymbol{\theta} = 0\,. \tag{1.6}$$

It will be convenient for further presentment if we introduce components of a tensor of curvature of the plate's median surface in its bent state by assuming

$$\chi_{xx} = -\frac{\partial^2 w}{\partial x^2}\,, \qquad \chi_{yy} = -\frac{\partial^2 w}{\partial y^2}\,, \qquad \chi_{xy} = -\frac{\partial^2 w}{\partial x \partial y}\,, \tag{1.7}$$

where χ_{xx} and χ_{yy} are the curvatures of the surface in the respective planes (X,Z) and (Y,Z), and χ_{xy} is its twist. The 'minus' sign in expressions (1.7) shows that the curvature of the surface is assumed to be positive if the surface's convexity looks toward increasing Z's. Taking relations (1.4) into account, we can rewrite the expressions of the curvature tensor's components as

$$\chi_{xx} = \theta_{y,x}\,, \qquad \chi_{yy} = -\theta_{x,y}\,, \qquad \chi_{xy} = -\theta_{x,x} = \theta_{y,y}\,. \tag{1.8}$$

Now we use relationships (1.3) to express strains ε_{xx}, ε_{yy} and γ_{xy} via the components of the curvature tensor. We have

$$\varepsilon_{xx} = \varepsilon_{0xx} + z\chi_{xx}\,, \quad \varepsilon_{yy} = \varepsilon_{0yy} + z\chi_{yy}\,, \quad \gamma_{xy} = \gamma_{0xy} + 2z\chi_{xy}\,. \tag{1.9}$$

So, we conclude from (1.9) that, under the assumptions of the thin plate bending theory, strains ε_{xx}, ε_{yy} and γ_{xy} together with stresses σ^{xx}, σ^{yy} and τ^{xy} vary over the thickness of the plate linearly.

Let a plate in the (X,Y)-plane occupies an area Ω with a boundary Γ. Then, according to the assumptions we have taken, we can write an expression of energy E accumulated in the deformed plate in the form of an integral,

$$\mathsf{E} = \int_{\Omega}\int_{-h/2}^{h/2} \frac{\sigma^{xx}\varepsilon_{xx} + \sigma^{yy}\varepsilon_{yy} + \tau^{xy}\gamma_{xy}}{2}\, dz d\Omega . \tag{1.10}$$

Here we deem it convenient to switch to a tensor notation where the designations of global coordinates such as X, Y are replaced by indexed coordinates such as X_1, X_2. It means that the indexes of the tensors take values from 1 to 2. Then (1.10) can be rewritten as

$$\mathsf{E} = \frac{1}{2}\int_{\Omega}\int_{-h/2}^{h/2} \sigma^{ij}\varepsilon_{ij}\, dz d\Omega . \tag{1.11}$$

Here and further we assume $\sigma^{11} = \sigma^{xx}$, $\sigma^{22} = \sigma^{yy}$, $\sigma^{12} = \sigma^{21} = \tau^{xy}$, $\varepsilon_{11} = \varepsilon_{xx}$, $\varepsilon_{22} = \varepsilon_{yy}$, $\varepsilon_{12} = \varepsilon_{21} = ½\, \gamma_{xy}$.

Now let's turn to physical relationships of the plate theory. According to the no-pressure assumption, $\sigma^{zz} = 0$, so in this case Hooke's law for an isotropic material will become

$$\varepsilon_{xx} = \frac{1}{E}(\sigma^{xx} - \nu\sigma^{yy}), \qquad \varepsilon_{yy} = \frac{1}{E}(\sigma^{yy} - \nu\sigma^{xx}), \qquad \gamma_{xy} = \frac{1}{G}\tau^{xy}, \tag{1.12}$$

or, inverted,

$$\sigma^{xx} = \frac{E}{1-\nu^2}(\varepsilon_{xx} + \nu\varepsilon_{yy}), \quad \sigma^{yy} = \frac{E}{1-\nu^2}(\varepsilon_{yy} + \nu\varepsilon_{xx}), \quad \tau^{xy} = G\gamma_{xy} . \tag{1.13}$$

In the general case of an anisotropic material, we will have this tensor form instead of (1.13)[1]

$$\sigma^{ij} = C^{ijkl}\varepsilon_{kl} , \qquad \varepsilon_{ij} = D_{ijkl}\sigma^{kl}, \tag{1.14}$$

where tensors $\{C^{ijkl}\}$ and $\{D_{ijkl}\}$ are mutually inverse.

For an isotropic material, relations (1.12) and (1.13) can be represented in the form of (1.14) if we assume

[1] Obviously, the stiffness coefficients of the material, C^{ijkl}, we use here are different from those used in the general three-dimensional elasticity. Here we do not give relationships between these two sets of coefficients; an interested reader can turn to literature on the subject where the bending of anisotropic plates is considered. See [2], for example.

$$C^{ijkl} = \frac{E}{1-\nu^2}\,[\nu\delta^{ij}\delta^{kl} + \frac{1-\nu}{2}\,(\delta^{ik}\delta^{jl} + \delta^{jk}\delta^{il})]\,,$$

$$D_{ijkl} = \frac{1}{E}\,[-\nu\delta_{ij}\delta_{kl} + (1+\nu)\,\delta_{ik}\delta_{jl}\,] \tag{1.15}$$

where δ^{ij} and δ_{ij} are Kronecher's deltas.

The formulas of the strains from (1.9) can be rewritten in the tensor form, too, so we have

$$\varepsilon_{ij} = \varepsilon_{0ij} + z\chi_{ij}\,. \tag{1.16}$$

Returning to the energy expression in (1.11) and taking (1.14) and (1.16) into account, we will derive this by integrating over z

$$\mathsf{E} = \frac{1}{2}\int_{\Omega}\int_{-h/2}^{h/2} C^{ijkl}\varepsilon_{ij}\varepsilon_{kl}\,dz d\Omega = \frac{1}{2}\int_{\Omega} hC^{ijkl}\varepsilon_{0ij}\varepsilon_{0kl}\,d\Omega \; +$$

$$+\frac{1}{2}\int_{\Omega}\frac{h^3}{12}C^{ijkl}\chi_{ij}\chi_{kl}\,d\Omega\,. \tag{1.17}$$

As strains ε_{0ij} at the level of the plate's median plane depend only on tangential displacements u_0 and v_0, and curvatures χ_{ij} depend only on lateral displacements w, hence energy E splits into a sum of two independent functionals, each depending on its own functional argument:

$$\mathsf{E} = \mathsf{E}_{\mathrm{P}}(u_0,v_0) + \mathsf{E}_{\mathrm{B}}(w) \tag{1.18}$$

where E_{B} is a bending energy, and E_{P} is a plane stress energy. Recalling the notion of a Lagrangian energy space, we can say that the states of a plate caused by plane stress and by bending are orthogonal in the energy metric.

As we are interested only with the bending in this section (the plane stress was considered earlier in Section **4**.3), further we will mean by energy E only the energy of bending (so the index $_{\mathrm{B}}$ will not be necessary). In other words, we will assume

$$\mathsf{E} = \frac{1}{2}\int_{\Omega}\frac{h^3}{12}C^{ijkl}\chi_{ij}\chi_{kl}\,d\Omega = \frac{1}{2}\int_{\Omega}\frac{h^3}{12}C^{ijkl}w_{,ij}w_{,kl}\,d\Omega\,.$$

Introducing these designations for the anisotropy tensor's coefficients, where the thickness of the plate is allowed for,

$$c^{ijkl} = \frac{h^3}{12} C^{ijkl}, \qquad d_{ijkl} = \frac{12}{h^3} D_{ijkl}, \tag{1.19}$$

we can rewrite the expression of the plate bending energy, E, as

$$\mathsf{E} = \frac{1}{2} \int_\Omega c^{ijkl} \chi_{ij} \chi_{kl} d\Omega = \frac{1}{2} \int_\Omega c^{ijkl} w_{,ij} w_{,kl} d\Omega . \tag{1.20}$$

This will give the following for an isotropic material of the plate:

$$\mathsf{E} = \frac{1}{2} \int_\Omega D \left[w_{,xx}^2 + 2\nu w_{,xx} w_{,yy} + w_{,yy}^2 + 2(1-\nu) w_{,xy}^2 \right] d\Omega , \tag{1.21}$$

where D is a so-called *cylindrical (flexural) rigidity of the plate,*

$$D = \frac{Eh^3}{12(1-\nu^2)} .$$

So, as long as we are interested with stresses in the plate, caused solely by its bending, we have the right to remove terms related to the tangential displacements from the expressions of the strain components in (1.3) and (1.9). That is, instead of (1.3) and (1.9) we assume

$$\varepsilon_{xx} = z\theta_{y,x} , \qquad \varepsilon_{yy} = -z\theta_{x,y} , \quad \gamma_{xy} = z(\theta_{y,y} - \theta_{x,x}) , \tag{1.22}$$

$$\varepsilon_{xx} = z\chi_{xx} , \qquad \varepsilon_{yy} = z\chi_{yy} , \qquad \gamma_{xy} = 2z\chi_{xy} . \tag{1.23}$$

The stresses in the sections of the plate will become as follows after applying Hooke's law from (1.14) or (1.13):

- for a general anisotropic material

$$\sigma^{ij} = zC^{ijkl} \chi_{kl} = -zC^{ijkl} w_{,kl} ; \tag{1.24}$$

- for an isotropic material

$$\sigma^{xx} = \frac{E}{1-\nu^2} (\chi_{xx} + \nu\chi_{yy}) z, \qquad \sigma^{yy} = \frac{E}{1-\nu^2} (\chi_{yy} + \nu\chi_{xx}) z,$$

$$\tau^{xy} = \frac{E}{1+\nu} \chi_{xy} z. \tag{1.25}$$

The bending moments per unit of length in the plate's sections will be defined as the respective integral characteristics of the stresses, by assuming

$$M^{xx} = \int_{-h/2}^{h/2} \sigma^{xx} z dz\,, \quad M^{yy} = \int_{-h/2}^{h/2} \sigma^{yy} z dz\,, \quad M^{xy} = M^{yx} = \int_{-h/2}^{h/2} \tau^{xy} z dz\,, \quad (1.26)$$

which, when integrated, will give

- for a general anisotropic material,

$$M^{ij} = c^{ijkl}\chi_{kl} = -c^{ijkl} w_{,kl} \quad (1.27)$$

- for an isotropic material,

$$M^{xx} = D(\chi_{xx} + \nu\chi_{yy}) = -D(w_{,xx} + \nu w_{,yy}),$$
$$M^{yy} = D(\chi_{yy} + \nu\chi_{xx}) = -D(w_{,yy} + \nu w_{,xx}),$$
$$M^{xy} = D(1-\nu)\,\chi_{xy} = -D(1-\nu)\,w_{,xy}\,. \quad (1.28)$$

We will think that external actions upon the plate which are distributed over area Ω and cause its bending consist of the following components:

- a load distributed over the area of the plate's median surface, of an intensity q, acting in the direction of the Z-axis;
- moment loads distributed over the area of the plate's median surface, of intensities m^x and m^y, with their positive directions defined by the right-hand screw convention.

Those distributed external moments are created by loads p^x and p^y applied to the top and bottom faces of the plate, as shown in Fig. 5.1 where the loads are presented as acting in their positive direction. The formal sign convention for those surface loads is as follows: we assume p^x to be positive if that load acts in the positive direction of the X-axis on a face with the positive z-coordinate (on the bottom face). The same convention of signs is applicable to the surface load, p^y, too.

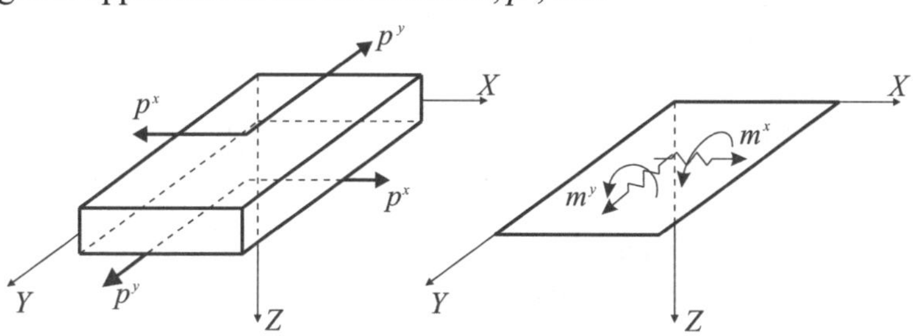

Fig. 5.1. Loads upon the external surfaces of a plate = moment loads at the level of the plate's median surface

Obviously, with this method of applying the moment load and with the sign convention defined above, the following formulas hold:

$$m^x = -hp^y, \qquad m^y = hp^x. \tag{1.29}$$

The last group of equations which we have not used yet is a group of equilibrium equations. Let us first write the simultaneous equations of equilibrium for three-dimensional elasticity. As we know specific external actions applied to the plate, we can write

$$\begin{aligned}
&\frac{\partial\sigma^{xx}}{\partial x}+\frac{\partial\tau^{xy}}{\partial y}+\frac{\partial\tau^{xz}}{\partial z}=0,\\
&\frac{\partial\tau^{yx}}{\partial x}+\frac{\partial\sigma^{yy}}{\partial y}+\frac{\partial\tau^{yz}}{\partial z}=0,\\
&\frac{\partial\tau^{zx}}{\partial x}+\frac{\partial\tau^{zy}}{\partial y}+\frac{q}{h}=0.
\end{aligned} \tag{1.30}$$

We should note that the third of the equilibrium equations does not have the $\partial\sigma^{zz}/\partial z$ term, which agrees with the no-pressure assumption. Also, the intensity of the volumetric load acting in the direction of the Z-axis is assumed to be q/h, which corresponds to a uniform distribution of that load over the plate's thickness.

We multiply the first two of those equations by z and integrate them over the thickness of the plate.. Taking into account (1.26), we derive

$$M^{xx}_{,x}+M^{xy}_{,y}+\int_{-h/2}^{h/2}\frac{\partial\tau^{xz}}{\partial z}z\,dz=0, \qquad M^{yx}_{,x}+M^{yy}_{,y}+\int_{-h/2}^{h/2}\frac{\partial\tau^{yz}}{\partial z}z\,dz=0.$$

We know that

$$\int_{-h/2}^{h/2}\frac{\partial\tau^{xz}}{\partial z}z\,dz=\tau^{xz}z\Big|_{-h/2}^{h/2}-\int_{-h/2}^{h/2}\tau^{xz}dz=p^xh-Q^x=m^y-Q^x,$$

$$\int_{-h/2}^{h/2}\frac{\partial\tau^{yz}}{\partial z}z\,dz=\tau^{yz}z\Big|_{-h/2}^{h/2}-\int_{-h/2}^{h/2}\tau^{yz}dz=p^yh-Q^y=-m^x-Q^y.$$

Here we denote by Q^x and Q^y the shear forces per unit of length, which act in the respective sections of the plate,

$$Q^x=\int_{-h/2}^{h/2}\tau^{xz}dz, \qquad Q^y=\int_{-h/2}^{h/2}\tau^{yz}dz, \tag{1.31}$$

and are integral (with respect to the plate's thickness) characteristics of tangential stresses τ^{xz} and τ^{yz}. Of course, involving these nonzero tangential stresses is in contradiction with the no-shear assumption. The contradiction reveals itself by violating the physical law $\tau^{xz} = G\gamma_{xz}$. This violation of the physical law for the tangential stresses is a tradeoff of the kinematic assumptions of the Kirchhoff–Love plate theory — first of all, of the straight-normals assumption.

As a result, the first two integrated equations of equilibrium are

$$M^{xx}_{,x} + M^{xy}_{,y} - Q^{x} = -m^{y}, \qquad M^{yy}_{,y} + M^{yx}_{,x} - Q^{y} = m^{x}. \tag{1.32}$$

As for the third equation of system (1.30), it will yield the following after being integrated over the thickness of the plate (without previously multiplying it by coordinate z):

$$Q^{x}_{,x} + Q^{y}_{,y} + q = 0. \tag{1.33}$$

After differentiating the first of equations (1.32) with respect to x and the second with respect to y, we add together what we have obtained and exclude the shear forces using equation (1.33). The result is a single equation of equilibrium that does not contain any shear forces,

$$-M^{xx}_{,xx} - 2M^{xy}_{,xy} - M^{yy}_{,yy} = \bar{q}, \tag{1.34}$$

where we denote

$$\bar{q} = q - m^{x}_{,y} + m^{y}_{,x}. \tag{1.35}$$

The function, $\bar{q}$, defined by formula (1.35) can be reasonably called a *generalized transverse load*.

It follows from (1.27) for a general anisotropic material or from (1.28) for an isotropic material that moments M^{ij} are linear aggregates built of second derivatives of deflection function w, therefore substituting these expressions to (1.34) will give a fourth-order differential equation that governs the thin plate bending. This equation contains a sole unknown function — a deflection function, w, and its physical sense is an equation of equilibrium in displacements, that is, a Lame-type equation which is alike to (**4**.1.7) in its operator form:

$$\boldsymbol{L}\mathbf{u} = \bar{\boldsymbol{X}}.$$

The vector of desirable displacements, **u**, and the vector of loads, $\bar{\boldsymbol{X}}$, are scalars or, if it is more to your liking, first-order vectors:

$$\mathbf{u} = \|[w]\| , \qquad \bar{X} = \|[\bar{q}]\| . \tag{1.36}$$

Having made the substitutes indicated above, we find that the governing equation of the bending analysis in the general case of an anisotropic material will be as follows (the tensor coordinate notation is used):

$$\left(c^{ijkl} w_{,kl}\right)_{,ij} = \bar{q} . \tag{1.37}$$

This equation becomes especially simple for a plate made of a homogeneous isotropic material and of a constant thickness. The flexural rigidity of the plate will be $D = Const.$ Substituting (1.28) to (1.34), we find that the Lame operator is a biharmonic operator multiplied by cylindrical rigidity D

$$\boldsymbol{L} = D\,\nabla^2\nabla^2 = D\left(\frac{\partial^4}{\partial x^4} + 2\frac{\partial^4}{\partial x^2 \partial y^2} + \frac{\partial^4}{\partial y^4}\right). \tag{1.38}$$

In the theory of elasticity, the equation of plate bending in the form

$$\nabla^2\nabla^2 w = \bar{q}\,/D \tag{1.39}$$

is sometimes referred to as a *Germain-Lagrange equation*[2].

5.1.1 Local basis in points of boundary Γ of area Ω

Let $(\boldsymbol{i}_x, \boldsymbol{i}_y, \boldsymbol{i}_z)$ be unit vectors of a right-hand Cartesian coordinate system, (X,Y,Z), introduced earlier. In addition, let $\boldsymbol{n}$ be a unit vector of an external normal to boundary Γ of area Ω occupied by the plate in bending, and let $\boldsymbol{t}$ be a vector tangential to the same boundary Γ and of the unit length. These vectors have the following matrix representation:

$$\boldsymbol{n} = \|[n_x , n_y]\|^{\mathsf{T}}, \qquad \boldsymbol{t} = \|[t_x , t_y]\|^{\mathsf{T}}$$

where n_x , n_y and t_x , t_y are components of those vectors with respect to axes X and Y.

For definiteness, we will assume the triple of unit vectors, $(\boldsymbol{n}, \boldsymbol{t}, \boldsymbol{i}_z)$, to be right-hand, that is,

---

[2] As S.P. Timoshenko notes ([22]), the priority in deriving a biharmonic equation for a thin plate bending problem belongs to Sophie Germain who, however, made a mistake in her paper submitted for a competition to the French Academy; the mistake was noticed and corrected by Joseph Lagrange who was in the jury.

$$\boldsymbol{i}_z = \boldsymbol{n} \times \boldsymbol{t} = \boldsymbol{i}_x \times \boldsymbol{i}_y . \tag{1.40}$$

Further we will treat n as a local coordinate along a normal to the plate's contour, zero on the contour itself and increasing toward the external normal, $\boldsymbol{n}$. The designation of s will be used for a so-called *arc coordinate* which means the length of the contour line, Γ, counted off from an arbitrarily chose point of the contour, $s = 0$, and increasing toward the direction indicated by vector $\boldsymbol{t}$.

We will say that unit vectors ($\boldsymbol{n}$, $\boldsymbol{t}$, $\boldsymbol{i}_z$) make up a *local basis* in each point of boundary Γ. It is easy to understand that the local basis is unambiguous everywhere in points of Γ where the boundary curve is smooth cnough (differentiable). However, in a breakpoint of the boundary there can be two local bases because the $\boldsymbol{n}$ and $\boldsymbol{t}$ unit vectors jump to alter their directions when they pass the breakpoint – these unit vectors have a discontinuity as vector functions of arc coordinate s.

Turning to Appendix F (Section F.4), we note that the set of parameters (n,s) introduced above can be treated as a system of orthogonal curvilinear coordinates on the (X,Y)-plane, defined in a certain vicinity of boundary Γ and associated with the closed curve of Γ. Consequently, we can use all formulas given in that Appendix. In particular, formula (F.36) defines a rotation matrix, $\boldsymbol{\omega}$, as

$$\boldsymbol{\omega} = \left\| \begin{bmatrix} 0 & -1 \\ 1 & 0 \end{bmatrix} \right\| ,$$

which causes a vector multiplied by it to rotate by the angle of $\pi/2$. The relation between the components of slope vector $\boldsymbol{\theta}$ and the derivatives of the plate's deflection function established by formulas (1.4) earlier can be also written in the following form using the rotation matrix $\boldsymbol{\omega}$:

$$\boldsymbol{\theta} = -\boldsymbol{\omega}\,\mathrm{grad}w. \tag{1.41}$$

Thus, the vector of the normal's slope, $\boldsymbol{\theta}$, is the gradient of function w rotated by angle $-\pi/2$.

We will need auxiliary formulas for further presentment, which help replace the differentiation of the functions with respect to global coordinates (X,Y) by the differentiation with respect to local coordinates (n,s) defined in every point of boundary Γ of area Ω. To do it, here we present formulas borrowed from Section F.4. We have

$$\begin{bmatrix} \partial/\partial n \\ \partial/\partial s \end{bmatrix} = \boldsymbol{\omega}_{\varphi}^{\mathsf{T}} \begin{bmatrix} \partial/\partial x \\ \partial/\partial y \end{bmatrix}, \qquad \begin{bmatrix} \partial/\partial x \\ \partial/\partial y \end{bmatrix} = \boldsymbol{\omega}_{\varphi} \begin{bmatrix} \partial/\partial n \\ \partial/\partial s \end{bmatrix}, \tag{1.42}$$

where the rotation matrix, $\boldsymbol{\omega}_{\varphi}$, is

$$\boldsymbol{\omega}_{\varphi} = \begin{bmatrix} n_x & -n_y \\ n_y & n_x \end{bmatrix} = \begin{bmatrix} n_x & t_x \\ n_y & t_y \end{bmatrix}, \qquad \boldsymbol{\omega}_{\varphi}^{-1} = \boldsymbol{\omega}_{\varphi}^{\mathsf{T}}. \tag{1.43}$$

Formulas (1.42) establish a relationship between first derivatives of an arbitrary function — for example, w — in the global and local coordinate systems.

Appendix F also presents a relation between second derivatives in these coordinate systems. The formulas are

$$w_{,kl} = n_k n_l w_{,nn} + (n_k t_l + n_l t_k) w_{,ns} + t_k t_l w_{,ss} + k t_k t_l w_{,n} - k(n_k t_l + n_l t_k) w_{,s} \,. \tag{1.44}$$

It should be reminded also that formulas (1.42) and (1.44) hold only for points of the plate which belong to curve Γ. The formulas are more complicated for the rest of points of area Ω, but we will not need those, so they are not presented.

Also, we will need two more matrices composed of the components of vectors $\boldsymbol{n}$ and $\boldsymbol{t}$

$$\boldsymbol{A}_n = \begin{bmatrix} n_x & 0 & n_y \\ 0 & n_y & n_x \end{bmatrix} = \begin{bmatrix} t_y & 0 & -t_x \\ 0 & -t_x & t_y \end{bmatrix}, \tag{1.45}$$

$$\boldsymbol{A}_t = \begin{bmatrix} t_x & 0 & t_y \\ 0 & t_y & t_x \end{bmatrix} = \begin{bmatrix} -n_y & 0 & n_x \\ 0 & n_x & -n_y \end{bmatrix}. \tag{1.46}$$

As we move along contour Γ, matrices $\boldsymbol{A}_n$ and $\boldsymbol{A}_t$ vary. Formulas (F.52) give

$$\frac{\partial \boldsymbol{A}_n}{\partial s} = k\boldsymbol{A}_t, \qquad \frac{\partial \boldsymbol{A}_t}{\partial s} = -k\boldsymbol{A}_n. \tag{1.47}$$

5.1.2 Matrix representation of basic relationships in the Kirchhoff–Love plate theory

Now we are going to convert all relationships of the thin plate theory that we have obtained into a general matrix form. For the plate bending analysis, the matrix representation of "stresses" $\boldsymbol{\sigma}$ and "strains" $\boldsymbol{\varepsilon}$ can be conveniently written as

$$\boldsymbol{\sigma} = \|[M^{xx}, M^{yy}, M^{xy}]\|^{\mathsf{T}}, \qquad \boldsymbol{\varepsilon} = \|[\chi_{xx}, \chi_{yy}, 2\chi_{xy}]\|^{\mathsf{T}}, \tag{1.48}$$

while displacement $\mathbf{u}$ and load $\bar{X}$ are determined by formulas (1.36).

If we assume

$$\boldsymbol{A} = \left\| \begin{bmatrix} -\dfrac{\partial^2}{\partial x^2} \\ -\dfrac{\partial^2}{\partial y^2} \\ -2\dfrac{\partial^2}{\partial x \partial y} \end{bmatrix} \right\|, \qquad \boldsymbol{A}^{\mathsf{T}} = \left\| \left[-\frac{\partial^2}{\partial x^2}, -\frac{\partial^2}{\partial y^2}, -2\frac{\partial^2}{\partial x \partial y} \right] \right\|, \tag{1.49}$$

then both geometric equations (1.7) and the equilbrium equations with respect to the moments, (1.34), will be satisfied. The same equations will acquire a general matrix form,

$$\boldsymbol{\varepsilon} = \boldsymbol{A}\mathbf{u}, \qquad \boldsymbol{A}^{\mathsf{T}}\boldsymbol{\sigma} = \bar{\boldsymbol{X}}.$$

Let us give also a matrix form of the physical relationship operators according to (1.27) and (1.28). For an isotropic material, we have

$$\boldsymbol{C} = \frac{Eh^3}{12(1-\nu^2)} \left\| \begin{bmatrix} 1 & \nu & 0 \\ \nu & 1 & 0 \\ 0 & 0 & \dfrac{1-\nu}{2} \end{bmatrix} \right\|, \qquad \boldsymbol{C}^{-1} = \frac{12}{Eh^3} \left\| \begin{bmatrix} 1 & -\nu & 0 \\ -\nu & 1 & 0 \\ 0 & 0 & 2(1+\nu) \end{bmatrix} \right\|. \tag{1.50}$$

For a general anisotropic material, matrices $\boldsymbol{C}$ and $\boldsymbol{C}^{-1}$ will be completely filled symmetric matrices of third order. A relation between the matrix and tensor representations of algebraic equations $\boldsymbol{C}$ and $\boldsymbol{C}^{-1}$ is given by formulas

$$\boldsymbol{C} = \left\| \begin{bmatrix} c^{1111} & c^{1122} & c^{1112} \\ c^{2211} & c^{2222} & c^{2221} \\ c^{1211} & c^{1222} & c^{1212} \end{bmatrix} \right\|, \quad \boldsymbol{C}^{-1} = \left\| \begin{bmatrix} d_{1111} & d_{1122} & d_{1112} \\ d_{2211} & d_{2222} & d_{2221} \\ d_{1211} & d_{1222} & d_{1212} \end{bmatrix} \right\|. \tag{1.51}$$

If we follow a general rule for building the Lame operator as in (**4**.1.6) and make formal matrix transformations, we will arrive, as expected, at the same equation (1.37) or (1.39).

The bending moment, M^{nn}, and the torque, M^{nt}, which act on the contour of the plate (Fig. 5.2), are defined as

$$M^{nn} = M^{xx} n_x n_x + M^{yy} n_y n_y + 2M^{xy} n_x n_y = M^{ij} n_i n_j = \boldsymbol{n}^{\mathsf{T}} \boldsymbol{A}_n \boldsymbol{\sigma},$$
$$M^{nt} = M^{xx} n_x t_x + M^{yy} n_y t_y + M^{xy} (n_x t_y + n_y t_x) = M^{ij} n_i t_j = \boldsymbol{t}^{\mathsf{T}} \boldsymbol{A}_n \boldsymbol{\sigma}. \tag{1.52}$$

These formulas can be easily validated by using the equilbrium equations with respect to the moments for an element of the plate cut out from the vicinity of the plate's contour as shown in Fig. 5.2. One should keep in mind, when deriving relationships (1.52), that the moments shown in Fig. 5.2 are not full moments acting on the respective faces of the extracted element but moments per unit of length. Note also a circumstance that the positive torque M^{nt} acting on the contour of the plate is opposite to unit vector $\boldsymbol{n}$, as can be seen in Fig. 5.2. At the same time, the positive bending moment, M^{nn}, has the same direction with unit vector $\boldsymbol{t}$ [3].

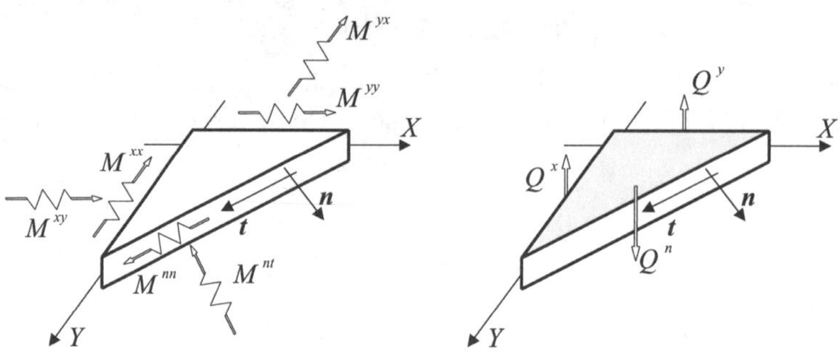

Fig. 5.2. Internal forces in sections of the plate near its contour

[3] The convention of signs for these moments is based on the following considerations. A positive bending moment, M^{nn}, is created by positive (tensioning) normal stresses σ^{nn} which act at those points of an oblique section in Fig. 5.2 which have a positive Z coordinate. A positive torque, M^{nt}, is created by positive tangential stresses τ^{nt} which act along the oblique plane in Fig. 5.2. In its turn, a positive tangential stress, τ^{nt}, which acts at a point with a positive Z coordinate has the same direction with the $\boldsymbol{t}$-axis.

We can also define the bending moment, M^{tt}, as a moment which acts on the contour in a cross-section the normal to which coincides with the $\boldsymbol{t}$-axis, so that

$$M^{tt} = M^{xx}\, t_x t_x + M^{yy}\, t_y t_y + 2M^{xy}\, t_x t_y = M^{ij}\, t_i t_j = \boldsymbol{t}^{\mathrm{T}} \boldsymbol{A}_t \boldsymbol{\sigma}\,. \tag{1.53}$$

Formulas (1.52), (1.53) define the components of the moment tensor in the ($\boldsymbol{n}$, $\boldsymbol{t}$)-axes via the components of the same tensor in the original global coordinate system. The latter are expressed via the derivatives of the deflection function by formulas (1.27) in the general case or by (1.28) for an isotropic material of the plate. However, for the analysis of plates with a curvilinear contour, it is useful to have expressions for quantities M^{nn}, M^{tt}, and M^{nt} also via the derivatives of the deflection along the normal to the contour and along the contour itself. Based on (1.52) and taking (1.27) and (1.44) into account, we have

$$M^{nn} = M^{ij}\, n_i n_j = -c^{ijkl} w_{,kl}\, n_i n_j =$$

$$= -c^{ijkl} n_i n_j [n_k n_l w_{,nn} + (n_k t_l + n_l t_k) w_{,ns} + t_k t_l w_{,ss} + k t_k t_l w_{,n} - k(n_k t_l + n_l t_k) w_{,s}]\,.$$

In particular, coefficients c^{ijkl} are defined by expressions (1.15) and (1.19) for an isotropic material, which gives the following after the substitutions:

$$M^{nn} = -D[w_{,nn} + \nu(w_{,ss} + k w_{,n})]. \tag{1.54}$$

These transformations use the index lifting/lowering rules known in tensor analysis and based on Kronecher's deltas, for example,

$$\delta^{ij} n_j = n^i\,,$$

as well as inequalities which can be easily validated:

$$n^k n_k = \boldsymbol{n}\cdot\boldsymbol{n} = 1, \quad t^k t_k = \boldsymbol{t}\cdot\boldsymbol{t} = 1, \quad n^k t_k = \boldsymbol{n}\cdot\boldsymbol{t} = 0.$$

Similarly, further we have

$$M^{tt} = -D[w_{,ss} + \nu w_{,nn} + k w_{,n})]\,. \tag{1.55}$$

$$M^{nt} = -D(1 - \nu)[w_{,ns} - k w_{,s})]\,. \tag{1.56}$$

Let us determine shear forces Q^x and Q^y which act in the plate's cross-sections as shown in Fig. 5.2 and Fig. 5.3, and shear force Q^n which acts on contour Γ.

From formulas (1.32) we derive

$$Q^x = M^{xx}_{,x} + M^{xy}_{,y} + m^y\,, \qquad Q^y = M^{yy}_{,y} + M^{yx}_{,x} - m^x \tag{1.57}$$

and

$$\boldsymbol{Q} = \|[Q^x, Q^y]\|^{\mathrm{T}}, \qquad Q^n = Q^x n_x + Q^y n_y = \boldsymbol{n}^{\mathrm{T}}\boldsymbol{Q}. \tag{1.58}$$

The formula for Q^n follows from an equation of equilibrium with respect to projections onto the Z-axis for an element cut off from the plate in the vicinity of its contour (Fig. 5.2).

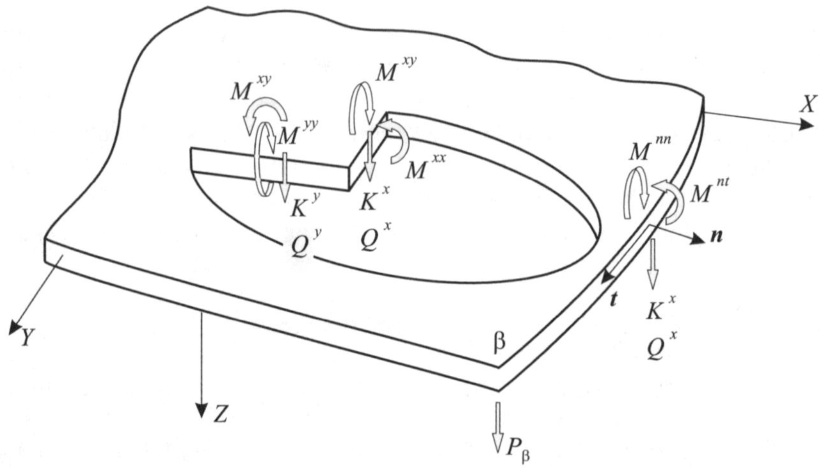

Fig. 5.3. Internal forces in the plate's cross-sections and forces on its contour

Relationships (1.57) can be represented also in the tensor notation as

$$Q^i = M^{ij}_{,j} - \omega^{ij} m_j . \tag{1.59}$$

Now let us consider a matrix notation for the shear forces. In order to obtain it, we introduce an auxiliary differentiation matrix, $\boldsymbol{A}_d$, and a vector of external moments, $\boldsymbol{m}$,

$$\boldsymbol{A}_d = \left\| \begin{bmatrix} \dfrac{\partial}{\partial x} & 0 & \dfrac{\partial}{\partial y} \\ 0 & \dfrac{\partial}{\partial y} & \dfrac{\partial}{\partial x} \end{bmatrix} \right\|, \qquad \boldsymbol{m} = \left\| \begin{bmatrix} m^x \\ m^y \end{bmatrix} \right\|. \tag{1.60}$$

Note that matrix $\boldsymbol{A}_d$ has the same structure as matrix $\boldsymbol{A}_n$. These matrices can be converted to each other by replacing the respective direction cosine of vector $\boldsymbol{n}$ with the differentiation with respect to the same coordinate; this can be symbolized as

$$\boldsymbol{A}_n \Leftrightarrow \boldsymbol{A}_d : \quad n_x \Leftrightarrow \partial/\partial x , \qquad n_y \Leftrightarrow \partial/\partial y .$$

As a result, (1.57) will acquire the following matrix form:

$$\boldsymbol{Q} = \boldsymbol{A}_d \boldsymbol{\sigma} - \boldsymbol{\omega} \boldsymbol{m},$$

so, consequently, $Q^n = \boldsymbol{n}^\mathrm{T}\boldsymbol{Q} = \boldsymbol{n}^\mathrm{T}\boldsymbol{A}_d\boldsymbol{\sigma} - \boldsymbol{n}^\mathrm{T}\boldsymbol{\omega}\boldsymbol{m}$. But then from (F.37) we derive $\boldsymbol{t}^\mathrm{T} = -\boldsymbol{n}^\mathrm{T}\boldsymbol{\omega}$, therefore

$$Q^n = \boldsymbol{n}^\mathrm{T}\boldsymbol{A}_d\boldsymbol{\sigma} + \boldsymbol{t}^\mathrm{T}\boldsymbol{m} = M^{ij}_{,i} n_j + m^t, \tag{1.61}$$

where m^t is a component of an external moment, distributed over area, with respect to the local axis $\boldsymbol{t}$,

$$m^t = m^x t_x + m^y t_y. \tag{1.62}$$

5.1.3 Basic integral identity for thin plate bending

Thus, what we have obtained above is a basic governing equation for the thin plate bending analysis within the limits of the Kirchhoff–Love theory – equation (1.37) for a general case, or biharmonic equation (1.39) for a particular case. To get a closed mathematical statement of the problem in the differential form, we need to formulate admissible boundary conditions. In order to define the boundary conditions correctly, we will need definitions of boundary displacements $\boldsymbol{u}$ and boundary stresses/forces $\boldsymbol{p}$; we require that a basic integral identity for the plate bending analysis hold:

$$\int_\Omega \boldsymbol{\sigma}^\mathrm{T} \boldsymbol{A}\mathbf{u} d\Omega = \int_\Omega \mathbf{u}^\mathrm{T} \boldsymbol{A}^\mathrm{T} \boldsymbol{\sigma} d\Omega + \oint_\Gamma \boldsymbol{u}^\mathrm{T} \boldsymbol{p} d\Gamma. \tag{1.63}$$

Now we drop the matrix notation for a while and again turn to a more natural form for this particular problem – a coordinate tensor notation. It is not hard to see that rewriting the left part of (1.63) in the tensor form and applying the Gauss–Ostrogradsky formula twice will give

$$\begin{aligned}\int_\Omega \boldsymbol{\sigma}^\mathrm{T} \boldsymbol{A}\mathbf{u} d\Omega &= -\int_\Omega M^{ij} w_{,ij} d\Omega = \int_\Omega M^{ij}_{,i} w_{,j} d\Omega - \oint_\Gamma M^{ij} w_{,j} n_i d\Gamma = \\ &= -\int_\Omega M^{ij}_{,ij} w d\Omega - \oint_\Gamma M^{ij} w_{,j} n_i d\Gamma + \oint_\Gamma M^{ij}_{,i} w n_j d\Gamma = \\ &= \int_\Omega \mathbf{u}^\mathrm{T} \boldsymbol{A}^\mathrm{T} \boldsymbol{\sigma} d\Omega - \oint_\Gamma M^{ij} w_{,j} n_i d\Gamma + \oint_\Gamma M^{ij}_{,i} w n_j d\Gamma. \end{aligned} \tag{1.64}$$

Now we will be interested with an overall contour integral, J, contained in the right part of formula (1.64), which can be represented as follows by using the symmetry $M^{ij} = M^{ji}$ and an insignificant transposition of umbral ("dummy") indexes

$$J = \oint_{\Gamma} (wM^{ij}_{,i} n_j - w_{,i} M^{ij} n_j) d\Gamma .$$

First of all, we note that formulas (1.42) imply

$$w_{,i} = w_{,n} n_i + w_{,s} t_i . \tag{1.65}$$

Using (1.65) and (1.52), we can easily validate the following useful identity[4],

$$w_{,i} M^{ij} n_j = w_{,n} M^{nn} + w_{,s} M^{nt} , \tag{1.66}$$

which we substitute to the integrand in J.

The aggregate $w_{,s} M^{nt}$ that appears in the integral can be replaced by expression $(wM^{nt})_{,s} - wM^{nt}_{,s}$, which follows from the formula of differentiation of a product. We have

$$J = \oint_{\Gamma} [-w_{,n} M^{nn} + w(M^{ij}_{,i} n_j + M^{nt}_{,s}) - (wM^{nt})_{,s}] d\Gamma . \tag{1.67}$$

Let a piecewise-continuous function $f(s)$ be defined on contour Γ, and let it have discontinuities in a certain finite number of singular points $s = s_\beta$ so that the function be differentiable on each segment of its continuity, that is, between two adjacent singular points. Then the integral of the derivative of this function with respect to arc coordinate s, taken over the whole closed contour Γ, will be equal to an alternated sum of leaps of the function in all singular points, that is,

$$\oint_{\Gamma} \frac{\partial f}{\partial s} d\Gamma = -\sum_{\beta} \left[f(s_\beta + 0) - f(s_\beta - 0) \right] . \tag{1.68}$$

We assume $f(s) = wM^{nt}$ in (1.68) and take into account the continuity of the deflection function, w, along the whole contour, to derive

$$-\oint_{\Gamma} (wM^{nt})_{,s} d\Gamma = \sum_{\beta} w(s_\beta) P_\beta , \tag{1.69}$$

where P_β denote leaps of the torque, M^{nt}, as we move along contour Γ,

---

[4] Note that here and below we do not use n and t as summation indexes; they just indicate the direction of a normal to the boundary, $\boldsymbol{n}$, and a tangential direction, $\boldsymbol{t}$.

$$P_\beta = M^{nt}(s_\beta + 0) - M^{nt}(s_\beta - 0)\,. \tag{1.70}$$

The sum in (1.69) is taken over all singular points of the contour, that is, over points in which moment M^n experiences a discontinuity. In particular, the discontinuity takes place in breakpoints of contour Γ. This becomes obvious when we turn to second formula in (1.52) where it can be seen clearly how torques M^{nt} experience discontinuities in all those points – the components of tangential vector $\boldsymbol{t}$ change in leaps.

It will be convenient for us to write the right part of (1.69) as a part of a contour integral, which can be dome by using Dirac delta functions, $\delta(s-s_\beta)$

$$\sum_\beta w(s_\beta)P_\beta = \oint_\Gamma w \sum_\beta P_\beta \delta(s - s_\beta)d\Gamma\,. \tag{1.71}$$

It is also easy to show that $-w_{,n} = \theta_t$. Indeed, this equality follows from a chain of formal transformations,

$$-w_{,n} = -\boldsymbol{n}^{\mathsf{T}}\mathrm{grad}w = -(\boldsymbol{\omega}^{\mathsf{T}}\boldsymbol{t})^{\mathsf{T}}\mathrm{grad}w = -\boldsymbol{t}^{\mathsf{T}}\boldsymbol{\omega}\mathrm{grad}w = \boldsymbol{t}^{\mathsf{T}}\boldsymbol{\theta} = \theta_t. \tag{1.72}$$

So, the final form of contour integral J is

$$J = \oint_\Gamma \left\{ \theta_t M^{nn} + w\left[M^{ij}_{,i} n_j + M^{nt}_{,s} + \sum_\beta P_\beta \delta(s - s_\beta) \right] \right\} d\Gamma\,. \tag{1.73}$$

Now we return to formula (1.64) and rewrite it as

$$\int_\Omega \boldsymbol{\sigma}^{\mathsf{T}} \boldsymbol{A}\mathbf{u}d\Omega =$$

$$= \int_\Omega \mathbf{u}^{\mathsf{T}} \boldsymbol{A}^{\mathsf{T}}\boldsymbol{\sigma}d\Omega + \oint_\Gamma \left\{ \theta_t M^{nn} + w\left[M^{ij}_{,i} n_j + M^{nt}_{,s} + \sum_\beta P_\beta \delta(s - s_\beta) \right] \right\} d\Gamma\,,$$

which is the same as the basic integral formula (1.63), if we take the following expressions for the vectors of boundary displacements $\boldsymbol{u}$ and boundary stresses $\boldsymbol{p}$,

$$\boldsymbol{u} = \|[w, \theta_t]\|^{\mathsf{T}}, \qquad \boldsymbol{p} = \|[\, M^{ij}_{,i} n_j + M^{nt}_{,s} + \sum_\beta P_\beta \delta(s - s_\beta)\,,\, M^{nn}]\|^{\mathsf{T}}. \tag{1.74}$$

The quantity of $M^{ij}_{,i} n_j + M^{nt}_{,s} + \sum_\beta P_\beta \delta(s - s_\beta)$, which we denote by K^n, was brought in the plate bending theory by Kirchhoff, as a generalized force conjugated by energy to the plate's deflection on the contour, $w \in \Gamma$. That is why K^n is called a *generalized* (*or Kirchhoff's*) *shear force*,

$$K^n = M^{ij}_{,i} n_j + M^{nt}_{,s} + \sum_{\beta} P_\beta \delta(s - s_\beta)\,. \tag{1.75}$$

Now we want to find out how the common shear force, Q^n, and Kirchhoff's shear force, K^n, relate to each other. In order to do it, we transform expression $M^{ij}_{,i} n_j$. From formula (1.61) we derive

$$M^{ij}_{,i} n_j = Q^n - m^t\,.$$

Consequently,

$$K^n = Q^n + M^{nt}_{,s} - m^t + \sum_{\beta} P_\beta \delta(s - s_\beta)\,. \tag{1.76}$$

Thus, Kirchhoff's contour shear force per unit of length differs from the common shear force, Q^n, by the derivative of torque $M^{nt}_{,s}$ and the value of m^t in all regular points of the contour; in addition, it differs by a shear force, concentrated in that point and equal to a leap of the torque, $P_\beta = M^{nt}(s_\beta + 0) - M^{nt}(s_\beta - 0)$, in a singular point of the contour where torque M^{nt} experiences a discontinuity.

Now we will define operators $\boldsymbol{H}_\mathrm{u}$ and $\boldsymbol{H}_\sigma$ which map the vector of displacements $\mathbf{u}$ and vector of stresses $\boldsymbol{\sigma}$ into the vector of contour displacements $\boldsymbol{u}$ and the vector of contour stresses $\boldsymbol{p}$, in accordance with general formulas (**1**.2.7)

$$\boldsymbol{u} = \boldsymbol{H}_\mathrm{u}\mathbf{u}\,, \qquad \boldsymbol{p} = \boldsymbol{H}_\sigma \boldsymbol{\sigma}\,.$$

As for operator $\boldsymbol{H}_\mathrm{u}$, it is clear from (1.72) and (1.42) that

$$\boldsymbol{H}_\mathrm{u} = \left\| \begin{bmatrix} 1 \\ -\dfrac{\partial}{\partial n} \end{bmatrix} \right\| = \left\| \begin{bmatrix} 1 \\ -n_x \dfrac{\partial}{\partial x} - n_y \dfrac{\partial}{\partial y} \end{bmatrix} \right\|. \tag{1.77}$$

It can be validated directly that aggregate $M^{ij}_{,i} n_j$ can be represented as

$$M^{ij}_{,i} n_j = \boldsymbol{n}^\mathrm{T} \boldsymbol{A}_d \boldsymbol{\sigma}\,. \tag{1.78}$$

Further,

$$M^{nt}_{,s} = \frac{\partial}{\partial s}\left(\boldsymbol{t}^\mathrm{T} \boldsymbol{A}_n \boldsymbol{\sigma}\right) = \frac{\partial(\boldsymbol{t}^\mathrm{T} \boldsymbol{A}_n)}{\partial s} \boldsymbol{\sigma} + \boldsymbol{t}^\mathrm{T} \boldsymbol{A}_n \frac{\partial}{\partial s} \boldsymbol{\sigma}\,.$$

Turning to Frenet formulas (F.52) and differentiation formulas (1.47), we have

$$\frac{\partial}{\partial s}\left(\boldsymbol{t}^{\mathsf{T}}\boldsymbol{A}_n\right) = -k\boldsymbol{n}^{\mathsf{T}}\boldsymbol{A}_n + k\boldsymbol{t}^{\mathsf{T}}\boldsymbol{A}_t = -k\boldsymbol{n}^{\mathsf{T}}(\boldsymbol{A}_n + \boldsymbol{\omega}\,\boldsymbol{A}_t)\,.$$

Thus, replacing vector $\boldsymbol{t}^{\mathsf{T}}$ by $-\boldsymbol{n}^{\mathsf{T}}\boldsymbol{\omega}$ will give

$$M^{nt}_{,s} = -\boldsymbol{n}^{\mathsf{T}}[k(\boldsymbol{A}_n + \boldsymbol{\omega}\,\boldsymbol{A}_t) + \boldsymbol{\omega}\boldsymbol{A}_n\frac{\partial}{\partial s}]\boldsymbol{\sigma}\,. \tag{1.79}$$

Now we introduce a symbolic operator, $\boldsymbol{\Gamma}_\beta$, that acts on a certain function, $a(s)$, and maps that function into the following expression: $\delta(s - s_\beta)[a(s_\beta + 0) - a(s_\beta - 0)]$.

Taking second formula from (1.74) and the relationships obtained above, we can present operator $\boldsymbol{H}_\sigma$ in its symbolic form:

$$\boldsymbol{H}_\sigma = \left\|\begin{bmatrix} \boldsymbol{n}^{\mathsf{T}}\boldsymbol{A}_d - \boldsymbol{n}^{\mathsf{T}}\left[k(\boldsymbol{A}_n + \boldsymbol{\omega}\boldsymbol{A}_t) + \boldsymbol{\omega}\boldsymbol{A}_n \partial/\partial s\right] + \sum_\beta \boldsymbol{\Gamma}_\beta \boldsymbol{t}^{\mathsf{T}}\boldsymbol{A}_n \\ \boldsymbol{n}^{\mathsf{T}}\boldsymbol{A}_n \end{bmatrix}\right\|\,; \tag{1.80}$$

this maps the vector of stresses $\boldsymbol{\sigma}$ into the vector of boundary stresses $\boldsymbol{p}$.

It will be also useful for further presentment to write an expression for the scalar product of the contour forces/stresses and the contour displacements, $(\boldsymbol{p}, \boldsymbol{u})_\Gamma$, in an explicit expanded form. We derive this from (1.74) and (1.75):

$$(\boldsymbol{p}, \boldsymbol{u})_\Gamma = \oint_\Gamma K^n w\, d\Gamma - \oint_\Gamma M^{nn} w_{,n}\, d\Gamma\,. \tag{1.81}$$

5.1.4 Boundary conditions for thin plate bending

The basic integral identity established above permits us to write correct boundary conditions for the plate bending analysis right now. As it follows from formula (1.74) for the contour displacements and contour stresses, two boundary conditions can be formulated in each point of the plate's contour, Γ.

According to the structure of the contour displacements and the contour stresses, the full boundary, Γ, of area Ω can be divided into two sets of boundaries such that

$$\Gamma = \Gamma_K \cup \Gamma_w\,,\quad \Gamma_K \cap \Gamma_w = \varnothing,\ \Gamma = \Gamma_M \cup \Gamma_\theta\,,\quad \Gamma_M \cap \Gamma_\theta = \varnothing, \tag{1.82}$$

where kinematic-type boundary conditions are stated on Γ_w and Γ_θ and static-type conditions on Γ_K and Γ_M,

$$w = \bar{w} \in \Gamma_w,\ \theta_t = \bar{\theta}_t \in \Gamma_\theta \qquad \text{kinematic boundary conditions,}$$

$$K^n = \bar{K} \in \Gamma_K,\ M^{nn} = \bar{M} \in \Gamma_M \qquad \text{static boundary conditions.} \qquad (1.83)$$

Here θ_t is a slope of the normal to the plate's median surface with respect to local axis $\boldsymbol{t}$, which can be represented as follows using (1.72):

$$\theta_t = -w,_n = -w,_x n_x - w,_y n_y\,. \qquad (1.84)$$

Overscored symbols in formulas (1.83) denote, as usual, quantities specified on the respective segments of contour Γ.

Obviously, the coresponding matrices of boundary condition extraction, $\boldsymbol{E}_u$ and $\boldsymbol{E}_p$, are

$$\boldsymbol{E}_u = \left\|\begin{bmatrix} e_w & 0 \\ 0 & e_\theta \end{bmatrix}\right\|, \quad \boldsymbol{E}_p = \left\|\begin{bmatrix} e_K & 0 \\ 0 & e_M \end{bmatrix}\right\|, \qquad (1.85)$$

where e_w, e_θ, e_K, e_M are characteristic functions of contour conditions; each of those is equal to one on a segment of the contour where a boundary condition corresponding to its subscript is stated and is equal to zero on the rest of the contour. It should be kept in mind that the matrices satisfy conditions (**1**.2.5).

So, we repeat that the Kirchhoff–Love theory of plates states two boundary conditions in each point of the contour, which is in accordance with all energy theorems of structural mechanics that follow from the basic integral identity.

It is a matter of history that the question of boundary conditions in the technical theory of plate bending was a subject of a long scientific discussion. One of creators of the plate bending theory, French mathematician and mechanician Simeon Denis Poisson arrived at the Germain–Lagrange biharmonic equation and formulated three boundary conditions on a fixation-free edge of a plate: (in our notation) for shear force Q^n, for bending moment M^{nn}, and for torque M^{nt}. As we have already told, mechanics of solids should thank another prominent scientist, Gustav Robert Kirchhoff, for his detecting a contradiction of this (redundant) set of boundary conditions to the energy laws of statics. Kirchhof used a variational formulation of the problem to show [11] that the natural conditions for the Lagrange functional in the plate bending theory include only two conditions on a fixation-free edge: for bending moment M^{nn} and for generalized (Kirchhoff's) shear force K^n. The publication of Kirchhoff's paper started a hot and one of longest scientific discussions in the history of mechanics. The participants of the discussion were numerous prominent scientists, and the very fact of an incorrect, from the standpoint of energy, formulation of the boundary conditions in the technical theory of plate bending was called a "*Poisson's mistake in the plate theory*".

In the course of that discussion, Kelvin and Tait suggested a mechanical model that explained, on a level of mathematical abstraction, how it was admissible (and even necessary!) to transit from the common shear force to Kirchhoff's generalized shear force on a fixation-free edge of the plate. The Kelvin–Tait transformation has been described many times in the literature, including various textbooks. We will not give a mechanical interpretation of the Kelvin–Tait transformation here because this is a bit of a deviation from the main course of the book; instead, we will refer to such books as [19] and [3] where this subject is addressed with a great carefulness and accuracy[5].

---

[5] It is curious that the said discussion was continued with a new passion at the end of twentieth century in Russia. Of course, particular issues and the whole subject were different in this new outbreak; the main interest of the participants was addressed to contradictions in the Kirchhoff–Love theory and its comparison with Reissner's theory of plates from the standpoint of matching the respective solutions to the solutions of three-dimensional elasticity. We will not participate in that discussion; instead, we suggest that the reader, if interested, made himself familiar with opinions of the discussion's participants and assessed them as best he could; it can be educational at the least. Here's a list of the relevant publications.

1. Goldenweiser AL (1992) A general theory of thin elastic bodies (shells, roofs, spacers) (in Russian). Bull. Acad. Sci., Mech. Deform. Solids, 3: 5 – 17
2. Vasiliev VV (1992) On a theory of thin plates (in Russian). Bull. Acad. Sci., Mech. Deform. Solids, 3: 26 – 46
3. Zhilin PA (1992) On theories of plates by Poisson and by Kirchhoff from the standpoint of the modern plate theory (in Russian). Bull. Acad. Sci., Mech. Deform. Solids, 3: 48 – 64
4. Alfutov NA (1992) Some paradoxes in the theory of thin elastic plates (in Russian). Bull. Acad. Sci., Mech. Deform. Solids, 3: 65 – 72
5. Darevsky VM (1995) On static boundary conditions in the classic theory of shells and plates (in Russian). Bull. Acad. Sci., Mech. Deform. Solids, 4: 129–132
6. Zhilin PA (1995) On the classic theory of plates and the Kelvin–Tait transformation (in Russian). Bull. Acad. Sci., Mech. Deform. Solids, 4: 133 – 139
7. Vasiliev VV (1995) To a discussion on classic theory of plates (in Russian). Bull. Acad. Sci., Mech. Deform. Solids, 4: 140 – 150
8. Goldenweiser AL (1997) On approximate methods of analysis of thin elastic shells and plates (in Russian). Bull. Acad. Sci., Mech. Deform. Solids, 3: 134–149
9. Vasiliev VV (1997) An asymptotic method of justification of the plate theory (in Russian). Bull. Acad. Sci., Mech. Deform. Solids, 3: 150 – 155

Here we just emphasize once again that Kirchhoff's boundary conditions are quite correct in the technical theory of plates by Kirchhoff–Love from the standpoint of variational formulations, which is not the case for Poisson's conditions.

5.1.5 Important functionals for thin plate bending

Lagrange functional

Earlier we have established an admissible set of external force actions on the plate, specified in area Ω, which includes a lateral load, q, and moments m^x and m^y distributed over the median surface of the plane.

To complete the picture, now we supplement these external forces with a load specified on the plate's contour, Γ. This contour load consists of contour forces $\bar{Q}$ per unit of length with the same positive direction as the Z-axis has and contour bending moments $\bar{M}^t$ and torques $\bar{M}^n$ per unit of length. Fig. 5.4 shows those external forces and mments as acting in their positive direction.

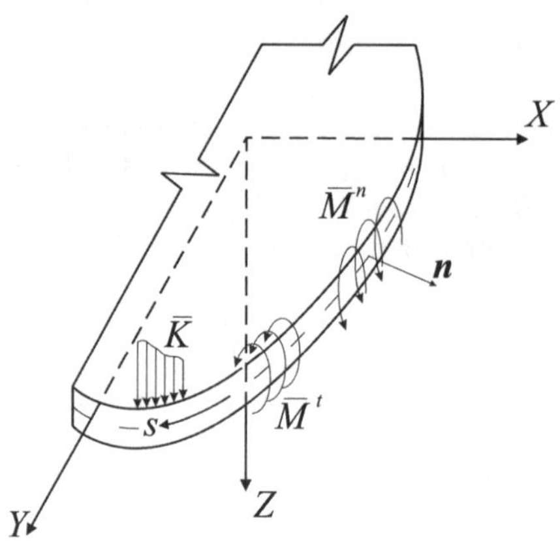

Fig. 5.4. External loads on the contour of the plate

Now we are able to write out an expression of force potential Π_s as a virtual work of all external forces on the respective displacements. We have

10. Goldenweiser AL (1997) Notes on the paper by V.V. Vasiliev "An asymptotic method of justification of the plate theory" (in Russian). Bull. Acad. Sci., Mech. Deform. Solids, 4: 150 – 158

$$\Pi_s = \int_\Omega (qw + m^x\theta_x + m^y\theta_y)d\Omega + \oint_\Gamma (\bar{Q}w + \bar{M}^t\theta_t + \bar{M}^n\theta_n)d\Gamma . \qquad (1.86)$$

Here θ_n and θ_t are components of the normal's slope vector, $\boldsymbol{\theta}$, in axes $\boldsymbol{n}$ and $\boldsymbol{t}$.

First of all, we want to transform an expression of the virtual work of loads distributed over area Ω by making substitutions according to (1.4) and using the Gauss–Ostrogradsky formula. We have

$$\int_\Omega (qw + m^x\theta_x + m^y\theta_y)d\Omega = \int_\Omega (qw + m^x w_{,y} - m^y w_{,x})d\Omega =$$

$$= \int_\Omega (q - m^x_{,y} + m^y_{,x})w d\Omega + \oint_\Gamma (m^x n_y - m^y n_x)w d\Gamma .$$

Now recall a designation introduced earlier by formula (1.62):

$$m^t = -m^x n_y + m^y n_x = m^x t_x + m^y t_y ,$$

and formula (1.35) for a generalized lateral load $\bar{q}$. Now we have

$$\int_\Omega (qw + m^x\theta_x + m^y\theta_y)d\Omega = \int_\Omega \bar{q} w d\Omega - \oint_\Gamma m^t w d\Gamma . \qquad (1.87)$$

Then, considering (1.4) and (1.46), we express slopes θ_t and θ_n via the derivatives of deflections w,

$$\theta_t = -w_{,n} , \qquad \theta_n = w_{,s} .$$

After substituting, we transform the expression of Π_s according to (1.86) into the following:

$$\Pi_s = \int_\Omega \bar{q} w d\Omega - \oint_\Gamma m^t w d\Gamma + \oint_\Gamma (\bar{Q}w + \bar{M}^t\theta_t + \bar{M}^n\theta_n)d\Gamma =$$

$$= \int_\Omega \bar{q} w d\Omega + \oint_\Gamma [\bar{Q}w - \bar{M}^t w_{,n} + (\bar{M}^n w)_{,s} - \bar{M}^n_{,s} w - m^t w]d\Gamma =.$$

$$= \int_\Omega \bar{q} w d\Omega + \oint_\Gamma (\bar{K}w - \bar{M}^t w_{,n})d\Gamma . \qquad (1.88)$$

Here we denote

$$\bar{K} = \bar{Q} - m^t - \bar{M}^n_{,s} - \sum_\beta \bar{P}_\beta \delta(s - s_\beta) , \qquad (1.89)$$

where $\bar{P}_\beta$ is a leap of the contour torque, $\bar{M}^n$, as we move along the contour coordinate, s:

$$\bar{P}_\beta = \bar{M}^n(s_\beta + 0) - \bar{M}^n(s_\beta - 0). \tag{1.90}$$

Function $\bar{K}(s)$ defined on contour Γ by formula (1.89) can be called a *generalized (or Kirchhoff's) contour load.* Note that the summation in (1.89) is performed over all singular points of the contour with the coordinates $s = s_\beta$, that is, over points where the given external contour moment, $\bar{M}^n$, experiences a discontinuity. We can think of two types of singular points:

- point $s = s_\beta$ is a singular one if function $\bar{M}^n$ specified on contour Γ has a discontinuity in that point, though the point itself may belong to a smooth part of the contour;
- point $s = s_\beta$ is a singular one if it is a breakpoint of the contour because the given external moment $\bar{M}^n$ has a leap of its direction.

Note that the integral in formula (1.88) is taken formally over the whole closed contour Γ. If we recall, however, that the external (active) contour actions $\bar{K}$ and $\bar{M}^t$ are nonzero only on parts of the contour where the respective characteristic functions e_K and e_M are equal to one, and if we introduce a simplified (index-free) designation of $\bar{M}$ for the contour bending moment, $\bar{M}^t$, that is, if we assume

$$\bar{M} = \bar{M}^t,$$

then we can rewrite expression (1.88) of force potential Π_s as

$$\Pi_s = \int_\Omega \bar{q} w d\Omega + \oint_\Gamma e_K \bar{K} w d\Gamma - \oint_\Gamma e_M \bar{M} w_{,n} d\Gamma. \tag{1.91}$$

If we introduce, as we did all the time, a common designation for a vector of external forces $\bar{\boldsymbol{p}}$ specified on the contour by assuming

$$\bar{\boldsymbol{p}} = \left\| \left[\bar{K}, \bar{M} \right] \right\|^{\mathsf{T}}, \tag{1.92}$$

then, according to (1.2.5) the expression of force potential Π_s will acquire a general operator form:

$$\Pi_s = \int_\Omega \bar{\boldsymbol{X}} \cdot \mathbf{u} d\Omega + \int_\Gamma \boldsymbol{E}_p \bar{\boldsymbol{p}} \cdot \boldsymbol{E}_p \boldsymbol{H}_u \mathbf{u} d\Gamma = (\bar{\boldsymbol{X}}, \mathbf{u}) + (\boldsymbol{E}_p \bar{\boldsymbol{p}}, \boldsymbol{E}_p \boldsymbol{H}_u \mathbf{u})_\Gamma . \tag{1.93}$$

Now, recalling the expression of the strain energy, E, from (1.20) and (1.21), we will present the Lagrange functional, $\mathsf{L} = \mathsf{E} - \Pi_s$, for a flexural Kirchhoff-Love plate. We have

$$\mathsf{L} = \frac{1}{2}\int_{\Omega} c^{ijkl} w_{,ij} w_{,kl} d\Omega - \int_{\Omega} \bar{q} w d\Omega - \oint_{\Gamma} (e_K \bar{K} w - e_M \bar{M} w_{,n}) d\Gamma =$$

$$= ½\,(\boldsymbol{CA}\mathbf{u}, \boldsymbol{A}\mathbf{u}) - (\bar{\boldsymbol{X}}, \mathbf{u}) - (\boldsymbol{E}_p \bar{\boldsymbol{p}}, \boldsymbol{E}_p \boldsymbol{H}_u \mathbf{u})_\Gamma \ . \tag{1.94}$$

The set, $\mathscr{R}_k$, of kinematically admissible stress-and-strain fields which we search for the minimum of Lagrangian functional L will consist of all fields of the type

$$F = \{\boldsymbol{CA}\mathbf{u}, \boldsymbol{A}\mathbf{u}, \mathbf{u}\} \quad \text{at} \quad \boldsymbol{E}_u \boldsymbol{u} = \bar{\boldsymbol{u}} \in \Gamma.$$

The expanded form of this can be written as follows for an isotropic plate:

$$F = \left\{ \left\| \begin{bmatrix} -D(w_{,xx} + \nu w_{,yy}) \\ -D(w_{,yy} + \nu w_{,xx}) \\ -D(1-\nu) w_{,xy} \end{bmatrix} \right\|, \left\| \begin{bmatrix} -w_{,xx} \\ -w_{,yy} \\ -2w_{,xy} \end{bmatrix} \right\|, w \right\} :$$

$$e_w(w - \bar{w}) = 0, \qquad e_\theta(-w_{,n} - \bar{\theta}_t) = 0 \quad \in \Gamma.$$

Earlier, in Chapter 2, we established that the condition of minimum of functional L on set $\mathscr{R}_k$ implies the Lame equation (a Euler equation for the L functional) in area Ω and static (natural) boundary conditions on contour Γ

$$\boldsymbol{A}^{\mathsf{T}}\boldsymbol{CA}\mathbf{u} - \bar{\boldsymbol{X}} \in \Omega, \qquad \boldsymbol{E}_p(\boldsymbol{p} - \bar{\boldsymbol{p}}) = \mathbf{0} \in \Gamma.$$

This gives the already known relationships when written in components:

$$\left(c^{ijkl} w_{,kl}\right)_{,ij} = \bar{q} \in \Omega, \qquad K^n - \bar{K} = 0 \in \Gamma_K, \qquad M^{nn} - \bar{M} = 0 \in \Gamma_M. \tag{1.95}$$

Castigliano functional

To derive the Castigliano functional, we use general formula (**1**.3.1) and assume $\boldsymbol{K} = \boldsymbol{O}$ – there is no elastic foundation.

An expression of the strain energy represented as a functional of the moment tensor's components is as follows for the general case of an anisotropic material:

$$\mathsf{E}(\boldsymbol{\sigma}) = ½(\boldsymbol{C}^{-1}\boldsymbol{\sigma}, \boldsymbol{\sigma}) = \frac{1}{2}\int_{\Omega} d_{ijkl} M^{ij} M^{kl} d\Omega. \tag{1.96}$$

This gives the following for an isotropic plate of a constant thickness:

$$\mathsf{E}(\boldsymbol{\sigma}) =$$

$$= \frac{1}{2D(1-\nu^2)} \iint_{\Omega} \left[(M^{xx})^2 - 2\nu M^{xx} M^{yy} + (M^{yy})^2 + 2(1+\nu)(M^{xy})^2 \right] d\Omega \tag{1.97}$$

Further, from (1.81) we derive this expression for a potential of external kinematical actions

$$\Pi_k(\boldsymbol{\sigma}) = (\boldsymbol{E}_u \boldsymbol{H}_\sigma \boldsymbol{\sigma}, \boldsymbol{E}_u\, \bar{\boldsymbol{u}})_\Gamma = \oint_{\Gamma} e_w K^n \overline{w} d\Gamma + \oint_{\Gamma} e_\theta M^{nn} \overline{\theta}_t d\Gamma . \tag{1.98}$$

A point of minimum of the Castigliano functional,

$$\mathsf{K}(\boldsymbol{\sigma}) = \mathsf{E}(\boldsymbol{\sigma}) - \Pi_k(\boldsymbol{\sigma}) =$$

$$= \frac{1}{2}\int_{\Omega} d_{ijkl} M^{ij} M^{kl} d\Omega - \oint_{\Gamma} e_w K^n \overline{w} d\Gamma - \oint_{\Gamma} e_\theta M^{nn} \overline{\theta}_t d\Gamma , \tag{1.99}$$

should be sought for in a set of statically admissible stress-and-strain fields.

As the Castigliano functional depends only on “stresses” M^{ij} when there is no elastic foundation, the “stresses” must satisfy the equilibrium equations in area Ω,

$$-M^{xx}_{,xx} - 2M^{xy}_{,xy} - M^{yy}_{,yy} = \overline{q} \quad \in \Omega, \tag{1.100}$$

and static boundary conditions on the plate’s contour, Γ,

$$K^n = \overline{K} \in \Gamma_K, \qquad M^{nn} = \overline{M} \in \Gamma_M . \tag{1.101}$$

As we established in Chapter **2** for the general case, Euler equations for the Castigliano functional are conditions of strain compatibility written in terms of stresses. For the thin plate bending analysis, the strain compatibility conditions consist of the two following relationships:

$$\frac{\partial \chi_{xx}}{\partial y} - \frac{\partial \chi_{xy}}{\partial x} = 0, \qquad \frac{\partial \chi_{yy}}{\partial x} - \frac{\partial \chi_{xy}}{\partial y} = 0 . \tag{1.102}$$

Actually, the strain compatibility conditions (1.102) can be obtained from the requirement that the third mixed derivatives of the deflection function should be independent from the order of differentiation. If we turn to the operator-based form of these equations as in (**4**.1.8), we will find easily that the compatibility operator, $\boldsymbol{S}$, is

$$\boldsymbol{S} = \left\|\begin{matrix} 0 & \dfrac{\partial}{\partial x} & -\dfrac{\partial}{2\partial y} \\ \dfrac{\partial}{\partial y} & 0 & -\dfrac{\partial}{2\partial x} \end{matrix}\right\|. \tag{1.103}$$

Here we take into account the fact that strain vector $\boldsymbol{\varepsilon}$ for the plate bending is defined by second formula in (1.48). Also, (1.49) helps make sure that operator $\boldsymbol{SA}$ is indeed an annihilating operator, as required by (**4**.1.10), that is,

$$\boldsymbol{SA} = \boldsymbol{O}.$$

Further, a general strain compatibility condition written in terms of stresses was already presented earlier by formula (**4**.1.12-*b*), which in our case is

$$\boldsymbol{SC}^{-1}\boldsymbol{\sigma} = \mathbf{0}.$$

In particular, the strain compatibility conditions in terms of stresses, written in components for a plate of a constant thickness made of a homogeneous isotropic material, are

$$M^{xx}_{,y} - \nu M^{yy}_{,y} - (1+\nu)M^{xy}_{,x} = 0\,,$$

$$M^{yy}_{,x} - \nu M^{xx}_{,x} - (1+\nu)M^{xy}_{,y} = 0\,. \tag{1.104}$$

Reissner functional

Here we present an expression of the Reissner functional for the bending of a Kirchhoff–Love plate. Writing in components, we have

$$(\boldsymbol{A}\mathbf{u},\boldsymbol{\sigma}) = -\int_{\Omega}[M^{xx}w_{,xx} + M^{yy}w_{,yy} + 2M^{xy}w_{,xy}]d\Omega = -\int_{\Omega}M^{ij}w_{,ij}d\Omega\,, \tag{1.105}$$

$$(\boldsymbol{E}_u\boldsymbol{u}, \boldsymbol{E}_u\boldsymbol{p})_\Gamma - \Pi_\text{k}(\boldsymbol{\sigma}) = \oint_\Gamma e_w K^n(w - \overline{w}d\Gamma + \oint_\Gamma e_\theta M^{nn}(-w_{,n} - \overline{\theta}_t)d\Gamma\,. \tag{1.106}$$

Now we derive an expression of the Reissner functional's first form from the general formula (**3**.1.4):

$$\mathsf{R}_1(\boldsymbol{\sigma}, \mathbf{u}) = \tfrac{1}{2}(\boldsymbol{C}^{-1}\boldsymbol{\sigma}, \boldsymbol{\sigma}) - (\boldsymbol{A}\mathbf{u}, \boldsymbol{\sigma}) + (\boldsymbol{E}_u\boldsymbol{p}, \boldsymbol{E}_u\boldsymbol{u})_\Gamma + \Pi_s - \Pi_k =$$

$$= \int\limits_\Omega \left[\frac{1}{2} d_{ijkl} M^{ij} M^{kl} + M^{ij} w_{,ij} + \overline{q} w\right] d\Omega + \oint\limits_\Gamma e_w K^n (w - \overline{w}) d\Gamma +$$

$$+ \oint\limits_\Gamma e_\theta M^{nn}(-w_{,n} - \overline{\theta}_t) d\Gamma + \oint\limits_\Gamma e_K \overline{K} w d\Gamma - \oint\limits_\Gamma e_M \overline{M} w_{,n} d\Gamma . \qquad (1.107)$$

The second form of the Reissner functional can be obtained in a similar way:

$$\mathsf{R}_2(\boldsymbol{\sigma}, \mathbf{u}) = \tfrac{1}{2}(\boldsymbol{C}^{-1}\boldsymbol{\sigma}, \boldsymbol{\sigma}) - (\boldsymbol{A}^{\mathsf{T}}\boldsymbol{\sigma}, \mathbf{u}) - (\boldsymbol{E}_p\boldsymbol{p}, \boldsymbol{E}_p\boldsymbol{u})_\Gamma + \Pi_s - \Pi_k =$$

$$= \int\limits_\Omega \left[\frac{1}{2} d_{ijkl} M^{ij} M^{kl} + M^{ij}_{,ij} w + \overline{q} w\right] d\Omega - \oint\limits_\Gamma e_w K^n \overline{w} d\Gamma -$$

$$- \oint\limits_\Gamma e_\theta M^{nn} \overline{\theta}_t d\Gamma + \oint\limits_\Gamma e_K (\overline{K} - K^n) w d\Gamma - \oint\limits_\Gamma e_M (\overline{M} - M^{nn}) w_{,n} d\Gamma . \qquad (1.108)$$

Herrmann functional

As we noted in Section **3**.1.1, there exists a third form of the Reissner functional for a number of applications where geometry operator $\boldsymbol{A}$ contains differential operations of an even order. This third form is derived from the Reissner functional's first or second form by formally moving some of the derivatives from the displacements/stresses onto the stresses/displacements using the Gauss–Ostrogradsky formula, to make the differentiation orders of the stresses and the displacements equal.

For example, let us consider the second form of the Reissner functional and try to transform in this way the $-(\boldsymbol{A}^{\mathsf{T}}\boldsymbol{\sigma}, \mathbf{u})$ term contained in the expression of $\mathsf{R}_2(\boldsymbol{\sigma},\mathbf{u})$ by moving half of the derivatives from the moments onto the displacements. We have

$$-(\boldsymbol{A}^{\mathsf{T}}\boldsymbol{\sigma}, \mathbf{u}) = \int\limits_\Omega M^{ij}_{,ij} w d\Omega = -\int\limits_\Omega M^{ij}_{,i} w_{,j} d\Omega + \oint\limits_\Gamma M^{ij}_{,i} n_j w d\Gamma .$$

As the condition $e_w + e_K = 1$ holds in any point of contour Γ, we can divide the contour integral in the above relationship into two integrals. This gives

$$-(\boldsymbol{A}^{\mathsf{T}}\boldsymbol{\sigma}, \mathbf{u}) = -\int\limits_\Omega M^{ij}_{,i} w_{,j} d\Omega + \oint\limits_\Gamma e_w M^{ij}_{,i} n_j w d\Gamma + \oint\limits_\Gamma e_K M^{ij}_{,i} n_j w d\Gamma .$$

Substituting this relation to (1.108) produces a component-based third form of the Reissner functional. Thus, we have

$$\mathsf{R}_3(\boldsymbol{\sigma},\mathbf{u}) = \iint_{\Omega}\left[\frac{1}{2}d_{ijkl}M^{ij}M^{kl} - M^{ij}_{,i}w_{,j} + \bar{q}w\right]d\Omega -$$

$$-\oint_{\Gamma} e_w(K^n\bar{w} - M^{ij}_{,i}n_j w)d\Gamma - \oint_{\Gamma} e_\theta M^{nn}\bar{\theta}_t d\Gamma +$$

$$+\oint_{\Gamma} e_K(\bar{K} - K^n + M^{ij}_{,i}n_j)wd\Gamma - \oint_{\Gamma} e_M(\bar{M} - M^{nn})w_{,n}d\Gamma . \qquad (1.109)$$

By using formulas (1.75) and (1.70) we derive

$$\oint_{\Gamma}(K^n - M^{ij}_{,i}n_j)wd\Gamma = \oint_{\Gamma}[M^{nt}_{,s} + \sum_{\beta}P_\beta\delta(s - s_\beta)]wd\Gamma =$$

$$= \oint_{\Gamma}[M^{nt}_{,s}w - (M^{nt}w)_{,s}]d\Gamma = -\oint_{\Gamma}M^{nt}w_{,s}d\Gamma . \qquad (1.110)$$

Substituting formula (1.110) to (1.109) and making some insignificant transformations, we produce a final functional which we denote by $\mathsf{H}(\boldsymbol{\sigma},\mathbf{u})$:

$$\mathsf{H}(\boldsymbol{\sigma},\mathbf{u}) = \iint_{\Omega}\left[\frac{1}{2}d_{ijkl}M^{ij}M^{kl} - M^{ij}_{,i}w_{,j} + \bar{q}w\right]d\Omega + \oint_{\Gamma}M^{nt}w_{,s}d\Gamma -$$

$$-\int_{\Gamma_w}K^n(\bar{w} - w)d\Gamma - \int_{\Gamma_\theta}M^{nn}\bar{\theta}_t d\Gamma + \int_{\Gamma_K}\bar{K}wd\Gamma -$$

$$-\int_{\Gamma_M}(\bar{M} - M^{nn})w_{,n}d\Gamma . \qquad (1.111)$$

Although this functional is nothing but the third form of the Reissner functional for the thin plate bending, nearly the same form of it was derived and used for the first time by Herrmann [10] as a variational basis for a mixed finite element method. Therefore we deem it reasonable to name the functional a *Herrmann functional* and give it a proper designation as indicated above.

An unquestionable advantage of the Herrmann functional is that all functions in its integrand are differentiated once at the most. This fact permits us to advance only minimal requirements of smoothness to the approximations of both the displacements and the stresses. In subsequent publications (see [1], [6], for example) the Herrmann functional was used many times to try and build various versions of mixed finite-element approaches to the plate bending, where it showed a good applicability and high efficiency.

5.2 Static-geometric analogy in the theory of plates

It is well-known that the problems of the bending and the plane stress in a plate are closely related, and their relation reveals itself in a so-called static-geometric analogy. Before we describe the static-geometric analogy in the theory of plates, we want to represent the statements of the plane stress and bending problems via a stress function.

5.2.1 A stress function vector in the theory of plates

Consider a plane stress state in a plate, taking an assumption that there is no volumetric load upon it. However, even if the original problem's statement contains external loads distributed over area Ω, the problem can always be reformulated in such way that there will be no more external forces and the equations of equilibrium in area Ω will be homogeneous. To do so, we represent a general solution of the equlibrium equations and the strain compatibility equations in terms of stresses, (**4**.1.12),

$$\boldsymbol{A}^{\mathsf{T}}\boldsymbol{\sigma} = \bar{\boldsymbol{X}}, \qquad \boldsymbol{SC}^{-1}\boldsymbol{\sigma} = \boldsymbol{0} \quad \in \Omega$$

as a sum of general solution $\boldsymbol{\sigma}_0$ of homogeneous equations

$$\boldsymbol{A}^{\mathsf{T}}\boldsymbol{\sigma}_0 = \boldsymbol{0}, \quad \boldsymbol{SC}^{-1}\boldsymbol{\sigma}_0 = \boldsymbol{0}$$

and some (arbitrary) partial solution $\boldsymbol{\tau}$ of inhomogeneous equations $\boldsymbol{A}^{\mathsf{T}}\boldsymbol{\tau} = \bar{\boldsymbol{X}}$, $\boldsymbol{SC}^{-1}\boldsymbol{\tau} = \boldsymbol{0}$, that is, we assume

$$\boldsymbol{\sigma} = \boldsymbol{\sigma}_0 + \boldsymbol{\tau}\,.$$

Usually, the construction of partial solution $\boldsymbol{\tau}$ is not especially difficult, therefore the requirement of no volumetric forces in the problem statement is not really restrictive. So we will assume for the future that the equilibrium equations are homogeneous — that is, $\bar{\boldsymbol{X}} = \boldsymbol{0}$ [6].

Before doing anything with the plane stress problem, we want to make an insignificant transformation by switching from stresses σ^{ij} to forces per unit of line, N^{ij}, that is, by assuming

$$N^x = h\sigma^x, \quad N^y = h\sigma^y, \quad N^{xy} = h\tau^{xy}, \qquad \boldsymbol{\sigma}^{\mathsf{T}} = \|[N^x, N^y, N^{xy}]\|\,. \tag{2.1}$$

[6] Some possibilities for finding a partial solution, $\boldsymbol{\tau}$, for the plane stress are considered in a well-known book by N.I. Muskhelishvili [18]. One of methods for the construction of a partial solution for the plate bending was indicated by A.I. Lurie [16].

If we turn to the equations of equlibrium for the plane stress analysis, we will derive the following from (**4**.3.2), where the conversion into the linear forces according to (2.1) should be taken into account:

$$-\frac{\partial N^x}{\partial x}-\frac{\partial N^{xy}}{\partial y}=0, \qquad -\frac{\partial N^{yx}}{\partial x}-\frac{\partial N^y}{\partial y}=0. \tag{2.2}$$

We introduce a stress function, Φ (an Airy function), defined in area Ω and related to the forces per unit of length, N^{ij}, via the formulas

$$N^x=\frac{\partial^2\Phi}{\partial y^2}, \qquad N^y=\frac{\partial^2\Phi}{\partial x^2}, \qquad N^{xy}=-\frac{\partial^2\Phi}{\partial x\partial y}. \tag{2.3}$$

By substituting (2.3) to (2.2), we make sure that the homogeneous equilibrium equations (2.2) are satisfied automatically.

To switch to a matrix form of the plane stress problem, it is convenient to introduce a one-dimensional vector of stress functions $\boldsymbol{\Phi}$ by assuming

$$\boldsymbol{\Phi}=\|[\Phi]\|. \tag{2.4}$$

Relationships (2.3) can be represented in the matrix form as

$$\boldsymbol{\sigma}=\boldsymbol{S}^{\mathsf{T}}\boldsymbol{\Phi}, \tag{2.5}$$

where the matrix differential operator $\boldsymbol{S}^{\mathsf{T}}$ is an operator conjugate in the Lagrangian sense to the Saint-Venant compatibility operator for the plane stress analysis:

$$\boldsymbol{S}=\left\|\left[\frac{\partial^2}{\partial y^2},\ \frac{\partial^2}{\partial x^2},\ -\frac{\partial^2}{\partial x\partial y}\right]\right\|.$$

It is not surprising because we established the following identity earlier by formula (**4**.1.11):

$$\boldsymbol{A}^{\mathsf{T}}\boldsymbol{S}^{\mathsf{T}}=\boldsymbol{O}.$$

Now we substitute an expression of vector $\boldsymbol{\sigma}$ from (2.5) to equation (**4**.1.12-*b*) and arrive at a governing equation of the problem in terms of the stress function. We have

$$\boldsymbol{M}\boldsymbol{\Phi}=\boldsymbol{0}, \tag{2.6}$$

where

$$\boldsymbol{M}=\boldsymbol{S}\boldsymbol{C}^{-1}\boldsymbol{S}^{\mathsf{T}}. \tag{2.7}$$

The matrix differential operator $\boldsymbol{M}$ defined by (2.7) will be further called a *compatiblity operator in stress functions*.

The $\boldsymbol{M}$ operator takes its simplest form in the case of a homogeneous isotropic material of a constant-thickness plate, $h = Const$. It is easy to see that the $\boldsymbol{M}$ operator for the plane stress analysis is just a common biharmonic operator (up to a constant factor), or

$$\boldsymbol{M} = \frac{1}{Eh}\nabla^2\nabla^2 . \tag{2.8}$$

Now let's return to the plate bending. Based on (1.103) and formal rules of conjugation, we have

$$\boldsymbol{S}^{\mathsf{T}} = \left\|\begin{bmatrix} 0 & -\dfrac{\partial}{\partial y} \\ -\dfrac{\partial}{\partial x} & 0 \\ \dfrac{\partial}{2\partial y} & \dfrac{\partial}{2\partial x} \end{bmatrix}\right\| . \tag{2.9}$$

It is now clear that the vector of stress functions $\boldsymbol{\Phi}$ in the plate bending analysis is a two-dimensional vector:

$$\boldsymbol{\Phi}^{\mathsf{T}} = \|[\Phi_x , \Phi_y]\| , \tag{2.10}$$

and the components of the moment tensor are expressed via the components of the vector of stress functions $\boldsymbol{\Phi}$ as follows:

$$M^{xx} = -\frac{\partial \Phi_y}{\partial y}, \quad M^{yy} = -\frac{\partial \Phi_x}{\partial x}, \quad M^{xy} = \frac{1}{2}\left(\frac{\partial \Phi_x}{\partial y} + \frac{\partial \Phi_y}{\partial x}\right). \tag{2.11}$$

And again we can see how the homogeneous equation of equlibrium that follows from (1.34),

$$-M^{xx}_{,xx} - 2M^{xy}_{,xy} - M^{yy}_{,yy} = 0 ,$$

is satisfied automatically by representations (2.11).

Based on the general formula (2.7) and making some necessary transformations, we conclude that the compatibility operator in stress functions, $\boldsymbol{M}$, for the plate bending is

$$\boldsymbol{M} = \frac{12}{Eh^3}\left\|\begin{bmatrix} \left(-\nabla^2 + \frac{1-\nu}{2}\frac{\partial^2}{\partial y^2}\right) & -\frac{1-\nu}{2}\frac{\partial^2}{\partial x \partial y} \\ -\frac{1-\nu}{2}\frac{\partial^2}{\partial x \partial y} & \left(-\nabla^2 + \frac{1-\nu}{2}\frac{\partial^2}{\partial x^2}\right) \end{bmatrix}\right\|. \tag{2.12}$$

Physical meaning of the stress function in plane stress

The stress function, Φ, can have a certain physical meaning, and it is useful to find out what that is. Let us begin with the plane stress.

Let a stress function, Φ, be an exact solution of the problem. This means that equation $\nabla^2\nabla^2\Phi = 0$ holds and all boundary conditions are met. We denote by P^x and P^y the forces per unit of length, which take place on contour Γ. Those values are predefined and equal to $P^x = h\overline{p}^x$, $P^y = h\overline{p}^y$, respectively, in locations of the contour where the original statement of the problem contains given static boundary conditions according to (**4**.3.7). In places where kinematic boundary conditions are formulated, P^x and P^y mean rcactivc forces which appear on the contour in response to an external load upon the plate. In any case, we have the following on the whole contour of the plate:

$$N^x n_x + N^{xy} n_y = P^x, \qquad N^{xy} n_x + N^y n_y = P^y \in \Gamma. \tag{2.13}$$

The samc boundary conditions, expressed via stress function Φ, will look like

$$\frac{\partial^2\Phi}{\partial y^2} n_x - \frac{\partial^2\Phi}{\partial x \partial y} n_y = P^x, \qquad -\frac{\partial^2\Phi}{\partial x \partial y} n_x + \frac{\partial^2\Phi}{\partial x^2} n_y = P^y \in \Gamma. \tag{2.14}$$

But we know that function Φ is defined up to linear terms because if Φ is a solution then the following stress function is also a solution:

$$\Phi + \alpha x + \beta y + \gamma\,,$$

where α, β, and γ are arbitrary constants. And indeed, adding the above terms to the stress function does not violate either the main equation, $\nabla^2\nabla^2\Phi = 0$, or boundary conditions (2.14) .

We can handle the constants in such way that we make stress function Φ itself and its first derivatives equal to zero in a certain predefined point O of contour Γ. This point O will be an origin of arc coordinate s counted off along contour Γ, so that

$$\Phi(0)=0, \qquad \frac{\partial\Phi}{\partial x}(0)=0\,, \qquad \frac{\partial\Phi}{\partial y}(0)=0\,.$$

Let a point S (a current point), located on contour Γ, have an arc coordinate s. According to formulas (F.57) from Appendix F, we have the following on the contour curve, Γ:

$$\frac{\partial}{\partial s}=t_i\frac{\partial}{\partial x_i}, \qquad \frac{\partial}{\partial n}=n_i\frac{\partial}{\partial x_i}; \qquad \frac{\partial}{\partial x_i}=n_i\frac{\partial}{\partial n}+t_i\frac{\partial}{\partial s}. \tag{2.15}$$

Using these equalities, expressions (2.14) for the boundary conditions can be rewritten as

$$\frac{\partial}{\partial s}\left(\frac{\partial\Phi}{\partial y}\right)=P^x, \qquad -\frac{\partial}{\partial s}\left(\frac{\partial\Phi}{\partial x}\right)=P^y \quad \in\Gamma\,.$$

By integrating the latter from the origin, O, to the current point, S, over the arc coordinate, we obtain

$$\frac{\partial\Phi}{\partial y}(s)=\int_0^s P^x ds=R^x, \qquad \frac{\partial\Phi}{\partial x}(s)=-\int_0^s P^y ds=-R^y, \tag{2.16}$$

where R^x and R^y are projections onto axes X and Y of the general vector, $\boldsymbol{R}$, of an external load applied to the piece of the contour from point O to point S. By differentiating function Φ with respect to s and using (2.15) and (2.16), we derive

$$\frac{\partial\Phi}{\partial s}=\frac{\partial\Phi}{\partial x}t_x+\frac{\partial\Phi}{\partial y}t_y = R^x n_x+R^y n_y=\boldsymbol{R}\cdot\boldsymbol{n}\,. \tag{2.17}$$

We can treat the piece of curve Γ from point O to point S formally as an axis of a planar curvilinear bar free from fixation in point O. Then the scalar product $\boldsymbol{R}\cdot\boldsymbol{n}$ can be treated as a shear force, Q, in the cross-section of that bar in point S, taken with an opposite sign, that is,[7]

$$\boldsymbol{R}\cdot\boldsymbol{n}=-Q.$$

Recalling Kirchhoff equations (**4**.7.30) for a curvilinear bar, we will have (when there is no moment load, $m=0$)

[7] The convention of signs for stresses in cross-sections of a curvilinear bar follows Fig. 4.7.

$$\frac{dM}{ds} = Q\,,$$

where M is a bending moment in the bar's cross-section. Thus,

$$\frac{\partial\Phi}{\partial s} = -\,Q = -\frac{dM}{ds}\,.$$

Integrating the latter relationship from zero to s, remembering that $\Phi(0) = 0$, and assuming $M(0) = 0$, we have the final result:

$$\Phi(s) = -M(s)\,. \tag{2.18}$$

Thus, the value of stress function $\Phi(s)$ in an arbitrary point of boundary Γ with coordinate s is equal to a negative moment of all contour load (both active and reactive forces) applied to the piece of the contour from origin O to the current point, with respect to that point.

We further differentiate stress function Φ along the normal to contour Γ and take into account equalities (2.15) and (2.16) to derive

$$\frac{\partial\Phi}{\partial n} = \frac{\partial\Phi}{\partial x}n_x + \frac{\partial\Phi}{\partial y}n_y = -R^x t_x - R^y t_y = -\boldsymbol{R}\cdot\boldsymbol{t}\,.$$

The scalar product $\boldsymbol{R}\cdot\boldsymbol{t}$ can be treated as a negative longitudinal force, N, that develops in the cross-section of the "bar" at point S, and thus[8]

$$\frac{\partial\Phi}{\partial n}(s) - N(s)\,. \tag{2.19}$$

An important conclusion from the above is that the mathematical aspect of the solution of the plane stress problem for a plate with static boundary conditions is to find a function Φ, biharmonic in area Ω, such that its values and the values of its normal derivative are specified on boundary Γ as in (2.18) and (2.19), respectively.

Physical meaning of the stress function in plate bending

Now let us find out what the physical meaning of the vector of stress functions $\boldsymbol{\Phi}$ is in the plate bending analysis. Formula (2.12) for operator $\boldsymbol{M}$ in the plate bending gives that the components of vector function $\boldsymbol{\Phi}$ satisfy the following homogeneous system of equations:

---

[8] We remind that a longitudinal force in a bar is positive when it causes a tension.

$$-\frac{\partial^2\Phi_x}{\partial x^2}-\frac{1+\nu}{2}\frac{\partial^2\Phi_x}{\partial y^2}-\frac{1-\nu}{2}\frac{\partial^2\Phi_y}{\partial x\partial y}=0,$$

$$-\frac{1-\nu}{2}\frac{\partial^2\Phi_x}{\partial x\partial y}-\frac{\partial^2\Phi_y}{\partial y^2}-\frac{1+\nu}{2}\frac{\partial^2\Phi_y}{\partial x^2}=0. \tag{2.20}$$

The plate is subject to the following on contour Γ: a generalized shear force, K^n, and a bending moment, M^{nn}, so that[9]

$$K^n = M^{ij}_{,i}n_j + M^{nt}_{,s}, \qquad M^{nn} = M^{ij}n_in_j\,. \tag{2.21}$$

Further, formulas (2.11), which establish a relation between the components of the stress functions vector and those of the moment tensor, can be written in the following tensor form:

$$M^{ij}=\frac{1}{2}\omega^{ik}\left(\frac{\partial\Phi_k}{\partial x_l}+\frac{\partial\Phi_l}{\partial x_k}\right)\omega^{lj}, \tag{2.22}$$

where ω^{ij} is an antisymmetric tensor introduced by formula (F.36) in Appendix F, which we called a rotation tensor. The equivalence between (2.11) and (2.22) can be easily checked by a direct substitution.

Seeing that, according to formula (F.39),

$$t_l=\omega^{lj}n_j \ \text{ and } \ t_k=-\omega^{ik}n_i\,,$$

now we want to represent boundary forces according to (2.21) via the components of the stress functions vector[10].

Let us begin with an expression for bending moment M^{nn} on the contour of the plate. We have

$$M^{nn}=\frac{1}{2}\omega^{ik}\left(\frac{\partial\Phi_k}{\partial x_l}+\frac{\partial\Phi_l}{\partial x_k}\right)\omega^{lj}n_in_j=-\frac{1}{2}\left(\frac{\partial\Phi_k}{\partial x_l}+\frac{\partial\Phi_l}{\partial x_k}\right)t_kt_l=-t_i\frac{\partial\Phi_i}{\partial s}. \tag{2.23}$$

Further,

[9] In order to simplify the formulas, we confine ourselves to considering only a smooth piece of contour Γ that does not contain any breakpoints.

[10] Here and in other places we use a formal lifting and/or lowering of indexes of vectors and/or tensors without additional comments, only to follow the convention of summation over repeating indexes placed on different levels. As known from tensor analysis, this operation is admissible for tensor components defined in a Cartesian coordinate system.

$$M^{nt}_{,s} = \frac{\partial}{\partial s}(M^{ij} n_i t_j) = \frac{1}{2}\frac{\partial}{\partial s}\left[\omega^{ik}\left(\frac{\partial \Phi_k}{\partial x_l} + \frac{\partial \Phi_l}{\partial x_k}\right)\omega^{lj} n_i t_j\right] =$$

$$= \frac{1}{2}\frac{\partial}{\partial s}\left[\left(\frac{\partial \Phi_k}{\partial x_l} + \frac{\partial \Phi_l}{\partial x_k}\right) n_l t_k\right] = \frac{1}{2}\frac{\partial}{\partial s}\left(t_i \frac{\partial \Phi_i}{\partial n} + n_i \frac{\partial \Phi_i}{\partial s}\right).$$

Also,

$$M^{ij}_{,i} n_j = \frac{1}{2}\frac{\partial}{\partial x_i}\left[\omega^{ik}\left(\frac{\partial \Phi_k}{\partial x_l} + \frac{\partial \Phi_l}{\partial x_k}\right)\omega^{lj}\right] n_j = \frac{1}{2}\omega^{ik}\left(\frac{\partial^2 \Phi_k}{\partial x_l \partial x_i} + \frac{\partial^2 \Phi_l}{\partial x_k \partial x_i}\right) t_l .$$

It is easy to notice that

$$\omega^{ik} \frac{\partial^2 \Phi_l}{\partial x_k \partial x_i} = 0 \quad (l = 1,2),$$

which can be validated by writing this aggregate out with a component-by-component summation. Consequently,

$$M^{ij}_{,i} n_j = \frac{1}{2}\omega^{ik} \frac{\partial^2 \Phi_k}{\partial s \partial x_i} = \frac{1}{2}\frac{\partial}{\partial s}\left[\omega^{ik}\left(n_i \frac{\partial \Phi_k}{\partial n} + t_i \frac{\partial \Phi_k}{\partial s}\right)\right] =$$

$$= \frac{1}{2}\frac{\partial}{\partial s}\left(-t_k \frac{\partial \Phi_k}{\partial n} + n_k \frac{\partial \Phi_k}{\partial s}\right).$$

Summing both terms for K^n yields a simple formula for Kirchhof's shear force on the contour via the components of the stress functions vector

$$K^n = M^{ij}_{,i} n_j + M^{nt}_{,s} = \frac{\partial}{\partial s}\left(n_i \frac{\partial \Phi_i}{\partial s}\right). \tag{2.24}$$

Now we will show that if some functions Φ_x and Φ_y satisfy both equation system (2.20) and boundary conditions (2.23) and (2.24) then the following functions F_x and F_y have the same properties:

$$F_x = \Phi_x + \alpha - \gamma y , \qquad F_y = \Phi_y + \beta + \gamma x ,$$

where α, β, and γ are arbitrary constants. And indeed, adding any linear term does not violate system of equations (2.20) because second derivatives of functions F_x and F_y annihilate the linear terms added to Φ_x and Φ_y. We have to check only that boundary conditions (2.23) and (2.24) hold. Let functions $x = X(s)$ and $y = Y(s)$ make up a parametric description of boundary Γ. This means that we have the following in any point of the boundary when moving along it:

$$\frac{dX}{ds}=t_x, \quad \frac{dY}{ds}=t_y.$$

So we have

$$t_i\frac{\partial \mathrm{F}_i}{\partial s}=t_x\frac{\partial \mathrm{F}_x}{\partial s}+t_y\frac{\partial \mathrm{F}_y}{\partial s}=t_i\frac{\partial \Phi_i}{\partial s}-t_x\gamma\frac{dY}{ds}+t_y\gamma\frac{dX}{ds}=t_i\frac{\partial \Phi_i}{\partial s}$$

and

$$\frac{\partial}{\partial s}\left(n_i\frac{\partial \mathrm{F}_i}{\partial s}\right)=\frac{\partial}{\partial s}\left(n_i\frac{\partial \Phi_i}{\partial s}\right)+\gamma\frac{\partial}{\partial s}(-n_xt_y+n_yt_x)=\frac{\partial}{\partial s}\left(n_i\frac{\partial \Phi_i}{\partial s}\right),$$

because $-n_xt_y+n_yt_x=-n_xn_x-n_yn_y=-1$. It is now clear that boundary conditions (2.23) and (2.24) hold for functions F_x and F_y, too.

By handling the α, β, and γ constants introduced above, we can always make the following conditions hold in a chosen point O of contour Γ:

$$\Phi_x(0)=0, \qquad \Phi_y(0)=0, \quad n_i(0)\frac{\partial \Phi_i}{\partial s}(0)=0.$$

This will be an origin of the arc coordinate, s, on contour Γ.

Integrating expression (2.24) of K^n over the arc coordinate from the origin, O, to the current point, S, will give

$$n_i\frac{\partial \Phi_i}{\partial s}=\int_0^s K^n ds=R_Z, \tag{2.25}$$

where R_Z is a projection onto the Z-axis of the general vector, $\boldsymbol{R}$, of an external load applied to the part of the contour between point O and point S.

Let us imagine that a piece of curve Γ from point O to point S is an axis of a planar curvilinear bar, free from fixation in point O and loaded by forces K^n which act out of the plane of the bar's axis and by torques M^{nn} distributed along the axis.

Now let's consider an elementary segment of that bar from a cross-section with coordinate s to a cross-section with coordinate $s+ds$ (Fig. 5.5) and denote by M_X and M_Y the moments with respect to axes X and Y, which act in the bar's cross-section with coordinate s. Equations of equilibrium of the moments with respect to axes X and Y for that segment of the bar will be

$$\frac{dM_X}{ds} = M^{nn}t_x - R_Z t_y, \qquad \frac{dM_Y}{ds} = M^{nn}t_y + R_Z t_x. \tag{2.26}$$

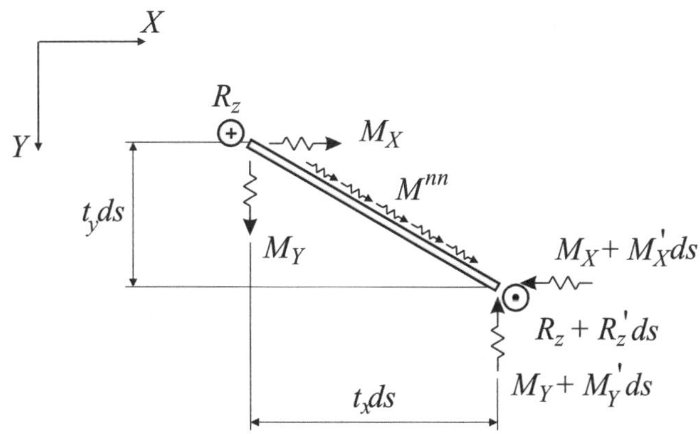

Fig. 5.5. An elementary fragment of a curvilinear bar

By integrating these equalities from zero to s and taking into account (2.23) and (2.25), we obtain

$$M_X = -\int_0^s \left(t_i t_x + n_i t_y\right)\frac{\partial \Phi_i}{\partial s} ds, \quad M_Y = -\int_0^s \left(t_i t_y - n_i t_x\right)\frac{\partial \Phi_i}{\partial s} ds.$$

But, as it can be easily seen,

$$t_i t_x + n_i t_y = \begin{cases} 1 & if\ i=1 \\ 0 & if\ i=2 \end{cases}, \qquad t_i t_y - n_i t_x = \begin{cases} 0 & if\ i=1 \\ 1 & if\ i=2 \end{cases},$$

hence the final result:

$$\Phi_x = -M_X, \qquad \Phi_y = -M_Y. \tag{2.27}$$

Thus, the vector of stress functions in every point S of contour Γ coincides (up to a sign) with the general moment of contour forces applied to a piece of the boundary between zero and s. Hence a conclusion: the mathematical aspect of the plate bending with static boundary conditions specified on the contour of the area is to find a vector function $\mathbf{\Phi}$ in area Ω such that its components satisfy the system of differential equations (2.20) and are known on boundary Γ according to (2.27).

Table 5.1

Bending	Plane stress
$\boldsymbol{\sigma}_{\mathrm{B}} = \|[M^{xx}, M^{yy}, M^{xy}]\|^{\mathsf{T}}$	$\boldsymbol{\sigma}_{\mathrm{P}} = \|[N^{x}, N^{y}, N^{xy}]\|^{\mathsf{T}}$
$\boldsymbol{\varepsilon}_{\mathrm{B}} = \|[\chi_{xx}, \chi_{yy}, 2\chi_{xy}]\|^{\mathsf{T}}$	$\boldsymbol{\varepsilon}_{\mathrm{P}} = \|[\varepsilon_{x}, \varepsilon_{y}, \gamma_{xy}]\|^{\mathsf{T}}$
$\mathbf{u}_{\mathrm{B}} = \|[w]\|^{\mathsf{T}}$	$\mathbf{u}_{\mathrm{P}} = \|[u, v]\|^{\mathsf{T}}$
$\boldsymbol{\Phi}_{\mathrm{B}} = \|[\Phi_{x}, \Phi_{y}]\|^{\mathsf{T}}$	$\boldsymbol{\Phi}_{\mathrm{P}} = \|[\Phi]\|^{\mathsf{T}}$
$\boldsymbol{A}_{\mathrm{B}} = \left\|\begin{bmatrix} -\frac{\partial^2}{\partial x^2} \\ -\frac{\partial^2}{\partial y^2} \\ -2\frac{\partial^2}{\partial x \partial y} \end{bmatrix}\right\|$	$\boldsymbol{A}_{\mathrm{P}} = \left\|\begin{bmatrix} \partial/\partial x & 0 \\ 0 & \partial/\partial y \\ \partial/\partial y & \partial/\partial x \end{bmatrix}\right\|$
$\boldsymbol{A}_{\mathrm{B}}^{\mathsf{T}} = \left\|\begin{bmatrix} -\frac{\partial^2}{\partial x^2}, -\frac{\partial^2}{\partial y^2}, -2\frac{\partial^2}{\partial x \partial y} \end{bmatrix}\right\|$	$\boldsymbol{A}_{\mathrm{P}}^{\mathsf{T}} = \left\|\begin{bmatrix} -\partial/\partial x & 0 & -\partial/\partial y \\ 0 & -\partial/\partial y & -\partial/\partial x \end{bmatrix}\right\|$
$\boldsymbol{C}_{\mathrm{B}} = \frac{Eh^3}{12(1-\nu^2)} \left\|\begin{bmatrix} 1 & \nu & 0 \\ \nu & 1 & 0 \\ 0 & 0 & (1-\nu)/2 \end{bmatrix}\right\|$	$\boldsymbol{C}_{\mathrm{P}} = \frac{Eh}{1-\nu^2} \left\|\begin{bmatrix} 1 & \nu & 0 \\ \nu & 1 & 0 \\ 0 & 0 & (1-\nu)/2 \end{bmatrix}\right\|$
$\boldsymbol{S}_{\mathrm{B}} = \left\|\begin{bmatrix} 0 & \frac{\partial}{\partial x} & -\frac{\partial}{2\partial y} \\ \frac{\partial}{\partial y} & 0 & -\frac{\partial}{2\partial x} \end{bmatrix}\right\|$	$\boldsymbol{S}_{\mathrm{P}} = \left\|\begin{bmatrix} \frac{\partial^2}{\partial y^2}, & \frac{\partial^2}{\partial x^2}, & -\frac{\partial^2}{\partial x \partial y} \end{bmatrix}\right\|$
$\boldsymbol{L}_{\mathrm{B}} = \boldsymbol{A}_{\mathrm{B}}^{\mathsf{T}} \boldsymbol{C}_{\mathrm{B}} \boldsymbol{A}_{\mathrm{B}} = \frac{Eh^3}{12(1-\nu^2)} \nabla^2\nabla^2$	$\boldsymbol{L}_{\mathrm{P}} = \boldsymbol{A}_{\mathrm{P}}^{\mathsf{T}} \boldsymbol{C}_{\mathrm{P}} \boldsymbol{A}_{\mathrm{P}} = \frac{Eh}{1-\nu^2} \times$ $\times \left\|\begin{bmatrix} -\nabla^2 + \frac{1+\nu}{2}\frac{\partial^2}{\partial y^2} & -\frac{1+\nu}{2}\frac{\partial^2}{\partial x \partial y} \\ -\frac{1+\nu}{2}\frac{\partial^2}{\partial x \partial y} & -\nabla^2 + \frac{1+\nu}{2}\frac{\partial^2}{\partial x^2} \end{bmatrix}\right\|$
$\boldsymbol{M}_{\mathrm{B}} = \boldsymbol{S}_{\mathrm{B}} \boldsymbol{C}_{\mathrm{B}}^{-1} \boldsymbol{S}_{\mathrm{B}}^{\mathsf{T}} = \frac{12}{Eh^3} \times$ $\times \left\|\begin{bmatrix} -\nabla^2 + \frac{1-\nu}{2}\frac{\partial^2}{\partial y^2} & -\frac{1-\nu}{2}\frac{\partial^2}{\partial x \partial y} \\ -\frac{1-\nu}{2}\frac{\partial^2}{\partial x \partial y} & -\nabla^2 + \frac{1-\nu}{2}\frac{\partial^2}{\partial x^2} \end{bmatrix}\right\|$	$\boldsymbol{M}_{\mathrm{P}} = \boldsymbol{S}_{\mathrm{P}} \boldsymbol{C}_{\mathrm{P}}^{-1} \boldsymbol{S}_{\mathrm{P}}^{\mathsf{T}} = \frac{1}{Eh} \nabla^2\nabla^2$

5.2.2 A static-geometric analogy in the theory of plates

Now let us define what is a static-geometric analogy. For convenience, we again write out basic matrix relationships of the plate bending (marked by subscript $_B$) and the plane stress (marked by subscript $_P$) in two parallel columns of Table 5.1. Also, please note that Table 5.1 describes the case of an isotropic material.

To begin with, notice that elasticity matrices $\boldsymbol{C}$ in these two problems are the same, up to factor c^2:

$$c^2\boldsymbol{C}_P = \boldsymbol{C}_B\,, \qquad c^2 = h^2/12. \tag{2.28}$$

Further, if we introduce matrix $\boldsymbol{\Lambda}$ as

$$\boldsymbol{\Lambda} = \left\|\begin{bmatrix} 0 & -1 & 0 \\ -1 & 0 & 0 \\ 0 & 0 & 2 \end{bmatrix}\right\|, \qquad \boldsymbol{\Lambda}^{-1} = \left\|\begin{bmatrix} 0 & -1 & 0 \\ -1 & 0 & 0 \\ 0 & 0 & 1/2 \end{bmatrix}\right\|, \tag{2.29}$$

we will find that:

- the operator of equlibrium, $\boldsymbol{A}^{\mathsf{T}}$, in one of the two problems becomes the operator of strain compatibility, $\boldsymbol{S}$, in the other problem and vice versa, by means of a transformation with the $\boldsymbol{\Lambda}$ matrix. Strictly speaking, the following relationships take place:

$$\boldsymbol{S}_P = \boldsymbol{A}_B^{\mathsf{T}}\boldsymbol{\Lambda}^{-1}, \qquad \boldsymbol{S}_B = \boldsymbol{A}_P^{\mathsf{T}}\boldsymbol{\Lambda}^{-1},$$

$$\boldsymbol{A}_P^{\mathsf{T}} = \boldsymbol{S}_B\boldsymbol{\Lambda}\,, \qquad \boldsymbol{A}_B^{\mathsf{T}} = \boldsymbol{S}_P\boldsymbol{\Lambda}\,. \tag{2.30}$$

  We can also say that the homogeneous equations of equilibrium for one problem turn into the strain compatibility equations for the other problem, and vice versa;
- the Lame operator, $\boldsymbol{L}_B$, for the plate bending coincides, up to a factor, with the compatibility operator in stress functions, $\boldsymbol{M}_P$, for the plane stress;
- the Lame operator, $\boldsymbol{L}_P$, coincides with the $\boldsymbol{M}_B$ operator up to a factor under an additional condition – the sign of Poisson ratio ν should be altered[11].

A reciprocity of the operators of the two mechanical problems as described above is known as a static-geometric analogy in the theory of plates. It is a particular manifestation of a more general static-geometric

[11] By the way, a negative value of the Poisson ratio does not violate the plane stress problem formulation because the matrix of elasticity, $\boldsymbol{C}$, is positive definite at any $|\nu|<1$, including negative values of ν.

analogy known in the classic theory of shells by Kirchhoff and Love [7], [15].

The reciprocity of the equations of the plane stress/strain and the plate bending makes it possible in some cases (when there is no load distributed over area Ω and under certain boundary conditions) to reduce one of the problems to the other.

Let us first consider a boundary-value problem for the plate bending with purely kinematical boundary conditions and a boundary-value problem for the plane stress with purely static boundary conditions[12]. These problems are formulated as follows:

$$\boldsymbol{A}_{\mathrm{B}}\mathbf{u}_{\mathrm{B}} = \boldsymbol{\varepsilon}_{\mathrm{B}}\,, \quad \boldsymbol{A}_{\mathrm{B}}^{\mathrm{T}}\boldsymbol{\sigma}_{\mathrm{B}} = \mathbf{0}\,, \quad \boldsymbol{C}_{\mathrm{B}}\boldsymbol{\varepsilon}_{\mathrm{B}} = \boldsymbol{\sigma}_{\mathrm{B}} \in \Omega\,; \quad \boldsymbol{u}_{\mathrm{B}} - \bar{\boldsymbol{u}}_{\mathrm{B}} = \mathbf{0} \in \Gamma\,, \quad (2.31\text{-}b)$$

$$\boldsymbol{A}_{\mathrm{P}}\mathbf{u}_{\mathrm{P}} = \boldsymbol{\varepsilon}_{\mathrm{P}}\,, \quad \boldsymbol{A}_{\mathrm{P}}^{\mathrm{T}}\boldsymbol{\sigma}_{\mathrm{P}} = \mathbf{0}\,, \quad \boldsymbol{C}_{\mathrm{P}}\boldsymbol{\varepsilon}_{\mathrm{P}} = \boldsymbol{\sigma}_{\mathrm{P}} \in \Omega\,; \quad \boldsymbol{p}_{\mathrm{P}} - \bar{\boldsymbol{p}}_{\mathrm{P}} = \mathbf{0} \in \Gamma\,. \quad (2.31\text{-}p)$$

It can be established that problems (2.31-*p*) and (2.31-*b*) are transformable into each other in the mathematical sense. And indeed, if we write the formulation of problem (2.31-*b*) in terms of displacements and the formulation of problem (2.31-*p*) in terms of the stress function, we will have, respectively,

$$\boldsymbol{L}_{\mathrm{B}}w = 0 \in \Omega\,; \qquad w = \bar{w}\,, \quad w_{,n} = -\bar{\theta}_t \in \Gamma\,,$$

$$\boldsymbol{M}_{\mathrm{P}}\Phi = 0 \in \Omega\,; \qquad \Phi = -\bar{M}\,, \quad \Phi_{,n} = \bar{N} \in \Gamma\,,$$

where $\bar{M}$ and $\bar{N}$ are a bending moment and a longitudinal force in a curvilinear bar depicted by curve Γ, cut apart in point O, and subjected to loads $\bar{\boldsymbol{p}}_{\mathrm{P}}$ defined on contour Γ. It is obvious that the two problems are identical from the standpoint of mathematics.

In the same way, we can write out an operator formulation of a boundary-value problem of the plate bending with purely static boundary

---

[12] Here we are conscious in our not using the terms of *first and second boundary-value problems* because this nomenclature, though frequently encountered, is not unified in the general use. For example, mathematicians, starting from S.G. Mikhlin [17], use the term of a *first boundary-value problem of elasticity* in their papers to refer to a problem with purely kinematical boundary conditions, while their term of a *second boundary-value problem* refers to purely static boundary conditions. The same nomenclature is used by L.A. Rozin [21]. In other papers, mainly by mechanicians (for example, see A.I. Lurie [16] or N.I. Muskhelishvili [18]), an exactly opposite terminology is used.

conditions and a formulation of the plane stress problem with purely kinematic boundary conditions,

$$\boldsymbol{A}_B\mathbf{u}_B = \boldsymbol{\varepsilon}_B\,, \quad \boldsymbol{A}_B^{\mathsf{T}}\boldsymbol{\sigma}_B = \mathbf{0}\,, \quad \boldsymbol{C}_B\boldsymbol{\varepsilon}_B = \boldsymbol{\sigma}_B \in\Omega\,; \quad \boldsymbol{p}_B - \bar{\boldsymbol{p}}_B = \mathbf{0} \in\varGamma\,, \tag{2.32-b}$$

$$\boldsymbol{A}_P\mathbf{u}_P = \boldsymbol{\varepsilon}_P\,, \quad \boldsymbol{A}_P^{\mathsf{T}}\boldsymbol{\sigma}_P = \mathbf{0}\,, \quad \boldsymbol{C}_P\boldsymbol{\varepsilon}_P = \boldsymbol{\sigma}_P \in\Omega\,; \quad \boldsymbol{u}_P - \bar{\boldsymbol{u}}_P = \mathbf{0} \in\varGamma\,. \tag{2.32-p}$$

As it follows from the above, problems (2.32-*p*) and (2.32-*b*) also turn out to be reducible to each other; however, in the latter case this reduction involves an additional condition: the Poisson ratio's sign in one of the two problems should be altered. To see this, we can write the formulation of problem (2.32-*b*) in terms of stress functions and the formulation of problem (2.31-*p*) in displacements, to derive

$$\boldsymbol{M}_B\boldsymbol{\Phi}_B = \mathbf{0} \in\Omega\,; \qquad \Phi_x = -\bar{M}_X\,, \quad \Phi_y = -\bar{M}_Y \in\varGamma\,,$$

$$\boldsymbol{L}_P\mathbf{u}_P = \mathbf{0} \in\Omega\,; \qquad u = \bar{u}\,, \quad v = \bar{v} \in\varGamma\,,$$

where $\bar{M}_X$ and $\bar{M}_Y$ are components (with respect to the respective axes X and Y) of a moment that acts in the cross-sections of a curvilinear bar depicted by curve $\varGamma$, cut apart in point O, and subjected to force loads $\bar{K}^n$ and moment loads $\bar{M}^{nn}$ defined on contour $\varGamma$.

This latter fact was established for the first time, apparently, in works by A.I. Lurie [16], [15] in a way different from what we use here. In essense, this is a coincidence between the operators $\boldsymbol{L}_P$ and $\boldsymbol{M}_B$ provided the sign of Poisson ratio ν is altered.

Proceeding from our static-geometric analogy, we can notice that Airy function Φ being determined in the plane stress problem up to a term $\alpha + \beta x + \gamma y$ can be interpreted as a calculation of the plate's deflection, w, up to a rigid displacement, $w_{\mathscr{R}} = \alpha + \beta x + \gamma y$, that does not affect the stress distribution in the plate. In a perfect resemblance to this, the vector of stress functions $\boldsymbol{\Phi} = \|[\Phi_x,\Phi_y]\|^{\mathsf{T}}$ is defined in the plate bending up to a vector

$$\|[\alpha - \gamma y, \beta + \gamma x]\|^{\mathsf{T}},$$

because the same vector is a vector of rigid displacements

$$\|[u_{\mathscr{R}}, v_{\mathscr{R}}]\|^{\mathsf{T}} = \|[\alpha - \gamma y,\ \beta + \gamma x]\|^{\mathsf{T}}$$

for the plane stress problem.

The static-geometric analogy is usually treated like this: if we make the following replacements/substitutes of functions in the equations of equlibrium and strain compatibility:

$$\boldsymbol{\varepsilon}_\mathrm{P} \to \boldsymbol{\Lambda}\boldsymbol{\sigma}_\mathrm{B}\,, \qquad \boldsymbol{\varepsilon}_\mathrm{B} \to \boldsymbol{\Lambda}\boldsymbol{\sigma}_\mathrm{P}\,,$$

$$\boldsymbol{\Lambda}^{-1}\boldsymbol{\varepsilon}_\mathrm{P} \leftarrow \boldsymbol{\sigma}_\mathrm{B}\,, \qquad \boldsymbol{\Lambda}^{-1}\boldsymbol{\varepsilon}_\mathrm{B} \leftarrow \boldsymbol{\sigma}_\mathrm{P}\,, \tag{2.33}$$

then equilibrium equations $\boldsymbol{A}^\mathsf{T}\boldsymbol{\sigma} = \mathbf{0}$ in one of the problems will become strain compatibility equations $\boldsymbol{S}\boldsymbol{\varepsilon} = \mathbf{0}$ in the other problem. And vice versa, the compatibility equations in any of the problems will become equilibrium equations in the other problem.

If we find a formal way to understand the substitution symbols in (2.33) as equalities, then we immediately arrive at relationships (2.30) established above. A corollary of substitutions (2.33) is a reciprocity (in a certain understanding) between the displacement function in one of the problems and the stress functions in the other,

$$\mathbf{u}_\mathrm{B} \leftrightarrow \boldsymbol{\Phi}_\mathrm{P}\,, \qquad \mathbf{u}_\mathrm{P} \leftrightarrow \boldsymbol{\Phi}_\mathrm{B}\,. \tag{2.34}$$

5.2.3 Boundary conditions for deformations in the theory of plates

As our previous presentment shows, the cornerstone of all constructions of variational formulations for problems of structural mechanics is the basic integral identity (1.2.19) which we will rewrite here as

$$(\boldsymbol{A}\mathbf{u}, \boldsymbol{\sigma}) = (\mathbf{u}, \boldsymbol{A}^\mathsf{T}\boldsymbol{\sigma}) + (\boldsymbol{H}_\sigma\boldsymbol{\sigma}, \boldsymbol{H}_u\mathbf{u})_\Gamma\,. \tag{2.35}$$

The contour scalar product, $(\boldsymbol{H}_\sigma\boldsymbol{\sigma}, \boldsymbol{H}_u\mathbf{u})_\Gamma$, also written as $(\boldsymbol{p}, \boldsymbol{u})_\Gamma$ in other designations, can be interpreted from the mechanical standpoint as a virtual work, A_Γ, of contour forces $\boldsymbol{p} = \boldsymbol{H}_\sigma\boldsymbol{\sigma}$ (both active and reactive) on the respective contour displacements $\boldsymbol{u} = \boldsymbol{H}_u\mathbf{u}$,

$$\mathsf{A}_\Gamma = (\boldsymbol{p}, \boldsymbol{u})_\Gamma\,. \tag{2.36}$$

As we showed earlier, the multipliers which participate in this scalar product also take part in the construction of boundary conditions.

Suppose we can compose this expression in another way, that is, construct different multipliers but maintain the final value of the contour scalar product, thus keeping the basic integral identity true, by assuming something like

$$(\boldsymbol{p}, \boldsymbol{u})_\Gamma = (\boldsymbol{a}, \boldsymbol{b})_\Gamma\,,$$

where vector $\boldsymbol{a}$ is not identical to $\boldsymbol{H}_\sigma\boldsymbol{\sigma}$ and vector $\boldsymbol{b}$ differs from $\boldsymbol{H}_u\mathbf{u}$. But then nothing restrains us from identifying vector $\boldsymbol{a}$ with a certain generalized contour force and vector $\boldsymbol{b}$ with a generalized contour displacement which is conjugate by energy to that force. Exactly speaking, we will suppose that

$$\boldsymbol{a} = \boldsymbol{H}_a \boldsymbol{\sigma}\,, \qquad \boldsymbol{b} = \boldsymbol{H}_b \mathbf{u}\,,$$

where operators $\boldsymbol{H}_a$ and $\boldsymbol{H}_b$ are different from operators $\boldsymbol{H}_\sigma$ and $\boldsymbol{H}_u$.

A different representation of the contour scalar product permits us to formulate the boundary conditions in another way, too, at the same time complying with all requirements of Kirchhoff's theorem of uniqueness of the solution. All that the boundary conditions have to do is to specify either of the following along i-th coordinate on any piece of contour Γ: a force component, a^i, or a kinematical component, b_i (following the exclusive "or" principle)[13]. We already encountered an example of an ambiguity of this kind in Chapter **4**, in a representation of a contour scalar product when we considered various formulations of boundary conditions for a curvilinear bar.

It turns out that the theory of plates always permits to represent the contour-based scalar product $(\boldsymbol{a}, \boldsymbol{b})_\Gamma$ in such way that the vector of generalized contour forces, $\boldsymbol{a}$, may have the meaning of a vector of forces in a curvilinear bar depicted by curve Γ. When we find out the mechanical meaning of vector $\boldsymbol{b}$, we conclude that its components are parameters which characterize a deformation of the curvilinear bar. A related conclusion can be made about the three-dimensional elasticity [20].

What follows from our consideration is a set of so-called *boundary conditions for deformations*.

We will consider contour Γ to be an axis of a closed curvilinear bar. We choose an arbitrary point O on contour Γ and assign it to be an origin for arc coordinate s.

Let a set of contour forces $\boldsymbol{p}$ applied to the curvilinear bar on a piece of contour Γ between the origin, O, and the current point, S, with arc coordinate s, form a general vector, $\boldsymbol{R}$, and a general moment, $\boldsymbol{M}$.

Boundary conditions for deformations in plane stress

For the plane stress analysis, general vector $\boldsymbol{R}$ is a two-component vector that belongs to the (X,Y)-plane,

$$\boldsymbol{R} = \|[R^x, R^y]\|^{\mathsf{T}}\,. \tag{2.37}$$

[13] There is a significant requirement: the respective algebraic operators of boundary condition extraction, $\boldsymbol{E}_a$ and $\boldsymbol{E}_b$, should be diagonal matrices with zeros and units in their main diagonal, and the matrices should be also related as $\boldsymbol{E}_a + \boldsymbol{E}_b = \boldsymbol{I}$, $\boldsymbol{E}_a\boldsymbol{E}_b = \boldsymbol{O}$.

We will not need general moment $\boldsymbol{M}$ itself for our plane stress analysis; instead, we will be interested with another moment M — a bending moment calculated in cross-section S of a planar curvilinear bar cut apart at point O. That will be a moment of all forces with respect to the current point, S, applied to a piece of the curvilinear bar between point O to point S.

The components of the general force vector, R^x, R^y, projected onto axes $\boldsymbol{n}$ and $\boldsymbol{t}$, will make up a shear force, Q, and a longitudinal force, N, in the same cross-section of the curvilinear bar:

$$Q = -\boldsymbol{R}\cdot\boldsymbol{n} = -R^x n_x - R^y n_y\,, \qquad N = -\boldsymbol{R}\cdot\boldsymbol{t} = -R^x t_x - R^y t_y\,. \tag{2.38}$$

Obviously,

$$\boldsymbol{R}(s) = \int_0^s \boldsymbol{p}ds + \boldsymbol{R}(0)\,. \tag{2.39}$$

Further, the work $\mathsf{A}_\Gamma(s)$ of contour forces distributed on a piece of the boundary between point O and point S can be represented as follows in the plane stress analysis:

$$\mathsf{A}_\Gamma(s) = \int_0^s \boldsymbol{p}\cdot\boldsymbol{u}ds = \int_0^s \frac{d\boldsymbol{R}}{ds}\cdot\boldsymbol{u}ds = \left[\boldsymbol{R}\cdot\boldsymbol{u}\right]\Big|_0^s - \int_0^s \boldsymbol{R}\cdot\frac{d\boldsymbol{u}}{ds}ds\,. \tag{2.40}$$

Now we represent the vector of contour displacements, $\boldsymbol{u}$, by its decomposition into components along the local basis' axes:

$$\boldsymbol{u} = u_n\boldsymbol{n} + u_t\boldsymbol{t}\,. \tag{2.41}$$

Then, taking into account the Frenet formulas (F.52), we will have

$$\frac{d\boldsymbol{u}}{ds} = \left(\frac{du_n}{ds} - ku_t\right)\boldsymbol{n} + \left(\frac{du_t}{ds} + ku_n\right)\boldsymbol{t}\,. \tag{2.42}$$

Now we use formulas (**4**.7.16) and (**4**.7.26) for longitudinal strain ε_0 on the axis of the curvilinear bar and for slope θ of its cross-section, replacing the earlier designations by those used here, $u_t \to v$ and $u_n \to w$, to obtain

$$\theta = -\frac{du_n}{ds} + ku_t\,, \quad \varepsilon_0 = \frac{du_t}{ds} + ku_n\,, \tag{2.43}$$

which gives

$$\frac{d\boldsymbol{u}}{ds} = -\theta\boldsymbol{n} + \varepsilon_0\boldsymbol{t}\,. \tag{2.44}$$

So we have

$$\mathsf{A}_\Gamma(s) = \left[\boldsymbol{R}\cdot\boldsymbol{u}\right]\Big|_0^s + \int_0^s \left(\theta\boldsymbol{R}\cdot\boldsymbol{n} - \varepsilon_0\boldsymbol{R}\cdot\boldsymbol{t}\right)ds =$$

$$= \left[\boldsymbol{R}\cdot\boldsymbol{u}\right]\Big|_0^s + \int_0^s \left(-\theta Q + \varepsilon_0 N\right)ds = \left[\boldsymbol{R}\cdot\boldsymbol{u}\right]\Big|_0^s + \int_0^s \left(-\theta\frac{dM}{ds} + \varepsilon_0 N\right)ds\,. \qquad (2.45)$$

Here we have used one of Kirchhoff's equations (**4**.7.30) for a curvilinear bar, which establishes a relationship between a shear force and a bending moment (in our case we should assume $m = 0$ in that equation).

We use the integration by parts once again:

$$\mathsf{A}_\Gamma(s) = \left[\boldsymbol{R}\cdot\boldsymbol{u}\right]\Big|_0^s - \left[\theta M\right]\Big|_0^s + \int_0^s \left(\kappa M + \varepsilon_0 N\right)ds\,, \qquad (2.46)$$

where κ is a parameter of flexural deformation of a curvilinear bar of a small curvature, which is a derivative of slope θ as can be seen from the respective formula in Table **4**.1:

$$\kappa = \frac{d\theta}{ds}\,. \qquad (2.47)$$

We supposed implicitly in the derivation of formula (2.46) for $\mathsf{A}_\Gamma(s)$ that curve Γ is smooth between point O and current point S, that is, it does not contain breakpoints.

Suppose that area Ω is simply connected; consequently, its bounding contour Γ is a single closed curve. Let the whole curve Γ of a length L consist of q smooth pieces separated by breakpoints with coordinates s_β, so the whole contour Γ is a combination of those smooth pieces,

$$\Gamma = \bigcup_{\beta=1}^{q} \Gamma_\beta\,. \qquad (2.48)$$

We assume that a piece of the curve, Γ_β, extends between a point with coordinate $s_{\beta-1}$ to a point with coordinate s_β. Thus, point O is also included formally in the set of breakpoints of the contour. The last piece of the contour starts at the point with the coordinate s_{q-1} and ends at the point with the coordinate $s_q = L$, that is, actually at point O.

Formula (2.46) can be applied now to each of q smooth pieces of the contour, so that the expression of the work of all contour forces, $\mathsf{A}_{\Gamma\beta}$, applied to a piece Γ_β of the contour curve, is written as

$$\mathsf{A}_{\Gamma\beta}=\int_{s_{\beta-1}}^{s_\beta}(\kappa M+\varepsilon_0 N)\,ds+\big[\boldsymbol{R}\cdot\boldsymbol{u}\big]\Big|_{s_{\beta-1}}^{s_\beta}-\big[\theta M\big]\Big|_{s_{\beta-1}}^{s_\beta}.$$

This formula uses designations of $\boldsymbol{R}$, M, and N for the respective quantities calculated for a curvilinear bar free from fixation at the point $s = 0$; that is, a build-up takes place when we move from one smooth piece of the contour to the next one.

By summing the work of the contour forces over all pieces and taking into account an obvious continuity of functions $\boldsymbol{R}$, M, and $\boldsymbol{u}$, we find[14]

$$\mathsf{A}_\Gamma(L)=\oint_\Gamma(\kappa M+\varepsilon_0 N)\,ds+[\boldsymbol{R}]_0\cdot\boldsymbol{u}(0)-[M]_0[\theta]_0-\sum_{\beta=1}^{q-1}M(s_\beta)[\theta]_\beta\,. \tag{2.49}$$

This formula uses the following designations:

- $[\boldsymbol{R}]_0 = \boldsymbol{R}(L) - \boldsymbol{R}(0)$ is a jump of vector function $\boldsymbol{R}$ at the zero point of contour as we follow the closed trajectory of Γ around the contour;
- $[\theta]_\beta = \theta(s_\beta+0) - \theta(s_\beta-0)$ is a jump of function $\theta(s)$ in the transition over angular point s_β;
- $[M]_0 = M(L) - M(0)$ is a jump of the moment after a full circle around contour Γ.

If we denote by symbol $\boldsymbol{P}$ a general vector of all external forces distributed over area Ω, and by symbol M_0 a moment of the same forces with respect to the point of the contour with the coordinate $s = 0$, then the integral equilibrium of the body requires that

$$[\boldsymbol{R}]_0 = \boldsymbol{P}\,,\qquad [M]_0 = M_0\,. \tag{2.50}$$

In particular, the jumps should be equal to zero in absence of volumetric forces:

$$[\boldsymbol{R}]_0 = \boldsymbol{0}\,,\qquad [M]_0 = 0\,. \tag{2.51}$$

So, the contour integral in the basic integral formula (2.35) can be written as follows in our case:

[14] The continuity of force quantities $\boldsymbol{R}$ and M in all points of the contour, with the only possible exception of the zero point, follows from an assumption that contour forces $\boldsymbol{p}$ do not contain concentrated force actions. The continuity of the vector of contour displacements, $\boldsymbol{u}$, is identical to a requirement that the contour curve, Γ, should be continuous during the deformation of the structure.

$$(\boldsymbol{H}_\sigma\boldsymbol{\sigma},\boldsymbol{H}_u\mathbf{u})_\Gamma=$$

$$=\oint_\Gamma(\kappa M+\varepsilon_0 N)\,ds+[\boldsymbol{R}]_0\cdot\boldsymbol{u}(0)-[M]_0[\theta]_0-\sum_{\beta=1}^{q-1}M(s_\beta)[\theta]_\beta. \qquad (2.52)$$

Hence we have immediately that the boundary conditions for the plane stress can be written in this form, too:

$$\varepsilon_0=\bar{\varepsilon}_0\in\Gamma_\varepsilon,\qquad \kappa=\bar{\kappa}\in\Gamma_\kappa,\qquad N=\bar{N}\in\Gamma_N,\qquad M=\bar{M}\in\Gamma_M, \qquad (2.53)$$

where a dash over a letter designates given external actions, as usual. Obviously, the division of the whole contour into pieces Γ_ε, Γ_κ, Γ_N, Γ_M must satisfy the conditions

$$\Gamma=\Gamma_\varepsilon\cup\Gamma_N,\qquad \Gamma_\varepsilon\cap\Gamma_N=\varnothing;\qquad \Gamma=\Gamma_\kappa\cup\Gamma_M,\quad \Gamma_\kappa\cap\Gamma_M=\varnothing. \qquad (2.54)$$

It should be remembered also that the formulation of boundary conditions according to (2.54) on smooth pieces of the contour requires a specification of the slope jumps, $[\theta]_\beta$, in breakpoints of the contour according to (2.52). Jumps like that should be specified only in points s_β which belong to the piece of the boundary Γ_κ. As for the pieces of the boundary Γ_M, their respective jumps are obtained by the solution of the problem.

For a multiply connected area Ω, the full boundary Γ consists of multiple closed curves $\Gamma_i\,(i=1,\ldots,m)$, where m is a number of connectivity of area Ω, and the formulation of boundary conditions for deformations becomes more complicated. The reasons for this are as follows:

- for a multiply connected area, the contour integral in (2.52) should be understood as a sum of integrals over all closed contours, and non-integral terms should be also written separately for each of the closed contours and then summed together. It is also obvious that $[\boldsymbol{R}]_0$ and $[M]_0$ in formulas (2.50) and (2.51) now mean sums of the respective jumps on all closed contours;
- mutuall displacements of separate contours as rigid bodies should be defined somehow for a multiply connected area. A mutual displacement of two contours even without a deformation of the contours themselves will create a strain in area Ω.

Boundary conditions in the form of (2.53) are called boundary conditions for deformations in the plane stress analysis. The mechanical meaning of these conditions is that they define a deformation of the area's contour via kinematical parameters ε_0 and κ, and they define force actions on the contour via static parameters N and M.

Boundary conditions for deformations in plate bending

In the plate bending analysis, contour Γ of a plate is subjected to a vector of edge loads, $\boldsymbol{p} = \|[K^n, M^{nn}]\|^{\mathsf{T}}$. However, here we will be interested mainly with a general moment, $\boldsymbol{M}$, of all contour forces applied to a piece of the curvilinear contour between the origin, O, and the current point, S. This moment can be represented as a two-component moment vector,

$$\boldsymbol{M} = \|[M_X, M_Y]\| \,. \tag{2.55}$$

The general vector, $\boldsymbol{R}$, of contour forces applied to the piece of the boundary between point O and point S is a two-component force vector directed along the Z-axis so that

$$\boldsymbol{R} = \|[R_Z]\|. \tag{2.56}$$

The components of the general moment vector, M_X and M_Y, when projected onto axes $\boldsymbol{n}$ and $\boldsymbol{t}$, make up bending moment M_n and torque M_t in the same cross-section of the curvilinear bar:

$$M_n = \boldsymbol{M} \cdot \boldsymbol{n} = M_X n_x + M_Y n_y \,, \qquad M_t = \boldsymbol{M} \cdot \boldsymbol{t} = M_X t_x + M_Y t_y \,. \tag{2.57}$$

Obviously,

$$R_Z(s) = \int_0^s K^n ds + R_Z(0) \,. \tag{2.58}$$

Also, we derive from (2.57) and (2.26) that

$$\frac{dM_t}{ds} = \frac{dM_X}{ds} t_x + \frac{dM_Y}{ds} t_y - kn_x M_X - kn_y M_Y = M^{nn} - kM_n \,,$$

$$\frac{dM_n}{ds} = \frac{dM_X}{ds} n_x + \frac{dM_Y}{ds} n_y + kt_x M_X + kt_y M_Y = -R_Z + kM_t \,.$$

Consequently,

$$M^{nn} = \frac{dM_t}{ds} + kM_n \,, \qquad R_Z = -\frac{dM_n}{ds} + kM_t \,. \tag{2.59}$$

Further, work $\mathsf{A}_\Gamma(s)$ of contour forces distributed over the piece of the boundary from point O to point S can be represented as follows for the plate bending:

$$\mathsf{A}_\Gamma(s) = \int_0^s \left(K^n w - M^{nn} w_{,n} \right) ds = \int_0^s \left[\frac{dR_Z}{ds} w - \frac{dM_t}{ds} w_{,n} - kM_n w_{,n} \right] ds \,,$$

which yields the following after integration by parts:

$$\mathsf{A}_{I}(s) = \left[R_Z w\right]\Big|_0^s - \left[M_t w_{,n}\right]\Big|_0^s - \int_0^s \left[R_Z w_{,s} - M_t w_{,sn} + k M_n w_{,n})\right] ds \, .$$

Substituting the expression of R_Z from (2.59) and integrating by parts again produces

$$\mathsf{A}_{I}(s) = \left[R_Z w\right]\Big|_0^s - \left[M_t w_{,n}\right]\Big|_0^s + \left[M_n w_{,s}\right]\Big|_0^s +$$

$$+ \int_0^s \left[-M_n (w_{,ss} + k w_{,n}) + M_t (w_{,sn} - k w_{,s})\right] ds \, . \tag{2.60}$$

Non-integral terms in this equation, which depend on moments M_n and M_t, will be transformed using (2.57) and (F.57). We have

$$M_n w_{,s} - M_t w_{,n} = (M_X n_x + M_Y n_y)(w_{,x} t_x + w_{,y} t_y) -$$

$$-(M_X t_x + M_Y t_y)((w_{,x} n_x + w_{,y} n_y) = M_X w_{,y} - M_Y w_{,x} , \tag{2.61}$$

and in this form the expression is obviously continuous around the whole contour, including transitions across breakpoints..

Reasoning in the same way as in the plane stress analysis, we can finally write out a full work of the contour forces on all pieces of the bar as follows:

$$\mathsf{A}_{I}(L) = \sum_{\beta - 0}^{q-1} [R_Z]_\beta w(s_\beta) + [M_X]_0 w_{,y}(0) - [M_Y]_0 w_{,x}(0) +$$

$$+ \oint_I (M_n \kappa_n + M_t \kappa_t) ds \, , \tag{2.62}$$

where the designations are

$$\kappa_n = -w_{,ss} - k w_{,n} , \qquad \kappa_t = w_{,sn} - k w_{,s} \, . \tag{2.63}$$

It can be shown that quantities κ_n and κ_t are parameters of flexural and torsional deformation of a planar curvilinear bar from its plane of curvature.

Further details are omitted – the consequent reasoning is exactly the same as that for the plane stress.

5.3 Bending of medium-thickness plates – Reissner's theory

The classic theory by Kirchhoff and Love is known to describe the stresses in a bent plate with a satisfactory accuracy in the case when the respective thickness of the plate, h/L, where L is a characteristic linear size of the plate's plane, is a value significantly less than one. As the plate's thickness h increases, the error of the theory, created chiefly by overly strict kinematical limitations of the Kirchhoff–Love straight-normals hypothesis, tends to grow. In this regard we have to reformulate the plate bending problem as a problem of three-dimensional elasticity. The bending of plates considered as a part of three-dimensional elasticity analysis is usually referred to as a theory of thick plates, though sometimes this term refers to an analysis not quite in the style of three-dimensional elasticity.

There are intermediate versions of the theory which consider so-called medium-thickness plates. Various authors built quite a few versions of the medium-thickness plate bending theory at various times[15]; most of those theories are intended to refine the classic Kirchhoff–Love theory somehow. A general feature of those theories is that all of them are two-dimensional – that is, all stress and displacement parameters depend on two coordinates only, which makes the problem generally much easier. Of course, it seems hardly possible to indicate exact borders between thin plates,, medium-thickness plates, and thick plates. As it is often the case for any qualitative classification, the borders between those concepts are vague and unclear.

We will consider in detail one of simplest (and, apparently, most popular) versions of the medium-thickness plate theory, developed by E. Reissner[16]. The version of Reissner's theory discussed here is about the same in comparison with the Kirchhoff–Love theory as the theory of beam bending by Timoshenko is in comparison with the classic beam theory by Bernoulli–Euler. To be more specific, the theory stated below will be based on the same principles of weakening the straight-normals hypothesis, which we used when we modified the Bernoulli–Euler beam theory to turn it into the Timoshenko theory.

---

[15] One should not think that the process of construction of the medium-thickness plate theories and theories of thick plates is finished; the process is still underway.

[16] It should be said that E. Reissner returned to the plate bending analysis many times, so his name is associated actually with more than one competing version of the theory.

The version of the Reissner theory discussed here is based on a refined expression of the strain energy of a bent plate, supplemented by an additional term to allow for the work of tangential stresses τ^{xz} and τ^{yz}. In other words, we can adopt the following instead of (1.10):

$$\mathsf{E} = \int_{\Omega}\int_{-h/2}^{h/2} \left(\frac{\sigma^{xx}\varepsilon_{xx} + \sigma^{yy}\varepsilon_{yy} + \tau^{xy}\gamma_{xy}}{2} + \frac{\tau^{xz}\gamma_{xz} + \tau^{yz}\gamma_{yz}}{2} \right) dz d\Omega . \tag{3.1}$$

First of all, we want to consider a distribution (along the thickness of the plate) of tangential stresses τ^{xz} and τ^{yz}. Recall that formulas (1.25) and (1.28) give the following representations of stresses in any point of a plate in bending, the point having the (x,y,z) coordinates:

$$\sigma^{xx} = \frac{12M^{xx}}{h^3} z , \quad \sigma^{yy} = \frac{12M^{yy}}{h^3} z , \quad \tau^{xy} = \frac{12M^{xy}}{h^3} z . \tag{3.2}$$

As for tangential stresses τ^{xz} and τ^{yz}, we should have assumed those to be equal to zero had we followed strictly the third of the hypotheses from the beginning of Section **5**.1 and Hooke's law:

$$\tau^{xz} = G\gamma_{xz} = 0, \qquad \tau^{yz} = G\gamma_{yz} = 0. \tag{3.3}$$

In this regard, keeping the expressions in (3.2) intact, we will adopt the straight-normals assumption in a weakened formulation by assuming that the normal remains rectilinear after the plate's deformation but it does not have to remain perpendicular to the deformed surface of the plate.

Treading nearly the same path as we did when we were deriving Zhuravsky's formula for tangential stresses in the beam bending analysis, we will try to extract expressions for our desirable tangential stresses from the equations of equilibrium:

$$\begin{aligned} &\frac{\partial \sigma^{xx}}{\partial x} + \frac{\partial \tau^{xy}}{\partial y} + \frac{\partial \tau^{xz}}{\partial z} = 0, \\ &\frac{\partial \tau^{yx}}{\partial x} + \frac{\partial \sigma^{yy}}{\partial y} + \frac{\partial \tau^{yz}}{\partial z} = 0. \end{aligned} \tag{3.4}$$

After substituting formulas (3.2) and integrating the equations over the z-coordinate, we will have

$$\tau^{xz} = -\frac{6M^{xx}_{,x}}{h^3} z^2 - \frac{6M^{xy}_{,y}}{h^3} z^2 + C_1 , \qquad \tau^{yz} = -\frac{6M^{yy}_{,y}}{h^3} z^2 - \frac{6M^{xy}_{,x}}{h^3} z^2 + C_2 ,$$

where C_1 and C_2 are constants of integration.

Taking into account the conditions of an equality between tangential stresses τ^{xz} and τ^{yz} on the faces of the plate (at $z = \pm h/2$) and the respective tangential loads p^x and p^y specified on those faces (Fig. 5.1), we will determine these constants and arrive at the following formulas:

$$\tau^{xz} = \left(M^{xx}_{,x} + M^{xy}_{,y}\right)\frac{3(h^2-4z^2)}{2h^3} + p^x, \ \tau^{yz} = \left(M^{yy}_{,y} + M^{xy}_{,x}\right)\frac{3(h^2-4z^2)}{2h^3} + p^y.$$

But from (1.32) and (1.29) we have

$$M^{xx}_{,x} + M^{xy}_{,y} = Q^x - m^y, \quad M^{yy}_{,y} + M^{yx}_{,x} = Q^y + m^x, \quad m^x = -hp^y, \quad m^y = hp^x,$$

so the final expressions of the tangential stresses are

$$\tau^{xz} = Q^x \frac{3(h^2-4z^2)}{2h^3} + m^y\left(\frac{6z^2}{h^3} - \frac{1}{2h}\right),$$

$$\tau^{yz} = Q^y \frac{3(h^2-4z^2)}{2h^3} - m^x\left(\frac{6z^2}{h^3} - \frac{1}{2h}\right). \tag{3.5}$$

Thus, according to what we have derived, tangential stresses τ^{xz} and τ^{yz} are distributed over the thickness of the plate as a quadratic parabola. Obviously, (3.5) contradicts relationships (3.3) that follow from Hooke's law, as we already mentioned. But it is also obvious that the contradiction is a tradeoff of the straight-normals hypothesis.

Formulas (3.5) show that tangential stresses τ^{xz} and τ^{yz} consist of a sum of two quantities one of which is proportional to the respective shear force and the other to the external moment load. But if we want to consider an integral characteristic of the tangential stresses over the height of the cross-section, h, then we will find that the second term (proportional to the moment load) makes a zero contribution to the integral, so nothing will change in the general form of the integrals in (1.31),

$$Q^x = \int_{-h/2}^{h/2} \tau^{xz} dz, \qquad Q^y = \int_{-h/2}^{h/2} \tau^{yz} dz,$$

if the integrands include only a part of tangential stresses τ^{xz} and τ^{yz} proportional to the shear forces.

Further, based on formula (1.2) and taking $u_0 = 0$, $v_0 = 0$, which is an exclusion of the plane stress, we will have

$$u = z\theta_y, \qquad v = -z\theta_x, \tag{3.6}$$

and thus the following expressions for the shear angles will take place:

$$\gamma_{xz} = \frac{\partial u}{\partial z} + \frac{\partial w}{\partial x} = \theta_y + w_{,x}, \qquad \gamma_{yz} = \frac{\partial v}{\partial z} + \frac{\partial w}{\partial y} = -\theta_x + w_{,y}. \tag{3.7}$$

In order to mitigate the above-mentioned contradiction, if not to eliminate it at all, we will think that formulas (3.7) define only certain shear strains, averaged over the plate's thickness, which we will denote by γ_x and γ_y without additional subscript $_z$. The geometrical sense of this is that the normal remains rectilinear after the plate is deformed, but it deviates from the right angle with respect to the median surface.

The shears averaged over the thickness of the plate will be found from the conditions of equivalence, in the sense of energy, between tangential stresses τ^{xz} and τ^{yz} and their integral characteristics Q^x and Q^y. In order to do that, we calculate a part of the strain energy per unit of area of the plate's median plane, which corresponds to the work of tangential stresses τ^{xz} and τ^{yz} denoted below by e_τ. Using the theorem of an average value from integral calculus, we have

$$\mathsf{e}_\tau = \frac{1}{2}\int_{-h/2}^{-h/2}\left(\tau^{xz}\gamma_{xz} + \tau^{yz}\gamma_{yz}\right)dz = \frac{1}{2}\gamma_x\int_{-h/2}^{-h/2}\tau^{xz}dz + \frac{1}{2}\gamma_y\int_{-h/2}^{-h/2}\tau^{yz}dz =$$

$$= \frac{1}{2}(Q^x\gamma_x + Q^y\gamma_y).$$

Further, it is natural to establish a proportional relation between average shears γ_x and γ_y over the plate's thickness and the integral characteristics of the tangential stresses, that is, shear forces

$$\gamma_x = \frac{Q^x}{Gh_x}, \qquad \gamma_y = \frac{Q^y}{Gh_y}, \tag{3.8}$$

where h_x and h_y are certain quantities of the dimensionality of length, which we will call *shear thicknesses*.

Substituting (3.8) to the expression of specific energy e_τ will yield

$$\mathsf{e}_\tau = \frac{1}{2}\left[\frac{(Q^x)^2}{Gh_x} + \frac{(Q^y)^2}{Gh_y}\right]. \tag{3.9}$$

At the same time,

$$\mathsf{e}_\tau = \frac{1}{2}\int_{-h/2}^{-h/2}\left(\tau^{xz}\gamma_{xz} + \tau^{yz}\gamma_{yz}\right)dz = \frac{1}{2}\int_{-h/2}^{-h/2}\left[\frac{(\tau^{xz})^2}{G} + \frac{(\tau^{yz})^2}{G}\right]dz =$$

$$=\frac{1}{2}\int_{-h/2}^{h/2}\left[\frac{(Q^x)^2}{G}\frac{9(h^2-4z^2)^2}{4h^6}+\frac{(Q^y)^2}{G}\frac{9(h^2-4z^2)^2}{4h^6}\right]dz=$$

$$=\frac{1}{2}\left(\frac{6(Q^x)^2}{5hG}+\frac{6(Q^y)^2}{5hG}\right).$$

Here we do not use a complete expression of the tangential stresses as in (3.5), but only a part of those which is proportional to the shear forces. The justification for this (approximate) replacement is a note made above that the additional terms can be omitted from formulas (3.5) for the tangential stresses averaged over the cross-section without making any harm to the integral equilibrium of the plate over its thickness.

Comparing this with formula (3.9), we conclude that the shear thicknesses should be taken as

$$h_x = h_y = \frac{5}{6}h. \tag{3.10}$$

Integrating the expression of specific energy e_τ over the whole area of the plate's median surface gives its strain energy, E_τ, generated by "vertical" tangential stresses,

$$\mathsf{E}_\tau = \frac{1}{2}\iint_\Omega\left(\frac{6(Q^x)^2}{5hG}+\frac{6(Q^y)^2}{5hG}\right)d\Omega. \tag{3.11}$$

When we sum the strain energies caused by stresses σ^{xx}, σ^{yy} and τ^{xy} as defined by formula (1.97) and energy E_τ that corresponds to the original formula (3.1), we will finally have

$$\mathsf{E}(Q^x, Q^y, M^{xx}, M^{yy}, M^{xy}) =$$

$$=\frac{1}{2D(1-\nu^2)}\iint_\Omega\left[(M^{xx})^2-2\nu M^{xx}M^{yy}+(M^{yy})^2+2(1+\nu)(M^{xy})^2\right]d\Omega+$$

$$+\frac{6(1+\nu)}{5Eh}\iint_\Omega\left[(Q^x)^2+(Q^y)^2\right]d\Omega. \tag{3.12}$$

Formula (3.12) represents an expression of the strain energy of a Reissner plate as a functional of stresses in the plate, so that the stress vector, $\boldsymbol{\sigma}$, of the Reissner plate is a five-component vector of the following form:

$$\boldsymbol{\sigma} = \|[Q^x, Q^y, M^{xx}, M^{yy}, M^{xy}]\|^{\mathsf{T}}. \tag{3.13}$$

Strains ε_{xx} , ε_{yy} and γ_{xy} were expressed in the Kirchhoff–Love thin plate theory via the curvatures of the bar's axis according to formulas (1.23). In the case of a Reissner plate, the role of the curvatures is played by derivatives of the normal's slopes. To see this, we derive the following from kinematical relationships (3.6):

$$\varepsilon_{xx} = z\theta_{y,x}\,, \qquad \varepsilon_{yy} = -z\theta_{x,y}\,, \qquad \gamma_{xy} = z(\theta_{y,y} - \theta_{x,x})\,. \tag{3.14}$$

As we can see, the Reissner theory, unlike the Kirchhoff–Love theory, does not define the whole set of displacements in a plate via a sole function of lateral deflections, w, because the slopes of the normal, θ_x and θ_y, are used in this theory as independent kinematical parameters not restricted by conditions (1.4). However, in the further presentment it will be convenient for us to use a vector $\boldsymbol{\varphi}$ rotated by 90° instead of the normal's slope vector $\boldsymbol{\theta}$ itself:

$$\boldsymbol{\varphi} = \boldsymbol{\omega\theta} \qquad \text{or} \qquad \varphi_x = -\theta_y\,, \quad \varphi_y = \theta_x\,. \tag{3.15}$$

Obviously, when there are no shears γ_{xz} and γ_{yz} , vector $\boldsymbol{\varphi}$ will be a gradient of deflection function w as can be derived from (3.7).

Now we introduce flexural strain parameters, χ_{xx}, χ_{yy}, χ_{xy}, by defining

$$\chi_{xx} = -\varphi_{x,x}\,, \quad \chi_{yy} = -\varphi_{y,y}\,, \quad \chi_{xy} = -\tfrac{1}{2}\,(\varphi_{x,y} + \varphi_{y,x})\,, \tag{3.16}$$

or, in a tensor component form,

$$\chi_{ij} = -\tfrac{1}{2}\,(\varphi_{i,j} + \varphi_{j,i})\,. \tag{3.17}$$

Then kinematic relationships (3.14) for the strain tensor components will be as follows:

$$\varepsilon_{xx} = z\chi_{xx}\,, \qquad \varepsilon_{yy} = z\chi_{yy}\,, \qquad \gamma_{xy} = 2z\chi_{xy}\,, \tag{3.18}$$

which are exactly of the same form as formulas (1.23) in the Kirchhoff–Love plate theory.

Accordingly, we will have the following formulas for the shear strains instead of (3.7):

$$\gamma_{xz} = w_{,x} - \varphi_x\,, \qquad \gamma_{yz} = w_{,y} - \varphi_y\,. \tag{3.19}$$

To use a matrix form, we assume the vector of displacements, $\mathbf{u}$, and the vector of strains, $\boldsymbol{\varepsilon}$, to be

$$\mathbf{u} = \|[w, \varphi_x\,, \varphi_y]\|^{\mathsf{T}}\,, \quad \boldsymbol{\varepsilon} = \|[\gamma_{xz}, \gamma_{yz}, \chi_{xx}\,, \chi_{yy}\,, 2\chi_{xy}]\|^{\mathsf{T}}\,. \tag{3.20}$$

A relation between the displacement vector, $\mathbf{u}$, and the strain vector, $\boldsymbol{\varepsilon}$, is based on the above formulas, (3.16) and (3.19), so we have the following operator of geometry, $\boldsymbol{A}$, in the conventional matrix form $\boldsymbol{\varepsilon}=\boldsymbol{A}\mathbf{u}$:

$$A = \left\| \begin{matrix} \frac{\partial}{\partial x} & -1 & 0 \\ \frac{\partial}{\partial y} & 0 & -1 \\ 0 & -\frac{\partial}{\partial x} & 0 \\ 0 & 0 & -\frac{\partial}{\partial y} \\ 0 & -\frac{\partial}{\partial y} & -\frac{\partial}{\partial x} \end{matrix} \right\| . \qquad (3.21)$$

But then the operator of equilibrium, conjugate to the operator of geometry, will be as follows, according to formal rules of construction of conjugate operators:

$$A^{\mathsf{T}} = \left\| \begin{matrix} -\frac{\partial}{\partial x} & -\frac{\partial}{\partial y} & 0 & 0 & 0 \\ -1 & 0 & \frac{\partial}{\partial x} & 0 & \frac{\partial}{\partial y} \\ 0 & -1 & 0 & \frac{\partial}{\partial y} & \frac{\partial}{\partial x} \end{matrix} \right\| . \qquad (3.22)$$

And indeed, equilibrium equations (1.32) and (1.33) do not depend on a currently adopted theory of plates, as can be easily seen. If we take into account the structure of the vector of "stresses", $\boldsymbol{\sigma}$, according to (3.13), then these equations will become a matrix equation, $A^{\mathsf{T}}\boldsymbol{\sigma} = \bar{X}$, provided we assume the vector of external distributed forces to be

$$\bar{X} = \|[q, -m^y, m^x]\|^{\mathsf{T}} . \qquad (3.23)$$

Now let us find out what the elasticity matrix, C, actually is. To construct this matrix, we can do all the reasoning and calculations in the same way as we did for the Kirchhoff–Love plate theory. However, we can do it in another way, too, because we already have an expression of the strain energy of a Reissner plate – formula (3.12). And indeed, strain energy E, being a quadratic functional of stresses $\boldsymbol{\sigma}$, can be represented as

$$\mathsf{E}(\boldsymbol{\sigma}) = \frac{1}{2}(\boldsymbol{\sigma}, C^{-1}\boldsymbol{\sigma}) = \frac{1}{2}\int_{\Omega}\sum_{i=1}^{n}\sum_{j=1}^{n} d_{ij}\sigma^i\sigma^j d\Omega = \int_{\Omega} f(\sigma^1,...,\sigma^5)d\Omega . \qquad (3.24)$$

Here n is a dimensionality of the stress vector (in the case we are considering, it is $n = 5$), and coefficients d_{ij} are components of matrix $C^{-1} = \|[d_{ij}]\|$.

The quadratic form in the integrand, denoted here by $f(\sigma^1,\ldots,\sigma^5) = f(Q^x,Q^y, M^{xx},M^{yy},M^{xy})$ for convenience, is defined by (3.12), so

$$f(Q^x, Q^y, M^{xx}, M^{yy}, M^{xy}) = \frac{6(1+\nu)}{5Eh}\left[(Q^x)^2 + (Q^y)^2\right] +$$

$$+\frac{1}{2D(1-\nu^2)}\left[(M^{xx})^2 - 2\nu M^{xx}M^{yy} + (M^{yy})^2 + 2(1+\nu)(M^{xy})^2\right]. \tag{3.25}$$

Formula (3.24) implies that the components of matrix C^{-1} are determined by double differentiation of function f, where

$$d_{ij} = \frac{\partial^2 f}{\partial\sigma^i \partial\sigma^j}. \tag{3.26}$$

By doing the said differentiation, we find

$$C^{-1} = \frac{1}{D(1-\nu^2)}\left\|\begin{bmatrix} \frac{(1+\nu)h^2}{5} & 0 & 0 & 0 & 0 \\ 0 & \frac{(1+\nu)h^2}{5} & 0 & 0 & 0 \\ 0 & 0 & 1 & -\nu & 0 \\ 0 & 0 & -\nu & 1 & 0 \\ 0 & 0 & 0 & 0 & 2(1+\nu) \end{bmatrix}\right\|. \tag{3.27}$$

Now, inverting the above matrix yields

$$C = D\left\|\begin{bmatrix} \lambda & 0 & 0 & 0 & 0 \\ 0 & \lambda & 0 & 0 & 0 \\ 0 & 0 & 1 & \nu & 0 \\ 0 & 0 & \nu & 1 & 0 \\ 0 & 0 & 0 & 0 & \frac{1-\nu}{2} \end{bmatrix}\right\|, \tag{3.28}$$

where we denote additionally

$$\lambda = \frac{5(1-\nu)}{h^2}. \tag{3.29}$$

The physical relationships of the Reissner theory of plates in the component form are:

$$Q^x = \lambda D\gamma_{xz} = \lambda D(w_{,x} - \varphi_x) = \lambda D(w_{,x} + \theta_y),$$

$$Q^y = \lambda D\gamma_{yz} = \lambda D(w_{,y} - \varphi_y) = \lambda D(w_{,y} - \theta_x),$$

$$M^{xx} = D(\chi_{xx} + \nu\chi_{yy}) = -D(\varphi_{x,x} + \nu\varphi_{y,y}) = -D(-\theta_{y,x} + \nu\theta_{x,y}),$$

$$M^{yy} = D(\chi_{yy} + \nu\chi_{xx}) = -D(\varphi_{y,y} + \nu\varphi_{x,x}) = -D(\theta_{x,y} - \nu\theta_{y,x}),$$

$$M^{xy} = D(1-\nu)\chi_{xy} =$$

$$= -\tfrac{1}{2}D(1-\nu)(\varphi_{x,y} + \varphi_{y,x}) = -\tfrac{1}{2}D(1-\nu)(-\theta_{y,y} + \theta_{x,x}). \tag{3.30}$$

Now we can write out an expression of the Lame operator for the theory of bending of Reissner plates. Based on the general formula (**4**.1.6), we have

$$\boldsymbol{L} = \boldsymbol{A}^{\mathrm{T}}\boldsymbol{C}\boldsymbol{A} =$$

$$= D\left\|\begin{bmatrix} -\lambda\left(\dfrac{\partial^2}{\partial x^2} + \dfrac{\partial^2}{\partial y^2}\right) & \lambda\dfrac{\partial}{\partial x} & \lambda\dfrac{\partial}{\partial y} \\ -\lambda\dfrac{\partial}{\partial x} & \left(\lambda - \dfrac{\partial^2}{\partial x^2} - \dfrac{1-\nu}{2}\dfrac{\partial^2}{\partial y^2}\right) & -\dfrac{1+\nu}{2}\dfrac{\partial^2}{\partial x \partial y} \\ -\lambda\dfrac{\partial}{\partial y} & -\dfrac{1+\nu}{2}\dfrac{\partial^2}{\partial x \partial y} & \left(\lambda - \dfrac{\partial^2}{\partial y^2} - \dfrac{1-\nu}{2}\dfrac{\partial^2}{\partial x^2}\right) \end{bmatrix}\right\|. \tag{3.31}$$

Returning from the components of vector $\boldsymbol{\varphi}$ to the original designations of the normal's slopes, θ_x and θ_y, as in (3.15), now we have the following governing set of differential equations in terms of displacements for the Reissner plate bending:

$$
\begin{aligned}
&-\lambda\nabla^2 w+\lambda\frac{\partial\theta_x}{\partial y}-\lambda\frac{\partial\theta_y}{\partial x}=\frac{q}{D},\\
&-\lambda\frac{\partial w}{\partial x}-\frac{1+\nu}{2}\frac{\partial^2\theta_x}{\partial x\partial y}-\lambda\theta_y+\frac{\partial^2\theta_y}{\partial x^2}+\frac{1-\nu}{2}\frac{\partial^2\theta_y}{\partial y^2}=-\frac{m^y}{D},\\
&-\lambda\frac{\partial w}{\partial y}+\lambda\theta_x-\frac{\partial^2\theta_x}{\partial y^2}-\frac{1-\nu}{2}\frac{\partial^2\theta_x}{\partial x^2}+\frac{1+\nu}{2}\frac{\partial^2\theta_y}{\partial x\partial y}=\frac{m^x}{D}\,.
\end{aligned}
\qquad (3.32)
$$

To make a closed mathematical statement of the problem in the differential form, we need to formulate also boundary conditions. In order to do that, we will use the basic integral identity in application to the Reissner plate bending.

However, before we turn to the basic integral identity, we would like to consider other forms of the simultaneous governing equations for problem (3.32).

5.3.1 A governing system of equations for a Reissner plate with respect to two unknown functions

In order to proceed with convenience, we will divide the system of equations (3.32) into two. The left part in each of the new systems will be inherited from (3.32), and we will assume $m^x - m^y - 0$ in thc right part of the first system and $q=0$ in the second system. In other words, the solution of the original problem will be divided into a sum of solutions of two problems: a homogeneous problem with respect to external moment actions and a homogeneous problem with respect to a lateral load. The general solution of the original problem can be represented as a sum of the general solution of the moment-free problem and *some* partial solution of the lateral-load-free problem[17]. As usual, a partial solution of the lateral-load-free (and, what is more important, not free from external moment actions) problem will be omitted. We will try to find a general solution of the problem where we assume $m^x = m^y = 0$.

Following [24], we introduce two functions Φ and Ψ by establishing the following relationship with slope vector $\boldsymbol{\theta}$:

$$
\boldsymbol{\theta} = -\omega\,\text{grad}\,\Phi + \text{grad}\,\Psi\,, \qquad (3.33)
$$

[17] Though this statement may sound a bit unusual, it is easily provable.

where rotation tensor $\boldsymbol{\omega}$ is defined by matrix $\boldsymbol{\omega}$ in the global coordinate system according to formula (F.36) from Appendix F. Paper [24] refers to functions Φ and Ψ as a *penetrating potential* and an *edge potential*, respectively. The meaning of these names will become clear later[18]. Representation (3.33) looks as follows in its component form:

$$\theta_x = \frac{\partial \Phi}{\partial y} + \frac{\partial \Psi}{\partial x}, \qquad \theta_y = -\frac{\partial \Phi}{\partial x} + \frac{\partial \Psi}{\partial y}. \tag{3.34}$$

We will show that the system of equations (3.32) with respect to three unknown functions is reducible to a system of two equations with respect to functions Φ and Ψ if we assume the following in addition to (3.34):

$$w = \Phi - \frac{1}{\lambda}\nabla^2\Phi. \tag{3.35}$$

To see this, we substitute (3.34) and (3.35) to the first of the equations of system (3.32) and find

$$\nabla^2\nabla^2\Phi = \frac{q}{D}. \tag{3.36}$$

Substituting the same to the second and third equation of system (3.32) (and recalling that the external moments are zero in this problem, $m^x = m^y = 0$) will produce the requirements that

$$\frac{\partial}{\partial x}\left(\nabla^2\Psi - c^2\Psi\right) = 0, \qquad \frac{\partial}{\partial y}\left(\nabla^2\Psi - c^2\Psi\right) = 0, \tag{3.37}$$

where we denote additionally

$$c^2 = \frac{2\lambda}{1-\nu} = \frac{10}{h^2}. \tag{3.38}$$

Equations (3.37) imply that the aggregate expression $(\nabla^2\Psi - c^2\Psi)$ is a constant, and this constant can be assumed to be zero because its value

---

[18] Strictly speaking, paper [24] deals with a slightly different version of a shear-based theory for medium-thickness plates, which is different from Reissner's theory. But that is insignificant for our reasoning because the structure of the problem's governing equations derived in [24] is exactly the same as the structure of equations (3.32). The difference lies in an expression of coefficient λ which is equal to $6(1-\nu)/h^2$ for an isotropic material in V.V. Vasiliev's theory of plates, unlike formula (3.29) for a Reissner plate

does not affect the components of the slope vector, $\boldsymbol{\theta}$, which we are interested with, because of (3.34). So,

$$\nabla^2\Psi - c^2\Psi = 0. \tag{3.39}$$

A system of two equations: an equation of fourth order, (3.36), with respect to function Φ and an equation of second order, (3.39), with respect to function Ψ fulfils our goal – it is a governing system of equations for the external-moment-free problem. Obviously, the overall order of the two equations is six[19]. The equations themselves keep functions Φ and Ψ separated, but, as we will see later, the solutions are still mutually dependent through boundary conditions.

Note that when the thickness of the plate, h, tends to zero, equality (3.35) implies that $w \to \Phi$, which shows an expected tendency of degeneration of system (3.36), (3.39) into one equation (3.36), a characteristic relationship of Kirchhoff's theory of plates.

When we have integrated equations (3.36) and (3.39) to determine functions Φ and Ψ, we can then find the internal forces in the cross-sections of the plate by formulas (3.30), which now become as follows after substituting expressions (3.34) and (3.35) to those:

$$Q^x = D\left(-\frac{\partial}{\partial x}\nabla^2\Phi + \lambda\frac{\partial\Psi}{\partial y}\right), \qquad Q^y = D\left(-\frac{\partial}{\partial y}\nabla^2\Phi - \lambda\frac{\partial\Psi}{\partial x}\right), \tag{3.40}$$

[19] Both types of equations are studied well by mathematics and are known in mechanics. Equation (3.36) is the same as the Germain–Lagrange equation for a classic Kirchhoff plate, and equation (3.39) is a Helmholtz equation. One of possible mechanical interpretations of the Helmholtz equation is a homogeneous equation of the deflection of a membrane supported by an elastic Winkler-type foundation. The general integral of the Germain–Lagrange equation (3.36) contains terms which vary across the area of the plate, Ω, at a comparatively slow rate. At the same time, the general integral of the Helmholtz equation (3.39) gives a solution with a comparatively quick variability. Equation (3.39) can be said also to describe a stress distribution in a plate similar to an edge effect. These were the considerations why function Φ was called a *penetrating potential* and function Ψ an *edge potential* in paper [24].

$$M^{xx} = -D\left[\frac{\partial^2\Phi}{\partial x^2} + \nu\frac{\partial^2\Phi}{\partial y^2} - (1-\nu)\frac{\partial^2\Psi}{\partial x\partial y}\right],$$

$$M^{yy} = -D\left[\frac{\partial^2\Phi}{\partial y^2} + \nu\frac{\partial^2\Phi}{\partial x^2} + (1-\nu)\frac{\partial^2\Psi}{\partial x\partial y}\right],$$

$$M^{xy} = -D(1-\nu)\left[\frac{\partial^2\Phi}{\partial x\partial y} + \frac{1}{2}\left(\frac{\partial^2\Psi}{\partial x^2} - \frac{\partial^2\Psi}{\partial y^2}\right)\right]. \tag{3.41}$$

The components of the strain vector, $\boldsymbol{\varepsilon}$, according to (3.20) will become as follows when expressed via the governing functions Φ and Ψ:

$$\gamma_{xz} = -\frac{1}{\lambda}\frac{\partial}{\partial x}\nabla^2\Phi + \frac{\partial\Psi}{\partial y}, \qquad \gamma_{yz} = -\frac{1}{\lambda}\frac{\partial}{\partial y}\nabla^2\Phi - \frac{\partial\Psi}{\partial x}$$

$$\chi_{xx} = -\frac{\partial^2\Phi}{\partial x^2} + \frac{\partial^2\Psi}{\partial x\partial y}, \qquad \chi_{yy} = -\frac{\partial^2\Phi}{\partial y^2} - \frac{\partial^2\Psi}{\partial x\partial y},$$

$$2\chi_{xy} = -2\frac{\partial^2\Phi}{\partial x\partial y} + \frac{\partial^2\Psi}{\partial y^2} - \frac{\partial^2\Psi}{\partial x^2}. \tag{3.42}$$

5.3.2 Basic integral identity in the Reissner plate theory

Based on earlier matrix relationships, we can write

$$(A\mathbf{u}, \boldsymbol{\sigma}) = \iint_\Omega \left[(w_{,x} - \varphi_x)Q^x + (w_{,y} - \varphi_y)Q^y\right]d\Omega +$$

$$+ \iint_\Omega \left[-\varphi_{x,x}M^{xx} - \varphi_{y,y}M^{yy} - (\varphi_{x,y} + \varphi_{y,x})M^{xy}\right]d\Omega .$$

By using the Gauss–Ostrogradsky formula, we remove the differentiation from the displacements in the integrand, to obtain

$$(A\mathbf{u}, \boldsymbol{\sigma}) =$$

$$= \iint_\Omega \left[w(-Q^x_{,x} - Q^y_{,y}) + \varphi_x(-Q^x + M^{xx}_{,x} + M^{xy}_{,y}) + \varphi_y(-Q^y + M^{yy}_{,y} + M^{xy}_{,x})\right]d\Omega +$$

$$+ \oint_\Gamma \left[w(Q^x n_x + Q^y n_y) - \varphi_x(M^{xx}n_x + M^{xy}n_y) - \varphi_y(M^{yy}n_y + M^{xy}n_x)\right]d\Gamma .$$

First of all, we would like to notice that the integral over area Ω is a scalar product, $(\mathbf{u}, A^{\mathsf{T}}\boldsymbol{\sigma})$. Further, we can easily check that the following identity is true:

$$\varphi_x(M^{xx}n_x + M^{xy}n_y) + \varphi_y(M^{yy}n_y + M^{xy}n_x) = \varphi_n M^{nn} + \varphi_t M^{nt},$$

where moments M^{nn} and M^{nt} are defined by formulas (1.52), and quantities φ_n and φ_t are projections of vector $\boldsymbol{\varphi}$ onto axes $\boldsymbol{n}$ and $\boldsymbol{t}$, where

$$\varphi_n = \varphi_x n_x + \varphi_y n_y = -\theta_y t_y - \theta_x t_x = -\theta_t,$$

$$\varphi_t = \varphi_x t_x + \varphi_y t_y = \theta_y n_y + \theta_x t_x = \theta_n. \tag{3.43}$$

As a result, the basic integral identity becomes

$$(A\mathbf{u}, \boldsymbol{\sigma}) = (\mathbf{u}, A^{\mathsf{T}}\boldsymbol{\sigma}) + \oint_{\Gamma}\left(wQ^n + \theta_t M^{nn} - \theta_n M^{nt}\right)d\Gamma. \tag{3.44}$$

Now it is clear that the vector of edge displacements, $\boldsymbol{u}$, and the vector of edge forces, $\boldsymbol{p}$, in the Reissner theory of plates should be

$$\boldsymbol{u} = \|[w, \theta_t, -\theta_n]\|^{\mathsf{T}}, \qquad \boldsymbol{p} = \|[Q^n, M^{nn}, M^{nt}]\|^{\mathsf{T}}. \tag{3.45}$$

Thus, unlike the Kirchhoff–Love theory, the Reissner plate theory states three boundary conditions on the contour of the plate, Γ, according to the dimensionality of vectors $\boldsymbol{u}$ and $\boldsymbol{p}$.

5.3.3 Important functionals for the Reissner plate

Lagrange functional

The strain energy stored in the Reissner plate can be represented as a quadratic functional of displacements, derived from the general formula

$$\mathsf{E}(w, \varphi_x, \varphi_y) = \frac{1}{2}(A\mathbf{u}, CA\mathbf{u}),$$

which yields the following after substitutions and matrix transformations:

$$\mathsf{E}(w, \varphi_x, \varphi_y) = \frac{\lambda D}{2}\iint_{\Omega}\left[(w_{,x} - \varphi_x)^2 + (w_{,y} - \varphi_y)^2\right]d\Omega +$$

$$+\frac{D}{2}\iint_{\Omega}\left[\varphi_{x,x}^2 + 2\nu\varphi_{x,x}\varphi_{y,y} + \varphi_{y,y}^2 + \frac{1-\nu}{2}(\varphi_{x,y} + \varphi_{y,x})^2\right]d\Omega. \tag{3.46}$$

Obviously, the first of the above integrals corresponds to a shear part of the energy and the second one to the bending energy itself.

As for the force potential, Π_s, an expression for it is given by the earlier formula (1.86), which can be represented as follows for a Reissner plate if we take into account relations (3.15) and (3.43):

$$\Pi_s =$$

$$= \int_\Omega (qw + m^x\theta_x + m^y\theta_y)d\Omega + \oint_\Gamma (e_Q\bar{Q}w + e_{Mt}\bar{M}^t\theta_t + e_{Mn}\bar{M}^n\theta_n)d\Gamma =$$

$$= \int_\Omega (qw - m^y\varphi_x + m^x\varphi_y)d\Omega + \oint_\Gamma (e_Q\bar{Q}w - e_{Mt}\bar{M}^t\varphi_n + e_{Mn}\bar{M}^n\varphi_t)d\Gamma . \quad (3.47)$$

Here we have already taken into account the fact that the external (contour) force actions are different from zero only on segments of the contour where the respective characteristic functions, e_Q, e_{Mt}, e_{Mn}, are equal to one. The expression of the Lagrangian functional, L, is constructed in the usual way, as $L = E - \Pi_s$.

As usual, kinematically admissible fields of displacements are searched for a minimum of the Lagrangian functional, which are displacement functions giving a finite value to integral (3.46) and satisfying kinematical boundary conditions

$$w = \bar{w} \ \in \Gamma_w , \qquad \varphi_n = -\bar{\theta}_t \ \in \Gamma_{\theta t} , \qquad \varphi_t = \bar{\theta}_n \ \in \Gamma_{\theta n} . \quad (3.48)$$

Euler equations for the Lagrangian functional is the system of differential equations obtained earlier, (3.32), and natural boundary conditions for it are static boundary conditions

$$Q^n = \bar{Q}^n \ \in \Gamma_Q , \qquad M^{nn} = \bar{M}^t \in \Gamma_{Mt} , \qquad M^{nt} = -\bar{M}^n \in \Gamma_{Mn} , \quad (3.49)$$

where

$$\Gamma_w \cup \Gamma_Q = \Gamma, \ \Gamma_w \cap \Gamma_Q = \varnothing; \qquad \Gamma_{\theta t} \cup \Gamma_{Mt} = \Gamma, \ \Gamma_{\theta t} \cap \Gamma_{Mt} = \varnothing;$$

$$\Gamma_{\theta n} \cup \Gamma_{Mn} = \Gamma, \qquad \Gamma_{\theta n} \cap \Gamma_{Mn} = \varnothing;$$

Castigliano functional

The strain energy, E, is already given by formula (3.12) in the form of a quadratic functional of “stresses”. Therefore all we have to do now is to write an explicit form of the kinematic potential. Based on (3.44) and (3.45), we have

$$\Pi_k(\boldsymbol{\sigma}) = \oint_{\Gamma} \left(e_w \overline{w} Q^n + e_{\theta t} \overline{\theta}_t M^{nn} - e_{\theta n} \overline{\theta}_n M^{nt} \right) d\Gamma \,. \tag{3.50}$$

A point of minimum of the Castigliano functional, $K(\boldsymbol{\sigma}) = E(\boldsymbol{\sigma}) - \Pi_k(\boldsymbol{\sigma})$, is searched for in a set of statically admissible stresses $\boldsymbol{\sigma}$. This means that stresses admitted to the comparison should satisfy equations of equlibrium (1.32), (1.33) over the whole area Ω and static boundary conditions (3.49). Now we want to derive strain compatibility conditions for a Reissner plate, which would be Euler equations for the Castigliano functional when formulated in terms of stresses.

As (3.20) shows, the displacement vector, **u**, is a three-component vector and the strain vector, $\boldsymbol{\varepsilon}$, has five components. Consequently, all five components of the strain vector cannot be specifed independently; they must be restricted by two additional relationships which would make it possible to restore the displacements from the strains. These additional relationships between the components of the strain vector are called Saint-Venant compatibility conditions.

As it is known, relationships of such a kind can be interpreted from the mathematical standpoint as conditions of independence of some mixed derivatives of the strain vector's components from the order of differentiation. In particular, two independent relationships of this type for the Reissner plate are

$$\frac{\partial^2 \gamma_{xz}}{2\partial y^2} - \frac{\partial^2 \gamma_{yz}}{2\partial x \partial y} + \frac{\partial \chi_{yy}}{\partial x} - \frac{\partial \chi_{xy}}{2\partial y} = 0 \,,$$

$$-\frac{\partial^2 \gamma_{xz}}{2\partial x \partial y} + \frac{\partial^2 \gamma_{yz}}{2\partial x^2} + \frac{\partial \chi_{xx}}{\partial y} - \frac{\partial \chi_{xy}}{2\partial x} = 0 \,. \tag{3.51}$$

The strain compatibility conditions from (3.51) can be validated immediately. Indeed, substituting the expressions of the strains via the displacements from (3.16) and (3.19) make relationship (3.51) an identity.

Another identity which seems obvious when we look at kinematical relationships (3.16),

$$\frac{\partial^2 \chi_{xx}}{2\partial y^2} + \frac{\partial^2 \chi_{yy}}{2\partial x^2} - \frac{\partial^2 \chi_{xy}}{\partial x \partial y} = 0$$

is not independent because it follows from (3.51): it can be obtained by differentiating the first equality there with respect to x, the second with respect to y, and summing the results.

Seeing that the strain vector's structure follows (3.20), we obtain the Saint-Venant operator for the Reissner plate:

$$\boldsymbol{S} = \left\| \begin{bmatrix} \dfrac{\partial^2}{2\partial y^2} & -\dfrac{\partial^2}{2\partial x \partial y} & 0 & \dfrac{\partial}{\partial x} & -\dfrac{\partial}{2\partial y} \\ -\dfrac{\partial^2}{2\partial x \partial y} & \dfrac{\partial^2}{2\partial x^2} & \dfrac{\partial}{\partial y} & 0 & -\dfrac{\partial}{2\partial x} \end{bmatrix} \right\| . \tag{3.52}$$

When written in terms of stresses, the strain compatibility conditions,

$$\boldsymbol{S}\boldsymbol{C}^{-1}\boldsymbol{\sigma} = \mathbf{0},$$

are Euler euqations for the Castigliano functional.

A stress function vector for the Reissner plate

As we already know the Saint-Venant operator, $\boldsymbol{S}$, we can derive its conjugate $\boldsymbol{S}^{\mathsf{T}}$. From formal rules of construction of Lagrange-conjugate operators we have

$$\boldsymbol{S}^{\mathsf{T}} = \left\| \begin{bmatrix} \dfrac{\partial^2}{2\partial y^2} & -\dfrac{\partial^2}{2\partial x \partial y} \\ -\dfrac{\partial^2}{2\partial x \partial y} & \dfrac{\partial^2}{2\partial x^2} \\ 0 & -\dfrac{\partial}{\partial y} \\ -\dfrac{\partial}{\partial x} & 0 \\ \dfrac{\partial}{2\partial y} & \dfrac{\partial}{2\partial x} \end{bmatrix} \right\| . \tag{3.53}$$

It is easy to check that

$$\boldsymbol{S}\boldsymbol{A} = \boldsymbol{O}, \qquad \boldsymbol{A}^{\mathsf{T}}\boldsymbol{S}^{\mathsf{T}} = \boldsymbol{O},$$

just the way the Saint-Venant operator should be.

Exactly as we did in Section 5.2 for a Kirchhoff–Love plate and for a plane stress, here we introduce a vector of stress functions $\boldsymbol{\Phi}$ related to stresses $\boldsymbol{\sigma}$ (in the case when there is no distributed load over area Ω – neither moments nor forces) via relationship (2.5) and an operator, $\boldsymbol{M}$, of strain compatibility in stress functions:

$$\boldsymbol{\sigma} = \boldsymbol{S}^{\mathrm{T}}\boldsymbol{\Phi}, \qquad \boldsymbol{M} = \boldsymbol{S}\boldsymbol{C}^{-1}\boldsymbol{S}^{\mathrm{T}}.$$

Obviously, the stress function vector $\boldsymbol{\Phi}$ is a two-component vector for both the Reissner plate and the Kirchhoff–Love plate, so

$$\boldsymbol{\Phi} = \|[\Phi_x, \Phi_y]\|^{\mathrm{T}}.$$

Making required matrix transformations and canceling out the insignificant common constant $h^2/[20D(1-\nu)]$ will yield an expression for the strain compatibility operator in stress functions,

$$\boldsymbol{M} = \left\| \begin{bmatrix} \left((\nabla^2 - c^2)\dfrac{\partial^2}{\partial y^2} - \dfrac{2c^2}{1+\nu}\dfrac{\partial^2}{\partial x^2} \right) & \left(-(\nabla^2 - c^2) - \dfrac{2c^2}{1+\nu} \right)\dfrac{\partial^2}{\partial x \partial y} \\ \left(-(\nabla^2 - c^2) - \dfrac{2c^2}{1+\nu} \right)\dfrac{\partial^2}{\partial x \partial y} & \left((\nabla^2 - c^2)\dfrac{\partial^2}{\partial x^2} - \dfrac{2c^2}{1+\nu}\dfrac{\partial^2}{\partial y^2} \right) \end{bmatrix} \right\|, \qquad (3.54)$$

where numerical parameter c^2 has been defined by formula (3.38).

A frightfully complicated structure of the matrix differential operator $\boldsymbol{M}$ is not actually as terrible as it seems, because the system of differential equations that corresponds to that operator can be reduced and simplified. To see this, we denote by M_{11}, M_{12}, M_{21}, M_{22} the elements of matrix operator $\boldsymbol{M}$ and find that each component of the stress function vector, Φ_x and Φ_y, satisfies a homogeneous differential equation with the operator

$$L = M_{22}M_{11} - M_{12}M_{21}.$$

Making some required operations, we have the final result[20]

$$\nabla^4(\nabla^2\Phi_x - c^2\Phi_x) = 0, \qquad \nabla^4(\nabla^2\Phi_y - c^2\Phi_y) = 0. \qquad (3.55)$$

However, there is another problem. The matter is that each of the two equations in (3.55) is a sixth-order equation. As we established earlier, we have only three boundary conditions. Obviously, an extra freedom was created by an additional differentiation added to (3.55) when we switched to operator L. This extra freedom can be eliminated if we recall that functions Φ_x and Φ_y must satisfy also the original system of differential equations with operator $\boldsymbol{M}$, hence two more conditions. Another degree of freedom can be eliminated by remembering the fact that actually we are interested with the stresses in the plate expressed via functions Φ_x and Φ_y rather than with the functions themselves. But functions Φ_x and Φ_y are

[20] If we do the same procedure for a Kirchhoff plate and recall formula (2.12), we will see that each of the components of the stress function vector should satisfy a biharmonic equation, that is, $\nabla^4\Phi_x = 0$, $\nabla^4\Phi_y = 0$.

differentiated in those expressions, which eliminates the third degree of freedom.

As we can see, this way of solving problems involves a lot of work. This is probably the reason why the method of stress functions is hardly ever used to solve practical problems for the Reissner plates. However, we deem it useful to present the method here so that the students do not have any suspicion about an incompleteness of this aspect of the Reissner plate theory, because the vector of stress functions is not even mentioned in many popular textbooks on the theory of plates.

A Reissner functional for a Reissner plate

We want to derive an expression of the Reissner functional's first form, based on the general formula (**3**.1.4):

$$\mathsf{R}_1(\boldsymbol{\sigma}, \mathbf{u}) = ½(\boldsymbol{C}^{-1}\boldsymbol{\sigma}, \boldsymbol{\sigma}) - (\boldsymbol{A}\mathbf{u}, \boldsymbol{\sigma}) + (\boldsymbol{E}_u\boldsymbol{p}, \boldsymbol{E}_u\boldsymbol{u})_\Gamma + \Pi_s - \Pi_k =$$

$$= \frac{1}{2D(1-\nu^2)} \iint_\Omega \left[(M^{xx})^2 - 2\nu M^{xx}M^{yy} + (M^{yy})^2 + 2(1+\nu)(M^{xy})^2 \right] d\Omega +$$

$$+ \frac{6(1+\nu)}{5Eh} \iint_\Omega \left[(Q^x)^2 + (Q^y)^2 \right] d\Omega - \iint_\Omega \left[(w_{,x} - \varphi_x)Q^x + (w_{,y} - \varphi_y)Q^y \right] d\Omega -$$

$$- \iint_\Omega \left[-\varphi_{x,x}M^{xx} - \varphi_{y,y}M^{yy} - (\varphi_{x,y} + \varphi_{y,x})M^{xy} \right] d\Omega +$$

$$+ \int_\Omega (qw - m^y\varphi_x + m^x\varphi_y) d\Omega + \oint_\Gamma (e_Q\bar{Q}w - e_{Mt}\bar{M}^t\varphi_n + e_{Mn}\bar{M}^n\varphi_t) d\Gamma +$$

$$+ \oint_\Gamma \left(e_w(w - \bar{w})Q^n + e_{\theta t}(-\varphi_n - \bar{\theta}_t)M^{nn} - e_{\theta n}(\varphi_t - \bar{\theta}_n)M^{nt} \right) d\Gamma \,. \qquad (3.56)$$

The Reissner functional's second form can be derived from the general formula (**3**.1.5):

$$\mathsf{R}_2(\boldsymbol{\sigma}, \mathbf{u}) = ½(\boldsymbol{C}^{-1}\boldsymbol{\sigma}, \boldsymbol{\sigma}) - (\boldsymbol{A}^\mathsf{T}\boldsymbol{\sigma}, \mathbf{u}) - (\boldsymbol{E}_p\boldsymbol{p}, \boldsymbol{E}_p\boldsymbol{u})_\Gamma + \Pi_s - \Pi_k =$$

$$= \frac{1}{2D(1-\nu^2)} \iint_\Omega \left[(M^{xx})^2 - 2\nu M^{xx}M^{yy} + (M^{yy})^2 + 2(1+\nu)(M^{xy})^2 \right] d\Omega +$$

$$+ \frac{6(1+\nu)}{5Eh} \iint_\Omega \left[(Q^x)^2 + (Q^y)^2 \right] d\Omega \; -$$

$$- \iint_\Omega \left[w(-Q^x_{,x} - Q^y_{,y}) + \varphi_x(-Q^x + M^{xx}_{,x} + M^{xy}_{,y}) + \varphi_y(-Q^y + M^{yy}_{,y} + M^{xy}_{,x}) \right] d\Omega +$$

$$+\int_{\Omega}(qw - m^y\varphi_x + m^x\varphi_y)d\Omega -$$

$$-\oint_{\Gamma}\left(e_Q(Q^n - \bar{Q})w - e_{Mt}(-M^{nn} + \bar{M}^t)\varphi_n + e_{Mn}(-M^{nt} + \bar{M}^n)\varphi_t\right)d\Gamma -$$

$$-\oint_{\Gamma}\left(e_w\bar{w}Q^n + e_{\theta t}\bar{\theta}_t M^{nn} - e_{\theta n}\bar{\theta}_n M^{nt}\right)d\Gamma . \tag{3.57}$$

5.4 Some examples

We present a few examples here in order to demonstrate solutions of problems from the theories of plates by Kirchhoff and by Reissner in some simple cases which allow an exact analysis. The same examples can be used as tests for debugging (and validating!) of software implementations declared in popular commercial software packages for structural analysis.

5.4.1 Example 1. A round plate loaded by a torque on its edge

Our first example is a round plate of a radius R, free from fixations and subjected to a contour torque action of an intensity m constant along the contour. We will use polar coordinates in this problem because, obviously, the solution depends only on the radial coordinate, $\rho = \sqrt{x^2 + y^2}$ and does not depend on the angular coordinate, α. Under these conditions, the Laplace operator ∇^2 is known to become unidimensional, as follows:

$$\nabla^2 = \frac{1}{\rho}\frac{d}{d\rho}\left(\rho\frac{d}{d\rho}\right) = \frac{d^2}{d\rho^2} + \frac{1}{\rho}\frac{d}{d\rho}.$$

First of all, we want to exclude rigid displacements by assuming that the plate is freely supported along its contour. This assumption does not violate the given boundary conditions for forces because, obviously, our external condition for this problem makes both shear forces $Q^n = Q^\rho$ and bending moments $M^{nn} = M^{\rho\rho}$ equal to zero. Moments $M^{\rho\rho}$ are equal to zero because such is the sense of the boundary conditions for a freely supported plate, and shear forces Q^ρ are zero because otherwise the global equilbrium of the plate in projection onto the Z-axis would be violated. Moreover, if we repeat the same consideration of equilibrium for any round part of the plate of a radius ρ, we will understand that shear forces

Q^{ρ} must be identical to zero in the whole plate, not only on its contour – that is, $Q^{\rho}(\rho) = 0$.

Furthermore, a general integral of the homogeneous biharmonic equation, $\nabla^2\nabla^2\Phi = 0$, for a penetrating potential with the above stated operator ∇^2 is known; it can be written as

$$\Phi = A_1\rho^2 \ln\rho + A_2 \ln\rho + A_3\rho^2 + A_4 .$$

This representation can be easily validated by direct differentiation.

On the other hand, a general integral of the homogeneous equation,

$$\nabla^2\Psi - c^2\Psi = 0,$$

is also known for a boundary-layer potential with this operator [14]; it can be represented as

$$\Psi = B_1 I_0(c\rho) + B_2 K_0(c\rho),$$

where I_0 and K_0 are modified Bessel functions of the first and second type, zero order. Here A_1, A_2, A_3, A_4, B_1, B_2 are constants of integration, yet to be found.

Using formula (3.35) gives

$$w = \Phi - \frac{1}{\lambda}\nabla^2\Phi = A_1\left[\rho^2 \ln\rho - \frac{4}{\lambda}(\ln\rho + 1)\right] + A_2 \ln\rho + A_3\left[\rho^2 - \frac{4}{\lambda}\right] + A_4 .$$

The limitedness of the deflection at the center of the plate gives $A_1 = A_2 = 0$. Furthermore, the deflection is equal to zero at the center of the circle (at $\rho = 0$) and on the contour (at $\rho = R$) – at the center, because of the symmetry, and on the contour, because of the boundary condition. Hence $A_3 = A_4 = 0$ too. Thus, the penetrating potential is identical to zero, $\Phi = 0$[21]. Seeing this and considering formulas (3.40), we have the shear forces in the plate:

$$Q^x = \lambda D\frac{\partial\psi}{\partial y}, \quad Q^y = -\lambda D\frac{\partial\psi}{\partial x},$$

so

---

[21] The fact that the penetrating potential is zero can be justified in another way in this problem, without having to find the constant of integration, $A_1, \ldots, A_4$. To see this, we should consider the symmetry and understand that the plate's deflection must be identical to zero. This means that the penetrating potential satisfies two conditions at once: $\nabla^2\nabla^2\Phi = 0$ and $\nabla^2\Phi - \lambda\Phi = 0$, hence $\Phi = 0$.

$$Q^{\rho} = Q^{n} = Q^{x} n_x + Q^{y} n_y = \lambda D \left(\frac{\partial \psi}{\partial y} t_y + \frac{\partial \psi}{\partial x} t_x \right) = \lambda D \frac{\partial \Psi}{R \partial \alpha} = 0,$$

which was to be expected.

Further, in our case (3.41) produces

$$M^{xx} = -M^{yy} = D(1-\nu) \frac{\partial^2 \Psi}{\partial x \partial y}, \qquad M^{xy} = -\frac{D(1-\nu)}{2} \left(\frac{\partial^2 \Psi}{\partial x^2} - \frac{\partial^2 \Psi}{\partial y^2} \right).$$

Consequently, formulas (1.52) will give

$$M^{nn} = M^{xx} n_x^2 + M^{yy} n_y^2 + 2M^{xy} n_x n_y =$$

$$= D(1-\nu) \left[\frac{\partial^2 \Psi}{\partial x \partial y} (n_x^2 - n_y^2) - \left(\frac{\partial^2 \Psi}{\partial x^2} - \frac{\partial^2 \Psi}{\partial y^2} \right) n_x n_y \right].$$

If we take a point on the X-axis, we will have for it

$$\frac{\partial}{\partial x} = \frac{\partial}{\partial \rho}, \quad \frac{\partial}{\partial y} = \frac{\partial}{R \partial \alpha}, \quad n_x = 1, \quad n_y = 0, \quad t_x = 0, \quad t_y = 1.$$

So finally we have $M^{nn} = 0$, on the contour of the plate too, which is in accordance with the free support boundary condition.

Now let us use the fact that the torque on the contour of the plate is equal to a given value, m. We have:

$$M^{nt} = M^{xx} n_x t_x + M^{yy} n_y t_y + M^{xy} (n_x t_y + n_y t_x) =$$

$$= D(1-\nu) \left[\frac{\partial^2 \Psi}{\partial x \partial y} (n_x t_x - n_y t_y) - \left(\frac{\partial^2 \Psi}{\partial x^2} - \frac{\partial^2 \Psi}{\partial y^2} \right) (n_x t_y + n_y t_x) \right].$$

Now we turn to points of the X axis again and have

$$M^{nt} = -D(1-\nu) \frac{\partial^2 \Psi}{\partial \rho^2}.$$

As function I_0 is singular in the infinity, and K_0 is singular at zero, the limitedness of the solution at zero requires that the constant of integration B_2 should be assumed equal to zero.

Let us present known relationships of the Bessel functions theory [14]

$$\frac{d}{d\rho} \left[\rho^k I_k (c\rho) \right] = c\rho^k I_{k-1} (c\rho).$$

This gives the following in our case:

$$\frac{d}{d\rho}[I_0(c\rho)] = cI_{-1}(c\rho), \qquad \frac{d^2}{d\rho^2}[I_0(c\rho)] = c\frac{d}{d\rho}\left(\rho\rho^{-1}I_{-1}(c\rho)\right) =$$

$$= c\frac{I_{-1}(c\rho)}{\rho} + c^2 I_{-2}(c\rho) = c^2\left(\frac{I_1(c\rho)}{c\rho} + I_2(c\rho)\right) = c^2\frac{c\rho I_0(c\rho) - I_1(c\rho)}{c\rho}.$$

Here we take into account two known facts from the theory of Bessel functions:

- $I_{-1}(c\rho) = I_1(c\rho)$ and $I_{-2}(c\rho) = I_2(c\rho)$;
- $c\rho I_2(c\rho) = c\rho I_0(c\rho) - 2\,I_1(c\rho)$.

The boundary condition for the torque gives the last, still undetermined, constant of integration B_1

$$B_1 = -\frac{mcR}{c^2 D(1-\nu)\left[cRI_0(cR) - I_1(cR)\right]}.$$

The final distribution of the torques along the radial coordinate of the plate is

$$M^{nt} = m\frac{R}{\rho}\left(\frac{c\rho I_0(c\rho) - I_1(c\rho)}{cRI_0(cR) - I_1(cR)}\right).$$

Graphs of the torque vs. the radial coordinate of the plate, based on the above formula, are shown in Fig. 5.6 for various values of the plate's respective thickness, h/R, where the abscissas are radial coordinates ρ/R and the ordinates are dimensionless torques M^{nt}/m.

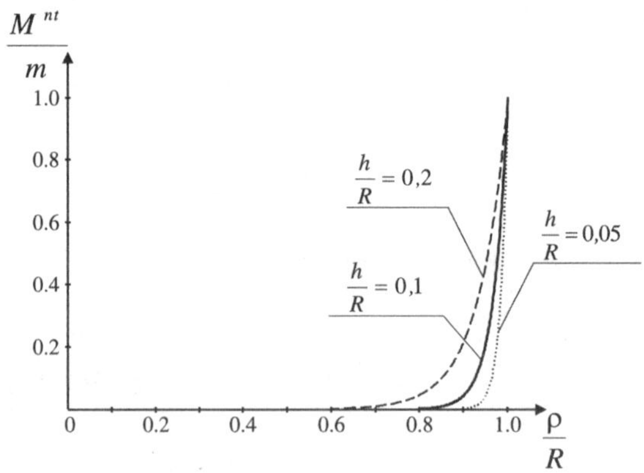

Fig. 5.6. Torque vs. the radial coordinate on the plate

The graphs show a clear edge effect, as was mentioned earlier, for the distribution of the torque. Even for a relatively thick plate (at $h/R = 0{,}2$) the M^{nt} moment falls quickly as we move away from the plate's edge.

To complete the presentment, we should find the slopes of the normal to the plate, which describe its strain state. Using formulas (3.34) and considering points which lie on the X-axis, we have

$$\theta_t = 0, \qquad \theta_n = \frac{\partial \Psi}{\partial \rho} = -\frac{m I_1(c\rho) R}{D(1-\nu)\left[cRI_0(cR) - I_1(cR)\right]}.$$

5.4.2 Example 2. A square plate loaded by torques on its edge

Our second example will be a square plate with its contour free from fixations. Let the plate be subjected to torques m evenly distributed along the contour and directed oppositely on each of the plate's sides as shown in Fig. 5.7.

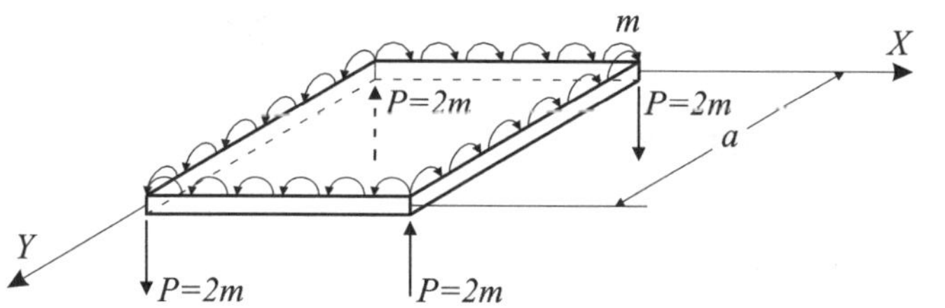

Fig. 5.7. A square plate subjected to torques in opposite directions on its contour

It can be checked directly that the solution of this simple problem according to Kirchhoff's theory is given by the following elementary formulas

$$w = -\frac{m}{D(1-\nu)} xy + \alpha x + \beta y + \gamma, \quad \theta_x = -\frac{m}{D(1-\nu)} x, \quad \theta_y = \frac{m}{D(1-\nu)} y,$$

$$Q^x = Q^y = 0, \quad K^x = K^y = 0, \qquad M^{xx} = M^{yy} = 0, \quad M^{xy} = m$$

with undetermined constants α, β, γ which characterize deflections w of the plate as a rigid body. Also, concentrated Kirchhoff's forces $P = 2m$ appear in the angles of the plate as shown in the same Fig. 5.7.

Now we want to solve the same problem according to the Reissner theory of plates. Obviously, the distribution of internal forces in a Reissner plate is the same as that in a Kirchhoff plate (except for generalized shear forces K^x and K^y which just do not exist in the Reissner plate).

And indeed, in this case all equations of equlibrium hold together with the strain compatibility conditions expressed via stresses, that is, $\boldsymbol{SC}^{-1}\boldsymbol{\sigma} = \mathbf{0}$. All the boundary conditions hold true, too. Then we find that the field of displacements for the Reissner plate coincides with that for the Kirchhoff plate. We can see that the shear forces are zero, therefore the shear angles are zero, too, wherefrom we have the relationships

$$\theta_y = -w_{,x}, \quad \theta_x = w_{,y},$$

which describe a Kirchhoff plate. After restoring the penetrating and edge potentials for this problem, we have

$$\Phi = -\frac{m}{D(1-\nu)} xy, \quad \Psi = 0.$$

Now we want to modify this problem by assuming the external contour moment, m, to be directed as in Fig. 5.8.

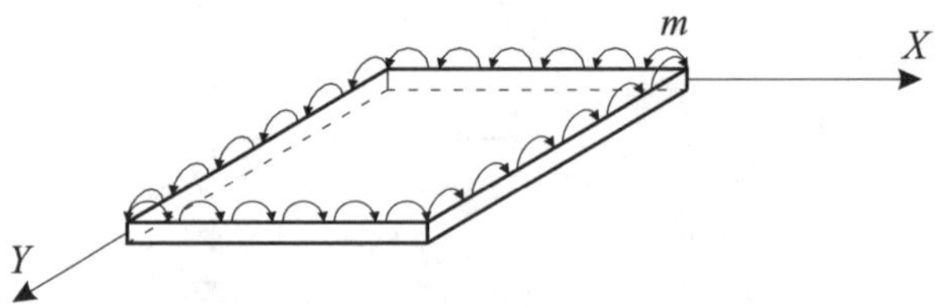

Fig. 5.8. A square plate subjected to codirectional torques on its contour

It turns out that there is no elementary solution for this problem, unlike the previous one. An approximate solution of the problem (or, rather, a solution for a linear combination of the problems from Fig. 5.7 and Fig. 5.8) can be found by the interested reader in paper [25]. Here we just want to notice that the very statement of the problem in Fig. 5.8 contains an incorrectness because the reciprocity of tangential stresses is violated in the angular points of the plate, which is $M^{xy} = M^{yx}$ in the theory of plates.

If we consider this problem within the limits of the Kirchhoff theory, we will understand easily that a Kirchhoff plate does not respond to that kind of load at all. All force and kinematical fields for the Kirchhoff plate are zero.

5.4.3 Example 3. A simply supported rectangular plate

Reissner plate

Consider a rectangular, simply supported plate of sizes $a \times b$ under a sinusoid-shaped load

$$q(x,y) = q_{mn} \sin\frac{m\pi x}{a} \sin\frac{n\pi y}{b} . \tag{3.58}$$

If we manage to find a solution for a plate subjected to load (3.58), we can always find a solution for the same plate under any arbitrary load $q(x,y)$ by expanding the latter into a double Fourier series and then summing the solutions over indexes m and n.

First of all, let us formulate boundary conditions carefully. It should be noted that, unlike a Kirchhoff plate, the Reissner plate requires more than just words to describe actual boundary conditions of "*simple support*" on the contour of the plate. Indeed, as we have noticed earlier, every point of the contour of the Reissner plate must have three boundary conditions. In particular, the boundary conditions for a simple support can be formulated in either of two competing forms:

- version "A" : $w = 0, \;\; M^{nn} = 0, \;\; \theta_n = 0 \;\; \in \Gamma -$ a cylindrical hinge;
- version "B" : $w = 0, \;\; M^{nn} = 0, \;\; M^{nt} = 0 \;\; \in \Gamma -$ a free support.

In these two versions, the first two conditions are the same and the third are different Version "A" has the plate's contour fixed against slopes around axis $\boldsymbol{n}$, and shear γ_s along the contour of the plate becomes zero automatically because

$$\gamma_s = w_{,s} - \varphi_t = w_{,s} - \theta_n = 0 ,$$

where it is obvious that $w_{,s} = 0$. On the contrary, version "B" permits the shear of the plate along its contour but eliminates the torque on the contour. It can be said that version "A" of the boundary conditions corresponds to an attachment of the flexural plate along its contour to some hypothetic wall that possesses an absolute flexibility from its plane and a perfect rigidity in its plate.

In order to focus the attention on differences in the problem statements for these two versions of boundary conditions for a simply supported Reissner plate, the authors of [9] even suggest to distinguish the conditions by their titles. For example, version "B" should be called a *free support*, while version "A" will refer to a *cylindrical hinge* on the edge of the plate[22].

Let us dwell on the solution for the boundary conditions of version "A", that is, for a cylindrical hinge rather than a free support on the plate's contour. Only in this case the solution can be obtained by elementary methods in a closed form (in double trigonometric series). We will seek the solution in the following form:

$$w = W_{mn} \sin\frac{m\pi x}{a} \sin\frac{n\pi y}{b},$$

$$\theta_x = X_{mn} \sin\frac{m\pi x}{a} \cos\frac{n\pi y}{b}, \qquad \theta_y = Y_{mn} \cos\frac{m\pi x}{a} \sin\frac{n\pi y}{b} \tag{3.59}$$

where amplitude factors of the displacements, that is, numerical coefficients W_{mn}, X_{mn} and Y_{mn} , are to be found. We note immediately that the solution in the form (3.59) satisfies all boundary conditions in version "A" but does not make the torque on the plate's contour equal to zero (which would be necessary for version "B") because from (3.30) we have

$$M^{xy} = D\frac{1-\nu}{2}(-\theta_{y,y} + \theta_{x,x}) = D\frac{1-\nu}{2}(-Y_{mn}\eta_n + X_{mn}\xi_m)\cos\frac{m\pi x}{a}\cos\frac{n\pi y}{b},$$

where the following designation is introduced for the sake of brevity:

$$\xi_m = \frac{m\pi}{a}, \qquad \eta_n = \frac{n\pi}{b}. \tag{3.60}$$

Substituting (3.59) to differential equations (3.32) produces a set of three simultaneous algebraic equations with respect to desirable parameters W_{mn}, X_{mn} and Y_{mn} :

---

[22] It should be said that the possibility of a dual formulation of the boundary conditions for a simply supported Reissner plate is not always mentioned explicitly in publications. Among works where this topic is addressed, we can mention a book by L.A. Gordon and L.A. Gotlif [9] and a paper by V.V. Vasiliev [24]. It is important because version "A", which is used commonly to build analytical solutions (see [8], for example), cannot be a basis for a comparison with solutions for a simply supported Kirchhoff plate, as we will see a bit later. Furthermore, version "B" of the boundary conditions for a simply supported Reissner plate should be adopted also for the purpose of comparison to a solution based on three-dimensional elasticity where the edge of the plate has fixations against lateral deflections only at points of the median surface of the plate (and no other fixations are allowed for the three-dimensional problem).

$$\begin{bmatrix} \left(\xi_m^2+\eta_n^2\right) & -\eta_n & \xi_m \\ -\lambda\xi_m & \frac{1+\nu}{2}\xi_m\eta_n & \left(-\lambda-\xi_m^2-\frac{1-\nu}{2}\eta_n^2\right) \\ -\lambda\eta_n & \left(\lambda+\eta_n^2+\frac{1-\nu}{2}\xi_m^2\right) & -\frac{1+\nu}{2}\xi_m\eta_n \end{bmatrix} \begin{bmatrix} W_{mn} \\ X_{mn} \\ Y_{mn} \end{bmatrix} = \begin{bmatrix} \frac{q_{mn}}{\lambda D} \\ 0 \\ 0 \end{bmatrix}.$$

A direct check helps make sure that the solution of the above equations is

$$W_{mn} = \frac{q_{mn}}{D}\frac{\lambda+\xi_m^2+\eta_n^2}{\lambda(\xi_m^2+\eta_n^2)^2},$$

$$X_{mn} = \frac{q_{mn}}{D}\frac{\eta_n}{(\xi_m^2+\eta_n^2)^2}, \qquad Y_{mn} = -\frac{q_{mn}}{D}\frac{\xi_m}{(\xi_m^2+\eta_n^2)^2}. \tag{3.61}$$

In a particular case when the plate is loaded by a constant load, $q = q_0$, the Fourier expansion gives

$$q = \sum_{m=1}\sum_{n=1} q_{mn}\sin\xi_m x\sin\eta_n y, \qquad \text{where } q_{mn} = \frac{16q_0}{\pi^2 mn},$$

and the summation invovles only odd indexes. Consequently, the solution for the constant load upon the Reissner plate with the cylindrical hinge on the contour is

$$w = \frac{16q_0}{\pi^2 D}\sum_{m=1}\sum_{n=1}\frac{1}{mn}\frac{\lambda+\xi_m^2+\eta_n^2}{\lambda(\xi_m^2+\eta_n^2)^2}\sin\xi_m x\sin\eta_n y,$$

$$\theta_x = \frac{16q_0}{\pi^2 D}\sum_{m=1}\sum_{n=1}\frac{1}{mn}\frac{\eta_n}{(\xi_m^2+\eta_n^2)^2}\sin\xi_m x\cos\eta_n y,$$

$$\theta_y = -\frac{16q_0}{\pi^2 D}\sum_{m=1}\sum_{n=1}\frac{1}{mn}\frac{\xi_m}{(\xi_m^2+\eta_n^2)^2}\cos\xi_m x\sin\eta_n y. \tag{3.62}$$

Now we can obtain all internal forces in the plate from (3.30). In particular, shear forces Q^x and Q^y, that develop on the contour and are equal to the respective reaction of the contour support, are

$$Q^x(0,y) = \lambda D(w_{,x} + \theta_y) = \frac{16q_0}{\pi^2}\sum_{m=1}\sum_{n=1}\frac{1}{mn}\frac{\xi_m}{(\xi_m^2 + \eta_n^2)}\sin\eta_n y,$$

$$Q^y(x,0) = \lambda D(w_{,y} - \theta_x) = \frac{16q_0}{\pi^2}\sum_{m=1}\sum_{n=1}\frac{1}{mn}\frac{\eta_n}{(\xi_m^2 + \eta_n^2)}\sin\xi_m x. \tag{3.63}$$

If we integrate these shear forces along the whole contour of the plate, we will have an overall reaction R that develops on the supported contour. Now let us determine reactions R_y and R_x that act along the separate faces of the plate:

$$R_y = \int_0^b Q^x(0,y)dy, \quad R_x = \int_0^a Q^y(x,0)dx.$$

Obviously, $R = 2(R_x + R_y)$, so a proper calculation gives

$$R = \frac{64q_0}{\pi^2}\sum_{m=1}\sum_{n=1}\frac{1}{mn}\frac{1}{\xi_m\eta_n} = q_0 ab\left(\frac{8}{\pi^2}\sum_{m=1}\frac{1}{m^2}\right)^2. \tag{3.64}$$

It is known [5] that the sum of this series (with odd indexes) is equal exactly to $\pi^2/8$, so we finally have $R = q_0ab$ as expected because the overall reaction must counterbalance the overall external load.

The bending moments and torques in the plate are

$$M^{xx} = -D(-\theta_{y,x} + \nu\theta_{x,y}) = \frac{16q_0}{\pi^2}\sum_{m=1}\sum_{n=1}\frac{1}{mn}\frac{\xi_m^2 + \nu\eta_n^2}{(\xi_m^2 + \eta_n^2)^2}\sin\xi_m x\sin\eta_n y,$$

$$M^{xy} = \frac{D(1-\nu)}{2}(\theta_{y,y} - \theta_{x,x}) =$$

$$= -\frac{16q_0}{\pi^2}(1-\nu)\sum_{m=1}\sum_{n=1}\frac{1}{mn}\frac{\xi_m\eta_n}{(\xi_m^2 + \eta_n^2)^2}\cos\xi_m x\cos\eta_n y.$$

Note the fact that slopes θ_x and θ_y together with forces M^{ij} and Q^i do not depend on parameter λ in this problem.

The solution obtained above can be a good test case for debugging computational algorithms and software intended for the analysis of Reissner plates.

Kirchhoff plate

A solution in double trigonometric series for a Kirchhoff plate simply supported on its contour was derived by Navier [23]. This solution for a separate term of the Fourier series can be also represented as

$$w = W_{mn} \sin\frac{m\pi x}{a} \sin\frac{n\pi y}{b}, \tag{3.65}$$

where the boundary conditions on the contour, $w = 0$, $M^{nn} = 0$, are satisfied.

Substituting (3.65) to the Germain–Lagrange equation (1.39) gives

$$W_{mn} = \frac{q_{mn}}{D(\xi_m^2 + \eta_n^2)^2}. \tag{3.66}$$

Thus, for a constant load q_0 we will have

$$w = \frac{16q_0}{\pi^2 D} \sum_{m=1} \sum_{n=1} \frac{1}{mn} \frac{1}{(\xi_m^2 + \eta_n^2)^2} \sin\xi_m x \sin\eta_n y,$$

$$\theta_x = w_{,y} = \frac{16q_0}{\pi^2 D} \sum_{m=1} \sum_{n=1} \frac{1}{mn} \frac{\eta_n}{(\xi_m^2 + \eta_n^2)^2} \sin\xi_m x \cos\eta_n y,$$

$$\theta_y = -w_{,x} = -\frac{16q_0}{\pi^2 D} \sum_{m=1} \sum_{n=1} \frac{1}{mn} \frac{\xi_m}{(\xi_m^2 + \eta_n^2)^2} \cos\xi_m x \sin\eta_n y. \tag{3.67}$$

When comparing solution (3.66) for the Kirchhoff plate with solution (3.62) for the Reissner plate, we notice that the deflections are different in these two models but the normal's slope prove to be the same.

When we have the expression of the deflection from (3.67), it is easy to derive expressions for internal forces in the Kirchhoff plate. Based on (1.28), we have

$$M^{xx} = -D(w_{,xx} + \nu w_{,yy}) = \frac{16q_0}{\pi^2} \sum_{m=1} \sum_{n=1} \frac{1}{mn} \frac{\xi_m^2 + \nu\eta_n^2}{(\xi_m^2 + \eta_n^2)^2} \sin\xi_m x \sin\eta_n y,$$

$$M^{xy} = -D(1-\nu)\, w_{,xy} = -\frac{16q_0}{\pi^2}(1-\nu) \sum_{m=1} \sum_{n=1} \frac{1}{mn} \frac{\xi_m \eta_n}{(\xi_m^2 + \eta_n^2)^2} \cos\xi_m x \cos\eta_n y.$$

Finally, using relationships (1.76), we obtain Kirchhoff's shear forces on the contour of the plate:

$$K^x(a,y) = M^{xx}_{,x} + 2M^{xy}_{,y} =$$

$$= -\frac{16q_0}{\pi^2}\sum_{m=1}\sum_{n=1}\frac{1}{mn}\frac{[\xi_m^2 + (2-\nu)\eta_n^2]\xi_m}{(\xi_m^2 + \eta_n^2)^2}\sin\eta_n y\,,$$

$$K^y(x,b) = M^{yy}_{,y} + 2M^{xy}_{,x} =$$

$$= -\frac{16q_0}{\pi^2}\sum_{m=1}\sum_{n=1}\frac{1}{mn}\frac{[\eta_n^2 + (2-\nu)\xi_m^2]\eta_n}{(\xi_m^2 + \eta_n^2)^2}\sin\xi_m x\,. \tag{3.68}$$

Note that the following concentrated forces P act in the angular points of the plate:

$$P = 2M^{xy} = \frac{32q_0}{\pi^2}(1-\nu)\sum_{m=1}\sum_{n=1}\frac{1}{mn}\frac{\xi_m\eta_n}{(\xi_m^2 + \eta_n^2)^2} \tag{3.69}$$

which are directed toward the growth of Z.

The overall reaction on the plate's contour (without taking into account the concentrated forces in the corners) is

$$2\int_0^b K^x(a,y)dy + 2\int_0^a K^x(x,b)dx = -\frac{64q_0}{\pi^2}\sum_{m=1}\sum_{n=1}\frac{1}{mn}\frac{\xi_m^4 + 2(2-\nu)\xi_m^2\eta_n^2 + \eta_n^4}{(\xi_m^2 + \eta_n^2)^2\xi_m\eta_n}\,.$$

And again, summing all the reactions on the contour of the plate together with the angular concentrated forces, we conclude that the reactions counterbalance the external load.

Now we want to calculate common shear forces Q^x and Q^y which develop on the plate's contour. We have

$$Q^x(a,y) = M^{xx}_{,x} + M^{xy}_{,y} = -\frac{16q_0}{\pi^2}\sum_{m=1}\sum_{n=1}\frac{1}{mn}\frac{\xi_m}{(\xi_m^2 + \eta_n^2)}\sin\eta_n y\,,$$

$$Q^y(x,b) = M^{yy}_{,y} + M^{xy}_{,x} = -\frac{16q_0}{\pi^2}\sum_{m=1}\sum_{n=1}\frac{1}{mn}\frac{\eta_n}{(\xi_m^2 + \eta_n^2)}\sin\xi_m x\,. \tag{3.70}$$

Let us see how reactions $r(x)$ are distributed along one of the sides of the square plate, that is, at $a = b$. Formulas (3.68) yield

$$r(x) = K^x(x,b) = -\frac{16q_0}{\pi^3}\sum_{m=1}\sum_{n=1}\frac{n^2 + (2-\nu)m^2}{m(m^2 + n^2)^2}\sin\frac{m\pi x}{a}\,. \tag{3.71}$$

For a Kirchhoff plate, these reactions distributed over the side of the plate should be supplemented by concentrated tearing forces at the beginning and at the end of the side, equal to half the concentrated angular reaction P:

$$P/2 = \frac{16q_0a^2}{\pi^4}(1-\nu)\sum_{m=1}\sum_{n=1}\frac{1}{(m^2+n^2)^2}.$$

The calculation gives (at $\nu = 0.3$)

$$P/2 = q_0a^2 = 0.032\,,$$

which accords with a result presented by Timoshenko [23].

A distribution of contour reactions $r(x)$ for the square Kirchhoff plate is shown in Fig. 5.9. The same figure presents a distribution of contour reactions calculated for the Reissner plate by (3.53).

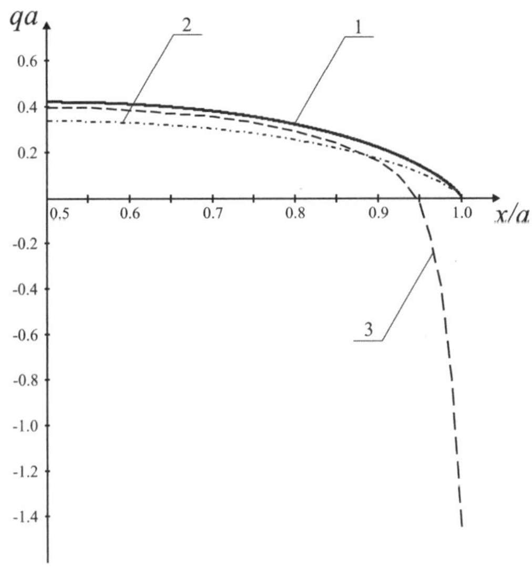

Fig. 5.9. A distribution of the reactive pressure on the contour along the side of a simply supported square plate

As we can see from the graphs, the fraction of the reaction of the supporting contour contributed by angular points is pretty big for a square Kirchhoff plate. At the same time, for the Reissner plate (version "A" of the boundary conditions) the reaction of the supporting contour in angular points is zero as follows from (3.53), so in this regard there is a significant difference between the two theories. An expectation that the analysis can be refined by using the Reissner theory of bending is deceived for several reasons – in particular, because the comparison should have involved the solution of a problem with the type "B" boundary conditions, that is, a free support of a Reissner plate along its contour. Results of calculations of that kind, done independently by various authors (such as Kromm [12], [13] and Donnell [4][23]), are shown in the same Fig. 5.9.

Three curves shown in Fig. 5.9 have the following meaning: curve 1 shows a distribution of Kirchhoff's contour forces (without the concentrated tearing force, $P = 0{,}032\, q_0 a^2$, in the corner of the plate). If we calculate the value of the tearing (negative) reaction on the quarter of the plate's contour, we will find that it is very close to $0{,}032 q_0 a^2$, that is, to the value of the concentrated tearing force in an angular point of Kirchhoff's plate per one side that joins the corner. Curve 2 shows a distribution of reactive contour forces Q calculated for the Reissner plate under the "cylindrical hinge" boundary conditions. Finally, curve 3 shows a distribution of the same reactive forces Q calculated for the Reissner plate but under the "free support" boundary conditions.

Note that a Kirchhoff plate cannot reproduce type "A" boundary conditions – a cylindrical hinge. The reason for this is an utter inability of Kirchhoff plates to transfer torques via their contours to a surrounding structure.

5.5 Final comments to Chapter 5

In general, when we take the Kirchhoff–Love hypotheses and use them to build a technical theory of plate bending, we produce contradictory relationships. Indeed, the basic assumptions of the theory are

$$\varepsilon_{zz} = 0, \quad \sigma^{zz} = 0\,,$$

and when we substitute those to an elastic relationship of the three-dimensional elasticity,

$$\varepsilon_{zz} = \frac{1}{E}(\sigma^{zz} - \nu\sigma^{xx} - \nu\sigma^{yy})\,,$$

we obtain the equality $\sigma^{xx} + \sigma^{yy} = 0$, thus $M^{xx} + M^{yy} = 0$, which cannot be true.

We also violate the elasticity relationships for the tangential stresses,

[23] We should say for the sake of justice that the calculations done by Kromm refer to a slightly different theory of plate bending that allows for shear deformations, was built by Kromm itself, and is different from Reissner's theory.

$$\gamma_{xz} = \frac{1}{G}\tau^{xz}, \quad \gamma_{yz} = \frac{1}{G}\tau^{yz},$$

because shears γ_{xz} and γ_{xz} are assumed to be zero and tangential stresses τ^{xz} and τ^{yz} taken from the equilibrium equations prove to be nonzero.

The contradictory relationships are an inevitable tradeoff of the Kirchhoff assumptions. These violations can be justified by the following. Fisrt, of all types of equations used in three-dimensional elasticity (geometric equations, equilibrium equations, physical equations), the elasticity relationships are least important. Second, the condition of $\sigma^{zz} = 0$ should not be treated as an absolutely exact equality; it is just an omission of a small stress component, σ^{zz}, from the elasticity relationships in comparison to the other two normal stress components σ^{xx} and σ^{yy}, that is,

$$\varepsilon_{xx} = \frac{1}{E}(\sigma^{xx} - \nu\sigma^{yy} - \nu\sigma^{zz}) \approx \frac{1}{E}(\sigma^{xx} - \nu\sigma^{yy}),$$

$$\varepsilon_{yy} = \frac{1}{E}(\sigma^{yy} - \nu\sigma^{xx} - \nu\sigma^{zz}) \approx \frac{1}{E}(\sigma^{yy} - \nu\sigma^{xx}).$$

This is how we treat the first two physical relationships from (1.12) in the Kirchhoff–Love theory of plates.

Sometimes a totally different logical justification is proposed: the plate of interest is assumed to be made of a transversally orthotropic material rather than an isotropic one, for which the third physical relationship can be written as

$$\varepsilon_{zz} = \frac{\sigma^{zz}}{E_1} - \nu_1 \frac{\sigma^{xx} + \sigma^{yy}}{E} \approx 0.$$

As the elasticity modulus E_1 is assumed to be big enough while coefficient ν_1 is assumed to be very small, ε_{zz} can be treated as small without requiring that $\sigma^{xx} + \sigma^{yy} \approx 0$ either exactly or approximately.

Similar physical relationships are adopted for the transverse tangential stresses,

$$\gamma_{xz} = \frac{1}{G_1}\tau^{xz}, \quad \gamma_{yz} = \frac{1}{G_1}\tau^{yz},$$

where the shear modulus, G_1, is very big, therefore we can think that shears γ_{xz} and γ_{yz} are approximately zero while tangential stresses τ^{xz} and τ^{yz} are not.

Furthermore, the Reissner plate theory presented in this chapter is a simplest version of a multitude of theories suggested for use with medium-thickness plates. This theory allows for a contribution made by lateral tangential stresses to the overall balance of the system's energy but ignores an energy accumulated by small stresses σ^{zz}. As the comparative analysis shows in most cases, taking into account a lateral shear introduces a much bigger correction to the solution of problems than taking into account a normal reduction of the plate. This holds true even for plates lying on an elastic foundation, if only that elastic foundation is not too stiff.

Discussions on Kirchhoff's theory of plates sometimes contain a statement that Kirchhoff's generalized shear forces K lack a "*physical sense*", and their concentrated component that appears in angular points of the plate is often called a "*fictitious force*" [24].

Moreover, some people say that Kirchhoff's shear forces do not counterbalance the normal load, q, applied to the plate (in projection onto the Z-axis) while common shear forces Q do. But the latter statement is based on a mere misunderstanding; its authors seem to forget to include concentrated Kirchhoff forces, acting in angular points of the plate's contour, in the overall Kirhhoff shear force to check the equilibrium of

---

[24] We can hardly do even as much as to give a comment on this cunning prestidigitation of words, at least until we define the meaning of the "*physical sense*" precisely. Of course, the concept of a concentrated force is an idealization, a model, or a strong abstraction. Still, doesn't mechanics use models of this kind extensively, including the model of a concentrated action? Why, then, is this classic and widely popular model quite admissible as an external load but protested against as an internal force in a Kirchhoff plate?

As a historical recollection, we remind that the adjective "*fictitious*" was already used earlier in mechanics in application to internal forces distributed according to a Dirac's delta function law. For example, there is a problem of a plate lying on a two-parametric foundation. The contour of the plate (and not even always angular points of it!) has an edge shear force which appears in any plate independently of what theory is used, including Reissner plates. If we adopt one of possible interpretations of the two-parametric foundation – a membrane in tension, supported by a Winkler-type layer of springs – we will understand that the derivative of the membrane's deflection function at contour points of the plate in the direction of a normal to that contour will have a discontinuity. But this discontinuity in the derivative can be caused by nothing but a concentrated action upon the membrane (in the direction of the normal to the plate's contour). This is how nonzero shear forces appear on the contour of a fixation-free plate lying on a two-parametric foundation as a result of an interaction between the membrane and the plate. The concept of *fictitious*, here used to emhasize that it is opposite to *real*, just expresses a disbelief of the authors of the term towards a new model or abstraction.

projections onto the Z-axis. Actually, those concentrated forces are a part of the K forces, distributed according to a Dirac's delta function law.

We want to show that the above said equation of equilibrium should not be checked in every particular problem because it holds for the general case. To see this, we extract a subarea $\omega \subseteq \Omega$ from area Ω occupied by the plate; the subarea is bounded by contour L which can be smooth or can have a certain number of breakpoints. In a particular case $\omega = \Omega$, then contour L coincides with boundary Γ or area Ω. A resultant of load q applied to the extracted subarea ω will be denoted by R, that is,

$$R = \int_{\omega} q d\Omega .$$

According to the equation of equlibrium (1.33), which holds for a Kirchhoff plate in the differential form, the same resultant can be also written as

$$R = -\int_{\omega}\left(Q^{x}_{,x} + Q^{y}_{,y}\right)d\Omega = -\int_{L}\left(Q^{x}n_{x} + Q^{y}n_{y}\right)dL = -\int_{L} Q^{n} dL ,$$

where we use the Gauss–Ostrogradsky formula and expression (1.58) for shear force Q^{n} on the contour. Thus, common shear forces Q^{n} counterbalance, in an integral sense, a load acting on any subarea ω extracted from area Ω. But the same holds true for the Kirchhoff shear forces, K. It suffices only to prove that

$$\int_{L}\left(K^{n} - Q^{n}\right)dL = 0.$$

By subsituting the expression of K^{n} from (1.76)[25], we transform the equality to be proved into

$$\int_{L}\left(M^{nt}_{,s} + \sum_{\beta} P_{\beta}\delta(s - s_{\beta}) \right) dL = 0,$$

which is true because by definition the P_{β} concentrated forces from (1.70) are jumps of torque M^{nt} at angular points of contour L.

The above discussion produces another, practically important, question which should be asked to the designers/analysts and, in a lesser degree, to the developers of software for structural analysis. What should we take for

[25] To simplify the formulas, we think that there is no external moment load. In this way we can set $m^{t} = 0$ in formula (1.76)

a force of interaction between a plate and a structure that supports the plate if our mechanical model for the plate is based on Kirchhoff's theory?

In response to this question, we can hear a choir of most ardent critics of the classic theory of plates: one should use common shear forces Q rather than Kirchhoff's shear forces K as integral characteristics of the respective lateral tangential stresses. But this recommendation is hardly reasonable; we prefer to agree with an exactly opposite answer as more logical.

And indeed, if we take the advice and use the common shear force as recommended above, we will eventually arrive at a contradiction. For example, if the contour of a plate is free from fixation and external loads, then nobody would mind that the mathematically justified condition for the contour is that Kirchhoff's shear force should be equal to zero. But then the common shear force, Q, becomes a nonzero quantity on the free egde of the plate and "hangs in the air" because it is not resisted by a surrounding structure.

Obviously, issues of this kind just do not appear when we do our analysis on the basis of a shear-aware theory of plates.

References

1. Allman JT, Key JE (1974) Triangular finite elements for the analysis of flexural plates under constant and linearly distributed bending moments (in Russian). In: Filin AP (ed) Analysis of elastic constructions using computers, Vol. 1. Sudpromgiz, Leningrad, pp 80–101
2. Ambartsumian SA (1967) A theory of anisotropic plates (in Russian). Nauka, Moscow
3. Biederman VL (1977) Mechanics of thin-walled constructions (in Russian). Mashinostroyeniye, Moscow
4. Donnell LG (1976) Beams, plates, and shells. McGraw-Hill
5. Dwight HB (1961) Tables of integrals and other mathematical data. The Macmillan Company, New York
6. Elsukova KP, Slivker VI (1973) On a finite element analysis of flexural plates (in Russian). In: Rozin LA (ed) Finite element method and structural mechanics. Coll. papers of Leningrad Polytechnic Inst. (LPI), Leningrad pp 45–59
7. Goldenweiser AL (1953) Theory of thin elastic shells (in Russian). Gostekhteorizdat, Moscow
8. Gordon LA, Skomorovsky YG, Friedman ES, Shoikhet BA (1971) Analysis and experimental investigations of medium-thickness plates (in Russian). Bulletin of B.Vedeneyev's All-Union Research Institute for Water Engineering, 95: 142–166

9. Gordon LA, Gotlif AA (1982) Static analysis of concrete and ferroconcrete structures in water engineering (in Russian). Energoizdat, Moscow
10. Herrmann RL (1967) Finite element bending analysis for plates. J. Eng. Mech. Div., No. EM5: 13–26.
11. Kirchhoff G (1850) Uber das Gleichgewicht und die Bewegung einer elastischen Scheibe. J. Reine Angew. Math. Bd. 40: 51–58
12. Kromm A (1953) Verallgemeinerte Theorie der Plattenstatik. Ingr. Arch. Bd. 21, 4: 266–286
13. Kromm A (1955). Über die Randquerkräfte bei gestützten Platten. Z. Angew. Für Math. Und Mech. Bd. 35, 6/7: 231–242
14. Lebedev NN (1963) Special functions and applications (in Russian). State Publ. House on Phys. & Math., Moscow Leningrad,
15. Lurie AI (1961) Static-geometric analogy in the theory of shells (in Russian). In: Problems of mechanics of continua. Coll. papers: In honour of 70th anniversary of Academician N.I. Muskhelishvili. Academy of Science of USSR, Moscow, pp 233-240
16. Lurie AI (1975) Static-geometric analogy in the theory of plates (in Russian). In: Advances in mechanics of deformable media. Coll. papers: In honour of 100th anniversary of Academician B.G. Galiorkin. Nauka, Moscow, pp 355-359
17. Mikhlin SG (1966) The Problem of the Minimum of a Quadratic Functional. Holden-Day
18. Muskhelishvili NI (1966) Some principal problems of the mathematical theory of elasticity (in Russian). Nauka, Moscow
19. Panovko YG (1985) Mechanics of deformable solids. Contemporary concepts, mistakes, and paradoxes (in Russian). Nauka, Moscow
20. Rozin LA (1964) Strain-based boundary conditions in elasticity (in Russian). Bull. Acad. Sci. USSR, Mech. & Mech. Engng., 4: 96–101
21. Rozin LA (1998) Problems of elasticity and numerical methods for their solution (in Russian). Publ. House of St.-Pet. State Tech. Univ., St.-Petersburg
22. Timoshenko SP (1953) History of strength of materials. McGraw-Hill, New York
23. Timoshenko SP, Voinovsky-Krieger S (1959) Theory of plates and shells. McGraw-Hill, New York Toronto London
24. Vasiliev VV (1992) On a theory of thin plates (in Russian). Bull. Acad. Sci., Mech. Deform. Solids, 3: 26–47
25. Zhilin PA (1995) On the classic theory of plates and the Kelvin–Tait transformation (in Russian). Bull. Acad. Sci., Mech. Deform. Solids, 4: 133 – 139

6 PARTICULAR CLASSES OF PROBLEMS IN STRUCTURAL MECHANICS – part 3

There are no rules for finding paths in the wilds of the nature. A creative thought is guided by generalized results of experiments, uses analogies, and brings old concepts revised in some way to newly born approaches. Here's when the flexibility and versatility of mathematical formulations of variational principles comes to help the researcher.

Polak L (1965) From a preface to the Russian translation of the book by C. Lanczos: Variational principles of mechanics. Mir, Moscow

6.1 Torsion of solid bars – Saint-Venant's theory

In the year when these words are being written (2003), one hundred and fifty years have passed since that significant day for the theory of elasticity when Barré de Saint-Venant submitted his famous memoir on torsion of prisms to the French Academy of Science [15], which became a basis for a theory of free (pure) torsion of prismatic bars; the theory was formulated in such details and in such manner that it has lived to our times in nearly the same form[1].

This theory can be derived from certain assumptions, or hypotheses, of a kinematic nature. These hypotheses play about the same role in the torsion of bars as the planar-sections hypothesis by Bernoulli–Euler does in the technical theory of bar bending[2]. But before we formulate those

[1] This moment indeed seems to be the beginning in the history of the theory of bar torsion, although the paper by Saint-Venant [12] itself was published only two years later, in 1855.

[2] When we build the theory of bar torsion according to Saint-Venant, we may omit the explicit introduction of any kinematical hypotheses; instead, we can start out from general equations of three-dimensional elasticity and use a so-called semi-inverse method [11], suggested and developed by that same Saint-Venant for solving problems of elasticity, in application to the bar torsion analysis.

assumptions, we deem it reasonable to discuss some conventions and a notation.

Let (X,Y,Z) be a right-hand Cartesian system of axes so that the X-axis is parallel to the generatrix of a prismatic bar and the Y-axis and Z-axis are principal central axes of inertia of the bar's cross-section[3]. The cross-section of the bar itself occupies an area Ω with a boundary Γ in the (X,Y)-plane (Fig. 6.1). It is convenient to associate the origin of the X-axis with the beginning section of the bar.

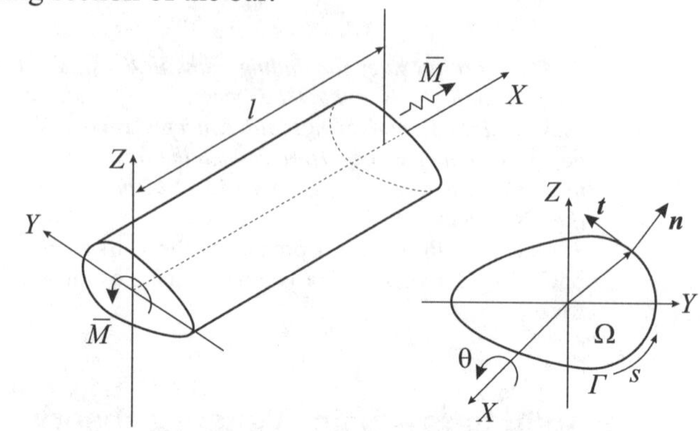

Fig. 6.1. A pure torsion of a prismatic bar

Let $\boldsymbol{n}$ be a unit vector of the external normal to boundary Γ of area Ω, and let $\boldsymbol{t}$ be a unit vector tangential to the same boundary, so that the mutual orientation of the vectors is defined by condition $\boldsymbol{n}\times\boldsymbol{t} = \boldsymbol{i}_x$ where $\boldsymbol{i}_x$ is the unit vector of axis X. The components of vectors $\boldsymbol{n}$ and $\boldsymbol{t}$ with respect to the (Y,Z) coordinate system will be denoted by n_y, n_z and t_y, t_z, respectively. Fig. 6.1 gives a hint that

$$n_y = t_z\,, \qquad n_z = -\,t_y\,. \tag{1.1}$$

We will assume the side surface of the bar to be free from any loads and all the external load upon the bar to consist of torques $\overline{M}$ applied to the butt cross-sections, that is, to the cross-sections with coordinates $x = 0$ and $x = l$ where l is the length of the bar in torsion[4]. According to the Saint-Venant principle, we deem the method of applying the torque insignificant when we want to know the distribution of stresses in the bar at a certain remote distance from the load application point.

---

[3] As we will see later, the requirement that the Y,Z axes be coincident with the principal central axes of inertia of the bar's cross-section can be omitted.

[4] These are exactly the conditions of loading that define a so-called *free* or *pure* torsion of bars.

So, we make the following kinematical assumptions (hypotheses):

- each cross-section of the bar does not change its profile in the course of deformation. In other words, the bar's cross-section behaves in its plane as if it were a rigid body rotating by a certain angle $\theta(x)$ around the X-axis.
- as the bar experiences the torsion, its cross-sections undergo a warping (a *deplanation*) such that the points of every cross-section can deviate from their original plane. Following Saint-Venant, we assume that all cross-sections of the bar experience the same deplanation. It means that the u displacement in the direction of X does not depend on the x-coordinate. Furthermore, the u deviation is assumed proportional to a rate of change of the twisting angle θ, that is, to the value of θ' (the stroke denotes the differentiation with respect to the longitudinal coordinate, x)[5]. The θ' parameter is often called a *twist* or a *twist factor* of the bar.

To exclude a rigid displacement of the body, we need to apply 6 external constraints to it. In particular, we can assume that the initial cross-section of the bar (that at $x = 0$) does not experience any rotation around the X-axis. We further assume that a point of the X-axis located at the origin of the (X,Y,Z) coordinate system does not experience any displacements at all. Finally, the end point of the bar on the X-axis with coordinates $(l,0,0)$ is restrained from displacements in the Y and Z directions. To put it another way, we assume

$$\theta_0 = \theta(0) = 0,$$

$$u_0 = u(0,0,0) = 0, \qquad v_0 = v(0,0,0) = 0, \qquad w_0 = w(0,0,0) = 0,$$

$$v_l = v(l,0,0) = 0, \quad w_l = w(l,0,0) = 0. \tag{1.2}$$

Obviously, these six external constraints just restrain the body from rigid displacements and do not affect the distribution of stresses in it. The values of the displacements can be calculated only up to their perfectly rigid components. It is easy to show that all points of the bar belonging to the X-axis do not move under these conditions, or

$$u(x,0,0) = 0, \qquad v(x,0,0) = 0, \qquad w(x,0,0) = 0, \tag{1.3}$$

[5] Further we will use the term *warping* more often than *deplanation* because it is more common in English-language technical literature.

where u, v, w are displacement components with respect to the corresponding axes of the (X,Y,Z) coordinate system, which are functions of the (x,y,z)-coordinates.

Indeed, to see this, let us suppose something like $v(x,0,0) \neq 0$ – and we will see immediately that the axis of the bar will have a curvature in the (X,Y)-plane during such a deformation. But then cross-sections of the bar which have a non-zero curvature must experience bending moments M_z. At the same time, external loads applied to the bar are not able to reproduce those bending moments. This contradiction proves that the equality $v(x,0,0) = 0$ is true. If we suppose that $u(x,0,0) \neq 0$, then we will have to admit that the external loads must reproduce a longitudinal force, N, in the bar, which cannot take place.

6.1.1 Saint-Venant torsion function. Lagrange functional

The mathematical treatment of the above kinematical hypotheses is obvious: they permit to express the displacement components u, v, w via the respective axes of the (X,Y,Z) system in the following simple form,

$$u = \theta'\varphi(y,z) - \theta'\varphi(0,0), \qquad v = -\theta z, \qquad w = \theta y, \tag{1.4}$$

where $\varphi(x,y)$ is a function called a *Saint-Venant torsion function* which is to be determined further[6].

According to second kinematical hypothesis, u does not depend on x, therefore twist θ' is a constant, and if we use (1.2) we arrive at the following linear relation between the torsion angle θ and the x coordinate:

$$\theta(x) = \theta' x. \tag{1.5}$$

Now, if we use general formulas **4**.2.1-*b* for the components of the strain tensor, we will find that all of them are zero, except for shear components ε_{yx} and ε_{zx}. However, it is more convenient to replace the strain tensor components ε_{yx} and ε_{zx} with shear angles $\gamma_{yx} = 2\varepsilon_{yx}$ and $\gamma_{zx} = 2\varepsilon_{zx}$, which we derive from (1.4) and (1.5) as follows:

$$\gamma_{yx} = v,_x + u,_y = \theta'(\varphi,_y - z), \qquad \gamma_{zx} = w,_x + u,_z = \theta'(\varphi,_z + y). \tag{1.6}$$

As follows from Hooke's law, all components of the stress tensor are zero, except for tangential stresses τ^{yx} and τ^{zx}. We have

---

[6] Here we assume the torsion angle of the cross-section, θ, to be small. This is what permits us to be sure that the displacement components v and w are proportional to angle θ by assuming $v = -\theta z$, $w = \theta y$ in formula (1.4).

$$\tau^{yx} = G\theta'(\varphi,_y - z), \qquad \tau^{zx} = G\theta'(\varphi,_z + y). \tag{1.7}$$

The boundary conditions for a prismatic bar in torsion are formulated as follows:

- on the butt $x = 0$: $\quad v = 0, \quad w = 0,$
- on the butt $x = l$: $\quad v = -\theta' lz, \quad w = \theta' ly,$
- on side faces: $\quad \tau^{xn} = \tau^{xy} n_y + \tau^{xz} n_z = 0.$

The latter condition follows from the requirement that there must be no load on the side faces of the bar, so that the component of the tangential stress in the direction of normal $\boldsymbol{n}$ to the section's contour Γ must be zero.

Taking account of (1.6) and (1.7), we can represent strain energy E of the bar in torsion as

$$\mathsf{E} = \frac{1}{2}\int_0^l \int_\Omega (\tau^{yx}\gamma_{yx} + \tau^{zx}\gamma_{zx}) d\Omega dx =$$

$$= \frac{G\theta'^2 l}{2} \int_\Omega [(\varphi_{,y} - z)^2 + (\varphi_{,z} + y)^2] d\Omega \,. \tag{1.8}$$

Based on (1.2), we can say that force potential Π_s is created solely by one external moment $\bar{M}$ applied to the butt $x = l$, and it can be written as

$$\Pi_s = \bar{M}\,\theta' l \,. \tag{1.9}$$

Thus, we have the following general expression of the full potential energy of the system

$$\mathsf{L}(\theta',\varphi) = \mathsf{E} - \Pi_s = \frac{G\theta'^2 l}{2} \int_\Omega [(\varphi_{,y} - z)^2 + (\varphi_{,z} + y)^2] d\Omega - \bar{M}\,\theta' l \,. \tag{1.10}$$

So, the Lagrange functional in the problem of bar torsion depends on two variables: the numerical parameter, θ', and the torsion function, φ. By varying the Lagrange functional with respect to those variables and using the Gauss–Ostrogradsky formula, we derive

$$\delta\mathsf{L} = G\theta' l \left(\int_\Omega [(\varphi_{,y} - z)^2 + (\varphi_{,z} + y)^2] d\Omega \right) \delta\theta' - \bar{M}\, l\delta\theta' +$$

$$+ G\theta'^2 l \int_\Omega (-\varphi_{,yy} - \varphi_{,zz}) \delta\varphi d\Omega +$$

$$+ G\theta'^2 l \oint_\Gamma [(\varphi_{,y} - z) n_y + (\varphi_{,z} + y) n_z] \delta\varphi d\Gamma \,. \tag{1.11}$$

Equaling δL to zero and considering the independence of variations $\delta\theta'$ and $\delta\varphi$ produces Euler equations and natural boundary conditions for functional L. The Euler equations are

$$\bar{M} = G\theta' \int_\Omega [(\varphi_{,y} - z)^2 + (\varphi_{,z} + y)^2] d\Omega, \tag{1.12}$$

$$\varphi_{,yy} + \varphi_{,zz} = 0 \ \in \Omega, \tag{1.13}$$

and the natural boundary conditions are

$$(\varphi_{,y} - z)n_y + (\varphi_{,z} + y)n_z = 0 \quad \in \Gamma, \tag{1.14}$$

which is equivalent to a static boundary condition, $\tau^{xn} = \tau^{xy}n_y + \tau^{xz}n_z = 0$, on the side faces of the bar.

Introducing a designation for the moment of inertia,

$$I_x = \int_\Omega [(\varphi_{,y} - z)^2 + (\varphi_{,z} + y)^2] d\Omega, \tag{1.15}$$

we can rewrite (1.12) in a form usual in the science of strength of materials:

$$\bar{M} = GI_x\theta'. \tag{1.16}$$

The geometric characteristic of a cross-section, I_x, is called its *torsional moment of inertia.*

Further, differential equation (1.13) defines a function φ, harmonic in area Ω. But the Laplace operator,

$$\nabla^2 = \frac{\partial^2}{\partial y^2} + \frac{\partial^2}{\partial z^2},$$

is invariant with respect to a Cartesian coordinate system. Relationships (1.1) help us rewrite boundary condition (1.14) for the desirable harmonic function, φ, also in a form invariant with respect to a rotation of the coordinate system, namely

$$\frac{\partial\varphi}{\partial n} = (\boldsymbol{r}, \boldsymbol{t}) \ \in \Gamma. \tag{1.17}$$

where $\boldsymbol{r} = y\boldsymbol{i}_y + z\boldsymbol{i}_z$ is a radius vector that defines an arbitrary point of boundary Γ (Fig. 6.1). Thus, we can omit the requirement that axes Y and Z should be principal axes of inertia, but we still keep the requirement that axis X should be coincident with the axis of centers of gravity of the bar's cross-sections because the scalar product $(\boldsymbol{r},\boldsymbol{t})$ depends generally on the location of the coordinate origin in the (Y,Z)-plane.

Mathematics puts it in such way that a combination of differential equation (1.13) and boundary condition (1.17) defines a *Neumann problem* for the Laplace equation.

The expression of moment of inertia I_x can be written also in another form, different from (1.15). In order to do that, we determine torque $\bar{M}$ from the condition of equilibrium,

$$\bar{M} = \int_\Omega (-\tau^{yx} z + \tau^{zx} y) d\Omega . \tag{1.18}$$

Substituting the expressions of the stresses from (1.7) gives

$$\bar{M} = G\theta' \int_\Omega [-(\varphi_{,y} - z)z + (\varphi_{,z} + y)y] d\Omega , \tag{1.19}$$

consequently,

$$I_x = \int_\Omega [-(\varphi_{,y} - z)z + (\varphi_{,z} + y)y] d\Omega . \tag{1.20}$$

Comparing between (1.15) and (1.20) gives the equality

$$\int_\Omega (\varphi_{,y}^2 + \varphi_{,z}^2) d\Omega = \int_\Omega (\varphi_{,y} z - \varphi_{,z} y) d\Omega , \tag{1.21}$$

which can be proved directly by using the Gauss–Ostrogradsky formula.

It can be checked (though we will not dwell on it) that the fields of displacements (1.4) and (1.5), strains (1.6), and stresses (1.7), which follow from the basic kinematical assumptions adopted earlier, satisfy accurately the equations of equilibrium in the volume of the bar and all boundary conditions on the side surface of the bar in torsion. As for boundary conditions on the butt surfaces of the bar, they are satisfied in an integral sense only – the stresses on those surfaces are statically equivalent to the given torque.

6.1.2 Prandtl stress function. Timoshenko functional

The torsion function, φ, can be replaced by a so-called *Prandtl stress function*, ψ, which is related to torsion function φ as follows:

$$\psi_{,z} = (\varphi_{,y} - z)G\theta', \qquad \psi_{,y} = -(\varphi_{,z} + y)G\theta'. \tag{1.22}$$

It follows directly from the definition that the Prandtl stress function satisfies the following differential equation (a Poisson equation) in area Ω:

$$\nabla^2\psi = -2G\theta' \quad \in\Omega. \tag{1.23}$$

Now we should formulate a boundary condition that the Prandtl function must satisfy. Formulas (1.7) for the tangential stresses in the cross-sections of a bar in torsion will become quite simple when (1.22) are substituted to express the stresses via function ψ:

$$\tau^{yx} = \psi_{,z}\,, \qquad \tau^{zx} = -\psi_{,y}\,. \tag{1.24}$$

Earlier we have said that the condition $\tau^{xn} = 0$ must be met on contour Γ, which gives

$$\tau^{xn} = \psi_{,z}\,n_y - \psi_{,y}\,n_z = \psi_{,y}\,t_y + \psi_{,z}\,t_z = \frac{\partial\psi}{\partial s} = 0 \quad \in\Gamma\,. \tag{1.25}$$

Here s is an arc coordinate counted off in the positive direction along the contour.

The cross-section in a bar may occupy a multiply connected area, Ω, in the general case. It means that boundary Γ may consist of multiple independent closed contours – say, it can include contours $\Gamma_0, \Gamma_1,\ldots, \Gamma_m$ for a $(m + 1)$-connected area, so that $\Gamma = \Gamma_0 \cup \Gamma_1 \cup \ldots \cup \Gamma_m$. We assume that Γ_0 designates an external boundary of area Ω, while $\Gamma_1,\ldots, \Gamma_m$ are boundaries of areas Ω_i (“holes”). If we denote by Ω_0 the whole area comprised by boundary Γ_0, then the original area Ω can be obtained by subtracting all the “holes” from it, that is, as

$$\Omega = \Omega_0 - \sum_{i=1}^{m}\Omega_i\,. \tag{1.26}$$

According to (1.25), the Prandtl function does not change on any of the contours, so this function obeys the following boundary condition on each contour:

$$\psi = C_i \quad \in\Gamma_i \quad (i = 0,1,\ldots, m), \tag{1.27}$$

where C_i is a constant, which is different on different closed contours. We could assume that $C_0 = 0$ without limiting the generality because adding an arbitrary constant to function ψ does not change the distribution of the stresses, and this is the way the problems are solved. However, we will not make an a priori assumption that $C_0 = 0$ for the sake of generality.

Here mathematics says that a combination of differential equation (1.23) and boundary conditions (1.27) defines a *Dirichlet problem* for the Poisson equation.

Potential energy E, accumulated in the twisted bar, can be expressed via the Prandtl function as follows:

$$\mathsf{E}(\psi)=\int_0^l\int_\Omega \frac{(\tau^{yx})^2+(\tau^{zx})^2}{2G}d\Omega dx = l\int_\Omega \frac{\psi_{,y}^2+\psi_{,z}^2}{2G}d\Omega\,. \tag{1.28}$$

Now it is convenient for us to replace the original problem for a bar under torques $\overline{M}$ with a problem where an external constraint is imposed to prevent the butt section $x = l$ from rotation. The external action upon the bar will be a kinematical action – a given nonzero angle of torsion, $\overline{\theta}_l$, of the bar's end section. In other words, six constraints from (1.2) are now supplemented by another one,

$$\theta(l) = \overline{\theta}_l\,. \tag{1.29}$$

In order to distinguish between the statement of the original problem by Saint-Venant and the statement of the problem of torsion of the same bar but with external constraint (1.29) imposed, we will refer to the original problem of a bar in torsion under a given moment $\overline{M}$ as problem (*a*), while the new problem of a bar subjected to torsion by the given angle, $\overline{\theta}_l$, will be called problem (*b*).

If we assume that the reactive torques at the beginning and end sections of the bar in problem (*b*) are equal to $\overline{M}$, then, obviously, the stress and strain distributions of problems (*a*) and (*b*) will be the same. It follows from this, in particular, that twist θ' will be also the same for problem (*a*) and problem (*b*). To put it another way,

$$\theta' = \frac{\overline{\theta}_l}{l}\,. \tag{1.30}$$

Now, turning to problem (*b*), we notice that there are no stresses on the side faces, and we know the displacements on the butt faces; in problem (*b*) those displacements are nonzero only at $x = l$ and are as follows:

$$\overline{v} = -\overline{\theta}_l\, z, \qquad \overline{w} = \overline{\theta}_l\, y.$$

Thus, we have the following for kinematic potential Π_{bk}, constructed for problem (*b*):

$$\Pi_{bk} = \int_{\Omega} (\tau^{yx}\overline{v} + \tau^{zx}\overline{w}) d\Omega = -\overline{\theta}_l \int_{\Omega} (\psi_{,z} z + \psi_{,y} y) d\Omega . \tag{1.31}$$

The integral participating in the above expression can be transformed using the Gauss–Ostrogradsky formula, which yields

$$\Pi_{bk} = 2\,\overline{\theta}_l \int_{\Omega} \psi d\Omega - \overline{\theta}_l \oint_{\Gamma} (\boldsymbol{r},\boldsymbol{n}) \psi d\Gamma . \tag{1.32}$$

We have established that the static boundary conditions on the side faces of the bar can be reduced to requirements (1.27) on each contour of the area. Consequently, after dividing the common contour integral in (1.32) into a sum of separate integrals over each of the closed contours and seeing that the ψ function is constant on each contour, we can rewrite (1.32) as

$$\Pi_{bk} = 2\,\overline{\theta}_l \int_{\Omega} \psi d\Omega - \overline{\theta}_l \left(C_0 \oint_{\Gamma_0} (\boldsymbol{r},\boldsymbol{n}) d\Gamma + \sum_{i=1}^{m} C_i \oint_{\Gamma_i} (\boldsymbol{r},\boldsymbol{n}) d\Gamma \right).$$

Furthermore, it is easy to note that

$$\oint_{\Gamma_0} (\boldsymbol{r},\boldsymbol{n}) d\Gamma = 2A_0 ,$$

where A_0 is the area of Ω_0[7]. Similarly, we have the following for the rest of the contours:

$$\oint_{\Gamma_i} (\boldsymbol{r},\boldsymbol{n}) d\Gamma = -2A_i ,$$

where A_i is the area of Ω_i (a positive value). The minus sign is present here because the direction of tracing around contour Γ_i is opposite to the positive direction for the external boundary of Ω_i. As a result, instead of (1.32) we have

$$\Pi_{bk} = 2\,\overline{\theta}_l \int_{\Omega} \psi d\Omega - 2\,\overline{\theta}_l \left(C_0 A_0 - \sum_{i=1}^{m} C_i A_i \right). \tag{1.33}$$

Now we can present an expression of the Castigliano functional in application to problem (*b*):

[7] This equality can be easily proved by using the Gauss–Ostrogradsky formula.

$$\mathsf{K}_b(\psi) = \mathsf{E} - \Pi_{bk} =$$

$$= l \iint_{\Omega} \left(\frac{\psi_{,y}^2 + \psi_{,z}^2}{2G} - 2\frac{\overline{\theta}_l}{l}\psi \right) d\Omega + 2\,\overline{\theta}_l \left(C_0 A_0 - \sum_{i=1}^{m} C_i A_i \right). \tag{1.34}$$

Thus, we have the following variational problem to solve, in order to determine function ψ:

minimize functional K_b on a set of functions ψ which satisfy static boundary conditions (1.27) on contours Γ_i ($i = 0,1,\ldots, m$).

From the standpoint of stationarity of the functional, the term

$$2\,\overline{\theta}_l \left(C_0 A_0 - \sum_{i=1}^{m} C_i A_i \right)$$

is an additive constant that has no influence on the variation of K_b with respect to ψ, so it can be omitted without affecting the search for a point of minimum of the Castigliano functional. When we do it, we arrive at the functional

$$\mathsf{T}(\psi) = \iint_{\Omega} \left(\frac{\psi_{,y}^2 + \psi_{,z}^2}{2G} l - 2\overline{\theta}_l \psi \right) d\Omega\,, \tag{1.35}$$

which was derived for the first time by S.P. Timoshenko for the Saint-Venant problem of torsion of prisms. The priority of S.P. Timoshenko is indicated by L.S. Leibenson [7]. Actually, as we have just seen, the Timoshenko functional is a part, subjected to variation, of the Castigliano functional which is constructed for problem (*b*) rather than for the original Saint-Venant problem – that is, for a problem perturbated by constraint (1.29).

Considering moment $\bar{M}$ to be an integral characteristics of the tangential stresses according to (1.18) and taking account of (1.31), we have the following:

$$\bar{M} = \int_{\Omega} (-\tau^{yx} z + \tau^{zx} y) d\Omega = -\int_{\Omega} (\psi_{,z} z + \psi_{,y} y) d\Omega = \frac{\Pi_{bk}}{\overline{\theta}_l}\,. \tag{1.36}$$

Substituting in the expression (1.33) for kinematical potential Π_{bk} will yield

$$\bar{M} = 2\int_{\Omega} \psi d\Omega - 2\left(C_0 A_0 - \sum_{i=1}^{m} C_i A_i \right). \tag{1.37}$$

Now take note again of the fact that the boundary conditions of (1.35) should be treated as main boundary conditions in the minimization of the Timoshenko functional (1.35), so the minimum $\mathsf{T}(\psi)$ must be sought for within a set of functions which are sure to satisfy conditions (1.27).

At the same time, the boundary conditions in the form of (1.27) are not convenient for multiply connected areas, because they contain unknown constants C_i. A standard way to overcome this difficulty is to make use of a well-known theorem by Bredt, which we are about to present.

Bredt theorem of the tangential stress circulation

Let L be a closed curve fully belonging to area Ω. A curvilinear integral taken over that curve,

$$T_L = \oint_L \tau^{sx} ds\,, \tag{1.38}$$

is called a *circulation* of the tangential stress over the curve. Here τ^{sx} is a component of the tangential stress in the direction of the tangent of the L curve.

As $\tau^{sx} = \tau^{yx} t_y + \tau^{zx} t_z = -\,\psi_{,y} n_y - \psi_{,z} n_z = -\,\psi_{,n}$, herefrom we derive

$$T_L = -\oint_L \frac{\partial \psi}{\partial n} ds\,. \tag{1.39}$$

Now we want to define the longitudinal displacement u via the Prandtl function. From (1.4) and (1.22) we derive

$$u_{,y} = \frac{\psi_{,z}}{G} + \theta' z, \qquad u_{,z} = -\frac{\psi_{,y}}{G} - \theta' y\,. \tag{1.40}$$

Obviously, a necessary condition for function u, defined in the two-dimensional area Ω by its partial derivatives, to be unambiguous is a requirement that

$$\oint_L \frac{\partial u}{\partial s} ds = 0, \tag{1.41}$$

which must be satisfied on every closed contour L fully belonging to area Ω. From (1.40) we derive

$$u_{,s} = u_{,y} t_y + u_{,z} t_z = -\frac{\psi_{,n}}{G} - \theta' (\boldsymbol{r}, \boldsymbol{n})_L\,, \tag{1.42}$$

which transforms the requirement of continuity of u according to (1.41) into

$$T_L = -\oint_L \frac{\psi_{,n}}{G} ds = 2\theta' A_L , \qquad (1.43)$$

where A_L is an area of a two-dimensional region belonging to the (X,Y) plane and bounded by the closed contour L.

Formula (1.43) is a mathematical representation of a well-known Bredt theorem of circulation of the tangential stress in torsion:

For any closed contour L, which lies completely within the cross-section of a bar, the circulation of the tangential stress is equal to area A_L, bounded by that contour, multiplied by $2G\theta'$.

Now, having established the Bredt theorem, we can get rid of the undetermined constants in the boundary conditions. As for the exterior contour itself, we have already said that we can assume $C_0 = 0$ without limiting the generality. The conditions on every interior contour according to (1.27) can be replaced now by linear relationships

$$-\oint_{\Gamma_i} \frac{\psi_{,n}}{G} ds = 2\theta' A_i \quad (i = 1, \ldots, m), \qquad (1.44)$$

which contain no undetermined constants whatsoever.

But then we have another problem: with the boundary conditions formulated as requirements (1.44), we deviate from the mathematical statement of the Dirichlet problem. The wish to stick to the Dirichlet problem leads to the following algorithm.

Instead of one problem for the $(m + 1)$-connected area, we will construct solutions for multiple Dirichlet problems in the number of $(m + 1)$. We denote by ψ_j the Prandtl function for problem j $(j = 0,1,\ldots, m)$, and assume

- for problem 0 : $\nabla^2\psi_0 = -2G\theta' \in\Omega$; $\psi_0 = C_i \in\Gamma_i$; $C_0 = 0$, $C_1 = 0$, $\ldots, C_m = 0$;
- for problem $j \neq 0$: $\nabla^2\psi_j = 0 \in\Omega$; $\psi_j = C_i \in\Gamma_i$; все $C_i = 0$ при $i \neq j$, $C_j = 1$.

Now it is obvious that the solution of the original problem with the Prandtl function ψ can be represented as a linear combination,

$$\psi = \psi_0 + \sum_{j=1}^{m} C_j \psi_j . \qquad (1.45)$$

The representation of the solution of the original problem as in (1.45) helps write out the Bredt theorem for each of the internal contours Γ_i ($i = 1,\dots, m$) as

$$-\oint_{\Gamma_i} \frac{\psi_{0,n} + \sum_{j=1}^{m} C_j \psi_{j,n}}{G} d\Gamma = 2\theta' A_i \quad (i = 1,\dots, m). \tag{1.46}$$

Unknown constants C_j ($j = 1,\dots, m$) will be determined from the solution of simultaneous algebraic equations (1.46)[8].

Center of twist

A center of twist is a point of the cross-section of a bar, which is not subjected to any displacements while the bar experiences a twist (torsion).

Earlier we imposed six external constraints (1.2) on a bar twisted by moments $\bar{M}$; the constraints prevent the bar from moving as a rigid body. We found that all points of the bar's axis (an axis of the centers of gravity of the bar's cross-sections) maintained their position in space — that is, the bar's axis proved to be at the same time an axis of twist, according to the above definition.

However, it is not hard to check that we could do without preliminarily excluding the rigid displacements, by assuming the following instead of (1.4):

$$v = -\theta' x(z - z_0), \qquad w = \theta' x(y - y_0),$$

$$u = \theta'[\varphi(y,z) - \varphi(y_0, z_0)] - \theta'(z_0 y - y_0 z), \tag{1.47}$$

where y_0, z_0 are arbitrary constants.

A direct check shows easily that the displacements from (1.47) are a solution of the Saint-Venant problem. It follows from here that the location

---

[8] The procedure described above can be shown to represent a mathematical analogue of a well-known method of structural mechanics of bar systems, the force (area-moment) method. The matrix of the system of equations (1.46) possesses all good qualities of the compliance matrix constructed for a principal system of the method of forces. The reader can derive this interpretation of the system of equations (1.46) for an exercise. Note that a convenient way of reasoning in this derivation is to refer to a so-called "*membrane analogy*" mentioned in many books on elasticity where the Saint-Venant problem is considered in sufficient detail – see [7], [3], for example.

of the center of twist cannot be determined unambiguously in the Saint-Venant problem[9].

6.1.3 Two-sided estimates of a section's torsion inertia moment

We introduce a functional, J_φ, as

$$\mathsf{J}_\varphi = \int_\Omega [(\varphi_{,y} - z)^2 + (\varphi_{,z} + y)^2] d\Omega\,, \tag{1.48}$$

which works on a set of functions φ defined in area Ω and possessing square-summable first partial derivatives. There are no requirements to boundary conditions for functions φ.

If now we denote by θ'_* and φ_* the respective twist and the torsion function, which solve the Saint-Venant problem exactly, then (1.15) produces

$$I_x = \mathsf{J}_\varphi(\varphi_*)\,. \tag{1.49}$$

Furthermore, because of (1.10) the Lagrange functional can be expressed via functional J_φ ,

$$\mathsf{L}(\theta',\varphi) = ½\, G(\theta')^2 l \mathsf{J}_\varphi(\varphi) - \bar{M}\,\theta' l\,. \tag{1.50}$$

But then, the Lagrangian functional takes a minimum value on the solution of the problem. In particular, the equality $\mathsf{L}(\theta'_*,\varphi_*) \le \mathsf{L}(\theta'_*,\varphi)$ holds, which is equivalent to

$$I_x \le \mathsf{J}_\varphi(\varphi) = \int_\Omega [(\varphi_{,y} - z)^2 + (\varphi_{,z} + y)^2] d\Omega\,, \tag{1.51}$$

and this is an estimate from above for the torsional moment of inertia of section сечения I_x. A combination of inequality (1.51) and equality (1.49) permits us to state the following variational problem of minimization with the intention to determine I_x,

---

[9] We notice this simple fact just because the literature sometimes gives inaccurate statements concerning the location of the center of twist in the cross-section of a bar in torsion. For example, a well-known monograph by A.N. Dinnik [3] says the following with regard to the Saint-Venant problem: "*In the case of asymmetric cross-sections or sections that have a single axis of symmetry, when the bar is twisted, its cross-sections turn around a point which may be different from the center of gravity and which is called a center of twist*".

$$I_x = \min_{\varphi} \mathsf{J}_\varphi(\varphi)\,. \tag{1.52}$$

To construct a guaranteed estimate from below, for the sake of simplicity we confine ourselves to the case of a singly connected cross-section of the bar[10]. We introduce the following functional J_ψ:

$$\mathsf{J}_\psi = 4\int_\Omega \psi d\Omega - \int_\Omega (\psi_{,y}^2 + \psi_{,z}^2) d\Omega\,, \tag{1.53}$$

which works on a set of functions ψ defined in area Ω and possessing square-summable first partial derivatives. The admissbile functions ψ must be identical to zero on contour Γ of the singly connected area Ω.

Now we will show that the determination of I_x can be based on a variational problem of the maximum of functional J_ψ, reciprocal to the variational problem (1.52). To put it another way,

$$I_x = \max_{\psi} \mathsf{J}_\psi(\psi)\,. \tag{1.54}$$

To see this relationship, we introduce the designation of

$$a = \frac{G\overline{\theta}_l}{l} \tag{1.55}$$

and understand that functionals $\mathsf{T}(\psi)$ and J_ψ are related through a simple dependence

$$\mathsf{T}(a\psi) = -\frac{a\overline{\theta}_l}{2}\mathsf{J}_\psi(\psi)\,. \tag{1.56}$$

We rewrite the Timoshenko functional as

$$\mathsf{T}(\psi) = \mathsf{E}(\psi) - 2\,\overline{\theta}_l \int_\Omega \psi d\Omega\;, \tag{1.57}$$

where $\mathsf{E}(\psi)$ is defined by (1.28). Let ψ_* be a Prandtl function that gives an exact solution to the Saint-Venant problem. Then the Clapeyron theorem and relationship (1.37) give

$$\mathsf{E}(\psi_*) = \frac{\overline{M}\overline{\theta}_l}{2}\,, \qquad 2\int_\Omega \psi_* d\Omega = \overline{M}\,,$$

[10] The reader who is interested with the estimation from below of the torsional moment of inertia for cross-sections with non-connected areas should take a look at the book by S.G. Mikhlin [9].

consequently,

$$\mathsf{T}(\psi_*) = -\frac{\overline{M}\overline{\theta}_l}{2} = -\frac{a\overline{\theta}_l}{2} I_x \,. \tag{1.58}$$

But the Castigliano functional $\mathsf{K}_b(\psi)$, and the Timoshenko functional $\mathsf{T}(\psi)$ along with it, takes its minimum value on the solution of the problem. This can be written also as an inequality,

$$\mathsf{T}(\psi_*) \le \mathsf{T}(a\psi), \tag{1.59}$$

which is true for any constant a and for any function ψ from the admissible set. Substituting (1.56) and (1.58), and seeing that parameters a and $\overline{\theta}_l$ are positive, we transform this inequality into

$$I_x \ \ge\ \mathsf{J}_\psi(\psi)\,. \tag{1.60}$$

Now, if we assume $a = 1$ and $\psi = \psi_*$ in relationship (1.56), we will get the equality

$$I_x \ =\ \mathsf{J}_\psi(\psi_*)\,. \tag{1.61}$$

A combination of inequality (1.60) and equality (1.61) is equivalent to the variational principle of maximum (1.54) being proved.

So, we have the following two-sided estimates for I_x:

$$\mathsf{J}_\psi(\psi) \le I_x \ \le \mathsf{J}_\varphi(\varphi)\,. \tag{1.62}$$

These inequalities can be made stronger in the following way, if we play with functions ψ and φ chosen for the estimation of I_x. We replace the functional argument of ψ with $\alpha\psi$ in the left-hand part of inequality (1.62), and choose parameter α in such way that the value of $\mathsf{J}_\psi(\alpha\psi)$ be maximum. Obviously,

$$\mathsf{J}_\psi(\alpha\psi) = 4\alpha \int_\Omega \psi\, d\Omega - \alpha^2 \int_\Omega (\psi_{,y}^2 + \psi_{,z}^2)\, d\Omega\,.$$

It should be clear that the maximum of $\mathsf{J}_\psi(\alpha\psi)$ is achieved at

$$\alpha = \frac{2\int_\Omega \psi\, d\Omega}{\int_\Omega (\psi_{,y}^2 + \psi_{,z}^2)\, d\Omega}, \quad \text{so that} \quad \max_\alpha \mathsf{J}(\alpha\psi) = \frac{\left(2\int_\Omega \psi\, d\Omega\right)^2}{\int_\Omega (\psi_{,y}^2 + \psi_{,z}^2)\, d\Omega}.$$

As a result, the left-hand inequality in (1.62) becomes stronger by turning into

$$\frac{4\left(\int_\Omega \psi d\Omega\right)^2}{\int_\Omega (\psi_{,y}^2 + \psi_{,z}^2) d\Omega} \le I_x . \tag{1.63}$$

It is worthwhile to remind that function ψ which we use in the above estimation should be zero on contour Γ of area Ω.

A similar idea works for making the upper estimate from (1.62) stronger, too. By reasoning in the same way as in the derivation of (1.63), we obtain the following after some simple transformations:

$$\min_\alpha \mathrm{J}_\varphi(\alpha\varphi) = I_r - \frac{\left(\int_\Omega (\varphi_{,y} z - \varphi_{,z} y) d\Omega\right)^2}{\int_\Omega (\varphi_{,y}^2 + \varphi_{,z}^2) d\Omega},$$

where I_r is a polar moment of inertia of the cross-section,

$$I_r = \int_\Omega (y^2 + z^2) d\Omega .$$

So the final estimates are

$$\frac{4\left(\int_\Omega \psi d\Omega\right)^2}{\int_\Omega (\psi_{,y}^2 + \psi_{,z}^2) d\Omega} \le I_x \le I_r - \frac{\left(\int_\Omega (\varphi_{,y} z - \varphi_{,z} y) d\Omega\right)^2}{\int_\Omega (\varphi_{,y}^2 + \varphi_{,z}^2) d\Omega} . \tag{1.64}$$

In the practical structural analysis, the engineers use a popular approximate formula recommended by Vlasov [17]

$$I_x \approx \frac{4 I_y I_z}{I_y + I_z}, \tag{1.65}$$

where I_y and I_z are moments of inertia of the cross-section with respect to the corresponding axes Y and Z. Based on inequality (1.64), this formula can be proved to always produce an upper (and generally rough) estimate of the torsional stiffness. To see this point, it suffices to assume $\varphi = yz$ and perform simple calculations in the right-hand part of inequality (1.64)[11].

6.1.4 A remark on Reissner-type functionals for the bar torsion analysis

First of all, we want to write out an expression of the second form of the Reissner functional for the general case of three-dimensional elasticity – by using formula (**4**.2.24) and removing terms from it which are related to the elastic foundation. In this way we have

$$\mathrm{R}_2(\boldsymbol{\sigma},\mathbf{u}) = \int_V (\frac{1}{2} D_{ijkl}\sigma^{ij}\sigma^{kl} + u_i\sigma^{ij}_{,j} + u_i\bar{X}^i)dV +$$

$$+\oint_S e_{pi}u_i(\bar{p}^i - \sigma^{ij}n_j)dS - \oint_S e_{ui}\bar{u}_i\sigma^{ij}n_j dS\,, \tag{1.66}$$

where the three-dimensional volume of the body is denoted by V and its bounding surface is re-denoted by S because the Γ symbol is already used in this section for another purpose.

Aggregate $\frac{1}{2}D_{ijkl}\sigma^{ij}\sigma^{kl}$ is a specific strain energy; in the case of the bar torsion we are dealing with, this aggregate is

$$\frac{(\tau^{yx})^2 + (\tau^{zx})^2}{2G}.$$

Further,

$$\sigma^{ij}_{,j}u_i = \tau^{xy}_{,y}u + \tau^{xz}_{,z}u + \tau^{yx}_{,x}v + \tau^{zx}_{,x}w = (\tau^{yx}_{,y} + \tau^{zx}_{,z})\theta'(\varphi - \varphi_0)\,, \tag{1.67}$$

where we denote $\varphi_0 = \varphi(0,0)$. Here we use the fact that (1.7) makes the tangential stresses independent of x.

Thus, the integral over volume V in formula (1.66) turns into

$$\int_V (\frac{1}{2} D_{ijkl}\sigma^{ij}\sigma^{kl} + u_i\sigma^{ij}_{,j} + u_i\bar{X}^i)dV =$$

$$= l\iint_\Omega \left[\frac{(\tau^{yx})^2 + (\tau^{zx})^2}{2G} + \theta'(\tau^{yx}_{,y} + \tau^{zx}_{,z})(\varphi - \varphi_0)\right]d\Omega\,. \tag{1.68}$$

---

[11] The reader is invited to make an exercise: show that Vlasov's formula produces a closest estimate when the Y and Z axes are principal central axes of inertia of the bar's cross-section.

Now let us consider the integrals over surface S in (1.66), meaning that we are going to construct them for problem (b). We note that the first of the integrals should be taken over the side surface of the bar only, because only that surface has the boundary conditions in problem (b) formulated in stresses (in zero ones), so only that surface has the characteristic functions for extracting the static boundary conditions, e_{pi}, different from zero. Thus,

$$\oint_S e_{pi} u_i (\bar{p}^i - \sigma^{ij} n_j) dS =$$

$$= -\int_0^l \oint_\Gamma (\tau^{yx} n_y + \tau^{zx} n_z) u \, d\Gamma dx = -\theta' l \oint_\Gamma \tau^{xn} (\varphi - \varphi_0) d\Gamma . \tag{1.69}$$

Finally, the second of the contour integrals in (1.66) should be taken over the butt section of the bar at $x = l$, because only in that place the characteristic functions of extraction of the kinematical boundary conditions, e_{ui}, and the given displacements are both different from zero. This integral is a kinematic potential, Π_{bk}, introduced by us earlier for problem (b) and represented via the torsion function, φ. To see this, we have

$$-\oint_S e_{ui} \bar{u}_i \sigma^{ij} n_j dS =$$

$$= -\int_\Omega (\tau^{yx} n_x (-\bar{\theta}_l z) + \tau^{zx} n_x \bar{\theta}_l y) d\Omega = \bar{\theta}_l \int_\Omega (\tau^{yx} z - \tau^{zx} y) d\Omega . \tag{1.70}$$

By summing expressions (1.68), (1.69), and (1.70) obtained for particular terms of functional (1.66), we arrive at a Reissner-type functional constructed for problem (b), which in our case depends on three functions τ^{yx}, τ^{zx}, φ. So,

$$\mathsf{R}_{b2}(\tau^{yx}, \tau^{zx}, \varphi) = -\bar{\theta}_l \oint_\Gamma \tau^{xn} (\varphi - \varphi_0) d\Gamma +$$

$$+ \iint_\Omega \left[\frac{(\tau^{yx})^2 + (\tau^{zx})^2}{2G} l + \bar{\theta}_l (\tau^{yx}_{,y} + \tau^{zx}_{,z})(\varphi - \varphi_0) + \bar{\theta}_l (\tau^{yx} z - \tau^{zx} y) \right] d\Omega . \tag{1.71}$$

Note that this functional is defined on a set of functions τ^{yx}, τ^{zx}, φ not subjected to any boundary conditions beforehand. Now, if we take the first variation of that functional and equal it to zero, we will obtain the following three Euler equations:

$$\frac{\tau^{yx}}{G}l-\bar{\theta}_l(\varphi_{,y}-z)=0\,,\quad \frac{\tau^{zx}}{G}l-\bar{\theta}_l(\varphi_{,z}+y)=0\,,\quad \tau^{yx}_{,y}+\tau^{zx}_{,z}=0\ \in\Omega \quad (1.72)$$

and the natural boundary conditions for the functional:

$$\tau^{xn}=0 \quad \in\Gamma. \qquad (1.73)$$

The Reissner functional from (1.71) can be transformed into its first form by removing the differentiation of the tangential stresses in the integrand over Ω. To do this, we use the Gauss–Ostrogradsky formula and have

$$\bar{\theta}_l\int_\Omega(\tau^{yx}_{,y}+\tau^{zx}_{,z})(\varphi-\varphi_0)d\Omega=$$
$$=-\bar{\theta}_l\int_\Omega(\tau^{yx}\varphi_{,y}+\tau^{zx}\varphi_{,z})d\Omega+\bar{\theta}_l\oint_\Gamma\tau^{xn}(\varphi-\varphi_0)d\Gamma\,.$$

Substituting the latter expression to (1.71) gives the first form of the Reissner functional,

$$\mathsf{R}_{b1}(\tau^{yx},\tau^{zx},\varphi)=$$
$$=\iint_\Omega\left[\frac{(\tau^{yx})^2+(\tau^{zx})^2}{2G}l-\bar{\theta}_l\tau^{yx}(\varphi_{,y}-z)-\bar{\theta}_l\tau^{zx}(\varphi_{,z}+y)\right]d\Omega\,. \qquad (1.74)$$

This ends our consideration of the Saint–Venant problem from the standpoint of its variational formulations.

6.1.5 A membrane analogy by Prandtl. Torsion of a narrow strip

It is known that the deflection, U, of a membrane stretched over contour Γ is determined by the solution of the following boundary-value problem:

$$U_{,yy}+U_{,zz}=-q/H \quad \in\Omega, \qquad U=0 \quad \in\Gamma,$$

which is mathematically identical to the bar torsion analysis if the latter uses the Prandtl stress function, ψ[12], and the tension in the membrane, H, and the lateral load, q, are chosen in such way that

$$2G\theta'=q/H\,.$$

[12] In this section we confine ourselves to singly connected areas Ω.

In that case the stress function, U, of the membrane will be identical to the stress function ψ. According to formula (1.37), volume V located between the surface of the membrane in its deformed state and its flat position before the load is applied will be equal numerically to half of the torque M_x. The slopes of the deformed surface of the membrane in the (X,Y) and (X,Z) planes will be equal to the respective components of the tangential stresses. Or, more exactly,

$$\tau^{yx} = U_{,z}\,, \qquad \tau^{zx} = -U_{,y}\,.$$

Furthermore, in the problem of torsion of a bar with its cross-section occupying the singly connected area Ω, the following statement holds[13]:

> *Torque $M^{(yx)}$ created by tangential stresses τ^{yx} is equal to torque $M^{(zx)}$ created by stresses τ^{zx}.*

In other words, the tangential stresses in the section, which are parallel to each of the Cartesian axes, create half of the full torque M_x each. This statement can be proved very easily. According to the defintion, we know that

$$M^{(yx)} = -\int_{\Omega} \tau^{yx} z d\Omega\,, \qquad M^{(zx)} = \int_{\Omega} \tau^{zx} y d\Omega\,.$$

Substituting in the expressions of the stresses via the Prandtl function according to (1.24) and using the Gauss–Ostrogradsky formula gives

$$M^{(yx)} = -\int_{\Omega} \psi_{,z} z d\Omega = \int_{\Omega} \psi d\Omega - \oint_{\Gamma} \psi n_z z d\Gamma\,,$$

$$M^{(zx)} = -\int_{\Omega} \psi_{,y} y d\Omega = \int_{\Omega} \psi d\Omega - \oint_{\Gamma} \psi n_y y d\Gamma\,.$$

But the Prandtl function is zero on the contour of the singly connected area Ω, so

$$M^{(yx)} = M^{(zx)} = \int_{\Omega} \psi d\Omega\,,$$

which is exactly half of the full torque. This result also corresponds to formula (1.37) obtained earlier; that was applied to a singly connected section of a twisted bar.

---

[13] As stated by S.P. Timoshenko and J.Goudier [16], this fact was established by Lord Kelvin.

Now let us show that the membrane analogy by Prandtl and the statement about the equality between moments $M^{(yx)}$ and $M^{(zx)}$ just proved above make it possible to obtain an approximate solution of the problem of torsion of a narrow strip with quite elementary means.

We consider a section in the shape of a strip with its characteristic sizes $b \times h$ where b is the strip's length and h is its width (Fig. 6.2).

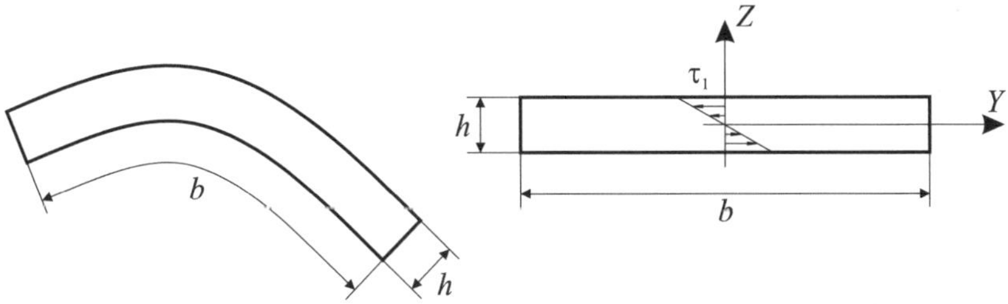

Fig. 6.2. Torsion of a bar the section of which is an elongated strip

The membrane analogy by Prandtl helps predict that the solution for torsion of the bar with the bent-strip-shaped section is not going to differ much from that for a rectilinear strip of the same sizes if the ratio $h/\rho \ll 1$, where ρ is a radius of curvature of the median line in the original bent strip. Furthermore, we can assume with a certain degree of accuracy that the deflection function of a membrane stretched over an elongated rectangular contour will be well approximated in the most part of area Ω by a square parabola, that is, $U = k(h^2 - 4z^2)$ with a certain coefficient k.

This approximation is apparently satisfactory along the b side of the rectangle, except for short pieces about h long near the ends of the rectangle. Therefore, in order to have a more accurate approximation of the end segments of area Ω, we set

$$U = \begin{cases} k(h^2 - 4z^2)\ , \ y \in [-(b/2 - h), (b/2 - h)] \\ k(h^2 - 4z^2)\left[1 - \dfrac{(y - b/2 + h)^2}{h^2}\right], \ y \in [(b/2 - h), b/2] \\ k(h^2 - 4z^2)\left[1 - \dfrac{(y + b/2 - h)^2}{h^2}\right], y \in [-b/2, -(b/2 - h)] \end{cases} . \quad (1.75)$$

The volume V located between the deflected membrane and the same membrane in its undeformed state is determined by integrating function U over the whole rectangle ($b \times h$), which gives

$$V = \frac{2}{3} kbh^3 \left(1 - \frac{2h}{3b}\right),$$

so we derive the following from Prandtl's analogy:

$$M_x = GI_x\theta' = \frac{4}{3} kbh^3 \left(1 - \frac{2h}{3b}\right).$$

According to the approximation we are using, we can think that stresses τ^{yx} vary little with the y coordinate, so $\tau^{yx}(y,z) \approx \tau^{yx}(b/2,z)$. To shorten the notation, we denote

$$\tau_1 = -\, \tau^{yx}(b/2,h/2).$$

The minus sign is used because the value of τ_1 is assumed to be positive while our convention of signs prescribes to have stress $\tau^{yx}(b/2,h/2)$ negative for a positive torque M_x. But then $\tau^{yx} = -2\tau_1 z/h$. Thus,

$$M^{(yx)} = -\int_\Omega \tau^{yx} z d\Omega = b \int_{-h/2}^{h/2} 2\tau_1 \frac{z^2}{h} dz = \tau_1 \frac{bh^2}{6}.$$

As we have just shown above, this value is half of the full torque M_x, therefore

$$\tau_1 = \frac{M_x}{W_x}, \qquad \text{where} \quad W_x = \frac{bh^2}{3}. \tag{1.76}$$

As for the torsional moment of inertia of the section, I_x, this parameter can be estimated from above by Vlasov's formula (1.65) which gives

$$I_x \le \frac{4b^4h^4}{12(bh^3 + b^3h)} = \frac{bh^3}{3(1 + h^2/b^2)}.$$

In our case we can obtain also an estimate from below if we use the left-hand part of formula (1.64) and, following the membrane analogy, replace the Prandtl function ψ defined by (1.75) with the deflection function of the membrane, U. Making necessary calculations and taking into account the estimate according to Vlasov's formula, we have the final result,

$$\frac{bh^3}{3}\left(1 - \frac{2h}{3b}\right) \le I_x \le \frac{bh^3}{3}\left(\frac{1}{1 + h^2/b^2}\right). \tag{1.77}$$

As we can see, for elongated strips where the dimensionless parameter $h/b \ll 1$, the two-sided estimates of the torsional moment of inertia according to (1.77) are fairly tight. In practice, a simplified formula is commonly used; that one gives a slightly overestimated torsional stiffness of the strip,

$$I_x = \frac{bh^3}{3}. \tag{1.78}$$

6.2 Thin-walled open-profile bars – a theory by Vlasov

This section deals with rectilinear thin-walled open-profile bars.

The structural mechanics uses the term of a *thin-walled rectilinear bar* to refer to a cylindrical shell (not necessarily a round cylinder) that has all three characteristic linear sizes substantially different in their orders of magnitude. The first characteristic size is the thickness of the bar, h, the second characteristic size is the extent of the profile of the shell's cross-section, l, and the third characteristic size is the bar's length, L. A thin-walled bar is supposed to have the following relations between the orders of magnitude of the said parameters:

$$\frac{h}{l} << 1, \quad \frac{l}{L} << 1. \tag{2.1}$$

The first of the above inequalities is used commonly with shells, while the second is a characteristic property of bars. This is why we can treat the thin-walled bar as a kind of a "child" element which partially inherits and thus combines the properties of its parents, both the bar and the shell.

Let us introduce a right-oriented Cartesian system of coordinates, (X,Y,Z), such that its X-axis coincides with the line of the centers of gravity of the bar's cross-sections and its Y and Z-axes are principal central axes of inertia of those sections. Further we will refer to the X-axis in the same way as we did in the theory of solid-section bars, that is, as a *longitudinal axis* of the thin-walled bar. Thus, the longitudinal axis of a thin-walled bar is parallel to the generatrix of the cylindrical shell that makes up the bar.

The theory of bars uses a most important notion of a *median surface*, being a surface that divides the shell into two equal parts in its thickness. The notion of a *profile* of a thin-walled bar, which was mentioned above and is quite intuitive, can be defined explicitly as a flat curve formed by the intersection between the bar's median surface and its cross-section's plane, that is, by the (Y,Z)-plane.

If the bar's profile does not contain any closed curves, then the profile is called an *open profile*. This is a kind of the thin-walled bars we will consider in this section of the book. Also, we will adopt a limitation according to which the cross-section of the bar may not vary along the X-axis. It follows from here, in particular, that the thickness of the wall, h, may not depend on the x-coordinate as well; however, it is a function of the arc coordinate s, i.e. $h = h(s)$.

In addition to the global coordinate system, every point of the median surface of a thin-walled bar will have its local coordinate system. To introduce such a system, we fixate a coordinate origin O on the profile of the bar and a certain direction in which the arc coordinate s will increase, so that the increment of the arc coordinate, ds, be an increment of the profile curve's length as we move along the profile in the assumed positive direction. The position of an arbitrary point M that belongs to the median surface of the shell will be defined by two coordinates, (x,s), as shown in Fig. 6.3. By definition, the beginning point of the profile, O, will have the zero arc coordinate and thus will be a coordinate origin.

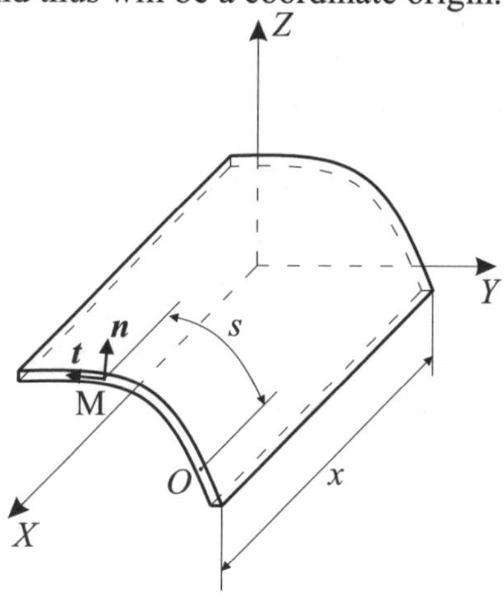

Fig. 6.3. Global and local coordinate systems for a thin-walled bar

Now let us define a unit vector $\boldsymbol{t}$ as a vector belonging to the plane of the bar's cross-section and tangential to the profile of the section in every point; the direction of the vector will coincide with the increasing direction of arc coordinate s. The above-mentioned local coordinate system is a curvilinear orthogonal system, with the vectors $(\boldsymbol{i}_x, \boldsymbol{n}, \boldsymbol{t})$ being unit vectors of its axes, where $\boldsymbol{n}$ is a unit vector of the normal to the bar's profile, the positive direction of which is defined by the requirement of the right-hand orientation of the triple of axes $(\boldsymbol{i}_x, \boldsymbol{n}, \boldsymbol{t})$, that is,

$$\boldsymbol{n} = \boldsymbol{t} \times \boldsymbol{i}_x \,. \tag{2.2}$$

Let a certain current point M of the profile become a point N through the deformation of the bar. The directed segment MN is a vector of a full displacement of the current point of the shell-bar's median surface. This vector can be expanded either over the global coordinate axes or over the local coordinates, which gives

$$u\boldsymbol{i}_x + v\boldsymbol{t} + w\boldsymbol{n} = U\boldsymbol{i}_x + V\boldsymbol{i}_y + W\boldsymbol{i}_z\,, \tag{2.3}$$

where (u,v,w) are components of the displacement vector of the M point with respect to the local coordinate system, and (U,V,W) are components of the same vector with respect to the global coordinates. The v local component is called a *tangential displacement* of the current point M, and the w component a *normal displacement* of the same point.

By multiplying the above equality scalarly first by vector $\boldsymbol{t}$ and then by vector $\boldsymbol{n}$, we can establish relationships between the components of the displacement vector, which have been introduced above, as

$$v = Vt_y + Wt_z\,, \qquad w = Vn_y + Wn_z\,, \qquad u = U,$$

and multiplying equality (2.3) scalarly by unit vectors $\boldsymbol{i}_y$ and $\boldsymbol{i}_z$ in order gives the inverse relationships

$$V = vt_y + wn_y, \qquad W = vt_z + wn_z\,.$$

The components of the basic unit vectors of the local coordinate system with respect to the global system of axes,

$$t_y = \boldsymbol{t}\cdot\boldsymbol{i}_y\,, \quad t_z = \boldsymbol{t}\cdot\boldsymbol{i}_z\,, \quad n_y = \boldsymbol{n}\cdot\boldsymbol{i}_y\,, \quad n_z = \boldsymbol{n}\cdot\boldsymbol{i}_z\,,$$

are not independent quantities. To see this, we make a scalar product of (2.2) with unit vector $\boldsymbol{i}_y$ and unit vector $\boldsymbol{i}_z$ and then use a known property [6] of the mixed vector product, $\boldsymbol{a}\cdot(\boldsymbol{b}\times\boldsymbol{c}) = \boldsymbol{b}\cdot(\boldsymbol{c}\times\boldsymbol{a})$, to obtain the following, after a chain of appropriate transformations:

$$n_y = \boldsymbol{i}_y\cdot\boldsymbol{n} = \boldsymbol{i}_y\cdot(\boldsymbol{t}\times\boldsymbol{i}_x) = \boldsymbol{t}\cdot(\boldsymbol{i}_x\times\boldsymbol{i}_y) = \boldsymbol{t}\cdot\boldsymbol{i}_z = t_z\,,$$

$$n_z = \boldsymbol{i}_z\cdot\boldsymbol{n} = \boldsymbol{i}_z\cdot(\boldsymbol{t}\times\boldsymbol{i}_x) = \boldsymbol{t}\cdot(\boldsymbol{i}_x\times\boldsymbol{i}_z) = -\,\boldsymbol{t}\cdot\boldsymbol{i}_y = -t_y\,, \tag{2.4}$$

which is identical to (1.1), as should be expected. Therefore we can re-write the formulas of the direct and inverse transformations between the displacement vector's components in the global and local coordinates in their final form that depends on the components of unit vector $\boldsymbol{t}$ only,

$$v = Vt_y + Wt_z\,, \qquad w = Vt_z - Wt_y\,,$$

$$V = vt_y + wt_z, \qquad W = vt_z - wt_y\,. \tag{2.5}$$

6.2.1 Basic assumptions in the theory of thin-walled open-profile bars

A technical theory of thin-walled open-profile bars can be based on the following set of hypotheses of the kinematic and static nature:

- *an unchanged-contour hypothesis*, according to which the cross-section of the bar does not change its shape in its plane;
- *a no-shear hypothesis*: there is no shear in the median surface of the thin-walled bar so that $\gamma_{xs} = 0$;
- *a no-pressure hypothesis*, according to which the longitudinal fibers of the thin-walled bar do not interact in their normal directions. In other words, we assume $\sigma^s = \sigma^n = 0$;
- *a moment-free-shell hypothesis*: there are no moments in the longitudinal direction, that is, the distribution of normal stresses σ^x over the thickness of the shell is assumed uniform, and tangential stresses τ^{xn} are believed negligibly small and therefore approximately equal to zero;
- *a hypothesis of a linear distribution of tangential stresses* τ^{xs} over the thickness of the shell.

Now we would like to give some comments on the above hypotheses and establish their corollaries.

The unchanged-contour hypothesis declares, actually, that the bar's profile behaves like a perfectly rigid body in the (Y,Z)-plane, that is, in the bar's cross-section plane. Thus, the profile has three degrees of freedom in the (Y,Z)-plane – two translational displacements of a selected point of the profile with respect to the (Y,Z) axes, and a rotation of the profile by a certain angle about that point (a slope). Of course, the applicability of this assumption is not unconditional; taking the assumption should be justified by structural features of the thin-walled bar in question, which prevent the deformation of the profile in the plane of the bar's cross-section. The thin-walled bar is usually assumed to have narrowly spaced transversal diaphragms installed throughout its length, the diaphragms being perfectly rigid *in* their planes and perfectly flexible *from* their planes. Those diaphragms are what keeps the profile's geometry unchanged when the thin-walled bar experiences the deformation.

Now let us drop for a time the longitudinal displacements $u = U$ of a current point, M, of the thin-walled bar's median surface and define a two-dimensional vector of displacements $\mathbf{v}$ of the same point in the plane of the bar's cross-section, that is, in the (Y,Z)-plane. The unchanged-profile hypothesis permits us to represent this vector as

$$\mathbf{v} = \mathbf{v}_P + \boldsymbol{\theta} \times \boldsymbol{R}_{PM}$$

where:

- $\mathbf{v}_P$ is a vector of displacements of a certain, arbitrarily chosen, point P of the section, which we call a *pole* and which has global coordinates y_P and z_P. The P pole itself does not have to belong to the section's profile, but it is assumed to have a rigid attachment to it;
- $\boldsymbol{\theta} = \theta \boldsymbol{i}_x$ is a vector of rotation of the section about an axis that goes through the pole, P, and is parallel to the X-axis. The positive direction of that vector is defined by right-hand screw and coincides with the direction of the X-axis. The absolute value of the rotation, $|\theta| = \|\boldsymbol{\theta}\|$, is assumed to be small, that is, $|\theta| \ll 1$;
- $\boldsymbol{R}_{PM} = (\boldsymbol{r} - \boldsymbol{r}_P) = (y - y_P)\boldsymbol{i}_y + (z - z_P)\boldsymbol{i}_z$ is a vector from the pole, P, to the current point of the profile, M, with coordinates y, z.

After expanding all separate components of the displacement vector $\mathbf{v}$ over the axes of the global coordinate system and then performing the vector multiplication, we have

$$\mathbf{v} = [\eta - \theta(z - z_P)]\, \boldsymbol{i}_y + [\zeta + \theta(y - y_P)]\, \boldsymbol{i}_z\,, \tag{2.6}$$

where $\eta = \mathbf{v}_P \cdot \boldsymbol{i}_y$, $\zeta = \mathbf{v}_P \cdot \boldsymbol{i}_z$ are components of the displacement of the pole P with respect to the global system of axes.

But the same vector $\mathbf{v}$ can be expanded over the axes of the local coordinate system,

$$\mathbf{v} = v\boldsymbol{t} + w\boldsymbol{n}, \tag{2.7}$$

where $v = \mathbf{v} \cdot \boldsymbol{t}$, $w = \mathbf{v} \cdot \boldsymbol{n}$ are components of the displacement of the current point, M, with respect to the local system of axes. By equaling expressions (2.6) and (2.7) for vector $\mathbf{v}$, we derive

$$[\eta - \theta(z - z_P)]\, \boldsymbol{i}_y + [\zeta + \theta(y - y_P)]\, \boldsymbol{i}_z = v\boldsymbol{t} + w\boldsymbol{n}.$$

Making the scalar product of this equality with unit vector $\boldsymbol{t}$ and considering that $t_y^2 + t_z^2 = 1$ gives

$$v = \eta t_y + \zeta t_z + \theta[(y - y_P)t_z - (z - z_P)t_y]\,. \tag{2.8}$$

Formula (2.8) expresses the *tangential displacement* v of the current point M of the profile via translational displacements η, ζ and the rotation anlge θ of the cross-section as a rigid body around the pole.

The hypothesis of no shear in the median surface of the shell-bar is a close analogy of the respective Bernoulli hypothesis; it states, actually, that

$$\gamma_{xs} = \frac{\partial u}{\partial s} + \frac{\partial v}{\partial x} = 0. \tag{2.9}$$

When we take the assumption of no shear, we can integrate (2.9) and derive an expression of the longitudinal displacements, u, as follows:

$$u = u_0(x) - \int_0^s \frac{\partial v}{\partial x} ds$$

where by u_0 we denote a longitudinal displacement of the beginning point of the profile, O, i.e. $u_0(x) = u(x,0)$.

Substituting the expression of tangential displacement v from formula (2.8) and denoting the arc coordinate of the current point M by s, we derive

$$u(x,s) = u_0(x) - \eta' \int_0^s t_y ds - \zeta' \int_0^s t_z ds - \theta' \int_0^s \rho ds,$$

where we use the notation of

$$\rho = (y - y_P)t_z - (z - z_P)t_y = (y - y_P)n_y + (z - z_P)n_z = \boldsymbol{R}_{PM} \cdot \boldsymbol{n}. \tag{2.10}$$

Here and further a stroke means a differentiation with respect to longitudinal coordinate x. Obviously, ρ is a projection of the $\boldsymbol{R}_{PM}$ vector onto the direction of normal $\boldsymbol{n}$ to the profile line at the current point M. It is obvious, too, that $t_y ds = dy$, $t_z ds = dz$. Also, the geometrical meaning of the $\rho ds = d\omega$ quantity can be represented as the doubled area of an elementary triangle hatched in Fig. 6.4-*a*. So, making the integration, we obtain

$$u = u_0 - \eta'(y - y_0) - \zeta'(z - z_0) - \theta'\omega \tag{2.11}$$

where ω is a so-called *sectorial area*, called also a *sectorial coordinate*, which is the doubled area of a figure swept by the radius vector $\boldsymbol{R}_{PM}$ when its end moves continuously along the profile from the origin O to the current point M – see Fig. 6.4-*b*. By y_0 and z_0 we denote global coordinates of the origin point of the profile, O.

The sectorial area is associated with a sign according to the following rule. At the zero point of the profile, it is equal to zero, $\omega(0) = 0$, and the increment of the sectorial coordinate, $d\omega = \rho ds$, is thought positive at a positive ρ and when we move toward bigger tangential coordinates s. For example, areas $\omega/2$ and $d\omega/2$ shown in Fig. 6.4 should get the minus sign because the convention of the right-hand orientation of the triple of axes

$(\boldsymbol{i}_x, \boldsymbol{n}, \boldsymbol{t})$ makes the ρ projection of vector $\boldsymbol{R}_{\mathrm{PM}}$ onto the direction of $\boldsymbol{n}$ negative.

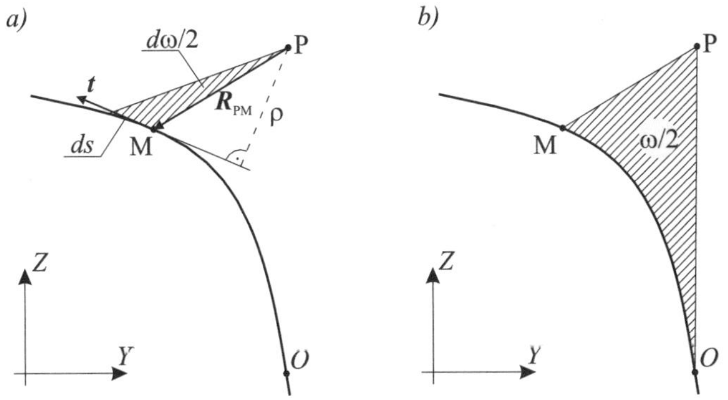

Fig. 6.4. Definition of the sectorial area, ω

Thus,

$$\omega(s) = \int_0^s \rho ds\,. \tag{2.12}$$

Now imagine that the pole P is bound to the zero point of the profile, O, by a perfectly rigid element PO (Fig. 6.5).

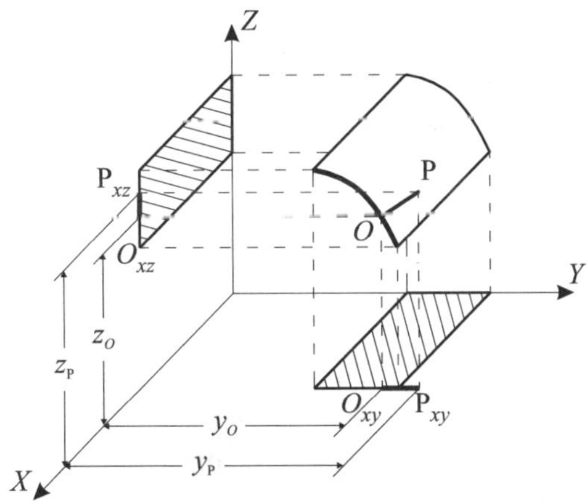

Fig. 6.5. A rigid binding of the pole P to the profile's zero point

Then the longitudinal displacement, u_O, of the zero point of the profile can be expressed via the longitudinal displacement, u_{P}, of the pole (that is, its displacement in the direction of axis X) and lateral displacements of the pole η, ζ as

$$u_0 = u_P + \eta'(y_P - y_0) + \zeta'(z_P - z_0). \tag{2.13}$$

This follows from apparent geometrical considerations if we notice that quantities η' and ζ' are slopes θ_z and θ_y of the perfectly rigid element, P*O*, in the respective planes (*X*,*Y*) and (*X*,*Z*). To be more particular,

$$\theta_y = -\zeta', \qquad \theta_z = \eta'. \tag{2.14}$$

Slopes θ_z and θ_y follow here the following convention of signs: a slope angle is assumed to be positive if the respective right-hand vector of rotation has the same direction as the respective axis of the global coordinate system.

By substituting the expression of u_0 thus obtained to (2.11), we find

$$u = u_P + \eta'(y_P - y) + \zeta'(z_P - z) - \theta'\omega.$$

We will assume that the pole P is, in its turn, bound rigidly to a point G of the bar's cross-section, which is the center of gravity of that cross-section. It means that the G point is located on the *X* axis. The relation between the longitudinal displacement, ξ, of the G point and the longitudinal displacement, u_P, of the pole P follows the law of plane, which gives

$$\xi = u_P + \eta' y_P + \zeta' z_P.$$

It follows from here and from the previous equality that

$$u = \xi - \eta' y - \zeta' z - \theta'\omega. \tag{2.15}$$

Formula (2.15) defines the longitudinal displacement, $u(x,s)$, of an arbitrary point M belonging to the bar's profile via:

- the longitudinal displacement ξ of the cross-section's center of gravity, as a function of x;
- the lateral displacements of the pole, (η, ζ), as functions of the x-coordinate;
- the angle of rotation (the slope), θ, of the bar's cross-section about the pole P, as a function of the section's coordinate, x;
- the global coordinates (y, z) of point M, as functions of coordinate s;
- the sectorial coordinate, ω, of point M, as a function of coordinate s.

Taking the derivative of kinematic relationship (2.15) with respect to coordinate x gives an expression of the relative longitudinal strain, $\varepsilon_x = du/dx$, at the current point of the profile, M,

$$\varepsilon_x = \xi' - \eta'' y - \zeta'' z - \theta'' \omega . \tag{2.16}$$

The no-pressure hypothesis is totally equivalent to its counterpart in the bar bending theory by Bernoulli–Euler. It states that normal stresses σ^s and σ^n can be neglected in the part of Hooke's law that relates to normal stresses σ^x, which gives

$$\sigma^x = E\varepsilon_x = E\frac{\partial u}{\partial x} .$$

Now, using the new form of Hooke's law, we can determine normal stresses σ^x, too:

$$\sigma^x = E\xi' - E\eta'' y - E\zeta'' z - E\theta'' \omega . \tag{2.17}$$

The moment-free-shell hypothesis that states the absence of longitudinal moments in the shell-bar seems intuitively reasonable, because a small thickness, h, of the bar's walls makes the variation of longitudinal normal stresses σ^x over the thickness hardly noticeable. This is confirmed also by a more accurate analysis of the stresses in long cylindrical shells.

The mathematical meaning of this hypothesis is that normal stresses σ^x are assumed to be functions of two coordinates only – longitudinal coordinate x and arc coordinate s,

$$\sigma^x = \sigma^x(x,s) .$$

Using this law and summing stresses σ^x over the whole area of the bar's cross-section, we can obtain the longitudinal force, N, in it. And if we integrate the same stresses over the bar's cross-section area after multiplying them by coordinate y or z, we will have overall bending moments in the respective sections, M_z or M_y. So,

$$N = \int_A \sigma^x dA , \qquad M_y = \int_A z\sigma^x dA , \qquad M_z = \int_A y\sigma^x dA , \tag{2.18}$$

where A is the area of the bar's cross-section.

Note that the longitudinal force, N, is assumed to be positive if it causes positive stresses σ^x. Bending moments M_y and M_z are assumed to be positive if they cause tension in fibers with positive coordinates z and with positive coordinates y, respectively. We would like to emphasize that the bending moments, according to their definition, are calculated with respect to the principal central axes of inertia.

As the element of area, dA, can be expressed via the thickness of the wall, h, and the element of the arc coordinate, ds, as $dA = hds$, we will have the following after substituting (2.17) to (2.18) and integrating:

$$N = EA\xi' - \theta'' E \int_l h\omega ds ,$$

$$M_y = -\zeta'' EI_y - \theta'' E \int_l h\omega z ds , \qquad M_z = -\eta'' EI_z - \theta'' E \int_l h\omega y ds . \qquad (2.19)$$

The integration takes account of the fact that the Y and Z axes are principal central axes of inertia of the bar's cross-section, and I_y and I_z denote, as usual, the respective moments of inertia of the bar's cross-section.

We have at our disposal an arbitrary choice of the coordinates of pole P and the origin point, O, of the bar's profile. So we choose those points in such way that the following three conditions should hold:

$$\int_l h\omega ds = 0 , \qquad \int_l h\omega y ds = 0 , \qquad \int_l h\omega z ds = 0 . \qquad (2.20)$$

The pole P and the origin O such that they satisfy conditions (2.20) are called a *principal pole* and a *principal origin* (*zero point*) of the profile, respectively. Appendix G shows how the second and the third of the conditions from (2.20) are sufficient to determine the coordinates of the principal pole. However, there can exist more than one (and even an infinite number of) origin of a particular profile that satisifies the first of conditions (2.20). To determine an unambiguous location of the principal origin among all possible origins of a particular profile which satisfy conditions (2.20), we can choose the nearest one to the principal pole, for example. This will be explained in detail in Appendix G.

As a result of selecting the pole and the origin on the basis of (2.19), we have

$$\xi' = \frac{N}{EA} , \qquad \zeta'' = -\frac{M_y}{EI_y} , \qquad \eta'' = -\frac{M_z}{EI_z} . \qquad (2.21)$$

The geometrical meaning of the ζ'' and η'' quantities is the respective inverted curvature, χ_z or χ_y, of the *pole axis* in the undeformed state of the bar. Obviously, the pole axis (the axis of poles) is an axis parallel to the longitudinal axis, X, to which the principal pole P belongs.

As for the expression of ξ', it can be treated as a *conditional relative strain*, ε_G, on the axis of the centers of gravity of the bar's sections. We mention the conditional, not actual, strain because the actual strain can be determined only for points of the cross-section which belong to the profile.

But neither the cross-section's center of gravity, G, nor the main pole P belong to the profile in the general case.

Thus, we have the following geometric relationships to relate displacement components ξ, η, ζ to strain parameters ε_G, χ_y, χ_z:

$$\varepsilon_G = \xi', \qquad \chi_z = -\zeta'', \qquad \chi_y = -\eta''. \tag{2.22}$$

Further, substituting (2.21) to formula (2.17) will give the following expression of the normal stresses:

$$\sigma^x = \frac{N}{A} + \frac{M_y}{I_y} z + \frac{M_z}{I_z} y - E\theta''\omega. \tag{2.23}$$

The first three terms in formula (2.23) are the same as those used to analyze a solid bar for bending in two planes and for tension. The last term is an additional one comparing to the classic theory of solid bars. This last term defines normal stresses caused by a non-uniform twist along the thin-walled bar. The stress itself,

$$E\theta''\omega,$$

is called a *normal stress of constricted torsion*. As we can see, the constricted-torsion normal stress varies over the cross-section according to the *law of sectorial areas* rather than the law of plane.

If we multiply equality (2.23) by ω and integrate the result over the whole area of the section, we will have this, considering (2.20):

$$\int_l \sigma^x h\omega ds = -E\theta'' \int_l h\omega^2 ds.$$

An integral characteristic of the stresses in the left-hand part of the formula is called a *bitorque* and denoted usually by letter B,

$$B = \int_l \sigma^x h\omega ds. \tag{2.24}$$

Also, a purely geometric property of the section, I_ω,

$$I_\omega = \int_l h\omega^2 ds, \tag{2.25}$$

is called a *sectorial moment of inertia of the section.*

Using these designations, we can write

$$\theta'' = -\frac{B}{EI_\omega}. \tag{2.26}$$

Therefore the final formula of the normal stresses in the bar's cross-section will be

$$\sigma^x = \frac{N}{A} + \frac{M_y}{I_y} z + \frac{M_z}{I_z} y + \frac{B}{I_\omega} \omega . \tag{2.27}$$

Note that a positive bitorque B causes tension in points of the bar which have a positive sectorial coordinate ω.

The hypothesis of a linear distribution of the tangential stresses over the thickness of the bar's walls suggests that the tangential stresses in the bar's cross-sections, τ^{xs}, along the tangent to the profile line can be represented as as sum,

$$\tau^{xs} = \tau + \tau_k, \tag{2.28}$$

where *the average tangential stress* τ is constant over the thickness of the wall and τ_k are *stresses of pure torsion* distributed linearly over the thickness (Fig. 6.6) with the value of zero on the profile line. The superscript ^{xs} of those components is not indicated because there are no other tangential stress components other than τ^{xs} in the bar's cross-sections, so there is not going to be any ambiguity.

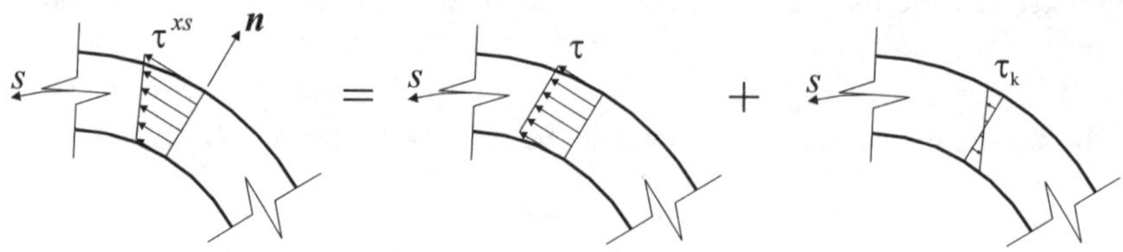

Fig. 6.6. A representation of tangential stresses τ^{xs} as a sum of average tangential stresses τ and pure-torsion tangential stresses τ_k.

We remind that, according to the moment-free hypothesis, the τ^{xn} component of the tangential stresses, orthogonal to the profile line, is assumed to be zero.

The pure-torsion tangential stresses, τ_k, reproduce the torque of the pure torsion according to Saint-Venant, which we will denote by H. Recall that the Saint-Venant theory, which determines the pure-torsion torque H, is already discussed in Section **6**.1. According to that theory, the pure-torsion torque, H, is calculated as

$$H = GI_x \theta' , \tag{2.29}$$

where I_x is the section's torsional moment of inertia.

We noted earlier in Section **6**.1.5 that the analysis of thin-walled open-profile sections can use the following assumption and still provide a practically sufficient accuracy of calculation:

$$I_x = \frac{1}{3}\int_l h^3 ds\,. \tag{2.30}$$

To determine the average tangential stress, τ, we will use the equilibrium equations. We extract an element from the thin-walled shell-bar — one limited by two longitudinal sections parallel to the generatrix and located at the distance *ds* from each other and by two cross-sections of the bar at the distance *dx* from each other – as shown in Fig. 6.7.

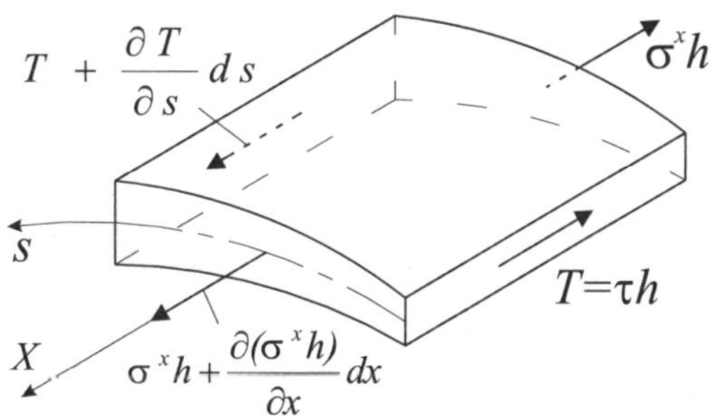

Fig. 6.7. A shell-bar element of the sizes *ds*×*dx*

Considering the equilibrium of this extracted element in the projections onto the *X* axes, we arrive at the differential equation

$$\frac{\partial(\tau h)}{\partial s} + \frac{\partial(\sigma^x h)}{\partial x} + q_x = 0$$

where $q_x(x,s)$ is the intensity of an external load in the direction of the *X*-axis per unit of area of the median surface of the bar.

By integrating this equation, we find

$$T = T_0 - \int_0^s \frac{\partial\sigma^x}{\partial x} h ds - \int_0^s q_x ds\,,$$

where $T(x,s) = \tau h$ is a tangential force per unit of length at the current point of the profile, which is also called sometimes a *flow of tangential stresses*. Here $T_0 = T(x,0)$ is a flow of the tangential stresses at the zero point of the profile.

By substituting expression (2.27) for normal stress σ^x, we obtain

$$T = T_0 - N'\frac{A(s)}{A} - M'_y\frac{S_y(s)}{I_y} - M'_z\frac{S_z(s)}{I_z} - B'\frac{S_\omega(s)}{I_\omega} - \int_0^s q_x ds\,. \qquad (2.31)$$

Here we use the designations of

$$A(s) = \int_0^s h ds\,,\;\; S_y(s) = \int_0^s hz ds\,,\;\; S_z(s) = \int_0^s hy ds\,,\;\; S_\omega(s) = \int_0^s h\omega ds\,. \qquad (2.32)$$

The right-hand part of formula (2.31) contains the still unknown stress T_0 at the zero point of the profile. To exclude this stress from formula (2.31), we use the known stress on one of the profile's edges, which should be obviously equal to an external shear force specified for that edge. However, we can assume the shear forces on the profile's edges to be zero, without limiting the generality of the consideration. This can be always achieved if we permit the longitudinal load, $q_x(x,s)$, to contain a Dirac delta function with respect to the arc coordinate, s.

For example, let one of the profile's edges defined by the coordinate of s_Γ have an external load in the form of a shear force per unit of length, equal to $\overline{T}(x)$, and let there be no other longitudinal loads upon the bar. Then it suffices to assume

$$q_x(x,s) = \overline{T}(x)\,\delta(s - s_\Gamma)$$

to be able to make $T(x,s_\Gamma) = 0$[14].

Now assuming $s = s_\Gamma$ in (2.31) and taking into account the equality of $T(x,s_\Gamma) = 0$, we have

$$T_0 = N'\frac{A(s_\Gamma)}{A} + M'_y\frac{S_y(s_\Gamma)}{I_y} + M'_z\frac{S_z(s_\Gamma)}{I_z} + B'\frac{S_\omega(s_\Gamma)}{I_\omega} + \int_0^{s_\Gamma} q_x ds\,,$$

which gives the following when substituted to formula (2.31):

$$T = -N'\frac{A_o(s)}{A} - M'_y\frac{S_{oy}(s)}{I_y} - M'_z\frac{S_{oz}(s)}{I_z} - B'\frac{S_{o\omega}(s)}{I_\omega} - \int_{s_\Gamma}^s q_x ds\,, \qquad (2.33)$$

where we use the designations of

---

[14] This can be interpreted as shifting the longitudinal load applied to the edge of the profile inside the profile by an infinitesimal distance.

$$A_o(s)=\int_{s_\Gamma}^{s} h ds\,,\quad S_{oy}(s)=\int_{s_\Gamma}^{s} hz ds\,,\quad S_{oz}(s)=\int_{s_\Gamma}^{s} hy ds\,,$$

$$S_{o\omega}(s)=\int_{s_\Gamma}^{s} h\omega ds\,. \tag{2.34}$$

The geometrical meaning of these parameters is obvious. To see this, we name a piece of the bar's cross-section between the edge of the profile, Γ, to the current point of the profile, M, with the coordinate s a *cut-off part of the cross-section*, and here's what we have:

- A_o is an area of the cut-off part of the cross-section;
- S_{oy} is a static moment of the cut-off part of the cross-section with respect to the Y-axis;
- S_{oz} is a static moment of the cut-off part of the cross-section with respect to the Z-axis;
- $S_{o\omega}$ is a sectorial static moment of the cut-off part of the cross-section.

Fig. 6.8 shows the cut-off part of the cross-section hatched.

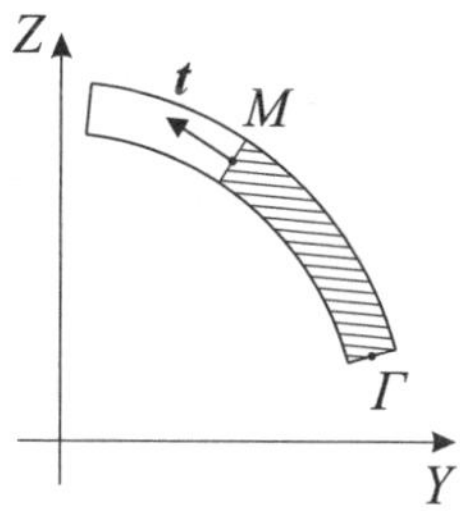

Fig. 6.8. A cut-off part of a cross-section

6.2.2 Equations of equilibrium of an open-profile thin-walled bar

Let us begin with discussing kinds of external loads that can be applied to a thin-walled bar. In addition to the longitudinal load introduced above, $q_x(x,s)$, the median surface of a shell-bar can be subjected also to loads $q_y(x,s)$, $q_z(x,s)$, which are lateral loads per unit of area of the median surface acting in the respective Y and Z directions. However, sometimes it proves more convenient to represent the lateral loads by expansions over local, rather than global, axes and operate their $q_t(x,s)$, $q_n(x,s)$ components. The mutual transition between the two methods of representation of the lateral loads is based on the simple formulas:

$$q_t = q_y t_y + q_z t_z\,, \qquad q_n = q_y t_z - q_z t_y\,,$$
$$q_y = q_t t_y + q_n t_z\,, \qquad q_z = q_t t_z - q_n t_y\,, \tag{2.35}$$

which become obvious if we take (2.4) into account.

As we will see later, the equations of equilibrium for a thin-walled bar will not include the actual loads, q_x, q_y and q_z, distributed over the median surface; instead, the equations will include their integral characteristics over the bar's cross-section such as:

$$p_x(x) = \int_l q_x(x,s)ds\,, \quad p_y(x) = \int_l q_y(x,s)ds\,, \quad p_z(x) = \int_l q_z(x,s)ds\,,$$

$$m_x(x) = \int_l \left[q_z(y-y_P) - q_y(z-z_P)\right]ds\,,$$

$$m_y(x) = \int_l q_x(x,s)zds\,, \quad m_z(x) = -\int_l q_x(x,s)yds\,,$$

$$m_B(x) = \int_l q_x(x,s)\omega ds\,. \tag{2.36}$$

Obviously, p_x, p_y, and p_z are loads per unit of length (of the bar's length) in the directions of the axes corresponding to the subscripts. Similarly, m_x, m_y and m_z are moment loads per unit of length. More exactly, m_x is an external moment per unit of length, which is created by loads q_y and q_z with respect to the pole axis, and m_y and m_z are external moments per unit of length created by load q_x, but these are taken with respect to the principal central axes of inertia of the bar's cross-section. We adopt a standard sign convention for the external moment loads, m_x, m_y and m_z – a positive right-hand moment vector has the same direction as the respective axis of the global coordinate system.

In addition to this, the last of formulas (2.36) defines a so-called *external bitorque per unit of length*, m_B, which is created by longitudinal load q_x in the same cross-section.

Further, we want to determine internal force factors acting in the cross-sections of the thin-walled bar, as integral characteristics of normal stresses σ^x and tangential stresses $\tau^{sx} = \tau + \tau_k$.

The integral characteristics of the normal stresses have been already established; they include:

- a normal force, N;
- a bending moment, M_y;
- a bending moment, M_z;
- a bitorque, B.

The tangential stresses, τ_k, create the pure-torsion moment, H. The average tangential stresses, τ, which act along the profile, also create a torque with respect to the pole P, which is called a *constricted-torsion moment* and denoted by symbol M_ω:

$$M_\omega = \int_l \tau h \rho ds = \int_l T d\omega . \tag{2.37}$$

The same average tangential stresses give, when projected onto the Y and Z axes, such integral characteristics as the respective lateral forces Q_y and Q_z,

$$Q_y = \int_l \tau h t_y ds - \int_l T dy , \qquad Q_z = \int_l \tau h t_z ds = \int_l T dz . \tag{2.38}$$

So, the integral characteristics of the tangential stresses are:

- a moment of pure torsion, H;
- a moment of constricted torsion, M_ω;
- a lateral (shear) force, Q_y;
- another lateral (shear) force, Q_z.

Let us consider an element of a thin-walled bar, cut out by two cross-sections with the distance between them denoted by dx (Fig. 6.9).

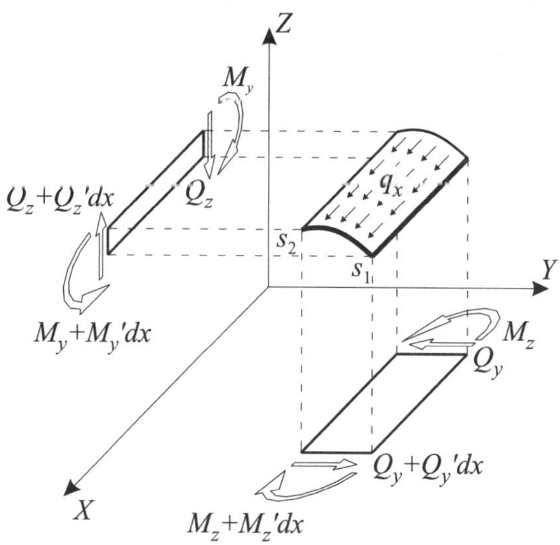

Fig. 6.9. The equilibrium of an element of a thin-walled bar

The conditions of equlibrium of the extracted element in terms of the moments with respect to the Y-axis and the Z-axis give

$$Q_y = M_z' - m_z, \qquad Q_z = M_y' + m_y. \tag{2.39}$$

Formulas (2.39) are counterparts of the well-known formula by Zhuravsky in the theory of bending of solid bars, in application to the theory of open-profile thin-walled bars.

By using the relationships of (2.39), formula (2.33) of the tangential force can be rewritten as

$$T = -N'\frac{A_o(s)}{A} -$$

$$-(Q_z - m_y)\frac{S_{oy}(s)}{I_y} - (Q_y + m_z)\frac{S_{oz}(s)}{I_z} - B'\frac{S_{o\omega}(s)}{I_\omega} - \int_{s_\Gamma}^{s} q_x ds. \tag{2.40}$$

Substituting this expression of T to (2.37) gives

$$M_\omega =. -\frac{N'}{A}\int_l A_o(s)d\omega - \frac{(Q_z - m_y)}{I_y}\int_l S_{oy}(s)d\omega - \frac{(Q_y + m_z)}{I_z}\int_l S_{oz}(s)d\omega -$$

$$-\frac{B'}{I_\omega}\int_l S_{o\omega}(s)d\omega - \int_l\left(\int_{s_\Gamma}^{s} q_x ds\right)d\omega.$$

We want to transform the integrals from this formula using integration by parts. Before doing that, we introduce some additional designations. Let $s = s_1$ be a minimum arc coordinate of the profile in question, and let $s = s_2$ be a maximum arc coordinate of the same profile. Obviously, both coordinates conform to ultimate points of the profile, so the lengths of the profile is $l = s_2 - s_1$. Now, using formulas (2.34) and assuming $s_\Gamma = s_1$, we have

$$\int_l A_o(s)d\omega = \int_l\left(\int_{s_1}^{s} hds\right)d\omega = \left[\omega\int_{s_1}^{s} hds\right]\Bigg|_{s_1}^{s_2} - \int_l \omega hds = \omega(s_2)A,$$

$$\int_l S_{oy}(s)d\omega = \int_l\left(\int_{s_1}^{s} hzds\right)d\omega = \left[\omega\int_{s_1}^{s} hzds\right]\Bigg|_{s_1}^{s_2} - \int_l \omega hzds = 0,$$

$$\int_l S_{oz}(s)d\omega = \int_l\left(\int_{s_1}^{s} hyds\right)d\omega = \left[\omega\int_{s_1}^{s} hyds\right]\Bigg|_{s_1}^{s_2} - \int_l \omega hyds = 0, \tag{2.41}$$

where we take account of the fact that the Y and Z axes are principal central axes of inertia of the bar's cross-section, and we also use equalities (2.20) which help choose the principal pole and the zero point of the profile.

In addition to that,

$$\int_l S_{o\omega}(s)d\omega = \int_l \left(\int_{s_1}^{s} h\omega ds \right) d\omega = \left[\omega \int_{s_1}^{s} h\omega ds \right]_{s_1}^{s_2} - \int_l \omega^2 h ds = -I_\omega , \qquad (2.42)$$

where I_ω is a sectorial moment of inertia of the cross-section, defined by (2.25).

Finally,

$$\int_l \left(\int_{s_1}^{s} q_x ds \right) d\omega = \left[\omega \int_{s_1}^{s} q_x ds \right]_{s_1}^{s_2} - \int_l q_x \omega ds = \omega(s_2) p_x - m_B .$$

Considering all the above, the formula of the constricted-torsion moment will become

$$M_\omega =. -(N' + p_x)\omega(s_2) + B' + m_B .$$

It is easy to notice that the following equality holds because of the equation of equilibrium of the bar's element shown in Fig. 6.9, in projections onto X:

$$N' + p_x = 0 . \qquad (2.43)$$

So the formula of M_ω becomes ultimately as simple as

$$M_\omega =. B' + m_B . \qquad (2.44)$$

If there is no longitudinal load, $q_x = 0$, the external bitorque also vanishes, and the formula acquires its simplest form possible:

$$M_\omega = B' . \qquad (2.45)$$

Let us return to the equations of equilibrium. Considering the conditions of equilibrium of the extracted element of the bar (Fig. 6.9) in terms of the projections onto the Y and Z axes, we have

$$Q_y' + p_y = 0 , \quad Q_z' + p_z = 0 , \qquad (2.46)$$

which gives the following together with equations (2.39) after excluding the lateral forces:

$$M_z'' = -p_y + m_z' , \quad M_y'' = -p_z - m_y' . \qquad (2.47)$$

Also, the overall torque M_x over the cross-section consists of the pure-torsion moment and the constricted-torsion moment, that is,

$$M_x = H + M_\omega . \tag{2.48}$$

We would like to note that M_x is understood as a torque created by tangential stresses τ^{xs} and calculated with respect to a longitudinal axis through the principal pole, P.

If we consider the last equation of equlibrium, which still remain unused, in terms of the moments with respect to a longitudinal exis through the bending center, then we can extract an element of the bar between two cross-sections at a distance of dx from each other and obtain

$$M_x' = -m_x . \tag{2.49}$$

It will be convenient for further presentment to introduce the notion of *generalized loads* $\overline{p}_y$, $\overline{p}_z$, $\overline{m}_x$ by defining them as

$$\overline{p}_y = p_y - m_z', \qquad \overline{p}_z = p_z + m_y', \qquad \overline{m}_x = m_x + m_B' . \tag{2.50}$$

Now the set of all differential equations of equilibrium gathered from (2.43), (2.44), (2.47) through (2.50) can be written in a final form:

$$-N' = p_x , \quad -M_z'' = \overline{p}_y , \quad -M_y'' = \overline{p}_z , \quad -H' - B'' = \overline{m}_x . \tag{2.51}$$

6.2.3 Center of twist, center of bending

Earlier in Section **6**.1 we gave a general definition of a *center of twist* as a point C_R of a cross-section which does not have any displacements when the bar is in torsion (twist). According to Saint-Venant's theory, the center of twist is undefined in the problem of pure torsion. In other words, in the pure torsion any point of the bar's cross-section can be assigned to be a center of twist. It is a different thing with the problem of constricted torsion. In that case, as we will show later, the position of the center of twist, C_R, is unambiguous.

Consider a cantilever bar with a concentrated force, P, applied to a certain point, C_B, on the free end of the bar in the plane of its cross-section. If that force does not cause any torsion of the bar, the point of its application is usually called a *bending center*[15]. Generally, the bending

[15] Sometimes the bending center is called a *shear center*. The English-language technical literature uses the latter term. Another usable term is a *center of stiffness of the cross-section*. The designations of C_B and C_R have been derived from the

center is a point of a cross-section of the bar such that lateral forces pass through it without causing any twist of the bar.

The locations of the two points can be calculated independently from their definitions. However, it suffices to find the position of one of the points only, because the following statement, a corollary of the unchanged-contour hypothesis, holds:

> *In the theory of thin-walled bars with an unchanged contour, the bending center and the center of twist coincide.*

This statement can be proved easily on the basis of Betty's work reciprocity theorem. To see the statement, we suppose an opposite thing: let the bending center, C_B, and the twist center, C_R, be two different points. Then we consider two states of the thin-walled bar. Let the bar be loaded by a torque M_x in State 1, and let it be loaded by a force P applied to the center of bending, C_B, perpendicular to the bar's axis, X, in State 2. The work of the forces of State 1 on the displacements of State 2 is equal to zero, $A_{12} = 0$, because the P force causes a bending of the bar without any torsion. On the other hand, the work of the forces of State 2 on the displacements of State 1 is different from zero, $A_{21} \neq 0$, because the displacement of the force's application point, C_B, in the torsion of the bar around the C_R point is nonzero. This is a contradiction with the Betty theorem which follows from our supposition of the two points not being coincident; this proves the original statement.

The following statement is also true in the theory of open-profile thin-walled bars:

> *The bending center,* C_B, *thus the center of twist,* C_R, *too, coincides with the principal pole of the profile,* P.

To see this point, we note that the position of the bending center, according to its definition, is characterized by the condition that tangential stresses τ_Q created by shear forces Q_y and Q_z (and by them only!) do not create a torque with respect to point C_B. According to formula (2.40), tangential stresses τ_Q caused by shear forces Q_y and Q_z are defined as

---

abbreviations of the *bending center* and *rotation center* (the latter is not quite correct but related and permits to distinguish between the centers easily – *Translator's note*). We prefer to use the 'bending center' term for reasons declared in Appendix G. The notion of the *shear center* denoted by C_S should be reserved for naming a totally different point which does not coincide, generally, with the bending center.

$$\tau_Q = -Q_z \frac{S_{oy}(s)}{hI_y} - Q_y \frac{S_{oz}(s)}{hI_z}. \tag{2.52}$$

Denoting the coordinates of the bending center by y_B and z_B, we formulate the condition of zero torque as

$$\int_l \tau_Q h \rho_{BM} ds = 0, \tag{2.53}$$

where ρ_{BM} is a projection of vector $\boldsymbol{R}_{BM}$ onto the direction of normal $\boldsymbol{n}$ to the profile at the current point, M, and vector $\boldsymbol{R}_{BM}$ goes from point C_B to point M. In other words, ρ_{BM} is a distance from C_B to the tangent of the profile at the current point M, taken with an appropriate sign.

Obviously, the bending center is defined unambiguously, that is, its location should not depend on a particular distribution of tangential stresses τ_Q over the profile. Therefore, in order to prove our statement, it suffices to show that (2.53) will hold if we place the principal pole at the bending center, which lets ul assume assume $\rho_{BM} = \rho$ or

$$\int_l \tau_Q h \rho ds = \int_l \tau_Q h d\omega = 0 .$$

But this equality is warranted because, as we have shown with formulas (2.41),

$$\int_l S_{oy}(s) d\omega = 0, \qquad \int_l S_{oz}(s) d\omega = 0.$$

This proves the statement.

6.2.4 Tangential stresses in the open-profile thin-walled bar theory

Earlier we established formula (2.40) as a general expression of the flow of tangential stresses, $T = \tau h$, at an arbitrary point of the profile with coordinates (x,s). Our next problem here is to simplify that formula and reduce it to a more convenient form both for calculation and for subsequent transformations.

First of all, we would like to note that the profile has only four degrees of freedom with regard to longitudinal displacements u. To see this, we use formulas (2.15) and (2.14) to derive

$$u = \xi - \theta_z y + \theta_y z - \theta' \omega . \tag{2.54}$$

The immediate consequence is that displacements u of an arbitrary point of the profile with its arc coordinate s, in an arbitrary cross-section of the bar with its longitudinal coordinate x, are defined unambiguously by four parameters:

- ξ, a longitudinal displacement of the cross-section's center of gravity which is rigidly connected to the principal pole P, and the latter is, in its turn, rigidly connected to the zero point of the profile, O;
- θ_y and θ_z, slopes of the rigid element OP - PG about the respective axes Y and Z;
- θ', a twist factor (an angle of twist of the bar per unit of its length).

Such are the parameters that can be treated as four degrees of freedom of the profile, as indicated above.

Each of the degrees of freedom corresponds to its own function of distribution of the displacements over the arc coordinate, s. As can be seen from (2.54), the functions are

$$1, \qquad y = y(s), \qquad z = z(s), \qquad \omega = \omega(s). \tag{2.55}$$

These four functions satisfy the following conditions of a generalized orthogonality, or, more particularly, the orthogonality with the weight of h throughout the length of the profile, l,

$$\int_l 1 \cdot yhds = 0\,, \qquad \int_l 1 \cdot zhds = 0\,, \qquad \int_l 1 \cdot \omega hds = 0\,,$$

$$\int_l y \cdot zhds = 0\,, \qquad \int_l y \cdot \omega hds = 0\,, \qquad \int_l z \cdot \omega hds = 0\,. \tag{2.56}$$

Obviously, the conditions of mutual orthogonality of the first three functions in (2.55) follow from our choice of the (Y,Z)-axes, which are principal central axes of inertia of the bar's cross-section by definition. As for the orthogonality of function $\omega(s)$ to the first three functions in (2.55), this is a consequence of our choice of the principal pole and the zero point of the profile, for which equalities (2.20) must hold.

Now let us turn to function $q_x = q_x(x,s)$ which characterizes a distribution of the longitudinal load upon the bar over the whole median surface of the bar. We represent this function as a series every term of which is a product of a function of x and a function of s, and the first four functions of s in that series will be the functions from (2.55) multiplied by $h(s)$. This gives

$$q_x(x,s) = a_1(x)h(s) + a_y(x)h(s)y(s) + a_z(x)h(s)z(s) + \\ + a_\omega(x)h(s)\omega(s) + \ldots \tag{2.57}$$

The ellipsis denotes the rest of the terms in the series.

However, the tail of the series can be (and should be!) omitted within the frames of the theory of thin-walled bars we are dealing with, because, as we have just shown, the first four terms of the series cover all the generalized forces which correspond to all degrees of freedom of the profile. Therefore the generalized external longitudinal force actions are only functions $a_1(x)$, $a_y(x)$, $a_z(x)$, $a_\omega(x)$.

We will determine those functions by multiplying (2.57) by 1, $y(s)$, $z(s)$, $\omega(s)$ in succession and integrating the results over the whole profile l. Considering the conditions of generalized orthogonality as in (2.56), we have

$$\int_l q_x ds = a_1 \int_l h ds = a_1 A\,, \qquad \int_l q_x y ds = a_y \int_l h y^2 ds = a_y I_z\,,$$

$$\int_l q_x z dx = a_z \int_l h z^2 ds = a_z I_y\,, \qquad \int_l q_x \omega\, ds = a_\omega \int_l h\omega^2 ds = a_\omega I_\omega\,. \tag{2.58}$$

Also, using formulas (2.36), now we can represent the desirable factors a_1, a_y, a_z, a_ω in the expansion of (2.57) as follows:

$$a_1 = \frac{p_x}{A}, \quad a_y = -\frac{m_z}{I_z}, \quad a_z = \frac{m_y}{I_y}, \quad a_\omega = \frac{m_B}{I_\omega}.$$

Substituting these coefficients to the expansion of load q_x in (2.57) and keeping only the first four terms will yield

$$q_x = \frac{p_x h}{A} - \frac{m_z h}{I_z} y + \frac{m_y h}{I_y} z + \frac{m_B h}{I_\omega}\omega\,. \tag{2.59}$$

If we integrate both parts of formula (2.59) by s from one edge of the profile with the arc coordinate s_Γ to the current coordinate, s, we will have

$$\int_{s_\Gamma}^{s} q_x ds = p_x \frac{A_o(s)}{A} + m_y \frac{S_{oy}(s)}{I_y} - m_z \frac{S_{oz}(s)}{I_z} + m_B \frac{S_{o\omega}(s)}{I_\omega}. \tag{2.60}$$

Now we can simplify formula (2.40) of the flow of tangential stresses. To see this, we substitute (2.43), (2.44) and (2.60) to (2.40), to find finally

$$T = -Q_z \frac{S_{oy}}{I_y} - Q_y \frac{S_{oz}}{I_z} - M_\omega \frac{S_{o\omega}}{I_\omega},$$

hence an expression for the tangential stresses which we were going to derive:

$$\tau = -Q_z \frac{S_{oy}}{hI_y} - Q_y \frac{S_{oz}}{hI_z} - M_\omega \frac{S_{o\omega}}{hI_\omega}. \tag{2.61}$$

This is exactly the formula we wanted to have.

6.2.5 A matrix form of basic relationships in the theory of open-profile thin-walled bars

Now we introduce a vector of "stresses" $\boldsymbol{\sigma}$ and a vector of given external force actions $\bar{\boldsymbol{X}}$, together with a vector of displacements $\mathbf{u}$ and a vector of strains $\boldsymbol{\varepsilon}$:

$$\boldsymbol{\sigma} = \left\|\left[N, H, M_y, M_z, B\right]\right\|^{\mathrm{T}}, \quad \bar{\boldsymbol{X}} = \left\|\left[p_x, \bar{p}_y, \bar{p}_z, \bar{m}_x\right]\right\|^{\mathrm{T}},$$

$$\mathbf{u} = \left\|\left[\xi, \eta, \zeta, \theta\right]\right\|^{\mathrm{T}}, \qquad \boldsymbol{\varepsilon} = \left\|\left[\varepsilon_G, \chi_x, \chi_z, \chi_y, \chi_B\right]\right\|^{\mathrm{T}}, \tag{2.62}$$

where we also use additional designations of the strain parameters, χ_x and χ_B,

$$\chi_x = \theta', \qquad \chi_B = -\theta''. \tag{2.63}$$

A relation between these vectors is written in the matrix form in a standard way, $\boldsymbol{\varepsilon} = \boldsymbol{A}\mathbf{u}$, where formulas (2.22) and (2.63) produce

$$\boldsymbol{A} = \left\|\begin{bmatrix} d/dx & 0 & 0 & 0 \\ 0 & 0 & 0 & d/dx \\ 0 & 0 & -d^2/dx^2 & 0 \\ 0 & -d^2/dx^2 & 0 & 0 \\ 0 & 0 & 0 & -d^2/dx^2 \end{bmatrix}\right\|, \quad \boldsymbol{A}\mathbf{u} = \left\|\begin{bmatrix} \xi' \\ \theta' \\ -\zeta'' \\ -\eta'' \\ -\theta'' \end{bmatrix}\right\| = \boldsymbol{\varepsilon}, \tag{2.64}$$

so, according to the formal rules of construction of conjugate operators, we have

$$A^{\mathsf{T}} = \begin{bmatrix} -\frac{d}{dx} & 0 & 0 & 0 & 0 \\ 0 & 0 & 0 & -\frac{d^2}{dx^2} & 0 \\ 0 & 0 & -\frac{d^2}{dx^2} & 0 & 0 \\ 0 & -\frac{d}{dx} & 0 & 0 & -\frac{d^2}{dx^2} \end{bmatrix}. \tag{2.65}$$

Obviously, the matrix differential operator $\boldsymbol{A}^{\mathsf{T}}$ is indeed an equilibrium operator because the equality $\boldsymbol{A}^{\mathsf{T}}\boldsymbol{\sigma} = \overline{\boldsymbol{X}}$ is just a matrix form of the equations of equilibrium from (2.51) with vectors of stresses $\boldsymbol{\sigma}$ and vectors of external force actions $\overline{\boldsymbol{X}}$ specified according to (2.62). Now let us turn to physical relationships which connect the strain vector and the displacement vector. According to (2.21) and (2.22), we can write the following:

$$N = \varepsilon_G EA, \qquad M_y = \chi_z EI_y, \qquad M_z = \chi_y EI_z, \tag{2.66}$$

and because of (2.26), (2.29) and (2.63)

$$H = \theta' GI_x, \qquad B = \chi_B EI_\omega . \tag{2.67}$$

This enables us to write an explicit form of the matrix of elasticity, $\boldsymbol{\sigma} = \boldsymbol{C\varepsilon}$:

$$\boldsymbol{C} = \begin{bmatrix} EA & 0 & 0 & 0 & 0 \\ 0 & GI_x & 0 & 0 & 0 \\ 0 & 0 & EI_y & 0 & 0 \\ 0 & 0 & 0 & EI_z & 0 \\ 0 & 0 & 0 & 0 & EI_\omega \end{bmatrix}. \tag{2.68}$$

Using the general formula, $\boldsymbol{L} = \boldsymbol{A}^{\mathsf{T}}\boldsymbol{CA}$, to build the Lame operator, we come up with a governing system of differential equations for the problem in terms of the displacements, $\boldsymbol{L}\mathbf{u} = \overline{\boldsymbol{X}}$. The expanded form of the system of equations is

$$EA\xi'' = p_x, \qquad EI_z\eta^{IV} = \overline{p}_y, \qquad EI_y\zeta^{IV} = \overline{p}_z,$$
$$EI_\omega\theta^{IV} - GI_x\theta'' = \overline{m}_x . \tag{2.69}$$

As we can see, the system of equations in terms of displacements for a thin-walled bar decomposes into four independent equations: one equation of second order and three equations of fourth order:

- the first of the equations, which contains the longitudinal displacement of the cross-section's center of gravity, $\xi(x)$, as an unknown function, describes a tension/compression of the thin-walled bar;
- the second and third equations describe a problem of bending. These equations are formally equivalent in their structure to the equations of bending of solid bars. However, there is a little difference between the two problems. To see this, take the second equation. For a solid bar, the sought-for function of displacements should have been a function that somehow characterizes the lateral displacements of the cross-section's center of gravity. For a thin-walled bar, the sought-for function is $\eta(x)$ which describes lateral dsiplacements of the principal pole P (the center of bending). As the center of bending does not belong to the material substance of the bar's cross-section in a general case (it lies outside the profile), η is actually a conditional displacement or, more exactly, a displacement of the pole which has an imaginable rigid connection to the principal zero point of the profile. It should be noted also that the relation between bending moment M_z and function η has a usual form $M_z = -EI_z\eta''$ though the bending moment itself is taken with respect to the principal central axis of inertia of the cross-section, Z, by definition according to (2.18);
- however, the most important equation for the whole theory of thin-walled bars is the last one in (2.69), which contains the cross-section's rotation angle, θ, as an unknown desirable function. It is the solution of this equation that determines the state of the thin-walled bar caused by the constriction of the torsion.

Each of the equations conforms to its specific set of boundary conditions; to formulate those, we will use the basic integral identity within the frames of the theory of open-profile thin-walled bars.

6.2.6 Basic integral identity in the theory of open-profile thin-walled bars

Now we can derive the following from the matrix relationships obtained above, using our usual approach of integration by parts:

$$(\mathbf{A}\mathbf{u},\boldsymbol{\sigma}) = \int_0^L \left[\xi' N + \theta' H - \zeta'' M_y - \eta'' M_z - \theta'' B\right] dx =$$

$$= \int_0^L \left[-\xi N' - \theta(B'' + H') - \zeta M_y'' - \eta M_z'' \right] dx + \left[\xi N \right]_0^L -$$

$$- \left[\theta' B \right]_0^L + \left[\theta(B' + H) \right]_0^L - \left[\zeta' M_y \right]_0^L + \left[\zeta \, M_y' \right]_0^L - \left[\eta' M_z \right]_0^L + \left[\eta \, M_z' \right]_0^L .$$

The matrix form of this formula is standard:

$$(A\mathbf{u}, \boldsymbol{\sigma}) = (\mathbf{u}, A^{\mathsf{T}}\boldsymbol{\sigma}) + (\boldsymbol{p}, \boldsymbol{u})_{\Gamma} ,$$

where the edge loads, $\boldsymbol{p}$, and the edge displacements, $\boldsymbol{u}$, are

$$\boldsymbol{p} = \left\| \left[nN, n(B' + H), -nM_y, -nM_z, -nB, nM_y', nM_z' \right] \right\|^{\mathsf{T}} ,$$

$$\boldsymbol{u} = \left\| \left[\xi, \theta, \zeta', \eta', \theta', \zeta, \eta \right] \right\|^{\mathsf{T}} . \qquad (2.70)$$

The components participating in the vectors are those that are directly used to construct the boundary condtions. Recall that the n parameter, a factor in the components of the $\boldsymbol{p}$ vector, is a projection onto the X-axis of the vector of the external normal to the bar's cross-section defined as follows: $n(0) = -1, \ n(L) = 1$.

6.2.7 Basic variational principles in the theory of open-profile thin-walled bars

First of all, we want to find an expression of the strain energy, E, of a thin-walled bar. Writing the expression of the energy as a functional of the displacement vector, **u**, we obtain $\mathsf{E}(\mathbf{u}) = ½(A\mathbf{u}, CA\mathbf{u})$ or

$$\mathsf{E}(\mathbf{u}) =$$

$$= \frac{1}{2} \int_0^L \left[EA(\xi')^2 + GI_x(\theta')^2 + EI_y(\zeta'')^2 + EI_z(\eta'')^2 + EI_\omega(\theta'')^2 \right] dx . \quad (2.71)$$

The same energy can be represented as a quadratic functional of the stresses, which gives $\mathsf{E}(\boldsymbol{\sigma}) = ½\,(\boldsymbol{\sigma}, C^{-1}\boldsymbol{\sigma})$ or

$$\mathsf{E}(\boldsymbol{\sigma}) = \frac{1}{2} \int_0^L \left[\frac{N^2}{EA} + \frac{H^2}{GI_x} + \frac{M_y^2}{EI_y} + \frac{M_z^2}{EI_z} + \frac{B^2}{EI_\omega} \right] dx . \qquad (2.72)$$

We will also give expressions of the force and kinematic potentials. Confining ourselves, for the sake of simplicity, to homogeneous boundary conditions for forces, we make the potential of external forces equal to the virtual work of external distributed loads. Obviously, this work can be written as a double integral,

$$\Pi_s(\mathbf{u}) = \int_0^L \left(\int_l (q_x u + q_t v + q_n w) ds \right) dx . \qquad (2.73)$$

Recall formula (2.8) which enables us to represent the tangential displacement, v, of an arbitrary point of the bar's profile via lateral displacements η and ζ of the principal pole and via the rotation angle, θ, of the bar's cross-section about the pole axis. Similarly, by making a scalar product of equality (2.7) with unit vector $\boldsymbol{n}$, we derive an expression of normal displacement w. Adding formula (2.15) to all this, we have finally:

$$u = \xi - \eta' y - \zeta' z - \theta' \omega ,$$

$$v = \eta t_y + \zeta\, t_z + \theta[(y - y_P)t_z - (z - z_P)t_y] ,$$

$$w = \eta t_z - \zeta\, t_y - \theta[(y - y_P)t_y + (z - z_P)t_z] . \qquad (2.74)$$

Here we take account of our earlier relations $n_y = t_z$, $n_z = -t_y$.

In addition to the above, we can use the representation of the components of the external actions, q_t and q_n, in the local coordinate system via components of the same actions, q_y and q_z, in the global coordinate system according to (2.35).

Substituting all these to (2.73), we transform the expression of the force potential into

$$\int_0^L \int_l \{ q_x [\xi - \eta' y - \zeta' z - \theta' \omega)] + q_y [\eta - \theta(z - z_P)] + q_z [\zeta + \theta(y - y_P)] \} ds dx .$$

Integrating over arc coordinate s and using the designations of external actions integral with respect to the bar's cross-section from (2.36), we simplify the above expression as follows:

$$\Pi_s(\mathbf{u}) = \int_0^L \left(\xi p_x + \eta p_y + \zeta p_z + \theta m_x + \eta' m_z - \zeta' m_y - \theta' m_B \right) dx . \qquad (2.75)$$

To transform this integral further, we will use the integration by parts and the designations of the generalized loads from (2.50). The result is

$$\Pi_s(\mathbf{u}) = \int_0^L \left(\xi p_x + \eta \bar{p}_y + \zeta\, \bar{p}_z + \theta \bar{m}_x \right) dx + \left[\eta m_z \right]_0^L - \left[\zeta m_y \right]_0^L - \left[\theta m_B \right]_0^L . \quad (2.76)$$

Please note that the expression of Π_s according to (2.76) comprises the virtual work only of loads specified along the bar, without the force actions at the ends. Note that this condition has been adopted to simplify the form of the formulas. Actually, this condition is not a limitation of the class of loads, because we can formally include the force actions at the ends in the distributed loads by using generalized functions such as Dirac's delta function.

If we use the general formula for the Lagrange functional in the form of

$$\mathsf{L}(\mathbf{u}) = \mathsf{E}(\mathbf{u}) - \Pi_s(\mathbf{u}),$$

then the conditions of minimum of the Lagrangian functional on kinematically admissible fields of displacements will consist of Euler equations for this same functional, the latter being equations (2.69), and of natural boundary conditions – in our case these are homogeneous conditions for the stresses at the begining and end sections of the bar,

$$\left[e_N EA\xi'\right]\Big|_0^L = 0, \qquad \left[e_{Mx}(-EI_\omega\theta''' + GI_x\theta' - m_B)\right]\Big|_0^L = 0,$$

$$\left[e_{Qy}(-EI_z\eta''' + m_z)\right]\Big|_0^L = 0, \quad \left[e_{Qz}(-EI_y\zeta''' - m_y)\right]\Big|_0^L = 0,$$

$$\left[e_{Mz} EI_z\eta''\right]\Big|_0^L = 0, \qquad \left[e_{My} EI_y\zeta''\right]\Big|_0^L = 0, \qquad \left[e_B EI_\omega\theta''\right]\Big|_0^L = 0. \qquad (2.77)$$

The physical meaning of these boundary conditions is quite clear:

- the first of them relates to longitudinal force N;
- the second to full torque M_x;
- the third and the fourth to shear forces Q_y and Q_z;
- the fifth and the sixth to bending moments M_y and M_z;
- the last, seventh, boundary condition defines bitorque B.

As usual, the characteristic functions for boundary condition extraction, e_N, e_{Mx}, e_{Qy}, e_{Qz}, e_{My}, e_{Mz}, e_B, are equal to one only at that end of the bar where an appropriate static boundary condition is formulated.

Now let us construct the potential of the kinematical external actions. First of all, it is useful to write out explicit expressions of operators $\boldsymbol{H}_\sigma$ and $\boldsymbol{H}_u$ which map stress vector $\boldsymbol{\sigma}$ and displacement vector $\mathbf{u}$ to the vectors of edge stresses, $\boldsymbol{p}$, and of edge displacements, $\boldsymbol{u}$, that is, $\boldsymbol{p} = \boldsymbol{H}_\sigma\boldsymbol{\sigma}$ and $\boldsymbol{u} = \boldsymbol{H}_u\mathbf{u}$. Based on formulas (2.70), we have

$$
\boldsymbol{H}_\sigma = \left\|\begin{matrix} n & 0 & 0 & 0 & 0 \\ 0 & n & 0 & 0 & n\dfrac{d}{dx} \\ 0 & 0 & -n & 0 & 0 \\ 0 & 0 & 0 & -n & 0 \\ 0 & 0 & 0 & 0 & -n \\ 0 & 0 & n\dfrac{d}{dx} & 0 & 0 \\ 0 & 0 & 0 & n\dfrac{d}{dx} & 0 \end{matrix}\right\|, \quad \boldsymbol{H}_u = \left\|\begin{matrix} 1 & 0 & 0 & 0 \\ 0 & 0 & 0 & 1 \\ 0 & 0 & \dfrac{d}{dx} & 0 \\ 0 & \dfrac{d}{dx} & 0 & 0 \\ 0 & 0 & 0 & \dfrac{d}{dx} \\ 0 & 0 & 1 & 0 \\ 0 & 1 & 0 & 0 \end{matrix}\right\| . \tag{2.78}
$$

Now, turning to the general formula (**2**.3.2), we are able to define the potential of kinematical external actions as

$$
\Pi_k(\boldsymbol{\sigma}) = (\boldsymbol{E}_u \boldsymbol{H}_\sigma \boldsymbol{\sigma},\, \boldsymbol{E}_u \bar{\boldsymbol{u}})_\Gamma =
$$

$$
= \left[e_\xi N \bar{\xi} \right]_0^L + \left[e_\theta (B' + H)\bar{\theta} \right]_0^L - \left[e_{\zeta'} M_y \bar{\zeta}' \right]_0^L - \left[e_{\eta'} M_z \bar{\eta}' \right]_0^L -
$$

$$
- \left[e_{\theta'} B \bar{\theta}' \right]_0^L + \left[e_\zeta M_y' \bar{\zeta} \right]_0^L + \left[e_\eta M_z' \bar{\eta} \right]_0^L . \tag{2.79}
$$

The characteristic functions of boundary condition extraction, e_ξ, e_θ, $e_{\zeta'}$, $e_{\eta'}$, $e_{\theta'}$, e_ζ, e_η, are equal to one only at that end of the bar where an appropriate kinematical boundary condition is formulated. Matrix $\boldsymbol{E}_u$ that extracts the kinematical boundary conditions is defined as a diagonal matrix,

$$
\boldsymbol{E}_u = \mathrm{diag}\left\| \left[e_\xi, e_\theta, e_{\zeta'}, e_{\eta'}, e_{\theta'}, e_\zeta, e_\eta \right] \right\| .
$$

In the similar way we can define the diagonal matrix $\boldsymbol{E}_p$ for extracting the static boundary conditions,

$$
\boldsymbol{E}_p = \mathrm{diag}\left\| \left[e_N, e_{Mx}, e_{My}, e_{Mz}, e_B, e_{Qz}, e_{Qy} \right] \right\| ,
$$

where the usual condition, $\boldsymbol{E}_u + \boldsymbol{E}_p = \boldsymbol{I}$, must hold where $\boldsymbol{I}$ is the identity matrix of seventh order.

Having the potential of kinematical external actions, $\Pi_k(\boldsymbol{\sigma})$, and the expression of strain energy $\mathsf{E}(\boldsymbol{\sigma})$, we can easily derive the Castigliano functional as

$$
\mathsf{K}(\boldsymbol{\sigma}) = \mathsf{E}(\boldsymbol{\sigma}) - \Pi_k(\boldsymbol{\sigma}),
$$

the minimum of which is to be sought for on statically admissible fields of stresses. Further details are omitted.

6.2.8 A remark on non-warped cross-sections in the open-profile thin-walled theory

The class of non-warped cross-sections occupies a special place among all possible kinds of cross-sections of open-profile thin-walled bars. This term denotes a cross-section for which the sectorial coordinate, ω, is zero in each point of the profile. This means that there is no warping component u_W of the longitudinal displacements u of the profile's point, that is,

$$u_W = -\theta'\omega = 0\,,$$

and thus the longitudinal displacements, u, of the profile's points should follow the law of plane because of (2.15):

$$u = \xi - \eta' y - \zeta' z\,. \tag{2.80}$$

The class of non-warped cross-sections consists of all possible so-called foil profiles (fig. 6.10) which consist of flat ribs intersected in one point. The simplest case of a foil-like profile is a standard angle section. For a foil profile, bending center P is at the intersection of the foil ribs (see Appendix G), so the sectorial coordinate ω is identical to zero.

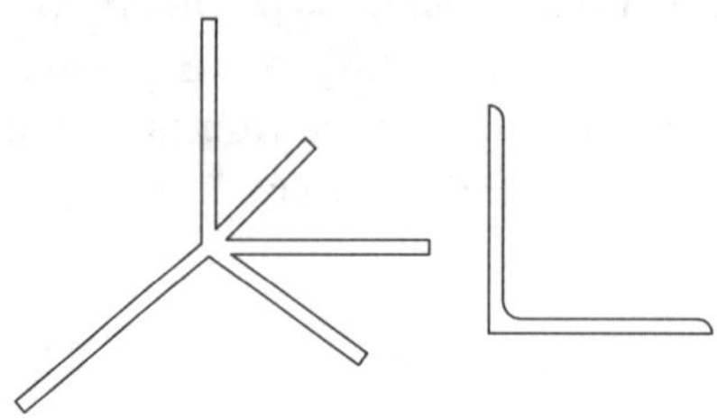

Fig. 6.10. Non-warped sections

For a non-warped section, formula (2.23) of stresses σ^x keeps only first three terms, so

$$\sigma^x = \frac{N}{A} + \frac{M_y}{I_y} z + \frac{M_z}{I_z} y\,. \tag{2.81}$$

If now we reproduce the process of derivation of formula (2.61) of tangential stresses τ, starting from (2.81) instead of (2.23), we will see that formula (2.61) should be replaced with the following for a non-warped section:

$$\tau = -Q_z \frac{S_{oy}}{hI_y} - Q_y \frac{S_{oz}}{hI_z}. \tag{2.82}$$

But if we started directly from the general formula (2.61) and assumed $\omega = 0$, we would have to deal with a singularity of the 0/0 type because $M_\omega \to 0,\ \ S_{o\omega} \to 0,\ I_\omega \to 0$ at $\omega \to 0$.

Bitorque B and the constricted-torsion moment M_ω are absent in a non-warped section (that is, they are identical to zero). As a result, the vector of stresses, $\boldsymbol{\sigma}$, and the vector of strains, $\boldsymbol{\varepsilon}$, will contain one component less, so formulas (2.62) will be replaced by

$$\boldsymbol{\sigma} = \left\| \left[N.H, M_y, M_z \right] \right\|^{\mathsf{T}}, \qquad \bar{\boldsymbol{X}} = \left\| \left[p_x, \bar{p}_y, \bar{p}_z, m_x \right] \right\|^{\mathsf{T}},$$

$$\mathbf{u} = \left\| \left[\xi, \eta, \zeta, \theta \right] \right\|^{\mathsf{T}}, \qquad \boldsymbol{\varepsilon} = \left\| \left[\varepsilon_G, \chi_x, \chi_z, \chi_y \right] \right\|^{\mathsf{T}}. \tag{2.83}$$

The sizes of the matrices of geometry, equilibrium, and elasticity will reduce accordingly. Obviously, matrix $\boldsymbol{A}$ in (2.64) should have its last row excluded, matrix $\boldsymbol{A}^{\mathsf{T}}$ in (2.65) should have its last column removed, and matrix $\boldsymbol{C}$ in (2.68) should lose its last row and its last column. As a result, we will have the following for a non-warped section:

$$\boldsymbol{A} = \left\| \begin{bmatrix} \dfrac{d}{dx} & 0 & 0 & 0 \\ 0 & 0 & 0 & \dfrac{d}{dx} \\ 0 & 0 & -\dfrac{d^2}{dx^2} & 0 \\ 0 & -\dfrac{d^2}{dx^2} & 0 & 0 \end{bmatrix} \right\|,$$

$$\boldsymbol{A}^{\mathsf{T}} = \left\| \begin{bmatrix} -\dfrac{d}{dx} & 0 & 0 & 0 \\ 0 & 0 & 0 & -\dfrac{d^2}{dx^2} \\ 0 & 0 & -\dfrac{d^2}{dx^2} & 0 \\ 0 & -\dfrac{d}{dx} & 0 & 0 \end{bmatrix} \right\|, \tag{2.84}$$

$$\boldsymbol{C} = \left\| \begin{matrix} EA & 0 & 0 & 0 \\ 0 & GI_x & 0 & 0 \\ 0 & 0 & EI_y & 0 \\ 0 & 0 & 0 & EI_z \end{matrix} \right\| . \tag{2.85}$$

We use our general formula, $\boldsymbol{L} = \boldsymbol{A}^{\mathrm{T}}\boldsymbol{CA}$, to build the Lame operator and arrive at a governing system of simultaneous differential equations in terms of displacements, $\boldsymbol{L}\mathbf{u} = \overline{\boldsymbol{X}}$. When expanded, this system becomes as follows:

$$\boxed{EA\xi'' = p_x, \quad EI_z\eta^{IV} = \overline{p}_y, \quad EI_y\zeta^{IV} = \overline{p}_z, \quad -GI_x\theta'' = m_x.} \tag{2.86}$$

As we can see, the system of the equilibrium equations in terms of the displacements for a thin-walled bar with a non-warped section is equivalent in its structure with the respective equations for solid bars. The last, fourth, equation in (2.86), which describes the torsion of the bar, is an equation of second order and corresponds to pure torsion. The fact that the differential equation has an order less by two than that of the counterpart equation from (2.69) shows how the problem of torsion *degenerates* for a non-warped profile.

The degeneration happens not only to the governing simultaneous equations but to the variational formulation of the problem, too. So instead of (2.71) and (2.72) we will have the following for strain energy E:

$$\mathsf{E}(\mathbf{u}) = \frac{1}{2}\int_0^L \left[EA(\xi')^2 + GI_x(\theta')^2 + EI_y(\zeta'')^2 + EI_z(\eta'')^2 \right] dx, \tag{2.87}$$

$$\mathsf{E}(\boldsymbol{\sigma}) = \frac{1}{2}\int_0^L \left[\frac{N^2}{EA} + \frac{H^2}{GI_x} + \frac{M_y^2}{EI_y} + \frac{M_z^2}{EI_z} \right] dx. \tag{2.88}$$

The expression of the force potential can be simplified, too. To make it simpler, we should get rid of the distributed bitorque action, m_B, in formulas (2.75) and (2.76). As a result, the following holds for the non-warped section:

$$\Pi_s(\mathbf{u}) = \int_0^L \left(\xi p_x + \eta\overline{p}_y + \zeta\overline{p}_z + \theta m_x \right) dx + \left[\eta m_z \right]_0^L - \left[\zeta m_y \right]_0^L . \tag{2.89}$$

6.3 Allowing for shearing in open-profile thin-walled bars

The theory of open-profile thin-walled bars was created by joint effort of a lot of scientists, though its most complete and general form should be attributed to V.Z. Vlasov, therefore the whole theory is usually associated with the name of this outstanding mechanician. One of fundamental postulates in this theory is an assumption that there should be no shearing in the median surface of a thin-walled bar.

As the theory was being developed, a lot of various suggestions were advanced in order to modify it; those were based on attempts to omit the no-shear hypothesis and to somehow take into account the effect of shear strains on the behavior of a thin-walled bar. In this regard we can refer to well-known publications by Adadurov [2], [1], Janelidze and Panovko [5], Goldenweiser [4], Vorobiev [18], Mescheriakov [8].

We will follow the general idea being developed in this book and try to take into account the effect of the shear strains by supplementing the expression of the strain energy of a thin-walled bar with a term caused by the work of the tangential stresses. This idea is essentially an application of the known procedure, which we have used to switch from the Bernoulli–Euler theory of beams to the theory by Timoshenko or from Kirchhoff's theory of plate bending to the theory by Reissner, to the case of the thin-walled bars. This reasoning seems to have been applied first to the shear strains in the thin-walled bar theory by L.N. Vorobiev in his paper [18], where it produces one of simplest versions of the shear theory of thin-walled bars.

Below we present, for the convenience of referencing, the expressions of the normal and tangential stresses in an open-profile thin-walled bar, taken from (2.27) and (2.61). Thus, we have:

$$\sigma^x = \frac{N}{A} + \frac{M_y}{I_y} z + \frac{M_z}{I_z} y + \frac{B}{I_\omega} \omega, \tag{3.1}$$

$$\tau = -Q_z \frac{S_{oy}}{hI_y} - Q_y \frac{S_{oz}}{hI_z} - M_\omega \frac{S_{o\omega}}{hI_\omega}. \tag{3.2}$$

Formula (3.2) defines only those tangential stresses τ which are distributed evenly over the thickness of the walls, h, of the bar's cross-section (see Fig. 6.6).

Exactly as in the Bernoulli–Euler beam theory, nonzero tangential stresses τ are in contradiction with the physical law,

$$\tau = G\gamma_{xs},$$

because, according to the no-shear hypothesis,

$$\gamma_{xs} = \frac{\partial u}{\partial s} + \frac{\partial v}{\partial x} = 0 .$$

If we recall the expression of strain energy $\mathsf{E}(\boldsymbol{\sigma})$,

$$\mathsf{E}(\boldsymbol{\sigma}) = \frac{1}{2}\int_0^L \left[\frac{N^2}{EA} + \frac{H^2}{GI_x} + \frac{M_y^2}{EI_y} + \frac{M_z^2}{EI_z} + \frac{B^2}{EI_\omega} \right] dx . \tag{3.3}$$

we will see easily that this energy shows the work of the normal stresses and of a part of tangential stresses τ_k distributed over the wall thickness linearly, with the zero value on the profile (Fig. 6.6). In other words, expression (3.3) includes the energy of normal stresses σ^x and the energy of tangential stresses τ_k of the pure (Saint-Venant's, as it is sometimes called) torsion.

Focusing on energy $\mathsf{E}(\tau)$ created by the tangential stresses τ which are distributed evenly over the profile's wall thickness, we can write

$$\mathsf{E}(\tau) = \frac{1}{2}\int_0^L \int_l \frac{\tau^2}{G} h ds dx , \tag{3.4}$$

which, after substituting (3.2) and integrating along the profile, gives the following expression of the strain energy consisting of eleven terms together with the energy from (3.3):

$$\mathsf{E}(\boldsymbol{\sigma}) = \frac{1}{2}\int_0^L \left[\frac{N^2}{EA} + \frac{H^2}{GI_x} + \frac{M_y^2}{EI_y} + \frac{M_z^2}{EI_z} + \frac{B^2}{EI_\omega} \right] dx +$$

$$+ \frac{1}{2}\int_0^L \left[\frac{Q_z^2}{GF_z} + \frac{Q_y^2}{GF_y} + \frac{M_\omega^2}{GF_\omega r^2} + \frac{2Q_z Q_y}{GF_{zy}} + \frac{2Q_z M_\omega}{GF_{z\omega} r} + \frac{2Q_y M_\omega}{GF_{y\omega} r} \right] dx . \tag{3.5}$$

Here we use the following designations for the new geometrical characteristics of the cross-section:

$$F_z = \frac{A}{\mu_{zz}}, \quad \mu_{zz} = \frac{A}{I_y^2}\int_l \frac{S_{oy}^2}{h}ds, \qquad F_y = \frac{A}{\mu_{yy}}, \quad \mu_{yy} = \frac{A}{I_z^2}\int_l \frac{S_{oz}^2}{h}ds,$$

$$F_{zy} = \frac{A}{\mu_{zy}}, \quad \mu_{zy} = \frac{A}{I_z I_y}\int_l \frac{S_{oz}S_{oy}}{h}ds, \quad F_\omega = \frac{A}{\mu_{\omega\omega}}, \quad \mu_{\omega\omega} = \frac{I_r}{I_\omega^2}\int_l \frac{S_{o\omega}^2}{h}ds,$$

$$F_{z\omega} = \frac{A}{\mu_{z\omega}}, \qquad \mu_{z\omega} = \frac{\sqrt{AI_r}}{I_y I_\omega}\int_l \frac{S_{oy}S_{o\omega}}{h}ds,$$

$$F_{y\omega} = \frac{A}{\mu_{y\omega}}, \qquad \mu_{y\omega} = \frac{\sqrt{AI_r}}{I_z I_\omega}\int_l \frac{S_{oz}S_{o\omega}}{h}ds, \tag{3.6}$$

where $I_r = I_y + I_z$ is a polar moment of inertia of the section, $r^2 = I_r/A$ is the square of the polar radius of inertia of the bar's cross-section.

Dimensionless factors μ will be called *cross-section shape factors*. Unlike the theory of beams by Timoshenko where only one cross-section shape factor is used, the theory of thin-walled bars makes use of six (dimensionless) factors of shape, which make up a symmetric *matrix*, **μ**, *of cross-section shape factors*,

$$\boldsymbol{\mu} = \left\|\begin{bmatrix} \mu_{zz} & \mu_{zy} & \mu_{z\omega} \\ \mu_{yz} & \mu_{yy} & \mu_{y\omega} \\ \mu_{\omega z} & \mu_{\omega y} & \mu_{\omega\omega} \end{bmatrix}\right\|, \qquad \mathbf{v} = \boldsymbol{\mu}^{-1} = \left\|\begin{bmatrix} \nu_{zz} & \nu_{zy} & \nu_{z\omega} \\ \nu_{yz} & \nu_{yy} & \nu_{y\omega} \\ \nu_{\omega z} & \nu_{\omega y} & \nu_{\omega\omega} \end{bmatrix}\right\|. \tag{3.7}$$

The matrix **v** introduced here is the inverse of **μ**, and its components will be useful further.

We introduce a new vector of stresses, **σ**, by assuming

$$\boldsymbol{\sigma} = \left\|\left[N, H, M_y, M_z, B, Q_z, Q_y, M_\omega\right]\right\|^{\mathsf{T}}. \tag{3.8}$$

In this formula, unlike (2.62) for the vector of stresses in the shear-free theory of thin-walled bars, the stress vector has eight components. Comparing to vector **σ** from (2.62), here we have three more components: Q_z, Q_y and M_ω. Obviously, the expression of strain energy E(**σ**) from (3.5) can be represented in a common matrix form, $\mathsf{E}(\boldsymbol{\sigma}) = ½\, \boldsymbol{\sigma}^{\mathsf{T}} \boldsymbol{C}^{-1} \boldsymbol{\sigma}$, if we adopt the following for matrix $\boldsymbol{C}^{-1}$:

$$
\boldsymbol{C}^{-1} = \left\|\begin{matrix}
\frac{1}{EA} & 0 & 0 & 0 & 0 & 0 & 0 & 0 \\
0 & \frac{1}{GI_x} & 0 & 0 & 0 & 0 & 0 & 0 \\
0 & 0 & \frac{1}{EI_y} & 0 & 0 & 0 & 0 & 0 \\
0 & 0 & 0 & \frac{1}{EI_z} & 0 & 0 & 0 & 0 \\
0 & 0 & 0 & 0 & \frac{1}{EI_\omega} & 0 & 0 & 0 \\
0 & 0 & 0 & 0 & 0 & \frac{\mu_{zz}}{GA} & \frac{\mu_{zy}}{GA} & \frac{\mu_{z\omega}}{rGA} \\
0 & 0 & 0 & 0 & 0 & \frac{\mu_{yz}}{GA} & \frac{\mu_{yy}}{GA} & \frac{\mu_{y\omega}}{rGA} \\
0 & 0 & 0 & 0 & 0 & \frac{\mu_{\omega z}}{rGA} & \frac{\mu_{\omega y}}{rGA} & \frac{\mu_{\omega\omega}}{r^2GA}
\end{matrix}\right\| . \tag{3.9}
$$

If we assume the kinematic boundary conditions to be homogeneous, then the strain energy, $\mathsf{E}(\boldsymbol{\sigma})$, will be a Castigliano functional which should be minimized under the condition that the equilibrium equations should hold in the volume of the bar and at its ends.

Now we want to write out a full set of all equations of equilibrium for the eight-component vector of stresses, $\boldsymbol{\sigma}$, introduced above; we refer to (2.39), (2.43), (2.44), (2.46), (2.48), (2.49) to do it. So, we have

$$
-N' = p_x , \quad -H' - M'_\omega = m_x , \quad -M'_y + Q_z = m_y , \quad -M'_z + Q_y = -m_z ,
$$

$$
-Q'_z = p_z , \quad -Q'_y = p_y , \quad -B' + M_\omega = m_B . \tag{3.10}
$$

These equations of equilibrium can be represented in the standard matrix form,

$$
\boldsymbol{A}^{\mathsf{T}}\boldsymbol{\sigma} = \bar{\boldsymbol{X}} ,
$$

if we compose the vector of loads, $\bar{\boldsymbol{X}}$, as

$$
\bar{\boldsymbol{X}} = \left\| \left[p_x , p_y , p_z , m_x , m_y , m_z , m_B \right] \right\|^{\mathsf{T}} . \tag{3.11}
$$

and construct the equilibrium operator, $\boldsymbol{A}^{\mathsf{T}}$, as

$$
\begin{array}{cccccccc}
N & H & M_y & M_z & B & Q_z & Q_y & M_\omega \\
\hline\hline
\end{array}
$$

$$
A^{\mathrm{T}} = \left\|\begin{matrix}
-\frac{d}{dx} & 0 & 0 & 0 & 0 & 0 & 0 & 0 \\
0 & 0 & 0 & 0 & 0 & 0 & -\frac{d}{dx} & 0 \\
0 & 0 & 0 & 0 & 0 & -\frac{d}{dx} & 0 & 0 \\
0 & -\frac{d}{dx} & 0 & 0 & 0 & 0 & 0 & -\frac{d}{dx} \\
0 & 0 & -\frac{d}{dx} & 0 & 0 & 1 & 0 & 0 \\
0 & 0 & 0 & \frac{d}{dx} & 0 & 0 & -1 & 0 \\
0 & 0 & 0 & 0 & -\frac{d}{dx} & 0 & 0 & 1
\end{matrix}\right\|. \tag{3.12}
$$

For the convenience of visual perception, the corresponding components of the stress vector, $\boldsymbol{\sigma}$, are indicated above the columns of the $\boldsymbol{A}^{\mathrm{T}}$ operator. As we can see, matrix $\boldsymbol{A}^{\mathrm{T}}$ consists of seven rows (according to the number of equilibrium equations) and of eight columns (according to the number of the components of the stress vector, $\boldsymbol{\sigma}$). Obviously, the geometry operator, $\boldsymbol{A}$, which is conjugate to the equilibrium operator, is a matrix of the dimensions 8×7,

$$
\boldsymbol{A} = \left\|\begin{matrix}
d/dx & 0 & 0 & 0 & 0 & 0 & 0 \\
0 & 0 & 0 & d/dx & 0 & 0 & 0 \\
0 & 0 & 0 & 0 & d/dx & 0 & 0 \\
0 & 0 & 0 & 0 & 0 & -d/dx & 0 \\
0 & 0 & 0 & 0 & 0 & 0 & d/dx \\
0 & 0 & d/dx & 0 & 1 & 0 & 0 \\
0 & d/dx & 0 & 0 & 0 & -1 & 0 \\
0 & 0 & 0 & d/dx & 0 & 0 & 1
\end{matrix}\right\|. \tag{3.13}
$$

Let us find out the physical meaning of the geometry operator, ***A***. We will use the kinematical relationships of the theory of thin-walled bars where the shear is taken into account.

It is easy to see that the longitudinal displacements, u, of points of the bar's cross-section, as defined by (2.54), in the form of

$$u = \xi - \theta_z y + \theta_y z - \theta' \omega, \tag{3.14}$$

ensure with their first three terms that the displacements vary over the cross-section according to the law of plane, while the last term of the formula defines a warping displacement. Earlier, with regard to formulas (2.14), we already gave a geometrical treatment of slopes θ_z and θ_y of the rigid cross-sections of the bar with respect to the corresponding axes, Z and Y, as the derivatives of displacements η and ζ:

$$\theta_y = -\zeta', \qquad \theta_z = \eta'.$$

These relationships should be omitted from the shear-based theory, and slopes θ_z and θ_y should be treated as independent kinematical parameters which characterize the bar's displacements integrally over its section. Also, we adopt the same law of sectorial areas over the bar's section for the warping component of longitudinal displacement u as before, but with a functional factor $\beta(x)$ different from twist θ'; we will call that factor a *warping measure function*. As a result, in the shear theory formula (3.14) should be replaced by

$$u = \xi - \theta_z y + \theta_y z - \beta\omega, \tag{3.15}$$

Let us present also an expression of tangential displacement v, deriving it from (2.8), (2.10):

$$v = \eta\, t_y + \zeta\, t_z + \theta\rho\,. \tag{3.16}$$

Now we are able to find shear strain γ_{xs} using (2.9) but not assuming the shear equal to zero:

$$\gamma_{xs} = \frac{\partial u}{\partial s} + \frac{\partial v}{\partial x} = (\eta' - \theta_z)t_y + (\zeta' + \theta_y)t_z + (\theta' - \beta)\rho\,. \tag{3.17}$$

Here we take account of the fact thast $t_y = dy/ds$, $t_z = dz/ds$, $\rho = d\omega/ds$.

We will denote particular components of the above formula by

$$\gamma_y = \eta' - \theta_z, \quad \gamma_z = \zeta' + \theta_y, \quad \gamma_\omega = \theta' - \beta, \tag{3.18}$$

which gives the following alternative representation of (3.17):

$$\gamma_{xs} = \gamma_y\, t_y + \gamma_z\, t_z + \gamma_\omega \rho. \tag{3.19}$$

The strain parameters γ_y, γ_z, γ_ω can be interpreted geometrically in this way:

- γ_y is a shear in the (X,Y)-plane, averaged over the cross-section;
- γ_z is a shear in the (X,Z)-plane, averaged over the cross-section;
- $\gamma_\omega \rho$ is a shear in each point of the profile, caused by the warping of the cross-section.

Parameter γ_ω, which is equal to the difference between twist θ' and warping measure β, characterizes a *warping strain of shear*. As we can see, this parameter is the same for the whole section.

Notice that displacement vector **u** will have seven components rather than four, as it used to have in (2.62) for the shear-free theory of thin-walled bars, because now we take account of the shear strains:

$$\mathbf{u} = \left\|\left[\xi, \eta, \zeta, \theta, \theta_y, \theta_z, -\beta\right]\right\|^{\mathrm{T}}, \tag{3.20}$$

These are the additional displacement components:

- θ_y, a slope (rotation angle) of the bar's cross-section about the Y-axis;
- θ_z, a slope (rotation angle) of the bar's cross-section about the Z-axis;
- $-\beta$, an inverted warping measure different from twist factor θ'.

But then the strain vector, $\boldsymbol{\varepsilon}$, will be

$$\boldsymbol{\varepsilon} = \left\|\left[\varepsilon_G, \chi_x, \chi_z, \chi_y, \chi_B, \gamma_z, \gamma_y, \gamma_\omega\right]\right\|^{\mathrm{T}}, \tag{3.21}$$

where we use the notation of

$$\varepsilon_G = \xi', \quad \chi_x = \theta', \quad \chi_z = \theta'_y, \quad \chi_y = -\theta'_z, \quad \chi_B = -\beta',$$

$$\gamma_z = \zeta' + \theta_y, \quad \gamma_y = \eta' - \theta_z, \quad \gamma_\omega = \theta' - \beta. \tag{3.22}$$

When written in the matrix form, $\boldsymbol{A}\mathbf{u} = \boldsymbol{\varepsilon}$, these relationships produce exactly the geometry matrix, $\boldsymbol{A}$, obtained earlier by us as a matrix operator conjugate to the equilibrium operator, $\boldsymbol{A}^{\mathrm{T}}$.

The stiffness matrix, $\boldsymbol{C}$, can be derived by formal inversion of compliance matrix $\boldsymbol{C}^{-1}$, as follows:

$$\boldsymbol{C} = \begin{bmatrix} EA & 0 & 0 & 0 & 0 & 0 & 0 & 0 \\ 0 & GI_x & 0 & 0 & 0 & 0 & 0 & 0 \\ 0 & 0 & EI_y & 0 & 0 & 0 & 0 & 0 \\ 0 & 0 & 0 & EI_z & 0 & 0 & 0 & 0 \\ 0 & 0 & 0 & 0 & EI_\omega & 0 & 0 & 0 \\ 0 & 0 & 0 & 0 & 0 & \nu_{zz}GA & \nu_{zy}GA & \nu_{z\omega}rGA \\ 0 & 0 & 0 & 0 & 0 & \nu_{yz}GA & \nu_{yy}GA & \nu_{y\omega}rGA \\ 0 & 0 & 0 & 0 & 0 & \nu_{\omega z}rGA & \nu_{\omega y}rGA & \nu_{\omega\omega}r^2GA \end{bmatrix} . \tag{3.23}$$

Now we are able to derive a governing system of equations in terms of displacements for the theory of thin-walled bars where the shear strains are taken into account. Based on the general formula of the Lame operator equation, we have

$$\boldsymbol{L}\mathbf{u} = \bar{\boldsymbol{X}} , \qquad \boldsymbol{L} = \boldsymbol{A}^{\mathrm{T}}\boldsymbol{C}\boldsymbol{A} ,$$

and, by making necessary matrix transformations, we derive the following system of seven differential equations with respect to seven components of displacement vector **u**,

$$\begin{gathered} -EA\xi'' = p_x , \\ -GA\left[\nu_{yz}(\zeta'' + \theta_y') + \nu_{yy}(\eta'' - \theta_z') + \nu_{y\omega}r(\theta'' - \beta')\right] = p_y , \\ -GA\left[\nu_{zz}(\zeta'' + \theta_y') + \nu_{zy}(\eta'' - \theta_z') + \nu_{z\omega}r(\theta'' - \beta')\right] = p_z , \\ -GI_x\theta'' - GAr\left[\nu_{\omega z}(\zeta'' + \theta_y') + \nu_{\omega y}(\eta'' - \theta_z') + \nu_{\omega\omega}r(\theta'' - \beta')\right] = m_x , \\ -EI_y\theta_y'' + GA\left[\nu_{zz}(\zeta' + \theta_y) + \nu_{zy}(\eta' - \theta_z) + \nu_{z\omega}r(\theta' - \beta)\right] = m_y , \\ -EI_z\theta_z'' - GA\left[\nu_{yz}(\zeta' + \theta_y) + \nu_{yy}(\eta' - \theta_z) + \nu_{y\omega}r(\theta' - \beta)\right] = m_z , \\ EI_\omega\beta'' + GAr\left[\nu_{\omega z}(\zeta' + \theta_y) + \nu_{\omega y}(\eta' - \theta_z) + \nu_{\omega\omega}r(\theta' - \beta)\right] = m_B . \end{gathered} \tag{3.24}$$

As we can see, each of the desirable seven components of displacement vector **u** is contained in the above system together with its major derivative of up to second order. Unlike the shear-free theory by Vlasov, the bending and torsion states of the system are not separated but coupled. The physical meaning of the equations is easy to see:

- first three equations are equations of equilibrium in terms of projections onto the respective axes X, Y and Z;
- the fourth, fifth, and sixth equations are equations of equilibrium in moments with respect to the corresponding axes;
- the last, seventh, equation is an equation of equilibrium of bitorques.

6.3.1 Basic integral identity in the shear theory of open-profile thin-walled bars

Based on the matrix relationships obtained earlier and using our common approach of integration by parts, we have

$$(A\mathbf{u},\,\boldsymbol{\sigma}) = \int_0^L [\xi' N + \theta' H + \theta_y' M_y - \theta_z' M_z - \beta' B + (\zeta' + \theta_y) Q_z +$$

$$+(\eta' - \theta_z) Q_y + (\theta' - \beta) M_\omega]dx = \int_0^L [-\xi N' - \eta Q_y' - \zeta Q_z' - \theta(H' + M_\omega') -$$

$$-\theta_y (M_y' - Q_z) + \theta_z (M_z' - Q_y) - \beta(M_\omega - B')]dx + [\xi N]\Big|_0^L + \left[\eta Q_y\right]\Big|_0^L +$$

$$+[\zeta Q_z]\Big|_0^L + [\theta(H + M_\omega)]\Big|_0^L + \left[\theta_y M_y\right]\Big|_0^L - [\theta_z M_z]\Big|_0^L - [\beta B]\Big|_0^L .$$

The matrix form of the above equation is standard:

$$(A\mathbf{u},\,\boldsymbol{\sigma}) = (\mathbf{u},\,A^{\mathrm{T}}\boldsymbol{\sigma}) + (\boldsymbol{p},\,\boldsymbol{u})_\Gamma ,$$

where we use the following vector of boundary loads, $\boldsymbol{p}$, and the vector of boundary displacements, $\boldsymbol{u}$:

$$\boldsymbol{p} = \left\|\left[nN, nQ_y, nQ_z, n(H + M_\omega), nM_y, -nM_z, nB \right]\right\|^{\mathrm{T}} ,$$

$$\boldsymbol{u} = \left\|\left[\xi, \eta, \zeta, \theta, \theta_y, \theta_z, -\beta \right]\right\|^{\mathrm{T}} .$$

As usual, n denotes an X-component of the vector of external normal to the bar's section. That is, $n = 1$ when $x = L$, and $n = -1$ when $x = 0$.

Let us derive explicit expressions of operators $\boldsymbol{H}_\sigma$ и $\boldsymbol{H}_u$ using the above components of the vector of boundary stresses, $\boldsymbol{p}$, and the vector of boundary displacements, $\boldsymbol{u}$, where $\boldsymbol{p} = \boldsymbol{H}_\sigma \boldsymbol{\sigma}$ and $\boldsymbol{u} = \boldsymbol{H}_u \mathbf{u}$. These are

$$\boldsymbol{H}_\sigma = \begin{Vmatrix} n & 0 & 0 & 0 & 0 & 0 & 0 & 0 \\ 0 & 0 & 0 & 0 & 0 & 0 & n & 0 \\ 0 & 0 & 0 & 0 & 0 & n & 0 & 0 \\ 0 & n & 0 & 0 & 0 & 0 & 0 & n \\ 0 & 0 & n & 0 & 0 & 0 & 0 & 0 \\ 0 & 0 & 0 & -n & 0 & 0 & 0 & 0 \\ 0 & 0 & 0 & 0 & n & 0 & 0 & 0 \end{Vmatrix}, \qquad \boldsymbol{H}_u = \boldsymbol{I}, \qquad (3.25)$$

where $\boldsymbol{I}$ is an identity matrix of seventh order.

6.3.2 Basic variational principles in the shear theory of open-profile thin-walled bars

Earlier we defined an expression of strain energy $\mathsf{E}(\boldsymbol{\sigma})$ by formula (3.5) as a quadratic functional of stresses $\boldsymbol{\sigma}$. The same energy but now represented as a quadratic functional of the displacements can be derived from the general formula, $\mathsf{E}(\mathbf{u}) = ½\,(\boldsymbol{A}\mathbf{u}, \boldsymbol{CA}\mathbf{u})$, or

$$\mathsf{E}(\mathbf{u}) = \frac{1}{2}\int_0^L \{EA\xi'^2 + GI_x\theta'^2 + EI_y\theta_y'^2 + EI_z\theta_z'^2 + EI_\omega\beta'^2 +$$

$$+GA[\nu_{zz}(\zeta'+\theta_y)^2 + \nu_{yy}(\eta'-\theta_z)^2 + \nu_{\omega\omega}r^2(\theta'-\beta)^2 +$$

$$+2\nu_{zy}(\zeta'+\theta_y)(\eta'-\theta_z) + 2\nu_{z\omega}r(\zeta'+\theta_y)(\theta'-\beta) +$$

$$+2\nu_{y\omega}r(\eta'-\theta_z)(\theta'-\beta)\}dx\,. \qquad (3.26)$$

Now let us deal with expressions of the force and kinematic potential. As earlier in Section 6.2.7, we use homogeneous boundary conditions for forces for the sake of simplicity. Then the potential of external forces will be a virtual work of the external distributed loads, written as a double integral:

$$\Pi_s(\mathbf{u}) = \int_0^L \left(\int_l (q_x u + q_t v + q_n w)\,ds \right) dx\,. \qquad (3.27)$$

Earlier we used (2.74) to transform this integral, but in our present case the expression of longitudinal displacement u should be replaced according to (3.15). Therefore, instead of (2.74), the shear theory uses

$$u = \xi - \theta_z y + \theta_y z - \beta\omega ,$$

$$v = \eta t_y + \zeta t_z + \theta[(y - y_P)t_z - (z - z_P)t_y] ,$$

$$w = \eta t_z - \zeta t_y - \theta[(y - y_P)t_y + (z - z_P)t_z] . \qquad (3.28)$$

Substituting (3.28) and (2.35) to (3.27) transforms the force potential into

$$\int_0^L \int_l \left\{ q_x[\xi - \theta_z y + \theta_y z - \beta\omega] + q_y[\eta - \theta(z - z_P)] + q_z[\zeta + \theta(y - y_P)] \right\} ds dx .$$

Integrating over arc coordinate s and using the notation of external loads integral over the bar's cross-section from (2.36) yields the final expression of the force potential:

$$\Pi_s(\mathbf{u}) = \int_0^L \left(\xi p_x + \eta p_y + \zeta p_z + \theta m_x + \theta_z m_z + \theta_y m_y - \beta m_B \right) dx . \qquad (3.29)$$

If we take the difference between this force potential and that for the shear-free theory defined by (2.75), we will have an increment of the force potential, $\Delta\Pi_s(\mathbf{u})$, caused by taking the shear into account:

$$\Delta\Pi_s(\mathbf{u}) = \int_0^L \left[(\theta_z - \eta')m_z + (\theta_y + \zeta')m_y + (\theta' - \beta)m_B \right] dx . \qquad (3.30)$$

As we can see, this increment of force potential $\Delta\Pi_s(\mathbf{u})$ is a virtual work of the external distributed moment and bitorque loads on the shear strains, which is quite expectable.

Following the formal rules formulated in Chapter two for the general case, it is easy to build also the potential of the kinematical external actions and to write out explicit expressions of the basic functionals, the Lagrange functional and the Castigliano functional.

The reader can make this derivation by himself, if need be.

6.3.3 A remark on the shear theory of open-profile thin-walled bars for non-warped cross-sections

As we established in Section **6**.2.8, the non-warped case makes the problem degenerate both in the differential form and in the variational form of the shear-free theory of open-profile thin-walled bars.

Now let us see what changes the non-warpedness introduces to the shear theory. Formulas (3.1) and (3.2) of the normal and tangential stresses will be replaced by (2.81) and (2.82). As a result, formula (3.5) of the strain energy transforms into

$$\mathsf{E}(\boldsymbol{\sigma}) = \frac{1}{2}\int_0^L \left[\frac{N^2}{EA} + \frac{H^2}{GI_x} + \frac{M_y^2}{EI_y} + \frac{M_z^2}{EI_z} + \frac{Q_z^2}{GF_z} + \frac{Q_y^2}{GF_y} + \frac{2Q_zQ_y}{GF_{zy}}\right]dx. \qquad (3.31)$$

The matrix of cross-section shape factors, $\boldsymbol{\mu}$, together with its inverse matrix $\mathbf{v}$, becomes a second-order matrix:

$$\boldsymbol{\mu} = \left\|\begin{bmatrix} \mu_{zz} & \mu_{zy} \\ \mu_{yz} & \mu_{yy} \end{bmatrix}\right\|, \qquad \mathbf{v} = \boldsymbol{\mu}^{-1} = \left\|\begin{bmatrix} \nu_{zz} & \nu_{zy} \\ \nu_{yz} & \nu_{yy} \end{bmatrix}\right\|. \qquad (3.32)$$

Further, the vector representations of "stresses" $\boldsymbol{\sigma}$, strains $\boldsymbol{\varepsilon}$, displacements $\mathbf{u}$ and external load $\bar{\boldsymbol{X}}$ will change, so we will have the following instead of (3.8), (3.11), (3.20) and (3.21):

$$\boldsymbol{\sigma} = \left\|\left[N, H, M_y, M_z, Q_z, Q_y\right]\right\|^{\mathsf{T}}, \quad \bar{\boldsymbol{X}} = \left\|\left[p_x, p_y, p_z, m_x, m_y, m_z\right]\right\|^{\mathsf{T}},$$

$$\mathbf{u} = \left\|\left[\xi, \eta, \zeta, \theta, \theta_y, \theta_z\right]\right\|^{\mathsf{T}}, \qquad \boldsymbol{\varepsilon} = \left\|\left[\varepsilon_{\mathrm{G}}, \chi_x, \chi_z, \chi_y, \gamma_z, \gamma_y\right]\right\|^{\mathsf{T}}. \qquad (3.33)$$

As we can see, the vector of stresses, $\boldsymbol{\sigma}$, has lost its bitorque B and its constricted-torsion moment M_ω, which are absent in a non-warped section. The respective strain components χ_B and γ_ω are also excluded from vector $\boldsymbol{\varepsilon}$. The vector of external force loads has lost its bimoment load m_B, and the displacement vector, $\mathbf{u}$, does not contain warping measure β anymore.

Following the changes made to vectors $\boldsymbol{\sigma}$ and $\boldsymbol{\varepsilon}$ for a non-warped section, we should exclude:

- sixth and eighth rows and seventh column from geometry matrix $\boldsymbol{A}$ as defined in (3.13);
- fifth and eighth columns and seventh row from equlibrium matrix $\boldsymbol{A}^{\mathsf{T}}$ as defined in (3.12);
- rows and columns number five and eight from elasticity matrix $\boldsymbol{C}$ as defined in (3.23).

The result is

$$\boldsymbol{A} = \begin{array}{c} \begin{array}{cccccc} \xi & \eta & \zeta & \theta & \theta_y & \theta_z \end{array} \\ \left\| \begin{array}{cccccc} \frac{d}{dx} & 0 & 0 & 0 & 0 & 0 \\ 0 & 0 & 0 & \frac{d}{dx} & 0 & 0 \\ 0 & 0 & 0 & 0 & \frac{d}{dx} & 0 \\ 0 & 0 & 0 & 0 & 0 & -\frac{d}{dx} \\ 0 & 0 & \frac{d}{dx} & 0 & 1 & 0 \\ 0 & \frac{d}{dx} & 0 & 0 & 0 & -1 \end{array} \right\| \end{array}, \tag{3.34}$$

$$\boldsymbol{A}^{\mathsf{T}} = \begin{array}{c} \begin{array}{cccccc} N & H & M_y & M_z & Q_z & Q_y \end{array} \\ \left\| \begin{array}{cccccc} -d/dx & 0 & 0 & 0 & 0 & 0 \\ 0 & 0 & 0 & 0 & 0 & -d/dx \\ 0 & 0 & 0 & 0 & -d/dx & 0 \\ 0 & -d/dx & 0 & 0 & 0 & 0 \\ 0 & 0 & -d/dx & 0 & 0 & 1 \\ 0 & 0 & 0 & d/dx & 0 & -1 \end{array} \right\| \end{array}, \tag{3.35}$$

and

$$\boldsymbol{C} = \left\| \begin{array}{cccccc} EA & 0 & 0 & 0 & 0 & 0 \\ 0 & GI_x & 0 & 0 & 0 & 0 \\ 0 & 0 & EI_y & 0 & 0 & 0 \\ 0 & 0 & 0 & EI_z & 0 & 0 \\ 0 & 0 & 0 & 0 & \nu_{zz}GA & \nu_{zy}GA \\ 0 & 0 & 0 & 0 & \nu_{yz}GA & \nu_{yy}GA \end{array} \right\|. \tag{3.36}$$

The governing system of equations in terms of displacements is derived from the general formula of the Lame operator equation:

$$\boldsymbol{L}\mathbf{u} = \bar{X}\,, \qquad \boldsymbol{L} = A^{\mathrm{T}}CA\,.$$

After making appropriate matrix transformations, we will have the following system of six simultaneous differential equations with respect to six components of the displacement vector, **u**,

$$\begin{gathered}
-EA\xi'' = p_x\,,\\
-GA\left[\nu_{yz}(\zeta'' + \theta_y') + \nu_{yy}(\eta'' - \theta_z')\right] = p_y\,,\\
-GA\left[\nu_{zz}(\zeta'' + \theta_y') + \nu_{zy}(\eta'' - \theta_z')\right] = p_z\,,\\
-GI_x\theta'' = m_x\,,\\
-EI_y\theta_y'' + GA\left[\nu_{zz}(\zeta' + \theta_y) + \nu_{zy}(\eta' - \theta_z)\right] = m_y\,,\\
-EI_z\theta_z'' - GA\left[\nu_{yz}(\zeta' + \theta_y) + \nu_{yy}(\eta' - \theta_z)\right] = m_z\,.
\end{gathered} \qquad (3.37)$$

As for the expression of strain energy E represented as a functional of the displacements, we will have the following for the non-warped section instead of (3.26) which was derived for a warped section:

$$\mathsf{E}(\mathbf{u}) = \frac{1}{2}\int_0^L \{EA\xi'^2 + GI_x\theta'^2 + EI_y\theta_y'^2 + EI_z\theta_z'^2 +$$

$$+ GA[\nu_{zz}(\zeta' + \theta_y)^2 + \nu_{yy}(\eta' - \theta_z)^2 + 2\nu_{zy}(\zeta' + \theta_y)(\eta' - \theta_z)]\}dx\,. \qquad (3.38)$$

6.3.4 Remark on a matrix of cross-section shape factors, μ

Earlier we have defined the components of matrix with formulas (3.6) – a matrix of dimensionless factors of the cross-section's shape. Also, according to (3.7), matrix **ν** is defined by formal inversion of matrix **μ**. The latter definition is based on the implicit assumption that the inversion is possible, i.e. that the **μ** matrix is nondegenerate.

The **μ** matrix can be shown to be positive definite – hence invertible. To see this, take note than formula (3.4) implies the strict positivity of energy $\mathsf{E}(\tau) > 0$ at $\tau \neq 0$. Hence follows, as is easy to see, the positive definiteness of the matrix:

$$\begin{bmatrix} \frac{\mu_{zz}}{GA} & \frac{\mu_{zy}}{GA} & \frac{\mu_{z\omega}}{rGA} \\ \frac{\mu_{yz}}{GA} & \frac{\mu_{yy}}{GA} & \frac{\mu_{y\omega}}{rGA} \\ \frac{\mu_{\omega z}}{rGA} & \frac{\mu_{\omega y}}{rGA} & \frac{\mu_{\omega\omega}}{r^2GA} \end{bmatrix} =$$

$$= \frac{1}{GA} \begin{bmatrix} 1 & 0 & 0 \\ 0 & 1 & 0 \\ 0 & 0 & 1/r \end{bmatrix} \cdot \begin{bmatrix} \mu_{zz} & \mu_{zy} & \mu_{z\omega} \\ \mu_{yz} & \mu_{yy} & \mu_{y\omega} \\ \mu_{\omega z} & \mu_{\omega y} & \mu_{\omega\omega} \end{bmatrix} \cdot \begin{bmatrix} 1 & 0 & 0 \\ 0 & 1 & 0 \\ 0 & 0 & 1/r \end{bmatrix}.$$

If we consider that $GA > 0$ and $r > 0$, we will understand that the positive definiteness of the matrix implies the same for matrix $\boldsymbol{\mu}$.

This kind of reasoning works only when we deal with a warped section. A non-warped section is a special degenerate case, for which the last two terms in (3.1) and (3.2) vanish. This makes the order of matrices $\boldsymbol{\mu}$ and $\mathbf{v}$ lower by one (see Section **6**.3.3 for more detail). At the same time, the positive definiteness of these truncated matrices is still maintained even in the degenerate case of the non-warped section.

The latter statement can be proved directly, without using any energy considerations. To see this, consider the $\boldsymbol{\mu}$ matrix for a non-warped cross-section:

$$\boldsymbol{\mu} = \begin{bmatrix} \mu_{zz} & \mu_{zy} \\ \mu_{yz} & \mu_{yy} \end{bmatrix} = \begin{bmatrix} \frac{A}{I_y^2} \int_l \frac{S_{oy}^2}{h} ds & \frac{A}{I_y I_z} \int_l \frac{S_{oy} S_{oz}}{h} ds \\ \frac{A}{I_z I_y} \int_l \frac{S_{oz} S_{oy}}{h} ds & \frac{A}{I_z^2} \int_l \frac{S_{oz}^2}{h} ds \end{bmatrix}. \tag{3.38}$$

As we know from linear algebra [14], a symmetric matrix is positive definite if and only if all its principal minors are positive. Because $\mu_{zz} > 0$, it becomes clear that the positive definiteness of the second-order matrix $\boldsymbol{\mu}$ will follow immediately from the positiveness of its determinant. We have

$$\det \boldsymbol{\mu} = \frac{A^2}{I_y^2 I_z^2} \left[\int_l \frac{S_{oy}^2}{h} ds \int_l \frac{S_{oz}^2}{h} ds - \left(\int_l \frac{S_{oz} S_{oy}}{h} ds \right)^2 \right]. \tag{3.39}$$

But the brackets in (3.39) contain a positive expression because of the well-known Buniakovsky inequality [13]:

$$\left(\int_l UVds\right)^2 \le \int_l U^2 ds \int_l V^2 ds ,$$

in which we should assume

$$U = S_{oy} / \sqrt{h} , \qquad V = S_{oz} / \sqrt{h} .$$

6.4 A semi-shear theory of open-profile thin-walled bars

Let us return to formula (3.2) of the tangential stresses in particular points of the cross-section of a thin-walled bar and rewrite it here as a sum of two terms:

$$\tau = \tau_Q + \tau_M, \qquad \tau_Q = -Q_z \frac{S_{oy}}{hI_y} - Q_y \frac{S_{oz}}{hI_z}, \qquad \tau_M = -M_\omega \frac{S_{o\omega}}{hI_\omega}. \tag{4.1}$$

Tangential stresses τ_Q caused by shear forces Q_z and Q_y will be called *tangential stresses of bending*. Then stresses τ_M caused by the constricted-torsion moment, M_ω, can be reasonably called *tangential stresses of torsion*.

As known [5], one of methods of building approximate (applied) theories is based on dividing the stresses and/or strains into *primary* (say, σ_o, ε_o) and *secondary* (say, σ_1, ε_1) and then excluding the secondary components from the governing equations because they are less significant than the primary ones. The energy-based approach to this division is to neglect the energy of the secondary components comparing to the energy of the primary stress and strain components. In other words, the general expression of the energy,

$$\mathsf{E}(\sigma) = \mathsf{E}(\sigma_o) + \mathsf{E}(\sigma_1) + \mathsf{A}(\sigma_o,\varepsilon_1) \approx \mathsf{E}(\sigma_o)$$

should lose its second and third terms because they are far less than the first. Here $\mathsf{A}(\sigma_o,\varepsilon_1)$ is a virtual work of the primary stresses on the strains caused by the secondary stresses (or vice versa, because the work reciprocity theorem holds).

If we use this approach to compare between the shear theory of thin-walled bars discussed in Section 6.3 and the shear-free theory by V.Z. Vlasov, we can see that Vlasov's theory adopts the normal stresses, σ^x, and the tangential stresses, τ_k, as the primary stress components. As for tangential stresses τ evenly distributed over the thickness, h, of the bar's cross-section walls, they are classified as secondary stress components – it suffices to compare the expressions of the energy from (3.5) and from (2.72).

However, we can also use an intermediate version of the theory where only a part of the tangential stresses, τ, is thought to be primary. Those are the tangential stresses of torsion, τ_M, while the tangential stresses of bending, τ_Q, are classified as secondary. A theory of this kind will be called a *semi-shear theory of thin-walled bar.*

The significance of the semi-shear theory, which is an intermediate one between the shear and shear-free versions of the theories, is as follows. First, as we can predict beforehand and will see later, the semi-shear theory is essentially simpler than the shear version. Second, we can expect that the energy of tangential stresses τ_Q and the energy of normal stresses σ^x are in about the same relation as are the additional terms in the expression of the energy in Timoshenko's theory of beam bending comparing to that in the Bernoulli–Euler theory. But, as we know well, there is a wide scope of practical problems where the refinement of the solutions achieved by using Timoshenko's theory can be neglected beforehand without introducing any substantial error.

So, according to what has been said, we adopt the following energy expression for the semi-shear theory instead of (3.5):

$$\mathsf{E}(\boldsymbol{\sigma}) = \frac{1}{2}\int_0^L \left[\frac{N^2}{EA} + \frac{H^2}{GI_x} + \frac{M_y^2}{EI_y} + \frac{M_z^2}{EI_z} + \frac{B^2}{EI_\omega} + \frac{M_\omega^2}{GF_\omega r_\omega^2} \right] dx\,, \tag{4.2}$$

where, as it is worth reminding,

$$F_\omega = \frac{A}{\mu_{\omega\omega}}\,, \qquad \mu_{\omega\omega} = \frac{I_r}{I_\omega^2}\int_l \frac{S_{o\omega}^2}{h}\,ds\,.$$

Obviously, in the semi-shear theory the shear forces should be omitted from stress vector $\boldsymbol{\sigma}$ in (3.8), and this gives the following six-component vector:

$$\boldsymbol{\sigma} = \left\| \left[N, H, M_y, M_z, B, M_\omega \right] \right\|^{\mathsf{T}}. \tag{4.3}$$

It is also obvious that matrix $\boldsymbol{C}^{-1}$ from (3.9) should be replaced by the following for the purposes of the semi-shear theory:

$$\boldsymbol{C}^{-1} = \begin{bmatrix} \frac{1}{EA} & 0 & 0 & 0 & 0 & 0 \\ 0 & \frac{1}{GI_x} & 0 & 0 & 0 & 0 \\ 0 & 0 & \frac{1}{EI_y} & 0 & 0 & 0 \\ 0 & 0 & 0 & \frac{1}{EI_z} & 0 & 0 \\ 0 & 0 & 0 & 0 & \frac{1}{EI_\omega} & 0 \\ 0 & 0 & 0 & 0 & 0 & \frac{\mu_{\omega\omega}}{r^2 GA} \end{bmatrix}. \tag{4.4}$$

As we can see, the elasticity matrix, $\boldsymbol{C}$, is diagonal in the semi-shear theory, and this is a simplification of the governing simultaneous equations of the problem in comparison to the equations of the full shear theory.

The set of the equilibrium equations (3.10) will become as follows, after the shear forces are excluded:

$$-N' = p_x, \quad -H' - M'_\omega = m_x, \quad -M''_y = \bar{p}_z, \quad -M''_z = \bar{p}_y,$$

$$-B' + M_\omega = m_B, \tag{4.5}$$

where the generalized loads $\bar{p}_y$ and $\bar{p}_z$ are defined by the first two formulas in (2.50).

We introduce the vector of loads, $\bar{\boldsymbol{X}}$,

$$\bar{\boldsymbol{X}} = \left\| \left[p_x, \bar{p}_y, \bar{p}_z, m_x, m_B \right] \right\|^{\mathsf{T}}, \tag{4.6}$$

to write the system of equilibrium equations (4.5) in the matrix form, $\boldsymbol{A}^{\mathsf{T}} \boldsymbol{\sigma} = \bar{\boldsymbol{X}}$, where we assume

$$
\begin{array}{c}
\begin{array}{cccccc} N & H & M_y & M_z & B & M_\omega \end{array} \\
=================== \\
A^{\mathrm{T}} = \left\| \begin{array}{cccccc}
-\dfrac{d}{dx} & 0 & 0 & 0 & 0 & 0 \\
0 & 0 & 0 & -\dfrac{d^2}{dx^2} & 0 & 0 \\
0 & 0 & -\dfrac{d^2}{dx^2} & 0 & 0 & 0 \\
0 & -\dfrac{d}{dx} & 0 & 0 & 0 & -\dfrac{d}{dx} \\
0 & 0 & 0 & 0 & -\dfrac{d}{dx} & 1
\end{array} \right\| .
\end{array}
\tag{4.7}
$$

Then the operator of geometry, $\boldsymbol{A}$, conjugate to the equilibrium operator, will become

$$
\begin{array}{c}
\begin{array}{ccccc} \xi_G & \eta_P & \zeta_P & \theta & -\beta \end{array} \\
=============== \\
\boldsymbol{A} = \left\| \begin{array}{ccccc}
\dfrac{d}{dx} & 0 & 0 & 0 & 0 \\
0 & 0 & 0 & \dfrac{d}{dx} & 0 \\
0 & 0 & -\dfrac{d^2}{dx^2} & 0 & 0 \\
0 & -\dfrac{d^2}{dx^2} & 0 & 0 & 0 \\
0 & 0 & 0 & 0 & \dfrac{d}{dx} \\
0 & 0 & 0 & \dfrac{d}{dx} & 1
\end{array} \right\| .
\end{array}
\tag{4.8}
$$

This operator produces the vector of strains, $\boldsymbol{\varepsilon}$, when applied to the vector of displacements, $\mathbf{u}$, where we denote

$$
\mathbf{u} = \left\| [\xi, \eta, \zeta, \theta, -\beta] \right\|^{\mathrm{T}}, \quad \boldsymbol{\varepsilon} = \left\| [\varepsilon_G, \chi_x, \chi_z, \chi_y, \chi_B, \gamma_\omega] \right\|^{\mathrm{T}}, \tag{4.9}
$$

and

$$\varepsilon_G = \xi', \quad \chi_x = \theta', \quad \chi_z = -\zeta'', \quad \chi_y = -\eta'', \quad \chi_B = -\beta', \quad \gamma_\omega = \theta' - \beta. \quad (4.10)$$

The final system of governing equations of equilibrium in terms of displacements,

$$\boldsymbol{L}\mathbf{u} = \overline{\boldsymbol{X}}, \qquad \boldsymbol{L} = \boldsymbol{A}^{\mathrm{T}}\boldsymbol{C}\boldsymbol{A},$$

looks like

$$\boxed{\begin{gathered} -EA\xi'' = p_x, \quad EI_z\eta^{IV} = \overline{p}_y, \quad EI_y\zeta^{IV} = \overline{p}_z, \\ -GI_x\theta'' - \frac{r^2GA}{\mu_{\omega\omega}}(\theta'' - \beta') = m_x, \quad EI_\omega\beta'' + \frac{r^2GA}{\mu_{\omega\omega}}(\theta' - \beta) = m_B. \end{gathered}} \quad (4.11)$$

As we can see, the bending and torsion problems are separate in the semi-shear theory. This is why the system of equations (4.11) is much simpler than its counterpart in the full shear theory, that is, simpler than (3.24).

By the way, the problem of torsion described by the last two differential equations in (4.11) can be easily reduced to one equation of third order with respect to function β. To see this, we differentiate the last equation in (4.11) with respect to x and sum the result with the fourth equation of system (4.11) to produce

$$-GI_x\theta'' + EI_\omega\beta''' = m_x + m_B'. \quad (4.12)$$

Now we exclude the value of θ'' from the fourth equation of (4.11) using (4.12) and get

$$EI_\omega\left(1 + \frac{I_r}{\mu_{\omega\omega}I_x}\right)\beta''' - \frac{GI_r}{\mu_{\omega\omega}}\beta' = \frac{I_r}{\mu_{\omega\omega}I_x}m_x + \left(1 + \frac{I_r}{\mu_{\omega\omega}I_x}\right)m_B'.$$

It is convenient to introduce a dimensionless geometrical parameter, ψ, by defining

$$\psi = 1 + \frac{\mu_{\omega\omega}I_x}{I_r}. \quad (4.13)$$

With this notation, the final governing differential equation of third order for the analysis of torsion with the semi-shear theory of open-profile thin-walled bars will be

$$\psi EI_{\omega}\beta''' - GI_{x}\beta' = m_{x} + \psi m_{B}' . \tag{4.14}$$

It is also useful to present an expression of strain energy E, which is treated here as a functional of the displacements in the semi-shear theory. It is easy to derive this:

$$\mathsf{E}(\mathbf{u}) = \frac{1}{2}\int_{0}^{L}[EA\xi'^{2} + EI_{y}\zeta''^{2} + EI_{z}\eta''^{2} +$$

$$+EI_{\omega}\beta'^{2} + GI_{x}\theta'^{2} + \frac{GI_{x}}{\psi - 1}(\theta' - \beta)^{2}]dx . \tag{4.15}$$

Obviously, the torsion corresponds to only a part of the integrand in (4.15), which is placed on the second line.

Confining ourselves with the constricted torsion (no bending considered), we can write the following expression of the Lagrange functional,

$$\mathsf{L}(\theta,\beta) =$$

$$= \frac{1}{2}\int_{0}^{L}\left[EI_{\omega}\beta'^{2} + \frac{GI_{x}}{\psi - 1}(\theta' - \beta)^{2} + GI_{x}\theta'^{2}\right]dx - \int_{0}^{L} m_{x}\theta dx + \int_{0}^{L} m_{B}\beta dx . \tag{4.16}$$

Here we would like to note that the structure of the above functional bears some resemblance to the Lagrange functional in the Timoshenko beam theory – see formula (**4**.6.14). To bring these two functionals to a complete consistency, we would need, however, to support the Timoshenko beam not by a usual Winkler-type foundation but by an elastic foundation which would resist to the rotation of the original ground surface with the factor of C_2 [10]. Formula (**4**.6.14), transformed for this purpose, will look like

$$\mathsf{L}(v,\varphi) = \frac{1}{2}\int_{0}^{L}[EI\varphi'^{2} + GF_{y}(v' - \varphi)^{2} + C_{2}v'^{2}]dx - \int_{0}^{L} qv dx - \int_{0}^{L} m\varphi dx . \tag{4.17}$$

In (4.17) we re-denote the slope of the Timosheko beam by φ because the symbol of θ is already used here to designate a torsion angle of the thin-walled bar's cross-section.

The comparison between (4.16) and (4.17) leads to an important conclusion: the torsion of an open-profile thin-walled bar can be modeled and analyzed within the frames of the semi-shear theory as the bending of a Timoshenko beam. To do it, one needs to replace the designations in one's equations according to Table 6.1 presented below.

Table 6.1

Bending of a Timoshenko beam		Torsion of a thin-walled bar	
deflection	v	*torsion angle*	θ
slope of section	φ	*warping measure*	β
flexural stiffness	EI	*sectorial stiffness*	EI_ω
shearing stiffness	GF_y		$GI_x/(\psi - 1)$
stiffness of foundation	C_2	*stiffness of pure torsion*	GI_x
lateral load	q	*external torque per unit of line*	m_x
moment load	m	*external bitorque per unit of line*	$-m_B$
bending moment in the beam's cross-section	$M = -EI\varphi'$	*bitorque*	$B = -EI_\omega\beta'$
shear force in the beam's cross-section	$Q = GF_y(v' - \varphi)$	*constricted-torsion moment*	$M_\omega = \dfrac{GI_x}{\psi - 1}(\theta' - \beta)$
moment reaction of elastic foundation	$C_2 v'$	*pure-torsion moment*	$H = GI_x\theta'$
generalized shear force in the beam	$K = Q + C_2 v'$	*full torque*	$M_x = M_\omega + H$

Further details are not discussed here.

References

1. Adadurov RA (1947) Determining tangential stresses in thin-walled structures near an immuring (in Russian). In: Proc. Centr. Aerohydr. Inst., N614: 1–13
2. Adadurov RA (1948) Stresses and strains in a cylindrical shell with rigid transversal constraints (in Russian). Rep. Acad. Sci., 62, N2: 183–186
3. Dinnik AN (1955) Buckling. Torsion (in Russian). USSR Acad. Sci. Publishing House, Moscow
4. Goldenweiser AL (1949) On a theory of thin-walled bars (in Russian). Appl. Math and Mech., Vol.13
5. Janelidze GY, Panovko YG (1948) Statics of elastic thin-walled bars (in Russian). Gostekhteorizdat, Moscow Leningrad
6. Kochin NE (1965) Vector calculus and basics of tensor calculus (in Russian). Nauka, Moscow
7. Leibenzon LS (1951) Variational methods for solving problems of elasticity (in Russian). In: Academician L.S. Leibenson, collected works, vol. 1: Theory of elasticity. Academy of Sciences of USSR, Moscow

8. Mescheriakov VB (1968) General equations of the theory of open-profile thin-walled bars where shear is taken into account (in Russian). Proc. Moscow Inst. Transp. Engngrs, Problems in applied mechanics, issue 260, pp 82–93
9. Mikhlin SG (1970) Variational methods in mathematical physics (in Russian). Nauka, Moscow
10. Perelmuter AV, Slivker VI (2003) Numerical Structural Analysis. Springer-Verlag, Berlin Heidelberg New York
11. Rozin LA (1998) Problems of elasticity and numerical methods for their solution (in Russian). St.-Petersburg: Publ. House of St.-Pet. State Tech. Univ., St.-Petersburg
12. Saint-Venant Barré (1855) Mém. acad. sci. savants étrangers, XIV: 233–560.
13. Smirnov VI (1974) A course on advanced mathematics (in Russian). Vol. 2. Physmathgiz, Moscow
14. Strang G (1976) Linear algebra and its applications. Academic press, New York San Francisko London
15. Timoshenko SP (1957) History of strength of materials. McGraw-Hill, New York
16. Timoshenko SP, Goudier JN (1951) Theory of elasticity. New York
17. Vlasov VZ (1963) Selected works. Volume 2 (in Russian). USSR Acad. Sci. Publishing House, Moscow
18. Vorobiev LN (1955) An effect of a shear in the median surface on the strains and stresses in open-profile thin-walled bars with an undeformed contour (in Russian). Proc. Novocherkassk Polytech. Inst., Vol.26(40), pp. 92–111

7 PARTICULAR CLASSES OF PROBLEMS IN STRUCTURAL MECHANICS – part 4

The theory of analysis of thin-walled constructions no earlier became a chapter of structural mechanics than its clear and convenient working hypotheses were formulated on the basis of experimental data and made it possible to construct practical methods of calculation.
Kan SN, Panovko YG (1952) Principles of structural mechanics of thin-walled constructions (in russian). Oborongiz, Moscow

7.1 Closed-profile thin-walled bars – a theory by Umanski

We consider a closed cylindrical (not necessarily round) shell of the length L, with the perimeter of its cross-section's median line being l and the thickness of the wall being h. We assume that, exactly as in the theory of open-profile thin-walled bars, those characteristic parameters are related in comparison to one another as (**6**.2.1). This is the case when we deal with a closed-profile thin-walled bar.

As before, we introduce a right-hand Cartesian coordinate system, (X,Y,Z). The X axis is a longitudinal axis of the bar, parallel to the generatrix of the cylindrical shell, and it passes through the center of gravity of any cross-section of the bar. The Y and Z axes are central principal axes of inertia of the bar's cross-section. The origin of the coordinates will be located in such way that the beginning and the end sections of the bar have the respective coordinates of $x = 0$ and $x = L$.

We choose an arbitrary point, O, on the bar's profile, from which to count off the arc coordinate, s — a distance, taken with a certain sign, between the origin O and the current point of the profile, M. The sign of s is defined by the positive direction along the contour. We assume that the positive direction of coordinate s conforms to a counter-clockwise movement around the contour, when looking from the positive side of the X-axis.

As earlier in Section **6**.2, a right-hand triple of curvilinear axes is introduced in every point M of the profile, with local unit vectors $(\boldsymbol{i}_x, \boldsymbol{n}, \boldsymbol{t})$.

Unit vector $\boldsymbol{t}$ is tangential to the arc coordinate s, towards its increasing values, and unit vector $\boldsymbol{n}$ is an external normal to the profile. Take note of the important fact that, in order to describe the material properties, we introduce two dimensionless functions of the arc coordinate, $e(s)$ and $g(s)$, and the modulus of elasticity, E, together with the modulus of shear, G, which help define a fictitious unified material for the section (and thus can be called the unification moduli). Functions e and g characterize the distribution of the material's physical properties along the profile. The actual elasticity modulus, $E(s)$, and the actual shear modulus, $G(s)$, in an arbitrary point of the profile are defined as

$$E(s) = Ee(s), \qquad G(s) = Gg(s). \tag{1.1}$$

The reason why sections not uniform with respect to their material are taken into consideration is the wish both to generalize the relevant formulas and to serve practical purposes. In particular, when a modern bridge is being built, the sections of the bridge's stiffening girder are often composed of steel boxes, with a ferroconcrete slab to serve for the carriageway as shown further in Fig. 8.1.

With regard to what has been said, please note that the geometrical characteristic values of the sections are actually their combined physical and geometrical characteristics. Particularly, the area of the section, A, and the moments of inertia, I_y and I_z, are defined as

$$A = \oint ehds\,, \qquad I_y = \oint ehz^2 ds\,, \qquad I_z = \oint ehy^2 ds\,.$$

It is also obvious that the position of the center of gravity and those of the central principal axes of inertia of the section depend on function $e(s)$.

7.1.1 Pure torsion of a closed-profile thin-walled bar

We begin our consideration of the theory of closed-profile thin-walled bars with a simplest analysis of pure (free, nonconstricted) torsion of a thin-walled bar with its cross-section containing a single closed contour. Sometimes cross-sections of this kind are called *singly closed.*

Let each end of the bar have a specific external moment, M_x, applied to it, as shown in Fig. 7.1. There are no other external actions, nor external constraints (fixations of the bar). But then we can, just as we did in Saint-Venant's problem, impose six external constraints on the bar in order to eliminate its rigid displacements without affecting the stress distribution.

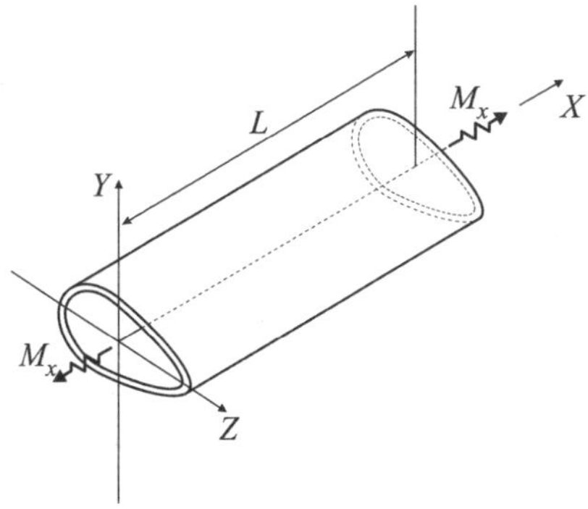

Fig. 7.1. Pure torsion of a thin walled bar with a singly closed cross-section

To solve this simple problem, we adopt the following set of assumptions, or hypotheses:

- *an unchangeable-contour hypothesis*, according to which the cross-section of the bar does not change its shape in its plane;
- *a uniform-tangential-stress hypothesis*, which states that tangential stresses τ^{xs} are distributed uniformly over the thickness of the shell;
- *an independent-longitudinal-displacement hypothesis*, which states that displacements u of the profile's points are independent of their longitudinal coordinate x.
- *a no-pressure hypothesis*, according to which the longitudinal fibers of the thin-walled bar do not interact with one another in their normal directions. In other words, we assume $\sigma^s = \sigma^n = 0$.

The unchangeable-contour hypothesis is an exact replica of its counterpart in the theory of open-profile thin-walled bars, and it has exactly the same sense.

It is the matter of course that the hypothesis is true just as long as the thin-walled bar is reinforced throughout its length by a set of sufficiently frequent transversal diaphragms, stiff in their planes and perfectly flexible from their planes.

The uniform-tangential-stress hypothesis states that the stresses τ^{xs} remain the same along the thickness of the shell, that is, they do not vary with the local coordinate n. Also, this hypothesis assumes that there are no tangential stresses other than τ^{xs}, directed tangentially to the profile, in the bar's cross-section. Therefore we can omit the ^{xs} superscript of these stresses and denote $\tau = \tau^{xs}$.

The independent-longitudinal-displacement hypothesis assumes that the displacements $u = u(s)$ of the profile's points are a function of the arc coordinate s only and do not depend on x. The geometrical meaning of this fact is an equal deformation of all cross-sections of the shell bar from their planes. This hypothesis implies the absence of the longitudinal strain, $\varepsilon_x = du/dx = 0$, in pure torsion.

The no-pressure hypothesis and the $\varepsilon_x = 0$ condition imply also the absence of the normal stress in pure torsion, that is, $\sigma^x = 0$.

We will show that the flow of tangential stresses $T = \tau h$ is the same in all points of the profile. To see this, we cut an element of the $dx \times ds$ sizes from the shell bar and check its equilibrium in projections onto X, to obtain immediately that $\partial T/\partial s = 0$. The similar condition of equilibrium of the same element in projection onto the tangential direction, $\boldsymbol{t}$, gives $\partial T/\partial x = 0$. So, the flow of the tangential stress T in pure torsion is a constant.

This constant, which is the only characteristic value of the stress distribution in the system, can be easily determined from the equation of equilibrium in terms of the moments with respect to the X-axis. If we take an arbitrary cross-section of the bar and select an arbitrary point P to be a pole in the plane of that section, then the total torque created by the T flow with respect to the pole will be equal to

$$\oint (\boldsymbol{R}_{\mathrm{PM}} \times T\boldsymbol{t}) \cdot \boldsymbol{i}_x ds = T\oint \boldsymbol{R}_{\mathrm{PM}} \cdot \boldsymbol{n} ds = T\oint \rho ds\,,$$

where ρ is a projection of vector $\boldsymbol{R}_{\mathrm{PM}}$ onto the direction of normal $\boldsymbol{n}$ to the profile of the bar at the current point M.

Parameter ρ, considered as a function of arc coordinate s, will be called an *arm function of pole* P, or just an *arm of pole* P.

It is clear that the integral of the arm function over the whole profile,

$$\oint \rho ds = \Omega, \tag{1.2}$$

has the geometrical meaning of the doubled area bounded by the bar's profile, and it does not depend on a particular location of pole P. Pole P can be chosen, for example, among the points of the profile itself, or even outside the area bounded by the profile. To be particular (although we do not have to), we can place pole P in the center of gravity of the bar's cross-section.

Equaling the moment of internal forces (stresses) from (1.1) to the external torque, M_x, yields immediately

$$T = \frac{M_x}{\Omega}. \tag{1.3}$$

Formula (1.3) is well-known in the science of strength of materials, where it is often called *the Bredt formula*.

Thus, all the components of the stress distribution in the bar are now determined, and we still have used nothing but the equilibrium equations to find them. Hence an important conclusion:

> *The problem of a pure torsion of a singly closed profile belongs to the category of statically determinate problems.*

Now let us determine the kinematical parameters. Of all strain components, the only one different from zero is the shear, γ_{xs}, which can be determined from Hooke's law as

$$\gamma_{xs} = \frac{T}{Ggh}. \tag{1.4}$$

On the other hand,

$$\gamma_{xs} = \frac{\partial u}{\partial s} + \frac{\partial v}{\partial x}, \tag{1.5}$$

where v is the tangential displacement of the current point of the profile, M.

We will use about the same above-mentioned six external constraints as we have used with the Saint-Venant problem. Strictly speaking, we assume that:

- the location of pole P at the beginning and end sections of the bar does not change in the bar crosss-section's plane – this gives us four constraints;
- the slope, θ, of the beginning section of the bar is kept equal to zero – this is the fifth constraint;
- the start point, O, of the beginning section of the bar is restrained from displacements toward X – this is the sixth constraint.

Clearly, by excluding only the rigid displacements of the body with these constraints, we make the axis of poles of all cross-sections of the bar keep its position in the (Y,Z)-plane. As a result, the tangential displacement, v, of the current point, M, of an arbitrary section of the bar can be determined as the product of the section's slope θ by arm ρ of pole P around which the rotation takes place[1]

$$v = \theta\rho, \tag{1.6}$$

wherefrom

---

[1] It should be obvious that we assume the slopes θ to be small (infinitesimal).

$$\frac{\partial v}{\partial x} = \theta' \rho . \tag{1.7}$$

Now, excluding shear γ_{xs} from (1.5) using (1.4), and then using (1.7) yields

$$\frac{du}{ds} = \frac{T}{Ggh} - \theta' \rho . \tag{1.8}$$

Note that we are entitled to change the partial derivative to the ordinary differentiation here, because function u is independent of coordinate x as we said earlier. The integration of (1.8) with respect to the arc coordinate from the start point of the profile, O, to the current point M will give

$$u = u_0 + \frac{T}{G} \int_0^s \frac{ds}{gh} - \theta' \int_0^s \rho ds , \tag{1.9}$$

where u_0 is a longitudinal displacement of the zero point of the profile.

The displacement continuity condition requires that the function $u(s)$ be periodic with a period equal to the perimeter of the profile, l. The mathematical condition of space periodicity is $u(s) = u(s + l)$; in particular, $u_0 = u(l)$. Therefore, by assuming $s = l$ in (1.9) we get this expression for twist θ':

$$\theta' = \frac{T}{G\Omega} \oint \frac{ds}{gh} = \frac{M_x}{G\Omega^2} \oint \frac{ds}{gh} . \tag{1.10}$$

Placing the result in equation (1.9) gives

$$u = u_0 + \frac{M_x}{G\Omega^2} \left(\Omega \frac{\int_0^s \frac{ds}{gh}}{\oint \frac{ds}{gh}} - \omega \right) \oint \frac{ds}{gh} , \tag{1.11}$$

where ω is a so-called *sectorial coordinate* of the current point of the profile, M, which is equal to the doubled area of the sector swept by radius vector $\boldsymbol{R}_{PM}$ as the end of this vector moves from origin O to the current point, M, that is,

$$\omega = \int_0^s \rho ds , \tag{1.12}$$

and

$$\omega(l) = \Omega .$$

It may seem that formula (1.11) includes a still unknown displacement, u_0. However, in our case this displacement has to be equal to zero because this was the way we imposed one of the above said constraints to exclude the rigid displacement of the bar. So, finally,

$$u = \frac{M_x}{G\Omega^2}\left(\Omega\frac{\int_0^s \frac{ds}{gh}}{\oint\frac{ds}{gh}} - \omega\right)\oint\frac{ds}{gh}. \quad (1.13)$$

Note that all values in the right-hand part of formula (1.10) of the twist do not depend on x. Consequently, the twist is a constant in the problem of pure torsion of the singly closed profile. Hence the slope of the bar's section, θ, should be a linear function of coordinate x:

$$\theta = \theta' x .$$

Returning to formula (1.10), we can rewrite it with standard designations from the science of strength of material:

$$\theta' = \frac{M_x}{GI_x}. \quad (1.14)$$

Comparing (1.10) and (1.14) gives immediately an expression of the torsional moment of inertia of the bar's cross-section,

$$I_x = \frac{\Omega^2}{\oint\frac{ds}{gh}}. \quad (1.15)$$

After introducing the above designations, we can rewrite formula (1.13) of longitudinal displacements $u(s)$ as

$$u = \frac{M_x}{GI_x}\alpha(s), \quad (1.16)$$

where

$$\alpha(s) = \frac{I_x}{\Omega}\int_0^s \frac{ds}{gh} - \omega . \quad (1.17)$$

Now we could consider the solution of the problem to be completed but for one circumstance, still unexplained. The matter is that parameter α, which is a factor in (1.16), is a function of the arc coordinate, $\alpha = \alpha(s)$, and is (together with sectorial coordinate ω) dependent on the location on the profile of pole P and of the start point, O, from which to count off the arc coordinate, s. It means that the choice of the pole and start point O affect also the longitudinal displacements, u. But in our pure torsion case this choice means just a certain way of eliminating the rigid displacements of the bar.

We are not going to dwell on the determination of the pole P location here; we will find this out later, when discussing a more general behavior of a thin-walled bar.

Note that function $\alpha(s)$ introduced earlier characterizes the distribution of the warp displacements along the profile of a thin-walled bar in pure torsion. That is why the $\alpha(s)$ function can and should be called a *warp function.*

7.1.2 A general behavior of a closed-profile thin-walled bar

To analyze the general behavior of a thin-walled singly closed bar, we introduce the following assumptions or hypotheses:

- *an unchangeable-contour hypothesis*, which assumes the cross-section of the bar does not change its profile in the section's plane;
- *a longitudinal-moment-free-shell hypothesis*, according to which the distribution of normal stresses σ^x over the thickness of the shell is assumed to be uniform, and tangential stresses τ^{xn} are thought negligible thus equivalent to zero;
- *a no-pressure hypothesis*, which assumes that the longitudinal fibers of the thin-walled bar do not interact in their common normal directions. In other words, it assumes $\sigma^s = \sigma^n = 0$;
- *a uniform-tangential-stress hypothesis*, which assumes tangential stress τ^{xs} to be uniformly distributed over the thickness of the shell;
- *Umanski's hypothesis* of warping in the general analysis of the behavior of a thin-walled bar.

Of all the hypotheses listed above, the new one (and the most important one, too) is Umanski's hypothesis [7], [8]. The main point of it follows.

In the general analysis of a closed-profile thin-walled bar, the longitudinal displacements of the points of the bar's cross-sections consist of displacements obeying the law of plane and of additional warp displacements. Or,

$$u = A + By + Cz + u_w\,, \tag{1.18}$$

where $u_w = u_w(x,s)$ are the longitudinal warp displacements which appear when the cross-sections leave their planes, while contours A, B and C are different for different cross-sections[2]. More accurately, the said values are actually not constants but parameters which are functions of the longitudinal coordinate, x.

The key idea that belongs to A.A. Umanski is simple and elegant. The meaning of Umanski's hypothesis is that the warp component of the longitudinal displacements in the general mechanical behavior of a thin-walled bar for each particular cross-section of it is inherited from the free torsion of the bar. Of course, the inheritance should not be understood as a direct borrowing; instead it means that a pattern of variation of that component in the bar's cross-section plane (that is, in the Y,Z-plane) is inherited. In essence, Umanski's hypothesis assumes that warp displacements u_w are the same as those in free torsion, up to a certain scaling factor.

Turning to formula (1.16), which defines the warping of a singly closed section in free torsion, we rewrite it in application to the general analysis of the thin-walled bar as

$$u_w = \beta(x)\alpha(s)\,, \tag{1.19}$$

where $\beta(x)$ is a function yet to be determined, which we will call a *warp measure function.*

So, according to Umanski's hypothesis, warp displacements u_w can be represented with a sufficient accuracy (approximated, if it is more to one's liking) as the product of the warp function, $\alpha(s)$, by the warp measure function, $\beta(x)$.

We should emphasize once again that the $\alpha(s)$ function is defined completely only when the P pole is chosen and a start point, O, is assigned on the contour of the section to be an origin of the arc coordinate, s. We will return to the definition of those points later, and now we want to notice that the meaning of the warp function, $\alpha(s)$, implies that the pole, P, should be the section's center of twist which, as was proved in Section **6**.2.3, is the same as its center of bending.

Obviously, the first three terms in formula (1.18), which define the longitudinal displacements of the bar's cross-section as a rigid body, can be treated from the standpoint of geometry as a translatory displacement of the section along the X-axis and as displacements of the section's points

---

[2] The subscript of $_w$ corresponds, quite obviously, to *warping*.

caused by the rotation of the section about the *Y* and *Z* axes. Thus, instead of (1.18) we can write

$$u(x,s) = \xi - \theta_z y + \theta_y z + \beta\alpha\,, \tag{1.20}$$

where

- $\xi = \xi(x)$ is a translatory displacement of the section along the *X* axis;
- θ_y and θ_z are slopes of the section with respect to the *Y* and *Z* axes, respectively.

Based on (1.20) and the no-pressure hypothesis, we have this for the normal stresses:

$$\sigma^x = Ee\frac{\partial u}{\partial x} = Ee\left(\xi' - \theta_z' y + \theta_y' z + \beta'\alpha\right). \tag{1.21}$$

The section-integral characteristics of normal stresses σ^x are thus determined as follows:

$$N = \oint \sigma^x h ds = EA\xi' + E\beta'\oint e\alpha h ds\,,$$

$$M_y = \oint z\sigma^x h ds = EI_y\theta_y' + E\beta'\oint e\alpha z h ds\,,$$

$$M_z = \oint y\sigma^x h ds = -EI_z\theta_z' + E\beta'\oint e\alpha y h ds\,. \tag{1.22}$$

To simplify the formulas, we introduce three values, ω_0, ω_y, ω_z, which are constants for the section as a whole:

$$\omega_0 = \frac{1}{A}\oint e\alpha h ds\,, \qquad \omega_y = \frac{1}{I_y}\oint e\alpha z h ds\,, \qquad \omega_z = \frac{1}{I_z}\oint e\alpha y h ds\,. \tag{1.23}$$

This enables us to rewrite (1.22) as

$$\xi' = \frac{N}{EA} - \beta'\omega_0\,, \qquad \theta_y' = \frac{M_y}{EI_y} - \beta'\omega_y\,, \qquad \theta_z' = -\frac{M_z}{EI_z} + \beta'\omega_z\,. \tag{1.24}$$

Substituting these relations to the general formula (1.21) will transform the formula of the normal stresses into

$$\sigma^x = \frac{N}{A}e + \frac{M_z}{I_z}ey + \frac{M_y}{I_y}ez - Ee\beta'\left(\varpi + \omega_z y + \omega_y z\right), \tag{1.25}$$

where

$$\varpi = \omega_0 - \alpha \,. \tag{1.26}$$

The function of the arc coordinate, $\varpi = \varpi(s)$, introduced by formula (1.26) is called a *generalized sectorial coordinate*, and it plays an important part in the whole theory of the closed-profile thin-walled bars. The role that belongs to this function in the closed-profile bar theory is about the same as that of the usual sectorial coordinate, $\omega = \omega(s)$, in the theory of open-profile bars. Note one essential property of the generalized sectorial coordinate $\varpi(s)$, such as:

> *The generalized sectorial coordinate, ϖ, does not depend on a particular origin of the arc coordinate s.*

And indeed, if we expand the expression from (1.26) for ϖ as a function of s, we will see how both $\alpha(s)$ and ω_0 do depend on the location of the arc coordinate's origin separately, but the variations of these functions as the O point is moving along the profile annihilate each other thus keeping the value of $\varpi(s)$ intact.

Also, it follows directly from (1.26) and (1.23) that the generalized sectorial coordinate, ϖ, as a function of s satisfies the condition

$$\oint \varpi ehds = 0 \,. \tag{1.27}$$

We will choose the pole, P, right now as a point with coordinates (y_P, z_P) such that it should meet two extra conditions of orthogonality for the generalized sectorial coordinate,

$$\oint \varpi ehyds = 0 \,, \qquad \oint \varpi ehzds = 0 \,. \tag{1.28}$$

Later we will see that meeting these requirements does indeed make pole P coincident with the center of bending/twist.

It is easy to check that conditions (1.28) imply that the two constants are equal to zero, $\omega_y = 0$ and $\omega_z = 0$. Therefore formula (1.25) becomes as simple as

$$\sigma^x = \frac{N}{A}e + \frac{M_z}{I_z}ey + \frac{M_y}{I_y}ez - Ee\beta'\varpi \,. \tag{1.29}$$

Once again we should call the reader's attention to the fact that the formula is true for any start point of the profile, O, whatsoever.

Finally, it should be noted that the first three terms in formula (1.29) are standard for the science of resistance of materials. They define the normal stresses in a section, caused by the longitudinal displacements of the section according to the law of plane. The last term, one proportional to function β', is a normal stress caused by the warping of the section.

Tangential stresses

Let us extract an element from the thin-walled bar by using two longitudinal cross-sections and two lateral cross-sections, as shown in Fig. 7.2.

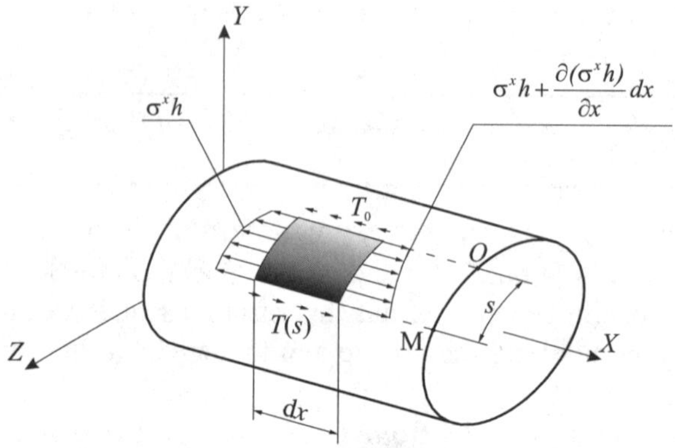

Fig. 7.2. The equilibrium of an extracted element of the bar

One of the longitudinal cross-sections will go through the zero point of the profile, O, and the second longitudinal section will go through the current point of the profile, M, with the arc coordinate of s. The two cross-sections are at the distance dx from each other.

We compose an equation of equilibrium of the extracted element of the bar in terms of projections onto X and thus have

$$\int_0^s \frac{\partial \sigma^x}{\partial x} h ds + T(s) - T_0 + \int_0^s q_x ds = 0 . \tag{1.30}$$

Substituting the expression of σ^x from (1.29) in (1.30) will give

$$T = T_0 - N' \frac{A_o}{A} - M_z' \frac{S_{oz}}{I_z} - M_y' \frac{S_{oy}}{I_y} + E\beta'' S_{o\varpi} - \int_0^s q_x ds , \tag{1.31}$$

where we use the designations of

$$A_o = \int_0^s ehds , \quad S_{oz} = \int_0^s yehds , \quad S_{oy} = \int_0^s zehds , \quad S_{o\varpi} = \int_0^s \varpi ehds . \tag{1.32}$$

Obviously, A_o, S_{oy} and S_{oz} are the material-adjusted area and static moments of the cut-off part of the section, that is, the part between the start point O and the current point of the coordinate s. As for the value of $S_{o\varpi}$, it can be called a *sectorial static moment* of the cut-off part of the section.

Exactly in the same way as we did with the open-profile thin-walled bars and similarly to (**6**.2.36), we introduce the following integral characteristics of the external actions with respect to the cross-section of the bar:

$$p_x(x) = \oint q_x(x,s)ds\,, \quad p_y(x) = \oint q_y(x,s)ds\,, \quad p_z(x) = \oint q_z(x,s)ds\,,$$

$$m_x(x) = \oint \left[q_z(y - y_P) - q_y(z - z_P) \right] ds\,,$$

$$m_y(x) = \oint q_x(x,s) z ds\,, \qquad m_z(x) = -\oint q_x(x,s) y ds\,,$$

$$m_B(x) = \oint q_x(x,s)\varpi ds\,. \tag{1.33}$$

As can be seen, the only difference between formulas (1.33) and (**6**.2.36) is the replacement of sectorial coordinate ω in the expression of the distributed bitorque load, m_B, by the generalized sectorial coordinate ϖ.

Further, it is easy to notice that the equation of equlibrium in projections onto X for a piece of the bar cut off by adjacent cross-sections at the distance of dx between them makes the following equality true:

$$-N' = p(x)\,, \tag{1.34}$$

and the conditions of equilibrium of that piece of the bar in terms of the moments with respect to the Y and Z axes yield

$$Q_y = M_z' - m_z\,, \qquad Q_z = M_y' + m_y\,, \tag{1.35}$$

where Q_y and Q_z are shear forces in the bar's cross-section, and

$$Q_y = \oint \tau t_y h ds\,, \qquad Q_z = \oint \tau t_z h ds\,. \tag{1.36}$$

Just as we did in the theory of open-profile bars, we will represent the longitudinal load, q_x, as an expansion over four degrees of freedom (solely!) which participate in the formation of the longitudinal displacements according to (1.20):

$$q_x(x,s) = a_1(x)\, eh + a_y(x)\, ehy + a_z(x)\, ehz + a_\omega(x)\, eh\alpha\,. \tag{1.37}$$

By multiplying this expansion successively by 1, $y(s)$, $z(s)$, $\varpi(s)$ with subsequent integration over the whole profile, we get

$$p_x = a_1 A + a_\omega \omega_0 A\,, \qquad -m_z = a_y I_z\,, \qquad m_y = a_z I_y\,, \qquad m_B = -a_\omega I_\varpi\,,$$

where I_ϖ is a sectorial moment of inertia of the material-adjusted section,

$$I_\varpi = \oint \varpi^2 ehds\,.$$

Having determined the coefficients in expansion (1.37), we obtain the final expression of the longitudinal load,

$$q_x(x,s) = p_x \frac{eh}{A} - m_z \frac{ehy}{I_z} + m_y \frac{ehz}{I_y} + m_B \frac{eh\varpi}{I_\varpi}\,, \tag{1.38}$$

which should be used for the analysis. Note that the value of ω_0, which the a_1 coefficient depends on directly, does not participate in the final expression of the longitudinal load in (1.38). This should be clear because the expression of q_x should not depend on the choice of the start point, O.

If we integrate the expression of the longitudinal load over the arc coordinate from zero to s, we will have

$$\int_0^s q_x ds = p_x \frac{A_\text{o}}{A} - m_z \frac{S_\text{oz}}{I_z} + m_y \frac{S_\text{oy}}{I_y} + m_B \frac{S_{\text{o}\varpi}}{I_\varpi}\,.$$

By substituting this in (1.31) using (1.34) and (1.35), we transform the expression of the flow of tangential stresses into

$$T = T_0 - Q_y \frac{S_\text{oz}}{I_z} - Q_z \frac{S_\text{oy}}{I_y} + (EI_\varpi \beta'' - m_B)\frac{S_{\text{o}\varpi}}{I_\varpi}\,. \tag{1.39}$$

According to its method of derivation, formula (1.39) should be true for any arbitrary start point O of the profile. However, particular terms in this formula are of course dependent on the choice of O.

To exclude the flow T_0 from (1.39), we use the condition of static equivalence of flow $T(s)$ over the whole bar's section to the torque, M_x, acting with respect to the center of twist, P. Thus we have

$$M_x = \oint T(s)\rho ds\,. \tag{1.40}$$

Substituting in the formula of $T(s)$ from (1.39) gives

$$M_x = T_0\Omega - \frac{Q_y}{I_z}\oint S_\text{oz}\rho ds - \frac{Q_z}{I_y}\oint S_\text{oy}\rho ds + \frac{EI_\varpi\beta'' - m_B}{I_\varpi}\oint S_{\text{o}\varpi}\rho ds\,,$$

so, consequently,

$$T_0 = \frac{M_x}{\Omega} + \frac{Q_y}{\Omega I_z}\oint S_{oz}\rho ds + \frac{Q_z}{\Omega I_y}\oint S_{oy}\rho ds - \frac{EI_{\varpi}\beta'' - m_B}{\Omega I_{\varpi}}\oint S_{o\varpi}\rho ds . \quad (1.41)$$

The final result is obtained by excluding the initial flow of T_0 from (1.39) using (1.41):

$$T = \frac{M_x}{\Omega} - \frac{Q_y}{I_z}\bar{S}_{oz} - \frac{Q_z}{I_y}\bar{S}_{oy} + \frac{EI_{\varpi}\beta'' - m_B}{I_{\varpi}}\bar{S}_{o\varpi} , \quad (1.42)$$

where we use the designations of

$$\bar{S}_{oz} = S_{oz} - \frac{1}{\Omega}\oint S_{oz}\rho ds , \qquad \bar{S}_{oy} = S_{oy} - \frac{1}{\Omega}\oint S_{oy}\rho ds ,$$

$$\bar{S}_{o\varpi} = S_{o\varpi} - \frac{1}{\Omega}\oint S_{o\varpi}\rho ds . \quad (1.43)$$

Properties of functions $\bar{S}_{oz}$, $\bar{S}_{oy}$, $\bar{S}_{o\varpi}$

The functions just introduced possess peculiar properties which we are going to prove right now and use when necessary in our subsequent presentment. The following five properties are worth noticing.

- **Property 1**:
  This is a pretty important feature of these functions.

  > *Neither function $\bar{S}_{oz}$, $\bar{S}_{oy}$, $\bar{S}_{o\varpi}$ depends on the choice of an origin for the arccoordinate s; each is an unambiguous characteristic of any point belonging to the profile.*

  This statement can be validated by directly using (1.43) and (1.32).

- **Property 2**:
  It follows immediately from the definition of (1.43) that

$$\oint \bar{S}_{oz}\rho ds = 0 , \qquad \oint \bar{S}_{oy}\rho ds = 0 , \qquad \oint \bar{S}_{o\varpi}\rho ds = 0 . \quad (1.44)$$

What the first two equalities in (1.44), if we take (1.42) into account, state is that tangential stresses τ_Q, i.e. tangential stresses caused by the shear forces and nothing else, do not create a torque with respect to a pole that satisfies conditions (1.28). This means we used such a condition for choosing a pole, when we were deriving the formula, that the pole then became the center of bending, and thus the center of twist, too.

- **Property 3**:

We show that the functions obey the following integration formulas with respect to the whole profile:

$$\oint \bar{S}_{oy} t_y ds = 0\,, \qquad \oint \bar{S}_{oz} t_y ds = -I_z\,, \qquad \oint \bar{S}_{o\varpi} t_y ds = 0\,,$$

$$\oint \bar{S}_{oy} t_z ds = -I_y\,, \qquad \oint \bar{S}_{oz} t_z ds = 0\,, \qquad \oint \bar{S}_{o\varpi} t_z ds = 0\,. \tag{1.45}$$

To prove those, we first notice that

$$\oint t_y ds = \oint \frac{dy}{ds} ds = 0\,, \qquad \oint t_z ds = \oint \frac{dz}{ds} ds = 0\,.$$

But then (1.43) and (1.32) yield

$$\oint \bar{S}_{oy} t_y ds = \oint S_{oy} t_y ds = \oint \left(\int_0^s zehds \right) t_y ds = \left[y \int_0^s zehds \right]_{s=0}^{s=l} - \oint yzehds = 0\,,$$

$$\oint \bar{S}_{oz} t_y ds = \oint S_{oz} t_y ds = \oint \left(\int_0^s yehds \right) t_y ds = \left[y \int_0^s yehds \right]_{s=0}^{s=l} - \oint y^2 ehds = -I_z\,,$$

$$\oint \bar{S}_{o\varpi} t_y ds = \oint S_{o\varpi} t_y ds = \oint \left(\int_0^s \varpi ehds \right) t_y ds = \left[y \int_0^s \varpi ehds \right]_{s=0}^{s=l} - \oint y\varpi ehds = 0.$$

The transformations presented here are based on the integration by parts, on the space periodicity $y(0) = y(l)$, and on the earlier formulas (1.27) и (1.28). The other three formulas in (1.45) are proved in a similar way.

- **Property 4**:

The following formulas hold for the integration over the section's profile:

$$\oint \frac{d\varpi}{ds} ds = 0\,,$$

$$\oint \bar{S}_{oy} \frac{d\varpi}{ds} ds = 0\,, \qquad \oint \bar{S}_{oz} \frac{d\varpi}{ds} ds = 0\,, \qquad \oint \bar{S}_{o\varpi} \frac{d\varpi}{ds} ds = -I_\varpi\,. \tag{1.46}$$

The first of the formulas in (1.46) is obvious because there is a periodicity in the generalized sectorial coordinate ϖ, i.e. $\varpi(0) = \varpi(l)$. Further, we use the integration by parts to obtain

$$\oint \bar{S}_{oy} \frac{d\varpi}{ds} ds = \left[\bar{S}_{oy} \varpi \right]_{s=0}^{s=l} - \oint \frac{d\bar{S}_{oy}}{ds} \varpi ds = -\oint \frac{dS_{oy}}{ds} \varpi ds = -\oint zeh\varpi ds = 0\,.$$

Here we have used (1.43) and (1.28). The third formula in (1.46) is proved in the same way. So we finally have

$$\oint \bar{S}_{o\varpi} \frac{d\varpi}{ds} ds = \left[\bar{S}_{o\varpi} \varpi \right] \Big|_{s=0}^{s=l} - \oint \frac{d\bar{S}_{o\varpi}}{ds} \varpi ds = -\oint \varpi eh \varpi ds = -I_{\varpi} ,$$

because (1.43) and (1.32) yield

$$\frac{d\bar{S}_{o\varpi}}{ds} = \frac{dS_{o\varpi}}{ds} = \varpi eh .$$

Thus all the formulas in (1.46) have been proved.

- **Property 5**:

It is useful to notice, to end this consideration, that the functions in question obey also the following integration formulas with respect to the section's profile:

$$\oint \frac{\bar{S}_{oy}}{gh} ds = 0 , \qquad \oint \frac{\bar{S}_{oz}}{gh} ds = 0 , \qquad \oint \frac{\bar{S}_{o\varpi}}{gh} ds = I_{\varpi} \frac{\Omega}{I_x} . \tag{1.47}$$

To see this, we take (1.17) and (1.26) to write

$$\frac{d\varpi}{ds} = -\frac{I_x}{\Omega gh} + \rho .$$

By multiplying this equality by $\bar{S}_{oy}$, integrating the result over the whole closed profile, and taking account of the already proved relationships (1.44) and (1.46), we establish that the first relation in (1.47) is true. The same technique is used with the second formula in (1.47). Applying this procedure to function $\bar{S}_{o\varpi}$, we arrive at the third formula in (1.47).

To conclude this section, we would like to note that multiplying both parts in formula (1.42) successively by t_y and t_z and integrating over the whole closed contour will yield, taking (1.45) and (1.46) into account:

$$Q_y = \oint T t_y ds , \qquad Q_z = \oint T t_z ds ,$$

which is, as expected, in full accord with (1.36).

7.1.3 General physical relationships in the theory of closed-profile thin-walled bars

Up to this point, we have derived formulas (1.29) for the normal stresses and formulas (1.42) for the tangential stresses.

We begin with making these formulas similar to (**6**.3.1) and (**6**.3.2), the latter being used in the open-profile bar theory. In other words, we assume

$$\sigma^x = \frac{N}{A}e + \frac{M_y}{I_y}ez + \frac{M_z}{I_z}ey + \frac{B}{I_\varpi}e\varpi\,, \tag{1.48}$$

$$\tau = \frac{M_x}{h\Omega} - Q_z\frac{\bar{S}_{oy}}{hI_y} - Q_y\frac{\bar{S}_{oz}}{hI_z} - M_\varpi\frac{\bar{S}_{o\varpi}}{hI_\varpi}\,, \tag{1.49}$$

where bitorque B and the constricted-torsion moment M_ϖ are related to the warp measure β as

$$B = -EI_\varpi\beta'\,, \qquad M_\varpi = -EI_\varpi\beta'' + m_B\,, \tag{1.50}$$

which relations follow from the comparison between (1.29) and (1.42), on one hand, and (1.48) and (1.49), on the other hand. By the way, formulas (1.50) imply also a relation between B and M_ϖ as follows:

$$M_\varpi = B' + m_B\,, \tag{1.51}$$

which is formally the same as (**6**.2.44) in the open-profile bar theory.

As for bitorque B, the first of formulas (1.50) defines its relation to the warp measure. However, the same value can be treated also as a section-integral static characteristic of the normal stresses. To see this, we multiply (1.48) by ϖ and integrate the result over the whole section, to obtain

$$B = \oint \sigma^x \varpi h ds\,. \tag{1.52}$$

And again we find a full accord with (**6**.2.24) for the open-profile bars, with the only exception that the sectorial coordinate, ω, is replaced by the generalized sectorial coordinate, ϖ. Formula (1.52) permits to introduce [4] a purely static definition of bitorque B, totally independent of all kinematic parameters, such as:

Bitorque B is a moment of the normal stresses, σ^x*, on the "arm" of* ϖ.

We can also establish a purely static characteristic for the constricted-torsion torque. To do this, we multiply both parts of formula (1.49) of the tangential stresses by $d\varpi/ds$ and integrate the result over the whole section.

Using the properties of functions $\overline{S}_{oz}$, $\overline{S}_{oy}$, $\overline{S}_{o\varpi}$ proved earlier, we obtain this by means of the integration:

$$M_\varpi = \oint \tau h \frac{d\varpi}{ds} ds \,. \tag{1.53}$$

Hence a purely static definition of the constricted-torsion torque, namely:

> *The constricted-torsion torque, M_ϖ, is a moment of the tangential stresses, τ, on the "arm" of $d\varpi/ds$.*

Note that the two definitions of the internal stress factors in the cross-sections of the thin-walled bars are true not only for the closed profiles but for the open profiles, too. And indeed, the static definition of bitorque B for an open-profile section is a verbal description of (**6**.2.24). On the other hand, for the open profiles we use the conventional sectorial coordinate, ω, instead of the generalized one, ϖ, and (**6**.2.12) gives $d\omega/ds = \rho$. Now, using formula (**6**.2.37), we obtain the above-cited verbal definition of the constricted-torsion torque.

For the closed profile, the derivative $d\varpi/ds$ can be written as follows on the basis of (1.26) and (1.17):

$$\frac{d\varpi}{ds} = -\frac{d\alpha}{ds} = \rho - \frac{I_x}{gh\Omega} \,.$$

Hence, taking account of (1.53) and (1.40), we have

$$M_\varpi = M_x - \frac{I_x}{\Omega} \oint \frac{\tau}{g} ds \,. \tag{1.54}$$

The relation between the components of the stress state in the bar, N, M_y, M_z, and the displacements of the section can be restored from (1.24) where we should assume $\omega_y = \omega_z = 0$ because we use the center of bending as the pole P. Also, conditions (1.28) define the location of the main pole, P, but those conditions are, together with the requirement of (1.27), invariant with respect to the location of the profile's zero point. That's why we are able to choose the zero point, O, to our liking. In particular, it is convenient to place it in such way that constant ω_o defined by the first formula in (1.23) should become zero, i.e. $\omega_o = 0$. So,

$$N = EA\xi' \,, \qquad M_y = EI_y\theta'_y \,, \qquad M_z = -EI_z\theta'_z \,. \tag{1.55}$$

Up to this point we did not establish any relation between the slopes of the section, θ_y and θ_z, and the η and ζ functions, and we did not do that intentionally. The geometrical meaning of functions η and ζ is the same as that in the open-profile thin-walled bar theory: these functions are displacements of the center of bending of the bar's cross-section in the respective directions of Y and Z. We remind that the whole section of the bar is treated as a perfectly rigid body with respect to displacements in the (Y,Z)-plane (this is what the unchangeable-contour hypothesis assumes).

This is where the theory bifurcates. Taking one of the approaches, we can think for the sake of generality that slopes θ_y and θ_z of the section, which we used in the original formula (1.20) and further, are independent kinematical variables not identified with the derivatives of the displacement functions, η and ζ. To put it another way, we can assume for the general case that

$$\theta_y \neq -\zeta' , \qquad \theta_z \neq \eta' . \tag{1.56-a}$$

The assumption that slopes θ_y and θ_z from (1.56) are fully independent of displacements η and ζ originates a full shear theory of closed-profile thin-walled bars.

There can be the second way to construct the theory, when slopes θ_y and θ_z are identified with the derivatives of functions η and ζ. In that case (1.56-*a*) is replaced by the following convention:

$$\theta_y \equiv -\zeta' , \qquad \theta_z \equiv \eta' . \tag{1.56-b}$$

We can predict beforehand that excluding the θ_y and θ_z slopes from the list of the independent kinematical parameters and adopting the convention of (1.56-*b*) will correspond to the semi-shear version of the theory.

However, even before dividing the theory into the shear and semi-shear versions, it is useful to introduce, at least formally so far, a moment (torque) H as a difference between the full torque, M_x, and the constricted-torsion torque, M_{ϖ},

$$H = M_x - M_{\varpi} . \tag{1.57}$$

We will see a bit later that this value can be treated as a pure-torsion torque, hence the designation.

Comparing (1.57) with (1.54) yields an expression of the pure-torsion torque, H, as an integral characteristics of the tangential stresses:

$$H = \frac{I_x}{\Omega} \oint \frac{\tau}{g} ds . \tag{1.58}$$

Thus, from (1.57) we have a statement that

the full torque, M_x, is a sum of the pure-torsion torque, H, and the constricted-torsion torque, M_ϖ .

Let us first pay some attention to the formulation of physical relationships in the shear theory of closed-profile thin-walled bars.

7.1.4 First (energy-based) version of the theory

The expression of the strain energy, E, represented as a quadratic functional of the stresses, can be written in the form of

$$\mathsf{E}(\sigma^x, \tau) = \mathsf{E}_\sigma + \mathsf{E}_\tau ,$$

where

$$\mathsf{E}_\sigma = \int_0^L \oint \frac{(\sigma^x)^2}{2Ee} h ds dx , \qquad \mathsf{E}_\tau = \int_0^L \oint \frac{\tau^2}{2Gg} h ds dx . \tag{1.59}$$

The first of the integrals up there conforms to an energy of the normal stresses while the second describes an energy stored by the tangential stresses. We express the normal and tangential stresses in the integrands via the components of stress vector $\boldsymbol{\sigma}$, i.e. by using formulas (1.48) and (1.49). Substituting (1.48) and (1.49) in (1.59) and taking note of (1.57) gives

$$\mathsf{E}_\sigma = \int_0^L \left[\frac{N^2}{2EA} + \frac{M_y^2}{2EI_y} + \frac{M_z^2}{2EI_z} + \frac{B^2}{2EI_\varpi} \right] dx ,$$

$$\mathsf{E}_\tau = \int_0^L \left[\oint \left(\frac{H + M_\varpi}{h\Omega} - Q_z \frac{\bar{S}_{oy}}{hI_y} - Q_y \frac{\bar{S}_{oz}}{hI_z} - M_\varpi \frac{\bar{S}_{o\varpi}}{hI_\varpi} \right)^2 \frac{h}{2Gg} ds \right] dx . \tag{1.60}$$

We transform the expression of energy E_τ in this way:

$$\mathsf{E}_\tau = \frac{1}{2G} \int_0^L \left[\frac{H^2}{\Omega^2} \oint \frac{ds}{gh} + \frac{Q_z^2}{I_y^2} \oint \frac{\bar{S}_{oy}^2}{gh} ds + \frac{Q_y^2}{I_z^2} \oint \frac{\bar{S}_{oz}^2}{gh} ds + \frac{M_\varpi^2}{I_\varpi^2} \oint \left(\bar{S}_{o\varpi} - \frac{I_\varpi}{\Omega} \right)^2 \frac{1}{gh} ds \right] dx +$$

$$+ \frac{1}{G} \int_0^L \left[-\frac{HQ_z}{\Omega I_y} \oint \frac{\bar{S}_{oy} ds}{gh} - \frac{HQ_y}{\Omega I_z} \oint \frac{\bar{S}_{oz} ds}{gh} - \frac{HM_\varpi}{\Omega I_\varpi} \oint \left(\bar{S}_{o\varpi} - \frac{I_\varpi}{\Omega} \right) \frac{ds}{gh} \right] dx +$$

$$+ \frac{1}{G} \int_0^L \frac{Q_z Q_y}{I_y I_z} \oint \frac{\bar{S}_{oz} \bar{S}_{oy} ds}{gh} dx +$$

$$+\frac{1}{G}\int_0^L\left[\frac{Q_z M_\varpi}{I_y I_\varpi}\oint\bar{S}_{oy}\left(\bar{S}_{o\varpi}-\frac{I_\varpi}{\Omega}\right)\frac{ds}{gh}+\frac{Q_y M_\varpi}{I_z I_\varpi}\oint\bar{S}_{oz}\left(\bar{S}_{o\varpi}-\frac{I_\varpi}{\Omega}\right)\frac{ds}{gh}\right]dx\,. \tag{1.61}$$

This is where we make use of property 5 of functions $\bar{S}_{oy}, \bar{S}_{oz}, \bar{S}_{o\varpi}$, which we established earlier. We take formulas (1.47) and use (1.15) to make sure that the second of the integrals in (1.61) vanishes.

It is appropriate at this point to introduce another designation for another geometric characteristic of the section,

$$\bar{\bar{S}}_{o\varpi}=\bar{S}_{o\varpi}-\frac{I_\varpi}{\Omega}=S_{o\varpi}-\frac{1}{\Omega}\oint S_{o\varpi}\rho ds-\frac{I_\varpi}{\Omega}\,. \tag{1.62}$$

Now, just as we did in the theory of open-profile thin-walled bars where shear was taken into account, we introduce a *matrix of shape factors of the section*, $\boldsymbol{\mu}$, by defining

$$\boldsymbol{\mu}=\left\|\begin{bmatrix}\mu_{zz} & \mu_{zy} & \mu_{z\varpi}\\ \mu_{yz} & \mu_{yy} & \mu_{y\varpi}\\ \mu_{\varpi z} & \mu_{\varpi y} & \mu_{\varpi\varpi}\end{bmatrix}\right\|=$$

$$=\left\|\begin{bmatrix}\frac{A}{I_y^2}\oint\frac{\bar{S}_{oy}^2}{gh}ds & \frac{A}{I_y I_z}\oint\frac{\bar{S}_{oy}\bar{S}_{oz}}{gh}ds & \frac{\sqrt{I_r A}}{I_y I_\varpi}\oint\frac{\bar{S}_{oy}\bar{\bar{S}}_{o\varpi}}{gh}ds\\ \frac{A}{I_z I_y}\oint\frac{\bar{S}_{oz}\bar{S}_{oy}}{gh}ds & \frac{A}{I_z^2}\oint\frac{\bar{S}_{oz}^2}{gh}ds & \frac{\sqrt{I_r A}}{I_z I_\varpi}\oint\frac{\bar{S}_{oz}\bar{\bar{S}}_{o\varpi}}{gh}ds\\ \frac{\sqrt{I_r A}}{I_\varpi I_y}\oint\frac{\bar{\bar{S}}_{o\varpi}\bar{S}_{oy}}{gh}ds & \frac{\sqrt{I_r A}}{I_\varpi I_z}\oint\frac{\bar{\bar{S}}_{o\varpi}\bar{S}_{oz}}{gh}ds & \frac{I_r}{I_\varpi^2}\oint\frac{\bar{\bar{S}}_{o\varpi}^2}{gh}ds\end{bmatrix}\right\|. \tag{1.63}$$

In these designations, formula (1.61) of energy E_τ becomes

$$\mathsf{E}_\tau=\int_0^L\left[\frac{H^2}{2GI_x}+\frac{Q_z^2\mu_{zz}}{2GA}+\frac{Q_y^2\mu_{yy}}{2GA}+\frac{M_\varpi^2\mu_{\varpi\varpi}}{2r^2GA}+\frac{Q_z Q_z\mu_{zy}}{GA}+\frac{Q_z M_\varpi\mu_{z\varpi}}{rGA}+\frac{Q_y M_\varpi\mu_{y\varpi}}{rGA}\right]dx\,.$$

Here I_r and r are the respective polar moment of inertia of the section and the polar radius of inertia. Summing E_τ and E_σ and switching to the matrix representation, we have the final relation, $\mathsf{E}(\boldsymbol{\sigma})=½\,\boldsymbol{\sigma}^{\mathrm{T}}C^{-1}\boldsymbol{\sigma}$, where the vector of stresses $\boldsymbol{\sigma}$ is

$$\boldsymbol{\sigma} = \left\| \left[N, H, M_y, M_z, B, Q_z, Q_y, M_\varpi \right] \right\|^{\mathrm{T}}. \tag{1.64}$$

It is easy to notice that this type of stress vector $\boldsymbol{\sigma}$ implies the following matrix $\boldsymbol{C}^{-1}$:

$$\boldsymbol{C}^{-1} = \left\| \begin{matrix} \frac{1}{EA} & 0 & 0 & 0 & 0 & 0 & 0 & 0 \\ 0 & \frac{1}{GI_x} & 0 & 0 & 0 & 0 & 0 & 0 \\ 0 & 0 & \frac{1}{EI_y} & 0 & 0 & 0 & 0 & 0 \\ 0 & 0 & 0 & \frac{1}{EI_z} & 0 & 0 & 0 & 0 \\ 0 & 0 & 0 & 0 & \frac{1}{EI_\varpi} & 0 & 0 & 0 \\ 0 & 0 & 0 & 0 & 0 & \frac{\mu_{zz}}{GA} & \frac{\mu_{zy}}{GA} & \frac{\mu_{z\varpi}}{rGA} \\ 0 & 0 & 0 & 0 & 0 & \frac{\mu_{yz}}{GA} & \frac{\mu_{yy}}{GA} & \frac{\mu_{y\varpi}}{rGA} \\ 0 & 0 & 0 & 0 & 0 & \frac{\mu_{\varpi z}}{rGA} & \frac{\mu_{\varpi y}}{rGA} & \frac{\mu_{\varpi\varpi}}{r^2GA} \end{matrix} \right\|. \tag{1.65}$$

An important point here is that the components of stress vector $\boldsymbol{\sigma}$ from (1.64) are exactly the same as the components of the same vector in the shear theory of open-profile bars – see formula (**6**.3.8).

Physical relationships in the first (energy-based) version of the semi-shear theory

Recalling what we did in the semi-shear theory of open-profile thin-walled bars, we decompose tangential stresses τ into a sum of the tangential stresses of bending, τ_Q, and the tangential stresses of torsion, τ_M. According to (1.49), we have

$$\tau = \tau_M + \tau_Q, \quad \tau_M = \frac{M_x}{h\Omega} - M_\varpi \frac{\bar{S}_{o\varpi}}{hI_\varpi}, \quad \tau_Q = -Q_z \frac{\bar{S}_{oy}}{hI_y} - Q_y \frac{\bar{S}_{oz}}{hI_z}. \tag{1.66}$$

Categorizing the tangential stresses of bending, τ_Q, into the secondary class, but keeping the energy contribution of stresses τ_M in the expression of energy E_τ, we have

$$\mathsf{E}_\tau = \int_0^L \left[\oint \left(\frac{H + M_\varpi}{h\Omega} - M_\varpi \frac{\overline{S}_{o\varpi}}{hI_\varpi} \right)^2 \frac{h}{2Gg} ds \right] dx .$$

After being integrated over the bar's section, this formula becomes

$$\mathsf{E}_\tau = \int_0^L \left[\frac{H^2}{2GI_x} + \frac{M_\varpi^2 \mu_{\varpi\varpi}}{2r^2 GA} \right] dx . \tag{1.67}$$

For the semi-shear version of the theory, we should deprive stress vector $\boldsymbol{\sigma}$ taken from (1.64) of the shear forces, thus turning it into

$$\boldsymbol{\sigma} = \left\| \left[N, H, M_y, M_z, B, M_\varpi \right] \right\|^{\mathsf{T}} , \tag{1.68}$$

which has the same form as stress vector $\boldsymbol{\sigma}$ from (**6**.4.3) used in the semi-shear theory of open-profile bars.

We present also equilibrium equations for a thin-walled bar, the expanded form of which is

$$-N' = p_x , \qquad -Q_y' = p_y , \qquad -Q_z' = p_z ,$$

$$-H' - M_\varpi' = m_x , \qquad -M_y' + Q_z = m_y , \qquad M_z' - Q_y = m_z ,$$

$$-B' + M_\varpi = m_B . \tag{1.69}$$

The physical meaning of the above equations is obvious:

- the first three equations are differential equations in terms of projections onto the X, Y, Z axes;
- the fourth to sixth equations are differential equations in terms of moments with respect to the X, Y, Z axes;
- as for the last equation in (1.69), it is a generalized equation of equilibrium that relates to a bitorque state of the bar. This equation was established earlier by (1.51).

If now we sum energies E_τ and E_σ and switch to the matrix form, we will have $\mathsf{E}(\boldsymbol{\sigma}) = ½\,\boldsymbol{\sigma}^{\mathsf{T}} \boldsymbol{C}^{-1} \boldsymbol{\sigma}$ with the diagonal matrix $\boldsymbol{C}^{-1}$ as follows:

$$C^{-1} = \begin{Vmatrix} 1/EA & 0 & 0 & 0 & 0 & 0 \\ 0 & 1/GI_x & 0 & 0 & 0 & 0 \\ 0 & 0 & 1/EI_y & 0 & 0 & 0 \\ 0 & 0 & 0 & 1/EI_z & 0 & 0 \\ 0 & 0 & 0 & 0 & 1/EI_{\varpi} & 0 \\ 0 & 0 & 0 & 0 & 0 & \mu_{\varpi\varpi}/r^2GA \end{Vmatrix}. \qquad (1.70)$$

Governing equations of the first version of the semi-shear theory

Remaning within the boundaries of the shear theory, we should keep all equations of equilibrium in (1.69) when building the operator of equilibrium, A^{T}, because (1.64) dictates so. In the matrix form, this produces operator A^{T} already established by (**6**.3.12) for the shear theory of the open-profile bars.

As it follows immediately, the respective operators of geometry, A, are coincident, too, because they are Lagrange-conjugate to the equilibrium operator.

The coincidence also takes place for load vectors $\bar{X}$, displacements $\mathbf{u}$ and strains $\boldsymbol{\varepsilon}$, for which formulas (**6**.3.11), (**6**.3.20), (**6**.3.21) hold as before.

Finally, the matrices of physical relationships, C^{-1}, of the two theories are also equivalent in their form of representation: (1.65) for the closed-profile bars and (**6**.3.9) for the open profiles.

Thus, the only difference in the equations of the shear theories of the closed and open bar profiles is solely in the calculation of the geometrical characteristics of the cross-sections. The two theories are totally equivalent in all the other aspects. Hence a conclusion: the final governing equations (**6**.3.24) are applicable to the closed-profile thin-walled bars. One should remember, of course, that the sectorial coordinate ω must be replaced everywhere, including the subscripts, by the generalized sectorial coordinate ϖ. It should be also obvious that matrix $\mathbf{v}$ is an inverse of matrix $\boldsymbol{\mu}$. The components of this matrix for the closed-profile bars are calculated by formulas (1.63).

It goes without saying that the respective functionals in the two theories have the same form of representation, too.

Now let us return to the interpretation of torque H as a pure-torsion torque which we promised to give earlier. We use the physical relationship of $\boldsymbol{\sigma} = \boldsymbol{CA}\mathbf{u}$. In particular, it will permit us to establish an expression of torque H via the displacements or, after making substitutions,

$$H = GI_x\theta' , \tag{1.71}$$

wherefrom we have the promised treatment of torque H as a pure-torsion torque.

Comparing this result with (1.58) gives

$$\oint \frac{\tau}{g} ds = G\Omega\theta' .$$

Recalling the Bredt theorem of the tangential stress circulation (see Section **6**.1.2) derived for bars of solid sections and made of a homogeneous material (that is, at $g = 1$), we can see that this theorem is applicable to the closed-profile thin-walled bars, too.

A semi-shear theory by Janelidze–Panovko

Reasoning similarly to the previous section, we find out that the semi-shear theories of closed- and open-profile thin-walled bars are identical in the mathematical sense, too.

Therefore all basic formulas of Section **6**.4 are applicable (with the stipulations of the previous section) also to the semi-shear theory of closed-profile thin-walled bars. We present here a governing differential equation (with respect to the warp measure, β) for the constricted-torsion analysis in the semi-shear theory of closed-profile thin-walled bars. Referring to (**6**.4.14) and (**6**.4.13), we rewrite those formulas here in application to bars which have closed profiles of their cross-sections:

$$\boxed{\psi EI_{\varpi}\beta''' - GI_x\beta' = m_x + \psi m_B' ,} \tag{1.72}$$

where

$$\psi = 1 + \frac{\mu_{\varpi\varpi} I_x}{I_r} .$$

It should be said that the theory suggested and developed by A. Umanski relates, strictly speaking, exclusively to the semi-shear theory, though this fact is never indicated explicitly or discussed in either the papers by A. Umanski himself or subsequent works of other people where this theory is presented.

The matter is that both Umanski himself and his followers make an implicit assumption when they begin to construct their theory: they assume the bending analysis and the torsion analysis of a thin-walled bar to be separated in the mathematical sense. But, as we just discussed, this separation takes place only in the semi-shear version of the theory. The key point that makes the separation of the two analyses possible is the diagonal structure of the $\boldsymbol{C}$ elasticity matrix.

Moreover, the original theory by A.A. Umanski is slightly different also from the energy-based version of the semi-shear theory presented above. Actually, equation (1.72) is exactly coincident (up to the notation) not with Umanski's equation but with a similar governing equation derived for the first time by Janelidze and Panovko (see equation (29), page 116 of [4]). This is not just a chance; the authors use similar assumptions in their book [4] and also base their consideration on an energy-related reasoning[3]. We believe what we said is a sufficiently good reason to name the above-stated semi-shear theory of closed-profile thin-walled bars a *Janelidze–Panovko theory*.

7.1.5 Second (physical) version of the theory

We will start from Hooke's law for tangential stresses τ, which states this for the general case:

$$\tau = Gg\left(\frac{\partial u}{\partial s} + \frac{\partial v}{\partial x}\right).$$

In the notation that we use, the vector $\mathbf{v}_P = \eta \boldsymbol{i}_y + \zeta \boldsymbol{i}_z$ describes a full displacement of the center of bending (point P) in the (Y,Z)-plane. But, according to the kinematical hypotheses that we use, the tangential displacement v of an arbitrary point of the profile consists of a projection of vector $\mathbf{v}_P$ onto the direction of the tangent to the profile at the point in question and of a tangential displacement caused by the rotation of the profile around point P as a pole. Or, to put it another way,

$$v = \mathbf{v}_P \cdot \boldsymbol{t} + \theta\rho = \eta t_y + \zeta t_z + \theta\rho, \tag{1.73}$$

wherefrom

[3] Janelidze and Panovko [4] derive the differential equation (1.72) using directly the Castigliano variational principle.

$$\frac{\partial v}{\partial x} = \eta' t_y + \zeta' t_z + \theta' \rho .$$

At the same time, according to (1.20)

$$\frac{\partial u}{\partial s} = -\theta_z t_y + \theta_y t_z + \beta \frac{d\alpha}{ds} .$$

Therefore the shear strain, $\gamma_{xs} = \partial u/\partial s + \partial v/\partial x$, becomes equal to

$$\gamma_{xs} = (\eta' - \theta_z) t_y + (\zeta' + \theta_y) t_z + \left(\theta' \rho + \beta \frac{d\alpha}{ds} \right) .$$

According to (1.17) and (1.12)

$$\frac{d\alpha}{ds} = \frac{I_x}{gh\Omega} - \rho ,$$

so, finally,

$$\gamma_{xs} = \gamma_y\, t_y + \gamma_z\, t_z + \gamma_\varpi \rho, \tag{1.74}$$

where

$$\gamma_y = \eta' - \theta_z , \qquad \gamma_z = \zeta' + \theta_y , \qquad \gamma_\varpi = (\theta' - \beta) + \beta \frac{I_x}{\rho g h \Omega} . \tag{1.75}$$

As we can see, formula (1.74) for the total shear γ_{xs} is exactly equivalent to (**6**.3.19) in its structure, the latter being used in the theory of open-profile thin-walled bars where shear is allowed for. Furthermore, the first two formulas in (1.75) coincide with the first two in (**6**.3.18). As for the formulas for the shear parameter γ_ϖ, there is a significant difference from the respective formula of γ_ω. The difference is in the additional term $(\beta I_x)/(\rho g h \Omega)$ which participates in the expression of the shear parameter γ_ϖ. We have

$$\gamma_\varpi = \gamma_\omega + \beta \frac{I_x}{\rho g h \Omega} .$$

It should be apparent because the closed section supplements the warp-caused shear strain, $\gamma_\omega \rho$, with the pure-torsion-caused shear strain equal to $(\beta I_x)/(g h \Omega)$.

If we equal shears γ_{xs} defined by (1.74) and by (1.49), we will have

$$(\eta' - \theta_z)t_y + (\zeta' + \theta_y)t_z + (\theta' - \beta)\rho + \beta\frac{I_x}{gh\Omega} =$$

$$= \frac{M_x}{Ggh\Omega} - Q_z\frac{\overline{S}_{oy}}{GghI_y} - Q_y\frac{\overline{S}_{oz}}{GghI_z} - M_\varpi\frac{\overline{S}_{o\varpi}}{GghI_\varpi}. \qquad (1.76)$$

However, as we have said, the physical law in the form of (1.76) cannot be thought, actually, to hold in every point of the bar's section. There is no reason to believe that the two expressions of the tangential stress (one following from Hooke's law and the other from the equations of equilibrium) will produce the same result. All we can demand is that the equality (1.76) should be satisfied at least in the integral, if not pointwise, sense. As we will see further, this way of reasoning leads to the original theory by Umanski (of course, within the boundaries of the semi-shear theory), although the approach of A.A. Umanski himself, presented also in [4], is different from what we use here.

For further presentment we will need some additional geometrical characteristics of the section. These characteristics were endenizened in the theory of thin-walled bars by R.A. Adadurov [2], [1] and used with the same notation by G.Y. Janelidze and Y.G. Panovko [4]. We present those characteristics here in their generalized form, for a heterogeneous material. So, we assume

$$I_\rho = \oint \rho^2 ghds, \qquad K_y = \oint t_z \rho ghds, \qquad K_z = \oint t_y \rho ghds,$$

$$L_{yz} = L_{zy} = \oint t_y t_z ghds, \quad L_y = \oint t_z^2 ghds, \quad L_z = \oint t_y^2 ghds. \qquad (1.77)$$

Note that the value of I_ρ is called, following A.A. Umanski, a *directed moment of inertia of the section.*

For the physical law of (1.76) to hold in average, we multiply this equality by $gh\rho$ and integrate the result over the profile. Hence a physical relationship, integral for the section, between torque M_x and the displacement components:

$$M_x = GK_z(\eta' - \theta_z) + GK_y(\zeta' + \theta_y) + GI_\rho(\theta' - \beta) + GI_x\beta. \qquad (1.78)$$

It is also useful to introduce a dimensionless value, μ, which is usually referred to as a *warp factor*,

$$\mu = 1 - \frac{I_x}{I_\rho}, \qquad (1.79)$$

and which is also one of the section's geometric characteristics.

In the designations thus introduced, the physical relationship (1.78) can be rewritten as

$$M_x = GK_z(\eta' - \theta_z) + GK_y(\zeta' + \theta_y) + GI_\rho(\theta' - \mu\beta) . \tag{1.80}$$

The following two integral relations of a physical nature, which follow from (1.76), are derived by multiplying both parts of (1.76) by ght_y and ght_z and integrating over the whole profile. These relations provide expressions of shear forces Q_y and Q_z via displacement components,

$$Q_y = GL_z(\eta' - \theta_z) + GL_{yz}(\zeta' + \theta_y) + GK_z(\theta' - \beta) ,$$

$$Q_z = GL_{yz}(\eta' - \theta_z) + GL_y(\zeta' + \theta_y) + GK_y(\theta' - \beta) . \tag{1.81}$$

Finally, the fourth physical relationship for the integral stress characteristics over the bar's section will be derived from (1.76) by multiplying by $ghd\varpi/ds$ and integrating over the whole profile. Using (1.46), we have the following result:

$$M_\varpi =$$

$$= (\eta' - \theta_z)G\oint t_y gh\frac{d\varpi}{ds}ds + (\zeta' + \theta_y)G\oint t_z gh\frac{d\varpi}{ds}ds + (\theta' - \beta)G\oint \rho gh\frac{d\varpi}{ds}ds .$$

We noticed earlier that

$$\frac{d\varpi}{ds} = -\frac{d\alpha}{ds} = -\frac{I_x}{gh\Omega} + \rho ,$$

therefore

$$\oint t_y gh\frac{d\varpi}{ds}ds = K_z , \quad \oint t_z gh\frac{d\varpi}{ds}ds = K_y , \quad \oint \rho gh\frac{d\varpi}{ds}ds = I_\rho - I_x = \mu I_\rho .$$

As a result, the expression of M_ϖ becomes

$$M_\varpi = GK_z(\eta' - \theta_z) + GK_y(\zeta' + \theta_y) + \mu GI_\rho(\theta' - \beta) . \tag{1.82}$$

The set consisting of the first equality from (1.50) and equalities (1.55), (1.80), (1.81) and (1.82) is exactly the set of the desirable physical relationships. All those can be written conveniently in a unified matrix form, so we have

$$
\begin{Vmatrix} N \\ H \\ M_y \\ M_z \\ B \\ Q_z \\ Q_y \\ M_\varpi \end{Vmatrix} = \begin{Vmatrix} EA & 0 & 0 & 0 & 0 & 0 & 0 & 0 \\ 0 & GI_x & 0 & 0 & 0 & 0 & 0 & 0 \\ 0 & 0 & EI_y & 0 & 0 & 0 & 0 & 0 \\ 0 & 0 & 0 & EI_z & 0 & 0 & 0 & 0 \\ 0 & 0 & 0 & 0 & EI_\varpi & 0 & 0 & 0 \\ 0 & 0 & 0 & 0 & 0 & GL_y & GL_{yz} & GK_y \\ 0 & 0 & 0 & 0 & 0 & GL_{zy} & GL_z & GK_z \\ 0 & 0 & 0 & 0 & 0 & GK_y & GK_z & \mu GI_\rho \end{Vmatrix} \cdot \begin{Vmatrix} \varepsilon \\ \chi_x \\ \chi_z \\ \chi_y \\ \chi_B \\ \gamma_z \\ \gamma_y \\ \gamma_\omega \end{Vmatrix}, \tag{1.83}
$$

where the pure-torsion torquc, H, is dcfined by (1.57) and

$$
\varepsilon = \xi', \quad \chi_x = \theta', \quad \chi_z = \theta_y', \quad \chi_y = -\theta_z', \quad \chi_B = -\beta',
$$

$$
\gamma_z = \zeta' + \theta_y, \qquad \gamma_y = \eta' - \theta_z, \qquad \gamma_\omega = \theta' - \beta. \tag{1.84}
$$

We introduce, as we did earlier, a vector of "stresses" $\boldsymbol{\sigma}$, a vector of displacements $\mathbf{u}$, and a vector of strains $\boldsymbol{\varepsilon}$ by assuming

$$
\boldsymbol{\sigma} = \left\| N, H, M_y, M_z, B, Q_z, Q_y, M_\varpi \right\|^{\mathsf{T}},
$$

$$
\mathbf{u} = \left\| \xi, \eta, \zeta, \theta, \theta_y, \theta_z, -\beta \right\|^{\mathsf{T}}, \quad \boldsymbol{\varepsilon} = \left\| \varepsilon, \chi_x, \chi_z, \chi_y, \chi_B, \gamma_z, \gamma_y, \gamma_\omega \right\|^{\mathsf{T}}. \tag{1.85}
$$

Comparing these formulas with (**6**.3.8), (**6**.3.20) and (**6**.3.21) from the theory of open-profile bars, we can see that the former and the latter are identical. But with these designations the formula of physical relationships (1.83) will havc a standard matrix form:

$$
\boldsymbol{\sigma} = \bar{\boldsymbol{C}}\, \boldsymbol{\varepsilon},
$$

if matrix $\bar{\boldsymbol{C}}$ is the linear transformation matrix from (1.83). Here we put a bar over the designation of the elasticity matrix intentionally, to separate the notation of this version of the theory from the notation of the first (energy-based) version. We would like to note that the operator of geometry, $\boldsymbol{A}$, which defines a differential relation between displacements $\mathbf{u}$ and strains $\boldsymbol{\varepsilon}$ according to (1.84), is our familiar matrix differential operator (**6**.3.13). It should be clear because the operator of equilibrium, $\boldsymbol{A}^{\mathsf{T}}$, conjugate to operator $\boldsymbol{A}$, may not change from one version of the theory to another: the equilibrium equations are the same in both versions.

It is convenient to switch from the designations by Adadurov as in (1.77) to dimensionless coefficients $\bar{\nu}$ by defining

$$\overline{\nu}_{yy} = \frac{1}{A}\oint t_y^2 ghds\,, \qquad \overline{\nu}_{zz} = \frac{1}{A}\oint t_z^2 ghds\,,$$

$$\overline{\nu}_{yz} = \overline{\nu}_{zy} = \frac{1}{A}\oint t_y t_z ghds\,, \qquad \overline{\nu}_{\varpi\varpi} = \frac{\mu}{I_r}\oint \rho^2 ghds = \mu\frac{I_\rho}{I_r}\,,$$

$$\overline{\nu}_{\varpi z} = \overline{\nu}_{z\varpi} = \frac{1}{\sqrt{I_r A}}\oint t_z \rho ghds\,, \qquad \overline{\nu}_{\varpi y} = \overline{\nu}_{y\varpi} = \frac{1}{\sqrt{I_r A}}\oint t_y \rho ghds\,. \qquad (1.86)$$

We remind that I_r denotes the polar moment of inertia of the section, i.e. $I_r = I_y + I_z$.

As can be seen at this point, the geometrical characteristics of the section according to Adadurov, introduced earlier by (1.77), together with the directed moment of inertia I_ρ, can be expressed via the dimensionless geometrical characteristics of the section, $\overline{\nu}$, as follows:

$$I_\rho = r^2 A\overline{\nu}_{\varpi\varpi}/\mu\,, \qquad K_y = rA\overline{\nu}_{z\varpi}\,, \qquad K_z = rA\overline{\nu}_{y\varpi}\,,$$

$$L_{yz} = L_{zy} = A\overline{\nu}_{yz}\,, \qquad L_y = A\overline{\nu}_{zz}\,, \qquad L_z = A\overline{\nu}_{yy}\,. \qquad (1.87)$$

Therefore the matrix of elasticity, $\overline{\boldsymbol{C}}$, will be as follows with the new designations:

$$\overline{\boldsymbol{C}} = \left\| \begin{matrix} EA & 0 & 0 & 0 & 0 & 0 & 0 & 0 \\ 0 & GI_x & 0 & 0 & 0 & 0 & 0 & 0 \\ 0 & 0 & EI_y & 0 & 0 & 0 & 0 & 0 \\ 0 & 0 & 0 & EI_z & 0 & 0 & 0 & 0 \\ 0 & 0 & 0 & 0 & EI_\varpi & 0 & 0 & 0 \\ 0 & 0 & 0 & 0 & 0 & GA\overline{\nu}_{zz} & GA\overline{\nu}_{zy} & rGA\overline{\nu}_{z\varpi} \\ 0 & 0 & 0 & 0 & 0 & GA\overline{\nu}_{yz} & GA\overline{\nu}_{yy} & rGA\overline{\nu}_{y\varpi} \\ 0 & 0 & 0 & 0 & 0 & rGA\overline{\nu}_{\varpi z} & rGA\overline{\nu}_{\varpi y} & r^2 GA\overline{\nu}_{\varpi\varpi} \end{matrix} \right\| . \qquad (1.88)$$

Exactly as in the theory of open-profile thin-walled bars which allows for shear, the symmetric matrix $\overline{\boldsymbol{\nu}}$ and its inverse are positive definite. But this fact is not true for a non-warped section. The non-warped sections require a separate consideration which we will do later.

The structure of matrix $\overline{\boldsymbol{\nu}}$ is defined by formula (1.89):

$$\overline{\nu} = \left\| \begin{bmatrix} \overline{\nu}_{zz} & \overline{\nu}_{zy} & \overline{\nu}_{z\varpi} \\ \overline{\nu}_{yz} & \overline{\nu}_{yy} & \overline{\nu}_{y\varpi} \\ \overline{\nu}_{\varpi z} & \overline{\nu}_{\varpi y} & \overline{\nu}_{\varpi\varpi} \end{bmatrix} \right\| =$$

$$= \left\| \begin{bmatrix} \frac{1}{A}\oint t_z^2 hds & \frac{1}{A}\oint t_z t_y hds & \frac{1}{\sqrt{I_r A}}\oint t_z \rho hds \\ \frac{1}{A}\oint t_y t_z hds & \frac{1}{A}\oint t_y^2 hds & \frac{1}{\sqrt{I_r A}}\oint t_y \rho hds \\ \frac{1}{\sqrt{I_r A}}\oint t_z \rho hds & \frac{1}{\sqrt{I_r A}}\oint t_y \rho hds & \frac{\mu}{I_r}\oint \rho^2 hds \end{bmatrix} \right\| . \tag{1.89}$$

Now that we know all matrix operators, we can find the governing equations for our problem, Lame equations, in the usual way. And again we obtain a system of equations totally equivalent in its structure to (**6**.3.24) but with coefficients ν replaced by the coefficients of matrix $\overline{\nu}$.

A semi-shear theory by Umanski

If we confine ourselves to the semi-shear theory when adopting the second approach to the derivation of physical relationships, then we will have the following instead of equation (1.72):

$$\boxed{\overline{\psi} E I_\varpi \beta''' - G I_x \beta' = m_x + \overline{\psi} m'_B ,} \tag{1.90}$$

where

$$\overline{\psi} = 1 + \frac{I_x}{\overline{\nu}_{\varpi\varpi} I_r} .$$

This is an equation derived in the theory which was suggested by A.A. Umanski himself. By replacing the $\overline{\nu}_{\varpi\varpi}$ coefficient with its expression from (1.86) and using (1.79), we have

$$\overline{\psi} = 1/\mu , \tag{1.91}$$

so (1.90) can be transformed into a form commonly used in the presentment of A. Umanski's theory[4]

---

[4] See [4], for example. It should be said that [4] does not consider a most general load upon the bar because all equations are derived there from the

$$EI_{\varpi}\beta''' - \mu GI_x\beta' = \mu m_x + m'_B\,. \tag{1.92-a}$$

Sometimes the basic governing equation of Umanski's theory contains slope θ rather than function β as an unknown. After making necessary transformations, we have

$$EI_{\varpi}\theta^{IV} - \mu GI_x\theta'' = \mu m_x - (1-\mu)\frac{EI_{\varpi}}{GI_x}m''_x + \mu m'_B\,. \tag{1.92-b}$$

7.1.6 A remark on non-warped cross-sections of closed-profile thin-walled bars

We would like to draw attention to a circumstance related to the warp factor, μ. The fact is that the warp factor μ is always nonnegative and strictly less than one:

$$0 \le \mu < 1\,. \tag{1.93}$$

The definition of the warp factor, μ, from (1.79) and the positiveness of the moments of inertia, I_x and I_ρ, imply that the estimates of (1.93) are equivalent to one inequality,

$$I_\rho \ge I_x\,,$$

which is, in its turn, equivalent to

$$\left(\oint \rho\, ds\right)^2 \le \oint \rho^2 gh\,ds \oint \frac{ds}{gh}\,. \tag{1.94}$$

But the latter is a corollary of an inequality well known in mathematics, one by Buniakovsky [6], for two functions U and V, which is

$$\left(\oint UV ds\right)^2 \le \oint U^2 ds \oint V^2 ds\,, \tag{1.95}$$

where we should assume

$$U = \rho\sqrt{gh}\,,\quad V = 1/\sqrt{gh}\,.$$

By the way, the equality takes place in (1.95) only when functions U and V are linearly dependent, that is, when $U = CV$ where C is a constant.

assumption that there is no bitorque load, i.e. $m_B = 0$. However, this fact does not have any significant effect on the subsequent reasoning.

Hence the warp factor μ is equal to zero if and only if the ρgh product is a constant.

Physically, the fact that the warp factor is equal to zero is equivalent to the requirement that the cross-section of the thin-walled bar should not be warped when in torsion. And indeed, the absence of the warp displacements according to (1.16) is equivalent to the warp function being equal to zero, $\alpha(s) = 0$, or, according to (1.17)

$$\frac{I_x}{\Omega}\int_0^s \frac{ds}{gh} - \int_0^s \rho ds = 0 . \tag{1.96}$$

We will show that the equality of (1.96) implies $\mu = 0$. To do it, we differentiate (1.96) with respect to s and obtain

$$\frac{I_x}{\Omega} = \rho gh .$$

But the left-hand part of this equality contains a constant independent of arc coordinate s, hence $\rho gh = Const$, and this, as we have just found out, implies the desirable equality $\mu = 0$.

The inverse statement is true, too: the condition that $\mu = 0$ implies the non-warpedness of the section, that is, $\alpha(s) = 0$. The reasoning is like this:

$$\mu = 0 \Rightarrow \rho gh = Const \Rightarrow \alpha(s) = \frac{I_x}{\Omega}\int_0^s \frac{\rho ds}{\rho gh} - \int_0^s \rho ds = \left(\frac{I_x}{\Omega \rho gh} - 1\right)\int_0^s \rho ds .$$

And further,

$$\mu = 0 \Rightarrow I_x = I_\rho \Rightarrow \frac{I_x}{\Omega \rho gh} = \frac{I_\rho}{\Omega \rho gh} = \frac{\oint \rho^2 ghds}{\Omega \rho gh} = \frac{\oint \rho ds}{\Omega} = 1 \Rightarrow \alpha(s) = 0.$$

A simplest case of a non-warped section is a ring of a constant thickness. This is, of course, not the only member in the category of non-warped sections.

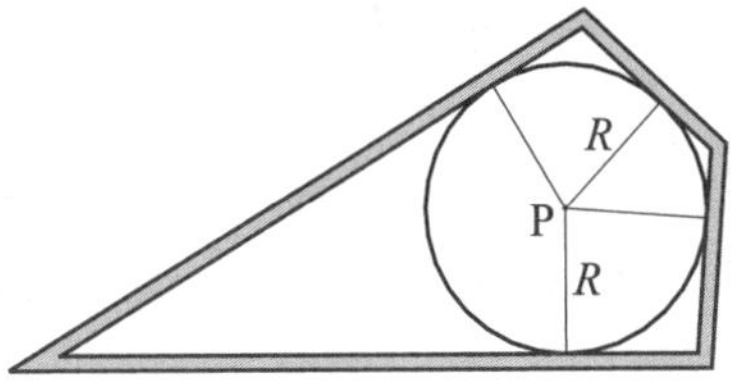

Fig. 7.3. A non-warped section with a constant thickness of the wall

In particular, the non-warped sections include any closed one made of a homogeneous material with a constant thickness of the walls, $h = Const$, provided its profile consists of straight line segments touching the same circle. An example is shown in Fig. 7.3.

To validate this statement, it suffices to show that the center of twist, P, for such a profile coincides with the center of the inscribed circle because $g = 1$ and $\rho gh = Rh = Const$ [5].

As we will see, considering the non-warped sections as a separate class is critically important for the theory of closed-profile thin-walled bars. From the standpoint of theory, this type of the sections is degenerate, and the very notion of the warp measure, β, makes no sense. The torsion of the non-warped sections occurs under any conditions without constriction effects, i.e. exactly in the same way as the pure (free) torsion does.

Basic relationships of the theory of closed-profile thin-walled bars for non-warped cross-sections

Just as we did in the shear theory of open-profile bars, we have to consider a non-warped section of a closed profile separately because this is a degenerate case.

Omitting detailed transformations in order to save space (an attentive reader can restore those easily if he reads this section carefully), we present only final results. Instead of formulas (1.48) and (1.49) for the stresses in the case of the non-warped profile, we have

$$\sigma^x = \frac{N}{A}e + \frac{M_y}{I_y}ez + \frac{M_z}{I_z}ey\,, \tag{1.97}$$

$$\tau = \frac{M_x}{h\Omega} - Q_z\frac{\overline{S}_{oy}}{hI_y} - Q_y\frac{\overline{S}_{oz}}{hI_z}\,, \tag{1.98}$$

and the physical relationships become

$$\sigma^x = Ee(\xi' + \theta'_y z - \theta'_z y)\,, \tag{1.99}$$

$$\tau = Gg\left[(\eta' - \theta_z)t_y + (\zeta' + \theta_y)t_z + \theta'\rho\right]. \tag{1.100}$$

These are the basic vectors for the non-warped profiles:

---

[5] The reader is invited to prove that the center of twist coincides with the center of the inscribed circle for exercise. Should there be any problems with this, additional guidelines can be taken from Appendix G.

$$\boldsymbol{\sigma} = \left\| \left[N, M_x, M_y, M_z, Q_z, Q_y \right] \right\|^{\mathsf{T}}, \quad \bar{\boldsymbol{X}} = \left\| \left[p_x, p_y, p_z, m_x, m_y, m_z \right] \right\|^{\mathsf{T}},$$

$$\mathbf{u} = \left\| \left[\xi, \eta, \zeta, \theta, \theta_y, \theta_z \right] \right\|^{\mathsf{T}}, \qquad \boldsymbol{\varepsilon} = \left\| \left[\varepsilon, \chi_x, \chi_z, \chi_y, \gamma_z, \gamma_y \right] \right\|^{\mathsf{T}}, \tag{1.101}$$

and these are operators $\boldsymbol{A}$ and $\boldsymbol{A}^{\mathsf{T}}$:

$$\boldsymbol{A} = \left\| \begin{bmatrix} \partial/\partial x & 0 & 0 & 0 & 0 & 0 \\ 0 & 0 & 0 & \partial/\partial x & 0 & 0 \\ 0 & 0 & 0 & 0 & \partial/\partial x & 0 \\ 0 & 0 & 0 & 0 & 0 & -\partial/\partial x \\ 0 & 0 & \partial/\partial x & 0 & 1 & 0 \\ 0 & \partial/\partial x & 0 & 0 & 0 & -1 \end{bmatrix} \right\|,$$

$$\boldsymbol{A}^{\mathsf{T}} = \left\| \begin{bmatrix} -\partial/\partial x & 0 & 0 & 0 & 0 & 0 \\ 0 & 0 & 0 & 0 & 0 & -\partial/\partial x \\ 0 & 0 & 0 & 0 & -\partial/\partial x & 0 \\ 0 & -\partial/\partial x & 0 & 0 & 0 & 0 \\ 0 & 0 & -\partial/\partial x & 0 & 1 & 0 \\ 0 & 0 & 0 & \partial/\partial x & 0 & -1 \end{bmatrix} \right\|.$$

As for the elasticity matrix, $\boldsymbol{C}$, it can be presented in two versions:

- first (energy-based) version:

$$\boldsymbol{C} = \left\| \begin{bmatrix} EA & 0 & 0 & 0 & 0 & 0 \\ 0 & GI_x & 0 & 0 & 0 & 0 \\ 0 & 0 & EI_y & 0 & 0 & 0 \\ 0 & 0 & 0 & EI_z & 0 & 0 \\ 0 & 0 & 0 & 0 & GA\nu_{zz} & GA\nu_{zy} \\ 0 & 0 & 0 & 0 & GA\nu_{yz} & GA\nu_{yy} \end{bmatrix} \right\|; \tag{1.102}$$

- second (physical) version:

$$\bar{C} = \left\| \begin{bmatrix} EA & 0 & 0 & 0 & 0 & 0 \\ 0 & GI_x & 0 & 0 & 0 & 0 \\ 0 & 0 & EI_y & 0 & 0 & 0 \\ 0 & 0 & 0 & EI_z & 0 & 0 \\ 0 & 0 & 0 & 0 & GA\bar{\nu}_{zz} & GA\bar{\nu}_{zy} \\ 0 & 0 & 0 & 0 & GA\bar{\nu}_{yz} & GA\bar{\nu}_{yy} \end{bmatrix} \right\| . \qquad (1.103)$$

Building the Lame operator according to the general rule, we arrive at a governing system of equations (1.104) in terms of the displacements of the non-warped sections.

Note that the equations are given for the first (energy-based) version of the theory in (1.104). Obviously, to switch to the second (physical) version of the theory, one needs only to replace the components of matrix $\boldsymbol{\nu}$ by the components of $\bar{\boldsymbol{\nu}}$ in these formulas.

$$-EA\xi'' = p_x ,$$

$$-GA\left[\nu_{yz}(\zeta'' + \theta_y') + \nu_{yy}(\eta'' - \theta_z')\right] = p_y ,$$

$$-GA\left[\nu_{zz}(\zeta'' + \theta_y') + \nu_{zy}(\eta'' - \theta_z')\right] = p_z ,$$

$$-GI_x\theta'' = m_x ,$$

$$-EI_y\theta_y'' + GA\left[\nu_{zz}(\zeta' + \theta_y) + \nu_{zy}(\eta' - \theta_z)\right] = m_y ,$$

$$-EI_z\theta_z'' - GA\left[\nu_{yz}(\zeta' + \theta_y) + \nu_{yy}(\eta' - \theta_z)\right] = m_z . \qquad (1.104)$$

7.2 Multiple-contour, closed-profile, thin-walled bars

7.2.1 Pure torsion of a multiple-contour profile

The preceding presentment assumed that the cross-section of the bar was a single closed contour. In that way we could analyze the stress distribution in the bar in pure torsion without using any strain compatibility conditions, by involving the equilibrium equations only.

However, if there are multiple closed contours, the static equations are no longer sufficient because the pure torsion problem becomes a statically indeterminate one.

To be particular, let us consider a cross-section of a thin-walled bar, the profile of which contains three closed base contours (Fig. 7.4-*a*).

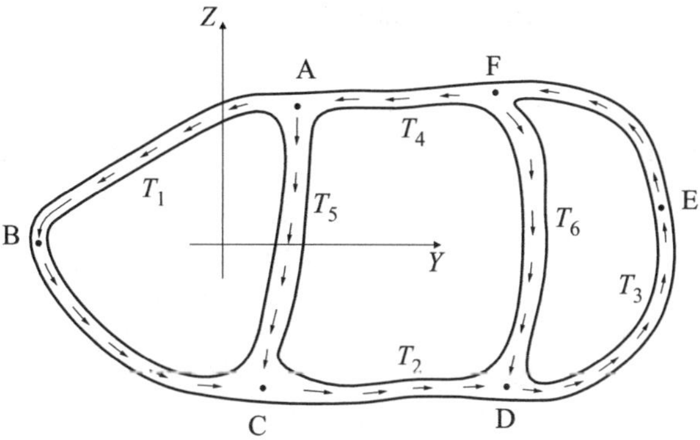

Fig. 7.4-*a*. Pure torsion of a three-contour profile – flows of tangential stresses along each of the non-bifurcating segments of the profile

The *base closed contours* will mean here and further a set of closed contours Γ_i $(i = 1,\dots, k)$ such that each of them comprises a two-dimensional area Ω_i and satisfies two conditions:

- no intersection – $\Omega_i \cap \Omega_j = \varnothing$ at $i \neq j$, where $\varnothing$ is an empty area;
- completeness – the union of all k areas Ω_i coincides with area Ω_0 bounded by the exterior closed contour Γ_0.

There are the following base contours for the profile shown in Fig. 7.4-*a*:

$$\Gamma_1 : \text{ABCA}\,; \qquad \Gamma_2 : \text{ACDFA}\,; \qquad \Gamma_2 : \text{DEFD}\,. \tag{2.1}$$

In the list of points which every closed contour passes, the first and the last points of the contour have the same names because they are coincident. We adopt the convention that the positive direction of tracing a contour will be the counter-clockwise direction, if we look from the positive direction of the longitudinal axis X. Note that the base contours defined in (2.1) have their points indicated in the order that corresponds to the positive direction of the contour tracing.

Each of the contours contains both independent segments not included in any other contour and segments which are common for it and another contour. For example, take a three-contour profile shown in Fig. 7.4-*a*; the common boundary for its first and second contours is $\Gamma_{12} = \Gamma_1 \cap \Gamma_2$, and for its second and third contours the common boundary occupies the profile's segment $\Gamma_{23} = \Gamma_2 \cap \Gamma_3$.

In addition to the base contours, the profile can be defined by such objects as non-bifurcating segments, i.e. continuous pieces between two points of bifurcation. In the example we are considering, the profile contains six non-bifurcating segments:

segment 1: ABC; segment 2: CD; segment 3: DEF;
segment 4: FA; segment 5: AC; segment 6: FD.

Reasoning the same way as we did with pure torsion of a singly closed profile, we find out that the flow of the tangential stresses along each of the non-bifurcating segment of the contour is a constant. In other words, we have six unknown flows, T_i ($i = 1,\ldots, 6$), according to the number of the non-bifurcating pieces of the profile (Fig. 7.4-*a*). Generally, each of the segments has its own constant, so

$$T_1 \neq T_2 \neq T_3 \neq T_4 \neq T_5 \neq T_6 \,.$$

These six flows are actually not independent but related through conditions of equilibrium.

To establish these relationships, we introduce the notion of a flow on each of the base contours, $\overline{T}_1$, $\overline{T}_2$, $\overline{T}_3$, as shown in Fig. 7.4-*b*. The flows meet one another on common segments of the base contours.

A contour flow is assumed positive if it moves in the positive direction around the contour of the section the exterior normal to which has the same direction as the *X*-axis. With this sign convention, the positive flows of adjacent base contours have the opposite directions when meeting on a common boundary (see Fig. 7.4-*b*).

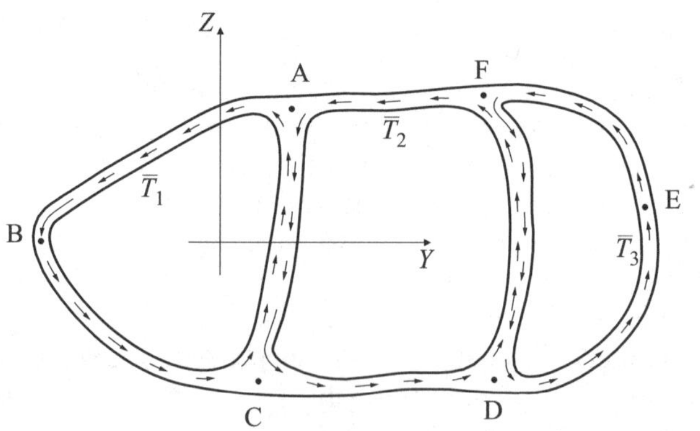

Fig. 7.4-*b*. Pure torsion of a three-contour profile – flows of tangential stresses along each of the base contours

The flows on the non-bifurcating segments of the profile can be proved to have the following relation to the flows on the contours:

$$T_1 = \overline{T}_1,\quad T_2 = \overline{T}_2,\quad T_3 = \overline{T}_3,\quad T_4 = \overline{T}_2,\quad T_5 = -\overline{T}_1 + \overline{T}_2,\quad T_6 = -\overline{T}_2 + \overline{T}_3 .$$

As the number of the base contours is less than the number of the non-bifurcating segments of the profile, it will be more convenient further to deal with the flows on the contours.

Obviously, the total torque in the section, created by contour flows $\overline{T}_1$, $\overline{T}_2$, $\overline{T}_3$, is equal to the sum of the torques created by each particular flow. Reasoning the same way as in Section **7**.1, we obtain the following instead of (1.3):

$$M_x = \overline{T}_1\Omega_1 + \overline{T}_2\Omega_2 + \overline{T}_3\Omega_3, \tag{2.2}$$

where Ω_i is the doubled area comprised by i-th contour.

As we can see, one equation of statics contains three unknown flows. This means our problem has the redundance of two, so the strain compatibility conditions should be involved to solve it.

Apparently, it is easiest to formulate the strain compatibility conditions on the variational basis, by considering the Castigliano functional. To do it, we should define the strain energy via the stresses, which will be exactly the Castigliano functional that we need for our analysis.

Up to an insignificant multiplier L, where L is the length of the bar in torsion, the Castigliano functional can be written as

$$\mathsf{K} = \oint_{\Gamma_1} \frac{\overline{\tau}_1^2}{2Gg} h\,ds + \oint_{\Gamma_2} \frac{\overline{\tau}_2^2}{2Gg} h\,ds + \oint_{\Gamma_3} \frac{\overline{\tau}_3^2}{2Gg} h\,ds - \int_{\Gamma_{12}} \frac{\overline{\tau}_1\overline{\tau}_2}{Gg} h\,ds - \int_{\Gamma_{23}} \frac{\overline{\tau}_2\overline{\tau}_3}{Gg} h\,ds ,$$

where $\overline{\tau}_i = \overline{T}_i / h$ is a tangential stress created by flow $\overline{T}_i$. The two last integrals account for the mutual work of the opposite flows of the tangential stresses on common boundaries of the contours. After expressing the stresses via the flows and allowing for the constancy of the flows on each of the contours, we can transform this formula into a more convenient relation,

$$\mathsf{K}(\overline{T}_1, \overline{T}_2, \overline{T}_3) =$$

$$= \frac{\overline{T}_1^2}{2G} \oint_{\Gamma_1} \frac{ds}{gh} + \frac{\overline{T}_2^2}{2G} \oint_{\Gamma_2} \frac{ds}{gh} + \frac{\overline{T}_3^2}{2G} \oint_{\Gamma_3} \frac{ds}{gh} - \frac{\overline{T}_1\overline{T}_2}{G} \int_{\Gamma_{12}} \frac{ds}{gh} - \frac{\overline{T}_2\overline{T}_3}{G} \int_{\Gamma_{23}} \frac{ds}{gh}, \tag{2.3}$$

which turns the Castigliano functional into a quadratic form of three scalar values, flows $\overline{T}_1$, $\overline{T}_2$, $\overline{T}_3$.

The minimum of this quadratic form should be sought for in a statically admissible set, i.e. in a set of flows which satisfy the only equation of equilibrium (2.2). We use the standard Lagrangian multipliers approach that helps reduce the problem of a conditional extremum of function K to the problem of unconditional stationarity of the modified function M,

$$\mathsf{M}(\overline{T}_1,\overline{T}_2,\overline{T}_3,\lambda)=\mathsf{K}(\overline{T}_1,\overline{T}_2,\overline{T}_3)-\lambda(M_x-\overline{T}_1\Omega_1-\overline{T}_2\Omega_2-\overline{T}_3\Omega_3)\,, \tag{2.4}$$

with the Lagrangian multiplier λ. The condition of stationarity of M produces the following system of simultaneous linear algebraic equations:

$$\left\|\begin{array}{cccc} \dfrac{1}{G}\oint\limits_{\Gamma_1}\dfrac{ds}{gh} & -\dfrac{1}{G}\int\limits_{\Gamma_{12}}\dfrac{ds}{gh} & 0 & \Omega_1 \\ -\dfrac{1}{G}\int\limits_{\Gamma_{12}}\dfrac{ds}{gh} & \dfrac{1}{G}\oint\limits_{\Gamma_2}\dfrac{ds}{gh} & -\dfrac{1}{G}\int\limits_{\Gamma_{23}}\dfrac{ds}{gh} & \Omega_2 \\ 0 & -\dfrac{1}{G}\int\limits_{\Gamma_{23}}\dfrac{ds}{gh} & \dfrac{1}{G}\oint\limits_{\Gamma_3}\dfrac{ds}{gh} & \Omega_3 \\ \Omega_1 & \Omega_2 & \Omega_3 & 0 \end{array}\right\| \cdot \left\|\begin{array}{c} \overline{T}_1 \\ \overline{T}_2 \\ \overline{T}_3 \\ \lambda \end{array}\right\| = \left\|\begin{array}{c} 0 \\ 0 \\ 0 \\ M_x \end{array}\right\|. \tag{2.5}$$

The solution of this system of equations resolves the problem of redundance.

It is of little avail to present the solution of this system of equations here in its general form because its analytic expression is too bulky. The only thing worth mentioning is that the solution can be represented as

$$\overline{T}_1=\overline{a}_1\frac{M_x}{\Omega_0}\,,\qquad \overline{T}_2=\overline{a}_2\frac{M_x}{\Omega_0}\,,\qquad \overline{T}_3=\overline{a}_3\frac{M_x}{\Omega_0}\,, \tag{2.6}$$

where the dimensionless coefficients $\overline{a}_1$, $\overline{a}_2$, $\overline{a}_3$ satisfy the condition

$$\overline{a}_1\,\Omega_1+\overline{a}_2\,\Omega_2+\overline{a}_3\,\Omega_3=\Omega_0, \tag{2.7}$$

and Ω_0 is a total area of the three contours,

$$\Omega_0=\Omega_1+\Omega_2+\Omega_3\,. \tag{2.8}$$

By the way, it is worth noticing that the values of those coefficients showing a relative distribution of the flows between the contours do not depend on the common shear modulus, G, nor they depend on the M_x torque.[6]

---

[6] It is important because it permits to formally assume $G = 1$ in the system of equations of the type (2.5) written for the coefficients of distribution of the flows

It is also obvious that the result will be similar in the general case for a k-contour profile:

$$\overline{T}_i = \overline{a}_i \frac{M_x}{\Omega_0} \quad (i = 1,\dots, k), \qquad \sum_{i=1}^{k} \overline{a}_i \Omega_i = \Omega_0 , \qquad \Omega_0 = \sum_{i=1}^{k} \Omega_i . \qquad (2.9)$$

A topologic structure of a multiple-contour profile

When we construct a governing system of equations for a general multiple-contour profile, there is such an essential thing as the topologic structure of the multiple-contour profile in question. This structure is characterized by the following three integer parameters:

- k, the number of base (non-intersecting, closed) contours which make up the profile;
- n, the number of non-bifurcating segments in the profile;
- t, the number of points of bifurcation of the profile.

It wiil be convenient for us to think topologically — in terms of some elementary concepts of the theory of directed graphs which we borrow mainly from [3][7]. We introduce new notions in application only to that (pretty narrow) class of graphs which will help us in our analysis.

A vertex or *a node of a graph* will refer to any point of bifurcation where the profile goes along more than one path. For example, the profile shown in Fig. 7.4-*a* and 7.4-*b* is associated with the graph shown in Fig. 7.5 that has four nodes, v_1, v_2, v_3, v_4. Those four nodes of the graph conform to four points of bifurcation of the profile, the respective A, C, F, and D.

An edge or *an arc of the graph* will refer to any non-bifurcating segment of the profile. The graph shown in Fig. 7.5 has six edges r_1, r_2, r_3, r_4, r_5, r_6 which conform to the respective non-bifurcating segments ABC, CD, DEF, FA, AC, FD of the profile, as shown in Fig. 7.4-*a* and 7.4-*b* .

A base loop of the graph will refer to a set of edges which make up any base contour of the profile. The graph in Fig. 7.5 has three base loops: $c_1 = (r_1, r_5)$, $c_2 = (r_5, r_2, r_6, r_4)$, $c_3 = (r_3, r_6)$.

---

over the contours – see the general formula (2.15) later. So it is worth doing some work in order to prove this statement.

[7] This is not a new idea to use techniques of the theory of graphs in application to the analysis of thin-walled bars. For example, [5] uses this approach to calculate geometrical characteristics of the sections of thin-walled bars.

A frame loop, c_0, *of the graph* will refer to a set of edges that make up the outermost bounding contour of the profile, Γ_0. In our case $c_0 = (r_1, r_2, r_3, r_4)$.

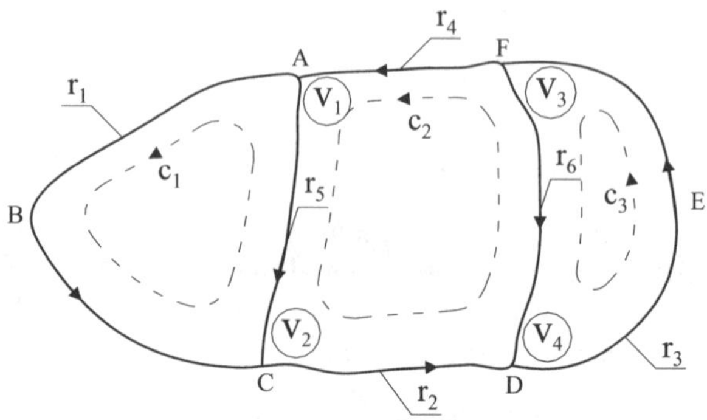

Fig. 7.5. An oriented graph of a profile

In terms of the graph theory, the three integral parameters introduced above can be interpreted as follows:

- k is the number of base loops in the graph;
- n is the number of edges in the graph;
- t is the number of vertices in the graph.

In the class of graphs that we are dealing with, each edge can be included in one base loop or two base loops. If an edge participates in one base loop only, we will call it a *frame edge.* Otherwise, such an edge will be called a *double edge.*

The direction of any frame edge (indicated by an arrow on the graph's picture) is predefined by the choice of a positive direction of the respective base contour. As for any double edge, its direction does not obey any rule and can be chosen arbitrarily.

The set of all frame edges listed in succession is exactly the frame loop of the graph.

The presence of a definite direction on each edge permits to associate each vertex (node) with a list of *incoming* edges and a list of *outgoing* edges. An edge r_i is said to be *incident* to a vertex v_α if that edge either comes to or goes from the v_α vertex. In the class of graphs that we are dealing with, each vertex has at least three edges incident to it.

It can be proved that the integer parameters introduced above, n, t and k, are not independent in this particular class of graphs. Instead, they are related as

$$n - t = k - 1 . \tag{2.10}$$

From the standpoint of structural mechanics, this number of $(k-1)$ is exactly the redundance of the multiple-contour profile in the pure torsion analysis. Indeed, if the profile contains k independent contours, then the unknown contour flows $\overline{T}_i\,(i=1,\ldots,k)$ obey exactly one static equation of the type (2.2):

$$\sum_{i=1}^{k} \overline{T}_i \Omega_i = M_x\,, \tag{2.11}$$

and the other $(k-1)$ equations have to have the sense of strain compatibility conditions. Note that the above-introduced parameters are as follows for the profile shown in Fig. 7.4:

$$k=3; \quad n=6; \quad t=4.$$

We introduce a so-called *incidence matrix*, **G**, which is used to describe the topologic structure of a directed graph by algebraic means. This matrix contains t rows and n columns, and its elements can take one of three values: 0, 1, or –1. Each i-th column of matrix $\boldsymbol{G}$ corresponds to a separate edge of the graph, and each α-th row to a separate vertex of it. All elements are zeros in every i-th column, except for two elements one of which is equal to one (this element's row number indicates a node which i-th edge of the graph goes out of) and the other to minus one (this element's row number indicates a node which i-th edge of the graph comes to).

Numbering the vertices and edges of the graph from our example as shown in Fig. 7.5, we obtain the following matrix $\boldsymbol{G}$:

$$\boldsymbol{G} =. \begin{matrix} \\ v_1 \\ v_2 \\ v_3 \\ v_4 \end{matrix} \begin{Vmatrix} r_1 & r_2 & r_3 & r_4 & r_5 & r_6 \\ 1 & 0 & 0 & -1 & 1 & 0 \\ -1 & 1 & 0 & 0 & -1 & 0 \\ 0 & 0 & -1 & 1 & 0 & 1 \\ 0 & -1 & 1 & 0 & 0 & -1 \end{Vmatrix}.$$

In addition to the incidence matrix, **G**, there is another important matrix used to describe the topologic structure of the profile; this is a *loop matrix*, ***F***. This matrix has the dimensions of $k \times n$. An arbitrary element, f_{ij}, of that matrix, $\boldsymbol{F} = \|[f_{ij}]\|$, is equal to:

- zero, if edge r_j is not included in the loop c_i;
- one, if edge r_j is included in the loop c_i, and the chosen direction on this edge coincides with the positive direction of tracing the contour that conforms to the loop;

- minus one, if edge r_j is included in the loop c_i, and the chosen direction on this edge is different from the positive direction of tracing the the contour that conforms to the loop.

For the profile that we use as an example, the loop matrix $\boldsymbol{F}$ will be

$$\boldsymbol{F} = \begin{matrix} \\ c_1 \\ c_2 \\ c_3 \end{matrix} \begin{matrix} r_1 & r_2 & r_3 & r_4 & r_5 & r_6 \\ \end{matrix} \left\| \begin{bmatrix} 1 & 0 & 0 & 0 & -1 & 0 \\ 0 & 1 & 0 & 1 & 1 & -1 \\ 0 & 0 & 1 & 0 & 0 & 1 \end{bmatrix} \right\| .$$

Obviously, edge r_j is a frame one if one of the elements in j-th column of the loop matrix is equal to one and all the others are zeros.

The two matrices, $\boldsymbol{G}$ and $\boldsymbol{F}$, are not thoroughly independent. For the class of graphs we are dealing with, the incidence matrix and the loop matrix are mutually orthogonal[8], that is,

$$\boldsymbol{G}\boldsymbol{F}^{\mathrm{T}} = \boldsymbol{O}, \tag{2.12}$$

where $\boldsymbol{O}$ is a zero matrix of the $t{\times}k$ dimensions.

It is further useful to associate each j-th edge of the graph with an *edge weight*, p_j, which is defined as

$$p_j = \int_{l_j} \frac{ds}{gh} \quad (j = 1,\ldots, n)\,, \tag{2.13}$$

where l_j is a geometrical length of the profile's segment which conforms to the edge of graph r_j. Similarly, we can associate each closed base contour c_i with a *contour weight*, $\overline{p}_i$, by defining

$$\overline{p}_i = \oint_{\Gamma_i} \frac{ds}{gh} = \sum_{j=1}^{n} |\, f_{ij} \,|\, p_j \quad (i = 1,\ldots, k). \tag{2.14}$$

We suggest that the reader think over the proof for what we present here: a general form of the governing system of equations for calculating the flow distribution coefficients, $\overline{a}_i$ $(i = 1,\ldots, k)$, in a multiple-contour profile in pure torsion:

---

[8] A formal proof of this statement will be an exercise for the reader.

$$\begin{Vmatrix} \overline{p}_1 & -p_{12} & \cdots & -p_{1j} & \cdots & -p_{1k} & \Omega_1 \\ -p_{21} & \overline{p}_2 & \cdots & -p_{2j} & \cdots & -p_{2k} & \Omega_2 \\ \vdots & \vdots & \ddots & \vdots & \vdots & \vdots & \vdots \\ -p_{j1} & -p_{j2} & \cdots & \overline{p}_j & \cdots & -p_{jk} & \Omega_j \\ \vdots & \vdots & \vdots & \vdots & \ddots & \vdots & \vdots \\ -p_{k1} & -p_{k2} & \cdots & -p_{kj} & \cdots & \overline{p}_k & \Omega_k \\ \Omega_1 & \Omega_2 & \cdots & \Omega_j & \cdots & \Omega_k & 0 \end{Vmatrix} \cdot \begin{Vmatrix} \overline{a}_1 \\ \overline{a}_2 \\ \vdots \\ \overline{a}_j \\ \vdots \\ \overline{a}_k \\ \lambda \end{Vmatrix} = \begin{Vmatrix} 0 \\ 0 \\ \vdots \\ 0 \\ \vdots \\ 0 \\ \Omega_0 \end{Vmatrix}, \qquad (2.15)$$

where:

- $p_{ij} = p_{ji} = 0$, if contours c_i and c_j do not have common edges;
- $p_{ij} = p_{ji} = p_\alpha$, if edge r_α is a common one of contours c_i and c_j.

The solution of this system of equations produces coefficients $\overline{a}_i$ ($i = 1,\ldots, k$) for each of the base contours, thus the values of flows $\overline{T}_i$ for each of the base contours, too:

$$\overline{T}_i = \overline{a}_i \frac{M_x}{\Omega_0} \ (i = 1,\ldots, k). \qquad (2.16)$$

Now we can determine flows T_j ($j = 1,\ldots, n$) on each of the non-bifurcating segments of the profile:

$$T_j = \sum_{i=1}^{k} f_{ij} \overline{T}_i \quad (j = 1,\ldots, n). \qquad (2.17)$$

We can associate each of the non-bifurcating segments with its particular dimensionless coefficient a_j by defining

$$a_j = \sum_{i=1}^{k} f_{ij} \overline{a}_i \quad (j = 1,\ldots, n). \qquad (2.18)$$

Obviously, with these designations introduced, the flows T_j ($j = 1,\ldots, n$) on non-bifurcating segments of the contour can be written as follows:

$$T_j = a_j \frac{M_x}{\Omega_0} \quad (j = 1,\ldots, n). \qquad (2.19)$$

We would like to note that it is appropriate to refer to coefficients a_j ($j = 1,\ldots, n$) as *coefficients of flow distribution over segments*, while coefficients $\bar{a}_i$ $(i = 1,\ldots, k)$ can be entitled *coefficients of flow distribution over contours*.

It is also convenient to introduce a piecewise-constant function, $a(s)$, which takes the values of a_j for the argument s from j-th piece of the contour. Function $a(s)$ will be called a *flow distribution function*, meaning the distribution along the profile. Using this function and following (2.19), the flow, $T(s)$, in any point of the profile can be represented as

$$T(s) = a(s)\frac{M_x}{\Omega_0} . \tag{2.20}$$

A warp function for a multiple-contour profile

Further we will deal with the warp function, $\alpha(s)$, constructed for a multiple-contour profile.

Exactly as we did in Section **7**.1, we will choose an arbitrary point of the (Y,Z)-plane to be the pole P and will impose the same six constraints on the bar, which will prevent the bar from moving as a rigid whole.

Formula (1.8), which can be more conveniently written as

$$\frac{du}{ds} = \frac{T(s)}{Ggh} - \theta'\rho , \tag{2.21}$$

will still remain true for each of the segments of the profile. However, here's what we should keep in mind about this formula:

- flows T take different values in different parts of the profile;
- the increase of coordinate s (that is, the sign of ds) and the sign of parameter ρ should be consistent in common parts of the contours.

Let us now consider the contour Γ_0, which is the outermost, bounding contour of the section. For example, take a profile shown in Fig. 7.4-*b*; the positive direction around its contour Γ_0 is to move successively between the points ABCDEFA of the profile. Integrating (2.21) along that contour in the positive direction gives this for an arbitrary point $\mathrm{M} \in \Gamma_0$:

$$u_\mathrm{M} = u_\mathrm{A} + \frac{1}{G}\int_\mathrm{A}^\mathrm{M}\frac{Tds}{gh} - \theta'\int_\mathrm{A}^\mathrm{M}\rho ds . \tag{2.22}$$

Flow T remains in the integrand because it is different in different parts of the profile. After tracing the Γ_0 contour, we return to point A, so the condition of unambiguity of the longitudinal displacement in that point yields a general expression of the twist:[9]

$$\theta' = \frac{1}{G\Omega_0} \oint_{\Gamma_0} \frac{Tds}{gh}, \tag{2.23}$$

where Ω_0 is the doubled area bounded by exterior contour Γ_0 which we introduced earlier:

$$\Omega_0 = \oint_{\Gamma_0} \rho ds = \Omega_1 + \Omega_2 + \Omega_3 .$$

Using the representation (2.20) of flow $T(s)$, we transform (2.23) into

$$\theta' = \frac{M_x}{G\Omega_0^2} \oint_{\Gamma_0} \frac{ads}{gh}. \tag{2.24}$$

After writing the twist expression in standard designations of the science of strength of materials,

$$\theta' = \frac{M_x}{GI_x},$$

we find an expression of the torsional moment of inertia of the multiple-contour section by comparing the previous relation with (2.24):

$$I_x = \frac{\Omega_0^2}{\oint_{\Gamma_0} \frac{ads}{gh}}. \tag{2.25}$$

Formula (1.15) for a singly closed section follows naturally from the general formula (2.25), when assuming $\Omega_0 = \Omega$ and $a(s) \equiv 1$ in the latter.

If we begin the movement from the A point of the contour and move constantly in the positive direction ofthe flow, we can eventually reach any point M of the profile. There can be multiple paths to some points M. For example, for a point M belonging to the FA segment, the potential paths by which it can be reached include ACDFM, ACDEFM, ABCDFM. Let us choose one of the paths to M and integrate (2.21) from point A to point M along that. The result is again (2.22) in which, however, the

---

[9] Note that (2.23) is actually nothing but a re-formulation of the Bredt theorem of the tangential stress circulation in application to the bounding contour Γ_0 for a thin-walled section with a closed multiple-contour profile under the condition of $g = 1$.

chosen path of integration may be different from contour Γ_0. Substituting the expression of the twist from (2.24) to (2.22) gives

$$u_{\mathrm{M}} = u_{\mathrm{A}} + \frac{M_x}{G\Omega_0}\int_{\mathrm{A}}^{\mathrm{M}}\frac{a\,ds}{gh} - \frac{M_x}{G\Omega_0^2}\oint_{\Gamma_0}\frac{a\,ds}{gh}\int_{\mathrm{A}}^{\mathrm{M}}\rho\,ds\,.$$

Of course, in this formula the value of each integral with the variable upper limit,

$$\int_{\mathrm{A}}^{\mathrm{M}}\frac{a\,ds}{gh} \quad \text{and} \quad \int_{\mathrm{A}}^{\mathrm{M}}\rho\,ds\,,$$

depends on what path of integration has been chosen, but the ultimate result for u_{M} should be invariant with respect to the path. Otherwise, the uniqueness of the displacements at point M would be violated.

Further, in our case we are entitled to assume constant u_{A} to be zero, which is equivalent to the fixation of the bar against rigid longitudinal displacements exactly at the A point. As a result, the final formula for the warp displacements in pure torsion of a multiple-contour profile becomes

$$u = \frac{M_x}{GI_x}\alpha(s)\,, \tag{2.26}$$

which is the same as (1.16) if we take the warp function, $\alpha(s)$, to be

$$\alpha(s) = \frac{I_x}{\Omega_0}\int_{\mathrm{A}}^{\mathrm{M}}\frac{a\,ds}{gh} - \int_{\mathrm{A}}^{\mathrm{M}}\rho\,ds \tag{2.27}$$

with an arbitrary path of integration from the start point, A, to the current point of the profile, M.

To conclude this section, we would like to present a useful validation identity which is required to hold for the function of distribution of the flows over the non-bifurcating segments of the profile:

$$\sum_{j=1}^{n} a_j^2 \int_{l_j}\frac{ds}{gh} = \oint_{\Gamma_0}\frac{a\,ds}{gh}\,. \tag{2.28}$$

The reader is kindly asked to prove this identity by himself.

7.2.2 A general behavior of a multiple-contour profile

By reasoning the same way as we did for a singly closed profile, we find that formulas (1.19) through (1.29) remain true for a multiply closed profile, too. In particular, formulas (1.23) and (1.26) define a generalized sectorial coordinate, ϖ, in each point of the multiply closed profile. The only thing to be particularly careful about when defining the generalized sectorial coordinate ϖ for the multiply closed profile is a bit different, comparing to (1.23), form of parameters ω_0, ω_y, ω_z:

$$\omega_0 = \frac{1}{A}\sum_{j=1}^{n}\int_{l_j}\alpha(s)ehds\,, \qquad \omega_y = \frac{1}{I_y}\sum_{j=1}^{n}\int_{l_j}\alpha(s)zehds\,,$$

$$\omega_z = \frac{1}{I_z}\sum_{j=1}^{n}\int_{l_j}\alpha(s)yehds\,, \tag{2.29}$$

though the meaning of the formulas does not change comparing to (1.23).

However, the formulas of the tangential stresses for a multiply-closed profile need refining.

Tangential stresses in a multiple-contour profile

On each of the non-bifurcating segments of the profile, formula (1.30) will be

$$\int_{S_i}^{M}\frac{\partial\sigma^x}{\partial x}hds + T_M - T_{iS} + \int_{S_i}^{M} q_x ds = 0\,, \tag{2.30}$$

where the symbol of S_i denotes a start point of the non-bifurcating i-th segment in question, and the symbol of M is, as usual, an arbitrary current point of that segment.

Substituting formulas (1.29) in here permits to express the value of the flow T_M in the current point M of the profile's segment i via its value T_{iS} in the start point of the same segment,

$$T_M = T_{iS} - N'\frac{A_{io}}{A} - M_z'\frac{S_{ioz}}{I_z} - M_y'\frac{S_{ioy}}{I_y} + E\beta'' S_{io\varpi} - \int_{S_i}^{M} q_x ds\,, \tag{2.31}$$

where instead of (1.32) we use this notation:

$$A_{io} = \int_{S_i}^{M} ehds\,, \quad S_{ioz} = \int_{S_i}^{M} yehds\,, \quad S_{ioy} = \int_{S_i}^{M} zehds\,, \quad S_{io\varpi} = \int_{S_i}^{M} \varpi ehds\,. \qquad (2.32)$$

Obviously, A_{io}, S_{ioy} and S_{ioz} are the respective area and static moments of the cut-off part of the current cross-section's segment, that is, a part of the section between the initial (start) point of i-th segment S_i and the current point M. Similarly, the value of $S_{io\varpi}$ is a *sectorial static moment* of the same cut-off part of the section.

Reasoning the same way as we did with a singly closed profile, we arrive also at (1.39), which is as follows on each non-bifurcating segment of the multiple-contour profile:

$$T_{\mathrm{M}} = T_{i\mathrm{S}} - Q_y \frac{S_{ioz}}{I_z} - Q_z \frac{S_{ioy}}{I_y} - M_\varpi \frac{S_{io\varpi}}{I_\varpi}\,, \qquad (2.33)$$

where we denote

$$M_\varpi = -EI_\varpi \beta'' + m_B\,. \qquad (2.34)$$

Clearly, there can be exactly n relationships like (2.33) – according to the number of the non-bifurcating pieces of the profile.

But relationship (2.33) is true for any point M of the non-bifurcating piece of the contour in question, thus for its "tail point" T_i, too. This permits us to write exactly n equations which have the meaning of equilibrium equations and establish a simple linear relation between the start and tail tangential stresses in each part of the profile:

$$T_{i\mathrm{T}} - T_{i\mathrm{S}} = -Q_y \frac{S_{iz}}{I_z} - Q_z \frac{S_{iy}}{I_y} - M_\varpi \frac{S_{i\varpi}}{I_\varpi} \quad (i = 1,\ldots, n)\,, \qquad (2.35)$$

where we simplify the designations in this way:

- $T_{i\mathrm{T}}$ denotes a flow of the tangential stresses on the "tail" of i-th segment;
- $T_{i\mathrm{S}}$ denotes a flow of the tangential stresses at the "start" of i-th segment;
- S_{iz}, S_{iz}, $S_{i\varpi}$ denote static moments and a sectorial moment of a part of the bar's section which is located between the start and tail points of i-th segment.

To put it another way,

$$S_{iz} = \int_{S_i}^{T_i} yehds\,, \quad S_{iy} = \int_{S_i}^{T_i} zehds\,, \quad S_{i\varpi} = \int_{S_i}^{T_i} \varpi ehds\,. \qquad (2.36)$$

So, we have $2n$ flows in total to be our unknowns: $T_{i\mathrm{T}}$ and $T_{i\mathrm{S}}$ ($i = 1,\ldots, n$), and we still have only n equations of equilibrium to calculate those (2.35).

We can compose t additional equations of equilibrium by extracting a vicinity of each bifurcation point and making an equation of equilibrium of the flows in projections onto the X-axis. For example, by extracting an element of the bar's section in the vicinity of point C (Fig. 7.6) and considering the equilibrium of the extracted elements in projections onto axis X, we will have

$$-T_{1\mathrm{T}} + T_{2\mathrm{S}} - T_{5\mathrm{T}} = 0\ .$$

There are exactly t equations of this kind: one for each bifurcation point.

To write the equations of this kind in the general case, we can use a matrix introduced earlier: the incidence matrix of directed graph $\boldsymbol{G}$ which describes the topology of a multiple-contour profile.

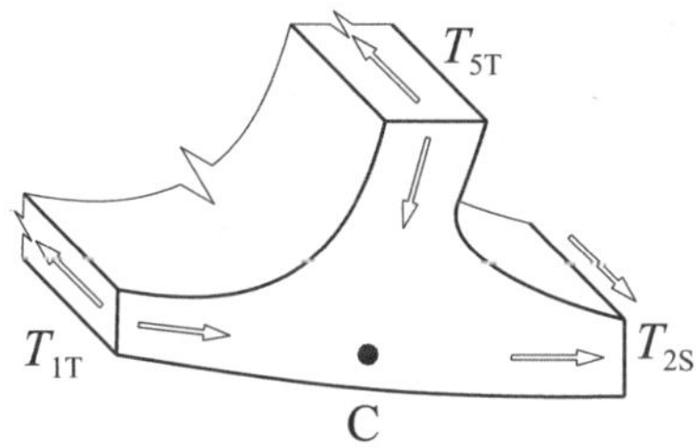

Fig. 7.6. An element of the bar's cross-section in the vicinity of point C

Now, having introduced matrix $\boldsymbol{G} = \|[g_{\alpha i}]\|$, we can write the equations of equilibrium for α-th point of bifurcation of the profile for the most general case,

$$\sum_{i=1}^{n} g_{\alpha i}\left[(g_{\alpha i}+1)T_{i\mathrm{S}} - (g_{\alpha i}-1)T_{i\mathrm{T}}\right] = 0 \qquad (\alpha = 1,\ldots, t). \tag{2.37}$$

It is easy to make sure that actually α-th equation will contain positive start values of the flows on segments the beginnings of which coincide with point α. The same equation will contain negative tail values of the flows on segments the ends of which are coincident with point α. The form (2.37) of the equations of equilibrium of the flows in α-th point of bifurcation is universal and therefore convenient for building practical algorithms and developing analytic software.

However, the set of equations (2.37), in the number of t, contains exactly one equation which is linearly dependent on the rest $(t-1)$ equations. This fact is equivalent to the rank of matrix $\boldsymbol{G}$ being $(t-1)$.

The inequality *rank* $\boldsymbol{G} < t$ can be easily proved. To see this, we add up all rows of $\boldsymbol{G}$ and get a zero row because the sum of the elements of each column is zero. This means that the last t-th equation in system (2.31) is a linear combination of first $(t-1)$ equations[10].

Finally, another equation of equilibrium can be derived from the condition of static equivalence between the torque created by flow $T(s)$ and the total torque, M_x, acting in the section:

$$M_x = \sum_{i=1}^{n} \int_{S_i}^{T_i} T_i(s)\rho(s)ds\,. \tag{2.38}$$

Here $T_i(s)$ means a value of the T_M flow in an arbitrary point of i-th segment, defined by (2.33). The symbol of ρ denotes, as usual, *an arm function*, that is, a projection of the PM vector onto the normal unit vector, $\boldsymbol{n}$, which comes from the current point M on the i-th segment of the contour. We remind that P is a center of twist of the profile.

Replacing $T_i(s)$ with the expression of the flow from (2.33) and integrating gives

$$M_x = \sum_{i=1}^{n}\left[T_{iS}\int_{S_i}^{T_i}\rho ds - \frac{Q_y}{I_z}\int_{S_i}^{T_i} S_{ioz}\rho ds - \frac{Q_z}{I_y}\int_{S_i}^{T_i} S_{ioy}\rho ds - \frac{M_{\varpi}}{I_{\varpi}}\int_{S_i}^{T_i} S_{io\varpi}\rho ds \right]. \tag{2.39}$$

So, we have the following to calculate $2n$ unknown flows T_{iS} and T_{iT}:

- n equilibrium equations (2.35);
- $(t-1)$ equilibrium equations (2.37);
- one equilibrium equation (2.39).

In total we have $(n+t)$ equations of equilibrium. Consequently, the problem of calculating the flows becomes statically indeterminate with its redundance equal to

$$2n - (n+t) = k - 1\,.$$

As we can see, this number is the same as the redundance in our earlier analysis of pure torsion, which is of course quite expectable.

Clearly, the lacking $(k-1)$ equations should be based on the strain compatibility conditions. To construct these lacking equations, we again turn to the Castigliano variational principle.

[10] It would be well to show that first $(t-1)$ equations in system (2.37) are linearly independent, hence the equality *rank* $\boldsymbol{G} = (t-1)$. How about another exercise in proving propositions?

However, before we use the Castigliano principle directly, it will be useful to simplify the set of equilibrium equations with respect to the unknown flows by avoiding the determination of the tail flows, $T_{i\mathrm{T}}$. A matrix form will be most convenient for both systems, (2.35) and (2.37). To arrive at it, we introduce a vector of the flows at the start points of the segments, $\boldsymbol{T}_\mathrm{S}$, and a vector of the flows at the end points of the segments, $\boldsymbol{T}_\mathrm{T}$, by defining

$$\boldsymbol{T}_\mathrm{S} = \left\| \left[T_{1\mathrm{S}}, \ldots, T_{n\mathrm{S}} \right] \right\|^\mathrm{T}, \qquad \boldsymbol{T}_\mathrm{T} = \left\| \left[T_{1\mathrm{T}}, \ldots, T_{n\mathrm{T}} \right] \right\|^\mathrm{T}. \tag{2.40}$$

With these designations, we can rewrite the system of equilibrium cquations (2.35) as

$$\boldsymbol{T}_\mathrm{T} = \boldsymbol{T}_\mathrm{S} - \Delta \boldsymbol{T}, \tag{2.41}$$

where (2.35) defines $\Delta\boldsymbol{T}$ as a vector of increments of the flows on each of the non-bifurcating segments of the profile,

$$\Delta\boldsymbol{T} = \frac{Q_y}{I_z}\boldsymbol{S}_z + \frac{Q_z}{I_y}\boldsymbol{S}_y + \frac{M_\varpi}{I_\varpi}\boldsymbol{S}_\varpi , \tag{2.42}$$

and vectors $\boldsymbol{S}_z, \boldsymbol{S}_y, \boldsymbol{S}_\varpi$ are

$$\boldsymbol{S}_z = \left\| \begin{bmatrix} S_{1z} \\ \vdots \\ S_{nz} \end{bmatrix} \right\|, \quad \boldsymbol{S}_y = \left\| \begin{bmatrix} S_{1y} \\ \vdots \\ S_{ny} \end{bmatrix} \right\|, \quad \boldsymbol{S}_\varpi = \left\| \begin{bmatrix} S_{1\varpi} \\ \vdots \\ S_{n\varpi} \end{bmatrix} \right\| \tag{2.43}$$

with the components defined according to (2.36).

If we supplement the incidence matrix, $\boldsymbol{G}$, with a truncated matrix $\bar{\boldsymbol{G}}$ from which the last row of $\boldsymbol{G}$ has been removed and with a matrix $|\bar{\boldsymbol{G}}|$ composed of absolute values of the $\bar{\boldsymbol{G}}$ components, so that

$$\bar{\boldsymbol{G}} = \left\| \begin{bmatrix} g_{11} & \cdots & g_{1n} \\ \vdots & \ddots & \vdots \\ g_{t-1,1} & \cdots & g_{t-1,n} \end{bmatrix} \right\|, \qquad |\bar{\boldsymbol{G}}| = \left\| \begin{bmatrix} |g_{11}| & \cdots & |g_{1n}| \\ \vdots & \ddots & \vdots \\ |g_{t-1,1}| & \cdots & |g_{t-1,n}| \end{bmatrix} \right\|, \tag{2.44}$$

then the system of first independent $(t-1)$ equations of equilibrium in (2.37) can be rewritten in the matrix form as

$$(|\bar{\boldsymbol{G}}| + \bar{\boldsymbol{G}})\boldsymbol{T}_\mathrm{S} - (|\bar{\boldsymbol{G}}| - \bar{\boldsymbol{G}})\boldsymbol{T}_\mathrm{T} = \boldsymbol{0}.$$

By excluding the tail flows $\boldsymbol{T}_\mathrm{T}$ using (2.41), we get the equations of equilibrium in the number of $(t-1)$, which contain only start flows $\boldsymbol{T}_\mathrm{S}$ for unknowns:

$$2\bar{\boldsymbol{G}}\boldsymbol{T}_\mathrm{S} + (|\,\bar{\boldsymbol{G}}\,| - \bar{\boldsymbol{G}})\Delta\boldsymbol{T} = \boldsymbol{0}\,. \tag{2.45}$$

It will be convenient for us also to rewrite the condition of equilibrium in terms of moments with respect to the X-axis – that is, equation (2.39) – as

$$M_x = \boldsymbol{\omega}^\mathrm{T}\boldsymbol{T}_\mathrm{S} - \frac{Q_y}{I_z}S_{\rho z} - \frac{Q_z}{I_y}S_{\rho y} - \frac{M_\varpi}{I_\varpi}S_{\rho\varpi}\,, \tag{2.46}$$

where we use these additional designations:

$$\boldsymbol{\omega} = \|[\omega_1,\ldots,\omega_n]\|^\mathrm{T}, \quad \omega_i = \int_{\mathrm{S}_i}^{\mathrm{T}_i}\rho\,ds\,,$$

$$S_{\rho z} = \sum_{i=1}^{n}\int_{\mathrm{S}_i}^{\mathrm{T}_i}S_{ioz}\rho\,ds\,, \quad S_{\rho z} = \sum_{i=1}^{n}\int_{\mathrm{S}_i}^{\mathrm{T}_i}S_{ioz}\rho\,ds\,, \quad S_{\rho\varpi} = \sum_{i=1}^{n}\int_{\mathrm{S}_i}^{\mathrm{T}_i}S_{io\varpi}\rho\,ds\,. \tag{2.47}$$

Assuming for simplification that the boundary conditions are homogeneous, we can establish an identity between the Castigliano functional and the expression of the strain energy via the stresses[11]. Thus,

$$\mathsf{K} = \int_0^L\left[\sum_{i=1}^{n}\left(\int_{l_i}\frac{\sigma^2}{2Ee}h\,ds + \int_{l_i}\frac{\tau^2}{2Gg}h\,ds\right)\right]dx\,. \tag{2.48}$$

The summation comprises all non-bifurcating parts of the profile.

Now we replace the normal stress, σ, with its representation from (1.48), and the tangential stress, τ, with its representation via the flow, $\tau = T/h$, where flow $T(s)$ is defined by (2.33) in its turn in every non-bifurcating part of the profile.

As a result, we have

---

[11] The assumption like this does not make the subsequent conclusions any narrower in this case. The matter is that all we need is Euler equations for functional K, which determine the strain compatibility conditions expressed via flows $T_{i\mathrm{S}}$. But the Euler equations for a quadratic functional are known to be independent from the boundary conditions.

$$\mathsf{K}=\frac{1}{2G}\int_0^L\sum_{i=1}^n\int_{l_i}\left[T_{i\mathrm{S}}^2-2T_{i\mathrm{S}}\frac{Q_y}{I_z}S_{ioz}-2T_{i\mathrm{S}}\frac{Q_z}{I_y}S_{ioy}-2T_{i\mathrm{S}}\frac{M_\varpi}{I_\varpi}S_{io\varpi}\right]\frac{ds}{gh}dx+(\ldots),$$

where (…) denotes all other terms not depending on flows $T_{i\mathrm{S}}$.

We introduce another series of additional designations by denoting

$$\boldsymbol{p}=\left\|\begin{bmatrix}p_1 & 0 & \cdots & 0\\ 0 & \ddots & \ddots & \vdots\\ \vdots & \ddots & \ddots & 0\\ 0 & \cdots & 0 & p_n\end{bmatrix}\right\|,\quad \boldsymbol{S}_{hz}=\left\|\begin{bmatrix}S_{1hz}\\ \vdots\\ S_{nhz}\end{bmatrix}\right\|,\quad \boldsymbol{S}_{hy}=\left\|\begin{bmatrix}S_{1hy}\\ \vdots\\ S_{nhy}\end{bmatrix}\right\|,$$

$$\boldsymbol{S}_{h\varpi}=\left\|\begin{bmatrix}S_{1h\varpi}\\ \vdots\\ S_{nh\varpi}\end{bmatrix}\right\| \tag{2.49}$$

with components

$$p_i=\int_{l_i}\frac{ds}{gh},\quad S_{ihz}=\int_{l_i}\frac{S_{ioz}}{gh}ds,\quad S_{ihy}=\int_{l_i}\frac{S_{ioy}}{gh}ds,\quad S_{ih\varpi}=\int_{l_i}\frac{S_{io\varpi}}{gh}ds. \tag{2.50}$$

The diagonal matrix $\boldsymbol{p}$ can be called (and for a good reason) a *weight matrix* of the non-bifurcating parts of the profile.

In terms of the designations just introduced, we can rewrite the K functional as

$$\mathsf{K}=\frac{1}{2G}\boldsymbol{T}_\mathrm{S}^\mathsf{T}\boldsymbol{p}\boldsymbol{T}_\mathrm{S}-\boldsymbol{T}_\mathrm{S}^\mathsf{T}\frac{Q_y}{GI_z}\boldsymbol{S}_{hz}-\boldsymbol{T}_\mathrm{S}^\mathsf{T}\frac{Q_z}{GI_y}\boldsymbol{S}_{hy}-\boldsymbol{T}_\mathrm{S}^\mathsf{T}\frac{M_\varpi}{GI_\varpi}\boldsymbol{S}_{h\varpi}+(\ldots). \tag{2.51}$$

Further we should minimize the K functional with respect to the components of the $\boldsymbol{T}_\mathrm{S}$ vector under additional conditions (2.45) and (2.46).

We use a standard Lagrange-multiplier-based approach to reduce this problem to the search for a stationary point of a modified functional, M, by assuming

$$\mathsf{M}=\mathsf{K}+\boldsymbol{\Lambda}^\mathsf{T}\left[2\bar{\boldsymbol{G}}\boldsymbol{T}_\mathrm{S}+(|\bar{\boldsymbol{G}}|-\bar{\boldsymbol{G}})\Delta\boldsymbol{T}\right]+$$

$$+\lambda_t\left[\boldsymbol{\omega}^\mathsf{T}\boldsymbol{T}_\mathrm{S}-\frac{Q_y}{I_z}S_{\rho z}-\frac{Q_z}{I_y}S_{\rho y}-\frac{M_\varpi}{I_\varpi}S_{\rho\varpi}-M_x\right], \tag{2.52}$$

where $\mathbf{\Lambda} = \|[\lambda_1, \ldots, \lambda_{t-1}]\|^{\mathrm{T}}$ is a vector of Lagrangian multipliers of the order $t-1$ and λ_t is another (additional) Lagrangian multiplier.

The stationarity of M yields the following set of simultaneous linear algebraic equations of the order $(n+t)$[12]:

$$\left\|\begin{matrix} \boldsymbol{p}/G & 2\bar{\boldsymbol{G}}^{\mathrm{T}} & \boldsymbol{\omega} \\ 2\bar{\boldsymbol{G}} & \mathbf{0} & \mathbf{0} \\ \boldsymbol{\omega}^{\mathrm{T}} & \mathbf{0} & 0 \end{matrix}\right\| \left\|\begin{matrix} \boldsymbol{T}_{\mathrm{S}} \\ \mathbf{\Lambda} \\ \lambda_t \end{matrix}\right\| = M_x \left\|\begin{matrix} \mathbf{0} \\ \mathbf{0} \\ 1 \end{matrix}\right\| +$$

$$+\frac{Q_y}{I_z}\left\|\begin{matrix} \boldsymbol{S}_{hz}/G \\ (\bar{\boldsymbol{G}}-|\bar{\boldsymbol{G}}|)\boldsymbol{S}_z \\ S_{\rho z} \end{matrix}\right\| + \frac{Q_z}{I_y}\left\|\begin{matrix} \boldsymbol{S}_{hy}/G \\ (\bar{\boldsymbol{G}}-|\bar{\boldsymbol{G}}|)\boldsymbol{S}_y \\ S_{\rho y} \end{matrix}\right\| + \frac{M_\varpi}{I_\varpi}\left\|\begin{matrix} \boldsymbol{S}_{h\varpi}/G \\ (\bar{\boldsymbol{G}}-|\bar{\boldsymbol{G}}|)\boldsymbol{S}_\varpi \\ S_{\rho\varpi} \end{matrix}\right\|. \quad (2.53)$$

The solution of this system defines the components of the vector of start flows, $\boldsymbol{T}_{\mathrm{S}}$, which we want to know. This vector can be represented as

$$\boldsymbol{T}_{\mathrm{S}} = M_x \boldsymbol{b}_x + \frac{Q_y}{I_z}\boldsymbol{b}_z + \frac{Q_z}{I_y}\boldsymbol{b}_y + \frac{M_\varpi}{I_\varpi}\boldsymbol{b}_\varpi . \quad (2.54)$$

---

[12] We anticipate a perplexity, if not a protest, of some (supposedly, not very numerous) readers educated by books on structural mechanics written in fifties of the last century or even much newer publications. Why, such a reader would ask, do you want me to solve a system of equations of a higher order, $(n+t)$, while I know that I can (and how I can) use a system of equations of the order $(n-t) = (k-1)$?

Here I count on the understanding of mostly younger professionals in structural mechanics, for which the calculation in our science can (and should) no longer be done without computer programming. Such readers would not need my explanation to understand that the principal criteria by which a computational algorithm is assessed in these latter days, since the second half of the twentieth century, are not ingenious tricks for reducing the orders of equations systems but, rather, such things as versatility, simple algorithmization etc. It is a wondrous thing how our attitude to "*good and bad*" has changed along with the tools of caclulation! From manual calculations on paper, through the now forgotten abacus, counting frame, and sliding rule, through the arithmometer, through first analog and digital electron-tube computers to the contemporary advanced personal computers – this is one impressive way, not only for structural mechanics. What then awaits us ahead?

Each of vectors $\boldsymbol{b}_x$, $\boldsymbol{b}_y$, $\boldsymbol{b}_z$, $\boldsymbol{b}_\varpi$ is a vector of the order n composed of coefficients which are taken from the solution of equations (2.53). More accurately, each of the vectors is a solution of the system of equations with the same **M** matrix,

$$\mathbf{M} = \left\| \begin{matrix} p/G & 2\bar{\boldsymbol{G}}^{\mathrm{T}} & \boldsymbol{\omega} \\ 2\bar{\boldsymbol{G}} & \mathbf{0} & \mathbf{0} \\ \boldsymbol{\omega}^{\mathrm{T}} & \mathbf{0} & 0 \end{matrix} \right\|, \tag{2.55}$$

but with different right-hand parts, where

$$\mathbf{M}\left\| \begin{matrix} \boldsymbol{b}_x \\ \boldsymbol{\Lambda} \\ \lambda_t \end{matrix} \right\| = \left\| \begin{matrix} \mathbf{0} \\ \mathbf{0} \\ 1 \end{matrix} \right\|, \qquad \mathbf{M}\left\| \begin{matrix} \boldsymbol{b}_z \\ \boldsymbol{\Lambda} \\ \lambda_t \end{matrix} \right\| = \left\| \begin{matrix} \boldsymbol{S}_{hz}/G \\ (\bar{\boldsymbol{G}} - |\bar{\boldsymbol{G}}|)\boldsymbol{S}_z \\ S_{\rho z} \end{matrix} \right\|,$$

$$\mathbf{M}\left\| \begin{matrix} \boldsymbol{b}_y \\ \boldsymbol{\Lambda} \\ \lambda_t \end{matrix} \right\| = \left\| \begin{matrix} \boldsymbol{S}_{hy}/G \\ (\bar{\boldsymbol{G}} - |\bar{\boldsymbol{G}}|)\boldsymbol{S}_y \\ S_{\rho y} \end{matrix} \right\|, \qquad \mathbf{M}\left\| \begin{matrix} \boldsymbol{b}_\varpi \\ \boldsymbol{\Lambda} \\ \lambda_t \end{matrix} \right\| = \left\| \begin{matrix} \boldsymbol{S}_{h\varpi}/G \\ (\bar{\boldsymbol{G}} - |\bar{\boldsymbol{G}}|)\boldsymbol{S}_\varpi \\ S_{\rho\varpi} \end{matrix} \right\|. \tag{2.56}$$

Of course, the values of Lagrangian multipliers $\lambda_1, \ldots, \lambda_t$ are different in the solution of each particular system of equations, but this is inessential for us because we are little interested with the multipliers at the moment.

Excluding the start flows from the general formula (2.33) of flows in an arbitrary point using (2.54) gives

$$T_i(s) = M_x b_{ix} - \frac{Q_y}{I_z}\bar{S}_{ioz} - \frac{Q_z}{I_y}\bar{S}_{ioy} - \frac{M_\varpi}{I_\varpi}\bar{S}_{io\varpi}, \tag{2.57}$$

where functions $\bar{S}_{ioz}, \bar{S}_{ioy}, \bar{S}_{io\varpi}$ are defined on each of the non-bifurcating parts of the contour as

$$\bar{S}_{ioz} = S_{ioz} - b_{iz}, \quad \bar{S}_{ioy} = S_{ioy} - b_{iy}, \quad \bar{S}_{io\varpi} = S_{io\varpi} - b_{i\varpi}. \tag{2.58}$$

Formula (2.57) is a counterpart of (1.49) which we established and used earlier for singly closed cross-sections. In exactly the same way, formulas (2.58) are counterparts of (1.43).

Now let us show that the components of vector $\boldsymbol{b}_x$ are not quite unfamiliar to us. They are related in a simple way to the components of the earlier vector, $\boldsymbol{a} = \|[a_1,\ldots,a_n]\|^{\mathrm{T}}$, where a_i are coefficients of distribution of the flows over the non-bifurcating parts of the profile in pure torsion – see (2.18). To see this, we assume we are dealing with a pure torsion. Then $M_x = H$, $Q_y = Q_z = M_\varpi = 0$, so (2.57) gives

$$T_i = Hb_{ix}\,.$$

But the flows obey formula (2.19) when in pure torsion, or

$$T_i = a_i \frac{H}{\Omega_0}\,.$$

Equaling these two expressions of the pure-torsion flows gives

$$\boldsymbol{a} = \Omega_0 \boldsymbol{b}_x\,. \tag{2.59}$$

Now we represent the full torque, M_x, as a sum of the pure-torsion torque, H, and the constricted-torsion torque, M_ϖ, to rewrite formula (2.57) in a slightly more convenient form:

$$T_i(s) = \frac{H}{\Omega_0} a_i - \frac{Q_y}{I_z}\overline{S}_{ioz} - \frac{Q_z}{I_y}\overline{S}_{ioy} - \frac{M_\varpi}{I_\varpi}\overline{\overline{S}}_{io\varpi}\,, \tag{2.60}$$

where we assume

$$\overline{\overline{S}}_{io\varpi} = \overline{S}_{io\varpi} - a_i \frac{I_\varpi}{\Omega_0}\,. \tag{2.61}$$

This ends our consideration of the tangential stresses in the theory of thin-walled bars having multiple-contour closed profiles.

Functions $\overline{S}_{ioz}, \overline{S}_{ioy}, \overline{S}_{io\varpi}$ just introduced are multiple-contour generalizations of functions $\overline{S}_{oz}, \overline{S}_{oy}, \overline{S}_{o\varpi}$ which we introduced earlier for a singly closed profile. These functions can be shown to have characteristic properties of functions $\overline{S}_{oz}, \overline{S}_{oy}, \overline{S}_{o\varpi}$. We will not spend our time giving formal grounds; we just want to note that the following formulas are true:

$$\sum_{i=1}^{n} a_i \int_{l_i} \frac{\overline{S}_{ioy}}{gh} ds = 0\,, \qquad \sum_{i=1}^{n} a_i \int_{l_i} \frac{\overline{S}_{ioz}}{gh} ds = 0\,, \qquad \sum_{i=1}^{n} a_i \int_{l_i} \frac{\overline{\overline{S}}_{io\varpi}}{gh} ds = 0\,; \tag{2.62}$$

these will be needed in the next section.

Governing equations for thin-walled bars having closed multiple-contour profiles

It is easy to notice that the structure of the governing equations for the shear and semi-shear theories, which we have established for a singly closed profile, is thoroughly applicable to a multiple-contour profile.

We confine ourselves to considering the first (energy-based) version of the theory and find that the only difference in the multiple-contour case lies in formulas for the elements of the shape factor matrix, $\boldsymbol{\mu}$. To obtain the $\boldsymbol{\mu}$ matrix using (2.59), we write the expression of the strain energy, E_τ, in the form of

$$\mathsf{E}_\tau = \int_0^L \sum_{i=1}^n \left[\int_{l_i} \left(\frac{H}{\Omega_0 h} a_i - \frac{Q_y}{hI_z}\bar{S}_{ioz} - \frac{Q_z}{hI_y}\bar{S}_{ioy} - \frac{M_\varpi}{hI_\varpi}\bar{\bar{S}}_{io\varpi} \right)^2 \frac{h}{2Gg} ds \right] dx \,. \qquad (2.63)$$

Now we transform the expression of energy E_τ. We have

$$\mathsf{E}_\tau = \frac{1}{2G}\int_0^L \sum_{i=1}^n \left[\frac{H^2}{\Omega_0^2} a_i^2 \int_{l_i} \frac{ds}{gh} + \frac{Q_z^2}{I_y^2}\int_{l_i}\frac{\bar{S}_{ioy}^2 ds}{gh} + \frac{Q_y^2}{I_z^2}\int_{l_i}\frac{\bar{S}_{ioz}^2 ds}{gh} + \frac{M_\varpi^2}{I_\varpi^2}\int_{l_i}\frac{\bar{\bar{S}}_{io\varpi}^2 ds}{gh} \right] dx +$$

$$+\frac{1}{G}\int_0^L \sum_{i=1}^n \left[-\frac{HQ_z}{\Omega_0 I_y} a_i \int_{l_i}\frac{\bar{S}_{ioy} ds}{gh} - \frac{HQ_y}{\Omega_0 I_z} a_i \int_{l_i}\frac{\bar{S}_{ioz} ds}{gh} - \frac{HM_\varpi}{\Omega_0 I_\varpi} a_i \int_{l_i}\frac{\bar{\bar{S}}_{io\varpi} ds}{gh} \right] dx +$$

$$+\frac{1}{G}\int_0^L \frac{Q_z Q_y}{I_y I_z} \sum_{i=1}^n \int_{l_i} \frac{\bar{S}_{ioz}\bar{S}_{ioy} ds}{gh} dx +$$

$$+\frac{1}{G}\int_0^L \sum_{i=1}^n \left[\frac{Q_z M_\varpi}{I_y I_\varpi}\int_{l_i}\frac{\bar{S}_{ioy}\bar{\bar{S}}_{io\varpi} ds}{gh} + \frac{Q_y M_\varpi}{I_z I_\varpi}\int_{l_i}\frac{\bar{S}_{ioz}\bar{\bar{S}}_{io\varpi} ds}{gh} \right] dx \,. \qquad (2.64)$$

The second line of this expression of E_τ vanishes because of the properties of (2.62). Note also that (2.25) and (2.28) imply

$$\frac{1}{2G}\int_0^L \sum_{i=1}^n \frac{H^2}{\Omega_0^2} a_i^2 \int_{l_i}\frac{ds}{gh} dx = \int_0^L \frac{H^2}{2GI_x} dx \,. \qquad (2.65)$$

Thus we finally have the following for the multiple-contour profile:

$$\mathsf{E}_\tau = \int_0^L \left[\frac{H^2}{2GI_x} + \frac{Q_z^2 \mu_{zz}}{2GA} + \frac{Q_y^2 \mu_{yy}}{2GA} + \frac{M_\varpi^2 \mu_{\varpi\varpi}}{2r^2 GA} + \frac{Q_z Q_z \mu_{zy}}{GA} + \frac{Q_z M_\varpi \mu_{z\varpi}}{rGA} + \frac{Q_y M_\varpi \mu_{y\varpi}}{rGA} \right] dx \,,$$

where numerical parameters μ_{zz}, μ_{zy}, $\mu_{z\varpi}$, μ_{yy}, $\mu_{y\varpi}$, $\mu_{\varpi\varpi}$ make up a matrix of the section's shape factors, $\boldsymbol{\mu}$, given below:

$$\boldsymbol{\mu} = \left\| \begin{bmatrix} \mu_{zz} & \mu_{zy} & \mu_{z\varpi} \\ \mu_{yz} & \mu_{yy} & \mu_{y\varpi} \\ \mu_{\varpi z} & \mu_{\varpi y} & \mu_{\varpi\varpi} \end{bmatrix} \right\| =$$

$$= \left\| \begin{bmatrix} \frac{A}{I_y^2}\sum_{i=1}^{n}\int_{l_i}\frac{\bar{S}_{ioy}^2}{gh}ds & \frac{A}{I_y I_z}\sum_{i=1}^{n}\int_{l_i}\frac{\bar{S}_{ioy}\bar{S}_{ioz}}{gh}ds & \frac{\sqrt{I_r A}}{I_y I_\varpi}\sum_{i=1}^{n}\int_{l_i}\frac{\bar{S}_{ioy}\bar{\bar{S}}_{io\varpi}}{gh}ds \\ \frac{A}{I_z I_y}\sum_{i=1}^{n}\int_{l_i}\frac{\bar{S}_{ioz}\bar{S}_{ioy}}{gh}ds & \frac{A}{I_z^2}\sum_{i=1}^{n}\int_{l_i}\frac{\bar{S}_{ioz}^2}{gh}ds & \frac{\sqrt{I_r A}}{I_z I_\varpi}\sum_{i=1}^{n}\int_{l_i}\frac{\bar{S}_{ioz}\bar{\bar{S}}_{io\varpi}}{gh}ds \\ \frac{\sqrt{I_r A}}{I_\varpi I_y}\sum_{i=1}^{n}\int_{l_i}\frac{\bar{\bar{S}}_{io\varpi}\bar{S}_{ioy}}{gh}ds & \frac{\sqrt{I_r A}}{I_\varpi I_z}\sum_{i=1}^{n}\int_{l_i}\frac{\bar{\bar{S}}_{io\varpi}\bar{S}_{ioz}}{gh}ds & \frac{I_r}{I_\varpi^2}\sum_{i=1}^{n}\int_{l_i}\frac{\bar{\bar{S}}_{io\varpi}^2}{gh}ds \end{bmatrix} \right\| . \quad (2.66)$$

As the operator of equilibrium, $\boldsymbol{A}^{\mathrm{T}}$, which we established earlier by (**6**.3.12), is the same for all kinds of profiles in the shear theory, its Lagrange-conjugate operator of geometry, $\boldsymbol{A}$, must not depend on the profile type, too. The latter operator is defined by (**6**.3.13).

The physical relationships are defined by matrix $\boldsymbol{C}^{-1}$ according to (1.65), where, of course, the geometrical characteristics of the section are understood as calculated for the multiple-contour profile.

Due to the said circumstances, the governing equations of the shear theory for the multiple-contour profile are (**6**.3.24), and those of the semi-shear theory are (**6**.4.11).

References

1. Adadurov RA (1947) Determining tangential stresses in thin-walled structures near an immuring (in Russian). Proc. Centr. Aerohydr. Inst., N614: pp 1–13
2. Adadurov RA (1948) Stresses and strains in a cylindrical shell with rigid transversal constraints (in Russian). Rep. Acad. Sci., 62, N2: pp 183–186
3. Christofides N (1975) Graph theory. An algorithmic approach. Academic Press, New York, London, San Francisko
4. Janelidze GY, Panovko YG (1948) Statics of elastic thin-walled bars. Gostekhteorizdat, Moscow Leningrad
5. Koniukhovsky VS, Khaletskaya OB (1995) Theory of Kirchhoff cycles in application to the bar torsion analysis (in Russian). In: Researches in

mechanics of structural constructions and materials. Collected papers of St.-Petersb. State Univ. of Civil Engng and Architecture, pp 26–44
6. Smirnov VI (1974) A course on advanced mathematics (in Russian). Vol. 2. Physmathgiz, Moscow
7. Umanski AA (1939) Bending and torsion of thin-walled aircraft constructions (in Russian). Oborongiz, Moscow
8. Umanski AA (1940) On normal stresses in torsion of an aircraft wing (in Russian). In: Tekhnika vozdushnogo flota (Aircraft engineering), N12, pp 48–65

8 PARTICULAR CLASSES OF PROBLEMS IN STRUCTURAL MECHANICS – part 5

...only those of all investigations are valuable which are initiated by applications... and only those theories are really useful which follow from the analysis of particular cases.
Liapunov AM. From an essay dedicated to P.L.Chebyshev.

8.1 Compound-profile thin-walled bars

The two preceding chapters are dedicated to two separate theories of the open-profile and the closed-profile thin-walled bars. However, a practical engineer often encounters thin-walled bars which cannot be unquestionably classifed into either category. A characteristic example is the cross-section of a box-shaped or even a two-box stiffening girder of a bridge's span (Fig. 8.1).

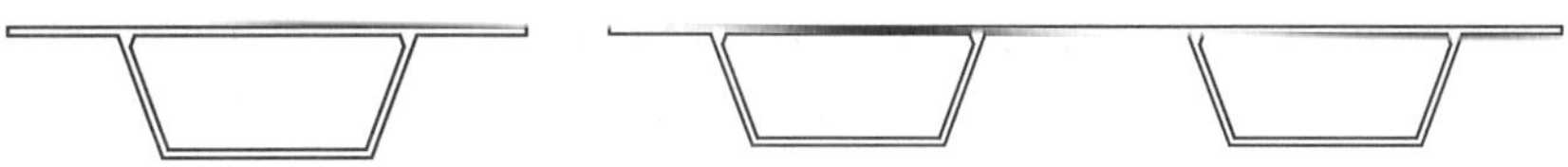

Fig. 8.1. Examples of compound thin-walled profiles

Another complication of the design model of a thin-walled bar, which is also often encountered in practice, is the presence of so-called elastic warping constraints in the cross-section. A structural prototype of such constraints can be something like transversal straps or struts (Fig. 8.2).

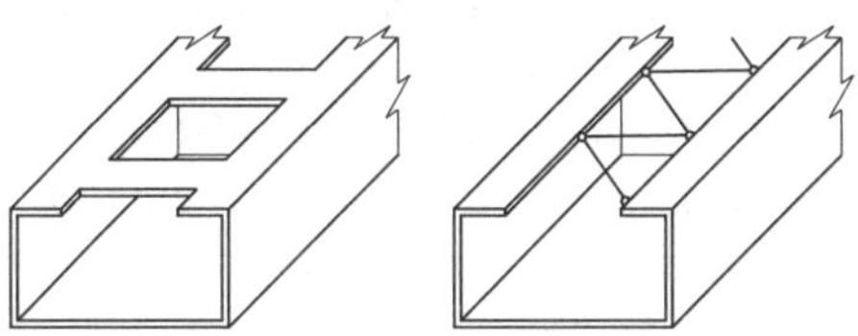

Fig. 8.2. Examples of thin-walled profiles with warping constraints

Of course, experts in the theory of thin-walled bars have always understood how pressing is the need of the engineering science for a unified practical theory free from any limitations of the profile type. There have been numerous attempts of developing such a theory. Not meaning to give a full review, we would like to mention works by D.G. Eliashvili [7], [8], I.V. Kruglenko [13], O.V. Luzhin [14], [15], and also by foreign authors [10], [9], [16].

It is important to notice that one of the hardest issues in the theory is a passing-to-limit requirement. This requirement demands that the theory of open-profile thin-walled bars should be implied by (follow from) the compound theory as a particular case, a result of passing to a limit. Similarly, the closed-profile thin-walled bar theory should be another particular case of the same compound theory. However, these two theories were no sooner combined into something more general than a lot of effort had been undertaken.

The matter is that the no-shear hypothesis in V.Z. Vlasov's theory stands in contradiction with the basic principles of A.A. Umansky's theory where the physical relationship $\tau^{xs} = G\gamma_{xs}$ is critical, so assuming no shear, $\gamma_{xs} = 0$, does not work for the closed-profile bars.

There is another circumstance that, according to opinions of some authors, is an obstacle for a union between the two theories. We mean the hypotheses concerning the distribution of the tangential stresses over the thickness of the walls of a thin-walled section. Umanski's theory uses the assumption that the tangential stresses are uniform throughout the thickness of the walls, while in Vlasov's theory the stresses are generally linear (see Fig. 6.6), so Vlasov assumes them to be $\tau^{xs} = \tau + \tau_k$. As it is not possible to omit the pure-torsion stresses τ_k in the theory by Vlasov, one could try to modify the theory by Umanski: by introducing stresses τ_k linearly distributed over the thickness of the walls and having zero values directly on the profile.

A similar reasoning led O.V. Luzhin to one of simplest compound theories, which is by far the best recognized theory among experts in the thin-walled bar analysis. E.A. Beilin [3], [6], [4] generalized the approach of the theory by Luzhin afterwards, by making modifications needed to allow for effects of the warping constraints. It is important to notice that Luzhin's theory deems it significant to formulate its physical relationships in the way Umanski does in his semi-shear theory of closed-profile thin-walled bars, rather than use the energy-based approach.

Of course, allowing for tangential stresses τ_k additionally in the closed part of the section introduces a certain refining correction to the general theory of thin-walled bars. However, as we will show later, this refinement is not substantial for a compound theory based on the energy approach.

It seems to us that the actual difficulty in the creation of a unified theory was how to connect the shear-free theory by Vlasov with the semi-shear theory by Umanski. However, as soon as we build the theory of open-profile thin-walled bars as a semi-shear one, there are no more difficulties of the unification. A similar unification of the full shear theories for the opne-profile and closed-profile bars does not meet any obstacles.

Having made these introductory notes, we begin to present the theory of compound-profile bars by using Luzhin's generalizing refinement (that allows for τ_k in the closed part of the section). As we will see further, ignoring this refinement will not make significant changes to the general theory.

For the sake of simplicity, we begin the consideration of the combined theory by assuming that the profile in question contains a single closed contour like one shown on the left in Fig. 8.1.

8.1.1 Pure torsion of a compound-profile thin-walled bar

The theory by Luzhin postulates for pure torsion of a compound-profile bar that the open parts of the contour have the tangential stresses τ_k only, while the closed contours have both the tangential stress τ and the tangential stress τ_k. The total torque created by all these stresses is

$$M_x = T\Omega + GI_k\theta' \qquad (1.1)$$

where

$$I_k = \frac{1}{3}\int_l gh^3 ds\,, \qquad (1.2)$$

and the integral is taken along the whole length of the compound contour[1]. The k subscript emphasizes that we mean the characteristics corresponding to the action of tangential stresses τ_k only.

[1] Formula (1.2), derived for the closed part of the profile by formally inheriting the respective formula for the open part, is of course a pretty rough approximation. However, the fact is that the moment of the tangential stresses τ_k over the closed part of the section is a small value comparing to the moment of the average tangential stresses, $T\Omega$. This means that the further refinement of the small term's value is not really practical, so the approximation given by (1.2) is quite usable in engineering applications for both the open and the closed parts of the section.

The first term in (1.1) follows from formula (7.1.3) for the closed-profile section. It defines a torque created by the flow, T, of tangential stresses τ which act along the closed part of the profile.

Making the same derivation for the closed part of the profile as we did in Section 7.1.1, we have that the twist θ' obeys the relation

$$\theta' = \frac{T}{G\Omega}\oint\frac{ds}{gh} = \frac{T\Omega}{GI_d}. \tag{1.3}$$

Here we follow Luzhin and Beilin [14], [4] in denoting

$$I_d = \frac{\Omega^2}{\oint\frac{ds}{gh}}. \tag{1.4}$$

In other words, I_d is a torsional moment of inertia of a section produced from the one in question by removing all segments of it not included in the closed part – see formula (7.1.15) for the purely closed-profile sections.

Please note that the integral of $\oint()ds$ means everywhere an integral taken along the segments of the profile which are included in its closed part only.

Replacing the twist θ' in (1.1) by using (1.3), we arrive at a formula that defines the flow T in the closed part of the profile,

$$T = \frac{M_x}{\Omega(1 + I_\text{k}/I_d)}. \tag{1.5}$$

Now, if we put the obtained expression of the flow in formula (1.3), we will have a relation of twist θ' vs. torque M_x. Thus,

$$\theta' = \frac{M_x}{G(I_d + I_\text{k})}. \tag{1.6}$$

It is now clear that the combined theory demands that the torsional moment of inertia, I_x, of the whole compound section should be calculated as

$$I_x = I_d + I_\text{k}. \tag{1.7}$$

Now let us construct the warp function, $\alpha(s)$, for the compound profile.

For the purpose of our further presentment, we would like to introduce the notion of a *characteristic function of the profile*, $\kappa(s)$, by defining

- $\kappa(s) = 1$, if the point of the profile with the arc coordinate s belongs to the closed part of the profile;
- $\kappa(s) = 0$, if the point of the profile with the arc coordinate s belongs to the open part of the profile only.

This function makes it possible to write the respective integrals over the whole closed part of the profile, over the whole open part of the profile, and over the whole profile as

$$\int_l (\cdot)\kappa\, ds = \oint ()ds\,, \qquad \int_l (\cdot)(1-\kappa)\, ds = \int_l (\cdot)\, ds - \oint ()ds\,, \qquad \int_l (\cdot)\, ds\,,$$

where $(\cdot)$ denotes an integrand, a function of the arc coordinate s.

Formula (1.2) of the geometric characteristic I_k can be conveniently represented as a sum of two terms,

$$I_k = I_{k0} + I_{k1}, \qquad I_{k0} = \frac{1}{3}\int_l gh^3(1-\kappa)ds\,, \qquad I_{k1} = \frac{1}{3}\int_l gh^3\kappa ds\,. \tag{1.8}$$

Obviously, I_{k0} characterizes the torsional rigidity of the section that corresponds to stresses τ_k in the open part of the profile. The value of I_{k1} characterizes a similar torsional rigidity which corresponds to stresses τ_k in the closed part of the profile only.

As for the longitudinal displacements, u, they still obey formula (7.1.9) within the closed part of the profile, where we can assume $u_0 = 0$ by means of choosing a location where a longitudinal constraint preventing the rigid displacements of the bar should be imposed. In other words, we have this within the closed part of the profile:

$$u = \frac{T}{G}\int_0^s \frac{ds}{gh} - \theta' \int_0^s \rho ds\,. \tag{1.9}$$

Replacing T and θ' by their expressions according to (1.5) and (1.6), we come up with the following formula for the longitudinal displacements, u, of points that belong to the closed part of the profile:

$$u = \frac{M_x}{GI_x}\left(\frac{I_d}{\Omega}\int_0^s \frac{ds}{gh} - \int_0^s \rho ds\right). \tag{1.10}$$

Let a point R of the profile be its bifurcating point. It means that the R point belongs simultaneously to both the open and the closed part of the profile. We consider an open segment of the profile between its end points R and F (Fig. 8.3).

Recall that we assume the no-shear hypothesis on the open segments of the profile; it says that

$$\frac{\partial u}{\partial s}+\frac{\partial v}{\partial x}=0,$$

wherefrom we have the following for an arbitrary point M of the segment in question:

$$u_{\mathrm{M}}=u_{\mathrm{R}}-\int_{\mathrm{R}}^{\mathrm{M}}\frac{\partial v}{\partial x}ds.$$

If we restrain the pole P from lateral displacements, thus satisfying one of the conditions of the rigid displacement prevention in the (Y,Z)-plane, we will have $v=\theta\rho$ and thus

$$u_{\mathrm{M}}=u_{\mathrm{R}}-\theta'\omega_{\mathrm{RM}}, \tag{1.11}$$

where ω_{RM} is an increment of the sectorial coordinate, ω, as we move from point R to the current point M, i.e.

$$\omega_{\mathrm{RM}}=\int_{\mathrm{R}}^{\mathrm{M}}\rho ds. \tag{1.12}$$

We can write a general expression for the longitudinal displacement, u_{M}, at an arbitrary current point of the profile, M, as follows, using the characteristic function of the profile, κ, which we have introduced above:

$$u=\frac{M_x}{GI_x}\left(\frac{I_d}{\Omega}\int_0^s\frac{\kappa ds}{gh}-\int_0^s\kappa\rho ds\right)-\theta'\int_0^s(1-\kappa)\rho ds$$

Here we integrate from a chosen origin (start point) of the profile, O, to the current point M with the arc coordinate s, i.e. along the O-R-M segment of the profile.

Substituting in the expression of the twist from (1.6) gives the final one,

$$u=\frac{M_x}{GI_x}\left(\frac{I_d}{\Omega}\int_0^s\frac{\kappa ds}{gh}-\int_0^s\rho ds\right)=\frac{M_x}{GI_x}\left(\frac{I_d}{\Omega}p-\omega\right), \tag{1.13}$$

where $p(s)$ is a so-called *weight function of the arc coordinate* s, and ω is the sectorial coordinate of the current point M,

$$p(s)=\int_0^s\frac{\kappa ds}{gh}, \qquad \omega=\int_0^s\rho ds. \tag{1.14}$$

We represent the longitudinal displacements u, as usual, via the warp function $\alpha(s)$ as

$$u = \frac{M_x}{GI_x}\alpha(s),$$

and thus we have this final relation for the compound profile:

$$\alpha = \frac{I_d}{\Omega}p - \omega. \qquad (1.15)$$

Note that formula (1.15) for the warp function α works in the limit cases, too: for a purely closed profile and for a purely open one. In the former case, the weight function p takes exactly the values it should take for the purely closed profile. In the latter case, the weight function p becomes identical to zero.

There is a pitfall worth noticing: a possibly erroneous treatment of formula (1.15) for points belonging to the open part of the profile. It is a mistake to think that the weight function p is always zero on the open part of a compound profile. If it were, it would make the warp function $\alpha(s)$ discontinuous which is absolutely inadmissible because the longitudinal displacements of the profile are continuous. A different thing about the weight function is true: $p(s)$ is a constant on the open part of the profile, and this constant is the value of the function at the bifurcation point on the closed segment of the profile, from which our open part in question protrudes. For example, this constant is $p(\mathrm{R})$ for the profile shown in Fig. 8.3, for a piece between the points R and F:

$$p(\mathrm{R}) = \int_O^{\mathrm{R}} \frac{ds}{gh}.$$

Obviously, the warp function, $\alpha(s)$, depends on the choice of the position of the pole, P, on which the arm function $\rho(s)$ is based, and on the position of the profile's origin or start point, O, from which the arc coordinate is counted. The mechanical interpretation of the pole P is a fixation point through which to restrain the bar in pure torsion from rigid displacements in the (Y,Z)-plane. In addition to that, choosing the start point of the profile, O, means choosing a method of fixation (by choosing a point on the profile) of this same bar against rigid translational displacements along X.

8.1.2 General behavior of a compound-profile thin-walled bar

Repeating the reasoning of Section 7.1.2, we can see that all formulas (7.1.18) through (7.1.29) work for a compound profile, too. However, it should be noted especially that the integration over a closed contour in those formulas is to be replaced by the integration over the whole length of the compound profile. For example, formulas (7.1.23), (7.1.27), and (7.1.28) are as follows for the compound profile:

$$\omega_0 = \frac{1}{A}\int_l \alpha(s)ehds\,,\quad \omega_y = \frac{1}{I_y}\int_l \alpha(s)zehds\,,\quad \omega_z = \frac{1}{I_z}\int_l \alpha(s)yehds\,, \tag{1.16}$$

$$\int_l \varpi ehds = 0\,,\quad \int_l \varpi ehyds = 0\,,\quad \int_l \varpi ehzds = 0\,, \tag{1.17}$$

where, as we should remind, the generalized sectorial coordinate ϖ is defined as

$$\varpi = \omega_0 - \alpha\,. \tag{1.18}$$

It can be checked directly that the independence of the generalized sectorial coordinate ϖ from the origin of the arc coordinate s takes place for the compound profile, too. Note also that second and third conditions in (1.17) define the location of the principal pole, P, of the profile.

Under these conditions, the formula of normal stresses σ^x,

$$\sigma^x = \frac{N}{A}e + \frac{M_z}{I_z}ey + \frac{M_y}{I_y}ez + \frac{B}{I_\varpi}e\varpi\,, \tag{1.19}$$

can be extended onto the compound profile without any modifications. However, things regarding average tangential stresses (or flows $T(s)$, which are nearly the same) should be revised.

Average tangential stresses

We will choose (arbitrarily) a certain point O of the profile to be its start point, i.e. its curvilinear coordinate origin. Moving from point O to the current point M gives the formula of the flows (7.1.39) which we represent here as

$$T_M = T_0 - Q_y\frac{S_{oz}}{I_z} - Q_z\frac{S_{oy}}{I_y} - M_\varpi\frac{S_{o\varpi}}{I_\varpi}\,, \tag{1.20}$$

where the integrals

$$S_{oz} = \int_{O}^{M} yehds\,, \quad S_{oy} = \int_{O}^{M} zehds\,, \quad S_{o\varpi} = \int_{O}^{M} \varpi ehds \tag{1.21}$$

should be understood as the respective static moments and the sectorial static moment of the cut-off part of the section located between the initial point O and the current point M of the profile.

To give an explanatory example, we show the cut-off parts of the section darkened in Fig. 8.3 for three different locations of the current point M.

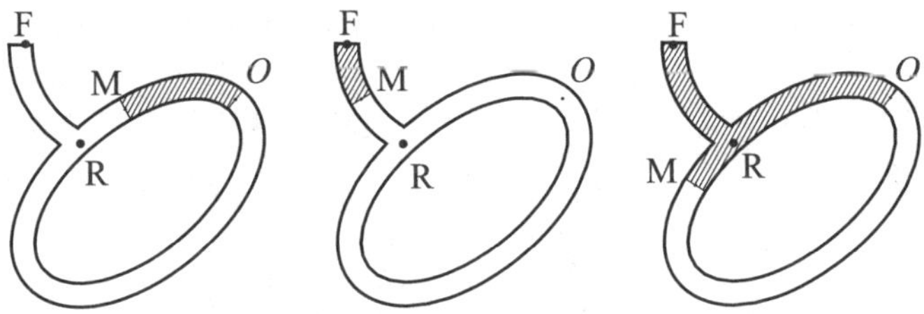

Fig. 8.3. Cut-off parts of the section for various locations of the M point

Obviously, formula (1.20) works when the current point of the profile M belongs to the closed part of the section (Fig. 8.3). In the case when the M point belongs to the open part of the section (which in our case lies between the bifurcating point of the profile, R, and the extreme point F), the expression of flow T_M will no longer contain the starting point O because the integration can be done by moving from the extreme point F to the current point M. However, as the flow T_F at the extreme point of the profile is guaranteed to equal zero, we obtain the following formula for the flows on the open segments of the profile:

$$T = -Q_y \frac{S_{oz}}{I_z} - Q_z \frac{S_{oy}}{I_y} - M_\varpi \frac{S_{o\varpi}}{I_\varpi}\,. \tag{1.22}$$

It is the matter of course that S_{oz}, S_{oy}, $S_{o\varpi}$ in the open part of the profile are understood as the following integrals different from (1.21):

$$S_{oz} = \int_{F}^{M} yehds\,, \quad S_{oy} = \int_{F}^{M} zehds\,, \quad S_{o\varpi} = \int_{F}^{M} \varpi ehds\,.$$

The overall torque M_x consists of the moment created by the flow of tangential stresses T and the moment of tangential stresses τ_k. This means

$$M_x = GI_k\theta' + \int_l T(s)\rho ds. \qquad (1.23)$$

Putting in the expressions of flow T from (1.20) in the closed part of the profile and from (1.22) in the open part of the profile yields an expression of flow T_0:

$$T_0 = \frac{M_x - GI_k\theta'}{\Omega} + \frac{Q_y}{\Omega I_z}\int_l S_{oz}\rho ds + \frac{Q_z}{\Omega I_y}\int_l S_{oy}\rho ds + \frac{M_\varpi}{\Omega I_\varpi}\int_l S_{o\varpi}\rho ds, \qquad (1.24)$$

where Ω is the doubled area comprised by the closed contour of the profile.

Next, substituting (1.24) to (1.20) gives the following for points belonging to the closed part of the profile:

$$T = \frac{M_x - GI_k\theta'}{\Omega} - \frac{Q_y}{I_z}\left(S_{oz} - \frac{1}{\Omega}\int_l S_{oz}\rho ds\right) -$$

$$-\frac{Q_z}{I_y}\left(S_{oy} - \frac{1}{\Omega}\int_l S_{oy}\rho ds\right) - \frac{M_\varpi}{I_\varpi}\left(S_{o\varpi} - \frac{1}{\Omega}\int_l S_{o\varpi}\rho ds\right). \qquad (1.25)$$

We will show that the general formula (1.25) for flow T works also for a limit case of a completely closed profile. To see this, we should assume the following for a completely closed section:

$$I_k = 0,$$

because the theory discussed in Chapter 7 is based on neglecting the tangential stresses τ_k. As a result, formula (1.25) becomes (**7**.1.49), after its left and right parts are divided by h, and now describes the tangential stresses in closed sections.

On the other hand, we have the already established relation (1.22) instead of (1.25) for points which belong to the open part of the profile. For a completely open profile, we can make sure that

$$\varpi = \omega.$$

Therefore formula (1.22) works also for a completely open profile because in that case it coincides with (**6**.2.61).

We deem it inconvenient, however, to have different formulas for the flows in the open and closed parts of the profile. It is easy to understand that two formulas (1.22) and (1.25) can be merged into one if we use our previously defined characteristic function of the profile, $\kappa(s)$. By multiplying (1.25) by κ and (1.22) by (1 – κ) and summing the results, we derive a general formula of the flows in any point of the compound profile,

$$T = \frac{\kappa(M_x - GI_k\theta')}{\Omega} - \frac{Q_y}{I_z}\bar{S}_{oz} - \frac{Q_z}{I_y}\bar{S}_{oy} - \frac{M_\varpi}{I_\varpi}\bar{S}_{o\varpi} \tag{1.26}$$

with the functions

$$\bar{S}_{oz} = S_{oz} - \frac{\kappa}{\Omega}\int_l S_{oz}\rho ds\,, \qquad \bar{S}_{oy} = S_{oy} - \frac{\kappa}{\Omega}\int_l S_{oy}\rho ds\,,$$

$$\bar{S}_{o\varpi} = S_{o\varpi} - \frac{\kappa}{\Omega}\int_l S_{o\varpi}\rho ds\,. \tag{1.27}$$

Properties of functions $\bar{S}_{oz}$, $\bar{S}_{oy}$, $\bar{S}_{o\varpi}$ and consequences of these properties

We show here that the three just defined functions of the arc coordinate s possess the properties established earlier in Section **7.1.2** for a purely closed profile. We would like to note also the consequences of the properties of the three functions which can be formulated in the theory of compound-profile thin-walled bars.

- **Property 1**:

Obviously, each of functions S_{oz}, S_{oy}, $S_{o\varpi}$ depends on the location of the start point O from which the arc coordinate is counted along the profile. However, if we trace carefully the way the $\bar{S}_{oz}$, $\bar{S}_{oy}$, $\bar{S}_{o\varpi}$ functions have been derived, we will find the following statement true:

> *Neither of functions $\bar{S}_{oz}$, $\bar{S}_{oy}$, $\bar{S}_{o\varpi}$ depends on a particular location of the origin of the arc coordinate s; instead, all three functions depend unambiguously on the points belonging to the profile.*

- **Property 2**:

It follows directly from the definition of (1.27) and from the obvious inequality

$$\int_l \kappa\rho ds = \oint \rho ds = \Omega$$

that

$$\int_l \bar{S}_{oz}\rho ds = 0, \qquad \int_l \bar{S}_{oy}\rho ds = 0, \qquad \int_l \bar{S}_{o\varpi}\rho ds = 0. \tag{1.28}$$

The first two equalities in (1.28), if we also recall (1.26), testify that the tangential stresses τ_Q, ones created solely by the shear forces, do not create any torque with respect to the pole that satisfies (1.17). This means the pole selection criterion of (1.17) which we used when deriving (1.26) makes the pole identical to the center of bending and thus to the center of twist, too.

Multiplying (1.26) by arm function ρ and integrating the result over the whole profile gives

$$\int_l T\rho ds = M_x - GI_k\theta',$$

which is, of course, identical to (1.23).

- **Property 3**:
  The following formulas of integration over the profile hold:

$$\int_l \bar{S}_{oy}\frac{d\varpi}{ds}ds = 0, \qquad \int_l \bar{S}_{oz}\frac{d\varpi}{ds}ds = 0, \qquad \int_l \bar{S}_{o\varpi}\frac{d\varpi}{ds}ds = -I_\varpi. \tag{1.29}$$

To begin, we prove the first one in (1.29). We split the integral to be calculated over the whole profile l into the sum of two integrals: the first one over the closed part and the second one over the open part of the profile.

For the convenience of reasoning, we take a profile shown in Fig. 8.3 and note that the generalized sectorial coordinate ϖ is continuous along the whole profile. At the same time, function $\bar{S}_{oy}$ has a discontinuity at the profile's bifurcation point, R, as we move along the closed part of the profile. We denote the leap of that function at point R by $[\bar{S}_{oy}]_R$. Fig. 8.3 and the definition of function $\bar{S}_{oy}$ make it clear that the value of this leap is the integral taken along the whole open part of the profile from point F to the profile's bifurcation point, R, that is,

$$[\bar{S}_{oy}]_R = \int_F^R zehds.$$

In other words, the leap in question is equal to the static moment, with respect to the Y-axis, of the whole open branch of the section that includes the point of bifurcation, R.

Now the formula of integration by parts will help us derive the following for the closed part of the profile:

$$\int_l \kappa \bar{S}_{oy} \frac{d\varpi}{ds} ds = \oint \bar{S}_{oy} \frac{d\varpi}{ds} ds = \int_{R+0}^{R-0} \bar{S}_{oy} \frac{d\varpi}{ds} ds = -\varpi(R)[\bar{S}_{oy}]_R - \int_{R+0}^{R-0} \frac{d\bar{S}_{oy}}{ds} \varpi ds =$$

$$= -\varpi(R) \int_F^R zeh\varpi ds - \int_{R+0}^{R-0} zeh\varpi ds =$$

$$= [1 - \varpi(R)] \int_F^R zeh\varpi ds - \int_l zeh\varpi ds = [1 - \varpi(R)] \int_F^R zeh\varpi ds .$$

Here we use the criterion of selection of the principal pole, (1.17). Also, we have this on the open part of the profile:

$$\int_F^R \bar{S}_{oy} \frac{d\varpi}{ds} ds = \int_F^R S_{oy} \frac{d\varpi}{ds} ds = \left[S_{oy} \varpi \right]_F^R - \int_F^R \frac{dS_{oy}}{ds} \varpi ds = \varpi(R) \int_F^R zehds - \int_F^R zeh\varpi ds .$$

By adding these two integrals, we obtain the first relation in (1.29). The second one in (1.29) is proved in the same way, as should be obvious.

Now let us justify the third formula in (1.29). We follow the same approach and obtain

$$[\bar{S}_{o\varpi}]_R = \int_F^R \varpi ehds .$$

Further,

$$\int_l \kappa \bar{S}_{o\varpi} \frac{d\varpi}{ds} ds = \oint \bar{S}_{o\varpi} \frac{d\varpi}{ds} ds = \int_{R+0}^{R-0} \bar{S}_{o\varpi} \frac{d\varpi}{ds} ds = -\varpi(R)[\bar{S}_{o\varpi}]_R - \int_{R+0}^{R-0} \frac{d\bar{S}_{o\varpi}}{ds} \varpi ds =$$

$$= -\varpi(R) \int_F^R eh\varpi^2 ds - \int_{R+0}^{R-0} eh\varpi^2 ds = [1 - \varpi(R)] \int_F^R eh\varpi^2 ds - \int_l eh\varpi^2 ds =$$

$$= [1 - \varpi(R)] \int_F^R eh\varpi^2 ds - I_\varpi .$$

The open part of the profile has

$$\int_F^R \bar{S}_{o\varpi} \frac{d\varpi}{ds} ds = \int_F^R S_{o\varpi} \frac{d\varpi}{ds} ds =$$

$$= \left[S_{o\varpi} \varpi \right]_F^R - \int_F^R \frac{dS_{o\varpi}}{ds} \varpi ds = \varpi(R) \int_F^R \varpi ehds - \int_F^R eh\varpi^2 ds .$$

Summing gives the third formula in (1.29). This proves the whole Property 3.

Now, after multiplying both parts of (1.26) by $d\varpi/ds$ and considering the following obvious equality

$$\int_l \kappa \frac{d\varpi}{ds} ds = 0,$$

that follows from the continuity of the generalized sectorial coordinate ϖ, we have

$$M_\varpi = \int_l \tau h \frac{d\varpi}{ds} ds. \tag{1.30}$$

This lets us extend the purely static definition of the constricted-torsion torque M_ϖ, presented in Section 7.1.3, onto the case of a compound profile, too.

Now let us turn to the general expression of the derivative $d\varpi/ds$; according to (1.14), (1.15), and (1.18), we have

$$\frac{d\varpi}{ds} = \rho - \frac{\kappa I_d}{gh\Omega}. \tag{1.31}$$

Putting (1.31) in the integrand of (1.30) and considering (1.23) gives

$$M_\varpi = M_x - GI_k\theta' - \frac{I_d}{\Omega}\int_l \frac{\kappa\tau}{g} ds. \tag{1.32}$$

We use the generalized Bredt formula of the tangential strain circulation (see Appendix D) and have

$$\int_l \frac{\kappa\tau}{g} ds = \oint \frac{\tau}{g} ds = G\Omega\theta'. \tag{1.33}$$

Combining (1.32) and (1.33) yields

$$M_x = H + M_\varpi, \tag{1.34}$$

where the pure-torsion torque H, following (1.6), is defined as

$$H = (GI_k + GI_d)\theta'. \tag{1.35}$$

Thus, the representation of torque M_x in the section of the bar as the sum of the pure-torsion torque, H, and the constricted-torsion torque, M_ϖ, is true also for a compound profile. This helps us rewrite (1.26), the formula of the flows, as

$$T = \frac{\kappa \bar{H}}{\Omega} - \frac{Q_y}{I_z}\bar{S}_{oz} - \frac{Q_z}{I_y}\bar{S}_{oy} - \frac{M_\varpi}{I_\varpi}\bar{\bar{S}}_{o\varpi}, \tag{1.36}$$

where we denote additionally

$$\bar{H} = M_x - GI_k\theta' - M_\varpi = GI_d\theta', \tag{1.37}$$

and function $\bar{\bar{S}}_{o\varpi}$ is defined as

$$\bar{\bar{S}}_{o\varpi} = \bar{S}_{o\varpi} - \frac{\kappa I_\varpi}{\Omega}. \tag{1.38}$$

Obviously, stress $\bar{H}$ is the part of the pure-torsion torque, H, which is created by the average tangential stresses only.

It is useful to define a dimensionless geometrical parameter, λ, by assuming

$$\lambda = \frac{I_d}{I_d + I_k}, \tag{1.39}$$

so we can represent torque $\bar{H}$ as a fraction of torque H:

$$\bar{H} = \lambda H. \tag{1.40}$$

Note that a fully open profile has λ equal to zero. For a fully closed profile, if we neglect the value of I_k in comparison to I_d, we can assume $\lambda = 1$.

- **Property 4**:

Finally, it is useful to note that the functions satisfy such formulas of integration over the profile as

$$\oint \frac{\bar{S}_{oy}}{gh} ds = 0, \qquad \oint \frac{\bar{S}_{oz}}{gh} ds = 0, \qquad \oint \frac{\bar{S}_{o\varpi}}{gh} ds = I_\varpi \frac{\Omega}{I_d}. \tag{1.41}$$

And indeed, by multiplying (1.31) by $\bar{S}_{oy}$, integrating the result over the profile, and considering the already proved relations of (1.28) and (1.29), we can easily establish that the first of (1.41) is true. The second one in (1.41) is proved in the same way. Making the same transformation on $\bar{S}_{o\varpi}$ gives the third formula in (1.41).

A corollary of the third formula in (1.41) and (1.38) is

$$\oint \frac{\bar{\bar{S}}_{o\varpi}}{gh} ds = I_\varpi \frac{\Omega}{I_d} - \frac{I_\varpi}{\Omega} \oint \frac{ds}{gh}.$$

Recalling the expression of (1.4) for the geometric characteristics I_d, we find out the following property of function $\overline{\overline{S}}_{o\varpi}$:

$$\oint \frac{\overline{\overline{S}}_{o\varpi}}{gh} ds = 0, \tag{1.42}$$

which we will need further.

Physical relationships of the theory of compound-profile thin-walled bars

Now we want to calculate the strain energy created by the tangential stresses. As we already noted, we deal with two types of the tangential stresses in a compound profile, τ and τ_k. Therefore the overall energy created by those can be represented as $E_\tau + E_{\tau k}$ where E_τ is the energy of average tangential stresses τ and $E_{\tau k}$ is that of tangential stresses τ_k[2]. Let us calculate $E_{\tau k}$ first. As the pure-torsion torque caused by the τ_k stresses only is equal to $(H - \overline{H})$, we have

$$E_{\tau k} = \int_0^L \frac{(H-\overline{H})^2}{2GI_k} dx = \int_0^L \frac{(1-\lambda)^2 H^2}{2GI_k} dx.$$

Further,

$$E_\tau = \frac{1}{2G} \int_0^L \int_l \frac{T^2}{gh} ds dx = \frac{1}{2G} \int_0^L \int_l \left(\frac{\kappa\lambda H}{\Omega} - \frac{Q_y}{I_z} \overline{S}_{oz} - \frac{Q_z}{I_y} \overline{S}_{oy} - \frac{M_\varpi}{I_\varpi} \overline{\overline{S}}_{o\varpi} \right)^2 \frac{ds}{gh} dx.$$

Summing these two expression gives

$$E_\tau + E_{\tau k} =$$

$$= \frac{1}{2G} \int_0^L \left[\frac{H^2}{I_x} + \frac{Q_z^2}{I_y^2} \int_l \frac{\overline{S}_{oy}^2 ds}{gh} + \frac{Q_y^2}{I_z^2} \int_l \frac{\overline{S}_{oz}^2 ds}{gh} + \frac{M_\varpi^2}{I_\varpi^2} \int_l \frac{\overline{\overline{S}}_{o\varpi}^2 ds}{gh} \right] dx +$$

$$+ \frac{1}{G} \int_0^L \left[-\frac{\lambda H Q_z}{I_y} \int_l \frac{\kappa \overline{S}_{oy} ds}{gh} - \frac{\lambda H Q_y}{I_z} \int_l \frac{\kappa \overline{S}_{oz} ds}{gh} - \frac{\lambda H M_\varpi}{I_\varpi} \int_l \frac{\kappa \overline{\overline{S}}_{o\varpi} ds}{gh} \right] dx +$$

[2] It is easy to understand, recalling the symmetric distribution of average stresses τ and the antisymmetric distribution of stresses τ_k over the thickness of the walls, h, that the virtual work of either stress on the displacements caused by the others is zero.

$$+\frac{1}{G}\int_0^L \frac{Q_z Q_y}{I_y I_z}\int_l \frac{\bar{S}_{oz}\bar{S}_{oy}\,ds}{gh}dx+$$

$$+\frac{1}{G}\int_0^L\left[\frac{Q_z M_\varpi}{I_y I_\varpi}\int_l \frac{\bar{S}_{oy}\bar{\bar{S}}_{o\varpi}\,ds}{gh}+\frac{Q_y M_\varpi}{I_z I_\varpi}\int_l \frac{\bar{S}_{oz}\bar{\bar{S}}_{o\varpi}\,ds}{gh}\right]dx. \tag{1.43}$$

The second line in the right part of (1.43) vanishes because of (1.41) and (1.42). The final energy of the tangential stress can be represented as

$$\mathsf{E}_\tau + \mathsf{E}_{\tau k} =$$

$$=\int_0^L\left[\frac{H^2}{2GI_x}+\frac{Q_z^2\mu_{zz}}{2GA}+\frac{Q_y^2\mu_{yy}}{2GA}+\frac{M_\varpi^2\mu_{\varpi\varpi}}{2r^2GA}+\frac{Q_z Q_z\mu_{zy}}{GA}+\frac{Q_z M_\varpi\mu_{z\varpi}}{rGA}+\frac{Q_y M_\varpi\mu_{y\varpi}}{rGA}\right]dx,$$

where numerical parameters μ_{zz}, μ_{zy}, $\mu_{z\varpi}$, μ_{yy}, $\mu_{y\varpi}$, $\mu_{\varpi\varpi}$ make up a matrix of the section's shape factors, $\boldsymbol{\mu}$, presented earlier in Chapter 7 – see (**7**.1.63).

As the formula (1.19) of the normal stresses is identical with its counterpart for open or closed profiles, the expression of the strain energy, E_σ, created by the normal stresses will not differ from its counterparts for the section types considered earlier. Now it is clear that the matrix of physical relationships, $\boldsymbol{C}^{-1}$, for the compound profile is identical in its form with (**6**.3.9) or (**7**.1.65).

This means that the system of governing equations of the full shear theory of closed-profile thin-walled bars is identical with (**6**.3.24). The similar identity takes place between the governing equations for the semi-shear theory.

8.1.3 Non-warped compound profiles

It is reasonable to define explicitly a class of non-warped compound profiles of thin-walled bars.

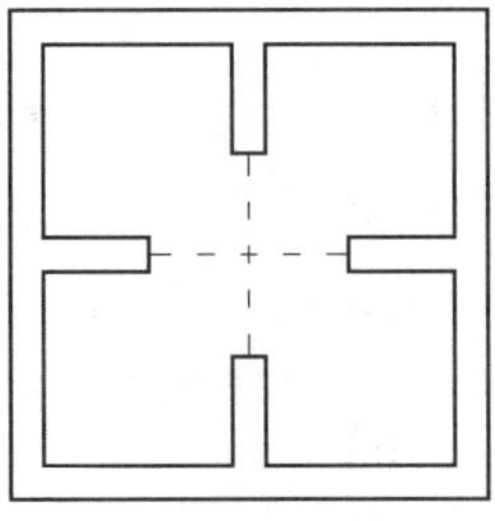

Fig. 8.4. An example of a non-warped compound profile

The meaning of the notion of a non-warped profile is such that the warp function, α, of such a profile should be identical to zero. Using formulas (1.14) and (1.15), we can derive a general expression of the warp function:

$$\alpha(s) = \frac{I_d}{\Omega}\int_0^s \frac{\kappa ds}{gh} - \int_0^s \rho ds .$$

When $\alpha \equiv 0$, we take the derivative of this identity with respect to s and reduce the non-warpedness condition to

$$\rho gh = \kappa \frac{I_d}{\Omega} . \tag{1.44}$$

As the characteristic function κ is equal by definition to one on the closed part and to zero on the open part of the profile, we can see that the compound profile does not warp if the following two conditions hold:

- on the closed part of the profile, $\rho gh = I_d / \Omega$;
- on the open part of the profile, $\rho gh = 0$, or, to put it another way, all walls of the open part of the profile should be straight and, when extended, should intersect the principal pole P.

A simplest example of a non-warped compound profile is shown in Fig. 8.4.

8.2 A multiple-contour compound profile

Let us discuss briefly the changes which have to be made to the basic formulas for the case when the closed part of the profile has multiple contours.

For the sake of generality, we will assume as earlier that different walls in a thin-walled section can be made of different materials. The modulus of elasticity, E(s), and the modulus of shear, G(s), at an arbitrary point of the profile are then defined as

$$E(\mathrm{s}) = Ee(s), \qquad G(\mathrm{s}) = Gg(s). \tag{2.1}$$

We will use the following limitations of the profile's type:

- Fisrt, we will assume the graph of the profile to be connected. It means that there is always at least one path on the profile from any arbitrary point to another arbitrary point of it.
- Second, we do not allow concentrated pointwise areas on the profile.
- And the last limitation: there are no warping constraints on the profile.

These limitations are adopted here solely to simplify the further formulas a little bit. When necessary, these limitations can be easily omitted. For example, the limitation of the absence of concentrated areas can be avoided by formally placing a short piece of the profile in a point with a concentrated area and by proportioning the thickness of the wall, h, as appropriate. The elimination of the two other limitations requires some rebuilding of the formulas, which we suggest that the reader do.

First of all, we want to discuss additions to be made in order to describe the topology of a generalized multiple-contour compound profile, comparing to the similar description from Section **7**.2.1.

8.2.1 Topology of a multiple-contour compound profile

For the sake of convenience, we take a profile example shown in the right part of Fig. 8.1. The graph of the profile is shown in Fig. 8.5.

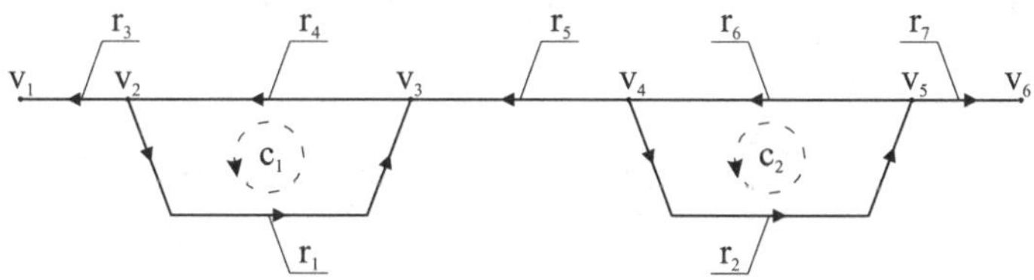

Fig. 8.5. The oriented graph of a two-contour compound profile

For the profile in question, the incidence matrix $\boldsymbol{G}$ looks like

$$\boldsymbol{G}=\begin{matrix} v_1 \\ v_2 \\ v_3 \\ v_4 \\ v_5 \\ v_6 \end{matrix}\left\|\overset{\begin{matrix} r_1 & r_2 & r_3 & r_4 & r_5 & r_6 & r_7 \end{matrix}}{\begin{bmatrix} 0 & 0 & -1 & 0 & 0 & 0 & 0 \\ 1 & 0 & 1 & -1 & 0 & 0 & 0 \\ -1 & 0 & 0 & 1 & -1 & 0 & 0 \\ 0 & 1 & 0 & 0 & 1 & -1 & 0 \\ 0 & -1 & 0 & 0 & 0 & 1 & 1 \\ 0 & 0 & 0 & 0 & 0 & 0 & -1 \end{bmatrix}}\right\|,$$

and the loop matrix $\boldsymbol{F}$ is

$$\boldsymbol{F}=\begin{matrix} c_1 \\ c_2 \end{matrix}\left\|\overset{\begin{matrix} r_1 & r_2 & r_3 & r_4 & r_5 & r_6 & r_7 \end{matrix}}{\begin{bmatrix} 1 & 0 & 0 & 1 & 0 & 0 & 0 \\ 0 & 1 & 0 & 0 & 0 & 1 & 0 \end{bmatrix}}\right\|.$$

The basic integer parameters of the graph take the values of $k=2$; $n=7$; $t=6$.

Now we will need a bit more detailed classification of the edges. As earlier, we will call an edge a *frame one* if it is included in one base loop only, or a *double one* if the edge is included in two base loops. However, a compound profile contains edges not included in any base loops. Edges of this kind will be called *free edges*. For example, the graph in Fig. 8.5 has the frame edges of r_1, r_2, r_4, r_6 and the free edges of r_3, r_5, r_7. There are no double edges.

Let ϑ_j be a classification identifier of the edge r_j in the sense that:

- $\vartheta_j = \ \ 1$ for a frame edge;
- $\vartheta_j = -1$ for a double edge;
- $\vartheta_j = \ \ 0$ for a single edge.

8.2.2 Pure torsion of a multiple-contour compound profile

Let k be the number of the base contours in the profile, and $k > 1$. In this case we will have the following instead of (1.1) for pure torsion:

$$M_x = GI_{\mathrm{k}}\theta' + \sum_{i=1}^{k} \overline{T}_i \Omega_i\,, \tag{2.2}$$

where $\overline{T}_i$ is a flow along i-th base contour, Ω_i is the doubled area comprised by this contour. Seeing that the material's properties may be variable over the profile, we assume

$$I_{\mathrm{k}} = \int_l \frac{gh^3}{3}\,ds\,. \tag{2.3}$$

Clearly, we have to use the strain compatibility conditions to be able to determine k unknown flows when we have only one equation of equilibrium. Reasoning the same way as we did in Section **7**.2.1, we construct the Castigliano functional,

$$\mathsf{K}(\overline{T}_1,\ldots,\overline{T}_k) = \sum_{i=1}^{k} \frac{\overline{T}_i^2}{2G} \oint_{\Gamma_i} \frac{ds}{gh} - \sum_{ij} \frac{\overline{T}_i \overline{T}_j}{G} \int_{\Gamma_{ij}} \frac{ds}{gh}\,, \tag{2.4}$$

where the symbol $\sum_{ij}$ means that the summation comprises all adjacent edges Γ_{ij} of any two adjacent base contours, so that each adjacent edge is included once in this sum. In other words, the summation comprises all double edges.

Considering the conditions of equilibrium (2.2) in terms of moments, the modified functional M with the Lagrangian multiplier $\overline{\lambda}$ becomes

$$\mathsf{M}(\overline{T}_1,...,\overline{T}_k)=\mathsf{K}(\overline{T}_1,...,\overline{T}_k)-\overline{\lambda}(M_x-GI_{\mathrm{k}}\theta'-\sum_{i=1}^{k}\overline{T}_i\Omega_i)\,. \tag{2.5}$$

The conditions of stationarity of functional M produce a system of linear algebraic equations the general solution of which can be represented as

$$\overline{T}_i=\overline{a}_i\frac{M_x-GI_{\mathrm{k}}\theta'}{\Omega_0}\quad(i=1,\dots,k),\qquad\qquad \Omega_0=\sum_{i=1}^{k}\Omega_i\,, \tag{2.6}$$

so that

$$\sum_{i=1}^{k}\overline{a}_i\Omega_i=\Omega_0\,. \tag{2.7}$$

Clearly, we can immediately build a system of simultaneous linear algebraic equations to determine coefficients $\overline{a}_i$ similar to (7.2.15). However, we should keep in mind that the weight factors of particular edges p_j and contours $\overline{p}_i$ are defined as follows, with the non-uniformity of the profile's material taken into account:

$$p_j=\int_{l_j}\frac{ds}{gh}\;(j=1,\dots,n),\qquad\qquad \overline{p}_i=\oint_{\Gamma_i}\frac{ds}{gh}\;(i=1,\dots,k). \tag{2.8}$$

Coefficients a_j of the distribution of flows T_j over non-bifurcated segments of the profile are defined by the earlier formula (7.2.18) in the pure torsion. It should be clear also that formulas (7.2.19) and (7.2.20) of the flows on the contours, T_j, and of flow $T(s)$ in an arbitrary point of the compound profile should undergo some slight revision:

$$T_j=a_j\frac{M_x-GI_{\mathrm{k}}\theta'}{\Omega_0}\quad(j=1,\dots,n),\qquad\qquad T(s)=a(s)\frac{M_x-GI_{\mathrm{k}}\theta'}{\Omega_0}\,. \tag{2.9}$$

Obviously, the function of flow distribution along the profile, $a(s)$, is assumed to be identical to zero on open segments of the profile.

A warp function for a multiple-contour compound profile

Let us consider an arbitrary closed base contour, Γ_0. Moving along it from a certain selected point $\mathrm{A}\in\Gamma_i$, we obtain the formula

$$u_{\mathrm{M}} = u_{\mathrm{A}} + \frac{1}{G}\int_{\mathrm{A}}^{\mathrm{M}} \frac{Tds}{gh} - \theta' \int_{\mathrm{A}}^{\mathrm{M}} \rho ds$$

which generalizes the formula (7.2.22) of the u_{M} displacement at any point M of the same contour. Having traced the Γ_i around, we return to the original point A. As a result, instead of (7.2.23) we will have this formula of the twist:

$$\theta' = \frac{1}{G\Omega_i} \oint_{\Gamma_i} \frac{Tds}{gh}. \tag{2.10}$$

Substituting the expression of flow $T(s)$ from (2.9) in here gives

$$\theta' = \frac{M_x \oint_{\Gamma_i} \frac{a(s)}{gh} ds}{G\left(\Omega_0 \Omega_i + I_{\mathrm{k}} \oint_{\Gamma_i} \frac{a(s)}{gh} ds\right)}. \tag{2.11}$$

Using the standard representation of $M_x = \theta' G I_x$, we produce the following expression of the torsional moment of inertia of the multiple-contour compound profile

$$I_x = I_{\mathrm{k}} + \Omega_0 \frac{\Omega_i}{\oint_{\Gamma_i} \frac{a(s)}{gh} ds}. \tag{2.12}$$

As the torsional moment of inertia of the section, I_x, is a numerical characteristic of the whole section, and (2.12) holds true for any index i from 1 to k, we have a useful identity:

$$\frac{\Omega_1}{\oint_{\Gamma_1} \frac{a(s)}{gh} ds} = \frac{\Omega_2}{\oint_{\Gamma_2} \frac{a(s)}{gh} ds} = \ldots = \frac{\Omega_k}{\oint_{\Gamma_k} \frac{a(s)}{gh} ds}, \tag{2.13}$$

which should be satisfied by function $a(s)$ that describes the distribution of the flows along the profile. The identity (2.13) can be also proved directly[3].

Formula (2.6) for flow $\overline{T}_i$ can be rewritten more conveniently using the substitution of $\theta' = M_x/GI_x$:

[3] This is our call to intellectual exercises.

$$\overline{T}_i = \overline{a}_i \frac{\lambda M_x}{\Omega_0} \quad (i = 1, \ldots, k), \qquad \lambda = 1 - \frac{I_k}{I_x}. \tag{2.14}$$

The same transformation takes place for (2.9) that describes flow T_j on each of the non-bifurcated segments of the profile,

$$T_j = a_j \frac{\lambda M_x}{\Omega_0} \quad (j = 1, \ldots, n). \tag{2.15}$$

This expression holds formally for all non-bifurcated segments of the profile from 1 to n, though actually the respective flows are equal to zero on open segments.

We take a fixed point A of the profile and restrain it from longitudinal displacements thus eliminating the rigid displacements of the bar lengthwise. Let a point M be the current point of the profile. We can move from point A to point M along more than one path, so let us choose one of the feasible ways in which we can reach point M. Then the longitudinal displacement of this point can be written as

$$u_{\mathrm{M}} = \frac{1}{G_0} \int_{\mathrm{A}}^{\mathrm{M}} \kappa \frac{T ds}{gh} - \theta' \int_{\mathrm{A}}^{\mathrm{M}} \rho ds$$

where κ(s) is the characteristic function of the profile introduced earlier.

Of course, each of the two integrals depends on what path of integration we choose, but the overall result should be independent. Otherwise we would have an ambiguity in the longitudinal displacements of the profile's points. Note also that the profile's characteristic function κ(s) can be actually omitted from the first integral because flow T is necessarily zero on single edges of the graph in the pure torsion analysis.

We replace θ′ with the ratio of M_x/GI_x and flow T with its expression from (2.15) to find

$$u_{\mathrm{M}} = \frac{M_x}{GI_x} \left(\frac{\lambda I_x}{\Omega_0} \int_{\mathrm{A}}^{\mathrm{M}} \kappa \frac{a ds}{gh} - \int_{\mathrm{A}}^{\mathrm{M}} \rho ds \right). \tag{2.16}$$

Now it is clear that the warp function, $\alpha(s)$, for the multiple-contour compound profile is

$$\alpha_{\mathrm{M}} = \frac{\lambda I_x}{\Omega_0} \int_{\mathrm{A}}^{\mathrm{M}} \kappa \frac{a ds}{gh} - \int_{\mathrm{A}}^{\mathrm{M}} \rho ds. \tag{2.17}$$

Also, we have the following instead of (1.4) for the geometric characteristic I_d of an inhomogeneous material:

$$I_d = \frac{\Omega_0 \Omega_i}{\oint\limits_{\Gamma_i} \frac{a(s)}{gh} ds}. \tag{2.18}$$

so the general formula (1.7) still holds.

8.2.3 A general behavior of a thin-walled multiple-contour compound profile

What we have to do here in essence is to repeat the reasoning of Section **7**.2.2 but in application to a compound multiple-contour profile. Therefore we indicate here only changes and additions in comparison to the formulas of Section **7**.2.2.

As for (**7**.2.29)-(**7**.2.37), they go unchanged for the case of the compound multiple-contour profile. The main change in the formulas comparing to Section **7**.2.2 is the necessity to replace the M_x torque by the expression

$$M_x - GI_k\theta' = M_x - GI_x\theta' \frac{I_k}{I_x} = M_x - (1-\lambda)H,$$

where we use our common designation of $H = GI_x\theta'$ – this is a pure-torsion torque. This is implied by the fact that the original formula (**7**.2.38) has to be represented as follows for a compound section, according to (1.23):

$$M_x - GI_k\theta' = \sum_{i=1}^{n} \int\limits_{\mathrm{S}_i}^{\mathrm{T}_i} T_i(s)\rho(s)ds.$$

Thus, the equation of equilibrium in terms of the torques will become as follows instead of (**7**.2.46):

$$M_x = \boldsymbol{\omega}^{\mathrm{T}} \boldsymbol{T}_{\mathrm{S}} + (1-\lambda)H - \frac{Q_y}{I_z} S_{\rho z} - \frac{Q_z}{I_y} S_{\rho y} - \frac{M_\varpi}{I_\varpi} S_{\rho\varpi}, \tag{2.19}$$

Formula (**7**.2.57) becomes

$$T_i(s) = \left[M_x - (1-\lambda)H\right] b_{ix} - \frac{Q_y}{I_z} \overline{S}_{ioz} - \frac{Q_z}{I_y} \overline{S}_{ioy} - \frac{M_\varpi}{I_\varpi} \overline{S}_{io\varpi}. \tag{2.20}$$

If we consider, particularly, the pure torsion, we should assume $M_x = H$, $Q_y = Q_z = M_\varpi = 0$, so (2.20) yields

$$T_i = \lambda H b_{ix} .$$

But the pure-torsion flows obey (2.9) or

$$T_i = a_i \frac{\lambda H}{\Omega_0} .$$

By equaling these two expressions of the pure-torsion flows, we find

$$\boldsymbol{a} = \Omega_0 \boldsymbol{b}_x, \tag{2.21}$$

which is the same as (7.2.59).

The representation of the full torque, M_x, as the sum of the pure-torsion torque, H, and the constricted-torsion torque, M_ϖ, lets us finally rewrite the formula (2.20) of the flows as

$$T_i(s) = \frac{\lambda H}{\Omega_0} a_i - \frac{Q_y}{I_z} \bar{S}_{ioz} - \frac{Q_z}{I_y} \bar{S}_{ioy} - \frac{M_\varpi}{I_\varpi} \bar{\bar{S}}_{io\varpi} , \tag{2.22}$$

where function $\bar{\bar{S}}_{io\varpi}$ is still defined by (7.2.61). As we can see, the only difference of formula (2.22) from its counterpart (7.2.60) is the presence of the λ factor at the pure-torsion torque, H.

Further, by composing an expression of the overall energy of the tangential stresses, we produce

$$\mathrm{E}_\tau + \mathrm{E}_{\tau k} =$$

$$= \int_0^L \left[\frac{H^2}{2GI_x} + \frac{Q_z^2 \mu_{zz}}{2GA} + \frac{Q_y^2 \mu_{yy}}{2GA} + \frac{M_\varpi^2 \mu_{\varpi\varpi}}{2r^2 GA} + \frac{Q_z Q_z \mu_{zy}}{GA} + \frac{Q_z M_\varpi \mu_{z\varpi}}{rGA} + \frac{Q_y M_\varpi \mu_{y\varpi}}{rGA} \right] dx ,$$

where numerical parameters μ_{zz}, μ_{zy}, $\mu_{z\varpi}$, μ_{yy}, $\mu_{y\varpi}$, $\mu_{\varpi\varpi}$ make up a matrix of shape factors for the section, $\boldsymbol{\mu}$, which is defined by the same formula (7.2.66).

8.3 Final comments to thin-walled bar theories

8.3.1 A remark on the energy-based comparison between Vlasov's shear-free theory and the semi-shear theory of open-profile thin-walled bars

Let us return to the open-profile thin-walled bars. In the constricted-torsion analysis according to Vlasov's theory (the shear-free theory), the Lagrange functional, $\tilde{\mathsf{L}}$, depends on the θ function only; its form in our designations is

$$\tilde{\mathsf{L}}(\theta) = \frac{1}{2}\int_0^L \left(EI_\omega \theta''^2 + GI_x \theta'^2\right)dx - \int_0^L \left(m_x\theta - m_B\theta'\right)dx\,. \tag{3.1}$$

At the same time, the Lagrange functional of the semi-shear theory, L, will depend on two unknown functions and can be represented as

$$\mathsf{L}(\theta,\beta) =$$

$$= \frac{1}{2}\int_0^L \left[EI_\omega \beta'^2 + GI_x \theta'^2 + \frac{GI_x}{\psi - 1}(\theta' - \beta)^2\right]dx - \int_0^L \left(m_x\theta - m_B\beta\right)dx\,. \tag{3.2}$$

Let us show that the exact values of these functionals, L_* and $\tilde{\mathsf{L}}_*$, on the solutions of the respective problems satisfy the inequality

$$\mathsf{L}_* \le \tilde{\mathsf{L}}_*\,. \tag{3.3}$$

To see this, note that functional L becomes equal to $\tilde{\mathsf{L}}$ if the twisted bar is subjected to an additional kinematic constraint:

$$\theta' - \beta = 0. \tag{3.4}$$

Herefrom we derive inequality (3.3) immediately as a particular manifestation of the general property of the Lagrange functional established in Section **2**.4 – see Table **2**.2 that demonstrates the influence of additional kinematic constraints on the value of the Lagrange functional.

It is easy to show by reasoning in the same way that the employment of the full shear theory also gives a lower value of the Lagrange functional on the solution of the problem, comparing to the solution that follows the semi-shear theory.

An example

Our example will be a simple problem of torsion of a cantilever bar of length L, loaded by torque $\bar{M}_x$ at its free end. The theory by Vlasov suggests the following differential equation for the angle of twist, θ:

$$EI_\omega \theta^{IV} - GI_x \theta'' = 0 \tag{3.5}$$

with the following boundary conditions:

$$\theta(0) = 0, \qquad \theta'(0) = 0,$$

$$B(L) = -EI_\omega \theta''(L) = 0, \qquad M_x(L) = GI_x\theta'(L) - EI_\omega\theta'''(L) = \bar{M}_x . \tag{3.6}$$

A direct check shows that the solution of the boundary-value problem (3.5), (3.6) is

$$\theta(x) = \frac{\bar{M}_x}{GI_x}\left[\frac{L}{k}\,\mathrm{th}k\left(\mathrm{ch}\frac{kx}{L} - 1\right) + x - \frac{L}{k}\,\mathrm{sh}\frac{kx}{L}\right], \tag{3.7}$$

where we use the designation of

$$k = L\sqrt{\frac{GI_x}{EI_\omega}} . \tag{3.8}$$

From (3.7) which expresses the angle of twist of the cantilever's end, $\theta_L = \theta(L)$, in Vlasov's theory, we derive

$$\theta_L = \frac{\bar{M}_x L}{GI_x}\left(1 - \frac{\mathrm{th}k}{k}\right). \tag{3.9}$$

The formulation of the same problem on the basis of the semi-shear theory produces a system of differential equations – see (**6**.4.11):

$$-GI_x\theta'' - \frac{I_r}{\mu_{\omega\omega}}(\theta'' - \beta') = 0 , \qquad EI_\omega\beta'' + \frac{I_r}{\mu_{\omega\omega}}(\theta' - \beta) = 0 , \tag{3.10}$$

with the boundary conditions

$$\theta(0) = 0, \qquad \beta(0) = 0, \qquad B(L) = -EI_\omega\beta'(L) = 0,$$

$$M_x(L) = GI_x\theta'(L) + \frac{GI_r}{\mu_{\omega\omega}}\big(\theta'(L) - \beta(L)\big) = \frac{GI_x}{\psi - 1}(\psi\theta' - \beta) = \bar{M}_x , \tag{3.11}$$

where, according to (**6**.4.13) and (**6**.3.6),

$$\psi = 1 + \frac{\mu_{\omega\omega} I_x}{I_r}, \qquad \mu_{\omega\omega} = \frac{I_r}{I_\omega^2} \int_l \frac{S_{\omega\omega}^2}{h} ds .$$

We present the solution of the boundary-value problem (3.10), (3.11) below. First of all, by making a linear combination of the two equations (3.10) we write the system in a more convenient form for our analysis:

$$\psi E I_\omega \beta''' - G I_x \beta' = 0 , \qquad \theta' = \beta - \frac{E I_\omega}{G I_x} (\psi - 1) \beta'' . \tag{3.12}$$

Now it is easy to see that the general solution of these simultaneous equations is

$$\beta = C_1 + C_2 \mathrm{ch} \frac{nx}{L} + C_3 \mathrm{sh} \frac{nx}{L},$$

$$\theta = C_1 x + C_2 \frac{L}{\psi n} \mathrm{sh} \frac{nx}{L} + C_3 \frac{L}{\psi n} \mathrm{ch} \frac{nx}{L} + C_4 , \tag{3.13}$$

where

$$n = L \sqrt{\frac{G I_x}{\psi E I_\omega}} = \frac{k}{\sqrt{\psi}} . \tag{3.14}$$

After determining the integration constants, $C_1, \ldots, C_4$, from the boundary conditions (3.11), we present the solution of this boundary-value problem in the form

$$\beta = \frac{\bar{M}_x}{G I_x} \left(1 - \mathrm{ch} \frac{nx}{L} + \mathrm{th}(n) \mathrm{sh} \frac{nx}{L} \right),$$

$$\theta = \frac{\bar{M}_x}{G I_x} \left[x - \frac{L}{\psi n} \mathrm{sh} \frac{nx}{L} + \frac{L}{\psi n} \mathrm{th} n \left(\mathrm{ch} \frac{nx}{L} - 1 \right) \right]. \tag{3.15}$$

Therefor the angle of twist of the cantilever's end in the semi-shear theory is

$$\theta_L = \frac{\bar{M}_x L}{G I_x} \left(1 - \frac{\mathrm{th} n}{\psi n} \right). \tag{3.16}$$

If we denote the ratio of the angle of twist from (3.9) to the angle of twist from (3.16) as ξ, we will have

$$\xi = \frac{k - \mathrm{th}k}{k - \frac{1}{\sqrt{\psi}}\mathrm{th}\frac{k}{\sqrt{\psi}}}, \tag{3.17}$$

and $\psi > 1$ implies $\xi < 1$. This is quite expectable because the stiffness of a bar in the shear-free theory is higher than that calculated according to the semi-shear theory.

It should be noted that the dimensionless parameter, ψ, is nearly equal to one often in the design practice. Therefore the refining contribution made by the semi-shear theory in comparison to the shear-free theory by Vlasov does not influence the final results really much. Apparently, the semi-shear theory does not give any valuable refinement of results of static analyses in application to the open-profile bars, therefore its virtue is based on something different: its letting us build a consistent theory of compound-profile thin-walled bars. Taking into account the full shear gives a yet less additional refinement of the numbers, which will certainly be ignored, too. These considerations allow us to omit the analysis of integration methods for the general system of equations of the type (**6**.3.24) though in principle it could be done. The semi-shear theory gives quite acceptable results from the practical accuracy standpoint, and it can be recommended as a reasonable engineering approach to the analysis of thin-walled bars.

8.3.2 A remark on the Luzhin equations for compound-profile thin-walled bars

As it should be obvious from the previous presentment, the traditional separation between the theories of thin-walled bars, based on the profile's type (open, closed, or compound) and used in nearly every book on the thin-walled bars (textbooks, scientific publications, reference manuals), is not really critical.

The form of the basic functionals in the variational formulation of the problem, thus the form of the governing differential equations and the respective boundary conditions, depends not on the profile's type but on whether certain stress components are included or not (or in what form they are) in the expression of the system's strain energy[4].

[4] In a different interpretation the origin of which can be found in the book by Janelidze and Panovko [11], applied theories of particular classes of elastic problems are classified by the way the stress components are divided into *primary* and *secondary* ones.

As we noted at the beginning of this chapter, all difficulties encountered in the attempts to unify the theories proceed mainly from the misunderstanding of this fact. Hence numerous earlier efforts of combining the shear-free theory by Vlasov and the semi-shear theory by Umanski; this approach was even considered to be the only right way to a general applied theory of compound-profile thin-walled bars.

We have already noted that this approach was used by O.V. Luzhin [14], [15] to construct a theory meant to take tangential stresses τ_k in the closed part of the profile into consideration. In our notation and our nomenclature, the reasoning by Luzhin can be described in this way.

First of all, the analysis of torsion and the analysis of bending in a thin-walled bar are assumed to be independent[5]. Based on the theory of constricted torsion of closed-profile thin-walled bars developed by A.A. Umanski, Luzhin considers the separate problem of constricted torsion of the thin-walled bar and advances a general relationship which binds together the torque M_x, the slope θ of the section with respect to the longitudinal axis X and the warp measure β:

$$\beta = A\theta' + BM_x \tag{3.18}$$

where A and B are certain constants which depend on the material and the geometry of the bar's section.

This relationship holds in Vlasov's theory of thin-walled bars if we assume

$$A = 1, \qquad B = 0. \tag{3.19}$$

At the same time, for relation (3.18) to be applicable in the theory by Umanski, we have to assume

$$A = \frac{1}{\mu}, \qquad B = -\frac{1}{\mu G I_\rho}, \tag{3.20}$$

where, as we should remind, μ is a warp factor defined for a closed profile by (7.1.79) as $\mu = 1 - I_d/I_\rho$, I_ρ being a directed moment of inertia of the section and I_d being defined by (1.4)[6]. And indeed, if we want to stick to Umanski's semi-shear theory, we should use (7.1.80) assuming $\eta_P' - \theta_z = 0$ and $\zeta_P' + \theta_y = 0$ to obtain

[5] As we know by now, this assumption is equivalent to dropping the full shear theory.

[6] Note that we designated the I_d geometric characteristic of a section as I_x in Chapter 7. This works because generally $I_x = I_d + I_k$, and for a purely closed profile we take $I_k = 0$.

$$M_x = GI_\rho(\theta' - \mu\beta),$$

and this is equivalent to (3.18) provided we assume (3.20).

Further, Luzhin extends the same relation of (3.18) formally onto the case of a compound profile by assuming

$$\beta = \left(1 + \frac{I_k}{I_\rho}\right)\frac{1}{\mu}\theta' - \frac{1}{\mu G I_\rho} M_x .$$

If we replace the warp factor, μ, with its expression via I_d and I_ρ, we will have this relation transformed into

$$\beta = \frac{I_\rho + I_k}{I_\rho - I_d}\theta' - \frac{1}{G(I_\rho - I_d)} M_x . \tag{3.21}$$

Relation (3.21) is not derived by Luzhin from any physical considerations; instead, he uses a formal selection of factors A and B in the linear relation of (3.18) to make this relation produce (3.19) and (3.20) in the respective limit cases of the purely open and purely closed profile. This means the relation of (3.21) is postulated by Luzhin[7]. Thus, the assumption of (3.21) is an auxiliary hypothesis in addition to ones adopted earlier.

By the way, proceeding from (3.21) to the limit case of the purely closed profile works if we assume $I_k = 0$, but this is true only when tangential stresses τ_k are assumed to be zero on the closed part of the profile. So when the τ_k tangential stresses are taken into consideration on the closed part of the profile, it is not really critical for the final equations constructed by Luzhin himself. There are, though, worse grounds for proceeding to the limit case of the purely open profile in the shear-free version of the theory. It is clear that the open profile corresponds to $I_d = 0$, and this makes Luzhin's relationship of (3.21) look like

$$\beta = \left(1 + \frac{I_k}{I_\rho}\right)\theta' - \frac{1}{G I_\rho} M_x . \tag{3.22}$$

But (3.22) is an additional relation for the open profile, which does not follow from any theoretical reasoning. Therefore Luzhin has to adopt an additional tendency, $I_\rho \to \infty$, to transit to a purely open profile according to Vlasov's theory and thus to be able to obtain the relationships (3.19) passing to the limit. Clearly, adopting this tendency is a bit of stretch

---

7 The fact that relation (3.21) should be treated as an additional postulate for a compound profile is recognized by O.V. Luzhin himself in [14].

because it actually does not take place. However, to be fair, we should note that the estimate of the orders of magnitude as $I_\rho >> I_k$ for the purely open profile can be at least considered.

If we adopt the semi-shear theory of thin-walled bars based on energy considerations (independently of the profile's type), then we have

$$H = G(I_k + I_d)\theta', \qquad M_\varpi = \frac{r^2 GA}{\mu_{\varpi\varpi}}(\theta' - \beta), \qquad M_x = H + M_\varpi,$$

which are also equivalent to (3.18) though with factors different from those in (3.21). To be particular, we have

$$\beta = \left(1 + \frac{I_k + I_d}{I_r}\mu_{\varpi\varpi}\right)\theta' - \frac{\mu_{\varpi\varpi}}{GI_r}M_x. \tag{3.23}$$

where $I_r = r^2 GA$ is a polar moment of inertia of the section and the dimensionless shape factor of the section, $\mu_{\varpi\varpi}$, is defined by relations

$$\mu_{\varpi\varpi} = \frac{I_r}{I_\varpi^2}\int_l \frac{\bar{\bar{S}}_{o\varpi}^2}{h}ds, \quad \bar{\bar{S}}_{o\varpi} = \bar{S}_{o\varpi} - \frac{\kappa I_\varpi}{\Omega}, \quad \bar{S}_{o\varpi} = S_{o\varpi} - \frac{\kappa}{\Omega}\int_l S_{o\varpi}\rho ds. \tag{3.24}$$

Quite expectedly, relation (3.23) works for the limit cases, too: the purely closed and purely open profiles. The characteristic function of the profile, κ, is identical to one in the former case and to zero in the latter case on the whole profile. Of course, this relation holds for a fully open profile in the semi-shear theory of open-profile bars, not in the shear-free theory by Vlasov, and instead of (3.22) we have

$$\beta = \left(1 + \frac{I_k}{I_\omega^2}\int_l \frac{S_{o\omega}^2}{h}ds\right)\theta' - \frac{1}{GI_\omega^2}\int_l \frac{S_{o\omega}^2}{h}ds\, M_x. \tag{3.25}$$

Now it becomes obvious that the relation (3.22) by Luzhin holds if and only if

$$I_\rho \int_l \frac{S_{o\omega}^2}{h}ds = I_\omega^2. \tag{3.26}$$

If (3.26) were true, this would mean the Luzhin equations produced the semi-shear theory for the limit case of the open profile. However, a direct calculation of practical examples shows easily that the requirement of (3.26) is not satisfied.

The things said above are already enough to drop the theory by Luzhin as one having no unquestionable logical justification. It makes sense to reject this theory also for another reason: the analytic relationships of the shear and semi-shear theories of compound-profile thin-walled bars presented above are provable to exactly the same extent as are the relationships of the theories of closed-profile and open-profile bars. There are no new hypotheses introduced for construction of those relationships in addition to those that form the fundamental basis of the thin-walled bar theory.

The fact that the attempts to unite the shear-free theory by Vlasov and the semi-shear theory by Umanski are illogical and doomed to failure can be explained by an analogy: imagine what happens if we try to unite the Bernoulli–Euler theory and the Timoshenko theory for the same beam subject to bending. Suppose a part of the section of a flexural beam (the analogue of the open part of a compound profile) is described by the Bernoulli-Euler theory (being a shear-free theory) and the rest of the section (the analogy of the closed part of the compound profile) obeys the Timoshenko theory that takes into consideration the contribution of the energy made by the flexural tangential stresses. Obviously, this kind of union between the two theories can be reasonable to any sensible extent only in very special circumstances (for example, a part of the section is made of a material which hardly resists to shear). But if the situation is usual and there are no special circumstances, this mixture of the two theories will look like an excruciating tumble of cunning tricks. And this is, essentially, the way of reasoning followed by all published and known papers which try to construct a unified theory of compound open/closed thin-walled profiles.

8.3.3 Classification criteria for separation between theories of thin-walled bars

We have already noted that the tradition in the theory of thin-walled bars separates the branches of the theory using the criterion of the profile type. There were the following reasons for this classification:

- the theories of open-profile bars (by Vlasov) and of closed-profile bars (by Umanski) are unrelated historically, i.e. they were constructed independently;
- the open-profile bars were identified with the shear-free theory while the closed-profile bars were associated with the semi-shear theory.

However, as we said earlier, when a mathematical formulation of a problem is being constructed using the energy-based (variational) approach, the type of the governing equations is not determined by the profile type; the criterion is how one divides the set of stress/strain components into primary and secondary components.

Table 8.1 presents a unified classification of the thin-walled bars which uses both classification criteria.

Table 8.1

	shear-free theory	**semi-shear theory**	**shear theory**
open profile	V.Z. Vlasov		L.N. Vorobiev
closed profile		G.Y. Janelidze, Y.G. Panovko - - - - - - - - A.A. Umanski	
compound profile		 - - - - - - - - O.V. Luzhin, E.A. Beilin	

As one can see from this table, the shear-free theory is constructed for the open profiles only; it should be obvious why. On the other hand, the total absence of the shear strains in bars the sections of which are based on closed contours is not admissible.

In the case of the semi-shear theory, the final governing equations for the closed and compound profiles differ by their method of establishing physical relationships[8]. We identify our opinion with that by Janelidze and Panovko [11]: the energy-based (variational) approach for the derivation of physical relationships in the theory is best justified and most convincing by its logic.

A more general version of the thin-walled bar theory should be recalled here, too. It was suggested by R.A. Adadurov [2], [1] who used a minimal set of hypotheses in his reasoning – he found it sufficient to assume the unchangeable-contour hypothesis only. The governing system of simultaneous equations thus derived is a fairly complex system of integral-differential equations. Therefore the theory created by Adadurov is usually excluded from the list of practical applied theories and treated as a theoretical sample that other applied theories can be compared with by estimating the accuracy of produced results. This is the reason why we do not discuss this theory here and do not include it in any of the cells of

---

[8] The said difference does not relate to the types of the differential equations; it relates solely to the values of the coefficients at functions which participate in the differential equations of the problem.

Table 8.1. For the interested reader, the theory created by Adadurov is presented in original papers by this author and also in [11].

8.3.4 Final remarks

The construction of any applied (and not only) theory seems to be finished when that theory permits a full solution of all problems from a sufficiently wide class where clear formulations have been established. However, the virtue of any theory lies not only in its completeness in the sense indicated above; it is equally important what prospects it opens for the further scientific research in the field and what attractions it has for new generations of researchers.

All that was said in Chapters 6 through 8 is related solely to the linear (first-order) static analysis of thin-walled bars. With primarily the interests of young scientists in mind, we deem it useful to present a short list of promising (in the author's opinion) directions of research on the thin-walled bars. Here they are:

- dynamics of thin-walled bars;
- stability/buckling of thin-walled bars;
- second-order (strain-based) analysis of thin-walled bars;
- curvilinear thin-walled bars (statics, dynamics, stability of those);
- thin-walled bars with the lengthwise-variable geometric parameters;
- computational implementation of a thin-walled bar as one of basic elements for structural design models created and analyzed by CAD software systems.

Of course, all these problems have been solved in various ways and to a various extent, and a lot of papers have been published on the respective subjects. When we say we invite new people to open the new phase of research concerning the techniques and approaches of the thin-walled bar analysis, we mean we would like to see the general energy-based approach we use in this book extended onto all of the listed parts of the thin-walled bar theory.

A book may appear one day in which the above-listed problems of engineering will be thoroughly addressed from the standpoint of the energy-based approach.

To conclude this section, we would like to call the reader's attention to a newly published textbook by E.A. Beilin [5] which presents principles of the theory of torsion of thin-walled bars on quite an elementary level. We recommend this textbook to people who wish to make a primary acquaintance with the theory of thin-walled bars. It is worth noticing that the presentment in the textbook is based on a traditional approach without involving any variational considerations.

References

1. Adadurov RA (1947) Determining tangential stresses in thin-walled structures near an immuring (in Russian). Proc. Centr. Aerohydr. Inst., N614, pp 1–13
2. Adadurov RA (1948) Stresses and strains in a cylindrical shell with rigid transversal constraints (in Russian). Rep. Acad. Sci., 62, 2: 183–186
3. Beilin EA (1987) Torsion of thin-walled bars of a closed profile in the presence of longitudinal shear seams (in Russian). In: Proc. XIV All-Union Conf. for Theory of Shells and Plates, Vol.1. Published by Tbilisi University, pp 182–187
4. Beilin EA (1991) A version of the unified theory of torsion of thin-walled bars of open, closed, and partially closed profiles (in Russian). In: Researches in mechanics of structural constructions and materials. Collected papers by Leningrad Institute of Civil Engineering, pp 57–74
5. Beilin EA (2003) Principles of the theory of torsion of thin-walled bars of an arbitrary profile (in Russian). St.-Petersburg Univ. of Civil Engng and Architec., St.-Petersburg
6. Beilin EA, Johnson RG (1991) Torsion of thin-walled bars with a partially closed contour of the section (in Russian). In: Structural mechanics and analysis, 3: 7–15
7. Eliashvili DG (1967) On a technical theory of constricted torsion of prismatic thin-walled bars (in Russian). Bull. Acad. Sci. USSR, Engineering Journal, Mech. Deform. Solids, 5: 149–154
8. Eliashvili DG (1967) On a unified applied theory of straight thin-walled bars (in Russian). In: Proc. of Leningrad Inst. Railway Engng. Issue 267, pp 3–14
9. Guo Zhong-heng (1981) A united theory of thinwalled elastic structures. J. Struct. Mech. Vol. 9; 2: 179–197
10. Hancook GJ, Harrison HB (1972) A general method of analysis of stresses in thin-walled sections with open and closed parts. Civ. Eng. Trans. Inst. Austral. Vol. 14; 2: 181–188
11. Janelidze GY, Panovko YG (1948) Statics of elastic thin-walled bars (in Russian). Gostekhteorizdat, Moscow Leningrad
12. Kan SN, Panovko YG (1952) Principles of structural mechanics of thin-walled constructions (in Russian). Oborongiz, Moscow
13. Kruglenko IV (1988) Bending and constricted torsion of thin-walled curvilinear bars of an arbitrary cross-section (in Russian). A thesis for the degree of Cand. Sci. (Engng). Leningrad Institute of Civil Engineering
14. Luzhin OV (1960) On an analogy in theories of constricted torsion of thin-walled bars (in Russian). Structural mechanics and analysis, 4: 13–14

15. Luzhin OV (1980) Torsion of thin-walled bars of a compound cross-section. – Problems in the analysis of spatial constructions (in Russian). Moscow Institute of Civil Engineering, pp 79–89
16. Usuki Tsueo (1983) Ein Beitrang zur Theorie dunnwandigen prismatischer Stabe mit offen-geschlossenem Profil aus vier Scheiben. Bautechnik, Vol. 60, 1: 14–23

9 THE RITZ METHOD AND ITS MODIFICATIONS

...the physical laws of nature are most appositely formulated in terms of the minimality principle which provides a natural approach to a more or less complete solution of particular problems.

Richard Courant, Herbert Robbins

9.1 The basic theorem of the Ritz method

We consider a quadratic functional, $\mathsf{J}(F)$, defined on a certain linear space $\mathcal{M}$,

$$\mathsf{J}(F) = \mathsf{E}(F) - \Pi(F)\,, \qquad F \in \mathcal{M}\,. \tag{1.1}$$

Here $\mathsf{E}(F)$ is a homogeneous quadratic functional which is, in its physical sense, the strain energy of an elastic system calculated on the elements of the stress-and-strain field F, and $\Pi(F)$ is a linear functional.

The linear space $\mathcal{M}$ can be, for example, any of the following energy spaces:

- $\mathcal{L}$ – the Lagrangian energy space;
- $\mathcal{K}$ – the Castiglianian energy space;
- $\mathcal{F}$ – the parametrized energy space.

As our subsequent reasoning relates equally to any of the three spaces and to the respective functionals defined on those spaces, we will use the general symbol of $\mathcal{M}$ to denote an energy space which to search for the point of minimum of the functional $\mathsf{J}(F)$.

As usual, we will denote by angular brackets the energy scalar product, $\langle F_1, F_2 \rangle$, so that

$$\mathsf{E}(F) = \langle F, F \rangle = \|F\|^2, \tag{1.2}$$

where symbol $\|F\|$ designates the norm of the stress-and-strain field F in the energy metric of the $\mathcal{M}$ space.

Let the functional in question take its minimum value at a point F_*:

$$\mathsf{J}(F_*) \le \mathsf{J}(F)\,. \tag{1.3}$$

The essence of the Ritz method which is used to search for the point of minimum of functional $\mathsf{J}(F)$ is as follows. Instead of searching the whole space $\mathcal{M}$ for the solution of the variational problem (1.3), we deal with a sequence of its finite-dimensional subspaces embedded in one another as the n index grows, the index itself being the dimensionality of its respective subspace[1]:

$$\mathcal{M}_1 \subset \mathcal{M}_2 \subset \ldots \subset \mathcal{M}_n \subset \ldots \subset \mathcal{M}\,. \tag{1.4}$$

An appropriate variational problem can be formulated on each of the subspaces, so that

$$F_{n*} = \arg \min_{F_n \in \mathcal{M}_n} \mathsf{J}(F)\ , \tag{1.5}$$

which is equivalent to

$$\mathsf{J}(F_{n*}) \le \mathsf{J}(F_n) \quad \forall\, F_n \in \mathcal{M}_n\,. \tag{1.6}$$

So, *an n-dimensional Ritz solution* of the original variational problem (1.3) is a solution of the finite-dimensional variational problem (1.5) or, which is the same, (1.6).

The technique of searching for a Ritz solution is very simple. It follows these steps. First of all, we have to build the subspace $\mathcal{M}_n \subset \mathcal{M}$. This can be done if we choose a sequence of linearly independent elements, $\Phi_1, \Phi_2, \ldots, \Phi_n, \ldots$, in $\mathcal{M}$. A linear envelope of the first n terms of the sequence is the desirable subspace, $\mathcal{M}_n$. In other words, subspace $\mathcal{M}_n$ consists of all possible elements F_n such that

$$F_n = \alpha_1\Phi_1 + \alpha_2\Phi_2 + \ldots + \alpha_n\Phi_n \tag{1.7}$$

with an arbitrary set of scalar factors $\alpha_1, \alpha_2, \ldots \alpha_n$.

[1] Actually, some versions of the Ritz method (such as the finite element method) may ignore the embeddedness of the subspaces as in (1.4). This happens when the sequence of finite element meshes obeys the different, a bit weaker condition $h \to 0$ where h is a so-called *mesh diameter*, that is, a minimum diameter of the sphere which can contain, in the geometrical sense, any of the finite elements. We can regard the diameter of the mesh, h, as a dimensionality, n, of the approximant space, $\mathcal{M}_n$, or, more exactly, $h = C/n$. Here C is a certain constant that depends on the used mesh type.

The set of n linearly independent elements, $\Phi_1, \Phi_2, \ldots, \Phi_n$, is said to make up a *basis* of the finite-dimensional space $\mathcal{M}_n$, and the elements themselves are often called *base* or *coordinate elements*.

As soon as the basis is known and fixed, the $\mathsf{J}(F_n)$ functional becomes a quadratic form of parameters $\alpha_1, \alpha_2, \ldots \alpha_n$,

$$\mathsf{J}(F_n) = \mathsf{J}(\alpha_1, \alpha_2, \ldots \alpha_n) = <\sum_{i=1}^{n}\alpha_i\Phi_i, \sum_{j=1}^{n}\alpha_j\Phi_j> - \Pi(\sum_{i=1}^{n}\alpha_i\Phi_i) =$$

$$= \sum_{i=1}^{n}\sum_{j=1}^{n}\alpha_i\alpha_j <\Phi_i, \Phi_j> - \sum_{i=1}^{n}\alpha_i\Pi(\Phi_i). \qquad (1.8)$$

This is how it looks in the matrix form:

$$\mathsf{J}(\boldsymbol{\alpha}) = \frac{1}{2}\boldsymbol{\alpha}^{\mathsf{T}}\mathbf{R}\boldsymbol{\alpha} - \boldsymbol{\alpha}^{\mathsf{T}}\mathbf{P}, \qquad (1.9)$$

where we use the designations of

$$\boldsymbol{\alpha} = \left\|\begin{bmatrix}\alpha_1\\ \vdots \\ \alpha_n\end{bmatrix}\right\|, \mathbf{R} = 2\left\|\begin{bmatrix}<\Phi_1,\Phi_1> & \cdots & <\Phi_1,\Phi_n>\\ \vdots & \cdots & \vdots \\ <\Phi_n,\Phi_1> & \cdots & <\Phi_n,\Phi_n>\end{bmatrix}\right\|, \mathbf{P} = \left\|\begin{bmatrix}\Pi(\Phi_1)\\ \vdots \\ \Pi(\Phi_n)\end{bmatrix}\right\|. \qquad (1.10)$$

The $\mathbf{R}$ matrix is a Gram matrix [18] built for linearly independent elements $\Phi_1, \Phi_2, \ldots, \Phi_n$. The course of linear algebra says that the Gram matrix possesses peculiar properties: it is nondegenerate and, moreover, positive definite. The problem of minimizing the quadratic form $\mathsf{J}(\boldsymbol{\alpha})$ from (1.9) consists of solving a set of simultaneous linear algebraic equations,

$$\mathbf{R}\boldsymbol{\alpha} = \mathbf{P}, \qquad (1.11)$$

wherefrom

$$\boldsymbol{\alpha} = \mathbf{R}^{-1}\mathbf{P}. \qquad (1.12)$$

The calculation of the components of vector $\boldsymbol{\alpha}$ gives an approximate (Ritz) solution, F_{n*}, of the original variational problem:

$$F_{n*} = \alpha_1\Phi_1 + \alpha_2\Phi_2 + \ldots + \alpha_n\Phi_n.$$

The key question with regard to the Ritz solution is: what relation is there between the exact solution of the original variational problem, F_*, and its Ritz approximation, F_{n*}?

The answer to this question is given by the basic theorem of the Ritz method formulated as follows.

The Ritz solution, F_{n*}, is a projection of the exact solution of the original problem, F_*, onto the finite-dimensional subspace $\mathcal{M}_n$.

It goes without saying that the projection we mention is based on the energy metric, i.e. one generated by the energy scalar product.

The basic theorem of the Ritz method states essentially that

$$F_* - F_{n*} \perp F_{n*} \quad \text{or, which is the same,} \quad < F_* - F_{n*} , F_{n*} > = 0 . \qquad (1.13)$$

An intuitive geometrical depiction of this statement is presented in Fig. 9.1 where the three-dimensional Euclidean space represents the whole energy space $\mathcal{M}$ and the darkened plane plays the role of its finite-dimensional subspace, $\mathcal{M}_n$.

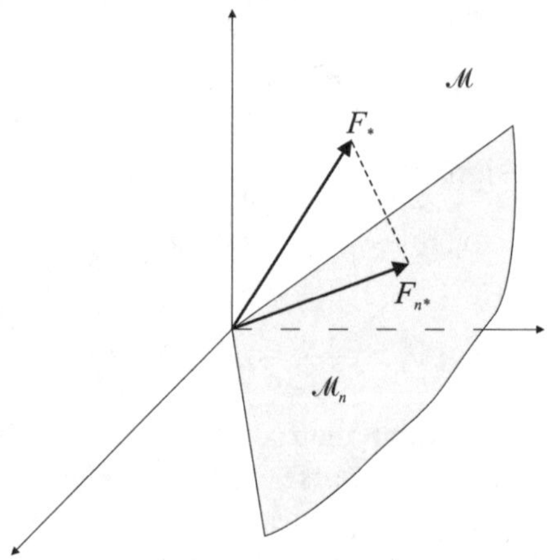

Fig. 9.1. A geometrical interpretation of the basic theorem of the Ritz method[2]

Before we start proving this important proposition, we would like to formulate its obvious corollary known as a generalized Pythagorean theorem:

$$\| F_* - F_{n*}\|^2 = \| F_*\|^2 - \| F_{n*}\|^2 . \qquad (1.14)$$

[2] Of course, pictures like that in Fig. 9.1 are not any proof by themselves. Their actual meaning and their educational importance are totally different. This is what such experts in teaching mathematics as R. Courant and H. Robbins say in this regard [3]:

"*The advantage of this geometrical method of describing mathematical facts is that it emphasizes certain aspects of the algebraic nature, which do not depend on the number of dimensions, n, and at the same time can be presented in a reader-friendly way for n ≤ 3*".

As the left-hand part is nonnegative here, we have a guaranteed estimate: the energy norm of the Ritz solution does not exceed the energy norm of the problem's exact solution,

$$||F_{n*}|| \leq || F_*|| \,. \tag{1.15}$$

Next goes the proof of the theorem.

As F_{n*} minimizes $\mathsf{J}(F)$ on $\mathcal{M}_n$, the following inequality holds for any number η and for any element $\varphi \in \mathcal{M}_n$:

$$\mathsf{J}(F_{n*}) \leq \mathsf{J}(F_{n*} + \eta\varphi) \,.$$

The right-hand part of this inequality is equivalent to the following expression (when expanded)

$$<F_{n*} + \eta\varphi, F_{n*} + \eta\varphi> - \Pi(F_{n*} + \eta\varphi) =$$

$$= \mathsf{J}(F_{n*}) + \eta[2<F_{n*}\,, \varphi> - \Pi(\varphi)] + \eta^2<\varphi, \varphi> .$$

Therefore

$$\eta[2<F_{n*}\,, \varphi> - \Pi(\varphi)] + \eta^2<\varphi, \varphi> \geq 0.$$

This inequality must hold for any number η of any sign, therefore the expression in the brackets must be zero:

$$2<F_{n*}\,, \varphi> - \Pi(\varphi) = 0. \tag{1.16}$$

Obviously, this equality must be true for any number n, and on the whole space $\mathcal{M}$ as well, that is,

$$2<F_*\,, \varphi> - \Pi(\varphi) = 0 \quad \forall\, \varphi \in \mathcal{M} \,. \tag{1.17}$$

Subtracting (1.16) from the above gives

$$<F_* - F_{n*}\,, \varphi> = 0. \tag{1.18}$$

As (1.18) holds true for any element $\varphi \in \mathcal{M}_n$, we can assume $\varphi = F_{n*}$, and this proves the proposition of (1.13).

This theorem has another equivalent formulation. It states that the Ritz solution, F_{n*}, is the one closest to the exact solution of the problem among all stress-and-strain fields which belong to a particular finite-dimensional space $\mathcal{M}_n$. To see this, we should make sure F_{n*} is a solution of the following minimum problem:

$$<F_* - F_{n*}, F_* - F_{n*}> = \min_{F_n \in \mathcal{M}_n} <F_* - F_n, F_* - F_n> \,. \tag{1.19}$$

We have

$$<F_* - F_{n*} - F_n, F_* - F_{n*} - F_n> =$$

$$= <F_* - F_{n*}, F_* - F_{n*}> - 2<F_* - F_{n*}, F_n> + <F_n, F_n> .$$

But due to (1.18) $2<F_* - F_{n*}, F_n> = 0$, therefore

$$<F_* - F_{n*}, F_* - F_{n*}> \leq <F_* - F_{n*} - F_n, F_* - F_{n*} - F_n>,$$

and the equality can take place only when $<F_n, F_n> = 0$, i.e. at $F_n = 0$. Consequently, F_{n*} is the only element in $\mathcal{M}_n$ which minimizes the functional $<F_* - F_n, F_* - F_n>$.

Following S.G. Mikhlin [7], we will call the sequence of Ritz elements $F_{1*}, F_{2*},\ldots$ a *minimizing sequence.*

As noted soundly by Strang and Fix [19], the proof of the theorem has only begun from the viewpoint of a mathematician (particularly, an expert in functional analysis) because the very existence of the F_* solution is not yet proved. The proof of existence is based on deeper mathematical considerations related to the completeness of the energy space $\mathcal{M}$, which is hardly the mathematical competence that can be required from a mechanical engineer. The interested reader can find a thorough mathematical discussion in the literature on the subject where the Ritz method is justified with all the strictness required by the mathematical style [7], [19], [14]. However, the proof above is quite sufficient for the engineer who is just sure the solution exists, therefore we finish it here[3]. As

---

[3] A discussion on the notion of "existence" and a strict separation between the meanings of this notion in pure and applied mathematics is presented in an interesting and educative book by I.I. Blekhman, A.D. Myshkis, and Y.G. Panovko [2]. In particular, as quoted from [2]:

"*…in general, we can say that a non-constructive proof of existence is not of any great value for an application-oriented analyst, and the lack of such a proof cannot unarm him in his search for an actual working solution*".

Nearly the same point of view towards the theorems of existence in elasticity has been advocated by V.V. Novozhilov. We would like to present a fairly long quotation from [9]:

"*However, proving the existence of a solution of an elasticity problem does not seem relevant from the physical point of view, because it is obvious that any solid body loaded and fixed in a certain admissible way must have at least one position of equilibrium (provided the strains appearing in it do not violate its continuity). Therefore, whenever we doubt the existence of a solution of elasticity equations, we essentially question their consistency with the physical problem they describe. The considerations used in Chapters I, II, III to give a mathematical formulation of the elasticity problems are based on a few thoroughly credible physical principles, therefore the equations thus obtained are in full accordance with the*

for the uniqueness of the solution, this issue is already resolved by the well-known Kirchhoff theorem discussed in Section **2**.2.2.

So, what conditions guarantee the convergence of the Ritz method, i.e. the fulfilment of the limit relationship for the following minimizing sequence?

$$||F_* - F_{n*}|| \to 0 \quad \text{as} \quad n \to \infty \,. \tag{1.20}$$

There are three such conditions:

- *Linear independence* – coordinate elements $\Phi_1, \Phi_2, \ldots, \Phi_n$ must be linearly independent for any n;
- *Belonging to the energy space* – all coordinate elements $\Phi_1, \Phi_2, \ldots, \Phi_n$ must be taken from the space $\mathcal{M}$;
- *Completeness* – the set of coordinate elements $\Phi_1, \Phi_2, \ldots, \Phi_n, \ldots$ must be complete in the $\mathcal{M}$ space.

Let us give a comment on each of the requirements above.

Linear independence

The requirement of linear independence is obvious. If there is an element (say, Φ_n) among coordinate elements $\Phi_1, \Phi_2, \ldots, \Phi_n$ which can be represented as a linear combination of the other $n-1$ elements, then the Ritz system of linear algebraic equations (1.11) becomes unsolvable because the Gram matrix, **R**, in (1.10) degenerates. This can be easily proved by algebraic methods [18].

Belonging to the energy space

The coordinate elements must belong to the energy space according to *sufficient* conditions of convergence for the Ritz method. This requirement was deemed for a long time a condition beyond all question, one of those which together ensure the convergence. However, the development of the practice and theory of computation[4] – first of all, of the finite element method – resulted in second thoughts about this requirements. A question: what if we do not require the $\Phi_1, \Phi_2, \ldots, \Phi_n$ elements to formally belong to the energy space $\mathcal{M}$; in other words, what if each separate finite-

---

problem and thus cannot produce absurd results. They are bound to give the solution of any particular problem that makes a real sense and has a correct formulation (meaning that the problem complies with the basic principles of the theory used)".

[4] In this exact order. The computational practice was ahead of the theory: it introduced the question, for the theory to answer, whether the inconsistent finite elements could be used justly.

dimensional space $\mathcal{M}_n$ is not necessarily a subspace of the original energy space $\mathcal{M}$? Can we then talk about the convergence of the Ritz solutions and, if we can, in what sense should we understand that convergence?

The answers to these questions are given by mathematical researches dedicated to the justification of the so-called inconsistent finite elements [19]. As it turned out, in a number of cases the requirement that the coordinate elements should belong to the energy space could be weakened. It has been found out that there exist such processes of construction of the minimizing sequences that each element of such a sequence does not belong to the original energy space $\mathcal{M}$ but is part of a bigger space $\mathcal{N}$ ($\mathcal{M} \subset \mathcal{N}$); at the same time, the limit element of this sequence *does* belong to $\mathcal{M}$. Here's the simplest geometric analogy: any point of the plane (being a two-dimensional energy space) can be approached as accurately as needed from the wider three-dimensional space, and no element of the approaching sequence is obliged to belong to the plane itself.

At the same time, we should notice the important fact that the basic theorem of the Ritz method holds true only for such approximant elements which do belong to the energy space. This is where the inequality (1.15) is ensured. The history of structural mechanics knows cases when ignoring this circumstance (or checking carelessly whether the coordinate elements belong to the energy space) produced fallacious deductions. An example is a known mistake made by Bernstein and discussed in detail by Panovko & Gubanova [12]. The example is fairly educative, so we present its short description below[5].

The subject of consideration is a flexural beam with hinged supports under a concentrated force P applied to the middle of the span. If we assume the flexural stiffness of the beam to vary according to a complicated law, then we have to use a computational numerical method to determine the deflection of the beam, Δ, in the middle of the span. We use the Ritz method to find the deflection function of the beam, $w(x)$. The Lagrangian functional for this problem is

$$\mathsf{L} = \frac{1}{2}\int_0^l EIw''^2 dx - Pw(l/2),$$

and note also that $\Delta = w(l/2)$.

[5] Actually, S.A. Bernstein [1] dealt with free oscillation frequencies rather than statics. The same problem is discussed in the mentioned book by Y.G. Panovko and I.I. Gubanova. However, this is not critical because the essence of the mistake made by Bernstein concerns the statics equally.

We take the only coordinate function, $\Phi_1(x)$, such that $\Phi_1(l/2) = 1$. Then, assuming $w(x) = \alpha\Phi_1(x)$, we will find the desirable factor, α, by minimizing the Lagrangian functional. An easy transformation gives

$$\alpha = \frac{P}{\int_0^l EIw''^2 dx} .$$

Apparently, the mechanical sense of (1.15) in this case is that the following estimate must hold true for sure:

$$\alpha \le \Delta .$$

However, if we take a "lid-shaped" function shown in Fig. 9.2 to be $\Phi_1(x)$, then the denominator in the formula of α becomes zero which leads to a formal violation of the inequality $\alpha \le \Delta$.

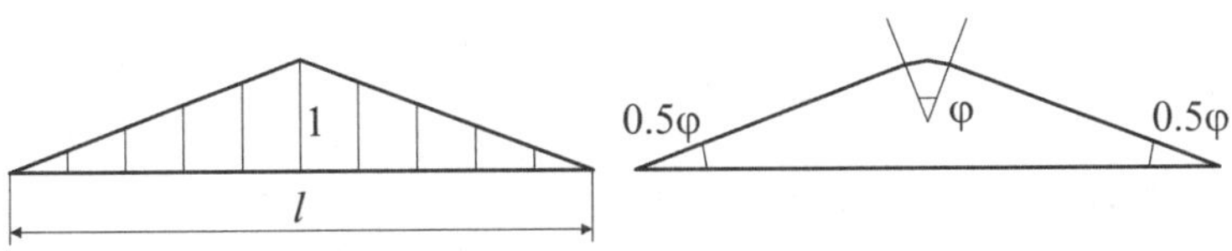

Fig. 9.2. An approximant

Bernstein explains the contradiction in this way, following intuitive considerations: "*The example has an obvious flaw (the violation of smoothness of the deflection curve), but it could be eliminated and the smoothness could be easily restored by assuming a finite length of the rounding, while the result of the calculation would remain the same*".

That's right: the violation of smoothness of the deflection curve! The matter is that the lid-shaped function does not belong to the Lagrangian energy space; the latter requires the existence of second square-summable generalized derivatives $\Phi_1''(x)$, and this existence is based on the existence of continuous first derivatives of the same function in the whole interval from 0 to l. This is exactly the minimum "smoothness of the deflection curve" mentioned by Bernstein. However, his speculation that *the result of the calculation would remain the same even after the smoothness of the curve was restored* is invalid.

We will present a correct treatment of this "paradox", borrowed from [12]. Let two straight segments of the approximant curve be mated by a circular arc of a radius ρ as shown in Fig. 9.2. We denote by a the length of a small part of the bar's longitudinal axis on which the approximant function is of the circular shape, and by φ the central angle; then we find that $a = 2\rho \sin(\varphi/2)$.

Choosing the approximant $\Phi_1(x)$ in this way makes the formula for α look like

$$\alpha = P \Big/ \int_{(l-a)/2}^{(l+a)/2} \frac{EI}{\rho^2} dx = P\rho^2 \Big/ \int_{(l-a)/2}^{(l+a)/2} EI dx \,.$$

If we maintain the value of φ and make the radius of the curve, ρ, tend to zero, then $a \to 0$ too. At a sufficiently small ρ we have

$$\alpha \to P\rho \Big/ 2EI_{x=l/2} \sin\frac{\varphi}{2} \to 0 \quad \text{as} \quad \rho \to 0 \,.$$

Consequently, the smoothed approximant $\Phi_1(x)$ shown in Fig. 9.2 on the right produces an infinitesimal rather than infinite displacement of the beam's central point as $\rho \to 0$. This is totally consistent with the basic theorem of the Ritz method.

Returning to the feasibility of building the minimizing sequence from a wider space than the energy one, we would like to notice again the violation of the conditions under which the inequality (1.15) holds; this fact was noticed in practice after the inconsistent finite elements came into usage.

The theoretical analysis of convergence and the estimation of errors of the respective sequences require finer mathematical approaches. Such an analysis has been done many times for one important class of problems where inconsistent finite elements are used. To mention some Russian-language mathematical publications dedicated to the convergence of inconsistent finite element approximations, we should refer to a series of papers by Kiev mathematicians I.D. Evzerov and V.S. Karpilovsky (see [4], [5] for examples).

Completeness

The last requirement, the completeness of the system of coordinate elements $\Phi_1, \Phi_2, \ldots, \Phi_n, \ldots$ in the $\mathcal{M}$ space, purports the possibility to approximate an arbitrary element $F \in \mathcal{M}$ with any given accuracy by a linear aggregate of the type $F \approx \alpha_1\Phi_1 + \alpha_2\Phi_2 + \ldots + \alpha_n\Phi_n$. It means that for any given $\varepsilon > 0$ one can find such integer N and such set of coefficients $\{\alpha_1, \alpha_2, \ldots, \alpha_n\}$ that the following will hold for $n > N$:

$$\|F - \alpha_1\Phi_1 - \alpha_2\Phi_2 - \ldots - \alpha_n\Phi_n\| \leq \varepsilon.$$

However, it is easy to prove that the requirement of completeness can be weakened. It suffices to require the possibility of approximating, at any given accuracy, not any arbitrary element of the space $\mathcal{M}$ but any element of a subspace, $\mathcal{K}$ ($\mathcal{M} \supset \mathcal{K}$), to which the problem's solution belongs for sure[6]. For example, the considerations of symmetry can give us a hint that the solution should be sought for in the class of symmetric functions without caring to approximate antisymmetric elements of the space $\mathcal{M}$.

9.2 The Ritz method in application to mixed functionals

The basic concept and the procedure of the Ritz method presented in the preceding section is essentially based in its proof part on the requirement that the functional of interest achieve its minimum at the point of its stationarity. However, the idea of replacing the original linear set $\mathcal{M}$, one to search for the solution, by its finite-dimensional approximation works, in the purely technical sense, also when the functional is not convex and thus does not generate a Hilbert space on the $\mathcal{M}$ linear set.

Of course, we cannot talk about projecting the exact solution, F_*, onto the finite-dimensional linear set because the very notion of projection loses its meaning when there is no proper metric. Nonetheless, a formal Ritz procedure of searching for an approximate solution can be constructed for this case, too.

9.2.1 The Ritz method in application to the Reissner functional

To be more particular, let us see what happens if we apply the Ritz method to the problem of finding a point of stationarity of the Reissner functional.

---

[6] The book already quoted earlier [2] presents different shades of understanding the role of the completeness of base functions in theoretical and applied mathematics. In particular, [2] says the following with regard to the completeness of the system of base functions in the Ritz or Galyorkin method:

"*Let us keep in mind that even if we work with a mathematically complete set of functions, we really take only a small subset of them for the purpose of our calculation. On the other hand, it is easy to see that any finite system of functions can be supplemented so as to become a complete system, i.e. the original finite set can be deemed formally a part of such a complete system. However, this does not mean the finite sets of functions we take for coordinate ones in the Galyorkin method are all the same from the standpoint of completeness. The finite set should posses the property of a practical completeness with respect to the problem in question, i.e. we must be sure (or at least cherish a hope) that the desirable solution can be approximated at an acceptable accuracy by a linear combination of our base functions*".

To simplify the formulas, we confine ourselves to homogeneous kinematical boundary conditions.

Recalling Chapter 3, we note that the original linear set $\mathcal{M}$ consists in our case of all possible stress-and-strain fields of the type $F = \{\boldsymbol{\sigma}, \boldsymbol{A}\mathbf{u}, \mathbf{u}\}$ which belong to the linear set $\mathcal{U}_{\mathrm{ko}}$. As for the requirements of smoothness for stresses $\boldsymbol{\sigma}$ and displacements $\mathbf{u}$, they are defined by the form of the Reissner functional we use, as discussed in Chapter 3. The original linear set $\mathcal{M}$ can be conveniently represented as the sum (union) of two linear sets $\mathcal{M} = \mathcal{M}_\sigma \cup \mathcal{M}_u$ where $\mathcal{M}_\sigma$ consists of all possible fields of the type $F_\sigma = \{\boldsymbol{\sigma}, \mathbf{0}, \mathbf{0}\}$ and $\mathcal{M}_u$ consists of all possible fields of the type $F_u = \{\mathbf{0}, \boldsymbol{A}\mathbf{u}, \mathbf{u}\}$ with homogeneously kinematically admissible displacements,

$$\boldsymbol{E}_u \boldsymbol{H}_u \mathbf{u} = \mathbf{0} \in \Gamma. \tag{2.1}$$

Let us build a system of linearly independent elements $\Psi_1, \Psi_2, \ldots, \Psi_m, \ldots$ on the set $\mathcal{M}_\sigma$. Similarly, we build a system of linearly independent elements $\Phi_1, \Phi_2, \ldots, \Phi_n, \ldots$ on the set $\mathcal{M}_u$. Obviously, the set of $(m+n)$ elements $\{\Psi_1, \Psi_2, \ldots, \Psi_m, \Phi_1, \Phi_2, \ldots, \Phi_n\}$ makes up a system of linearly independent elements in the original set $\mathcal{M}$.

Now let us approximate an arbitrary element $F \in \mathcal{M}$ by element F_{m+n} which is a linear combination of the selected *base elements*. In other words, we assume

$$\begin{aligned} F \approx F_{m+n} &= (X_1\Psi_1 + \ldots + X_m\Psi_m) + (Z_1\Phi_1 + \ldots + Z_n\Phi_n) = \\ &= \boldsymbol{X}^{\mathsf{T}}\boldsymbol{\Psi} + \boldsymbol{Z}^{\mathsf{T}}\boldsymbol{\Phi}, \end{aligned} \tag{2.2}$$

with $(m+n)$ coefficients $X_i\,(i = 1,\ldots, m)$, $Z_j\,(j = 1,\ldots, n)$ yet to be determined. Here we use the following designations of vectors:

$$\begin{aligned} \boldsymbol{X} &= \|[X_1, \ldots, X_m]\|^{\mathsf{T}}, & \boldsymbol{Z} &= \|[Z_1, \ldots, Z_n]\|^{\mathsf{T}}, \\ \boldsymbol{\Psi} &= \|[\Psi_1, \ldots, \Psi_m]\|^{\mathsf{T}}, & \boldsymbol{\Phi} &= \|[\Phi_1, \ldots, \Phi_n]\|^{\mathsf{T}}. \end{aligned} \tag{2.3}$$

For convenience, we repeat the general expression of the Reissner functional's first form from (**3**.1.4) here:

$$\mathsf{R}_1(\boldsymbol{\sigma}, \mathbf{u}) = \tfrac{1}{2}(\boldsymbol{C}^{-1}\boldsymbol{\sigma}, \boldsymbol{\sigma}) - \tfrac{1}{2}(\boldsymbol{K}\mathbf{u}, \mathbf{u}) - (\boldsymbol{A}\mathbf{u}, \boldsymbol{\sigma}) + (\boldsymbol{E}_u\boldsymbol{p}, \boldsymbol{E}_u\boldsymbol{u})_\Gamma + \Pi_{\mathrm{s}} - \Pi_{\mathrm{k}}\,.$$

According to (**3**.1.3), this holds for the homogeneously kinematically admissible fields:

$$\Pi_{\mathrm{s}}(\mathbf{u}) = (\bar{\boldsymbol{X}}, \mathbf{u}) + (\boldsymbol{E}_p\bar{\boldsymbol{p}}, \boldsymbol{E}_p\boldsymbol{u})_\Gamma, \qquad \Pi_{\mathrm{k}}(\boldsymbol{\sigma}) = 0\,, \quad (\boldsymbol{E}_u\boldsymbol{p}, \boldsymbol{E}_u\boldsymbol{u})_\Gamma = 0,$$

therefore for the linear set $\mathcal{M}$

$$\mathsf{R}_1(\boldsymbol{\sigma}, \mathbf{u}) = \tfrac{1}{2}(\boldsymbol{C}^{-1}\boldsymbol{\sigma}, \boldsymbol{\sigma}) - \tfrac{1}{2}(\boldsymbol{K}\mathbf{u}, \mathbf{u}) - (\boldsymbol{A}\mathbf{u}, \boldsymbol{\sigma}) + \\ + (\bar{\boldsymbol{X}}, \mathbf{u}) + (\boldsymbol{E}_p\bar{\boldsymbol{p}}, \boldsymbol{E}_p\boldsymbol{u})_\Gamma. \tag{2.4}$$

Fixing the integral values of m and n and subsituting approximations (2.2) in (2.4) makes the Reissner functional a quadratic form of $(m+n)$ desirable parameters:

$$\mathsf{R}_1(\boldsymbol{X}, \boldsymbol{Z}) = \frac{1}{2}\boldsymbol{X}^{\mathsf{T}}(\Psi, \boldsymbol{C}^{-1}\Psi)\boldsymbol{X} - \frac{1}{2}\boldsymbol{Z}^{\mathsf{T}}(\Phi, \boldsymbol{K}\Phi)\boldsymbol{Z} - \boldsymbol{Z}^{\mathsf{T}}(\boldsymbol{A}\Phi, \Psi)\boldsymbol{X} + \\ + \boldsymbol{Z}^{\mathsf{T}}(\bar{\boldsymbol{X}}, \Phi) + \boldsymbol{Z}^{\mathsf{T}}(\boldsymbol{E}_p\bar{\boldsymbol{p}}, \boldsymbol{E}_p\boldsymbol{H}_u\Phi)_\Gamma. \tag{2.5}$$

This is how it looks in the matrix form:

$$\mathsf{R}_1(\boldsymbol{X}, \boldsymbol{Z}) = \frac{1}{2}\left\|\left[\boldsymbol{X}^{\mathsf{T}}, \boldsymbol{Z}^{\mathsf{T}}\right]\right\| \left\|\begin{bmatrix} \mathbf{G} & -\mathbf{H} \\ -\mathbf{H}^{\mathsf{T}} & -\mathbf{K} \end{bmatrix}\right\| \left\|\begin{bmatrix} \boldsymbol{X} \\ \boldsymbol{Z} \end{bmatrix}\right\| + \left\|\left[\boldsymbol{X}^{\mathsf{T}}, \boldsymbol{Z}^{\mathsf{T}}\right]\right\| \left\|\begin{bmatrix} \mathbf{0} \\ \mathbf{P} \end{bmatrix}\right\|, \tag{2.6}$$

where the following designations are used: $<\Psi_i, \Psi_j> = \tfrac{1}{2}(\Psi_i, \boldsymbol{C}^{-1}\Psi_j)$,

$$\mathbf{G} = (\Psi, \boldsymbol{C}^{-1}\Psi) = 2\left\|\begin{bmatrix} <\Psi_1, \Psi_1> & \cdots & <\Psi_1, \Psi_m> \\ \vdots & \cdots & \vdots \\ <\Psi_m, \Psi_1> & \cdots & <\Psi_m, \Psi_m> \end{bmatrix}\right\|,$$

$$\mathbf{H} = (\boldsymbol{A}\Phi, \Psi) = \left\|\begin{bmatrix} (\Psi_1, \boldsymbol{A}\Phi_1) & \cdots & (\Psi_1, \boldsymbol{A}\Phi_n) \\ \vdots & \cdots & \vdots \\ (\Psi_m, \boldsymbol{A}\Phi_1) & \cdots & (\Psi_m, \boldsymbol{A}\Phi_n) \end{bmatrix}\right\|,$$

$$\mathbf{K} = (\boldsymbol{K}\Phi, \Phi) = \left\|\begin{bmatrix} (\Phi_1, \boldsymbol{K}\Phi_1) & \cdots & (\Phi_1, \boldsymbol{K}\Phi_n) \\ \vdots & \cdots & \vdots \\ (\Phi_n, \boldsymbol{K}\Phi_1) & \cdots & (\Phi_n, \boldsymbol{K}\Phi_n) \end{bmatrix}\right\|, \quad \mathbf{P} = \left\|\begin{bmatrix} \Pi_s(\Phi_1) \\ \vdots \\ \Pi_s(\Phi_n) \end{bmatrix}\right\|. \tag{2.7}$$

The conditions of stationarity of $\mathsf{R}_1(\boldsymbol{X}, \boldsymbol{Z})$ produce this set of simultaneous linear algebraic equations:

$$\left\|\begin{bmatrix} \mathbf{G} & -\mathbf{H} \\ -\mathbf{H}^{\mathsf{T}} & -\mathbf{K} \end{bmatrix}\right\| \left\|\begin{bmatrix} \boldsymbol{X} \\ \boldsymbol{Z} \end{bmatrix}\right\| + \left\|\begin{bmatrix} \mathbf{0} \\ \mathbf{P} \end{bmatrix}\right\| = \left\|\begin{bmatrix} \mathbf{0} \\ \mathbf{0} \end{bmatrix}\right\| \tag{2.8}$$

with respect to unknown vectors $\boldsymbol{X}$ and $\boldsymbol{Z}$.

The primary question with regard to the system of equations (2.8) is whether it can be solved unambiguously. First of all, we would like to notice the following two properties of blocks of the common matrix that makes up the system (2.8).

Property 1. The square matrix **G** of the order m is a positive definite matrix. This statement follows directly from the definition of matrix **G** according to which **G** can be looked at as a Gram matrix of the system of linearly independent elements $\Psi_1, \Psi_2, \ldots, \Psi_m$ with the Castiglianian energy metric.

Property 2. The square matrix **K** of the order n is a positive semi-definite matrix. This also follows from the definition of the matrix and the earlier condition (**1**.2.3) that the ***K*** operator must obey.

These two properties make the following theorem true:

The system of equations (2.8) can be solvable unambiguously with any right part and any positive definite operator ***K*** (including the zero operator ***K*** = ***O***) if and only if the following requirement is satisfied:

$$\text{rank}\, \mathbf{H} = n\,, \tag{2.9}$$

where n is the dimensionality of the finite-dimensional space of displacements.

First we prove the sufficiency (the "if" part) of condition (2.9). As the **G** matrix is positive definite, first m equations in (2.8) yield

$$\boldsymbol{X} = \mathbf{G}^{-1}\mathbf{H}\boldsymbol{Z},$$

which gives the following when substituted in the last n equations of (2.8):

$$(\mathbf{H}^{\mathsf{T}}\mathbf{G}^{-1}\mathbf{H} + \mathbf{K})\boldsymbol{Z} = \mathbf{P}. \tag{2.10}$$

The square matrix $\mathbf{H}^{\mathsf{T}}\mathbf{G}^{-1}\mathbf{H}$ of the order n can be interpreted as a Gram matrix for n column vectors of the **H** matrix specified in the Euclidean space of the dimensionality m with the scalar product $(\boldsymbol{x}, \boldsymbol{y})$ defined by the formula $(\boldsymbol{x}, \boldsymbol{y}) = \boldsymbol{x}^{\mathsf{T}}\mathbf{G}^{-1}\boldsymbol{y}$. As the rank of the **H** matrix is equal to the number of its columns according to (2.9), all columns of the matrix are linearly independent, hence the positive definiteness of the Gram matrix $\mathbf{H}^{\mathsf{T}}\mathbf{G}^{-1}\mathbf{H}$. This means that the following strict inequality holds for any nonzero vector $\boldsymbol{Z}$:

$$\boldsymbol{Z}^{\mathsf{T}}(\mathbf{H}^{\mathsf{T}}\mathbf{G}^{-1}\mathbf{H} + \mathbf{K})\boldsymbol{Z} > 0,$$

because, according to Property 2, $\boldsymbol{Z}^{\mathsf{T}}\mathbf{K}\boldsymbol{Z} \geq 0$. Thus the matrix $(\mathbf{H}^{\mathsf{T}}\mathbf{G}^{-1}\mathbf{H} + \mathbf{K})$ is positive definite, hence invertible.

Now we prove the necessity (the "only if" part) of condition (2.9). We suppose that $\text{rank}\, \mathbf{H} = t < n$. The theory of matrices says [18] that the rank of the product of matrices does not exceed the least rank of the matrix cofactors. This means rank $\mathbf{H}^{\mathrm{T}}\mathbf{G}^{-1}\mathbf{H} \leq t < n$, hence the square matrix $\mathbf{H}^{\mathrm{T}}\mathbf{G}^{-1}\mathbf{H}$ of the order n is degenerate and the system (2.8) is not necessarily solvable with any arbitrary vector $\mathbf{P}$ (at $\boldsymbol{K} = \boldsymbol{O}$, at the least).

Note an obvious corollary to the theorem just proved. Clearly, for the condition of (2.9) to hold, the following inequality must take place:

$$m \geq n\,. \tag{2.11}$$

The mechanical interpretation of the requirement (2.11) can be formulated as follows:

> *The Ritz system of equations which corresponds to the conditions of stationarity of the Reissner functional can be solved only if the dimensionality m of the finite-dimensional space of the stresses is not less than the dimensionality n of the finite-dimensional space of the displacements.*

9.2.2 The Ritz method in application to the generalized mixed functional Φ

We use the formula (**3**.5.12) for the generalized mixed functional, $\Phi(\boldsymbol{\sigma}, \mathbf{u})$, and have

$$\Phi = \frac{1}{2\kappa - 1}\left[\kappa \mathsf{L} + (1-\kappa)\mathsf{R}\right].$$

As we found out in Chapter 3, when the inequality ½ < κ < 1 holds, the functional Φ is convex and generates a metric, so the basic theorem of the Ritz method is applicable to this functional in these particular conditions.

Expanding the expression of the functional for the linear set $\mathscr{U}_{\mathrm{ko}}$ produces the following, up to the unimportant factor of $1/(2\kappa - 1)$:

$$\Phi(\boldsymbol{\sigma}, \mathbf{u}) = \frac{1-\kappa}{2}(\boldsymbol{C}^{-1}\boldsymbol{\sigma}, \boldsymbol{\sigma}) + \frac{2\kappa-1}{2}(\boldsymbol{K}\mathbf{u}, \mathbf{u}) + \frac{\kappa}{2}(\boldsymbol{A}\mathbf{u}, \boldsymbol{C}\boldsymbol{A}\mathbf{u}) -$$

$$- (1-\kappa)(\boldsymbol{A}\mathbf{u}, \boldsymbol{\sigma}) + (\overline{\boldsymbol{X}}, \mathbf{u}) + (\boldsymbol{E}_p\overline{\boldsymbol{p}}, \boldsymbol{E}_p\boldsymbol{u})_\Gamma\,. \tag{2.12}$$

Using the approximations of (2.2), we change this functional into a quadratic form,

$$\Phi(X, Z) = \frac{1-\kappa}{2} X^{\mathrm{T}}(\Psi, C^{-1}\Psi)X + \frac{2\kappa-1}{2} Z^{\mathrm{T}}(\Phi, K\Phi)Z +$$

$$+\frac{\kappa}{2} Z^{\mathrm{T}}(A\Phi, CA\Phi)Z - (1-\kappa)Z^{\mathrm{T}}(A\Phi, \Psi)X +$$

$$+ Z^{\mathrm{T}}(\bar{X}, \Phi) + Z^{\mathrm{T}}(E_p\,\bar{p}, E_p H_u \Phi)_{\Gamma}. \tag{2.13}$$

The matrix form for this is

$$\Phi(X,Z) = \frac{1}{2}\left\|X^{\mathrm{T}}, Z^{\mathrm{T}}\right\| \left\|\begin{matrix} (1-\kappa)\mathbf{G} & -(1-\kappa)\mathbf{H} \\ -(1-\kappa)\mathbf{H}^{\mathrm{T}} & \kappa\mathbf{R} + (2\kappa-1)\mathbf{K} \end{matrix}\right\| \left\|\begin{matrix} X \\ Z \end{matrix}\right\| +$$

$$+\left\|X^{\mathrm{T}}, Z^{\mathrm{T}}\right\| \left\|\begin{matrix} \mathbf{0} \\ \mathbf{P} \end{matrix}\right\|.$$

The result is the following Ritz system of equations:

$$\left\|\begin{matrix} (1-\kappa)\mathbf{G} & -(1-\kappa)\mathbf{H} \\ -(1-\kappa)\mathbf{H}^{\mathrm{T}} & \kappa\mathbf{R} + (2\kappa-1)\mathbf{K} \end{matrix}\right\| \left\|\begin{matrix} X \\ Z \end{matrix}\right\| = \left\|\begin{matrix} \mathbf{0} \\ -\mathbf{P} \end{matrix}\right\|, \tag{2.14}$$

where

$$\mathbf{R} = (A\Phi, CA\Phi). \tag{2.15}$$

As we can see, the system of equations (2.14) contains the system (1.11) as a particular case at $\kappa = 1$ and the system (2.8) at $\kappa = 0$.

It should be noticed that the requirement (2.11) at $\frac{1}{2} < \kappa < 1$ does not apply to functional $\Phi(\boldsymbol{\sigma}, \mathbf{u})$. The reason for this is that the Φ functional yields the following instead of (2.10):

$$[-(1-\kappa)\mathbf{H}^{\mathrm{T}}\mathbf{G}^{-1}\mathbf{H} + \kappa\mathbf{R} + (2\kappa-1)\mathbf{K}]Z = -\mathbf{P}. \tag{2.16}$$

Obviously, (2.16) becomes (2.10) in the particular case of $\kappa = 0$. At the same time, when the Φ functional is convex,, we can show that the matrix of the system of equations (2.16) of the order n is guaranteed to be positive definite[7] at any value of the integer parameter m. The meaning of it is that the Ritz system of equations is solvable regardless of a relationship between the dimensionalities in the approximations of the stresses and those of the displacements.

[7] The reader is invited to prove this.

An example

Our illustrative example will consist of an analysis of a beam which lies on an elastic bed with a reaction factor k; the beam is simply supported on its ends.

Following approximations (2.2), we introduce two systems of coordinate functions $\{\Phi_\alpha\}$ and $\{\Psi_\alpha\}$, both complete in the sense that the respective deflection w and moment M can be approximated as accurately as needed by linear aggregates composed of the said functions. We assume $m = n$ and

$$w = \sum_{\alpha=1}^{n} Z_\alpha \Phi_\alpha , \qquad M = \sum_{\alpha=1}^{n} X_\alpha \Psi_\alpha ,$$

so that

$$\Phi_\alpha(0) = \Phi_\alpha(l) = 0, \qquad \Psi_\alpha(0) = \Psi_\alpha(l) = 0,$$

where l is the length of the beam's span.

It is easy to see that the approximations thus introduced satisfy both the kinematic and the static boundary conditions for the beam with simple (hinged) supports on its ends.

The components of the respective blocks of the matrix of the Ritz equation system and the components of the absolute-term vector are defined as

$$\mathrm{G}_{\alpha\beta} = \int_0^l \frac{\Psi_\alpha \Psi_\beta}{EI} dx, \quad \mathrm{H}_{\alpha\beta} = \int_0^l \Psi'_\alpha \Phi'_\beta dx, \quad \mathrm{R}_{\alpha\beta} = \int_0^l EI \Phi''_\alpha \Phi''_\beta dx,$$

$$\mathrm{K}_{\alpha\beta} = \int_0^l k \Phi_\alpha \Phi_\beta dx, \qquad \mathrm{P}_\alpha = \int_0^l q \Phi_\alpha dx . \tag{2.17}$$

In addition to the already known designations, we introduce the following: EI is a bending stiffness of the beams' section, k is a subgrade reaction coefficient, q is a lateral load on the beam.

Assuming $\kappa = 1$, we get the classic version of the Ritz method based on the minimization of the Lagrangian functional. The system of equations (2.14) gets simpler to become

$$(\mathbf{R} + \mathbf{K})\boldsymbol{Z} = -\mathbf{P}$$

and does not contain the components of the X vector as unknowns. But then the bending moments M are defined by the relationship $M = -EIw''$ or

$$M = -EI\sum_{\alpha=1}^{n} Z_\alpha \Phi''_\alpha .$$

On the other hand, at $\kappa \neq 1$ the bending moments M will be calculated directly from solution of the Ritz equations, or

$$M = \sum_{\alpha=1}^{n} X_\alpha \Psi_\alpha$$

This approach does not require any differentiation of the coordinate functions, which may deteriorate the accuracy of the results. In particular, if the EI function is discontinuous (piecewise continuous, for example), then the bending moments calculated by the minimization of the Lagrangian functional will experience a discontinuity, too.

Assuming

$$\Phi_\alpha = \Psi_\alpha = \sin\frac{(2\alpha-1)\pi x}{l} \quad (\alpha = 1, 2, \ldots, n),$$

we can derive explicit expressions of the elements of all matrices needed for our calculation from formulas (2.17).

Without dwelling on the intermediate calculation, we present below the results for the beam with the stepwise bending stiffness EI under the concentrated force P applied to the middle of the beam's span (Fig. 9.3).

Table 9.1 gives the calculated values of the respective bending moment M^o,

$$M^\circ = 100\frac{M}{Pl}\frac{EI_0}{EI_1},$$

in the cross-section of the beam right under the load vs. the number of terms, n, kept in the expansion of the deflections and of the moments.

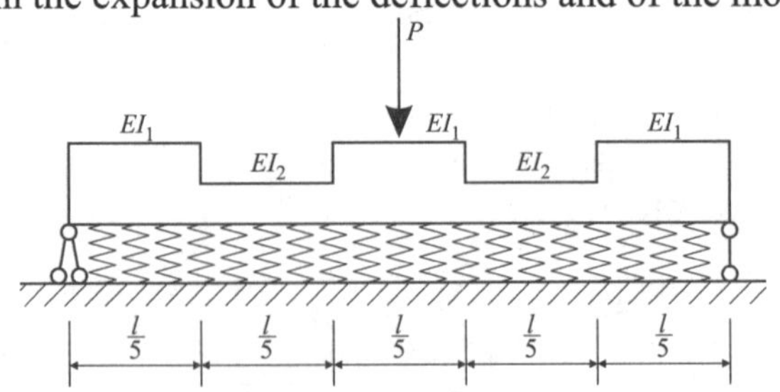

Fig. 9.3. The bending of a beam on an elastic bed

The following values of the parameters were used in the calculation:

$$\frac{EI_1}{EI_0}=\frac{2}{3}, \qquad \frac{EI_2}{EI_0}=\frac{1}{12}, \qquad l\sqrt[4]{k/EI_0}=6.25\,.$$

The exact values of M^{o} in the middle of the span is 7.37.

Table 9.1

n		2	3	4	5	6
$M_{\mathsf{L}}^{\text{o}}$	$\kappa=1$	4.25	6.14	6.04	5.43	5.85
$M_{\mathsf{R}}^{\text{o}}$	$\kappa=0$	2.68	4.40	5.39	5.87	6.11
M_{Φ}^{o}	$\kappa=2/3$	1.43	2.64	4.49	5.97	6.58

Fig. 9.4 presents curves of moments M^{o} at $n-6$. Curve 1 depicts the exact solution of the problem, curve 2 corresponds to M_{Φ}^{o}, and the dash line (curve 3) depicts $M_{\mathsf{L}}^{\text{o}}$. Here $M_{\mathsf{J}}^{\text{o}}$ is a relative bending moment calculated from the condition of stationarity of the functional J ($\mathsf{J}=\mathsf{L,R,\Phi}$).

It is interesting to track the variation of the accuracy at which the maximum bending moment is calculated vs. the parameter κ. The data presented in Table 9.2 help us watch this variation.

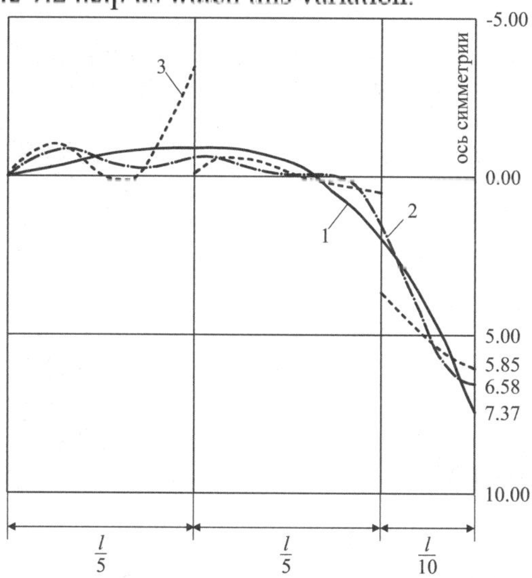

Fig. 9.4. Moment curves for a flexural beam

Table 9.2

κ	0.55	0.60	0.65	0.70	0.75	0.80	0.85	0.90	0.95
M_{Φ}^{o}	6.53	6.60	6.59	6.58	6.55	6.53	6.51	6.50	6.49

9.3 Method of two functionals

The Ritz method based on the minimization of the full potential energy functional (a Lagrangian functional) of an elastic system on a finite-dimensional subspace of a certain admissible space of displacement functions (a Lagrangian energy space) produces approximate solutions more accurate by displacements than by stresses. The finite element method (FEM), being a variation of the Ritz method with particular coordinate functions, features stress fields discontinuous on the boundaries between the elements where the displacement fields are continuous. This shortcoming of the Ritz method (FEM, in particular) vanishes if the variational formulation of the problem is based on the Castigliano functional. However, there is a great deal of difficulties in building systems of coordinate functions for the stress fields, which would satisfy the equilibrium equations in the volume and on the surface of the body. There is an alternative approach related to the mixed-type functionals where the fields of stresses and displacements are approximated independently. The Reissner functionals, being the best known representatives of the mixed-type functionals, posses two serious shortcomings:

- the lack of extreme properties (the stationarity point is a not a point of extremum for the Reissner functional) creates computational difficulties;
- the order of the mixed system of equations is the sum of the orders of both the displacement and stress approximations.

The first shortcoming in the above list can be eliminated by replacing the Reissner functional with the mixed one Φ which, as we have shown, has its minimum in the stationarity point for $\frac{1}{2} < \kappa < 1$. However, no simple reconstruction of the mixed functional is enough to eliminate the second shortcoming.

The paper [17] suggested a scheme for constructing Ritz approximations with the following properties:

- independent approximations of the stress and displacement fields are used;
- two Ritz systems of equations are constructed, each having a positive definite matrix;
- the boundary conditions (both static and kinematic) are satisfied precisely.

Unlike the traditional approach where the stationarity of one functional is sought for, our current method associates two functionals with the differential formulation of the problem; the conditions of minimum for those in finite-dimensional spaces yield the desirable Ritz systems of equations.

9.3.1 Weak and strong solutions with respect to stresses

As is known [13], [15], a *generalized* or *weak* solution of a problem where the strains and stresses in an elastic body should be found is a kinematically and statically admissible stress-and-strain field, $F_* = \{\boldsymbol{\sigma}_*, \boldsymbol{\varepsilon}_*, \mathbf{u}_*\} \in \mathscr{P}_k$, for which the following integral identity holds:

$$(\boldsymbol{A}^{\mathsf{T}}\boldsymbol{\sigma}_* + \boldsymbol{K}\mathbf{u}_* - \overline{\boldsymbol{X}}\,,\, \mathbf{u}_{ko}) + (\boldsymbol{E}_p\boldsymbol{H}_\sigma\boldsymbol{\sigma}_* - \boldsymbol{E}_p\,\overline{\boldsymbol{p}}\,,\, \boldsymbol{E}_p\,\boldsymbol{u}_{ko})_\Gamma = 0\,, \tag{3.1}$$

true for any homogeneously kinematically admissible field of displacements, $\mathbf{u}_{ko} \in \mathscr{U}_{ko}$.

As we already know, the mechanical interpretation of this integral identity is called the principle of virtual displacements – see (**1**.4.12). Further, the results of Chapter 2 imply that the above definition of the generalized solution is equivalent to a variational formulation where the Lagrangian functional should be minimized:

$$\mathsf{L}(\mathbf{u}) = \mathsf{E}(\mathbf{u}) - \Pi_s(\mathbf{u}) \tag{3.2}$$

where E is the potential strain energy, Π_s is the force potential. Note that the force potential can be treated as a potential of all *active* volumetric and surface forces.

Usually, as soon as we have determined the vector of displacements, $\mathbf{u}_*$, which minimizes the Lagrangial functional, finding the stress distribution in the system (i.e. finding the true vector of stresses $\boldsymbol{\sigma}_*$) is quite trivial because

$$\boldsymbol{\sigma}_* = \boldsymbol{C}\boldsymbol{A}\mathbf{u}_*\,. \tag{3.3}$$

Further we will refer to stresses $\boldsymbol{\sigma}_*$ determined by (3.3) as a *weak solution with respect to stresses*. This definition supposes that $\mathbf{u}_*$ in (3.3) is a weak solution with respect to displacements in the sense of the integral identity (3.1).

For reasons discussed later, it makes sense to re-formulate the problem of finding stresses $\boldsymbol{\sigma}$ so that it become variational.

Let $\mathscr{U}_{s/2}$ be a set of statically semi-admissible stress-and-strain fields, and let $\mathscr{U}_{so/2}$ be a set of homogeneously statically semi-admissible stress-

and-strain fields. To put it another way: $\boldsymbol{\sigma} \in \mathscr{U}_{s/2}$ if stresses $\boldsymbol{\sigma}$ satisfy the static boundary conditions and $\boldsymbol{\sigma}_0 \in \mathscr{U}_{so/2}$ if stresses $\boldsymbol{\sigma}_o$ satisfy the homogeneous static boundary conditions. Also, for these stresses the following integrals are supposed to exist and take finite values:

$$\int_\Omega \boldsymbol{\sigma}^T \boldsymbol{\sigma} d\Omega, \quad \int_\Omega \boldsymbol{\sigma}_o^T \boldsymbol{\sigma}_o d\Omega, \quad \int_\Omega (\boldsymbol{A}^T\boldsymbol{\sigma})^T (\boldsymbol{A}^T\boldsymbol{\sigma}) d\Omega, \quad \int_\Omega (\boldsymbol{A}^T\boldsymbol{\sigma}_o)^T (\boldsymbol{A}^T\boldsymbol{\sigma}_o) d\Omega .$$

For any vector $\mathbf{u}_k \in \mathscr{U}_k$ and any vector $\boldsymbol{\sigma}_0 \in \mathscr{U}_{so/2}$, the basic integral identity (1.2.17) holds, which can be written as follows:

$$(\boldsymbol{A}\mathbf{u}_k, \boldsymbol{\sigma}_o) = (\mathbf{u}_k, \boldsymbol{A}^T\boldsymbol{\sigma}_o) + (\boldsymbol{H}_\sigma \boldsymbol{\sigma}_o, \boldsymbol{H}_u \mathbf{u}_k)_\Gamma . \tag{3.4}$$

Definition.

We will call a *strong solution of a problem with respect to stresses* a vector $\boldsymbol{\sigma} \in \mathscr{U}_{s/2}$ that satisfies the integral identity

$$\int_\Omega (\boldsymbol{C}^{-1}\boldsymbol{\sigma})^T \boldsymbol{\sigma}_o d\Omega = \int_\Omega (\boldsymbol{A}^T\boldsymbol{\sigma}_o)^T \mathbf{u}_* d\Omega + \oint_\Gamma (\boldsymbol{E}_u \boldsymbol{H}_\sigma \boldsymbol{\sigma}_o)^T \boldsymbol{E}_u \bar{\boldsymbol{u}} d\Gamma \tag{3.5}$$

for any homogeneously statically semi-admissible vector $\boldsymbol{\sigma}_o \in \mathscr{U}_{so/2}$ where $\mathbf{u}_*$ is the weak solution with respect to displacements.

If $\mathbf{u}_*$ is a sufficiently smooth vector, then the weak stresses are identical with the strong stresses. To see this, we make the scalar product of $\boldsymbol{C}^{-1}\boldsymbol{\sigma}_*$ with $\boldsymbol{\sigma}_o$ and take (3.3) and (3.4) into account to obtain the integral identity (3.5).

Now let us show that if the strong stresses exist $\boldsymbol{\sigma}$ then they are also the weak stresses. From (3.5) and (3.4) we derive

$$(\boldsymbol{C}^{-1}\boldsymbol{\sigma} - \boldsymbol{A}\mathbf{u}_*, \boldsymbol{\sigma}_o) = 0,$$

and the arbitrariness of $\boldsymbol{\sigma}_o \in \mathscr{U}_{so/2}$ gives $\boldsymbol{\sigma} = \boldsymbol{C}\boldsymbol{A}\mathbf{u}_*$.

We introduce a functional D as follows on the set $\mathscr{U}_{s/2}$:

$$\mathsf{D}(\boldsymbol{\sigma}) = \mathsf{E}(\boldsymbol{\sigma}) - \Pi_r(\boldsymbol{\sigma}) \tag{3.6}$$

where $\mathsf{E}(\boldsymbol{\sigma}) = ½(\boldsymbol{C}^{-1}\boldsymbol{\sigma},\boldsymbol{\sigma})$ is the strain energy represented as a quadratic functional of the stresses and $\Pi_r(\boldsymbol{\sigma})$ is the potential of *reactive* volumetric and surface forces, which is equal to the following by definition:

$$\Pi_r(\boldsymbol{\sigma}) = (\boldsymbol{A}^T\boldsymbol{\sigma}, \mathbf{u}_*) + (\boldsymbol{E}_u \boldsymbol{H}_\sigma \boldsymbol{\sigma}, \boldsymbol{E}_u \bar{\boldsymbol{u}})_\Gamma . \tag{3.7}$$

It should be said that the expression $\mathsf{E}(\boldsymbol{\sigma}) = ½(\boldsymbol{C}^{-1}\boldsymbol{\sigma},\boldsymbol{\sigma})$ is, strictly speaking, a physical strain energy of only the deformable body itself, i.e. without the energy accumulated in the elastic medium surrounding the body.

We take the variation of functional D and write out its stationarity conditions:

$$\delta \mathsf{D} = (\boldsymbol{C}^{-1}\boldsymbol{\sigma}, \delta\boldsymbol{\sigma}) - (\boldsymbol{A}^{\mathsf{T}}\delta\boldsymbol{\sigma}, \mathbf{u}_*) - (\boldsymbol{E}_u\boldsymbol{H}_\sigma\delta\boldsymbol{\sigma}, \boldsymbol{E}_u\bar{\boldsymbol{u}})_\Gamma = 0.$$

By identifying variation $\delta\boldsymbol{\sigma}$ with $\boldsymbol{\sigma}_o$, we derive the integral identity (3.5). Thus the search for a strong solution with respect to stresses and the search for a stationarity point of functional D on the set of statically semi-admissible stresses, $\mathscr{U}_{s/2}$, are two equivalent problems.

The following statement is true: if a strong solution with respect to stresses, $\boldsymbol{\sigma}$, exists, then the D functional has its minimum in its point of stationarity. To see this, we write out the following for any statically semi-admissible field of stresses $\boldsymbol{\sigma}_{s/2} \in \mathscr{U}_{s/2}$:

$$\mathsf{D}(\boldsymbol{\sigma}_{s/2}) - \mathsf{D}(\boldsymbol{\sigma}) =$$

$$= \tfrac{1}{2}(\boldsymbol{C}^{-1}\boldsymbol{\sigma}_o,\boldsymbol{\sigma}_o) + (\boldsymbol{C}^{-1}\boldsymbol{\sigma},\boldsymbol{\sigma}_o) - (\boldsymbol{A}^{\mathsf{T}}\boldsymbol{\sigma}_o,\mathbf{u}_*) - (\boldsymbol{E}_u\boldsymbol{H}_\sigma\boldsymbol{\sigma}_o, \boldsymbol{E}_u\bar{\boldsymbol{u}})_\Gamma, \quad (3.8)$$

where

$$\boldsymbol{\sigma}_o = \boldsymbol{\sigma}_{s/2} - \boldsymbol{\sigma} \in \mathscr{U}_{so/2}\,.$$

As $\boldsymbol{\sigma}$ satisfies the integral identity (3.5), this annuls the last three terms in the right-hand part of (3.8), therefore

$$\mathsf{D}(\boldsymbol{\sigma}_{s/2}) - \mathsf{D}(\boldsymbol{\sigma}) = \tfrac{1}{2}(\boldsymbol{C}^{-1}\boldsymbol{\sigma}_o,\boldsymbol{\sigma}_o).$$

As the $\boldsymbol{C}$ matrix is positive definite, we have

$$\mathsf{D}(\boldsymbol{\sigma}_{s/2}) \geq \mathsf{D}(\boldsymbol{\sigma})\,,$$

which proves the proposition.

9.3.2 A remark on existence of a strong solution with respect to stresses

Let us now consider the question of existence of a minimum point for functional D[8]. For the beginning, we re-formulate the variational problem

---

[8] The author is aware that the very appearance of this section in the book discords from his attitude towards the theorems of existence uttered in footnote number [2]. An excuse for this inconsistency follows. First, the reader competent in mathematics can find the justification of principal theorems of existence in theory of elasticity in many well-known mathematical papers where, however, the goal of searching for a strong solution with respect to stresses itself is not addressed. Second, the reader oriented primarily at engineering applications is suggested just to skip this section and start reading the next one.

for $\mathsf{D}(\boldsymbol{\sigma})$ on the set $\mathscr{U}_{s/2}$ into an equivalent problem of minimization of $\mathsf{D}(\boldsymbol{\sigma})$ on the linear set $\mathscr{U}_{so/2}$.

Let stresses $\boldsymbol{\tau} \in \mathscr{U}_{s/2}$ (the $\mathscr{U}_{s/2}$ set is supposed not to be empty). Then for any vector $\boldsymbol{\sigma}_{s/2} \in \mathscr{U}_{s/2}$ such a vector $\boldsymbol{\tau}_o \in \mathscr{U}_{so/2}$ can be found that

$$\boldsymbol{\sigma}_{s/2} = \boldsymbol{\tau} + \boldsymbol{\tau}_o\,.$$

Functional D can be defined now on the set $\mathscr{U}_{so/2}$:

$$\mathsf{D}(\boldsymbol{\tau}_o) =$$

$$= \tfrac{1}{2}\,(\boldsymbol{C}^{-1}\boldsymbol{\tau}_o,\boldsymbol{\tau}_o) + (\boldsymbol{C}^{-1}\boldsymbol{\tau},\boldsymbol{\tau}_o) - (\boldsymbol{A}^{\mathsf{T}}\boldsymbol{\tau}_o,\mathbf{u}_*) - (\boldsymbol{E}_u\boldsymbol{H}_\sigma\boldsymbol{\tau}_o,\,\boldsymbol{E}_u\,\overline{\boldsymbol{u}}\,)_\Gamma + C_\tau \qquad (3.9)$$

where C_τ is a scalar value that depends solely on the fixed vector of $\boldsymbol{\tau}$.

The following bilinear form will be introduced on the linear set $\mathscr{U}_{so/2}$:

$$b(\boldsymbol{\sigma}_o,\boldsymbol{\tau}_o) = \tfrac{1}{2}(\boldsymbol{C}^{-1}\boldsymbol{\sigma}_o,\boldsymbol{\tau}_o)\,,$$

which satisfies all axioms of scalar product due to the $\boldsymbol{C}$ matrix's being positive definite. Completing $\mathscr{U}_{so/2}$ in the metric thus introduced yields a Hilbert space, $\mathscr{H}_\sigma$, and the metric itself will be called an energy metric[9]. Also, as the $\boldsymbol{C}$ matrix is positive definite, the energy metric in $\mathscr{H}_\sigma$ is equivalent to the metric of $L_2(\Omega)$:

$$\|\boldsymbol{\tau}_o\|^2_{L_2(\Omega)} = \int_\Omega \boldsymbol{\tau}_o^{\mathsf{T}}\boldsymbol{\tau}_o\,d\Omega\,, \qquad \|\boldsymbol{\tau}_o\|^2_{\mathscr{H}_\sigma} = \tfrac{1}{2}(\boldsymbol{C}^{-1}\boldsymbol{\tau}_o,\boldsymbol{\tau}_o)\,,$$

because the conditions of (**1**.2.3) give

$$\frac{m_d}{2}\,\|\boldsymbol{\tau}_o\|^2_{L_2(\Omega)} \le \|\boldsymbol{\tau}_o\|^2_{\mathscr{H}_\sigma} \le \frac{M_d}{2}\,\|\boldsymbol{\tau}_o\|^2_{L_2(\Omega)}\,. \qquad (3.10)$$

The linear part, $l(\boldsymbol{\tau}_o)$, of functional $\mathsf{D}(\boldsymbol{\tau}_o)$ is defined on the set of statically semi-admissible variations of stress vectors, $\mathscr{U}_{so/2}$, and is bounded in $\mathscr{H}_\sigma$. To see this,

$$|l(\boldsymbol{\tau}_o)| \le |(\boldsymbol{C}^{-1}\boldsymbol{\tau},\boldsymbol{\tau}_o)| + |(\boldsymbol{A}^{\mathsf{T}}\boldsymbol{\tau}_o,\mathbf{u}_*) + (\boldsymbol{E}_u\boldsymbol{H}_\sigma\boldsymbol{\tau}_o,\,\boldsymbol{E}_u\,\overline{\boldsymbol{u}}\,)_\Gamma\,| \le$$

---

[9] You should not think the Hilbert space $\mathscr{H}_\sigma$ just introduced coincides with the Castiglianian energy space $\mathscr{K}$ defined earlier in Section **2**.3.2. There is no such coincidence even if the elastic medium is absent, $\boldsymbol{K} = \boldsymbol{O}$. The matter is that the elements of the Castiglianian energy space, $\mathscr{K}$, are obliged to satisfy *both* the static boundary conditions *and* the homogeneous equations of equilibrium in area Ω. Therefore an inclusion takes place: $\mathscr{K} \subset \mathscr{H}_\sigma$. However, for $\boldsymbol{K} = \boldsymbol{O}$ the metric defined on $\mathscr{H}_\sigma$ is formally the same as that on $\mathscr{K}$. Thus we can refer to this metric as a Castiglianian energy metric.

$$\leq 2\|\boldsymbol{\tau}\|_{\mathscr{H}_\sigma}\|\boldsymbol{\tau}_o\|_{\mathscr{H}_\sigma} + \left|\int_\Omega \boldsymbol{\tau}_o^{\mathsf{T}} A\mathbf{u}_* d\Omega\right| \leq$$

$$\leq 2\|\boldsymbol{\tau}\|_{\mathscr{H}_\sigma}\|\boldsymbol{\tau}_o\|_{\mathscr{H}_\sigma} + \|A\mathbf{u}_*\|_{L_2}\|\boldsymbol{\tau}_o\|_{L_2} \leq$$

$$\leq \|\boldsymbol{\tau}_o\|_{\mathscr{H}_\sigma}\left(2\|\boldsymbol{\tau}\|_{\mathscr{H}_\sigma} + \sqrt{2/m_d}\,\|A\mathbf{u}_*\|_{L_2(\Omega)}\right).$$

This inequality's derivation is based on the Cauchy–Buniakovsky inequality, the basic integral identity (3.4), and the inequality (3.10).

According to the Hahn–Banach theorem [6], the linear bounded functional, $l(\boldsymbol{\tau}_o)$, defined on the linear set, $\mathscr{U}_{so/2}$, of the Hilbert space $\mathscr{H}_\sigma$ can be extended by continuity onto the whole space $\mathscr{H}_\sigma$ keeping the norm of the functional. The $\mathsf{D}(\boldsymbol{\tau}_o)$ functional thus extended will reach its lower bound on $\mathscr{H}_\sigma$ [8].

All that was said above implies that the minimum of functional D is guaranteed on $\mathscr{H}_\sigma$, and at the same time the "minimum point" does not necessarily belong to set $\mathscr{U}_{so/2}$. The condition for a strong stress state to exist is the inclusion

$$\boldsymbol{\sigma}_* = CA\mathbf{u}_* \in \mathscr{U}_{s/2}.$$

9.3.3 A remark on consistent approximations for minimization of functional D

To minimize the D functional by the Ritz method, we choose a sequence of finite-dimensional subspaces $\mathscr{H}_{m\sigma} \subset \mathscr{H}_\sigma$ with base vectors $\boldsymbol{\tau}_1, \boldsymbol{\tau}_2, \ldots, \boldsymbol{\tau}_m$. To ensure the convergence of the Ritz method (as $m \to \infty$), we need to be able to approximate any element $\boldsymbol{\sigma} \in \mathscr{H}_\sigma$ with any given accuracy by an element of the type

$$X_1\boldsymbol{\tau}_1 + X_2\boldsymbol{\tau}_2 + \ldots + X_m\boldsymbol{\tau}_m$$

with an appropriate finite dimensionality m and proper coefficients X_i $(i = 1, \ldots, m)$. As $\mathscr{H}_{m\sigma}$ is finite-dimensional, the D functional achieves its lower bound on it, and the problem of minimizing D on $\mathscr{H}_{m\sigma}$ has a unique solution.

Together with functional $\mathsf{D} = \mathsf{D}(\boldsymbol{\sigma}, \mathbf{u}_*)$, we consider a perturbed functional, $\tilde{\mathsf{D}}(\boldsymbol{\sigma}, \tilde{\mathbf{u}})$, where $\tilde{\mathbf{u}}$ is a distorted vector of displacements, which is close to the displacement vector $\mathbf{u}_*$ in some sense. In particular, $\tilde{\mathbf{u}}$ can be vector $\mathbf{u}_n$ that makes the Lagrangian functional L take a minimum value in the finite-dimensional subspace $\mathscr{L}_n$ of the Lagrangian energy space, $\mathscr{L}$. Let $\tilde{\boldsymbol{\sigma}}_*$ be a point of minimum of $\tilde{\mathsf{D}}(\boldsymbol{\sigma},\tilde{\mathbf{u}})$ in $\mathscr{H}_\sigma$, and let $\tilde{\boldsymbol{\sigma}}_m$ be a point of minimum of the same functional in $\mathscr{H}_{m\sigma}$. Note that

$$\tilde{\boldsymbol{\sigma}}_* = \boldsymbol{CA}\mathbf{u}_n .$$

It makes sense to pose and solve the problem of minimization of $\tilde{\mathsf{D}}(\boldsymbol{\sigma},\tilde{\mathbf{u}})$ in $\mathscr{H}_{m\sigma}$ if the approximate (Ritz) solution $\tilde{\boldsymbol{\sigma}}_m$ for the distorted functional is "*more exact than its exact solution* $\tilde{\boldsymbol{\sigma}}_*$" meaning how close they are to the exact solution, $\boldsymbol{\sigma}_*$, for the exact functional D. This strangely looking phrase is actually a verbal form of the following inequality:

$$e_2 = \| \tilde{\boldsymbol{\sigma}}_m - \boldsymbol{\sigma}_* \| \leq \| \boldsymbol{CA}\mathbf{u}_n - \boldsymbol{\sigma}_* \| = e_1 , \tag{3.11}$$

which is supposed to hold in terms of a certain common norm.

Error e_1 depends solely on the selection of subspace $\mathscr{L}_n$, and error e_2 depends on $\mathscr{L}_n$ and $\mathscr{H}_{m\sigma}$.

Let $\mathscr{L}_n$ be chosen in such way that the following estimate of the approximation order takes place:

$$e_1 = O(h^{\mathrm{p}}), \quad \mathrm{p} > 0 . \tag{3.12}$$

Here h is a parameter that has the dimension of length. This parameter is thought to be proportional to the maximum size of area Ω occupied by the elastic body. Also, $h \to 0$ when $n \to \infty$. In particular, the finite element method usually means by parameter h the minimum diameter of a sphere that can contain any of finite elements of a particular mesh division of area Ω.

We will call the $\mathscr{H}_{m\sigma}$ subspace *consistent with* $\mathscr{L}_n$ *by order* α if the following estimate holds:

$$e_2 = O(h^{\mathrm{p}+\alpha}), \quad \alpha > 0 . \tag{3.13}$$

Fig. 9.5 shows a qualitative picture of relationships between the consistent sets of $\mathscr{L}_n$ and $\mathscr{H}_{m\sigma}$, where the abscissa axis simulates the energy space $\mathscr{H}_\sigma$ and the ordinate axis shows values of the D functional (solid line) and the perturbed functional, $\tilde{\mathsf{D}}$ (dash line).

Clearly, if error e_1 has the order of h^{p}, then for error e_2 to have a higher order of smallness, the error e_3

$$e_3 = \| \tilde{\boldsymbol{\sigma}}_m - \tilde{\boldsymbol{\sigma}}_* \| \tag{3.14}$$

is required to have exactly the same order of h^p. Thus, in order to achieve the consistence, the Ritz solution for the distorted functional should be neither too “good” nor too “rough”.

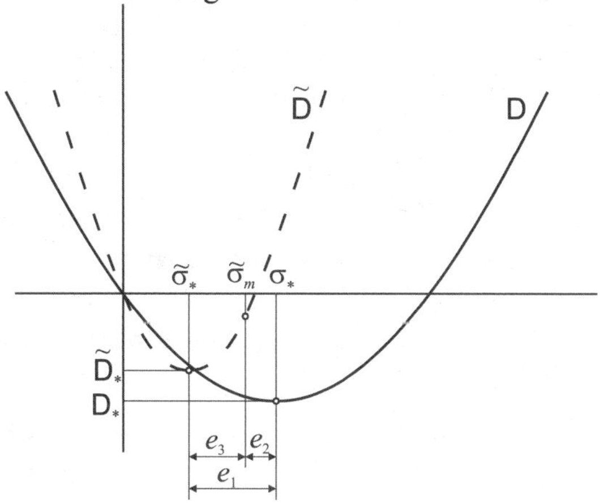

Fig. 9.5. Approximations of stresses and displacements consistent in their order

The question under what conditions the consistent approximations exist and how to estimate the respective errors is a subject of special mathematical studies[10], however, the very fact of existence of the consistent approximations has been validated in computational experiments.

9.3.4 A remark on the connection between functional D and the Reissner functional

Let us establish a relation between functional D and the Reissner functional R which is represented here in its second form:

$$\mathsf{R}_2(\boldsymbol{\sigma}, \mathbf{u}) = \tfrac{1}{2}(\boldsymbol{C}^{-1}\boldsymbol{\sigma}, \boldsymbol{\sigma}) - \tfrac{1}{2}(\boldsymbol{K}\mathbf{u}, \mathbf{u}) - (\boldsymbol{A}^{\mathsf{T}}\boldsymbol{\sigma}, \mathbf{u}) + (\overline{\boldsymbol{X}}, \mathbf{u}) -$$
$$- (\boldsymbol{E}_u \overline{\boldsymbol{u}}, \boldsymbol{E}_u \boldsymbol{p})_\Gamma + (\boldsymbol{E}_p(\overline{\boldsymbol{p}} - \boldsymbol{H}_\sigma\boldsymbol{\sigma}), \boldsymbol{E}_p \boldsymbol{u})_\Gamma . \quad (3.15)$$

If we fix the displacements, $\mathbf{u}$, in the above expression by assuming $\mathbf{u} = \mathbf{u}_*$ and consider only statically semi-admissible stresses, i.e assume $\boldsymbol{\sigma} \in \mathscr{U}_{s/2}$, then $\boldsymbol{E}_p(\overline{\boldsymbol{p}} - \boldsymbol{H}_\sigma\boldsymbol{\sigma}) = \mathbf{0} \in \Gamma$. Extracting a part of the resulting expression for $\mathsf{R}_2(\boldsymbol{\sigma}, \mathbf{u})$ which depends only on the variable stresses, we notice its coincidence with functional D.

[10] Actually, so special that they are far beyond both the subject of this book and the mathematical competence of its author.

In this regard, we can give the method of two functionals the following simple interpretation. The approximate Ritz solution is found in two phases. First, the minimization of the Lagrangian functional produces an approximate solution for displacements $\mathbf{u}_n$. Next, this vector of displacements is substituted in the Reissner functional which becomes a functional dependent on the stresses only, because the vector of displacements is now fixed. The second phase is to find a stationary point, $\tilde{\boldsymbol{\sigma}}_m$, of the functional just obtained. The set of statically semi-admissible stress fields is searched for this point. This makes the stationarity conditions of the Reissner functional its minimality conditions.

The method of two functionals and the method of conjugate approximations by Oden

If we classify the preceding two-phase procedure as belonging to the finite element method, which is nothing but a variation of the Ritz method, then it turns out that the procedure can be also a fairly general method for constructing continuous fields of stresses in the course of finite element analysis.

The foreign literature makes extensive references to a so-called *method of conjugate approximations* by Oden–Reddy [11], which is included even in textbooks [16] as a recipe (with no attempts of explaining). A detailed presentment of the Oden–Reddy method can be found in the book [10] by Oden himself. However, that particular presentment is much complicated by bulky index transformations and insignificant details which make it difficult to see the main idea.

The following interpretation can be given to the Oden–Reddy scheme after tracking through it carefully and establishing a relation between it and the variational problem of searching for stresses given a known field of displacements. The stress field is sought for as an expansion over the same base functions which have been used to minimize the Lagrangian functional. Further we can compose a functional, which can be called an Oden functional for convenience (though Oden himself does not treat his method as variational):

$$\mathsf{O}(\boldsymbol{\sigma}) = \tfrac{1}{2}(\boldsymbol{\sigma}, \boldsymbol{\sigma}) - (\boldsymbol{\sigma}, C\!A\mathbf{u})\,. \tag{3.16}$$

The conditions of stationarity of functional $\mathsf{O}(\boldsymbol{\sigma})$ produce physical equations

$$\boldsymbol{\sigma} = C\!A\mathbf{u}\,.$$

Let the set of vector functions $\Phi_1,\ldots,\Phi_n$ make up a system of base functions used to minimize the Lagrangian functional in the finite element analysis of displacements.

By expanding the stresses into finite sums over the base functions and substituting the expansions in (3.16), we make a quadratic form out of the functional. The condition of stationarity of the form produces a system of equations to calculate the coefficients of expansion of the stresses over the system of the same base functions.

The method of conjugate approximations by Oden yields continuous stress (force) fields, yet has a number of disadvantages:

- to determine the coefficients in the expansions of the stresses over functions $\Phi_1,\ldots,\Phi_n$, one has to compose and solve an additional set of simultaneous linear algebraic equations;
- the static boundary conditions are still not satisfied precisely in the Oden method;
- the approximation of the stresses is bound strictly to that of the displacements and does not leave any free choice of the coordinate functions;
- the variational formulation of the Oden method shows that the governing equations of this method follow from the stationarity of the $\mathsf{O}(\boldsymbol{\sigma})$ functional which has no clear energy-based meaning; this fact becomes obvious if we consider the dimensions of quantities contained in the integrands in (3.16);
- The method is oriented at obtaining continuous stress fields and cannot allow for discontinuities caused by such reasons as jump-like variations of the material constants in the integration domain of a two-dimensional or three-dimensional problem.

The method of two functionals is similar to the Oden–Reddy method in a certain sense (in its procedure) because it also requires the construction and solution of an additional system of equations. Therefore the first of the shortcomings of the Oden–Reddy conjugate approximations (a two-phase solution) remains in the method of two functionals, too. However, the other listed shortcomings vanish. The reason is that the method of two functionals can take the same functions $\Phi_1,\ldots,\Phi_n$ as those in the method of conjugate approximations to be base functions for minimizing $\mathsf{D}(\boldsymbol{\sigma})$, but it can use also a different, specifically chosen, system of coordinate functions $\Psi_1,\ldots,\Psi_m$.

9.3.5 Examples of application of the method of two functionals

Example 1

Our first example will be a Lame problem about the distribution of the stresses in a circular ring of an exterior radius b and an interior radius a under an internal pressure p (Fig. 9.6).

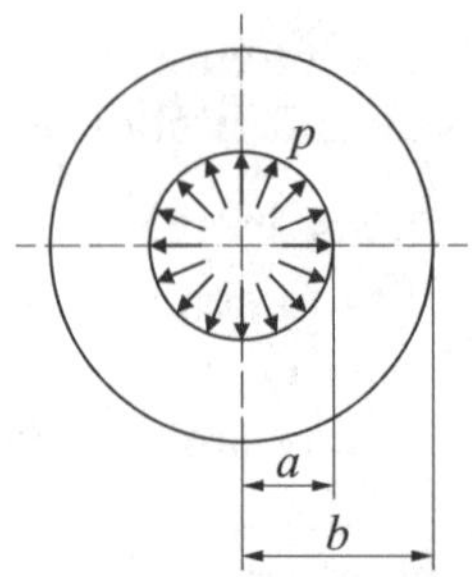

Fig. 9.6. The Lame problem

Let

- u be the radial displacement;
- σ^r, σ^φ be the respective radial and tangential stresses;
- E, ν be the respective modulus of elasticity and the Poisson ratio of the ring's material.

We assume

$$t = \frac{a}{b}, \qquad x = \frac{r}{b}$$

where r is the radial coordinate of a current point of the ring.

With these designations, the solution of the Lame problem is given by the formulas [20]

$$u_* = \frac{pbt^2}{E(1-t^2)}\left[(1-\nu)x + \frac{1+\nu}{x}\right], \ \sigma_*^r = \frac{pt^2}{1-t^2}\left(1-\frac{1}{x^2}\right), \ \sigma_*^\varphi = \frac{pt^2}{1-t^2}\left(1+\frac{1}{x^2}\right).$$

The L and D functionals for the Lame problem look like (up to the insignificant factor 2π)

$$\mathsf{L} = \frac{E}{2(1-\nu^2)}\int_1^t\left(\frac{u^2}{x} + 2\nu uu' + u'^2 x\right)dx - pbtu(t),$$

$$\mathsf{D} = \frac{b^2}{2E}\int_t^1 (\sigma^r\sigma^r - 2\nu\sigma^r\sigma^\varphi + \sigma^\varphi\sigma^\varphi)xdx - b\int_t^1\left(\sigma^\varphi - \sigma^r - x(\sigma^r)'\right)u_* dx. \quad (3.17)$$

Here and further the stroke means the differentiation with respect to coordinate x.

We divide the integration interval $[t,1]$ into n equal parts and try to minimize the above functionals using the Ritz method with a piecewise linear approximation of the desirable functions, $u(x)$, $\sigma^r(x)$, $\sigma^\varphi(x)$, thus making up the sets of $\mathscr{U}_{kn}$ and $\mathscr{U}_{(s/2)m}$. The $\mathscr{U}_{kn}$ set will be an n-dimensional set of kinematically admissible displacements, and the $\mathscr{U}_{(s/2)m}$ set will be an m-dimensional set of statically semi-admissible stresses. In our case we assume $m = n$.

With this method of construction of the coordinate functions, the Ritz approach actually becomes the finite element method where the maximum diameter of the finite element mesh, h, is the parameter

$$h = \frac{(1-t)b}{n}.$$

Without dwelling on details of the calculation, we present the results for $t = 0.5$. Fig. 9.7 shows curves of stresses σ^φ (upper curves) and σ^r (lower curves) multiplied by the factor $(1-t^2)/(pt^2)$ for $n = 4$.

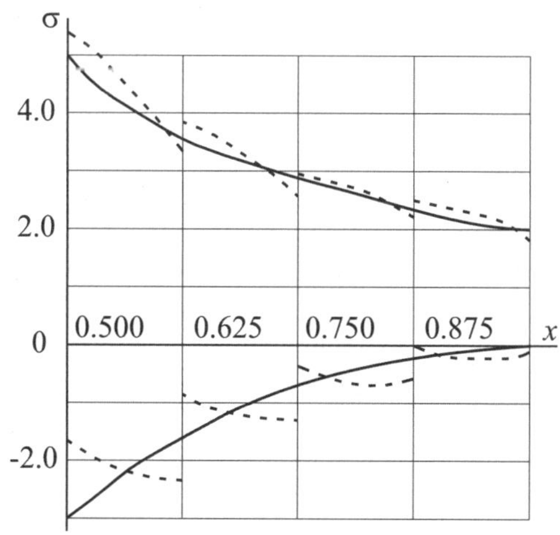

Fig. 9.7. Tangential and radial stresses in the Lame problem

The solid line depicts the exact stresses σ_*^φ and σ_*^r, the dash line depicts stresses σ_h^φ and σ_h^r calculated as

$$\sigma_h^r = \frac{E}{(1-\nu^2)b}\left(u_h' + \nu\frac{u_h}{x}\right), \qquad \sigma_h^\varphi = \frac{E}{(1-\nu^2)b}\left(\frac{u_h}{x} + \nu u_h'\right),$$

where u_h is a displacement function calculated by minimizing the Lagrangian functional.

If we calculate the $\tilde{\sigma}_h^\varphi$ and $\tilde{\sigma}_h^r$ stresses by the minimization of the $\tilde{\mathsf{D}}$ functional and put these on the graphs as in Fig.. 9.7, then the respective curves will nearly coincide with those depicting the exact stresses σ_*^φ and σ_*^r.

It is interesting to estimate the rate of convergence of the finite element method with respect to the stresses as $h \to 0$. Fig. 9.8 shows a relation (in percents) between the relative error of the calculated stresses (in the metric of the energy space $\mathscr{H}_\sigma$) and the number of the finite elements used, n.

The error ratio e_1/e_2 (Fig. 9.8) varies linearly vs. n, hence the finite-dimensional sets $\mathscr{U}_{kn}$ and $\mathscr{U}_{(s/2)m}$ are consistent by the order $\alpha = 1$.

The finite element analysis theory says that $e_1 = O(h)$ [19]. Consequently, $e_2 = O(h^2)$, which corresponds to the rate of convergence of displacements u_h to u_* in average.

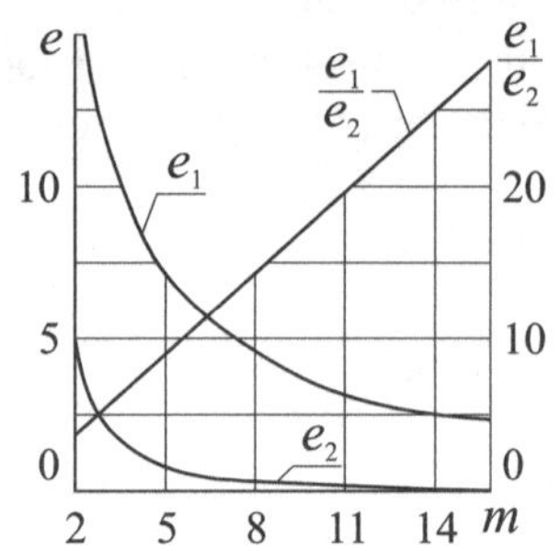

Fig. 9.8. Errors in the stresses for the Lame problem

Here

$$e_1 = \frac{\|\sigma_* - \sigma_h\|_{\mathscr{H}_\sigma}}{\|\sigma_*\|_{\mathscr{H}_\sigma}}100, \qquad e_2 = \frac{\|\sigma_* - \tilde{\sigma}_h\|_{\mathscr{H}_\sigma}}{\|\sigma_*\|_{\mathscr{H}_\sigma}}100. \tag{3.18}$$

Example 2

Let us illustrate the techniques of the method of two functionals (as a variation of the finite element method) by considering a problem of finding the stresses and strains in a cone-shaped structural part one end of which is fixed and the other subjected to an axial load of 42000 N (Fig. 9.9). This problem is discussed in the book by Oden [10] as an example of application of the method of conjugate approximations.

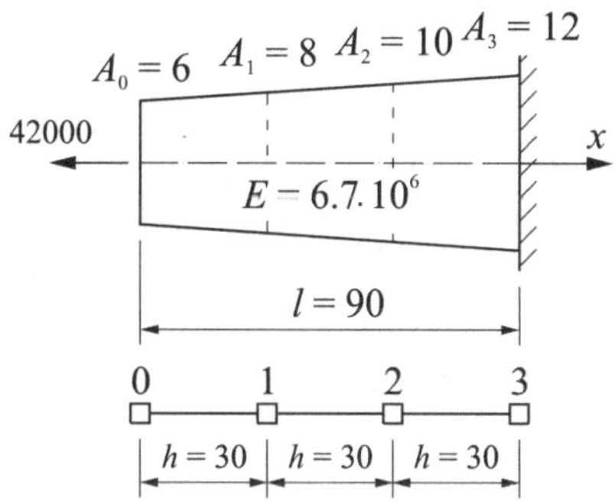

Fig. 9.9. Oden's problem

The area of the part's cross-section varies linearly between 6 cm^2 on its left end and 12 cm^2 on its right end. Also, the part is heated up by temperature $\Delta t = 20°$ evenly throughout its length. The linear thermal expansion factor is $\alpha = 7 \times 10^{-6}$. All sizes in Fig. 9.9 are in centimeters.

The solution of this problem by the finite element method using piecewise linear approximations of the displacements and the division of the part into three elements (Fig. 9.9) is presented in the book [16]. Here we repeat the calculated nodal values of the displacements from that book — Lagrangian displacements,

$$u_{0h} = -0.0753 \text{ cm}, \quad u_{1h} = -0.0450 \text{ cm}, \quad u_{2h} = -0.0207 \text{ cm}, \quad u_{3h} = -0,$$

where u_{ih} is the displacement in node i obtained from the finite element solution of the problem. At the same time, the exact values of the nodal displacements are equal to

$$u_{0*} = -0.0780 \text{ cm}, \quad u_{1*} = -0.0460 \text{ cm}, \quad u_{2*} = -0.0210 \text{ cm}, \quad u_{3*} = -0.$$

Let us calculate the stress field through the procedure of minimization of the $\tilde{\mathsf{D}}$ functional. We have

$$\tilde{D} = \int_0^l \left(\frac{S^2}{2EA} + u_h S' + \alpha \Delta t S \right) dx = \int_0^l \left(\frac{\sigma^2 A}{2E} + u_h (\sigma A)' + \alpha \Delta t \sigma A \right) dx \qquad (3.19)$$

where $S = \sigma A$ is the longitudinal stress in the cross-section of the part, A is its cross-section's area.

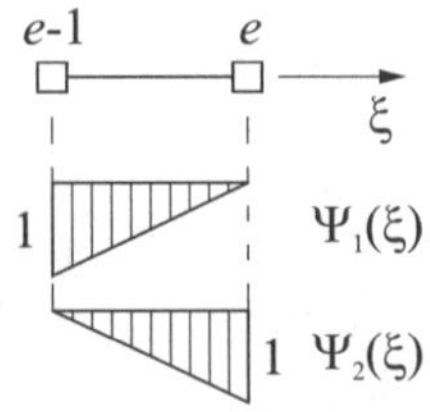

Fig. 9.10. Approximant functions on *e*-th finite element

We use a piecewise linear approximation of the stresses. On *e*-th finite element (Fig. 9.10) we have[11]

$$\sigma(\xi) = \sigma_{e-1}\Psi_1(\xi) + \sigma_e\Psi_2(\xi), \qquad A(\xi) = \frac{A_0}{3}(2 + e + \xi / h),$$

$$u_h(\xi) = u_{e-1,h}\Psi_1(\xi) + u_{e,h}\Psi_2(\xi)$$

where $\xi \in [0, h]$, $h = l/3$ is the finite element's length,

$$\Psi_1(\xi) = 1 - \xi/h, \qquad \Psi_2(\xi) = \xi/h .$$

Components $\{g_{ij}^{(e)}\}$ $(i,j = 1,2)$ of the compliance matrix, $\mathbf{G}_e$, for *e*-th finite element are determined by the formula

$$g_{ij}^{(e)} = \frac{A_0}{3E} \int_0^h \Psi_i(\xi)\Psi_j(\xi) a(\xi) d\xi \qquad (3.20)$$

where

$$a(\xi) = 2 + e + \xi/h .$$

Substitutions and an integration gives

[11] Never mistake the approximant functions of a particular finite element shown in Fig. 9.10 for the base functions of the Ritz method! As is known, because the finite element method is a special variation of the Ritz method, the base functions are constructed from element-by-element approximants for all finite elements meeting in one node (as it is said, "*on the star of the node*"). The element-by-element approximants used in the finite element method are often called *shape functions* in the literature on the subject.

$$\mathbf{G}_e = \frac{A_0 h}{36E}\left\|\begin{bmatrix} 9+4e & 5+2e \\ 5+2e & 11+4e \end{bmatrix}\right\|. \tag{3.21}$$

Components $\Delta_{pi}^{(e)}$ of the load vector $\mathbf{G}_{pe}$ for e-th finite element are determined as

$$\Delta_{pi}^{(e)} = \frac{A_0}{3}\int_0^h \left[(u_{e-1,h}\Psi_1 + u_{e,h}\Psi_2)\frac{d(\Psi_{,i}a)}{d\xi} + \alpha\Delta t\Psi_{,i}a \right] d\xi\,, \tag{3.22}$$

hence

$$\mathbf{G}_{pe} = \frac{A_0}{18}\left\|\begin{bmatrix} -5-3e & -7-3e \\ 8+3e & 10+3e \end{bmatrix}\right\| \cdot \left\|\begin{bmatrix} u_{e-1,h} \\ u_{e,h} \end{bmatrix}\right\| + \frac{A_0}{18}\alpha\Delta th\left\|\begin{bmatrix} 4+3e \\ 8+3e \end{bmatrix}\right\|. \tag{3.23}$$

The calculation yields the following.

- for element 1:

$$\mathbf{G}_1 = \frac{A_0 h}{36E}\left\|\begin{bmatrix} 13 & 7 \\ 7 & 15 \end{bmatrix}\right\|,$$

$$\mathbf{G}_{p1} = \frac{A_0}{18}\left\|\begin{bmatrix} -8 & -10 \\ 11 & 15 \end{bmatrix}\right\| \cdot \left\|\begin{bmatrix} u_{0,h} \\ u_{1,h} \end{bmatrix}\right\| + \frac{A_0}{18}\left\|\begin{bmatrix} 0.0294 \\ 0.0462 \end{bmatrix}\right\| = \frac{A_0}{18}\left\|\begin{bmatrix} 1.0813 \\ -1.3671 \end{bmatrix}\right\|$$

- for element 2

$$\mathbf{G}_2 = \frac{A_0 h}{36E}\left\|\begin{bmatrix} 17 & 9 \\ 9 & 19 \end{bmatrix}\right\|,$$

$$\mathbf{G}_{p2} = \frac{A_0}{18}\left\|\begin{bmatrix} -11 & -13 \\ 14 & 16 \end{bmatrix}\right\| \cdot \left\|\begin{bmatrix} u_{1,h} \\ u_{2,h} \end{bmatrix}\right\| + \frac{A_0}{18}\left\|\begin{bmatrix} 0.0420 \\ 0.0588 \end{bmatrix}\right\| = \frac{A_0}{18}\left\|\begin{bmatrix} 0.8061 \\ -0.9024 \end{bmatrix}\right\|$$

- for element 3

$$\mathbf{G}_3 = \frac{A_0 h}{36E}\left\|\begin{bmatrix} 21 & 11 \\ 11 & 23 \end{bmatrix}\right\|,$$

$$\mathbf{G}_{p3} = \frac{A_0}{18}\left\|\begin{bmatrix} -14 & -16 \\ 17 & 19 \end{bmatrix}\right\| \cdot \left\|\begin{bmatrix} u_{2,h} \\ u_{3,h} \end{bmatrix}\right\| + \frac{A_0}{18}\left\|\begin{bmatrix} 0.0546 \\ 0.0714 \end{bmatrix}\right\| = \frac{A_0}{18}\left\|\begin{bmatrix} 0.3444 \\ -0.2805 \end{bmatrix}\right\|$$

Let us merge, as the common finite element approach dictates, the compliance matrices of the elements into a single compliance matrix of the whole system and the load vectors into a common one. We have

$$\begin{matrix} \tilde{\sigma}_{0,h} & \tilde{\sigma}_{1,h} & \tilde{\sigma}_{2,h} & \tilde{\sigma}_{3,h} \end{matrix}$$

$$\mathbf{G}=\frac{A_0 h}{36E}\left\|\begin{bmatrix} 13 & 7 & 0 & 0 \\ 7 & 32 & 9 & 0 \\ 0 & 9 & 40 & 11 \\ 0 & 0 & 11 & 23 \end{bmatrix}\right\|, \qquad \mathbf{G}_p=\frac{A_0}{18}\left\|\begin{bmatrix} 1.0818 \\ -0.5610 \\ -0.5580 \\ -0.2805 \end{bmatrix}\right\|. \tag{3.24}$$

But for functional $\tilde{\mathsf{D}}$, the boundary condition on the free end of the part,

$$\tilde{\sigma}_{0,h}=7000 \text{ N/cm}^2$$

is the principal boundary condition because the minimum of that functional is sought for among statically semi-admissible fields of stresses. If this boundary condition is taken into consideration beforehand, our final governing system of equations with respect to the nodal displacement values will be

$$\left\|\begin{bmatrix} 32 & 9 & 0 \\ 9 & 40 & 11 \\ 0 & 11 & 23 \end{bmatrix}\right\| \cdot \left\|\begin{bmatrix} \tilde{\sigma}_{1,h} \\ \tilde{\sigma}_{2,h} \\ \tilde{\sigma}_{3,h} \end{bmatrix}\right\| = \left\|\begin{bmatrix} 201580 \\ 249340 \\ 125290 \end{bmatrix}\right\|. \tag{3.25}$$

Solution of this system of equations gives

$$\tilde{\sigma}_{1,h}=5141, \quad \tilde{\sigma}_{2,h}=4118, \quad \tilde{\sigma}_{3,h}=3478.$$

Table 9.3 below present comparative results; values of the stresses calculated by Oden's method of conjugate approximations [10] are borrowed from [16].

Table 9.3

No. of node	Exact stress	Stress by Oden's method	Stress by the method of two functionals	Constant stress on the element
0	7000	6132	7000	
				5829
1	5250	5222	5141	
				4489
2	4200	3935	4118	
				3685
3	3500	3558	3478	

Fig. 9.11 shows a visual picture of the obtained results. This is what we can see there:

- curve 1 depicts the exact solution, σ_*;
- stair-step line 2 presents the constant stresses on the elements, σ_h;
- polygonal line 3 shows the stresses calculated by Oden's method of conjugate approximations;
- piecewise linear curve 4 conforms to the stresses calculated by the method of two functionals.

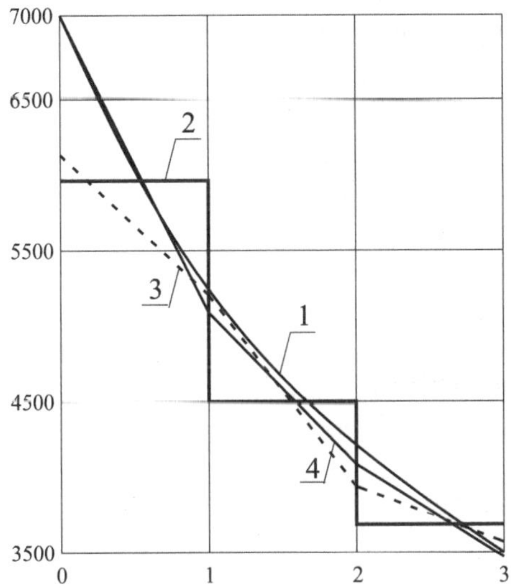

Fig. 9.11. Stresses found for the Oden problem

Example 3

Let us consider another interesting example. The interest follows from the fact that the solution for it obtained by the method of two functionals is a closed analytical expression, so the error estimation can be done explicitly.

P. Tong [21] proved the following simple but useful theorem for the one-dimensional analysis. Let

- the base functions for the FEM-minimization of the Lagrangian functional be functions that satisfy the respective homogeneous differential equation on each of the finite elements;
- each of the finite elements have two nodes and let every node have k degrees of freedom, so that $2k$ is the order of the problem's differential equation (the Euler equations for the Lagrange functional).

Then, independently of the mesh (the number and the relative sizes of its finite elements) and of the form of the partial solution (that is, of the load), the finite element solution u_h that describes the displacements in the nodes of the mesh will coincide with the exact solution in those points. In other words, u_h interpolates u_*:

$$\frac{d^j u_h}{dx^j}(x_i) = \frac{d^j u_*}{dx^j}(x_i), \qquad (j = 0, 1, \ldots, k-1), \tag{3.26}$$

where x_i is the coordinate of any nodal point of the mesh.

Now we consider a boundary-value problem,

$$EAu'' + q_0 \sin\frac{\pi x}{2l} = 0, \quad u(0) = u'(l) = 0, \quad x \in [0,l], \tag{3.27}$$

that modeles the tension of a cantilever bar of a constant cross-section under a longitudinal load

$$q(x) = q_0 \sin\frac{\pi x}{2l}.$$

The exact solution of the problem is

$$u_* = \frac{q_0}{EA}\left(\frac{2l}{\pi}\right)^2 \sin\frac{\pi x}{2l},$$

and the longitudinal force, N_*, is

$$N_* = EA\frac{du_*}{dx} = q_0 \frac{2l}{\pi}\cos\frac{\pi x}{2l}.$$

We divide the integration interval $[0,l]$ into n equal parts so that

$$h = l/n$$

and number the nodes from 0 to n by associating i-th node with coordinate x_i,

$$x_i = ih \quad (i = 0, \ldots, n).$$

Next we use a piecewise linear approximation of the displacements. The Tong theorem says that on i-th finite element, i.e. at $x \in [x_{i-1}, x_i]$,

$$u_h = \frac{q_0}{EA}\left(\frac{2l}{\pi}\right)^2 \left[\left(-\frac{x}{h} + i\right)\sin(i-1)\alpha + \left(\frac{x}{h} - i + 1\right)\sin i\alpha\right], \tag{3.28}$$

where we denote

$$\alpha = \frac{\pi}{2n} = \frac{\pi h}{2l}.$$

Using (3.28) to calculate the constant value of the longitudinal force on i-th finite element, we have

$$N_h = EA\frac{du_h}{dx} = q_0 \frac{4l}{\pi\alpha}\sin\frac{\alpha}{2}\cos\frac{(2i-1)\alpha}{2}. \tag{3.29}$$

Now we use the piecewise linear approximation of the longitudinal stress in the bar by assuming

$$\tilde{N}_h = X_0\Psi_0 + X_1\Psi_1 + \ldots + X_{n-1}\Psi_{n-1}.$$

The X_n coefficient at Ψ_n is assumed to be zero because of the boundary condition $u'(l) = 0$ on the right end of the bar, which is equivalent to $N(l) = 0$. The rest of the coefficients at the piecewise linear approximants Ψ_i $(i = 0, 1,\ldots, n)$ will be calculated by minimizing the functional $\tilde{\mathsf{D}}$,

$$\tilde{\mathsf{D}} = \int_0^l \left(\frac{N^2}{2EA} + N'u_h \right) dx. \tag{3.30}$$

The result is the following system of simultaneous linear algebraic equations with a three-diagonal matrix, which determines the X_i $(i = 0, 1,\ldots, n-1)$ coefficients — actual values of the desirable function $\tilde{N}_h$ in the nodes of the finite element mesh:

$$\frac{h}{6EA}\left\|\begin{bmatrix} 2 & 1 & & & & & \\ 1 & 4 & 1 & & & & \\ & \ddots & \ddots & \ddots & & & \\ & & 1 & 4 & 1 & & \\ & & & \ddots & \ddots & \ddots & \\ & & & & 1 & 4 & 1 \\ & & & & & 1 & 4 \end{bmatrix}\right\| \cdot \left\|\begin{bmatrix} X_0 \\ X_1 \\ \vdots \\ X_i \\ \vdots \\ X_{n-2} \\ X_{n-1} \end{bmatrix}\right\| = \frac{q_0}{EA}\left(\frac{2l}{\pi}\right)^2 \sin\alpha \left\|\begin{bmatrix} 0.5 \\ \cos\alpha \\ \vdots \\ \cos i\alpha \\ \vdots \\ \cos(n-2)\alpha \\ \cos(n-1)\alpha \end{bmatrix}\right\|.$$

The functions Ψ_i $(i = 0, 1,\ldots, n-1)$ themselves are shown in Fig. 9.12 for the convenience of visual perception.

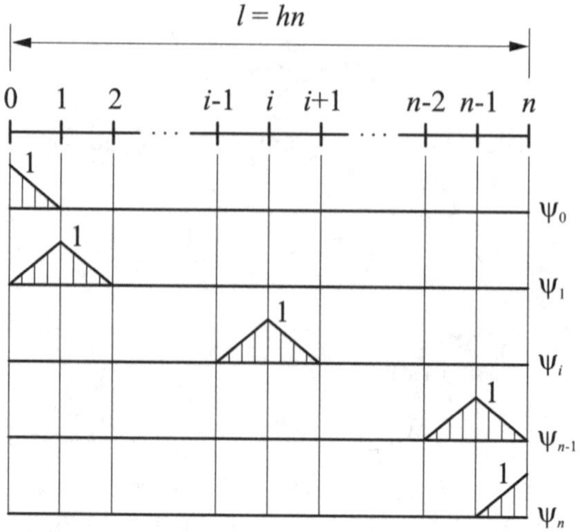

Fig. 9.12. Piecewise linear base functions Ψ_i

A direct substitution shows that the solution of the above system of equations is

$$X_i = q_0 \frac{2l}{\pi} \frac{3\sin\alpha}{\alpha(2+\cos\alpha)} \cos i\alpha\,, \qquad (i = 0, 1, \ldots, n-1)\,.$$

Thus, we have the following on i-th finite element:

$$\tilde{N}_h = q_0 \frac{2l}{\pi} \frac{3\sin\alpha}{\alpha(2+\cos\alpha)} \left[\left(-\frac{x}{h} + i \right) \cos(i-1)\alpha + \left(\frac{x}{h} - i + 1 \right) \cos i\alpha \right]. \quad (3.31)$$

Note that $\tilde{N}_h$ coincides with the linear interpolant of the exact solution, N_*, multiplied by coefficient $k(\alpha)$,

$$k(\alpha) = \frac{3\sin\alpha}{\alpha(2+\cos\alpha)}\,.$$

If we expand this coefficient as a function of α into a Taylor series, we will have a pointwise (not even an average) error of the stresses in the nodes of the finite element mesh. The result is

$$\max_{(i=0,1,\ldots,n)} \frac{|N_* - \tilde{N}_h|}{|N_*|} = O(\alpha^2)\,.$$

As we can see, in this problem the pointwise error of the stresses $\tilde{N}_h$ calculated by the method of two functionals in the nodes of the finite element mesh decreases at the rate of the order of h^2. At the same time, the root-mean-square error by stresses N_h is estimated as (intermediate calculations are omitted)

$$\| N_* \|_{L_2}^2 = \frac{2q_0^2 l^3}{\pi^2}, \qquad \| N_* - N_h \|_{L_2}^2 = \frac{2q_0^2 l^3}{\pi^2}\left[1 - \frac{2(1-\cos\alpha)}{\alpha^2}\right].$$

We can derive a simple asymptotic estimate of the error:

$$\frac{\| N_* - N_h \|_{L_2}}{\| N_* \|_{L_2}} = O(\alpha),$$

which shows the convergence of N_h to the exact solution N_* by metric L_2 at the rate of the order of h.

References

1. Bernstein SA (1941) Principles of structural dynamics (in Russian). Gosstroyizdat, Moscow
2. Blekhman II, Myshkis AD, Panovko YG (1976) Applied mathematics: subject, logic, peculiar approaches (in Russian). Naukova dumka, Kiev
3. Courant R, Robbins H (1967) What is mathematics? (in Russian). Prosvescheniye, Moscow
4. Evzerov ID (1981) The convergence of FEM for coordinate functions not belonging to the energy space. In: Computations with sparse matrices. Computing Center of Siberian Division of USSR Acad. Sci., pp 54–61
5. Karpilovsky VS (1982) Using and constructing some types of finite elements for problems in structural mechanics (in Russian). Synopsis of thesis for the Cand. Sci. (Engng) degree. Kiev
6. Liusternik LA, Sobolev VI (1965) Elements of functional analysis (in Russian). Nauka, Moscow
7. Mikhlin SG (1964) Variational Methods in Mathematical Physics. Macmillan, New York
8. Mikhlin SG (1966) The Problem of the Minimum of a Quadratic Functional. Holden-Day
9. Novozhilov VV (1958) Theory of elasticity (in Russian). Sudpromgiz, Leningrad
10. Oden JT (1972) Finite Elements of Nonlinear Continua. McGrow-Hill Publ.
11. Oden JT, Reddy JN (1973) Note on an approximate method for computing consistent conjugate stresses in elastic finite elements. Int. J. Numer. Meth. Eng., V.6, N1: 55–61.
12. Panovko YG, Gubanova II (1987) Stability and oscillations of elastic systems (in Russian). Nauka, Moscow

13. Pobedrya BE (1979) Some general theorems of mechanics of deformable solids (in Russian). Appl. Math and Mech., Vol. 43; 3: 531–541
14. Rektorys K (1985) Variational methods in mathematical physics and engineering (in Russian). Mir, Moscow
15. Rozin LA (1982) Development of the Bubnov–Galyorkin method in problems of structural mechanics (in Russian). Structural mechanics and analysis, 6: 10–15
16. Segerlind LJ (1976) Applied Finite Element Analysis. J.Wiley & Sons, New York
17. Slivker VI (1982) The Ritz method in problems of elasticity, based on sequentially minimizing two functionals (in Russian). Trans. Acad. Sci. USSR, Mech. of Solids, 2: 57–65
18. Strang G (1976) Linear algebra and its applications. Academic press, New York San Francisko London
19. Strang G, Fix GJ (1973) An analysis of the finite element method. Prentice-Hall Inc., Englewood Cliffs, N.Y.
20. Timoshenko SP (1932) A course of strength of materials (in Russian). Gostechizdat, Moscow, Leningrad
21. Tong P (1969) Exact solution of certain problems by finite element method. AIAA Journal, 1: 178–180

10 VARIATIONAL PRINCIPLES IN SPECTRAL PROBLEMS

It is the hardest thing for a physicist, to have a correct proportion of mathematical strictness. Or, it would be more correct to say this: he needs to know how to determine that proportion

Mandelshtam LI (1972) Lectures on the oscillation theory (in Russian). Nauka, Moscow

10.1 Basic concepts. Terminology

Up to this point, we have been dealing with static problems solely. New circumstances in the formulations of dynamical problems include the following.

First, given external force actions $\bar{\boldsymbol{X}}, \bar{\boldsymbol{p}}$ and external kinematical actions $\bar{\boldsymbol{u}}$ are assumed to vary with time according to a known law, that is,

$$\bar{\boldsymbol{X}} = \bar{\boldsymbol{X}}(t), \qquad \bar{\boldsymbol{p}} = \bar{\boldsymbol{p}}(t), \qquad \boldsymbol{u} = \boldsymbol{u}(t).$$

Second, the dynamical problems have to deal with inertia forces which can be represented as follows according to the second Newton's law:

$$-\boldsymbol{\rho}\ddot{\mathbf{u}}(t),$$

where the usual notation of two dots above the letter means the differentiation with respect to time *t*, and $\boldsymbol{\rho}$ denotes an algebraic operator that describes inertial properties of the mechanical system (a distribution of the system's inertial characteristics over the Ω area).

In the case when the mechanical system in question does not contain any additional weights except for the construction's weight itself, the $\boldsymbol{\rho}$ operator is an algebraic operator proportional to the scalar function, ρ, where ρ is the density of the material of the system as a function of the points of area Ω. The $\boldsymbol{\rho}$ operator is assumed to be a symmetric positive definite operator.

As a result, the problem of equilibrium turns into a problem concerning the motion of the mechanical system because all stress and strain fields we want to know will, generally speaking, vary with time.

Of all the governing set of equations of problem (**1**.2.2) and (**1**.2.4), only the equations of equilibrium (**1**.2.2-*a*) will be modified because the dynamics implies the presence of inertia forces:

$$\boldsymbol{A}^{\mathsf{T}}\boldsymbol{\sigma} + \boldsymbol{K}\mathbf{u} + \rho\ddot{\mathbf{u}} = \bar{\boldsymbol{X}} \,. \tag{1.1}$$

Sometimes the equations of motion (1.1) are called *equations of dynamic equilibrium*.

We will not deal with integration of the equations of motion in this section; instead, we will consider a particular but practically and theoretically important issue. With that in mind, we will assume that

- there is no external load $\bar{X}$ distributed over the area of integration;
- both static and kinematic boundary conditions are homogeneous.

Having adopted these limitations, we arrive at a problem of *free* (*natural*) *oscillations* of the mechanical system. The free oscillations imply that the time history of the motion is harmonic and all stress and strain components to be found can be represented as

$$\boldsymbol{\sigma} = \boldsymbol{\sigma}_0 \sin(\omega t + \mu), \quad \boldsymbol{\varepsilon} = \boldsymbol{\varepsilon}_0 \sin(\omega t + \mu), \quad \mathbf{u} = \mathbf{u}_0 \sin(\omega t + \mu), \tag{1.2}$$

where the zero subscript denotes amplitude values of the respective physical parameters.

As a result, the problem of free oscillations of a mechanical system is defined by the following set of equations:

$$\boldsymbol{A}^{\mathsf{T}}\boldsymbol{\sigma}_0 + \boldsymbol{K}\mathbf{u}_0 = \omega^2\boldsymbol{\rho}\mathbf{u}_0 \quad \text{– equations of dynamic equilibrium,} \tag{1.3-a}$$

$$\boldsymbol{A}\mathbf{u}_0 = \boldsymbol{\varepsilon}_0 \quad \text{– geometric equations,} \tag{1.3-b}$$

$$\boldsymbol{\sigma}_0 = \boldsymbol{C}\boldsymbol{\varepsilon}_0 \text{ или } \boldsymbol{\varepsilon}_0 = \boldsymbol{C}^{-1}\boldsymbol{\sigma}_0 \quad \text{– physical equations.} \tag{1.3-c}$$

The set of governing equations for problem (1.3) must be supplemented with boundary conditions that look like the following in the operator form:

$$\boldsymbol{E}_p\boldsymbol{H}_\sigma\boldsymbol{\sigma}_0 = \mathbf{0} \quad \text{– static boundary conditions,} \tag{1.4-a}$$

$$\boldsymbol{E}_u\boldsymbol{H}_u\mathbf{u}_0 = \mathbf{0} \quad \text{– kinematic boundary conditions,} \tag{1.4-b}$$

and are specified on boundary Γ of area Ω.

The scalar parameter, ω, for which a nonzero solution of problem (1.3) – (1.4) exists, is called a *natural oscillation frequency* of the mechanical system. It is also referred to as an *eigenfrequency*; both terms are equivalent and refer to the free oscillations at a frequency immanent in the system. All natural frequencies together make up a *frequency spectrum* of the mechanical system.

Parameter μ which participates as an additional term in the argument of sine in (1.2) is called a *phase shift* or an *initial phase* of the oscillatory motion. Note that the amplitude values of the stress and strain components in the system that performs free oscillations can be determined only up to a scalar multiplier.

A nonzero displacement, $\mathbf{u}_0$, which conforms to the frequency of natural oscillations, ω, is a vector function of spatial coordinates of the system and determines a *natural oscillation mode* (an *eigenvector*) of the system. Different natural frequencies conform to different natural modes of oscillation.

Although the tradition understands the oscillation mode as a displacement function, $\mathbf{u}_0$, of the mechanical system, we can also use this term in reference to the respective stresses, $\boldsymbol{\sigma}_0 = \boldsymbol{CA}\mathbf{u}_0$, or the respective strains, $\boldsymbol{\varepsilon}_0 = \boldsymbol{A}\mathbf{u}_0$.

If we switch from the general formulation (1.3) – (1.4) to the statement of the problem in terms of displacements by using the Lame operator, we will have this:

$$\boldsymbol{L}\mathbf{u}_0 = \omega^2 \boldsymbol{\rho}\mathbf{u}_0 \quad \in \Omega\,, \tag{1.5}$$

$$\boldsymbol{E}_p\boldsymbol{H}_\sigma\boldsymbol{CA}\mathbf{u}_0 = \mathbf{0}\,, \qquad \boldsymbol{E}_u\boldsymbol{H}_u\mathbf{u}_0 = \mathbf{0} \quad \in \Gamma. \tag{1.6}$$

where the Lame operator is defined as

$$\boldsymbol{L} = \boldsymbol{A}^{\mathsf{T}}\boldsymbol{CA} + \boldsymbol{K}\,. \tag{1.7}$$

Another term will be useful for the presentment: an *eigenload*, $\boldsymbol{X}_0$. A load $\boldsymbol{X}_0$ will be called an eigenload that corresponds to eigenfrequency ω if the static problem,

$$\boldsymbol{L}\mathbf{u}_0 = \boldsymbol{X}_0 \quad \in \Omega\,,$$

$$\boldsymbol{E}_p\boldsymbol{H}_\sigma\boldsymbol{CA}\mathbf{u}_0 = \mathbf{0}\,, \qquad \boldsymbol{E}_u\boldsymbol{H}_u\mathbf{u}_0 = \mathbf{0} \quad \in \Gamma,$$

produces a solution in terms of displacements, $\mathbf{u}_0$, which coincides (up to a scalar multiplier) with the respective natural frequency mode of the system.

The square of the system's natural frequency, ω^2, is also called an *eigenvalue* of operator $\boldsymbol{L}$ which corresponds to the inertia operator $\boldsymbol{\rho}$; it is denoted usually by $\lambda = \omega^2$.

The problem that requires us to calculate the eigenfrequencies and natural modes of oscillation of a mechanical system is called a *spectral problem*[1].

10.2 The spectral problem as a variational problem

10.2.1 The spectrum of a mechanical system with a finite number of degrees of freedom

Let us start our consideration from a simplest problem formulation where a mechanical system of interest will possess a finite number of degrees of freedom (DOFs). In this case we will deal only with algebraic operators described by certain matrices rather than differential operators and boundary conditions.

Let the position of a mechanical system be fully defined by n generalized coordinates $q_1, q_2,\ldots, q_n$ which make up a vector of generalized coordinates, $\boldsymbol{q} = \|[q_1,q_2,\ldots, q_n]\|^{\mathsf{T}}$. The state of the system in the course of its motion is also defined by a vector of generalized velocities,

$$\dot{\boldsymbol{q}} = \|[\dot{q}_1,\ldots,\dot{q}_n]\|^{\mathsf{T}}. \tag{2.1}$$

Suppose the system is in equilibrium in a fixed state of interest; it means the velocities of all its material particles are zeros, $\dot{\boldsymbol{q}} = \mathbf{0}$. Now we force the system somehow to leave its state of equilibrium and suppose that this forced deviation is limited to a small vicinity of the system's equilibrium state. By releasing the deviated system, we let it return to the initial equilibrium by itself. As a result, the mechanical system starts moving, and its motion will be a linear combination of n harmonic oscillations.

The physical characterisitcs of the system are fully defined by one function E which is a potential energy of the system. To simplify the reasoning without limiting its generality, we can assume that all generalized coordinates are equal to zero in the equilibrium state of the system around which the free oscillatory motion occurs; that is, $\boldsymbol{q} = \mathbf{0}$. If it is not so, then we can always switch to a new set of generalized coordinates by subtracting the values of the old generalized coordinates in the equilibrium state of the system from those coordinates, which validates our assumption. Note that this shift of the system's generalized coordinates

---

[1] As noted by R. Bellman [1], the words "spectrum" and "spectral" were introduced in mathematical studies by D. Hilbert.

does not affect the generalized velocities because time derivatives of the constants are equal to zero.

With the above-said small deviations from the state of equilibrium, the principal term of the expansion of potential energy $\mathsf{E} = \mathsf{E}(\boldsymbol{q})$ of the deviated state into a power series will be a homogeneous quadratic form of the generalized coordinates. To put it another way,

$$\mathsf{E} = \mathsf{E}(\boldsymbol{q}) = \frac{1}{2}\sum_{i=1}^{n}\sum_{j=1}^{n} r^{ij} q_i q_j + \ldots, \tag{2.2}$$

where coefficients r^{ij} are defined as

$$r^{ij} = \frac{\partial^2 \mathsf{E}}{\partial q_i \partial q_j}(0,\ldots,0). \tag{2.3}$$

The ellipsis in (2.2) denotes terms containing the generalized coordinates raised to powers higher than second, so our assumption of smallness of the initial deviation makes those values effective zeros.

The representation of (2.2) can be easily validated using formal rules for expanding a function of n coordinates into a Taylor series. To see this,

$$\mathsf{E}(\boldsymbol{q}) = \mathsf{E}(0,\ldots,0) + \sum_{i=1}^{n} \frac{\partial \mathsf{E}}{\partial q_i}(0,\ldots,0) q_i + \frac{1}{2}\sum_{i=1}^{n}\sum_{j=1}^{n} \frac{\partial^2 \mathsf{E}}{\partial q_i \partial q_j}(0,\ldots,0) q_i q_j + \ldots. \tag{2.4}$$

But, as we have agreed, the zero position of the system conforms to its state of equilibrium, and the potential energy has its stationary value in this state. It means

$$\frac{\partial \mathsf{E}}{\partial q_i}(0,\ldots,0) = 0 \quad (i = 1,\ldots, n).$$

Assuming $\mathsf{E}(0,\ldots,0) = 0$, which is nothing more than a relative reference level for the potential energy, we obtain (2.2), (2.3).

Further we will assume that the state of equilibrium of the system, $\boldsymbol{q} = \boldsymbol{0}$, is a stable equilibrium. According to the known Lagrange–Dirichlet theorem [15], this means the quadratic form (2.2), along with the matrix $\boldsymbol{r} = \|[r^{ij}]\|$ composed of the coefficients of the same quadratic form, is positive definite[2]. It means

---

[2] Actually, the Lagrange–Dirichlet theorem says just that the conditions of minimum of the potential energy are sufficient conditions for the equilibrium to be stable. The theorem says nothing about the necessary conditions of stability.

$$\mathsf{E}(\boldsymbol{q}) > 0 \text{ for } \boldsymbol{q} \neq \boldsymbol{0}\,. \tag{2.5}$$

In addition to potential energy E, the motion of the system is described also by its kinetic energy, T, which is, as known from theoretical mechanics [14], a homogeneous positive definite quadratic form of the generalized velocities. To put it another way,

$$\mathsf{T} = \mathsf{T}(\dot{\boldsymbol{q}}) = \frac{1}{2}\sum_{i=1}^{n}\sum_{j=1}^{n} m^{ij}\dot{q}_i\dot{q}_j\,, \tag{2.6}$$

where matrix $\boldsymbol{m} = \|[m^{ij}]\|$ is called a *matrix of inertial characteristics of the system*, or, shorter, a *mass matrix*. The mass matrix $\boldsymbol{m}$ is also supposed to be symmetric and positive definite which ensures the positiveness of the system's kinetic energy at any nonzero velocities of its material particles.

The matrix representations of the potential energy, E, and of the kinetic energy, T, look like

$$\mathsf{E} = \frac{1}{2}\boldsymbol{q}^{\mathsf{T}}\boldsymbol{r}\boldsymbol{q}\,, \quad \mathsf{T} = \frac{1}{2}\dot{\boldsymbol{q}}^{\mathsf{T}}\boldsymbol{m}\dot{\boldsymbol{q}}\,. \tag{2.7}$$

It is quite clear that the equations of motion of the above system in the absence of any exterior perturbations look like

$$\boldsymbol{m}\ddot{\boldsymbol{q}} + \boldsymbol{r}\boldsymbol{q} = \boldsymbol{0}\,. \tag{2.8}$$

The d'Alembert principle lets us treat this equation, exactly as the earlier equation (1.1), as an equation of mechanical equilibrium of a mechanical system that has a finite number of DOFs and where only inertia forces $-\boldsymbol{m}\ddot{\boldsymbol{q}}$ are taken for an external force action.

Note that the matrix differential equation of motion (2.8) can be easily derived in a formal way from (2.7) by using the known second-kind Lagrange equations which say[3]

Therefore the assumption of the quadratic form (1.9) being positive definite is a bit stronger condition. See [15] for more details on this.

[3] We do not have to use the second-kind Lagrange equations to derive the matrix equation (2.8); instead, we could use the energy conservation law which says $d(\mathsf{E}+\mathsf{T})/dt = 0$. Substituting in the expressions of the potential and kinetic energies from (2.7) and seeing that matrices $\boldsymbol{r}$ and $\boldsymbol{m}$ are symmetric, we transform this condition into $\dot{\boldsymbol{q}}^{\mathsf{T}}(\boldsymbol{m}\ddot{\boldsymbol{q}} + \boldsymbol{r}\boldsymbol{q}) = 0$. The velocity vector, $\dot{\boldsymbol{q}}$, must not be identical to zero, hence equation (2.8).

$$\frac{d}{dt}\frac{d\mathsf{T}}{d\dot{\boldsymbol{q}}}+\frac{d\mathsf{E}}{d\boldsymbol{q}}=\mathbf{0}\,.$$

The solution of the set of differential equations (2.8) is sought for in a form similar to (1.2):

$$\boldsymbol{q}=\boldsymbol{z}\sin(\omega t+\mu) \tag{2.9}$$

where $\boldsymbol{z}=\|[z_1,\ldots,z_n]\|^{\mathsf{T}}$ is a vector of displacement amplitudes which corresponds to the generalized coordinates q_i $(i = 1,\ldots, n)$.

Putting (2.9) into (2.8) and canceling out the sinusoid multiplier gives a homogeneous linear algebraic set of equations,

$$(\boldsymbol{r}-\lambda\boldsymbol{m})\boldsymbol{z}=\mathbf{0}\,, \tag{2.10}$$

where we want to be consistent with the traditional notation of mathematics and denote by λ the square of the oscillation frequency,

$$\lambda=\omega^2\,. \tag{2.11}$$

Now we will formulate briefly some definitions and propositions of the theory of symmetric matrices which we are going to use and which we deem known to the reader.

A generalized eigenvalue/eigenvector problem for a couple of matrices $\boldsymbol{r}$ and $\boldsymbol{m}$ is a problem that requires to find nonzero vectors $\boldsymbol{z}$ and their respective numbers λ which satisfy the homogeneous equation (2.10). If a pair consisting of a scalar λ and a vector $\boldsymbol{z}$ satisfies that equation then λ is called an *eigenvalue* (or a *characteristic value*) *of the **r** matrix with respect to **m** matrix*, and vector $\boldsymbol{z}$ which corresponds to the eigevnalue is called an *eigenvector*[4].

It is known that:

a) There are at most n eigenvalues for any symmetric matrices $\boldsymbol{r}$ and $\boldsymbol{m}$;

b) If matrix $\boldsymbol{m}$ is not degenerate, then there exist precisely n eigenvalues (taking into account their multiplicity) and precisely n linearly independent eigenvectors;

c) If at least one of the matrices $\boldsymbol{r}$ or $\boldsymbol{m}$ is positive definite, then all eigenvalues λ take real values;

d) If both $\boldsymbol{r}$ and $\boldsymbol{m}$ are positive definite, the eigenvalues are strictly positive;

e) An eigenvalue of multiplicity k corresponds to exactly k linearly independent vectors;

[4] We would like to emphasize that the definition above treats the *eigenvalue* and the *characteristic value* terms as mutually synonymous.

f) A positive definite matrix can be decomposed into a product of two mutually transposed square matrices of the same order. Note that such a decomposition is not unique, which is demonstrated by the following simple example:

$$\left\|\begin{bmatrix} 2 & 1 \\ 1 & 2 \end{bmatrix}\right\| = \left\|\begin{bmatrix} \sqrt{2} & 0 \\ \frac{1}{\sqrt{2}} & \sqrt{\frac{3}{2}} \end{bmatrix}\right\| \cdot \left\|\begin{bmatrix} \sqrt{2} & \frac{1}{\sqrt{2}} \\ 0 & \sqrt{\frac{3}{2}} \end{bmatrix}\right\| = \left\|\begin{bmatrix} \frac{1}{\sqrt{2}} & \sqrt{\frac{3}{2}} \\ -\frac{1}{\sqrt{2}} & \sqrt{\frac{3}{2}} \end{bmatrix}\right\| \cdot \left\|\begin{bmatrix} \frac{1}{\sqrt{2}} & -\frac{1}{\sqrt{2}} \\ \sqrt{\frac{3}{2}} & \sqrt{\frac{3}{2}} \end{bmatrix}\right\| .$$

Further we will confine ourselves to free oscillations only of mechanical systems which have positive definite matrices $\boldsymbol{r}$ and $\boldsymbol{m}$. As we mentioned earlier, the positive definiteness of those matrices takes place in our problems because the kinetic energy of the system is positive by definition and the potential energy is positive, too, because we consider oscillations of the system in the vicinity of its state of equilibrium which is assumed to be stable[5]. It is useful to index all eigenvalues by putting them in the ascending order:

$$0 < \lambda_1 \le \lambda_2 \le \ldots \le \lambda_n . \tag{2.12}$$

The same indexing can be used with their respective eigenvectors, so i-th eigenvalue λ_i and i-th eigenvector z_i for (2.10) will relate as follows:

$$(\boldsymbol{r} - \lambda_i \boldsymbol{m}) z_i = \boldsymbol{0} \quad (i = 1, \ldots, n). \tag{2.13}$$

Note that the inequalities in the sequence of (2.12) are not necessarily strict (except for $0 < \lambda_1$ which *is* strict). But then admitting the presence of equalities in (2.12) means that the full set of n eigenvalues can include equal eigenvalues under different indices. If a value λ is repeated k times in the set (2.12), then we are dealing with a *k-fold eigenvalue* (or an *eigenvalue of multiplicity k).*

One of the most important theorems in the theory of real-value symmetric matrices, which we will call a theorem of multiplicity, was formulated above as Proposition (*e*).

[5] A lot of books can be used to refresh the knowledge of the above propositions and statements from the theory of symmetric matrices (or, at worst, to get a primary acquaintance with them). We recommend any of the books [1], [25], [8].

Metric ideas in the spectral eigenvalue probem

Obviously, any eigenvector can be multiplied by any number without violating the inequality (2.13). We use this freedom to make the length of each eigenvector equal to 1 (the length is defined in the Euclidean metric)

$$\|z_i\|^2 = (z_i, z_i) = z_i^{\mathsf{T}} z_i = 1 \quad (i = 1, \ldots, n). \tag{2.14}$$

The requirement of (2.14) is essentially a method to choose a normalization of the eigenvectors. Any other normalization could be used, too, as long as there is a meaningful concept of the vector's length.

Let λ_i and λ_j be different eigenvalues, i.e $\lambda_i \neq \lambda_j$. Then it is easy to prove that the following rclationships of generalized orthogonality between the eigenvectors hold true:

$$z_i^{\mathsf{T}} \boldsymbol{m} z_j = 0, \qquad z_i^{\mathsf{T}} \boldsymbol{r} z_j = 0 \quad (i \neq j). \tag{2.15}$$

The first of the relationships in (2.15) is also referred to as an *orthogonality of the vectors with the mass matrix* ***m***. The second relationship in (2.15) is pronounced as an *orthogonality with the stiffness matrix* ***r***.

And indeed, the symmetry of matrix ***r*** gives

$$z_i^{\mathsf{T}} \boldsymbol{r} z_j = z_j^{\mathsf{T}} \boldsymbol{r} z_i \,.$$

By replacing $\boldsymbol{r} z_j$ with $\lambda_j \boldsymbol{m} z_j$ and $\boldsymbol{r} z_i$ with $\lambda_i \boldsymbol{m} z_i$ we turn this equality into

$$\lambda_j z_i^{\mathsf{T}} \boldsymbol{m} z_j = \lambda_i z_j^{\mathsf{T}} \boldsymbol{m} z_i , \tag{2.16}$$

which gives the following seeing that ***m*** is symmetric:

$$(\lambda_j - \lambda_i) z_i^{\mathsf{T}} \boldsymbol{m} z_j = 0 ,$$

hence the first of the generalized orthogonality relationships in (2.15) at $\lambda_i \neq \lambda_j$. The second relationship follows from the first one if we replace $\boldsymbol{m} z_j$ with $\boldsymbol{r} z_j / \lambda_j$ and take into account that $\lambda_j \neq 0$.

Note that the generalized orthogonality relationships (2.15) can be interpreted mechanically as a manifestation of the work reciprocity theorem by Betty. To see this, think of the displacement vector z_i as a solution of a quasi-static problem in which the mechanical system is subjected to statically applied inertia forces $\omega_i^2 \boldsymbol{m} z_i$ because $\boldsymbol{r} z_i = \omega_i^2 \boldsymbol{m} z_i$. But then two different modes of oscillation, z_i and z_j, each being caused by its particular set of forces, can be treated as two states of the mechanical system: state i state j, respectively. The Betty theorem gives

$$z_i^{\mathrm{T}}(\omega_j^2 \boldsymbol{m} z_j) = z_j^{\mathrm{T}}(\omega_i^2 \boldsymbol{m} z_i),$$

which is just another form of the same relation (2.16), hence (2.15).

The generalized orthogonality relationships have been derived with the limitation $\lambda_i \neq \lambda_j$. But if an eigenvalue, say λ_i, is k-fold then the above-stated theorem of multiplicity says this eigenvalue has exactly k linearly independent eigenvectors. Let

$$\lambda_i = \lambda_{i+1} = \ldots = \lambda_{i+k-1},$$

and let all eigenvectors $z_i, z_{i+1}, \ldots, z_{i+k-1}$, which we consider to be normalized to the length one as in (2.14), be linearly independent. A straightforward check shows that any linear combination of the vectors which is also a unit-length vector can be treated as a normalized eigenvector that conforms to eigenvalue λ_i. Of all possible linear combinations, we choose only vectors orthogonal with the mass matrix as candidates for our further consideration[6]. This makes it possible for us to extend the first of the generalized orthogonality relationships in (2.15) onto all eigenvectors independently of the presence or absence of multiple eigenvalues in the spectrum. The second of the orthogonality relationships in (2.15) will follow naturally from here.

To summarize, we would like to note that the orthogonality of the eigenvectors, both with the mass matrix and with the stiffness matrix, follows automatically for the different eigenvalues while for the multiple eigenvalues it can be proved by choosing a special system of linearly independent eigenvectors.

Having chosen such a system of linearly independent eigenvectors, we can write the following for <u>any</u> two unit-length eigenvectors instead of (2.15):

$$z_i^{\mathrm{T}} \boldsymbol{m} z_j = M_i \delta_{ij}, \qquad z_i^{\mathrm{T}} \boldsymbol{r} z_j = R_i \delta_{ij}, \tag{2.17}$$

where δ_{ij} is Kronecker's delta, $M_i = z_i^{\mathrm{T}} \boldsymbol{m} z_i$, $R_i = z_i^{\mathrm{T}} \boldsymbol{r} z_i$. Values M_i and R_i can be treated as the squares of the norms of eigenvector z_i in the metrics generated by the respective matrices $\boldsymbol{m}$ and $\boldsymbol{r}$.

---

[6] To build an orthogonal system of vectors out of any given linearly independent set of vectors, one can use a well-known orthogonalization process by Gram–Schmidt [25].

The original system of ordinary differential equations (2.8) contains all the unknown functions q_i $(i = 1,\ldots, n)$ in each of its equations in the general case, therefore it is not very convenient to analyze. There is a simple method of transforming this system in such way that each of the equations will contain only one unknown function.

We introduce a matrix **V** composed of the eigenvectors in such way that first column of the matrix contains first eigenvector, second column contains second eigenvector etc. To put it another way,

$$\mathbf{V} = \|[z_1,\ldots, z_n]\| = \left\|\begin{bmatrix} z_{11} & z_{12} & \cdots & z_{1n} \\ z_{21} & z_{22} & \cdots & z_{2n} \\ \vdots & \vdots & \vdots & \vdots \\ z_{n1} & z_{n2} & \cdots & z_{nn} \end{bmatrix}\right\| \tag{2.18}$$

where z_{ij} is i-th component of j-th eigenvector. Using this matrix as a matrix of linear transformation of the generalized coordinates helps us to switch to a new set of variables, $\boldsymbol{y} = \|[y_1,\ldots, y_n]\|^{\mathrm{T}}$, by defining

$$\boldsymbol{q} = \mathbf{V}\boldsymbol{y}\,. \tag{2.19}$$

Putting (2.19) in the main differential equation (2.8) and then multiplying the result by matrix $\mathbf{V}^{\mathrm{T}}$ on the left makes the following out of that equation:

$$\mathbf{M}\,\ddot{\boldsymbol{y}} + \mathbf{R}\,\boldsymbol{y} = \mathbf{0} \tag{2.20}$$

where eigenvectors z_i follow the conditions of generalized orthogonality as in (2.17) and thus matrices **M** and **R** are diagonal matrices[7]

$$\mathbf{V}^{\mathrm{T}}\boldsymbol{r}\mathbf{V} = \mathbf{R} = \mathrm{diag}\|\lfloor R_1,\ldots,R_n\rfloor\|\,, \qquad \mathbf{V}^{\mathrm{T}}\boldsymbol{m}\mathbf{V} = \mathbf{M} = \mathrm{diag}\|\lfloor M_1,\ldots,M_n\rfloor\|\,, \tag{2.21}$$

and the positive definiteness of matrices $\boldsymbol{r}$ and $\boldsymbol{m}$ make their the diagonal elements strictly positive.

As matrices **R** and **M** are diagonal, the set of equations (2.20) decomposes into n independent equations so that

[7] The nomenclature of the theory of quadratic forms which is closely related to the theory of symmetric matrices includes a beautiful proposition (theorem) about the possibility to reduce two quadratic forms simultaneously (at least one has to be positive definite) to a diagonal form using the same linear transformation of the coordinates [25]. Relationships (2.21) known in the theory as *congruence transformations* are corollaries to that proposition.

$$M_i \frac{d^2 y_i}{dt^2} + R_i y_i = 0 \quad (i = 1, \dots, n) .$$

The general solution of these equations can be written as

$$y_i = A_i \sin(\omega_i t + \mu_i) \tag{2.22}$$

where:

- A_i is an amplitude value of the new generalized displacement, y_i;
- ω_i is a circular frequency of the system's free oscillations in i-th mode;
- μ_i is an initial phase shift.

Obviously, (2.22) is a solution of the ordinary differential equation derived above if

$$\omega_i = \sqrt{R_i / M_i} . \tag{2.23}$$

Values y_i are different from the original generalized coordinates q_i, and they are usually called *principal* or *normal* coordinates. Parameters $R_1, R_2, \dots, R_n$ and $M_1, M_2, \dots, M_n$ are called *effective stiffnesses* and *effective masses*, respectively (or *reduced masses/stiffnesses*, meaning that they are reduced to the principal coordinates). As we have said before, the effective stiffnesses and effective masses have the geometrical meaning of squared lengths of the eigenvectors in the respective metrics generated by matrices $\boldsymbol{r}$ and $\boldsymbol{m}$. It follows from here and from (2.23) that the free-oscillation frequency, ω_i, of i-th mode is the ratio of the norms of eigenvector z_i in those two metrics.

What we have said above means knowing the eigenvectors is a much better knowledge that knowing the eigenvalues. Indeed, the eigenvectors determine the transformation matrix $\mathbf{V}$ unambiguously, and the latter permits to calculate the effective stiffnesses and masses. Formula (2.23) gives us a tool to calculate the eigenvalues.

Returning to the original generalized coordinates q_i, assuming $\mathbf{V} = \|[z_{ij}]\|$ and using (2.19) gives

$$q_i = \sum_{j=1}^{n} z_{ij} A_j \sin(\omega_j t + \mu_j) .$$

This result means the motion of the system with respect to each generalized coordinate is a superposition of n harmonic oscillations.

All these things are well-known from the simplest theory of free linear harmonic oscillations of finite-dimensional systems, and they are not directly related to the variational definition of the system's spectral properties. That's why the preceding text of this section should be understood as a reminder and an introduction to the further presentment which makes the reader familiar with the necessary nomenclature and the notation.

10.2.2 A variational description of eigenvalues

Assuming that we know the natural (free) oscillation frequencies and modes of a mechanical system that has n degrees of freedom, we have a chance to track the changes in basic integral characteristics of the system as it moves. The integral characteristics which we want to know and track are the potential energy and the kinetic energy.

Based on (2.7) and (2.9), we calculate these values at any arbitrary moment of time:

$$\mathsf{E} = \frac{1}{2}\boldsymbol{q}^{\mathsf{T}}\boldsymbol{r}\boldsymbol{q} = \frac{1}{2}\boldsymbol{z}^{\mathsf{T}}\boldsymbol{r}\boldsymbol{z}\sin^2(\omega t + \mu)\ , \qquad \mathsf{T} = \frac{1}{2}\dot{\boldsymbol{q}}^{\mathsf{T}}\boldsymbol{m}\dot{\boldsymbol{q}} = \frac{\omega^2}{2}\boldsymbol{z}^{\mathsf{T}}\boldsymbol{m}\boldsymbol{z}\cos^2(\omega t + \mu)\,.$$

The above expressions and the equality $\boldsymbol{r}\boldsymbol{z} = \omega^2\boldsymbol{m}\boldsymbol{z}$ produce immediately the fact that the total energy of the system at any moment of time is the same and equal to

$$\mathsf{E} + \mathsf{T} - \frac{1}{2}\boldsymbol{z}^{\mathsf{T}}\boldsymbol{r}\boldsymbol{z}\,, \tag{2.24}$$

which is quite expectable from the standpoint of the energy conservation law. However, each separate term in (2.24) varies with time.

From this point on, we will use the designations of E and T to denote *amplitude* values of the respective energies, so we can write

$$\mathsf{E} = \frac{1}{2}\boldsymbol{z}^{\mathsf{T}}\boldsymbol{r}\boldsymbol{z}\,, \qquad \mathsf{T} = \frac{\omega^2}{2}\boldsymbol{z}^{\mathsf{T}}\boldsymbol{m}\boldsymbol{z}\,. \tag{2.25}$$

The above equalities should be understood as follows. If a mechanical system carries out a free oscillatory movement at a frequency ω and amplitude values of the generalized coordinates $\boldsymbol{z}$, then its amplitude values of the potential energy and the kinetic energy in the course of that oscillation can be calculated by (2.25).

Of course, these values must be equal for the true displacement vectors $\boldsymbol{z}$ which satisfy the principal differential equation (2.8). Hence immediately

$$\omega^2 = \frac{\boldsymbol{z}^{\mathsf{T}} \boldsymbol{r z}}{\boldsymbol{z}^{\mathsf{T}} \boldsymbol{m z}}. \tag{2.26}$$

Assuming the z vector in the right-hand part of (2.26) equal to one of the eigenvectors, say z_i, we can clearly see that the ratio above is equal to the respective eigenvalue λ_i, that is,

$$\lambda_i = \frac{\boldsymbol{z}_i^{\mathsf{T}} \boldsymbol{r z}_i}{\boldsymbol{z}_i^{\mathsf{T}} \boldsymbol{m z}_i}. \tag{2.27}$$

However, we will be more interested with the two quadratic forms in the numerator and in the denominator of (2.26) as functionals of the displacement vector z. The ratio of the two quadratic forms can be also treated as a functional of vector z.

This functional is called a *Rayleigh functional*, r, or sometimes a *Rayleigh ratio*[8]. Thus,

$$\mathsf{r} = \frac{\boldsymbol{z}^{\mathsf{T}} \boldsymbol{r z}}{\boldsymbol{z}^{\mathsf{T}} \boldsymbol{m z}}. \tag{2.28}$$

Actually, we are dealing with three methods of introducing a metric in the space of the system's generalized displacements (this is a finite-dimensional space in our case). We will work with three spaces the properties of which will depend on what metric is introduced:

- the basic (Euclidean) space $\mathbb{R}_n$ with the scalar product $(\boldsymbol{x}, \boldsymbol{y}) = \boldsymbol{x}^{\mathsf{T}} \boldsymbol{y}$;
- the E-space with the scalar product $(\boldsymbol{x}, \boldsymbol{y})_{\mathsf{E}} = \boldsymbol{x}^{\mathsf{T}} \boldsymbol{r y}$;
- the T-пространство with the scalar product $(\boldsymbol{x}, \boldsymbol{y})_{\mathsf{T}} = \boldsymbol{x}^{\mathsf{T}} \boldsymbol{m y}$.

We have mentioned before that effective stiffnesses R_i and effective masses M_i have the geometrical meaning of the squares of the respective E-length and T-length of i-th eigenvector z_i which has the Euclidean length one. To put it another way,

$$R_i = \| z_i \|_{\mathsf{E}}^2, \qquad M_i = \| z_i \|_{\mathsf{T}}^2, \qquad \text{at} \qquad \boldsymbol{r} z_i = \lambda_i z_i, \quad \| z_i \| = 1. \tag{2.29}$$

In the same terms, the Rayleigh functional is a ratio of the squared lengths of the same vector in two metrics, namely,

[8] An excuse: we use the lowercase letter r to denote the Rayleigh functional in order to avoid the confusion. The matter is that the capital letter R is already used as a symbol for the Reisner functional.

$$\mathsf{r}(z) = \frac{(z,z)_{\mathsf{E}}}{(z,z)_{\mathsf{T}}} = \frac{\| z \|_{\mathsf{E}}^2}{\| z \|_{\mathsf{T}}^2}. \tag{2.30}$$

Obviously, the Rayleigh ratio does not change when vector z is multiplied by any scalar k, i.e. $\mathsf{r}(kz) = \mathsf{r}(z)$ for any $k \neq 0$. This means it suffices to consider the functional only with vectors z the T-length of which is equal to one, $\|z\|_{\mathsf{T}} = 1$. The locus of all vectors of this kind is an n-dimensional sphere of the unit radius in the T-space with its center at the coordinate origin.

A remark on expansion of an arbitrary vector over a basis of eigenvectors

Let us take an arbitrary vector $\boldsymbol{x}$ of the T-length one in the T-space, which does not have to be any eigenvector of the problem in question. Let $\{\boldsymbol{x}_1, \boldsymbol{x}_2, \ldots, \boldsymbol{x}_n\}$ be a set of all eigenvectors of the T-length one.

We expand this vector over the basis of $\boldsymbol{x}_1, \boldsymbol{x}_2, \ldots, \boldsymbol{x}_n$, that is, we assume

$$\boldsymbol{x} = \alpha_1 \boldsymbol{x}_1 + \alpha_2 \boldsymbol{x}_2 + \ldots + \alpha_n \boldsymbol{x}_n, \tag{2.31}$$

and because $\|\boldsymbol{x}\|_{\mathsf{T}} = 1$ these coefficients satisfy the condition

$$\alpha_1^2 + \alpha_2^2 + \ldots + \alpha_n^2 = 1. \tag{2.32}$$

Coefficients α_i are calculated by a standard technique: making scalar products of the equality (2.31) with all the basis vectors in succession. Taking into account the T-orthogonality of those vectors gives

$$\alpha_i = (\boldsymbol{x}, \boldsymbol{x}_i)_{\mathsf{T}} \quad (i = 1, \ldots, n)$$

Thus, the expansion (2.31) can be written as

$$\boldsymbol{x} = (\boldsymbol{x}, \boldsymbol{x}_1)_{\mathsf{T}} \boldsymbol{x}_1 + (\boldsymbol{x}, \boldsymbol{x}_2)_{\mathsf{T}} \boldsymbol{x}_2 + \ldots + (\boldsymbol{x}, \boldsymbol{x}_n)_{\mathsf{T}} \boldsymbol{x}_n. \tag{2.33}$$

In exactly the same way, we can take any vector $\boldsymbol{y}$ of the unit E-length in the E-space, such that $\|\boldsymbol{y}\|_{\mathsf{E}} = 1$, and expand this vector over the set of eigenvectors $\{\boldsymbol{y}_1, \boldsymbol{y}_2, \ldots, \boldsymbol{y}_n\}$. Each of eigenvectors $\boldsymbol{y}_i$ is considered to be a unit E-length vector, i.e. $\|\boldsymbol{y}_i\|_{\mathsf{E}} = 1$. Then the expansion of vector $\boldsymbol{y}$ over the basis of the eigenvectors will be

$$\boldsymbol{y} = (\boldsymbol{y}, \boldsymbol{y}_1)_{\mathsf{E}} \boldsymbol{y}_1 + (\boldsymbol{y}, \boldsymbol{y}_2)_{\mathsf{E}} \boldsymbol{y}_2 + \ldots + (\boldsymbol{y}, \boldsymbol{y}_n)_{\mathsf{E}} \boldsymbol{y}_n. \tag{2.34}$$

A recursive variational definition of eigenvalues and eigenvectors

The following proposition, usually attributed to Rayleigh only, though we will refer to it as a Rayleigh–Weber principle[9], holds:

> *Functional* $\mathsf{r}(z)$ *gets its minimum from its first eigenvector* z_1, *and this minimum is equal to the least eigenvalue* λ_1 *of matrix* $\boldsymbol{r}$ *with respect to matrix* $\boldsymbol{m}$, *that is,*
>
> $$\min \mathsf{r}(z) = \mathsf{r}(z_1) = \lambda_1 .$$

The Rayleigh–Weber principle can be easily validated for the first eigenvalue by the following consideration. Let us find all vectors z of the T-length one such that the Rayleigh functional $\mathsf{r}(z) = z^{\mathsf{T}}\boldsymbol{r}z$ takes a stationary value on them. As the varied vector z must satisfy the additional condition $z^{\mathsf{T}}\boldsymbol{m}z = 1$, what we are dealing with is a conditional variational problem. We use a standard approach of reducing a conditional variational problem to an unconditional one by introducing a modified Lagrangian functional

$$\mathsf{p}(z) = z^{\mathsf{T}}\boldsymbol{r}z + \mu(1 - z^{\mathsf{T}}\boldsymbol{m}z)$$

with the Lagrangian multiplier μ. The necessary conditions of stationarity for this functional produce these equations:

$$\boldsymbol{r}z - \mu\boldsymbol{m}z = \boldsymbol{0}, \qquad z^{\mathsf{T}}\boldsymbol{m}z = 1.$$

But the first of the equations has a nonzero solution only if vector z is one of the eigenvectors and the Lagrangian multiplier μ is the respective eigenvalue.

Thus, all stationary values of the Rayleigh functional are taken on eigenvectors, and according to (2.27),

$$\mathsf{r}(z_i) = \frac{z_i^{\mathsf{T}}\boldsymbol{r}z_i}{z_i^{\mathsf{T}}\boldsymbol{m}z_i} = \lambda_i \qquad (i = 1,\dots,n). \tag{2.35}$$

The numbering of the eigenvalues satisfies the order defined in (2.12), hence the least of the stationary values of $\mathsf{r}(z)$ is $\min \mathsf{r}(z) = \mathsf{r}(z_1) = \lambda_1$.

---

[9] Before Rayleigh, this variational principle for the eigenvalues was known to Weber whose paper [28] presents a recursive calculation of the eigenvalues in application to an oscillating membrane. This principle was then employed extensively by Rayleigh in the determination of various physical properties of oscillating mechanical systems [21].

The Rayleigh–Weber principle formulated as above concerns only the first (minor) eigenvalue. However, the Rayleigh functional permits to develop a generalized variational description of the eigenvalues without confining ourselves to the first one only.

Now let us represent the general Rayleigh–Weber variational principle as the following statement:

> *Functional* $\mathsf{r}(z)$ *gets its minimum from i-th eigenvector* z_i, *and this minimum is equal to i-th eigenvalue* λ_i *of matrix* $\boldsymbol{r}$ *with respect to matrix* $\boldsymbol{m}$ *only when the minimum is sought for among vectors* z *such that they are* T*-orthogonal to the first* $(i-1)$ *eigenvectors* $z_1,\ldots, z_{i-1}$, *i.e.*
>
> $$\min \mathsf{r}(z) = \mathsf{r}(z_i) = \lambda_i$$
>
> *under the conditions* $z^{\mathsf{T}}\boldsymbol{m}z_1 = 0,\ldots, z^{\mathsf{T}}\boldsymbol{m}z_{i-1} = 0$.

Indeed, we know that all stationary values of the Rayleigh functional are taken on eigenvectors. But we have to exclude first $(i-1)$ eigenvectors from our consideration because any one of those (say, z_j such that $1 \le j \le i-1$) violates the requirement $z^{\mathsf{T}}\boldsymbol{m}z_j = 0$ at $z = z_j$. As for the other vectors admitted to the comparison, it is vector z_i that gives the minimum to $\mathsf{r}(z)$ according to (2.35), which proves the proposition.

The Rayleigh–Weber variational principle can be treated as an essentially new (variational) definition of the eigenvalues and the eigenvectors. As we have shown above, Rayleigh's variational definition is recursive: first eigenvalue and first eigenvector are calculated independently, then these data are used to find second eigenvalue and its respective eigenvector, and so on.

These considerations allow us to treat the general Rayleigh–Weber principle as a variational and recursive technique for finding the eigenvalues and eigenvectors.

A remark on the effect of constraints on first eigenvalue

The variational definition of the eigenvalues gives us a powerful tool for estimating the effect of constraints on the natural frequencies of a mechanical system.

If we take the variational definition of the eigenvalues instead of the original one, we will make it much easier to prove the well-known Routh theorem that states an increase (more exactly, no decrease) in the natural frequencies of a mechanical system when some new constraints are imposed on the system.

Let us first consider the Routh theorem for one additionally imposed constraint. The formulation of it follows. Let a mechanical system S have n

degrees of freedom, and let its spectrum of the squares of its natural frequencies be

$$0 < \lambda_1 \leq \lambda_2 \leq \ldots \leq \lambda_n .$$

We make a new system out of it by incorporating an additional constraint; the new system will be denoted by $\tilde{S}$. We suppose the system remains linear after the new constraint is imposed; this means the constraint is actually a requirement that all the generalized coordinates of the system should satisfy a linear homogeneous equation of the type

$$a_1 z_1 + a_2 z_2 + \ldots + a_n z_n = 0 \tag{2.36}$$

with at least some of its coefficients a_i not equal to zero; this requirement is equivalent to

$$a_1^2 + a_2^2 + \ldots + a_n^2 \neq 0 .$$

Equation (2.36) is usually called a constraint equation[10].

Obviously, incorporating the constraint (2.36) reduces the number of independent generalized coordinates by one. Therefore the system with the new constraint will actually have $(n-1)$ DOFs, and its frequency spectrum will be

$$0 < \tilde{\lambda}_1 \leq \tilde{\lambda}_2 \leq \ldots \leq \tilde{\lambda}_{n-1} .$$

The theorem by Routh states that the following mutual estimate of the frequencies of the two systems takes place:

$$\lambda_1 \leq \tilde{\lambda}_1 \leq \lambda_2, \quad \lambda_2 \leq \tilde{\lambda}_2 \leq \lambda_3, \ \ldots \ , \ \lambda_{n-1} \leq \tilde{\lambda}_{n-1} \leq \lambda_n , \tag{2.37}$$

It can be said that the frequency spectra of the original system and of the additionally constrained one are intermittent.

In this section we will prove only the first pair of inequalities in (2.37), $\lambda_1 \leq \tilde{\lambda}_1 \leq \lambda_2$. The others will be validated afterwards as a particular case of the more general Routh theorem about the effect of constraints on the frequency spectrum of a mechanical system.

To prove the above proposition, first of all we should re-formulate the constraint equation (2.36). It can be represented in the matrix form as $\boldsymbol{a}^{\mathsf{T}}\boldsymbol{z} = 0$ where vector $\boldsymbol{a} = \|[a_1, a_2, \ldots, a_n]\|^{\mathsf{T}}$ is not a zero vector. We assume

[10] The mechanical meaning of it is a *perfectly rigid kinematical constraint*; each constraint of this kind reduces the number of dynamical DOFs of a mechanical system by one. The effect of additional elastic constraints on the shift in the frequency spectrum will be discussed a bit later.

$\boldsymbol{b} = \boldsymbol{m}^{-1}\boldsymbol{a}$ which is admissible because the mass matrix is non-degenerate. Then vector $\boldsymbol{b}$ is also nonzero. The symmetry of matrix $\boldsymbol{m}$ helps represent the constraint equation in an equivalent form,

$$\boldsymbol{b}^{\mathsf{T}}\boldsymbol{m}\boldsymbol{z} = (\boldsymbol{b}, \boldsymbol{z})_{\mathsf{T}} = 0 . \tag{2.38}$$

This means the constraint condition can be treated from the mathematical standpoint as a requirement that any admissible displacement vector $\boldsymbol{z}$ should be T-orthogonal to the constraint vector $\boldsymbol{b}$.

The Rayleigh–Weber variational principle gives $\lambda_1 = \min \mathsf{r}(\boldsymbol{z})$, while $\tilde{\lambda}_1 = \min \mathsf{r}(\boldsymbol{z})$ under the condition $(\boldsymbol{b}, \boldsymbol{z})_{\mathsf{T}} = 0$. Hence immediately $\lambda_1 \le \tilde{\lambda}_1$ because no truncation of the admissible set of vectors can decrease the minimum of the functional. Thus, the least eigenvalue increases when a new constraint is imposed. Or, a more correct formulation would be "does not decrease" instead of "increases". The equality $\lambda_1 = \tilde{\lambda}_1$ can take place, and it becomes quite clear if we take any of the eigenvectors $\boldsymbol{z}_i$ at $i = 2, \ldots, n$ for the constraint vector $\boldsymbol{b}$.

Our next step of reasoning will be to validate the inequality $\tilde{\lambda}_1 \le \lambda_2$. In the particular case $\boldsymbol{b} = \boldsymbol{z}_1$ all admissible displacements must be T-orthogonal to the first eigenvector. The recursive variational definition of the second eigenvalue of the original system gives $\lambda_2 = \min \mathsf{r}(\boldsymbol{z})$ for all $\boldsymbol{z}$ which meet the condition $(\boldsymbol{z}_1, \boldsymbol{z})_{\mathsf{T}} = 0$. But the same variational definition holds for the first eigenvalue $\tilde{\lambda}_1$ of the additionally constrained system. Therefore, if $\boldsymbol{b} = \boldsymbol{z}_1$ then the equality $\tilde{\lambda}_1 = \lambda_2$ takes place. Now let us show that no constraint vector $\boldsymbol{b}$ can make the first eigenvalue $\tilde{\lambda}_1$ greater than λ_2.

We consider an arbitrary vector $\boldsymbol{x}$ of the T-length one such that it meets the constraint condition and at the same time is a linear combination of two first eigenvectors of the T-length one, $\boldsymbol{x}_1$ and $\boldsymbol{x}_2$, i.e.

$$(\boldsymbol{b}, \boldsymbol{x})_{\mathsf{T}} = 0 \qquad \text{and} \qquad \boldsymbol{x} = (\boldsymbol{x}, \boldsymbol{x}_1)_{\mathsf{T}}\boldsymbol{x}_1 + (\boldsymbol{x}, \boldsymbol{x}_2)_{\mathsf{T}}\boldsymbol{x}_2 . \tag{2.39}$$

Such a vector exists necessarily; this is obvious at least for dimensionality reasons if nothing else. To see this, consider that the number of dimensions in the subspace of vectors that satisfy the constraint condition is $(n-1)$, while the number of dimensions in the subspace made up by a linear combination of the two first eigenvectors is 2. Consequently, a subspace created by intersection of the above two subspaces can by no means have zero number of dimensions.

Now we have

$$\tilde{\lambda}_1 = \min \mathsf{r}(y) \leq \mathsf{r}(x) \tag{2.40}$$

where the minimum is taken with respect to all vectors y of the $(n-1)$-dimensional subspace defined above. On the other hand, for vector x of unit T-length (2.39) gives

$$\mathsf{r}(x) = (x,x)_{\mathsf{E}} = (x,x_1)_{\mathsf{T}}^2 \,\|\, x_1 \|_{\mathsf{E}}^2 + (x,x_2)_{\mathsf{T}}^2 \,\|\, x_2 \|_{\mathsf{E}}^2 = (x,x_1)_{\mathsf{T}}^2 \lambda_1 + (x,x_2)_{\mathsf{T}}^2 \lambda_2 .$$

Here we take into account the fact of T-orthogonality of vectors x_1 and x_2. Further, $\lambda_1 \leq \lambda_2$, so we immediately have the following estimate

$$\mathsf{r}(x) \leq [(x,x_1)_{\mathsf{T}}^2 + (x,x_2)_{\mathsf{T}}^2]\lambda_2 = \| x \|_{\mathsf{T}}^2 \lambda_2 = \lambda_2 .$$

Thus, $\mathsf{r}(x) \leq \lambda_2$, which proves the inequality $\tilde{\lambda}_1 \leq \lambda_2$ when taken together with (2.40).

Before we formulate and validate the general Routh theorem about the effect of constraints on the frequency spectrum of a mechanical system, we would like to discuss a variational technique for determining the eigenvalues independently; this approach will be called a Fischer–Courant maximin principle, or, shorter, a maximin principle[11].

A variational technique to find the eigenvalues independently from one another

The recursive variational definition of the eigenvalues that the Rayleigh–Weber principle establishes has an inconvenience: in order to find eigenvalue λ_i it supposes the previous $(i-1)$ eigenvectors are already known.

The variational principle by Fischer–Courant does not have this shortcoming because it gives a definition of an eigenvalue under any number, which is independent of the preceding elements of the spectrum. The formulation of the principle is

[11] The literature on the subject is not unanimous about this entitlement. For example, L. Collatz [3] attributes this principle to Courant only. An excellent book by S. Gould [9] indicates in its bibliographic notes that the principle was formulated clearly for the first time by Fischer [7] as early as in 1905, though Fischer used it just as an auxiliary tool. R. Courant was apparently the first to realize the importance of the principle from the standpoint of variational formulations of physical problems and to employ it extensively in problems of an inifinite number of dimensions described by equations of mathematical physics [4].

> *An eigenvalue λ_i is equal to a maximum value that the minimum of the Rayleigh functional,* $\mathsf{r}(z)$*, can take under the condition that the system is subjected to* $(i-1)$ *additional perfectly rigid constraints.*

To clear the things up, we would like to emphasize that the maximization here means an exhaustive search (by varying) of all possible constraints in the number of $(i-1)$; for any particular fixed set of constraints, a minimum of the Rayleigh functional is sought for, so the final result is a maximum of minima.

Suppose the said $(i-1)$ constraints are defined by vectors $\boldsymbol{b}_1,\ldots,\boldsymbol{b}_{i-1}$. Obviously, first eigenvalue $\tilde{\lambda}_1$ of the constrained system will depend on the choice of the constraints. In other words, $\tilde{\lambda}_1 = \tilde{\lambda}_1(\boldsymbol{b}_1,\ldots,\boldsymbol{b}_{i-1})$ can be treated as a functional of the set of constraints. According to the Rayleigh–Weber variational principle, $\tilde{\lambda}_1$ is a minimum of $\mathsf{r}(z)$ under the condition that vector z is orthogonal in the T-metric to all constraint vectors $\boldsymbol{b}_1,\ldots,\boldsymbol{b}_{i-1}$. Our intention is to choose such a set of the constraint vectors that gives $\tilde{\lambda}_1$ its largest possible value.

If we choose the constraint vectors as the first $(i-1)$ eigenvectors by assuming $\boldsymbol{b}_1 = z_1,\ldots,\boldsymbol{b}_{i-1} = z_{i-1}$, then the recursive variational definition gives

$$\tilde{\lambda}_1(z_1,\ldots,z_{i-1}) = \lambda_i \tag{2.41}$$

where λ_i is i-th eigenvalue of the system without the constraints. So, these $(i-1)$ constraints can be always put in such way that the first eighenvalue of the constrained system be equal to i-th eigenvalue of the unconstrained system.

Now we need to show that any other method of choosing the constraints would not make $\tilde{\lambda}_1$ larger, that is, $\tilde{\lambda}_1(\boldsymbol{b}_1,\ldots,\boldsymbol{b}_{i-1}) \leq \tilde{\lambda}_1(z_1,\ldots,z_{i-1})$. We construct two subspaces of the basic space $\mathbb{R}_n$. The first subspace, $\mathbb{R}_i$, is defined as a linear capsule on the first i eigenvectors $\boldsymbol{x}_1,\ldots,\boldsymbol{x}_i$ of the T-length one. The second subspace, $\mathbb{R}_{n-i+1}$, is a set of all vectors from $\mathbb{R}_n$ which are T-orthogonal to the constraint vectors. The sum of the dimensionalities of the subspaces is $(n+1)$ which is more than n in the original subspace. It means the intersection $\mathbb{R}_i \cap \mathbb{R}_{n-i+1}$ is not empty. We take an arbitrary vector $\boldsymbol{x}$ of the T-length one from that intersection and represent it as an expansion over the basis of $\boldsymbol{x}_1,\ldots,\boldsymbol{x}_i$:

$$\boldsymbol{x} = (\boldsymbol{x},\boldsymbol{x}_1)_{\mathsf{T}}\boldsymbol{x}_1 + (\boldsymbol{x},\boldsymbol{x}_2)_{\mathsf{T}}\boldsymbol{x}_2 + \ldots + (\boldsymbol{x},\boldsymbol{x}_i)_{\mathsf{T}}\boldsymbol{x}_i\,.$$

Then we find the Rayleigh functional's value on this vector:

$$\mathsf{r}(\boldsymbol{x}) = (\boldsymbol{x},\boldsymbol{x})_{\mathsf{E}} = (\boldsymbol{x},\boldsymbol{x}_1)_{\mathsf{T}}^2\lambda_1 + (\boldsymbol{x},\boldsymbol{x}_2)_{\mathsf{T}}^2\lambda_2 + \ldots + (\boldsymbol{x},\boldsymbol{x}_i)_{\mathsf{T}}^2\lambda_i\,.$$

Obviously, $\mathsf{r}(\boldsymbol{x}) \le \lambda_i$. Also, for any choice of the constraint vectors

$$\tilde{\lambda}_1(\boldsymbol{b}_1,\ldots,\boldsymbol{b}_{i-1}) \le \mathsf{r}(\boldsymbol{x})$$

because the left-hand part here is a minimum of the Rayleigh ratio over all vectors which are T-orthogonal to the constraint vectors, and the right-hand part has the argument of vector $\boldsymbol{x}$ which is T-orthogonal to all the constrain vectors and, moreover, obeys the additional limitation of belonging to the above introduced space $\mathbb{R}_i$. Merging the two inequalities gives

$$\tilde{\lambda}_1(\boldsymbol{b}_1,\ldots,\boldsymbol{b}_{i-1}) \le \lambda_i\,. \tag{2.42}$$

Establishing this relationship and the possibility for (2.41) to become an equality finishes the proof of equivalence between the recursive and the independent variational definitions for i-th eigenvalue.

To complete this section, we would like to note that the independent variational definition of i-th eigenvalue according to Fischer–Courant can be formulated as both the above maximin principle and, inversely, as a minimax principle. Here's the formulation:

Eigenvalue λ_i is equal to a minimum value which the maximum of the Rayleigh functional $\mathsf{r}(z)$ can take in conditions when the mechanical system has $(i-1)$ perfectly rigid constraints imposed on it.

We will not dwell on the proof of the minimax principle; the reader can do it by herself because all the needed information and approaches for such a proof have been already presented.

A general Routh theorem about the effect of constraints on the eigenvalues of a mechanical system

Having been established once and for all, the principle of maximin permits us to greatly simplify future considerations of various spectral properties of a mechanical system. In particular, this principle helps simplify the validation of the well-known Routh theorem of constraints with the following formulation[12]:

---

[12] The literature on the theory of mechanical oscillations often refers to this theorem as a *theorem of frequency separation*, and sometimes it is also called a

Eigenvalues $\tilde{\lambda}_i$ $(i = 1,\ldots, n-k)$ of a mechanical system subjected to k linearly independent constraints separate eigenvalues λ_i $(i = 1,\ldots, n)$ of the original mechanical system with n degrees of freedom in the sense that

$$\lambda_i \le \tilde{\lambda}_i \le \lambda_{i+k}$$

for any index i such that $i \le n-k$.

In particular, at $k = 1$ this theorem provides a proof for all inequalities in (2.37).

The validation of this theorem is a fairly simple matter if we use the maximin principle by Fischer–Courant[13]. We would like to present a simple statement as a corollary to the Routh theorem, which is often used in writings on mechanics:

If a mechanical system receives a perfectly rigid constraint, it spectrum, a discrete set of numbers on the axis of real numbers, can shift only to the right. Imposing each following constraint will shift the new spectrum again to the right, and so on.

The Routh theorem establishes, actually, something bigger. It dictates the widest limits up to which the mechanical system permits its spectrum to be shifted in such a way.

Kinematic constraints which do not alter the number of dynamic degrees of freedom in a mechanical system

It should be understood clearly that the Routh theorem considers only such mechanical constraints which reduce the number of dynamic degrees of freedom in the system. By *dynamic degrees of freedom* we understand a set of independent coordinates which determine the positions of all the masses of the system unambiguously. It is the masses not mass-free points that matter.

Rayleigh theorem. However, Rayleigh himself admits that a full proof of the theorem was given by Routh in [22]. Notes to the book by S. Gould [9] made by its Russian translation's editor, V.B. Lidsky, say that the essense of the theorem can be found in an earlier memoir by Poincaré [20]. At the same time, a book by Lancaster [13] on the theory of matrices says that even Cauchy already knew the theorem in his time. All this is an illustrative example of how vague and contradictory can be the opinions of experts when it comes to scientific priorities.

[13] A successful attempt at the proof of the Routh theorem can be the best evidence that the discussed material has been digested perfectly by the reader. If the attempt fails, something like [8] can help.

Now let us imagine that we deal with free flexural oscillations of a beam simply supported at its ends, which has a certain finite amount of masses placed on it (Fig. 10.1-*a*).

Together with the original system, we want to consider a system with a constraint $\tilde{S}$ which prohibits the motion of one mass — say, mass 3 as shown in Fig. 10.1-*b*. The mathematical description of the motion of the mechanical model in Fig. 10.1-*b* differs from that in Fig. 10.1-*a* by an additional requirement, $q_3 = 0$, where q_3 is a deviation of mass m_3 from the state of equilibrium in the oscillation process. Obviously, this constraint reduces the number of dynamic DOFs by one and therefore satisfies the conditions of applicability of the Routh theorem.

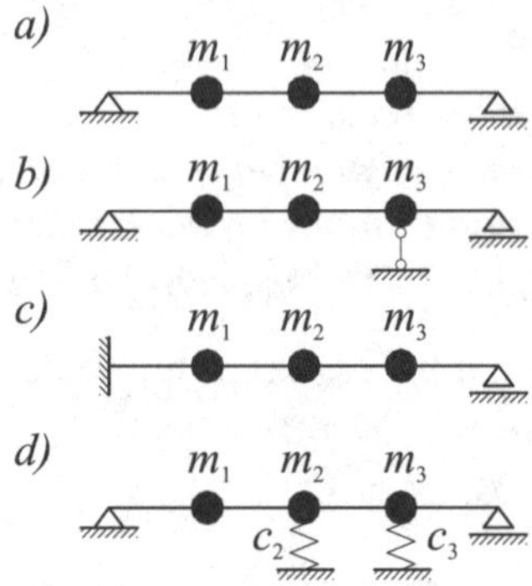

Fig. 10.1. Various kinds of constraints

There is another constraint type when we prevent the beam from rotating in its beginning section (Fig. 10.1-*c*). The number of dynamic DOFs in the system is not reduced by the additional constraint, so the Routh theorem is not applicable, though the constraint is a perfectly rigid one from the mechanical standpoint.

Another kind of constraints that does not change the number of dynamic DOFs is shown in Fig. 10.1-*d*. Here masses m_2 and m_3 are reinforced by springs, therefore the constraints like these can be called elastic. The elastic constraint category includes also constraints which increase the stiffness of all or some of the elements of a mechanical system. In our example problem it can be an increase in the flexural rigidity of the beam, *EI*, by making the material's elasticity modulus, *E*, or the moment of inertia of the beam's cross-section, *I*, larger.

For all constraint types listed above which do not reduce the number of dynamic DOFs, any changes in the mechanical system affect solely its potential energy counted off from its state of equilibrium.

Let $\mathsf{E} = \mathsf{E}(q_1,\ldots, q_n) = ½\boldsymbol{q}^{\mathsf{T}}\boldsymbol{rq}$ be an increment of the potential energy of a particular system in its small deviation from its state of equilibrium; the latter is assumed to take place at $\boldsymbol{q} = \mathbf{0}$. Similarly, let $\tilde{\mathsf{E}} = \tilde{\mathsf{E}}(q_1,\ldots,q_n) = ½\boldsymbol{q}^{\mathsf{T}}\,\tilde{\boldsymbol{r}}\,\boldsymbol{q}$ be an increment of the potential energy of the constrained system, where the constraint does not alter the number of DOFs in the system. First of all, for any deviation of the system from its state of equilibrium the following inequality holds:

$$\tilde{\mathsf{E}}(\boldsymbol{q}) \geq \mathsf{E}(\boldsymbol{q})\,. \tag{2.43}$$

According to their construction/definition, the values of E and $\tilde{\mathsf{E}}$ are the energies of the two closely related systems under a kinematic action — the $\boldsymbol{q}$ vector of given deviations of the masses from the systems' common state of equilibrium. But then we meet the conditions of a perturbed system considered earlier in Chapter 2. In particular, Table 2.2 says that if the system is subjected to a perfectly rigid constraint like that in Fig. 10.1-*c* and to a purely kinematic action, then the inequality (2.43) is sure to hold. If the constrain is elastic like one in Fig. 10.1-*d*, then a similar inequality takes place according to the central (framed) proposition of Section **2**.4.4.

For the purpose of further analysis, the inequality (2.43) can be represented in a convenient equivalent form by assuming

$$\tilde{\boldsymbol{r}} = \boldsymbol{r} + \Delta\boldsymbol{r} \tag{2.44}$$

where the matrix of stiffness increments, $\Delta\boldsymbol{r}$, is positively semi-definite, i.e for any vector $\boldsymbol{q}$ the equality holds:

$$\boldsymbol{q}^{\mathsf{T}}\Delta\boldsymbol{r}\,\boldsymbol{q} \geq 0\,. \tag{2.45}$$

Writing out an expression of the Rayleigh ratio in application to the original system and to the one perturbed by the constraint gives the following, after we take (2.45) into account:

$$\mathsf{r}(\boldsymbol{z}) = \frac{\boldsymbol{z}^{\mathsf{T}}\boldsymbol{rz}}{\boldsymbol{z}^{\mathsf{T}}\boldsymbol{mz}} \leq \tilde{\mathsf{r}}(\boldsymbol{z}) = \frac{\boldsymbol{z}^{\mathsf{T}}\boldsymbol{rz}}{\boldsymbol{z}^{\mathsf{T}}\boldsymbol{mz}} + \frac{\boldsymbol{z}^{\mathsf{T}}\Delta\boldsymbol{rz}}{\boldsymbol{z}^{\mathsf{T}}\boldsymbol{mz}}\,. \tag{2.46}$$

Obviously, the frequency spectrum of the system with the constraints imposed as shown above, will shift to the right on the real axis. It is also obvious that the major frequencies of the system can be made arbitrarily large by this shifting. To see this, imagine matrix $\Delta\boldsymbol{r}$ proportional to a certain numerical parameter β. The perturbation in the stiffness matrix $\beta\Delta\boldsymbol{r}$ can be made arbitrarily large by varying this parameter, and the respective maximum value of the Rayleigh ratio can also increase arbitrarily.

Of course, the qualitative effect of this type of constraints on the perturbation of the frequency spectrum is quite expectable. At the same time, we have a feeling that there must be tighter estimates of the type the Routh theorem provides.

And this feeling is right — the following theorem holds.

Let λ_i $(i=1,\ldots,n)$ be eigenvalues of the original mechanical system, and let values $\tilde{\lambda}_i$ $(i=1,\ldots,n)$ be eigenvalues for the mechanical system with constraints that do not decrease the number of its dynamic DOFs. Let the rank of matrix $\Delta\boldsymbol{r}$ *(a stiffness increment matrix of the system) be equal to k* $(1 \le k \le n)$*. Under these conditions, the following estimates take place:*

$$\lambda_i \le \tilde{\lambda}_i \quad (i=1,\ldots,n); \qquad \tilde{\lambda}_i \le \lambda_{i+k} \quad (i=1,\ldots,n-k).$$

The first part of the proposition is apparent and essentially follows from (2.46). We have to prove just that $\tilde{\lambda}_i \le \lambda_{i+k}$ $(i = 1,\ldots, n-k)$, which is not too hard using the Fischer–Courant variational principle[14].

Returning to the example shown in Fig. 10.1, we can see that for the constrained system in Fig. 10.1-*d* the following takes place:

$$\Delta\boldsymbol{r} = \left\| \begin{bmatrix} 0 & 0 & 0 \\ 0 & c_2 & 0 \\ 0 & 0 & c_3 \end{bmatrix} \right\|$$

where c_2 and c_3 are stiffnesses of the added springs which should be positive. Obviously, $k=2$. Therefore no calculation is required to immediately write out the following mutual estimates of the frequencies of the systems in Fig. 10.1-*a* and 10.1-*d*

$$\lambda_{1a} \le \lambda_{1b} \le \lambda_{3a}\,, \qquad \lambda_{2a} \le \lambda_{2b}, \qquad \lambda_{3a} \le \lambda_{3b}\,.$$

Knowing how to construct such estimates helps much in the solution of practical problems.

An exercise: show that imposing a constraint of the type shown in Fig. 10.1-*c* is mathematically equivalent to building up the stiffness matrix $\boldsymbol{r}$ according to (2.44). Find out the rank k of the matrix $\Delta\boldsymbol{r}$ for this case without making detailed calculations.

---

[14] Let it be another exercise. A full proof can be found in [8].

10.2.3 A geometric description of eigenvalues and eigenvectors. A Rayleigh ellipsoid

In Section **10**.2.1 we noted (see proposition *f*) that a symmetric positive definite matrix can be always represented as a product of two mutually transposed square matrices of the same order.

Now let us use this proposition to represent the inertia matrix, $\boldsymbol{m}$, as

$$\boldsymbol{m} = \mathbf{W}^{\mathsf{T}}\mathbf{W} . \tag{2.47}$$

After introducing a linear coordinate transformation,

$$\boldsymbol{x} = \mathbf{W}\boldsymbol{z} , \tag{2.48}$$

the spectral problem for the couple of matrices $\boldsymbol{rz} = \lambda \boldsymbol{mz}$ will become a standard spectral problem for matrix $\boldsymbol{p}$:

$$\boldsymbol{px} = \lambda \boldsymbol{x} , \qquad \boldsymbol{p} = \mathbf{W}^{-\mathsf{T}}\boldsymbol{r}\mathbf{W}^{-1} . \tag{2.49}$$

To shorten the notation, we denote by $^{-\mathsf{T}}$ a double operation: inversion plus transposition of the inverted matrix.

The eigenvalues are not changed by this replacement, and eigenvectors $\boldsymbol{z}_i$ of the original problem are related to eigenvectors $\boldsymbol{x}_i$ of the transformed problem through expressions that follow from (2.48),

$$\boldsymbol{x}_i = \mathbf{W}\boldsymbol{z}_i , \qquad \boldsymbol{z}_i = \mathbf{W}^{-1}\boldsymbol{x}_i . \tag{2.50}$$

Matrix $\boldsymbol{p}$ inherits its positive definiteness from matrix $\boldsymbol{r}$; this follows from a chain of identity transformations:

$$0 < (\boldsymbol{z},\boldsymbol{rz}) = (\mathbf{W}^{-1}\boldsymbol{x},\boldsymbol{r}\mathbf{W}^{-1}\boldsymbol{x}) = (\boldsymbol{x},\mathbf{W}^{-\mathsf{T}}\boldsymbol{r}\mathbf{W}^{-1}\boldsymbol{x}) = (\boldsymbol{x},\boldsymbol{px}) .$$

It is also obvious that eigenvectors $\boldsymbol{x}_1,\ldots,\boldsymbol{x}_n$ of the transformed problem are orthogonal both in the conventional (Euclidean) metric and in the metric generated by matrix $\boldsymbol{p}$. Normalizing these vectors in the usual way (by assuming $(\boldsymbol{x}_i,\boldsymbol{x}_i) = 1$) will give the following instead of (2.17):

$$(\boldsymbol{x}_i, \boldsymbol{x}_j) = \delta_{ij} , \quad (\boldsymbol{x}_i, \boldsymbol{px}_j) = \lambda_i\delta_{ij} . \tag{2.51}$$

The Rayleigh ratio will become as follows in terms of the new variables:

$$\mathsf{r} = \frac{\boldsymbol{z}^{\mathsf{T}}\boldsymbol{rz}}{\boldsymbol{z}^{\mathsf{T}}\boldsymbol{mz}} = \frac{\boldsymbol{z}^{\mathsf{T}}\boldsymbol{rz}}{\boldsymbol{z}^{\mathsf{T}}\mathbf{W}^{\mathsf{T}}\mathbf{W}\boldsymbol{z}} = \frac{\boldsymbol{x}^{\mathsf{T}}\mathbf{W}^{-\mathsf{T}}\boldsymbol{r}\mathbf{W}^{-1}\boldsymbol{x}}{\boldsymbol{x}^{\mathsf{T}}\boldsymbol{x}} = \frac{\boldsymbol{x}^{\mathsf{T}}\boldsymbol{px}}{\boldsymbol{x}^{\mathsf{T}}\boldsymbol{x}} \tag{2.52}$$

and now it can be treated as a functional of vector $\boldsymbol{x}$.

Now we know that the eigenvalues we want to find are equal to values of the Rayleigh functional in its points of stationarity. Without limiting the generality, this functional can be considered only with such arguments $\boldsymbol{x}$ the Euclidean norm of which is equal to one.

Suppose the denominator in the Rayleigh ratio is fixed equal to one; then we take the numerator and equal it to one, too, to produce the equation

$$(\boldsymbol{x}, \boldsymbol{p}\boldsymbol{x}) = 1. \tag{2.53}$$

Analytical geometry tells us that a set of unit vectors $\boldsymbol{x}$ which satisfy the equation (2.53) describes a second-order surface. We will show that the surface is an ellipsoid (at least in the three-dimensional space). To see this, we expand an arbitrary vector $\boldsymbol{x}$ of unit length over the basis of eigenvectors $\boldsymbol{x}_1, \boldsymbol{x}_2, \boldsymbol{x}_3$:

$$\boldsymbol{x} = (\boldsymbol{x},\boldsymbol{x}_1)\boldsymbol{x}_1 + (\boldsymbol{x},\boldsymbol{x}_2)\boldsymbol{x}_2 + (\boldsymbol{x},\boldsymbol{x}_3)\boldsymbol{x}_3 \ ,$$

so

$$\boldsymbol{p}\boldsymbol{x} = (\boldsymbol{x},\boldsymbol{x}_1)\boldsymbol{p}\boldsymbol{x}_1 + (\boldsymbol{x},\boldsymbol{x}_2)\boldsymbol{p}\boldsymbol{x}_2 + (\boldsymbol{x},\boldsymbol{x}_3)\boldsymbol{p}\boldsymbol{x}_3 = \lambda_1(\boldsymbol{x},\boldsymbol{x}_1)\boldsymbol{x}_1 + \lambda_2(\boldsymbol{x},\boldsymbol{x}_2)\boldsymbol{x}_2 + \lambda_3(\boldsymbol{x},\boldsymbol{x}_3)\boldsymbol{x}_3 \,.$$

Then the equation of our surface from (2.53) becomes

$$\lambda_1(\boldsymbol{x},\boldsymbol{x}_1)^2 + \lambda_2(\boldsymbol{x},\boldsymbol{x}_2)^2 + \lambda_3(\boldsymbol{x},\boldsymbol{x}_3)^2 = 1. \tag{2.54}$$

Recall that the equation of an ellipsoid in the three-dimensional space, its principal axes coincident with unit vectors $\boldsymbol{x}_1, \boldsymbol{x}_2, \boldsymbol{x}_3$ and its semi-axis lengths being a, b, c, is written as

$$\frac{(\boldsymbol{x},\boldsymbol{x}_1)^2}{a^2} + \frac{(\boldsymbol{x},\boldsymbol{x}_2)^2}{b^2} + \frac{(\boldsymbol{x},\boldsymbol{x}_3)^2}{c^2} = 1$$

This proves that (2.54) defines an ellipsoid, thus (2.53) does so too.

The ellipsoid thus constructed will be called a *Rayleigh ellipsoid.* Obviously, the principal axes of the Rayleigh ellipsoid are codirectional with the eigenvectors of the auxiliary problem, and the lengths of its semi-axes are equal to

$$a = \lambda_1^{-1/2}, \ b = \lambda_2^{-1/2}, \ c = \lambda_3^{-1/2} \,.$$

That is, the longest semi-axis a is directed along the first eigenvector, $\boldsymbol{x}_1 = \mathbf{W}z_1$, and the shortest one c along the major eigenvector $\boldsymbol{x}_3 = \mathbf{W}z_3$. By generalizing this "visible" three-dimensional geometric shape onto the n-dimensional space we arrive at an abstract geometric representation of an n-dimensional Rayleigh ellipsoid[15].

---

[15] Sometimes the direct Rayleigh ellipsoid is replaced with a reverse one, with its semi-axes equal to the square roots of the problem's eigenvalues. However, this choice is not critical.

An illustrative two-dimensional geometrical drawing is shown in Fig. 10.2.

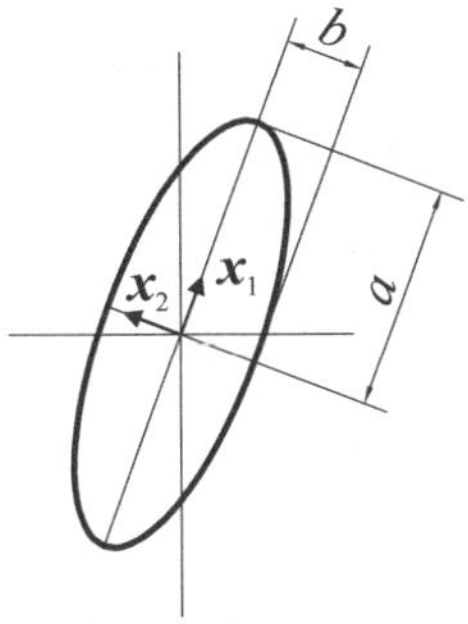

Fig. 10.2. A Rayleigh ellipsoid

Strictly speaking, there is nothing the Rayleigh ellipsoid can serve for in the analytical studies. However, as we have mentioned many times, a geometric representation of a problem permits to get a better understanding of it and to predict a result long before it is confirmed by accurate analysis.

For example, the effect of a constraint which is imposed on the system and reduces its number of dynamic DOFs can be represented as an intersection between the Rayleigh ellipsoid and a plane which goes through the coordinate origin and is orthogonal to the constraint vector. An ellipsoid in the $(n-1)$-dimensional space formed by this intersection is exactly the Rayleigh ellipsoid for the constrained system. The reader is invited to make up the rest of the geometric interpretation for the Routh theorem.

A notion of a maximum-rigidity constraint

The preceding presentment and the geometric analogy discussed above make it quite clear that if the constraint vector is $\boldsymbol{x}_1$, i.e. if all displacements allowed by the constraint are required to be orthogonal to the first eigenvector, then the first eigenvalue of the perturbed system, $\tilde{\lambda}_1$, is equal to the second eigenvalue λ_2 of the original problem. The geometrical meaning of it is that the length $\tilde{a}$ of the longest semi-axis of the $(n-1)$-dimensional ellipsoid will be equal to the length b of the second axis of the n-dimensional ellipsoid.

The Routh theorem says this shift of the first eigenvalue is a maximum achievable with one constraint. However, another question arises immediately. Does the above means that a constraint orthogonal to the first eigenvector is the only kind of constraint that provides a maximum growth of the first eigenvalue? In other words, may we suggest that a constraint of the following type:

$$(\boldsymbol{\beta},\boldsymbol{x}) = 0\,, \tag{2.55}$$

where vector $\boldsymbol{\beta}$ is different from the first eigenvector $\boldsymbol{x}_1$ of the original problem, though might increase $\tilde{\lambda}_1$ in comparison to λ_1, cannot let the equality $\tilde{\lambda}_1 = \lambda_2$ take place?

The answer to this question may seem to be undoubted yes. At least, the geometric analogy for $n = 2$ shows clearly this *is* the case (Fig. 10.2) because when a two-dimensional ellipse is intersected by a ray coming from the coordinate origin, the ellipse will cut off a piece of this ray the half-length of which is sure to be greater than b if the ray is not orthogonal to the long semi-axis of the ellipse.

However, this conclusion is a mistake even in three dimensions, let alone a larger dimensionality. Imagine a Rayleigh ellipsoid in three dimensions with the lengths of all its three semi-axes different, i.e. $a > b > c$. Now imagine the geometric picture of a constraint vector of the type $\boldsymbol{\beta} = \|[\cos\alpha,\, 0,\, \sin\alpha]\|^{\mathrm{T}}$ with a sufficiently small angle α[16]. The constraint plane $(\boldsymbol{\beta},\boldsymbol{x}) = 0$ cuts the three-dimensional Rayleigh ellipsoid apart and creates a two-dimensional ellipse with semi-axes $\tilde{a} \geq \tilde{b}$. This plane is created by turning the plane of axes $(\boldsymbol{x}_2, \boldsymbol{x}_3)$ about axis $\boldsymbol{x}_2$ by an angle α. If the α angle is small enough, it is obvious from geometrical considerations that the longest semi-axis $\tilde{a}$ of the two-dimensional ellipse remains equal to the second longest semi-axis b of the original three-dimensional ellipsoid. If we vary the α angle continuously, the equality $\tilde{a} = b$ will hold until the two-dimensional ellipse becomes a circle so that $\tilde{b} = \tilde{a}$. Increasing the α angle further would switch the eigenvalues in their ascending order, and the condition $\tilde{a} = b$ will be violated, becoming the inequality $\tilde{a} > b = \tilde{b}$.

Any constraint that increases maximally the first eigenvalue of a mechanical system is called a *maximum-rigidity constraint*. The notion of the maximum-rigidity constraint can be generalized in two ways. First, we can consider multiple (say, k) constraints instead of one. Second, we can deal also with constraints which do not reduce the number of dynamic DOFs in the system. In particular, we can switch from rigid to elastic constraints. Third, the notion of the maximum-rigidity constraints can be generalized onto infinite-dimensional systems.

[16] We assume the constraint vector's components to be specified with respect to the principal axes.

Seemingly the first formulated and correctly solved problem about the maximum-rigidity constraints in history was a problem of stability of a bar held by a set of elastic supports, not a free oscillations problem. In that problem, now known as Bubnov's problem [27], I.G. Bubnov found out that there is a certain limiting stiffness of the elastic supports which increases the first critical force of a compressed bar maximally. No further growth of the stiffness of the supports affects the value of the critical load.

We confine ourselves to the above brief introduction to the notion of the maximum-rigidity constraints and do not dwell on it any longer. It should be said that the problem of detecting all maximum-rigidity constraints for a given mechanical system subjected to a particular admissible set of constraints is one of most exciting topics studied by a branch of structural mechanics which is sometimes called a theory of qualitative methods for stability and dynamics. An interested reader can find more about this in the literature on the subject; there is much relevant information in Nudelman [17] and Dolberg [5], [6]. A fairly detailed presentment on the maximum-rigidity constraints can be found also in the above mentioned book by Gould [9].

10.3 A general spectral problem

Let us return from systems with a finite number of dynamic DOFs to a general problem which we started in Section **10**.1.

First of all, we rewrite the statement of the problem (1.5) – (1.6) that concerns the frequency spectrum of a mechanical system as

$$(\boldsymbol{A}^{\mathrm{T}}\boldsymbol{C}\boldsymbol{A} + \boldsymbol{K})\mathbf{u} = \lambda\boldsymbol{\rho}\mathbf{u} \qquad \in \Omega\,, \tag{3.1}$$

$$\boldsymbol{E}_p\boldsymbol{H}_\sigma\boldsymbol{C}\boldsymbol{A}\mathbf{u} = \mathbf{0}\,, \qquad \boldsymbol{E}_u\boldsymbol{H}_u\mathbf{u} = \mathbf{0} \qquad \in \Gamma. \tag{3.2}$$

Here and further we deal exclusively with amplitude values of the displacements, the stresses, and the strains, therefore we can omit the zero in the subscripts for the sake of brevity. Also, comparing to (1.5), the square of the frequency, ω^2, is re-denoted by λ.

Let us introduce expressions for the amplitude values of the potential and kinetic energy. Obviously, (2.25) should be replaced by the following in the general case, which follows from (**2**.1.13):

$$\mathsf{E} = \tfrac{1}{2}\,(\boldsymbol{C}\boldsymbol{A}\mathbf{u}, \boldsymbol{A}\mathbf{u}) + \tfrac{1}{2}\,(\boldsymbol{K}\mathbf{u}, \mathbf{u})\,. \tag{3.3}$$

At the same time, the general expression of the amplitude value of the kinetic energy becomes

$$\mathsf{T} = \tfrac{1}{2}\,\omega^2(\boldsymbol{\rho}\mathbf{u}, \mathbf{u})\,. \tag{3.4}$$

By generalizing the expression of the Rayleigh functional onto the most general spectral problem, we obtain

$$\mathsf{r}(\mathbf{u}) = \frac{(\boldsymbol{C}\boldsymbol{A}\mathbf{u}, \boldsymbol{A}\mathbf{u}) + (\boldsymbol{K}\mathbf{u}, \mathbf{u})}{(\boldsymbol{\rho}\mathbf{u}, \mathbf{u})}\,. \tag{3.5}$$

Obviously, it is sufficient to look for stationary values of the Rayleigh functional in the fields of displacements **u** which satisfy the mass normalization condition, $(\boldsymbol{\rho}\mathbf{u},\mathbf{u}) = 1$. To meet the mass normalization condition, we just have to introduce the multiplier into the function **u** because no such multiplier can change the value of the Rayleigh functional.

The result is a conditional variational problem reducible in standard way to an unconditional problem for the functional

$$\mathsf{p}(\mathbf{u},\mu) = (\boldsymbol{C}\boldsymbol{A}\mathbf{u}, \boldsymbol{A}\mathbf{u}) + (\boldsymbol{K}\mathbf{u},\mathbf{u}) + \mu[1 - (\boldsymbol{\rho}\mathbf{u},\mathbf{u})] \tag{3.6}$$

with the Lagrangian multiplier μ. The conditions of stationarity of $\mathsf{p}(\mathbf{u},\mu)$ on arbitrary homogeneously kinematically admissible fields of displacements **u** are as follows[17]:

$$\boldsymbol{A}^{\mathsf{T}}\boldsymbol{C}\boldsymbol{A}\mathbf{u} + \boldsymbol{K}\mathbf{u} = \mu\boldsymbol{\rho}\mathbf{u}, \qquad (\boldsymbol{\rho}\mathbf{u},\mathbf{u}) = 1. \tag{3.7}$$

Comparing (3.7) with (3.1) makes us conclude that the points of stationarity of the Rayleigh functional coincide with the eigenfunctions of the problem (3.1) - (3.2), and the Lagrangian multiplier μ is equal to the respective eigenvalue. It is obvious also that the *value* of the Rayleigh functional on any eigenfunction **u** will be equal to the respective eigenvalue, that is,

$$\lambda_i = \frac{(\boldsymbol{C}\boldsymbol{A}\mathbf{u}_i, \boldsymbol{A}\mathbf{u}_i) + (\boldsymbol{K}\mathbf{u}_i, \mathbf{u}_i)}{(\boldsymbol{\rho}\mathbf{u}_i, \mathbf{u}_i)} \quad (i = 1, 2, \ldots). \tag{3.8}$$

---

[17] Notice we do not require that the homogeneous static boundary conditions should be met when searching for a point of stationarity of the Rayleigh functional or, which is the same, of functional $\mathsf{p}(\mathbf{u},\mu)$. The static boundary conditions belong to the category of natural boundary conditions which are satisfied automatically at points of stationarity of $\mathsf{p}(\mathbf{u},\mu)$. The reasoning in the derivation of the governing differential equation (3.7) is the same as that used by us to formulate conditions of stationarity for the Lagrangian functional. Therefore we do not have to repeat it here.

Here and further we presume all eigenvalues λ_i and eigenvectors $\mathbf{u}_i$ are numbered in the eigenvalue ascending order, or

$$\lambda_1 \le \lambda_2 \le \ldots \le \lambda_i \le \lambda_{i+1} \le \ldots . \tag{3.9}$$

Return to Chapter **2** and recall that symbol $\mathscr{R}_0$ was used there to denote a set of homogeneously kinematically admissible rigid displacements of a mechanical system. Suppose first that $\mathscr{R}_0$ is an empty set so that the mechanical system in question is kinematically stable (its geometry does not change). We will show that all eigenvalues of the problem (3.1) – (3.2) are strictly positive in such a case.

To see this, we assume λ_i to be an eigenvalue of the problem (3.1) - (3.2) and $\mathbf{u}_i$ to be its respective eigenfunction. The denominator of (3.8) is a strictly positive value because earlier we specified that the inertia operator, $\boldsymbol{\rho}$, is a positive definite algebraic operator. The physical meaning of it is the positivity of the system's kinetic energy. On the other hand, the numerator in (3.8) is a doubled potential energy of strain, 2E, accumulated in the mechanical system with displacements $\mathbf{u}_i$. Energy E, as we found out in Chapter **2**, is strictly positive on any displacements of the system dfferent from perfectly rigid ones. The positivity of the numerator and denominator in (3.8) entails the strict inequality, $\lambda_i > 0$.

Now let $\mathscr{R}_0$ be non-empty so the dimensionality of this set be $k \ge 1$. According to a definition introduced in Chapter **2**, the k number is called a degree of kinematic instability of the system. It is quite clear that any function from $\mathscr{R}_0$ is an eigenfunction of the problem, and the respective eigenvalue of that eigenfuction is zero. This fact follows from the very definition of the rigid displacements according to which the numerator in (3.8) is equal to zero if $\mathbf{u}_i \in \mathscr{R}_0$. Thus,

> *A mechanical system that has the degree of kinematic instability k, has exactly the same number k of zero eigenvalues, and any rigid displacement is an eigenfuction of the problem that conforms to a zero eigenvalue. All the other eigenvalues of the system are strictly positive.*

What should *not* be forgotten: the above formulation lacks an explicit indication of the fact that we work with states of stable equilibrium only.

Let us show that all eigenfunctions of the problem (3.1) – (3.2) are orthogonal with respect to both mass and energy. First we should prove the orthogonality of eigenfunctions which correspond to different eigenvalues. We have

$$(\boldsymbol{A}^{\mathsf{T}}\boldsymbol{C}\boldsymbol{A} + \boldsymbol{K})\mathbf{u}_i = \lambda_i \boldsymbol{\rho}\mathbf{u}_i \ , \quad (\boldsymbol{A}^{\mathsf{T}}\boldsymbol{C}\boldsymbol{A} + \boldsymbol{K})\mathbf{u}_j = \lambda_j \boldsymbol{\rho}\mathbf{u}_j \ , \quad \lambda_i \ne \lambda_j \,. \tag{3.10}$$

Making a scalar product of the first of these equalities with $\mathbf{u}_j$ and of the second with $\mathbf{u}_i$ and subtracting one from the other gives

$$(\boldsymbol{A}^{\mathsf{T}}\boldsymbol{CA}\mathbf{u}_i,\mathbf{u}_j) + (\boldsymbol{K}\mathbf{u}_i,\mathbf{u}_j) - (\boldsymbol{A}^{\mathsf{T}}\boldsymbol{CA}\mathbf{u}_j,\mathbf{u}_i) - (\boldsymbol{K}\mathbf{u}_j,\mathbf{u}_i) = (\lambda_i - \lambda_j)(\boldsymbol{\rho}\mathbf{u}_i,\mathbf{u}_j)\ .$$

But the left-hand part here is zero, therefore at $\lambda_I \neq \lambda_j$ we have $(\boldsymbol{\rho}\mathbf{u}_i,\mathbf{u}_j) = 0$. This means eigenfunctions $\mathbf{u}_i$ and $\mathbf{u}_j$ are orthogonal with respect to mass.

The orthogonality with respect to mass implies the orthogonality with respect to energy, too. This can be proved by simply multiplying the first of the equalities in (3.7) by $\mathbf{u}_j$ and seeing that $(\boldsymbol{\rho}\mathbf{u}_i,\mathbf{u}_j) = 0$ gives

$$(\boldsymbol{A}^{\mathsf{T}}\boldsymbol{CA}\mathbf{u}_i,\mathbf{u}_j) = (\boldsymbol{CA}\mathbf{u}_i,\ \boldsymbol{A}\mathbf{u}_j) = 0 \ \text{при}\ \lambda_i \neq \lambda_j\ . \tag{3.11}$$

If an eigenvalue such as λ_i conforms to multiple (say, k) linearly independent eigenfunctions, the the eigenvalue is said to have a *multiplicity* of k.

In this case all the eigenfunctions can be orthogonalized by the Gram–Schmidt orthogonalization process. Therefore further we can think of the set of all eigenfunctions as an an orthogonal system both in the metric generated by the mass of the system and in the energy metric.

10.3.1 An expansion of an arbitrary function over the eigenfunctions

Let $\mathbf{u}$ be an arbitrary function of displacements which satisfies the homogeneous boundary conditions (3.2). And let $\{\mathbf{u}_1, \mathbf{u}_2,\ldots\}$ be a system of eigenfunctions, each conforming to its particular eigenvalue $\{\lambda_1, \lambda_2,\ldots\}$. To be definite without limiting the generality, we can think of each eigenfunction as having the T-length one, i.e. $(\boldsymbol{\rho}\mathbf{u}_i,\mathbf{u}_i) = 1$.

We expand the selected function $\mathbf{u}$, its T-norm being 1, over the basis made up of eigenfunctions $\{\mathbf{u}_1, \mathbf{u}_2,\ldots\}$:

$$\mathbf{u} = \alpha_1\mathbf{u}_1 + \alpha_2\mathbf{u}_2 + \ldots\ , \tag{3.12}$$

and because $\|\mathbf{u}\|_{\mathsf{T}} = 1$ or $(\boldsymbol{\rho}\mathbf{u},\mathbf{u}) = 1$, coefficients α_i obey the condition

$$\alpha_1^2 + \alpha_2^2 + \ldots. = 1\ . \tag{3.13}$$

We determine coefficients α_i in the expansion (3.12) by making scalar products of (3.12) with all eigenfunctions $\mathbf{u}_i$ in succession:

$$\alpha_i = (\boldsymbol{\rho}\mathbf{u},\mathbf{u}_i) = (\boldsymbol{\rho}\mathbf{u}_i,\mathbf{u}), \qquad \mathbf{u} = \sum_{i=1}^{\infty}(\boldsymbol{\rho}\mathbf{u},\mathbf{u}_i)\mathbf{u}_i\ . \tag{3.14}$$

As we know, all eigenfunctions $\mathbf{u}_i$ are mutually orthogonal with respect to both mass and energy. Therefore, for a sufficiently smooth displacement function $\mathbf{u}$ such that energy $\mathsf{E}(\mathbf{u})$ in (3.3) exists and is finite, we obtain the following from the expansion (3.14):

$$\mathsf{E}(\mathbf{u}) = \sum_{i=1}^{\infty} (\rho\mathbf{u},\mathbf{u}_i)^2 \mathsf{E}(\mathbf{u}_i). \tag{3.15}$$

Thus, the energy accumulated by a mechanical system on displacements $\mathbf{u}$ can be represented by a sum of energies taken with their coefficients and accumulated by the same system on each of the natural modes (eigenfunctions).

10.4 The Ritz method in the spectral problem

Up to this point we have established the fact that the Rayleigh functional takes a central place in the variational formulation of the spectral problem for a mechanical system. It seems now quite logical to use this functional directly and try to extend the Ritz method discussed in Chapter **9** onto the spectral problem[18]. This is the way it is usually done — see [24], for example.

However, there is another (and quite equivalent, generally speaking) opportunity for employing the Ritz approach in the solution of spectral problems. We will use the latter. Let us try to formulate a variational spectral problem without using the Rayleigh functional. There is a way to do it: we can use a quasi-static statement of the problem by formally classifying the inertia force, $\lambda\rho\mathbf{u}$, as an external action; this force appears in the system during its harmonic oscillation described by equations (3.1), (3.2) where the frequency is $\omega = \sqrt{\lambda}$.

Let us write out the Lagrangian and Reissnerian functionals (the latter in two forms) for this quasi-static problem. Based on results obtained in Chapter 2 and Chapter 3, we have

$$\mathsf{L}(\mathbf{u}) = \tfrac{1}{2}(\boldsymbol{C}\boldsymbol{A}\mathbf{u}, \boldsymbol{A}\mathbf{u}) + \tfrac{1}{2}(\boldsymbol{K}\mathbf{u}, \mathbf{u}) - \tfrac{1}{2}\lambda(\rho\mathbf{u},\mathbf{u}), \tag{4.1}$$

$$\mathsf{R}_1(\boldsymbol{\sigma}, \mathbf{u}) =$$

$$= \tfrac{1}{2}(\boldsymbol{C}^{-1}\boldsymbol{\sigma}, \boldsymbol{\sigma}) - \tfrac{1}{2}(\boldsymbol{K}\mathbf{u}, \mathbf{u}) - (\boldsymbol{A}\mathbf{u}, \boldsymbol{\sigma}) + (\boldsymbol{E}_u \boldsymbol{p}, \boldsymbol{E}_u \boldsymbol{u})_\Gamma + \tfrac{1}{2}\lambda(\rho\mathbf{u},\mathbf{u}), \tag{4.2}$$

---

[18] When the method is applied to spectral problems, it is sometimes called a Rayleigh–Ritz method, which is quite fair.

$$\mathrm{R}_2(\boldsymbol{\sigma}, \mathbf{u}) =$$

$$= \tfrac{1}{2}(\boldsymbol{C}^{-1}\boldsymbol{\sigma}, \boldsymbol{\sigma}) - \tfrac{1}{2}(\boldsymbol{K}\mathbf{u}, \mathbf{u}) - (\boldsymbol{A}^{\mathsf{T}}\boldsymbol{\sigma}, \mathbf{u}) - (\boldsymbol{E}_p\boldsymbol{p}, \boldsymbol{E}_p\,\boldsymbol{u})_\Gamma + \tfrac{1}{2}\lambda(\boldsymbol{\rho}\mathbf{u},\mathbf{u}). \tag{4.3}$$

Two following circumstances are taken into account. First, the homogeneous boundary conditions from (3.2) and the relationships of (**3**.1.3) give

$$\Pi_{\mathrm{k}}(\boldsymbol{\sigma}) = (\boldsymbol{E}_u\boldsymbol{p}, \boldsymbol{E}_u\,\overline{\boldsymbol{u}}\,)_\Gamma = 0 \;\text{ and }\; \Pi_{\mathrm{s}}(\mathbf{u}) = (\,\overline{\boldsymbol{X}}\,, \mathbf{u}) + (\boldsymbol{E}_p\,\overline{\boldsymbol{p}}\,, \boldsymbol{E}_p\,\boldsymbol{u})_\Gamma = (\,\overline{\boldsymbol{X}}\,, \mathbf{u}).$$

Second, in the expression of the potential of inertial forces $(\overline{\boldsymbol{X}},\mathbf{u})$ we cannot just assign formally $\overline{\boldsymbol{X}} = \lambda\boldsymbol{\rho}\mathbf{u}$. The multiplier of ½ must be added because the inertia forces, though conservative, are not dead. They depend on the displacements linearly, and the result is $\Pi_{\mathrm{s}}(\mathbf{u}) = \tfrac{1}{2}\lambda(\boldsymbol{\rho}\mathbf{u},\mathbf{u})$.

The stationarity conditions for the Lagrangian functional on kinematically admissible fields of displacements produce the following Euler equations and natural boundary conditions:

$$(\boldsymbol{A}^{\mathsf{T}}\boldsymbol{C}\boldsymbol{A} + \boldsymbol{K})\mathbf{u} = \lambda\boldsymbol{\rho}\mathbf{u} \quad \in\Omega\,, \qquad \boldsymbol{E}_p\boldsymbol{H}_\sigma\boldsymbol{C}\boldsymbol{A}\mathbf{u} = \mathbf{0} \;\; \in\Gamma, \tag{4.4}$$

and the kinematical boundary conditions,

$$\boldsymbol{E}_u\boldsymbol{H}_u\mathbf{u} = \mathbf{0} \;\; \in\Gamma, \tag{4.5}$$

are the principal conditions for the functional because the fields of displacements we admit to the consideration are kinematically admissible.

As we can see, the boundary-value problem (4.4), (4.5) is the same as the original problem (3.1), (3.2).

Using the stationarity conditions for the Reissner functional gives – see (**3**.1.10) and (**3**.1.11) – the following:

$$\boldsymbol{C}^{-1}\boldsymbol{\sigma} - \boldsymbol{A}\mathbf{u} = \mathbf{0}, \qquad \boldsymbol{A}^{\mathsf{T}}\boldsymbol{\sigma} + \boldsymbol{K}\mathbf{u} - \lambda\boldsymbol{\rho}\mathbf{u} = \mathbf{0} \;\; \in\Omega\,, \tag{4.6}$$

$$\boldsymbol{E}_p\boldsymbol{H}_\sigma\boldsymbol{\sigma} = \mathbf{0}, \qquad \boldsymbol{E}_u\boldsymbol{H}_u\mathbf{u} = \mathbf{0} \quad \in\Gamma. \tag{4.7}$$

Both types of boundary conditions in (4.7) are natural conditions for the Reissner functional. We use the first relation in (4.6) to exclude the stresses from the other equations, and again arrive at the original problem (3.1), (3.2).

Now that we have introduced the Lagrange and Reissner functionals into the spectral problem, we can use the general procedure of the Ritz method.

We start by constructing the Ritz method for the Lagrangian functional. To do it, we consider a system of linearly independent and kinematically admissible vectors of displacements $\{\boldsymbol{\varphi}_1,\boldsymbol{\varphi}_2,\ldots,\boldsymbol{\varphi}_n,\ldots\}$ – these will be coordinate functions for the Ritz method. This set can be considered complete in the sense that any vector of displacements $\mathbf{u} \in \mathcal{U}_{\mathrm{ko}}$ can be approximated as accurately as needed (in the Lagrangian energy metric) by a linear combination of the type

$$\mathbf{u} = Z_1\boldsymbol{\varphi}_1 + Z_2\boldsymbol{\varphi}_2 + \ldots + Z_n\boldsymbol{\varphi}_n \tag{4.8}$$

if we choose the number n of the terms in expansion (4.8) and coefficients Z_i. properly.

In addition to the linear independence and kinematical admissibility, we require one more thing from the coordinate functions. Namely, we want to assume that the set of the coordinate functions does not contain rigid displacements of the mechanical system. The mathematical equivalent of this requirement is that each coordinate function $\boldsymbol{\varphi}_i$ should be orthogonal in the main metric to any kinematically admissible rigid displacement $\mathbf{u}_j \in \mathcal{R}_{\mathrm{o}}$, that is,

$$(\mathbf{u}_j, \boldsymbol{\varphi}_i) = 0 \,. \tag{4.9}$$

As we already know from Chapter 2, meeting this requirement leads us to the Lagrangian energy space $\mathcal{L}$the elements of which are coordinate functions $\{\boldsymbol{\varphi}_1, \boldsymbol{\varphi}_2, \ldots, \boldsymbol{\varphi}_n, \ldots\}$. The mechanical treatment of the requirement (4.9) is just that there must be a statically determinate fixation of the system that prevents it from any rigid displacements.

Putting (4.8) in (3.18) makes the Lagrangian functional a quadratic form of parameters Z_i,

$$\mathsf{L}(\boldsymbol{Z}) = \tfrac{1}{2}\,\boldsymbol{Z}^{\mathsf{T}}(\mathbf{R} - \lambda\mathbf{M})\boldsymbol{Z} \,, \tag{4.10}$$

where we denote

$$\mathbf{R} = (\boldsymbol{CA\Phi}, \boldsymbol{A\Phi}) + (\boldsymbol{K\Phi}, \boldsymbol{\Phi}), \qquad \mathbf{M} = (\boldsymbol{\rho\Phi}, \boldsymbol{\Phi}),$$

$$\boldsymbol{Z} = \|[Z_1, \ldots, Z_n]\|^{\mathsf{T}} \,, \qquad \boldsymbol{\Phi} = \|[\boldsymbol{\varphi}_1, \ldots, \boldsymbol{\varphi}_n]\|^{\mathsf{T}}. \tag{4.11}$$

The conditions of stationarity of $\mathsf{L}(\boldsymbol{Z})$ produce the following finite-dimensional spectral problem:

$$(\mathbf{R} - \lambda\mathbf{M})\boldsymbol{Z} = \mathbf{0}. \tag{4.12}$$

Matrix **R** fits well the name of a *stiffness matrix* of a finite-dimensional system derived from the given infinite-dimensional system by the Ritz approximation of the displacements. Similarly, matrix **M** lends itself to the name of a *mass matrix*. Both matrices are sure to be positive definite because they are Gram matrices for a system of linearly independent coordinate functions,

$$\mathbf{R} = 2\left\|\begin{bmatrix} <\varphi_1,\varphi_1> & \cdots & <\varphi_1,\varphi_n> \\ \vdots & \vdots & \vdots \\ <\varphi_n,\varphi_1> & \cdots & <\varphi_n,\varphi_n> \end{bmatrix}\right\|, \quad \mathbf{M} = \left\|\begin{bmatrix} (\rho\varphi_1,\varphi_1) & \cdots & (\rho\varphi_1,\varphi_n) \\ \vdots & \vdots & \vdots \\ (\rho\varphi_n,\varphi_1) & \cdots & (\rho\varphi_n,\varphi_n) \end{bmatrix}\right\|.$$

We denote by $\lambda_{\mathrm{L}1}^{(n)}, \lambda_{\mathrm{L}2}^{(n)}, \ldots, \lambda_{\mathrm{L}n}^{(n)}$ a set of eigenvalues of the finite-dimensional problem (4.12) in the ascending order:

$$0 < \lambda_{\mathrm{L}1}^{(n)} \le \lambda_{\mathrm{L}2}^{(n)} \le \ldots \le \lambda_{\mathrm{L}n}^{(n)}. \tag{4.13}$$

Here the $^{(n)}$ superscript in parentheses indicates that the eigenvalues have been obtained by solving the n-dimensional spectral problem. The $_\mathrm{L}$ subscript emphasizes that the eigenvalues have been derived from the stationarity conditions for the Lagrangian functional, therefore we will call them *Lagrangian eigenvalues*.

Now imagine that our set of coordinate functions will not include all n functions $\{\varphi_1, \varphi_2, \ldots, \varphi_n\}$; instead, we will use only some of them. To be definite, we keep first $(n-k)$ functions of the set by assuming

$$\mathbf{\Phi} = \|[\varphi_1, \ldots, \varphi_{n-k}]\|^{\mathrm{T}}.$$

We arrive at the problem (4.12) again, but the dimensionality of all matrices and vectors is now reduced to $(n-k)$. The set of eigenvalues of this reduced problem is

$$0 < \lambda_{\mathrm{L}1}^{(n-k)} \le \lambda_{\mathrm{L}2}^{(n-k)} \le \ldots \le \lambda_{\mathrm{L}\,n-k}^{(n-k)}. \tag{4.14}$$

How do the sets of eigenvalues of the n-dimensional problem and of the $(n-k)$-dimensional problem relate to each other? Actually, we already know the answer. It suffices to understand that our $(n-k)$-dimensional problem is derived from the n-dimensional one by imposing k constraints on the latter; the constraints are of the type

$$\varphi_{n-k+1} = \mathbf{0}, \ldots, \varphi_n = \mathbf{0}. \tag{4.15}$$

Now we can use the Routh theorem of frequency separation; in our designations it gives

$$\lambda_{\mathrm{L}i}^{(n)} \le \lambda_{\mathrm{L}i}^{(n-k)} \le \lambda_{\mathrm{L}\,i+k}^{(n)} \qquad (i = 1, 2, \ldots, n-k). \tag{4.16}$$

The relation (4.16) implies an important conclusion: the sequence of i-th eigenvalues $\lambda_{\mathrm{L}i}^{(n)}$ decreases monotonously as the dimensionality n of the finite-dimensional Ritz approximation grows. However, as this monotonously decreasing sequence is bounded from below by zero, it will tend to a certain limit when n tends to infinity, which follows from a well-known theorem of mathematical analysis. Further, methods of functional analysis help prove that certain, fairly general conditions make these limits equal to the eigenvalues of the original infinite-dimensional problem[19]. For us, the most important qualitative conclusion that follows from the results presented above is:

The Lagrangian eigenvalues always estimate the true eigenvalues λ_i of the original mechanical system from above:

$$\lambda_i \le \lambda_{\mathsf{L}i}^{(n)} .$$

Let us introduce a Rayleigh ratio

$$\mathsf{r}_{\mathsf{L}} = \frac{\boldsymbol{Z}^{\mathsf{T}}\mathbf{R}\boldsymbol{Z}}{\boldsymbol{Z}^{\mathsf{T}}\mathbf{M}\boldsymbol{Z}} . \tag{4.17}$$

As the $\mathbf{M}$ matrix is positive definite, it can be represented as the product $\mathbf{M} = \mathbf{W}^{\mathsf{T}}\mathbf{W}$, which makes it possible to rewrite (4.17) as

$$\mathsf{r}_{\mathsf{L}} - \frac{\boldsymbol{Y}^{\mathsf{T}}\mathbf{W}^{-\mathsf{T}}\mathbf{R}\mathbf{W}\boldsymbol{Y}}{\boldsymbol{Y}^{\mathsf{T}}\boldsymbol{Y}}, \qquad \text{где } \boldsymbol{Y} = \mathbf{W}\boldsymbol{Z}. \tag{4.18}$$

The equation

$$\boldsymbol{Y}^{\mathsf{T}}\mathbf{W}^{-\mathsf{T}}\mathbf{R}\mathbf{W}\boldsymbol{Y} = 1, \tag{4.19}$$

defines, as we already know, an n-dimensional ellipsoid which we will call a *Rayleigh* L*-ellipsoid.* The sizes of the main semi-axes of the L-ellipsoid in the descending ordcr arc

$$1\Big/\sqrt{\lambda_{\mathsf{L}1}^{(n)}} \ge 1\Big/\sqrt{\lambda_{\mathsf{L}2}^{(n)}} \ge \ldots \ge 1\Big/\sqrt{\lambda_{\mathsf{L}n}^{(n)}} . \tag{4.20}$$

10.4.1 The Ritz method in the spectral problem, applied to the Reissner functional

Exactly as we did in the static problems, we should supplement the set of coordinate functions for the displacements, $\{\varphi_1,\varphi_2,\ldots,\varphi_n\}$, with a set of coordinate vector functions for the stresses, $\{\psi_1,\psi_2,\ldots,\psi_m\}$. We have noted that all boundary conditions are natural for the Reissnerian

[19] See, for example, [9] or [24].

functional, therefore we could avoid imposing any boundary requirements on the new coordinate functions. However, for the sake of simplicity we will think of functions $\{\varphi_1,\varphi_2,\ldots,\varphi_n\}$ as belonging to the Lagrange energy space and of functions $\{\psi_1,\psi_2,\ldots,\psi_m\}$ as being homogeneously statically semi-admissible, i.e. ones that satisfy homogeneous static boundary conditions. The next step is to use independent approximations:

$$\mathbf{u} = Z_1\varphi_1 + Z_2\varphi_2 + \ldots + Z_n\varphi_n\,, \qquad \boldsymbol{\sigma} = X_1\psi_1 + X_2\psi_2 + \ldots + X_n\psi_m\,. \tag{4.21}$$

Substituting the approximations (4.21) in the expression of the first form of the Reissner functional and seeing that $\varphi_i \in \mathscr{U}_{\text{ko}}$ produces a quadratic form,

$$\mathsf{R}_1(\boldsymbol{X},\boldsymbol{Z}) = \frac{1}{2}\boldsymbol{X}^{\mathsf{T}}(\Psi, C^{-1}\Psi)\boldsymbol{X} - \frac{1}{2}\boldsymbol{Z}^{\mathsf{T}}(\Phi, K\Phi)\boldsymbol{Z} - \boldsymbol{Z}^{\mathsf{T}}(A\Phi,\Psi)\boldsymbol{X} + \\ + \frac{\lambda}{2}\boldsymbol{Z}^{\mathsf{T}}(\rho\Phi,\Phi)\boldsymbol{Z}\,. \tag{4.22}$$

Recalling the designations from (**9**.2.7), we rewrite the above formula in the matrix form:

$$\mathsf{R}_1(\boldsymbol{X},\boldsymbol{Z}) = \frac{1}{2}\left\|\left[\boldsymbol{X}^{\mathsf{T}},\boldsymbol{Z}^{\mathsf{T}}\right]\right\| \left\|\begin{bmatrix}\mathbf{G} & -\mathbf{H}\\ -\mathbf{H}^{\mathsf{T}} & -\mathbf{K}+\lambda\mathbf{M}\end{bmatrix}\right\| \left\|\begin{bmatrix}\boldsymbol{X}\\ \boldsymbol{Z}\end{bmatrix}\right\|. \tag{4.23}$$

The stationarity conditions for this quadratic form give the following two homogeneous algebraic equations:

$$\mathbf{G}\boldsymbol{X} - \mathbf{H}\boldsymbol{Z} = \mathbf{0}, \qquad \mathbf{H}^{\mathsf{T}}\boldsymbol{X} + (\mathbf{K} - \lambda\mathbf{M})\boldsymbol{Z} = \mathbf{0}. \tag{4.24}$$

We use the first of equations (4.24) to exclude the $\boldsymbol{X}$ vector from the second equation and thus arrive at an eigenvalue problem for the couple of matrices:

$$(\mathbf{H}^{\mathsf{T}}\mathbf{G}^{-1}\mathbf{H} + \mathbf{K})\boldsymbol{Z} = \lambda\mathbf{M}\boldsymbol{Z}. \tag{4.25}$$

Let us denote by $\lambda_{\mathsf{R}1}^{(n,m)}, \lambda_{\mathsf{R}2}^{(n,m)}, \ldots, \lambda_{\mathsf{R}n}^{(n,m)}$ the set of eigenvalues of the finite-dimensional problem (4.25) in the ascending order,

$$0 < \lambda_{\mathsf{R}1}^{(n,m)} \le \lambda_{\mathsf{R}2}^{(n,m)} \le \ldots \le \lambda_{\mathsf{R}n}^{(n,m)}\,. \tag{4.26}$$

These numbers lend themselves obviously to the name of *Reissnerian eigenvalues*. The R subscript emphasizes that the eigenvalues are obtained from the stationarity conditions for the Reissnerian functional. The double superscript in parentheses, $^{(n,\,m)}$, indicates the dimensionalities involved: the first index n is the dimensionality of the displacement approximation space, and the second index m is the dimensionality of the stress approximation space.

From results of Section **9**.2.1 we already know that the solvalibity of the Ritz system of equations for a static problem (at least where there is no elastic medium, $\boldsymbol{K} = \boldsymbol{O}$) requires that $m \geq n$. As we will see a bit later, this key inequality is also important for the spectral problems. At the same time, we are interested how the Lagrangian and Reissnerian frequencies relate to one another. To get to know this, first we introduce a Rayleigh ratio for the problem (4.25):

$$\mathsf{r}_{\mathsf{R}} = \frac{\boldsymbol{Z}^{\mathsf{T}}(\mathbf{H}^{\mathsf{T}}\mathbf{G}^{-1}\mathbf{H}+\mathbf{K})\boldsymbol{Z}}{\boldsymbol{Z}^{\mathsf{T}}\mathbf{M}\boldsymbol{Z}} = \frac{\boldsymbol{Y}^{\mathsf{T}}\mathbf{W}^{-\mathsf{T}}(\mathbf{H}^{\mathsf{T}}\mathbf{G}^{-1}\mathbf{H}+\mathbf{K})\mathbf{W}\boldsymbol{Y}}{\boldsymbol{Y}^{\mathsf{T}}\boldsymbol{Y}}. \tag{4.27}$$

The equation

$$\boldsymbol{Y}^{\mathsf{T}}\mathbf{W}^{-\mathsf{T}}(\mathbf{H}^{\mathsf{T}}\mathbf{G}^{-1}\mathbf{H}+\mathbf{K})\mathbf{W}\boldsymbol{Y} = 1 \tag{4.28}$$

defines an n-dimensional ellipsoid which we will further call a *Rayleigh* R-*ellipsoid.* The lengths of the main semi-axes of the R-ellipsoid in the descending order are

$$1\Big/\sqrt{\lambda_{\mathsf{R}1}^{(n,m)}} \geq 1\Big/\sqrt{\lambda_{\mathsf{R}2}^{(n,m)}} \geq \ldots \geq 1\Big/\sqrt{\lambda_{\mathsf{R}n}^{(n,m)}}. \tag{4.29}$$

For the n-dimensional R-ellipsoid to exist, the quadratic form $\boldsymbol{Y}^{\mathsf{T}}\mathbf{W}^{-\mathsf{T}}(\mathbf{H}^{\mathsf{T}}\mathbf{G}^{-1}\mathbf{H}+\mathbf{K})\mathbf{W}\boldsymbol{Y}$ needs to be positive definite. Otherwise the numerator of the Rayleigh ratio in (4.27) could become negative or zero for a nonzero vector $\boldsymbol{Y}$. This would mean the equation (4.28) described a second-order surface which was not an ellipsoid.

In Chapter **9** the $\mathbf{H}^{\mathsf{T}}\mathbf{G}^{-1}\mathbf{H}$ matrix was proved to be positive definite if and only if rank $\mathbf{H} = n$. This implies the requirement (for $\boldsymbol{K} = \boldsymbol{O}$) of $m \geq n$, i.e the dimensionality of the stress approximation space should not be less than that of the displacement approximation space. If we assume the following for the approximations (4.21):

$$s = n - m > 0, \tag{4.30}$$

then we will have rank $\mathbf{H} = m$, and the n-dimensional R-ellipsoid will degenerate into an m-dimensional ellipsoid. In other words, s main semi-axes of it will be zero.

The components of matrix $\mathbf{H} = \|[\mathrm{H}_{ij}]\|$ $(i = 1,\ldots, m; j = 1,\ldots, n)$ are defined as

$$\mathrm{H}_{ij} = (\psi_i, \boldsymbol{A}\varphi_j) = (\boldsymbol{A}^{\mathsf{T}}\psi_i, \varphi_j)\,. \tag{4.31}$$

Both representations of H_{ij} in formula (4.31) produce the same result because, according to our condition, $\boldsymbol{\varphi}_j$ satisfies homogeneous kinematical boundary conditions and $\boldsymbol{\psi}_i$ homogeneous static boundary conditions. But then the two representations of H_{ij} according to (4.31) are true simply as corollaries to the basic integral identity (**1**.2.19) where the displacements **u** are $\boldsymbol{\varphi}_j$ and the stresses $\boldsymbol{\sigma}$ are $\boldsymbol{\psi}_i$.

We denote by L_φ a linear hull of the set of vectors $\{\boldsymbol{\varphi}_1,\boldsymbol{\varphi}_2,\ldots,\boldsymbol{\varphi}_n\}$ and by N_χ a linear hull of the set of vectors $\{\boldsymbol{\chi}_1,\boldsymbol{\chi}_2,\ldots,\boldsymbol{\chi}_m\}$ where $\boldsymbol{\chi}_i = \boldsymbol{A}^{\mathsf{T}}\boldsymbol{\psi}_i$. The following <u>theorem</u> holds:

In order for matrix **H** to have the rank of n, it is necessary and sufficient that the projection of N_χ onto L_φ, in the sense of a scalar product in the main metric, coincide with L_φ.

Sufficiency. The linear independence of the vectors $\{\boldsymbol{\varphi}_1,\boldsymbol{\varphi}_2,\ldots,\boldsymbol{\varphi}_n\}$ makes the L_φ set a finite-dimensional space of the dimensionality n. According to the condition of the theorem, $m \ge n$ because the dimensionality of no projection can exceed that of the projected set. Let $\boldsymbol{\chi}_i \in N_\chi$, then

$$\boldsymbol{\chi}_i = \boldsymbol{\chi}_{i1} + \boldsymbol{\chi}_{i2}, \qquad \boldsymbol{\chi}_{i1} \in L_\varphi, \qquad \boldsymbol{\chi}_{i2} \perp L_\varphi, \tag{4.32}$$

and there must be n linearly independent vectors among $\boldsymbol{\chi}_{i1}$. Let them be the first n vectors $\{\boldsymbol{\chi}_1,\boldsymbol{\chi}_2,\ldots,\boldsymbol{\chi}_n\}$. We introduce a square matrix **B** composed of first n rows of the **H** matrix,

$$\mathbf{B} = \|[(\boldsymbol{\chi}_{i1}, \boldsymbol{\varphi}_j)]\| = \|[(\boldsymbol{\chi}_i, \boldsymbol{\varphi}_j)]\| \quad (i = 1,\ldots, n;\ j = 1,\ldots, n).$$

This matrix is non-degenerate, $\det \mathbf{B} \neq 0$, because otherwise its rows would be linearly dependent,

$$\sum_{i=1}^{n} c_i B_{ij} = 0, \quad c_1^2 + c_2^2 + \ldots + c_n^2 \neq 0, \qquad (j = 1,\ldots, n),$$

or

$$\sum_{i=1}^{n} c_i (\boldsymbol{\chi}_{i1}, \boldsymbol{\varphi}_j) = 0, \qquad (j = 1,\ldots, n).$$

This means vector $\boldsymbol{\chi}$

$$\boldsymbol{\chi} = \sum_{i=1}^{n} c_i \boldsymbol{\chi}_{i1}$$

is orthogonal to all $\boldsymbol{\varphi}_j$ and at the same time it belongs to L_φ, i.e. $\boldsymbol{\chi} = \mathbf{0}$, which contradicts to the linear independence of vectors $\boldsymbol{\chi}_{i1}$ ($i = 1,\ldots, n$).

As the rows of matrix **B** coincide with the first n rows of matrix **H**, the condition det $\mathbf{B} \neq 0$ implies that rank $\mathbf{H} = n$.

Necessity. Obviously,

$$\operatorname{rank} \mathbf{H} = n \quad \Rightarrow \quad m \geq n.$$

By representing vector χ_i as an expansion (4.32), we can make sure that vectors χ_{i1} $(i = 1,\ldots, m)$ include n linearly independent ones because otherwise there would not be n linearly independent rows in matrix **H**. The linear hull of the n linearly indepdendent vectors χ_{i1} coincides with L_φ. This completes the proof of the theorem.

Further we will asume that the approximations we use meet the conditions of the above theorem.

Now let us carry out a comparison between the Lagrangian and Reissnerian eigenvalue problems. We are going to prove the following simple but important qualitative theorem.

An L-*ellipsoid is fully contained within the respective* R-*ellipsoid.*

Indeed, the surfaces of the L-ellipsoid and R-ellipsoid are swept by the respective radius vectors

$$\frac{\boldsymbol{Y}}{\|\boldsymbol{Y}\|} \mathrm{r}_{\mathrm{L}}^{-1/2}, \qquad \frac{\boldsymbol{Y}}{\|\boldsymbol{Y}\|} \mathrm{r}_{\mathrm{R}}^{-1/2}.$$

Therefore the condition of nesting of the ellipsoids is equivalent to the inequality

$$\mathrm{r}_{\mathrm{R}} \leq \mathrm{r}_{\mathrm{L}}. \tag{4.33}$$

For any two vectors of stresses $\boldsymbol{\sigma}$ and strains $\boldsymbol{\varepsilon} = \boldsymbol{A}\mathbf{u}$ which have a finite energy, the following inequality holds:

$$(\boldsymbol{\sigma}, \boldsymbol{\varepsilon}) \leq (\boldsymbol{C}^{-1}\boldsymbol{\sigma}, \boldsymbol{\sigma})(\boldsymbol{C}\boldsymbol{\varepsilon}, \boldsymbol{\varepsilon}), \tag{4.34}$$

which can be seen to follow from the Cauchy–Buniakovsky inequality if we notice the positive definiteness of the algebraic operator $\boldsymbol{C}$.

By using the approximations of (4.21), we transform inequality (4.34) into

$$(\boldsymbol{X}^{\mathsf{T}}\mathbf{H}\boldsymbol{Z})^2 \leq (\boldsymbol{X}^{\mathsf{T}}\mathbf{G}\boldsymbol{X})(\boldsymbol{Z}^{\mathsf{T}}\mathbf{R}\boldsymbol{Z}) \tag{4.35}$$

which holds for any m-dimensional vector $\boldsymbol{X}$ and any n-dimensional vector $\boldsymbol{Z}$. In particular, (4.35) is not violated if we assume

$$\boldsymbol{X} = \mathbf{G}^{-1}\mathbf{H}\boldsymbol{Z}.$$

The result is

$$(\boldsymbol{Z}^{\mathsf{T}}\mathbf{H}^{\mathsf{T}}\mathbf{G}^{-1}\mathbf{H}\boldsymbol{Z})^2 \le (\boldsymbol{Z}^{\mathsf{T}}\mathbf{H}^{\mathsf{T}}\mathbf{G}^{-1}\mathbf{H}\boldsymbol{Z})(\boldsymbol{Z}^{\mathsf{T}}\mathbf{R}\boldsymbol{Z}). \qquad (4.36)$$

Matrix $\mathbf{G}^{-1}$ is positive definite, therefore

$$\boldsymbol{Z}^{\mathsf{T}}\mathbf{H}^{\mathsf{T}}\mathbf{G}^{-1}\mathbf{H}\boldsymbol{Z} \ge 0$$

and it follows from (4.36) that

$$\boldsymbol{Z}^{\mathsf{T}}\mathbf{H}^{\mathsf{T}}\mathbf{G}^{-1}\mathbf{H}\boldsymbol{Z} \le \boldsymbol{Z}^{\mathsf{T}}\mathbf{R}\boldsymbol{Z} \qquad (4.37)$$

for any n-dimensional vector $\boldsymbol{Z}$. The inequality (4.37) is equivalent to (4.33), which proves the nesting theorem.

10.4.2 The method of two functionals in the spectral problem

A scheme of computing which we discussed in Section **9**.3 and called the *method of two functionals* is applicable to the spectral problems, too [24].

We begin with formula (4.17) of the Rayleigh ratio that conforms to the Lagrangian functional. Obviously, if the $\boldsymbol{Z}$ vector in that formula is replaced by i-th eigenvector $\boldsymbol{Z}_{\mathsf{L}\,i}$ of problem (4.12), we will have i-th Lagrangian eigenvalue. Or,

$$\lambda_{\mathsf{L}\,i}^{(n)} = \frac{\boldsymbol{Z}_{\mathsf{L}\,i}^{\mathsf{T}}\mathbf{R}\boldsymbol{Z}_{\mathsf{L}\,i}}{\boldsymbol{Z}_{\mathsf{L}\,i}^{\mathsf{T}}\mathbf{M}\boldsymbol{Z}_{\mathsf{L}\,i}}. \qquad (4.38)$$

The numerator is the doubled potential strain energy $2\mathsf{E}$ accumulated by a discrete n-dimensional elastic system as it moves in i-th mode of oscillations.

According to the general procedure of the method of two functionals, knowing the displacements (at least approximately) lets us pose a variational problem of finding the stresses from the condition of minimum of functional D. Based on formula (**9**.3.6) and the homogeneous kinematical boundary conditions, we can write

$$\mathsf{D} = \tfrac{1}{2}\,(\boldsymbol{C}^{-1}\boldsymbol{\sigma},\boldsymbol{\sigma}) - (\boldsymbol{A}^{\mathsf{T}}\boldsymbol{\sigma},\,\mathbf{u})\,. \qquad (4.39)$$

Let us insert the approximations (4.21) in here taking i-th eigenvector $\boldsymbol{Z}_{\mathsf{L}\,i}$ of problem (4.12) as vector $\boldsymbol{Z}$. This gives

$$\mathsf{D} = \tfrac{1}{2}\,\boldsymbol{X}^{\mathsf{T}}\mathbf{G}\boldsymbol{X} - \boldsymbol{X}^{\mathsf{T}}\mathbf{H}\boldsymbol{Z}_{\mathsf{L}\,i}\,. \qquad (4.40)$$

Minimization of the quadratic form (4.40) yields a stress vector $\boldsymbol{X}_{\mathsf{D}\,i}$ that corresponds to the eigenvector $\boldsymbol{Z}_{\mathsf{L}\,i}$:

$$X_{\mathsf{D}\,i} = \mathbf{G}^{-1}\mathbf{H}Z_{\mathsf{L}\,i}\,. \tag{4.41}$$

The doubled potential energy 2E can be expressed now via the "eigen" stresses $X_{\mathsf{D}\,i}$ by assuming

$$2\mathsf{E} = X_{\mathsf{D}i}^{\mathsf{T}}\mathbf{G}X_{\mathsf{D}i} = Z_{\mathsf{L}i}^{\mathsf{T}}\mathbf{H}^{\mathsf{T}}\mathbf{G}^{-1}\mathbf{H}Z_{\mathsf{L}\,i}\,. \tag{4.42}$$

After we replace the numerator in (4.38) by the energy expression from (4.42), we will get a formula of $\lambda_{\mathsf{D}\,i}$ – the square of the natural oscillation frequency of the system for i-th mode,

$$\lambda_{\mathsf{D}\,i}^{(n)} = \frac{Z_{\mathsf{L}i}^{\mathsf{T}}\mathbf{H}^{\mathsf{T}}\mathbf{G}^{-1}\mathbf{H}Z_{\mathsf{L}\,i}}{Z_{\mathsf{L}i}^{\mathsf{T}}\mathbf{M}Z_{\mathsf{L}\,i}}\,. \tag{4.43}$$

We already know that the spectrum of the Lagrangian frequencies is shifted on the numerical axis to the right with respect to the spectrum of true frequencies of the system, therefore refining the frequencies by using formula (4.43) instead of (4.38) makes sense only under the conditions

$$\lambda_{\mathsf{D}\,i}^{(n)} \le \lambda_{\mathsf{L}\,i}^{(n)} \qquad (i = 1,\ldots, n). \tag{4.44}$$

Seeing that we use the Rayleigh ratio (4.27) for the Reissner functional, the Reissnerian eigenvalues $\lambda_{\mathsf{R}\,i}^{(n,m)}$ are expressed as follows (provided that $\boldsymbol{K} = \boldsymbol{O}$):

$$\lambda_{\mathsf{R}\,i}^{(n,m)} = \frac{Z_{\mathsf{R}i}^{\mathsf{T}}\mathbf{H}^{\mathsf{T}}\mathbf{G}^{-1}\mathbf{H}Z_{\mathsf{R}\,i}}{Z_{\mathsf{R}i}^{\mathsf{T}}\mathbf{M}Z_{\mathsf{R}\,i}} = \frac{Y_{\mathsf{R}i}^{\mathsf{T}}\mathbf{W}^{-\mathsf{T}}\mathbf{H}^{\mathsf{T}}\mathbf{G}^{-1}\mathbf{H}\mathbf{W}Y_{\mathsf{R}\,i}}{Y_{\mathsf{R}i}^{\mathsf{T}}Y_{\mathsf{R}\,i}} \tag{4.45}$$

where

$$Y_{\mathsf{R}\,i} = \mathbf{W}Z_{\mathsf{R}\,i}\,.$$

The $Y_{\mathsf{R}\,i}$ is codirectional with i-th principal axis of the R-ellipsoid. Comparing (4.45) with (4.43) leads to a conclusion that

> *Values* $1/\sqrt{\lambda_{\mathsf{D}\,i}^{(n)}}$ *are equal to the lengths of segments cut off on the semi-axes of the* L*-ellipsoid by the surface of the* R*-ellipsoid.*

A geometry of the relationships between the Lagrangian and Reissnerian Rayleigh ellipsoids, where also the properties of the eigenvalues in the method of two functionals are depicted, is shown in Fig. 10.3 for the two-dimensional case $n = 2$.

The semi-axes of two ellipsoids shown in Fig. 10.3 are related to the eigenvalues as follows:

$$OA = \frac{1}{\sqrt{\lambda_{\mathsf{L}1}}}, \qquad OC = \frac{1}{\sqrt{\lambda_{\mathsf{R}1}}}, \qquad OE = \frac{1}{\sqrt{\lambda_{\mathsf{D}1}}},$$

$$OB = \frac{1}{\sqrt{\lambda_{\mathsf{L}2}}}, \qquad OD = \frac{1}{\sqrt{\lambda_{\mathsf{R}2}}}, \qquad OF = \frac{1}{\sqrt{\lambda_{\mathsf{D}2}}}. \tag{4.46}$$

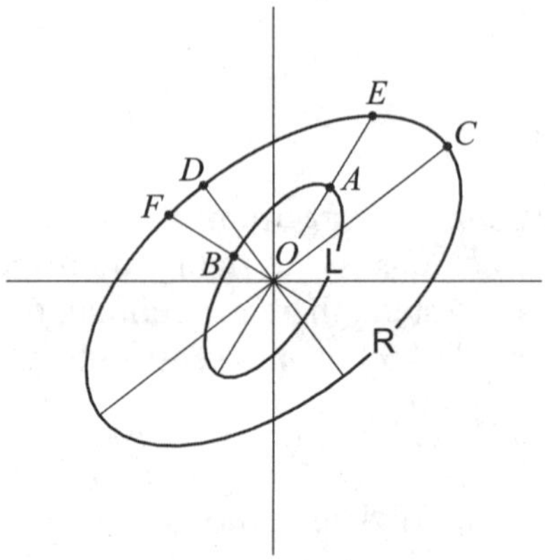

Fig. 10.3. A Rayleigh L-ellipsoid and a Rayleigh R-ellipsoid

The theorem of nesting the L-ellipsoid in the R-ellipsoid implies the estimates

$$\lambda_{\mathsf{R}1} \le \lambda_{\mathsf{D}1} \le \lambda_{\mathsf{L}1}, \quad \lambda_{\mathsf{D}i} \le \lambda_{\mathsf{L}i}, \quad \lambda_{\mathsf{R}i} \le \lambda_{\mathsf{L}i} \ (i = 1, \ldots, n),$$

$$\lambda_{\mathsf{D}n} \le \lambda_{\mathsf{R}n} \le \lambda_{\mathsf{L}n}. \tag{4.47}$$

Example 1.

Consider a problem of longitudinal oscillations of a bar. Let l be the length of the bar, EA its longitudinal stiffness, ρ its mass per unit of length, u a displacement along the bar's axis, N a longitudinal stress. The problem of free longitudinal oscillations of the bar is described by the differential equation

$$(EAu')' + \omega^2 \rho u = 0 \tag{4.48}$$

where each end of the bar has one of two possible boundary conditions specified on it: either $u = 0$ or $N = 0$, so that $N = EAu'$.

The functionals that correspond to this problem are as follows:

$$\mathsf{L}(u) = \frac{1}{2}\int_0^l \left[EA\left(\frac{du}{dx}\right)^2 - \omega^2 \rho A u^2 \right] dx,$$

$$\mathsf{D}(N) = \frac{1}{2}\int_0^l \left(\frac{N^2}{EA} + 2u\frac{dN}{dx} \right) dx,$$

$$\mathsf{R}(u, N) = \frac{1}{2}\int_0^l \left[\frac{N^2}{EA} + 2u\frac{dN}{dx} + \omega^2 \rho A u^2 \right] dx. \tag{4.49}$$

Assuming

$$u = \sum_{i=1}^{n} Z_i \varphi_i \in \mathscr{U}_{\text{ko}}, \qquad N = \sum_{i=1}^{m} X_i \psi_i \in \mathscr{U}_{\text{so/2}}, \tag{4.50}$$

we obtain the following expressions of the components of matrices **R**, **M**, **G**, **H**:

$$\mathsf{R}_{ij} = \int_0^l EA\varphi_i'\varphi_j' dx, \quad \mathrm{M}_{ij} = \int_0^l \rho\varphi_i\varphi_j dx,$$

$$\mathrm{G}_{ij} = \int_0^l \frac{\psi_i \psi_j}{EA} dx, \quad \mathrm{H}_{ij} = -\int_0^l \psi_i'\varphi_j dx. \tag{4.51}$$

Let us take as an example the Weinstein–Ficker problem [16] where the eigenvalues of the following equation should be determined:

$$u'' + \lambda(1 + \sin x)u = 0, \quad u(0) = 0, \quad u(\pi) = 0 \tag{4.52}$$

for eigenfunctions symmetric with respect to the point $x = \pi/2$, i.e. $u(\pi/2 + x) = u(\pi/2 - x)$.

We divide the integration interval $[0, \pi/2]$ into n equal pieces and, assuming $m = n$ in (4.50), use piecewise linear "lid" functions as coordinate ones:

$$\varphi_i(x) = \begin{cases} x/h + 1 - i, & (i-1)h \le x \le ih \\ -x/h + 1 + i, & ih \le x \le (i+1)h \\ 0 & \textit{otherwise} \end{cases}, \qquad \psi_i(x) = \begin{cases} x/h + 2 - i, & (i-2)h \le x \le (i-1)h \\ -x/h + i, & (i-1)h \le x \le ih \\ 0 & \textit{otherwise} \end{cases}.$$

Here $h = \pi/(2n)$ is the mesh's spacing, $i = 1, \ldots, n$.

Omitting the details, we present the first seven calculated eigenvalues of the Weinstein–Ficker problem at $n = 15$ – see Table 10.1.

Table 10.1

i	λ_i	$\lambda_{\mathrm{L}\,i}$	$\lambda_{\mathrm{R}\,i}$	$\lambda_{\mathrm{D}\,i}$
1	0.54031884	0.54081742	0,54031807	0.54031811
2	5.4486361	5.4945998	4.7965807	5.4480117
3	15.312608	15.674429	5.447986	15.298727
4	30.114984	31.520968	15.298160	30.006716
5	49.853254	53.734343	30.001928	49.343253
6	74.526762	83.267328	43.004475	72.740792
7	104.13529	121.30172	49.317578	98.991497

The error of calculation of the first three eigenvalues vs. the number n of the involved coordinate functions φ_i and ψ_i is shown in Table 10.2.

Table 10.2

n	$e_{\mathrm{L}1}$	$e_{\mathrm{L}2}$	$e_{\mathrm{L}3}$	$e_{\mathrm{R}1}$	$e_{\mathrm{R}2}$	$e_{\mathrm{R}3}$	$e_{\mathrm{D}1}$	$e_{\mathrm{D}2}$	$e_{\mathrm{D}3}$
2	5.273	33.23		0.558	54.66		0.464	54.67	
3	2.324	21.80	31.81	0.097	36.69	66.28	0.089	9.136	76.09
4	1.303	12.21	33.05	0.030	26.19	65.45	0.028	2.616	24.97
5	0.833	7.748	21.91	0.012	21.01	64.80	0.011	1.016	9.241
6	0.578	5.347	15.19	0.006	18.03	64.60	0.005	0.475	4.149
7	0.424	3.912	11.10	0.003	16.17	64.51	0.003	0.252	2.136
8	0.325	2.987	8.448	0.002	14.94	64.47	0.002	0.146	1.212
9	0.256	2.355	6.646	0.001	14.08	64.46	0.001	0.090	0.740
10	0.208	1.904	5.365	0.001	13.46	64.44	0.001	0.059	0.477
11	0.172	1.572	4.422	-	13.00	64.43	-	0.040	0.322
12	0.144	1.320	3.707	-	12.64	64.43	-	0.028	0.225
13	0.123	1.124	3.154	-	12.37	64.42	-	0.020	0.162
14	0.106	0.969	2.716	-	12.15	64.42	-	0.015	0.120
15	0.092	0.843	2.363	-	11.97	64.42	-	0.011	0.091

The following designations are used in the tables:
λ_i is i-th eigenvalue calculated as a half-sum of its upper and lower estimates; the latter are taken from results by A. Weinstein [16],

$$e_{\mathrm{L}\,i} = 100(\lambda_i - \lambda_{\mathrm{L}\,i})/\lambda_i\,, \qquad e_{\mathrm{R}\,i} = 100(\lambda_i - \lambda_{\mathrm{R}\,i})/\lambda_i\,, \quad e_{\mathrm{D}\,i} = 100(\lambda_i - \lambda_{\mathrm{D}\,i})/\lambda_i\,.$$

Note, by the way, that all calculated errors are indicated in Table 10.2 with the positive sign. Actually, all the calculated eigenvalues $\lambda_{\mathrm{L}\,i}$ are, as they should be, a bit higher than the true values of λ_i. At the same time, all the eigenvalues $\lambda_{\mathrm{R}\,i}$ and $\lambda_{\mathrm{D}\,i}$ are a little lower than the true values.

Note that the error in the second and third Reissnerian eigenvalues is very large. An explanation of this will be given later in Section **10**.4.3.

Example 2.

Our second example will be a problem that deals with axisymmetric oscillations of a round membrane of radius l, described by the same equation (4.48). We should assume

$$EA(x) = x,\ \ \lambda = \omega^2/S,\ \ \rho(x) = x\eta(x),\ \ u(l) = 0,\ \ u'(0) = 0,$$

where x is a radial coordinate, $\eta(x)$ is a mass per unit of area of the membrane, S is a tension in the membrane.

We use the coordinate functions

$$\varphi_i(x) = \sin\frac{i\pi x}{l} - \frac{i\pi x}{l} + i\pi\,, \qquad \psi_i(x) = \frac{i\pi}{l}\left(\cos\frac{i\pi x}{l} - 1\right)$$

and assume $m = n$. In our case

$$\mathrm{R}_{ij} = 16ij\int_0^{\pi/2} \xi\sin^2 i\xi\sin^2 j\xi\, d\xi\,,$$

$$\mathrm{M}_{ij} = \left(\frac{l}{\pi}\right)^2 \int_0^{\pi}(\sin i\xi - i\xi + i\pi)(\sin j\xi - j\xi + j\pi)\xi\eta\left(\frac{l\xi}{\pi}\right)d\xi\,,$$

$$\mathrm{G}_{ij} = 4ij\left(\frac{\pi}{l}\right)^2 \int_0^{\pi/2}\frac{\sin^2 i\xi\sin^2 j\xi}{\xi}d\xi,\ \ \mathrm{H}_{ij} = \frac{i^2\pi}{l}\int_0^{\pi}\sin i\xi(\sin j\xi - j\xi + j\pi)d\xi\,.$$

The first and second frequencies calculated by the method of two functionals for the round membrane with its inertial characteristic following a linear law of distribution,

$$\eta(x) = \frac{\eta_2 - \eta_1}{l}x + \eta_1\,,$$

are shown in Table 10.3. Note that to achieve the same accuracy of calculation with the same type of the coordinate functions in [23] required keeping 20 to 40 functions in the expansion of the displacements; this is about 10 times bigger than the number required by the method of two functionals.

Table 10.3

	$\omega_1 l\sqrt{\eta_1 / S}$			$\omega_2 l\sqrt{\eta_1 / S}$		
n	η_2/η_1			η_2/η_1		
	0.5	1.0	2.0	0.5	1.0	2.0
2	2.705	2.405	2.011	6.373	5.342	4.308
3				6.363	5.522	4.558
4				6.366	5.521	4.553
5				6.362	5.520	4.556
exact solution	2.405			5.520		

10.4.3 A remark on an effect which arises when the finite element method in its mixed form is applied to the spectral problems

Algorithms of the finite element method have become widely popular not only in the static analysis of elastic systems but also in the spectral problems. However, a formal application of the algorithms without a proper validation thereof can be unpredictably dangerous and can lead to pitfalls that are hard to detect.

There is a simple example for a one-dimensional problem in [11] where closed solutions of the mixed-form FEM equations are known and where, much to an analyst's chagrin, the spectrum of the frequencies can be distorted qualitatively by the approximate solution. In particular, we find that the FEM scheme for longitudinal oscillations of a bar described by a second-order differential equation, which is based on the stationarity conditions for the Reissner functional and an independent approximation of the displacements and the stresses, produces double series of frequencies which converge to second, fifth, eighth etc. natural frequencies.

A modification of the algorithm based on a specific choice of the coordinate functions (on a shifted mesh [12]) does not produce the extra series of the frequencies[20].

The known exact solutions of the FEM equations can be used as benchmarks for developing new software; they are also useful for

[20] The paper [12] uses one mesh for approximating static parameters and another for approximating kinematical values. Those two meshes are shifted with respect to each other. The coordinate functions in [12] are piecewise constant. At the same time, [11] uses only the idea of shifting the two meshes but keeps a continuous (piecewise linear) approximation.

validating theoretical estimates of the meshing error introduced by the finite elements.

So, let us consider a problem of longitudinal oscillations of a bar one end of which is fixed and the other free. Let ρ be a density of the bar's material; E be an elasticity modulus; A be an area of cross-section; l be a length of the bar.

The Lagrangian and Reissnerian functionals for the problem of free longitudinal oscillations of the bar are as shown in (4.49).

The stationarity conditions for the Lagrangian functional, L, on functions that satisfy the boundary condition $u(0) = 0$ and the stationarity conditions for the Reissner functional, R, on functions that satisfy the boundary conditions $u(0) = 0$ and $N(l) = 0$ produce the following equivalent Sturm–Liouville problems:

$$\left.\begin{array}{l} \dfrac{d}{dx}\left(EA\dfrac{du}{dx}\right)+\omega^2\rho A u=0 \\ u(0)=0,\ EA\dfrac{du}{dx}(l)=0 \end{array}\right\} \tag{4.53}$$

and

$$\left.\begin{array}{l} \dfrac{N}{EA}-\dfrac{du}{dx}=0,\ \dfrac{dN}{dx}+\omega^2\rho A u=0 \\ u(0)=0,\ EA\dfrac{du}{dx}(l)=0 \end{array}\right\}. \tag{4.54}$$

For the constant coefficients E =*Const*, A =*Const*, ρ =*Const*, the exact solution for the frequencies and the modes is known and looks like

$$\omega_i=\frac{(2i-1)\pi c}{2l},\quad u_i=A_u\sin\frac{\omega_i x}{c},\quad N_i=A_N\cos\frac{\omega_i x}{c} \tag{4.55}$$

where

$$c^2 = E/\rho .$$

We divide the integration interval $[0,l]$ into n equal parts by assuming

$$h = l/n$$

and switch to dimensionless variables as follows:

$$\xi = {}^{x}\!/_{h}, \quad v = {}^{u}\!/_{h}, \quad S = {}^{N}\!/_{EA}, \quad \lambda = \left({}^{\omega h}\!/_{c}\right)^2.$$

The Lagrangian and Reissnerian functionals in these designations become

$$\frac{\mathsf{L}}{EAh} = \frac{1}{2}\int_0^n \left(v'^2 - \lambda v^2\right) d\xi, \quad \frac{\mathsf{R}}{EAh} = \frac{1}{2}\int_0^n \left(S^2 + 2vS' + \lambda v^2\right) d\xi. \tag{4.56}$$

Here and further a stroke denotes the differentiation with respect to the dimensionless coordinate ξ.

We take piecewise linear "lid" functions $\varphi_i(\xi) = \psi_i(\xi)$ $(i = 1,\ldots, n)$ as the FEM coordinate functions

$$\varphi_i(\xi) = \psi_i(\xi) = \begin{cases} \xi + 1 - i, & \xi \in [i-1, i] \\ -\xi + 1 + i, & \xi \in [i, i+1], \\ 0 & otherwise \end{cases}$$

and use the following approximations:

$$v = \sum_{i=1}^{n} Z_i \varphi_i(\xi), \qquad N = \sum_{i=0}^{n-1} X_i \psi_i(\xi). \tag{4.57}$$

Substituting the approximations (4.57) in (4.56) turns the Lagrangian and Reissnerian functionals into the quadratic forms (4.10) and (4.23) with respect to variables Z_i and X_j $(i = 1,\ldots, n; j = 0,\ldots, n-1)$. The square matrices **R**, **M**, **G** and **H** of the order n are banded three-diagonal matrices:

$$\mathbf{R} = \begin{bmatrix} 2 & -1 & & & \\ -1 & 2 & -1 & & \\ & \ddots & \ddots & \ddots & \\ & & \ddots & \ddots & \ddots \\ & & -1 & 2 & -1 \\ & & & -1 & 1 \end{bmatrix}, \quad \mathbf{M} = \frac{1}{6}\begin{bmatrix} 4 & 1 & & & \\ 1 & 4 & 1 & & \\ & \ddots & \ddots & \ddots & \\ & & \ddots & \ddots & \ddots \\ & & 1 & 4 & 1 \\ & & & 1 & 2 \end{bmatrix},$$

$$\mathbf{G} = \frac{1}{6}\begin{bmatrix} 2 & 1 & & & \\ 1 & 4 & 1 & & \\ & \ddots & \ddots & \ddots & \\ & & \ddots & \ddots & \ddots \\ & & 1 & 4 & 1 \\ & & & 1 & 4 \end{bmatrix}, \quad \mathbf{H} = \frac{1}{2}\begin{bmatrix} 1 & & & & \\ 0 & 1 & & & \\ -1 & 0 & 1 & & \\ & \ddots & \ddots & \ddots & \\ & & -1 & 0 & 1 \\ & & & -1 & 0 & 1 \end{bmatrix}.$$

Treating the system of equations (4.12) that follows from the stationarity conditions for the Lagrangian functional as simultaneous equations in finite differences [2], we seek for its solution in the form

$$Z_\alpha = Z\sin(\alpha\nu)\,. \tag{4.58}$$

Putting (4.58) in (4.12) for equations No. $\alpha = 2,\ldots,n-1$ gives

$$\left[(1-\cos\nu)-\lambda\frac{2+\cos\nu}{6}\right]Z\sin\alpha\nu = 0,$$

wherefrom the condition of existence of a non-trivial solution yields

$$\lambda = 6\frac{1-\cos\nu}{2+\cos\nu}\,. \tag{4.59}$$

A direct check shows that the first of equations (4.12) is satisfied by any value of parameter ν. Substituting (4.58) in the last equation of the system (4.12) gives, taking (4.59) into account,

$$\cos n\nu = 0,$$

wherefrom

$$\nu_i = \frac{(2i-1)\pi}{2n}\,. \tag{4.60}$$

At $i = 1,\ldots, n$ we have different solutions for the natural frequencies and modes of oscillation. At $i = n+1,\ n+2,\ldots$ the solutions repeat those already found. Thus, we have find all n frequencies of oscillation of the discrete system with n degrees of freedom which will be Lagrangian frequencies $\omega_{\mathrm{L}\,i}$ so that

$$\omega_{\mathrm{L}\,i} = \frac{\omega_i}{\nu_i}\sqrt{\frac{6(1-\cos\nu_i)}{2+\cos\nu_i}}\,. \tag{4.61}$$

The ratio

$$\frac{\omega_{\mathrm{L}\,i}}{\omega_i} = \frac{\sqrt{\lambda_i}}{\nu_i}$$

characterizes the accuracy of the FEM computational algorithm. At a fixed number i and an increasing number of DOFs n we have $\nu_i \to 0$ and

$$\frac{\omega_{\mathrm{L}\,i}}{\omega_i} = 1+\frac{\nu_i^2}{24}+O(\nu_i^4)\,, \tag{4.62}$$

so that

$$\lim_{n\to\infty}\frac{\omega_{\mathrm{L}\,i}}{\omega_i}=1$$

and, expectedly, $\omega_{\mathrm{L}\,i} > \omega_i$. Note that $\omega_{\mathrm{L}\,1} < \omega_{\mathrm{L}\,2} <\ldots< \omega_{\mathrm{L}\,n}$.

Now let us determine the Reissnerian natural frequencies. The system of equations (4.24) that follows from the sationarity conditions for the Reissner functional can be treated in our case as a set of simultaneous equations in finite differences. We will seek a solution of this system in the form

$$Z_\alpha = Z\sin(\alpha\nu)\,,\quad X_{\alpha-1} = X\cos(\alpha-1)\nu\ ,\ (\alpha = 1,\ldots, n). \tag{4.63}$$

Putting (4.63) in (4.24) gives

$$[-X(2+\cos\nu)+3Z\sin\nu]\cos(\alpha-1)\nu = 0,$$

$$[3X\sin\nu-\lambda\,Z(2+\cos\nu)]\sin\alpha\nu = 0.$$

The condition of existence of a nontrivial solution for this system gives a relation between parameters λ and ν:

$$\lambda=\left(3\frac{\sin\nu}{2+\cos\nu}\right)^2. \tag{4.64}$$

Analysis of the rest of the equations in (4.24) shows that the ν value is defined by the condition $\cos n\nu = 0$ and thus ν_i is defined by (4.60). At $i = 1,\ldots, n$ we have various solutions for the natural frequencies and modes. At $i = n + 1,\ n + 2,\ldots$ the solutions repeat those we have already found. Thus, for the Reissnerian frequencies we have

$$\omega_{\mathrm{R}\,i}=\frac{\omega_i}{\nu_i}\frac{3\sin\nu_i}{2+\cos\nu_i}. \tag{4.65}$$

At a fixed number i and an increasing number of finite elements n, the asymptotic estimate $\nu_i \to 0$ holds; also,

$$\frac{\omega_{\mathrm{R}\,i}}{\omega_i}=1-\frac{\nu_i^4}{180}+O(\nu_i^6)\,, \tag{4.66}$$

which is much better (in terms of the rate of convergence) than the estimate (4.62) for the Lagrangian frequencies. Then the following is obvious for the Reissnerian frequencies:

$$\lim_{n\to\infty} \frac{\omega_{Ri}}{\omega_i} = 1,$$

however, unlike the Lagrangian frequencies, the Reissnerian frequencies are not put in the ascending order according to the standard numbering. For example, at $n = 4$ the calculation by the above formulas gives

$$\omega_{R1} = \frac{0.999\pi c}{2l} < \omega_{R4} = \frac{2.716\pi c}{2l} < \omega_{R2} = \frac{2.962\pi c}{2l} < \omega_{R3} = \frac{4.364\pi c}{2l}. \quad (4.67)$$

If we fix an integer number k ($k = 1, 2, \ldots$), we will find easily that

$$\lim_{n\to\infty} \frac{\omega_{R(n-k+1)}}{\omega_{3k-1}} = \lim_{n\to\infty} \frac{3\sin\nu_{n-k+1}}{\nu_{3k-1}(2+\cos\nu_{n-k+1})} = \lim_{n\to\infty} \frac{3\sin\nu_k}{\nu_{3k-1}(2-\cos\nu_k)} = 1.$$

Thus, the Reissnerian frequencies include additional series which converge to the accurate values of the frequencies with the numbers $3k - 1$.

This circumstance can be a serious difficulty for the use of the Reissner functional in the Ritz method when we want to find natural frequencies of an elastic system. Indeed, the only reasonable criterion for establishing a sequence of the frequencies in a discrete system during a numerical solution is based on putting all the frequencies in the ascending order. As we have just shown, there is a chance that "extra" or *spurious* frequencies stalk into the spectrum and distort it dramatically. Note that in the problem we are dealing with, the modes of oscillation that conform to the spurious frequencies feature many sign alterations, which is a usual indication of high-frequency oscillations.

To overcome the said difficulties, we need to develop a criterion that allows to filter the spurious frequencies out of the discrete system's spectrum, or we need to use consistent coordinate functions for approximating force and kinematic parameters which would give no chance to the spurious frequencies to get into the spectrum of the system.

In particular, in our problem the consistent approximation can be achieved by using coordinate functions on a so-called *shifted mesh*. The shifted meshes were suggested in [12]; the essence is that functions $\varphi_i(x)$ are based on one mesh and functions $\psi_i(x)$ on another, and the nodes of the two meshes are shifted with respect to one another. We are not going to dwell on an analysis of consistency of the shifted-mesh approximations; we just refer to [11] where this analysis is already done. Here we are going to show that one of effective methods for filtering the spurious frequencies out is the method of two functionals.

It turns out that in our problem the frequencies of the method of two functionals can be calculated in the closed form. First of all, we determine the Lagrangian form of the discrete system's natural oscillations. We take formula (4.58) and write out the Lagrangian eigenvector, $\boldsymbol{Z}_{\mathrm{L}i}$. Thus,

$$\boldsymbol{Z}_{\mathrm{L}i} = \|[\sin \nu_i, \sin 2\nu_i, \ldots, \sin n\nu_i]\|^{\mathrm{T}},$$

where parameter ν_i is defined by formula (4.60).

Calculating the matrix product $\mathbf{M}\boldsymbol{Z}_{\mathrm{L}i}$ gives

$$\mathbf{M}\boldsymbol{Z}_{\mathrm{L}i} = \frac{2+\cos\nu_i}{6} \left\| \begin{bmatrix} 2\sin\nu_i \\ \vdots \\ 2\sin(n-1)\nu_i \\ \sin n\nu_i \end{bmatrix} \right\|.$$

Consequently,

$$\boldsymbol{Z}_{\mathrm{L}i}^{\mathrm{T}}\mathbf{M}\boldsymbol{Z}_{\mathrm{L}i} = \frac{2+\cos\nu_i}{6}\left(2\sum_{k=1}^{n}\sin^2 k\nu_i - \sin^2 n\nu_i\right).$$

The sum of squares of sines that participates in the above formula can be reduced to (see (1.351.1) in [10])

$$\sum_{k=1}^{n}\sin^2 k\nu_i = \frac{n}{2} - \frac{\cos(n+1)\nu_i \sin n\nu_i}{2\sin\nu_i}.$$

Seeing that $\cos n\nu_i = 0$, we can expand $\cos(n+1)\nu_i$ and get

$$\sum_{k=1}^{n}\sin^2 k\nu_i = \frac{n}{2} + \frac{\sin^2 n\nu_i}{2}.$$

Therefore

$$\boldsymbol{Z}_{\mathrm{L}i}^{\mathrm{T}}\mathbf{M}\boldsymbol{Z}_{\mathrm{L}i} = n\frac{2+\cos\nu_i}{6}.$$

Now let us establish an expression of the numerator in the right part of the general formula (4.43) for the eigenvalues of the method of two functionals. By calculating the product $\mathbf{H}\boldsymbol{Z}_{\mathrm{L}i}$ and denoting it by $\mathbf{V}$ for the sake of brevity, we find

$$\mathbf{V} = \mathbf{H}\boldsymbol{Z}_{\mathrm{L}i} = \sin\nu_i \,\|[\tfrac{1}{2}, \cos\nu_i, \ldots, \cos(n-1)\nu_i]\|^{\mathrm{T}}.$$

In our designations, the numerator that we seek is

$$\mathbf{V}^{\mathrm{T}}\mathbf{G}^{-1}\mathbf{V}.$$

The **G** matrix cannot be inverted in its general form; however, we can calculate the desired product without inverting **G**. We denote

$$\boldsymbol{Y} = \mathbf{G}^{-1}\mathbf{V},$$

hence the $\boldsymbol{Y}$ vector is a solution of the following set of simultaneous algebraic equations:

$$\begin{Vmatrix} 2 & 1 & & & \\ 1 & 4 & 1 & & \\ & \ddots & \ddots & \ddots & \\ & & 1 & 4 & 1 \\ & & & 1 & 4 \end{Vmatrix} \cdot \begin{Vmatrix} Y_1 \\ Y_2 \\ \vdots \\ Y_{n-1} \\ Y_n \end{Vmatrix} = 6\sin\nu_i \begin{Vmatrix} 0.5_1 \\ \cos\nu_i \\ \vdots \\ \cos(n-2)\nu_i \\ \cos(n-1)\nu_i \end{Vmatrix}.$$

A direct substitution confirms that the solution of the system is the vector $\boldsymbol{Y}$ with its components being

$$Y_k = \frac{3\sin\nu_i}{2+\cos\nu_i}\cos(k-1)\nu_i \qquad (k=1,\ldots,n).$$

Thus,

$$\mathbf{V}^{\mathrm{T}}\mathbf{G}^{-1}\mathbf{V} = \mathbf{V}^{\mathrm{T}}\boldsymbol{Y} = \frac{3\sin^2\nu_i}{2+\cos\nu_i}\left(\frac{1}{2}+\sum_{k=2}^{n}\cos^2(k-1)\nu_i\right).$$

Using formula (1.351.2) from [10] gives

$$\sum_{k=2}^{n}\cos^2(k-1)\nu_i = \sum_{j=1}^{n-1}\cos^2 j\nu_i = \frac{n-1}{2} + \frac{\cos n\nu_i \sin(n-1)\nu_i}{2\sin\nu_i} = \frac{n-1}{2}.$$

Consequently,

$$\mathbf{V}^{\mathrm{T}}\mathbf{G}^{-1}\mathbf{V} = \frac{n}{2}\frac{3\sin^2\nu_i}{2+\cos\nu_i}.$$

So, according to formula (4.43) we have

$$\lambda_{\mathrm{D}\,i} = \left(\frac{3\sin\nu_i}{2+\cos\nu_i}\right)^2. \tag{4.68}$$

When comparing formula (4.68) for $\lambda_{\mathrm{D}\,i}$ with formula (4.64) for $\lambda_{\mathrm{R}\,i}$, we can see a total coincidence between the frequencies of the method of two functionals and the Reissnerian frequencies. This coincidence is not accidental; it accords with results presented in Section **10**.4.2. Indeed, by comparing (4.58) and (4.63) we find that eigenvectors $\boldsymbol{Z}_{\mathrm{L}\,i}$ and $\boldsymbol{Z}_{\mathrm{R}\,i}$ are coincident. This means the L-ellipsoid and R-ellipsoid are coaxial. Hence, due to the theorem proved in Section 1**0**.4.2, we have

$$\lambda_{\mathrm{D}\,i} = \lambda_{\mathrm{R}\,i}\,.$$

Note, however, that in spite of the totally coincident eigenvalues $\lambda_{\mathrm{D}\,i}$ and $\lambda_{\mathrm{R}\,i}$ the method of two functionals has certain advantages in its algorithm and approach:

1) Starting from the stationarity conditions for the Reissnerian functional, we ultimately have to solve a generalized eigenvalue problem for matrices of the order $2n$ – see formula (4.23),

$$\left(\left\|\begin{bmatrix} \mathbf{G} & -\mathbf{H} \\ -\mathbf{H}^{\mathrm{T}} & \boldsymbol{O} \end{bmatrix}\right\| - \lambda \left\|\begin{bmatrix} \boldsymbol{O} & \boldsymbol{O} \\ \boldsymbol{O} & -\mathbf{M} \end{bmatrix}\right\|\right) \cdot \left\|\begin{bmatrix} \boldsymbol{X} \\ \boldsymbol{Z} \end{bmatrix}\right\| = \left\|\begin{bmatrix} \boldsymbol{O} \\ \boldsymbol{O} \end{bmatrix}\right\|,$$

which are not of fixed sign, by the way, and this creates additional difficulties for the computation. Of course, we could reduce the problem to a generalized problem of eigenvalues of positive definite matrices of the order n,

$$(\mathbf{H}^{\mathrm{T}}\mathbf{G}^{-1}\mathbf{H} - \lambda\mathbf{M})\boldsymbol{Z} = \mathbf{0},$$

but

a) this would require an additional computational burden for inverting matrix **G** and calculating the matrix product $\mathbf{H}^{\mathrm{T}}\mathbf{G}^{-1}\mathbf{H}$;

b) matrix $\mathbf{H}^{\mathrm{T}}\mathbf{G}^{-1}\mathbf{H}$ becomes fully filled, so one of the advantages of the finite element method is lost – a band-like shape (or, generally, a sparse population) of the stiffness matrix.

When following the method of two functionals, we have first to solve the generalized eigenvalue problem for the n-dimensional problem

$$(\mathbf{R} - \lambda\mathbf{M})\boldsymbol{Z} = \mathbf{0}$$

wherefrom we obtain $\lambda_{\mathrm{R}\,i}$, $\boldsymbol{Z}_{\mathrm{L}\,i}$ $(i = 1,\ldots, p \le n)$, and then to calculate $\lambda_{\mathrm{D}\,i}$ using (4.43). In practice we need to find some rather than all eigenvalues: p first ones of the discrete problem where $p \ll n$. Anyway, we can think of p as a number never exceeding $n/2$ [19], [26] because the higher eigenvalues of the discrete problem have no sense for us — they are very far from the respective eigenvalues of the counterpart continuous problem (due to both the meshing error and to an inaccuracy of determining higher eigenvalues of a finite-dimensional system). Therefore we can avoid inverting matrix **G** and confine ourselves to the solution of a system of equations with p right-hand parts,

$$\mathbf{G}\|[Y_1,\ldots,Y_p]\| = \mathbf{H}\|[Z_{\mathrm{L}\,1},\ldots,Z_{\mathrm{L}\,p}]\| .$$

To be fair, we should note a shortcoming: the method of two functionals dictates to search for the eigenvectors $Z_{\mathrm{L}\,i}$ of the discrete problem even in cases when the original problem does not demand that in any way.

2) However, the most important advantage of the algorithm for searching $\lambda_{\mathrm{D}\,i}$ in comparison to other algorithms that calculate $\lambda_{\mathrm{R}\,i}$ is that the method of two functionals permits to detect the presence of spurious frequencies and filter them out automatically. This useful feature of the method of two functionals is based on the fact that the $\lambda_{\mathrm{D}\,i}$ frequencies are not numbered in the ascending order of their numerical values; instead, their numbering has been determined at an earlier stage of the algorithm by the ascending order of the $\lambda_{\mathrm{L}\,i}$ values.

For example, if we treat this problem numerically rather than analytically, we will get (for $n = 8$) the following results presented in Table 10.4.

Table 10.4

	$i = 1$	$i = 2$	$i = 3$	$i = 4$
$\omega_{\mathrm{L}\,i}/\omega_i$	1.0016	1.00145	1.0405	1.0792
$\omega_{\mathrm{R}\,i}/\omega_i$	0.9999	0.9749	0.5996	0.7102
$\omega_{\mathrm{D}\,i}/\omega_i$	0.9999	0.9993	0.9942	0.9753
$\omega_i\, 2l/\pi c$	1.0	3.0	5.0	7.0

It is this circumstance (an intrusion to the spectrum of spurious natural frequencies) that explains the low accuracy of the calculated numbers $\lambda_{\mathrm{R}\,i}$ in the Weinstein–Ficker problem (see Tables 10.1 and 10.2) comparing to the accuracy of $\lambda_{\mathrm{D}\,i}$.

10.4.4 A generalized mixed functional in the spectral problem

We still have one opportunity unused, and now we are going to discuss it here briefly. It will be useful to find out what we can do by employing the generalized mixed functional Φ introduced in Section **3**.5 for the calculation of natural frequencies of elastic systems.

We use the formula (**3**.5.11),

$$\Phi = \kappa \mathsf{L} + (1-\kappa)\mathsf{R}, \tag{4.69}$$

and put in the epxressions of functional $\mathsf{L}(\mathbf{u})$ from (4.1) and functional $\mathsf{R}_2(\boldsymbol{\sigma}, \mathbf{u})$ from (4.3) for the free harmonic oscillations of elastic systems. We confine ourselves to considering these functionals on kinematically admissible fields of displacements and statically semi-admissible fields of stresses; at $\boldsymbol{K} = \boldsymbol{O}$ we have

$$\Phi(\boldsymbol{\sigma}, \mathbf{u}) = \frac{\kappa}{2}(\boldsymbol{C}\boldsymbol{A}\mathbf{u}, \boldsymbol{A}\mathbf{u}) +$$

$$+\frac{1-\kappa}{2}(\boldsymbol{C}^{-1}\boldsymbol{\sigma}, \boldsymbol{\sigma}) - (1-\kappa)(\boldsymbol{A}^{\mathsf{T}}\boldsymbol{\sigma}, \mathbf{u}) + \lambda\frac{1-2\kappa}{2}(\boldsymbol{\rho}\mathbf{u}, \mathbf{u}). \tag{4.70}$$

Introducing an independent approximation of the displacement and stress fields according to (4.21) gives a quadratic form,

$$\Phi(\boldsymbol{X}, \boldsymbol{Z}) = \frac{1}{2}\left\|\boldsymbol{X}^{\mathsf{T}}, \boldsymbol{Z}^{\mathsf{T}}\right\| \left\|\begin{matrix} (1-\kappa)\mathbf{G} & -(1-\kappa)\mathbf{H} \\ -(1-\kappa)\mathbf{H}^{\mathsf{T}} & \kappa\mathbf{R} + \lambda(1-2\kappa)\mathbf{M} \end{matrix}\right\| \left\|\begin{matrix} \boldsymbol{X} \\ \boldsymbol{Z} \end{matrix}\right\|, \tag{4.71}$$

where matrices **R**, **H**, **G**, **M** coincide with the matrices denoted the same way and introduced earlier, provided the approximating functions of the displacements are homogeneously kinematically admissible and those of the stresses are homogeneously statically semi-admissible.

The stationarity conditions for the quadratic form $\Phi(\boldsymbol{X}, \boldsymbol{Z})$ bring about the following algebraic eigenvalue problem:

$$\mathbf{G}\boldsymbol{X} - \mathbf{H}\boldsymbol{Z} = \mathbf{0}, \qquad -\mathbf{H}^{\mathsf{T}}\boldsymbol{X} + \frac{\kappa}{1-\kappa}\mathbf{R}\boldsymbol{Z} + \lambda\frac{1-2\kappa}{1-\kappa}\mathbf{M}\boldsymbol{Z} = \mathbf{0}, \tag{4.72}$$

or, after the $\boldsymbol{X}$ vector is excluded from this system,

$$(\boldsymbol{\Phi} - \lambda\mathbf{M})\boldsymbol{Z} = \mathbf{0}. \tag{4.73}$$

Matrix $\boldsymbol{\Phi}$ is defined as

$$\boldsymbol{\Phi} = \frac{1}{2\kappa - 1}\left[\kappa\mathbf{R} - (1-\kappa)\mathbf{H}^{\mathsf{T}}\mathbf{G}^{-1}\mathbf{H}\right]. \tag{4.74}$$

Earlier in Section **3**.5.2 we considered a static problem and proved a theorem that the condition

$$½ < \kappa < 1 \tag{4.75}$$

makes the $\Phi(\boldsymbol{\sigma}, \mathbf{u})$ functional convex.

Now let us show that the condition (4.75) is sufficient also to ensure the positive definiteness of matrix $\mathbf{\Phi}$, hence the positiveness of all n eigenvalues of problem (4.73). And indeed, the inequality (4.37) produces the following estimate:

$$\boldsymbol{Z}^{\mathrm{T}}\mathbf{\Phi}\boldsymbol{Z} = \frac{1}{2\kappa - 1}\left[\kappa \boldsymbol{Z}^{\mathrm{T}}\mathbf{R}\boldsymbol{Z} - (1-\kappa)\boldsymbol{Z}^{\mathrm{T}}\mathbf{H}^{\mathrm{T}}\mathbf{G}^{-1}\mathbf{H}\boldsymbol{Z}\right] \geq$$

$$\geq \frac{1}{2\kappa - 1}\left[\kappa \boldsymbol{Z}^{\mathrm{T}}\mathbf{R}\boldsymbol{Z} - (1-\kappa)\boldsymbol{Z}^{\mathrm{T}}\mathbf{R}\boldsymbol{Z}\right] = \boldsymbol{Z}^{\mathrm{T}}\mathbf{R}\boldsymbol{Z} > 0\,, \tag{4.76}$$

where the inequality takes place for any nonzero vector $\boldsymbol{Z}$ because the $\mathbf{R}$ matrix is positive definite[21].

We introduce the Rayleigh ratio

$$\mathsf{r}_{\Phi} = \frac{\boldsymbol{Z}^{\mathrm{T}}\mathbf{\Phi}\boldsymbol{Z}}{\boldsymbol{Z}^{\mathrm{T}}\mathbf{M}\boldsymbol{Z}} \tag{4.77}$$

which generates an n-dimensional ellipsoid (an Φ-ellipsoid) in the $\mathbb{R}_n$ space, similar to the L-ellipsoid and R-ellipsoid.

Now let us find estimates to establish relations between the Rayleigh ratio r_{Φ}, on one hand, and the ratios r_{L} and r_{R}, on the other hand. As the denominators in the formulas for these three values are the same (and positive), it suffices to compare only their numerators. The estimate (4.76) yields immediately

$$\mathsf{r}_{\Phi} \geq \mathsf{r}_{\mathsf{L}}\,.$$

Further,

$$\boldsymbol{Z}^{\mathrm{T}}\mathbf{\Phi}\boldsymbol{Z} \leq \frac{\kappa}{2\kappa - 1}\boldsymbol{Z}^{\mathrm{T}}\mathbf{R}\boldsymbol{Z} \quad \text{or} \quad \mathsf{r}_{\Phi} \leq \frac{\kappa}{2\kappa - 1}\mathsf{r}_{\mathsf{L}}\,,$$

so our final estimate is a two-side one,

$$\mathsf{r}_{\mathsf{L}} \leq \mathsf{r}_{\Phi} \leq \frac{\kappa}{2\kappa - 1}\mathsf{r}_{\mathsf{L}}\,. \tag{4.78}$$

If we construct another ellipsoid (called a $\tilde{\Phi}$-ellipsoid) by extending the Φ-ellispoid with the extension factor k

[21] We still assume the approximations of (4.21) to be such that rank $\mathbf{H} = n$.

$$k = \sqrt{\frac{\kappa}{2\kappa - 1}}, \tag{4.79}$$

then the estimates (4.78) make the following statement true:

> *The Φ-ellipsoid is nested in the L-ellipsoid, and the L-ellipsoid is nested in the $\tilde{\Phi}$-ellipsoid.*

The respective two-dimensional geometrical picture is presented in Fig. 10.4. When the parameter κ tends to one, the Φ-ellipsoid and $\tilde{\Phi}$-ellipsoid will contract to the L-ellipsoid together.

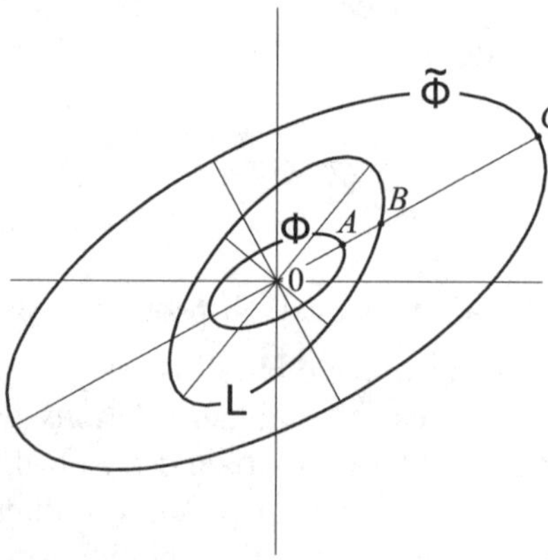

Fig. 10.4. An L-ellipsoid, a Φ-ellipsoid and Rayleigh's $\tilde{\Phi}$-ellipsoid

As the eigenvalues $\lambda_{\mathsf{L}\,i}$ are shifted along the semi-axis $(0,\infty)$ to the right with respect to the true frequency spectrum, i.e. $\lambda_{\mathsf{L}\,i} \ge \lambda_i$, an approximate (Ritz) solution for eigenvalues based on the stationarity of the $\mathbf{\Phi}(\boldsymbol{\sigma}, \mathbf{u})$ functional can consist of $\tilde{\lambda}_{\Phi i}$ rather than $\lambda_{\Phi\, i}$, where

$$\tilde{\lambda}_{\Phi i} = \frac{2\kappa - 1}{\kappa} \lambda_{\Phi i} \le \lambda_{\mathsf{L} i} \le \lambda_{\Phi i}\,. \tag{4.80}$$

We can also derive a mutual estimate of the Rayleigh ratios r_Φ and r_R. Definition (4.74) gives

$$\boldsymbol{Z}^\mathsf{T} \mathbf{\Phi} \boldsymbol{Z} \ge \frac{1-\kappa}{2\kappa - 1} \boldsymbol{Z}^\mathsf{T} \mathbf{H}^\mathsf{T} \mathbf{G}^{-1} \mathbf{H} \boldsymbol{Z}\,,$$

hence

$$r_\Phi \geq \frac{1-\kappa}{2\kappa-1} r_R . \tag{4.81}$$

Further details are omitted.

10.5 Final comments to Chapter 10

Practical engineers use various approximate formulas extensively which permit to estimate the fundamental natural frequency of a mechanical system roughly but in a simplest way possible. In particular, the well-known Rayleigh formula serves the same purpose. The Rayleigh formula can be interpreted as a result of applying the Ritz method to the Lagrange functional with a single-point approximation of the displacements.

To see this, we take $n = 1$ in the approximations of (4.8), i.e. we assume an approximate relationship

$$\mathbf{u} = Z\varphi ,$$

and get the following expression for the fundamental-tone natural frequency of the mechanical system:

$$\omega_{L1}^2 = \frac{(\boldsymbol{CA}\varphi, \boldsymbol{A}\varphi)}{(\rho\varphi,\varphi)} . \tag{5.1}$$

Formula (5.1) is essentially the same famous Rayleigh formula included in most textbooks and manuals on structural dynamics, taken in its particular form of representation. For example, in the case of natural oscillations of a flexural bar we should assume

$$(\boldsymbol{CA}\varphi, \boldsymbol{A}\varphi) = \int_0^l EI\left(\varphi''\right)^2 dx , \quad (\rho\varphi,\varphi) = \int_0^l m\varphi^2 dx ,$$

where EI is the bar's flexural rigidity, l is the bar's length, m is the bar's weight per unit of its length, and φ is a function that approximates the deflection of the bar and satisfies the kinematical boundary conditions.

As a result, the Rayleigh formula (5.1) becomes quite usual for the eye of an engineer:

$$\omega_{L1}^2 = \frac{\int_0^l EI\left(\varphi''\right)^2 dx}{\int_0^l m\varphi^2 dx} . \tag{5.2}$$

There are several known modifications of this formula which are presented and commented in detail in [18]. Here we just want to indicate another version of the approximate formula for the fundamental tone of oscillations because, as far as we know, it has not been published in literature yet.

The same idea which was used to validate the Rayleigh formula (5.1) or (5.2) works for an approximate formula based on the stationarity condition for the Reissner functional. To see this, we use an independent single-point approximations for both the displacements *and* the stresses, i.e. we assume the following instead of (4.21):

$$\mathbf{u} = Z\varphi , \qquad \boldsymbol{\sigma} = X\psi \ .$$

Using the Rayleigh ratio (4.27) with the single-term approximations gives

$$\omega_{\mathrm{R1}}^2 = \frac{(\boldsymbol{A}^{\mathrm{T}}\psi,\varphi)^2}{(\boldsymbol{C}^{-1}\psi,\psi)(\rho\varphi,\varphi)} . \tag{5.3}$$

In particular, the stress $\boldsymbol{\sigma}$ in a flexural bar is the bending moment, M, and formula (5.3) becomes

$$\omega_{\mathrm{R1}}^2 = \frac{\left(\int_0^l \psi'\varphi' dx\right)^2}{\int_0^l \frac{\psi^2}{EI} dx \int_0^l m\varphi^2 dx} . \tag{5.4}$$

Here we use the equality

$$\int_0^l \psi''\varphi dx = -\int_0^l \psi'\varphi' dx = \int_0^l \psi\varphi'' dx$$

that follows from the integration by parts formula provided function φ satisfies kinematical boundary conditions and function ψ satisfies static conditions.

Let us see, for example, what the Rayleigh formula (5.2) and formula (5.4) give for flexural oscillations of a cantilever bar. The simplest approximations can be

$$\varphi = x^2, \quad \psi = (l-x)^2 .$$

For $EI = Const, \quad m = Const$ we have

$$\int_0^l m\varphi^2 dx = \frac{ml^5}{5}, \ \int_0^l EI\,(\varphi'')^2\, dx = 4lEI\ , \ \int_0^l \frac{\psi^2}{EI} dx = \frac{l^5}{5EI}, \ \int_0^l \psi'\varphi' dx = -\frac{2l^3}{3} .$$

Substituting in (5.2) and (5.4) gives

$$\omega_{L1}^2 = 20\frac{EI}{ml^4}, \qquad \omega_{R1}^2 = \frac{100}{9}\frac{EI}{ml^4}.$$

The exact value of the fundamental-tone frequency is

$$\omega_1^2 = 12.36\frac{EI}{ml^4}.$$

References

1. Bellman R (1960) Introduction to matrix analysis. McGraw-Hill, New York
2. Bleich F, Melan E (1936) Finite-difference equations of structural statics (in Russian). Kharkov
3. Collatz L (1948) Eigenwertprobleme und ihre numerische Behandlung. (Mathematik und ihre Anwendungen in Physik und Technik. Reihe A). Chelsea Pub. Co.
4. Courant R, Gilbert D (1953) Methods of Mathematical Physics. Vol. I.. New York, Interscience
5. Dolberg MD (1951) On a generalization of Bubnov's problem (in Russian). Ukrainian Math. Journal, 3: 433–447
6. Dolberg MD (1957) On maximum-rigidity constraints (in Russian). Proc. of Kharkov State Univ. and Kharkov Math. Society, **25**, series 4
7. Fischer E (1905) Über quadratische Formen mit reellen Koeffizienten. – Monatshefte für Math. und Phys., 16: 234–249
8. Gantmakher FR (1967) Theory of matrices (in Russian). Nauka, Moscow
9. Gould S (1966) Variational methods for eigenvalue problems. Univ. of Toronto Pr, 2nd ed.
10. Gradshtein IS, Ryzhik IM (1971) Tables of integrals, series, sums and products (in Russian). Nauka, Moscow
11. Ispolov YG, Slivker VI (1984) On an effect that appears in applications of the mixed form of the finite element method (in Russian). Structural mechanics and analysis. 1: 43–48
12. Kukishev VL, Friedman VM (1976) A variational difference method in the theory of elastic oscillations based on the Reissner principle (in Russian). Trans. Acad. Sci. USSR, Mech. of solids, 5: 112–119
13. Lancaster P (1969) Theory of matrices. Acad. Press
14. Loitsiansky LG, Lurie AI (1955) A course of theoretical mechanics. Vol. 2 (in Russian). Gostekhteorizdat, Moscow
15. Mandelshtam LI (1972) Lectures on theory of oscillations (in Russian). Nauka, Moscow

16. Mikhlin SG (1964) Variational Methods in Mathematical Physics. Macmillan, New York
17. Nudelman YL (1949) Methods for determining natural frequencies and critical forces in bar systems (in Russian). State Publ. House of Tech. and Teor. Literature, Moscow Leningrad
18. Panovko YG, Gubanova II (1987) Stability and oscillations of elastic systems (in Russian). Nauka, Moscow
19. Parlett BN (1980) The Symmetric Eigenvalue Problem. Prentice-Hall, Inc., Englewood Cliffs, NJ
20. Poincaré H (1890) Sur les équations aux dérivées partielles de la physique mathematique. Amer. J. Math., 12: 211–294
21. Rayleigh JWS (1976) The theory of sound. Dover Publications
22. Routh EJ (1905) Dynamics of a system of rigid bodies. New York, 6 ed.
23. Sakato Y (1979) Eigenvalue Problem of Second-Order Differential Equation with variable coefficients. Mem. Chubu Inst. Technol., A 15: 113–116
24. Slivker VI (1984) A method of two functionals for frequency analysis of elastic systems (in Russian). In: Bulletin of B.Vedeneyev's All-Union Research Institute for Water Engineering, Vol.169. Energiya, Leningrad pp 20–27
25. Strang G (1976) Linear algebra and its applications. Academic press, New York San Francisko London
26. Strang G, Fix GJ (1973) An analysis of the finite element method. Prentice-Hall Inc., Englewood Cliffs, N.Y.
27. Timoshenko SP, Gere JM (1961) Theory of elastic stability. McGraw-Hill, New York
28. Weber H. (1869) Über die Integration der partiellen Differentialgleichungen. Math. Ann., 1: 1–366.

11 VARIATIONAL PRINCIPLES IN STABILITY/BUCKLING ANALYSIS

The structure of the world is indeed perfect, being built by our wisest Creator, therefore no such thing happens in the world in which we could not see the meaning of a certain maximum or minimum

Euler L.

The stability of equilibrium of elastic systems is one of most interesting branches of structural mechanics. One can often hear that the problems of stability/buckling are equivalent to eigenvalue problems and therefore identical in the mathematical sense to the analysis of natural oscillations of elastic systems[1]. But this is only partially true. The fact is that the mathematical basis of the stability analysis is much wider and more multifarious hence more complicated than its formal counterpart, the natural oscillation analysis.

Many a time the insidious peculiarity of the equilibrium stability problems (even in the linearized and purely static formulation) was a cause for errors both in solutions of particular practical problems and in various fundamental statements of the structural stability theory. It is not for nothing that the history of the foundation and development of this branch of structural analysis, more than of any other branch of this science, is overwhelmed by numerous mistakes, paradoxes, misbeliefs, incorrect or inaccurate formulations, treatments etc. Quite a few celebrities of the science were not able to avoid being vexatiously caught by the tricky net of the stability problems; the first of them seems to be the very founder of the static stability analysis – the famous author of the epigraph to this chapter.

---

[1] The essential point here is that we are talking about the equilibrium stability analysis in a linearized formulation. As we noted in the preface to the book, full treatment of the geometrically nonlinear problems is beyond the scope of our book.

A torrent of mistakes, many of which are brilliantly and educatively discussed in a well-known book by Y.G. Panovko and I.I. Gubanova [13], does not seem to dry up yet. An evidence of this is a commonly popular error in advanced computing software the detection and analysis of which is described in [15] or [16].

This is the reason why the author, in spite of the existence of many books (some of them outstanding) on the theory of structural stability, believes it useful and even necessary to do justice to this branch of structural mechanics and treat it from the standpoint of variational formulations. The present chapter is thoroughly dedicated to this kind of analysis.

11.1 Stability of systems with a finite number of degrees of freedom

Let us consider an elastic mechanical system with a finite number of degrees of freedom (DOFs). To be definite, we will think that the position of the mechanical system is defined by n generalized coordinates $q_1,\dots, q_n$ which together make up a vector of the generalized coordinates, $\boldsymbol{q}$:

$$\boldsymbol{q} = \|[q_1,\dots,q_n]\|^{\mathsf{T}}.$$

Let $\mathsf{E} = \mathsf{E}(\boldsymbol{q})$ be a potential energy of strain of the mechanical system in quesiton, and let $\Pi_s = \Pi_s(\boldsymbol{P},\boldsymbol{q})$ be a potential of external forces. This form of the force potential imports that Π_s depends both on the displacements and, parametrically, on the external forces defined by vector $\boldsymbol{P}$.

We assume the level of the load to be characterized by a single scalar parameter λ and to depend on this parameter proportionally. In other words, we assume $\boldsymbol{P} = \lambda\overline{\boldsymbol{P}}$ where $\overline{\boldsymbol{P}}$ is a fixed known vector which lends itself to the name of a *unit load vector.*

The above conditions of loading permit us to write the external force potential as

$$\Pi_s = \Pi_s(\lambda,\boldsymbol{q}) = \lambda\overline{\Pi}_s(\boldsymbol{q}) \qquad \text{where} \quad \overline{\Pi}_s(\boldsymbol{q}) = \Pi_s(1,\boldsymbol{q})\,. \tag{1.1}$$

The value of $\overline{\Pi}_s(\boldsymbol{q})$ will be entitled a *unit force potential.* Note immediately that when there is no external force, the force potential becomes identical to zero, $\Pi_s(0,\boldsymbol{q}) \equiv 0$.

In the system's position of equilibrium, its full potential energy L is

$$\mathsf{L} = \mathsf{E}(\boldsymbol{q}) - \lambda\overline{\Pi}_s(\boldsymbol{q})\,, \tag{1.2}$$

which is treated as a function of generalized coordinates $\boldsymbol{q}$ and takes a stationary value. Or, to put it another way, in the position of equilibrium we have

$$\frac{\partial \mathsf{L}}{\partial q_i} = \frac{\partial \mathsf{E}}{\partial q_i} - \lambda \frac{\partial \bar{\Pi}_s}{\partial q_i} = 0 \qquad (i = 1,\ldots, n)\,. \tag{1.3}$$

Any solution of the system of equations (1.3) with respect to generalized coordinates $\boldsymbol{q}$ defines a certain position of equilibrium. The following three options are available:

- the equilibrium equation system (1.3) has a unique solution;
- the equilibrium equation system (1.3) has multiple solutions;
- the equilibrium equation system (1.3) has no solution at all.

The no-solution case is of no interest because it means the mechanical system is unable to counterbalance its external loads. Let $\boldsymbol{q}_o$ be a solution of the set of equations (1.3). It goes without saying that $\boldsymbol{q}_o$ is a vector function of the load parameter, λ:

$$\boldsymbol{q}_o = \boldsymbol{q}_o(\lambda)\,. \tag{1.4}$$

Further we will be interested with the quality of this equilibrium state of the system, i.e. we will want to know whether this state is stable or not. We should remind a known definition of the stable equilibrium of a mechanical system given in courses on theoretical mechanics [11]:

"*The definition of a stability of equilibrium is based on considering* <u>*motions*</u> *that the system would perform after being unbalanced when its points receive fairly small initial deviations from its position of equilibrium and fairly small initial velocities. If the system would, after its equilibrium has been thus violated, deviate in its subsequent motion by only a little from the equilibrium position in question then this position of equilibrium is stable*".

The mathematical language suggests the following translation of the above qualitative definition into the terms of ε-δ. Let $\boldsymbol{q}_o$ be a displacement vector for the equilibrium position. This position is called stable if for any arbitrarily small number ε there is a respective small number δ such that for any perturbations of the generalized coordinates, $\delta\boldsymbol{q}$, and of the initial velocities, $\delta\dot{\boldsymbol{q}}$, which satisfy the conditions

$$\|\, \delta\boldsymbol{q} \,\| < \delta\,, \quad \|\, \delta\dot{\boldsymbol{q}} \,\| < \delta\,,$$

the following inequalities are assured at any moment of time t:

$$\| \boldsymbol{q}(t) \| < \varepsilon, \qquad \| \dot{\boldsymbol{q}}(t) \| < \varepsilon.$$

We do not specify any particular norm. The usual approach is to use a maximum absolute value of the generalized coordinate and that of the generalized velocity, that is,

$$\| \boldsymbol{q} \| = \max_i | q_i |, \qquad \| \dot{\boldsymbol{q}} \| = \max_i | \dot{q}_i |,$$

although any other can be used because all kinds of norms are equivalent in a finite-dimensional space.

In order to find out whether a certain state of equilibrium is stable or not, we use the Lagrange–Dirichlet theorem. It involves an analysis of the system's full potential energy for extremum in the vicinity of the equilibrium state of interest. If a stationary point of L is a point of minimum, then the Lagrange–Dirichlet theorem says the state of equilibrium is stable in the sense of the above definition[2].

As is known from calculi, the local minimum of L for a system with multiple DOFs will be ensured if the Hessian matrix $\mathbf{H}$ of function L is positive definite at the point of stationarity,

$$\mathbf{H} = \left\| \left[\frac{\partial^2 \mathsf{L}}{\partial q_i \partial q_j} \right] \right\|. \tag{1.5}$$

Clearly, the components of the Hessian matrix should be calculated at $\boldsymbol{q} = \boldsymbol{q}_o$, therefore they are functions of parameter λ. If at a certain fixed λ the $\mathbf{H}$ matrix is positive definite, then the equilibrium is stable.

The equilibrium of a mechanical system is assumed to be stable a priori for sufficiently small values of the external load. It means the equilibrium is stable at $\lambda = 0$. By changing λ we find its minimum value $\lambda = \lambda_{cr}$ at which the $\mathbf{H}$ matrix loses its positive definiteness. This value of $\lambda = \lambda_{cr}$ is called a *critical value* (*in the Lagrange–Dirichlet sense*).

Another, and a quite typical reasoning for the equilibrium stability analysis originates from Leonard Euler's works and is still taught by tradition in most textbooks on structural mechanics. It is based on a partly intuitive treatment of the stability rather than on the Lagrange–Dirichlet theorem.

The Eulerian idea of stability of equilibrium is essentially based on a totally different definition in which the concept of motion does not

---

[2] We remind that, generally speaking, the Lagrange–Dirichlet theorem gives only the sufficient conditions for the equilbrium of a system to be stable. The proof of the Lagrange–Dirichlet theorem can be found, for example, in [11].

participate at all[3]. According to Euler, the stability of a mechanical system is stable if the system tends to return to its original position after being deviated from the equilibrium by a small perturbation of its generalized coordinates. This approach also includes a gradual increase of the external load and supposes that at a certain moment of time another state of equilibrium may become feasible in addition to the original one. This second state will be adjacent to the original state in the sense that it will differ from the latter by an infinitesimal perturbation in the values of the generalized coordinates. In other words, the equilibrium state of the system ceases being unique, and one of two possibilities is realized: either the equilibrium modes will bifurcate or the curves of the equilibrium states in the "load vs. displacements" $(n + 1)$-dimensional space will have so-called *limit points*[4]. A value of the load parameter, $\lambda = \lambda_e$, at which this ambiguity of the equilibrium mode is permitted by the mechanical system, is called *critical* (*in the Euler sense*). A state of equilibrium is thought to be stable when the load is lower than the Euler critical value, and the equilibrium is unstable when the load is higher than the Euler critical value λ_e.

Euler's formulation of the stability problem seems to contain no allusion to the equlibrium stability analysis based on the Lagrange–Dirichlet theorem that requires a minimum of the system's full potential energy in its stable position. Strangely, the traditional engineering education, a great part of which still exists in our time, not only teaches the Euler approach to the engineer but forcedly trains him to use it exclusively in the elastic equilibrium stability analysis, leaving the more consistent Lagrange–Dirichlet analysis only in an earlier course of theoretical mechanics without employing it in structural analysis. Of course, under certain conditions both approaches produce the same results. However, the latter statement needs to be verified because otherwise the methodologies of the stability presentment in the courses of theoretical and structural mechanics would be inconsistent and a close relation between the two sciences would be lost.

---

[3] It was V.V. Bolotin who in 1965 noticed both this fact and the general shortcomings of the structural mechanics courses that teach the structural stability according to Euler's approach solely [4].

[4] Points where the equilibrium modes bifurcate are naturally called *bifurcation points*. It was H. Poincaré who introduced the notion of limit points into the theory of stability of mechanical equilibrium. No bifurcation of the equilibrium mode occurs at a limit point of the curve of equilibrium states, but the tangent to this curve is orthogonal to the loading axis, i.e. parallel to the n-dimensional plane of displacements.

So, now our intention is to validate the Euler method for a fairly general case of stability of elastic systems with a finite number of DOFs. The validation will consist of a proof of equivalence between the Eulerian critical load and that according to Lagrange–Dirichlet. Of course, the equivalence will obey certain, though sufficiently general conditions.

We will expand, purely formally for now, the expression of the potential strain energy, $\mathsf{E} = \mathsf{E}(\boldsymbol{q})$, and the expression of the force potential, $\Pi_s = \lambda\bar{\Pi}_s(\boldsymbol{q})$, into Taylor series over powers of the generalized coordinates in the vicinity of the zero vector, $\boldsymbol{q} = \mathbf{0}$:

$$\mathsf{E} = \mathsf{E}(\mathbf{0}) + \sum_{i=1}^{n} \frac{\partial \mathsf{E}}{\partial q_i} q_i + \frac{1}{2} \sum_{i=1}^{n} \sum_{j=1}^{n} \frac{\partial^2 \mathsf{E}}{\partial q_i \partial q_j} q_i q_j +$$

$$+ \frac{1}{6} \sum_{i=1}^{n} \sum_{j=1}^{n} \sum_{k=1}^{n} \frac{\partial^3 \mathsf{E}}{\partial q_i \partial q_j \partial q_k} q_i q_j q_k + \ldots$$

$$\Pi_s = \lambda\bar{\Pi}_s(\mathbf{0}) + \lambda \sum_{i=1}^{n} \frac{\partial \bar{\Pi}_s}{\partial q_i} q_i + \frac{\lambda}{2} \sum_{i=1}^{n} \sum_{j=1}^{n} \frac{\partial^2 \bar{\Pi}_s}{\partial q_i \partial q_j} q_i q_j + \ldots . \qquad (1.6)$$

Note that the derivatives of the functions E and Π_s in the right-hand parts of the above formulas are taken at the zero point $\boldsymbol{q} = \mathbf{0}$.

We suppose that the force potential vanishes when there are no displacements, i.e. $\bar{\Pi}_s(\mathbf{0}) = \Pi_s(1,\mathbf{0}) = 0$. Similarly, the potential strain energy is zero in the undeformed state of the system. This means $\mathsf{E}(\mathbf{0}) = 0$.

Finally, when there are no external forces (i.e. at $\lambda = 0$) the undeformed state of the system is postulated to be self-balanced and stable. The state's being self-balanced means in the mathematical language that when there is no external action on the system, we have

$$\frac{\partial \mathsf{L}}{\partial q_i} = \frac{\partial \mathsf{E}}{\partial q_i}(\mathbf{0}) = 0 \;\; (i = 1,\ldots, n) .$$

Thus, the formulas (1.6) of the expansions of E and Π_s into the Taylor series in the vicinity of zero become as simple as

$$\mathsf{E} = \frac{1}{2} \sum_{i=1}^{n} \sum_{j=1}^{n} \frac{\partial^2 \mathsf{E}}{\partial q_i \partial q_j} q_i q_j + \frac{1}{6} \sum_{i=1}^{n} \sum_{j=1}^{n} \sum_{k=1}^{n} \frac{\partial^3 \mathsf{E}}{\partial q_i \partial q_j \partial q_k} q_i q_j q_k + \ldots,$$

$$\Pi_s = \lambda \sum_{i=1}^{n} \frac{\partial \bar{\Pi}_s}{\partial q_i} q_i + \frac{\lambda}{2} \sum_{i=1}^{n} \sum_{j=1}^{n} \frac{\partial^2 \bar{\Pi}_s}{\partial q_i \partial q_j} q_i q_j + \ldots \quad . \qquad (1.7)$$

In order to investigate the quality of the equilibrium state, we need to use the expansions of E and Π_s in the vicinity of vector $\boldsymbol{q}_o$ rather than zero; we assume

$$\boldsymbol{q} = \boldsymbol{q}_o + \delta\boldsymbol{q}.$$

So, instead of (1.6) we have

$$\mathsf{E}(\boldsymbol{q}_o + \delta\boldsymbol{q}) = \mathsf{E}(\boldsymbol{q}_o) + \sum_{i=1}^{n} \frac{\partial \mathsf{E}}{\partial q_i} q_i + \frac{1}{2}\sum_{i=1}^{n}\sum_{j=1}^{n} \frac{\partial^2 \mathsf{E}}{\partial q_i \partial q_j} q_i q_j + \ldots,$$

$$\Pi_s(\boldsymbol{q}_o + \delta\boldsymbol{q}) = \lambda\bar{\Pi}_s(\boldsymbol{q}_o) + \lambda\sum_{i=1}^{n} \frac{\partial \bar{\Pi}_s}{\partial q_i} q_i + \frac{\lambda}{2}\sum_{i=1}^{n}\sum_{j=1}^{n} \frac{\partial^2 \bar{\Pi}_s}{\partial q_i \partial q_j} q_i q_j + \ldots, \qquad (1.8)$$

where all derivatives are taken at $\boldsymbol{q} = \boldsymbol{q}_o$. Recalling that vector $\boldsymbol{q}_o$ corresponds to the self-balanced state of the system, we find in accordance with (1.3) that

$$\frac{\partial \mathsf{L}}{\partial q_i}(\boldsymbol{q}_o) = \frac{\partial \mathsf{E}}{\partial q_i}(\boldsymbol{q}_o) - \lambda\frac{\partial \bar{\Pi}_s}{\partial q_i}(\boldsymbol{q}_o) = 0 \qquad (i = 1,\ldots, n), \qquad (1.9)$$

consequently,

$$\mathsf{L}(\boldsymbol{q}_o + \delta\boldsymbol{q}) - \mathsf{L}(\boldsymbol{q}_o) =$$

$$= \frac{1}{2}\sum_{i=1}^{n}\sum_{j=1}^{n} \frac{\partial^2 \mathsf{E}}{\partial q_i \partial q_j} \delta q_i \delta q_j - \frac{\lambda}{2}\sum_{i=1}^{n}\sum_{j=1}^{n} \frac{\partial^2 \bar{\Pi}_s}{\partial q_i \partial q_j} \delta q_i \delta q_j + \ldots \qquad (1.10)$$

The last expression shows that $\boldsymbol{q}_o$ is a point of minimum of $\mathsf{L}(\boldsymbol{q})$ if increment $\delta\mathsf{L}$ of functional L equal to $\delta\mathsf{L} = \mathsf{L}(\boldsymbol{q}_o + \delta\boldsymbol{q}) - \mathsf{L}(\boldsymbol{q}_o)$ is positive for any arbitrarily small variations $\delta\boldsymbol{q}$, and this is equivalent in its turn to the requirement that the following quadratic form should be positive definite:

$$\frac{1}{2}\sum_{i=1}^{n}\sum_{j=1}^{n}\left(\frac{\partial^2 \mathsf{E}}{\partial q_i \partial q_j} - \lambda\frac{\partial^2 \bar{\Pi}_s}{\partial q_i \partial q_j}\right)\delta q_i \delta q_j = \frac{1}{2}\delta\boldsymbol{q}^{\mathsf{T}} \boldsymbol{r}_\lambda \delta\boldsymbol{q}.$$

The elements of matrix $\boldsymbol{r}_\lambda$ of this quadratic form which is actually the Hessian matrix, $\mathbf{H}$, are

$$\boldsymbol{r}_\lambda = \left\|\left[\frac{\partial^2 \mathsf{E}}{\partial q_i \partial q_j}(\boldsymbol{q}_o) - \lambda\frac{\partial^2 \bar{\Pi}_s}{\partial q_i \partial q_j}(\boldsymbol{q}_o)\right]\right\| \qquad (1.11)$$

and are functions of parameter λ.

The $\boldsymbol{r}_\lambda$ matrix will be called a *full tangential stiffness matrix* of the system. This term is based on the consideration that the matrix is a generalized characteristic of the instantaneous stiffness at a certain point of the system's balanced state, i.e. the stiffness that corresponds to infinitesimal increments of the generalized coordinates, $\delta\boldsymbol{q}$.

We would like to emphasize again that the second derivatives are taken at the n-dimensional "point" $\boldsymbol{q} = \boldsymbol{q}_o$ the components of which depend on parameter λ according to (1.4). It is important to understand that parameter λ participates *both* as a factor at the second term in the representation of matrix $\boldsymbol{r}_\lambda$ in (1.11) *and* as an argument in the second derivatives of the strain energy and of the force potential. As it was said before, the value of parameter λ at which matrix $\boldsymbol{r}_\lambda$ ceases to be positive definite is called *critical*, λ_{cr} (in the Lagrange–Dirichlet sense).

Now let us determine the Euler critical load in the general case of a finite-dimensional system. Euler's approach is to find out a critical value of the load parameter *together* with the respective equilibrium state of the system determined by the equation (1.9), therefore we start by taking a perturbed state characterized by vector $\boldsymbol{q} = \boldsymbol{q}_o + \delta\boldsymbol{q}$, which is also assumed to be an equilibrium, that is,

$$\frac{\partial \mathsf{L}}{\partial q_i}(\boldsymbol{q}_o + \delta\boldsymbol{q}) = \frac{\partial \mathsf{E}}{\partial q_i}(\boldsymbol{q}_o + \delta\boldsymbol{q}) - \lambda\frac{\partial \overline{\Pi}_s}{\partial q_i}(\boldsymbol{q}_o + \delta\boldsymbol{q}) = 0\,. \tag{1.12}$$

Subtracting the equations (1.9) from the equations (1.12) of equilibrium of the perturbed state gives

$$\left[\frac{\partial \mathsf{E}}{\partial q_i}(\boldsymbol{q}_o + \delta\boldsymbol{q}) - \frac{\partial \mathsf{E}}{\partial q_i}(\boldsymbol{q}_o)\right] - \lambda\left[\frac{\partial \overline{\Pi}_s}{\partial q_i}(\boldsymbol{q}_o + \delta\boldsymbol{q}) - \frac{\partial \overline{\Pi}_s}{\partial q_i}(\boldsymbol{q}_o)\right] = 0 \quad (i = 1,\ldots, n)\,.$$

As the variations δq_j are assumed to be infinitesimal values, the expressions in the brackets can be treated as increments of first derivatives of the given functions E and $\overline{\Pi}_s$ between point $\boldsymbol{q}_o$ and the immediately adjacent point $\boldsymbol{q}_o + \delta\boldsymbol{q}$. They can be written equivalently using second partial derivatives of the same functions,

$$\sum_{j=1}^{n}\left(\frac{\partial^2 \mathsf{E}}{\partial q_i \partial q_j}(\boldsymbol{q}_o) - \lambda\frac{\partial^2 \overline{\Pi}_s}{\partial q_i \partial q_j}(\boldsymbol{q}_o)\right)\delta q_j = 0 \quad (i = 1,\ldots, n)\,. \tag{1.13}$$

Recalling the definition of matrix $\boldsymbol{r}_\lambda$ from (1.11), we can rewrite the same system of equations in the matrix form:

$$\boldsymbol{r}_\lambda \delta \boldsymbol{q} = \boldsymbol{0}\,. \tag{1.14}$$

The condition for a nontrivial solution of the homogeneous system of equations (1.14) with respect to variations δq_j to exist is that the determinant of the matrix of this system should be zero, which gives the Euler load λ_e. Thus, the analytical representation of the Euler load is given by the equation

$$\det\, \boldsymbol{r}_\lambda(\lambda_e) = 0\,. \tag{1.15}$$

But the same condition (1.15) is a criterion for achieving the Lagrange–Dirichlet critical load because, according to our convention of stability of a self-balanced undeformed state, matrix $\boldsymbol{r}_\lambda(0) = \boldsymbol{r}_\text{o}$ should be positive definite. Therefore, as the λ parameter is changing continuously, matrix $\boldsymbol{r}_\lambda(\lambda)$ can lose its positive definiteness only when the equality (1.15) begins to be true.

Thus, under the above stated conditions the following proposition is true:

The critical Euler load λ_e is equivalent to the Lagrange–Dirichlet critical load λ_{cr}.

As we can see, the fact that the Euler critical load and the Lagrange–Dirichlet one are equal has been established for a general geometrically nonlinear (second-order) analysis.

11.1.1 A functional of stability – Bolotin's functional

The system of homogeneous equations of stability (1.14) with respect to variations $\delta\boldsymbol{q}$ can be treated as a condition of stationarity of a certain homogeneous quadratic functional which we will call a *stability functional* and denote by S for the obvious reason. It is apparent that

$$\mathsf{S} = \frac{1}{2}\delta\boldsymbol{q}^\mathsf{T}\boldsymbol{r}_\lambda\delta\boldsymbol{q}\,.$$

Replacing matrix $\boldsymbol{r}_\lambda$ with its representation from (1.11) gives

$$\mathsf{S} = \frac{1}{2}\delta\boldsymbol{q}^\mathsf{T}\left\|\left[\frac{\partial^2\mathsf{L}}{\partial q_i \partial q_j}(\boldsymbol{q}_\text{o})\right]\right\|\delta\boldsymbol{q}\,, \tag{1.16}$$

and now we can see that the stability functional S can be interpreted mechanically as a *second special variation of the full potential strain energy of the system.* The '*special*' word in this statement reflects the fact that we do not mean to vary the full potential energy arbitrarily; instead, we take the variation in the vicinity of the equilibrium state of interest – the second derivatives of L are taken at point $\boldsymbol{q}_o$.

This critical theoretical proposition was clearly established by V.V. Bolotin [5], not for systems with a finite number of DOFs but for the general three-dimensional elasticity. He also derived a general stability functional for the general three-dimensional elasticity analysis.

11.1.2 Linear analysis and a linearized formulation of the stability problem

We are going to introduce a few notions to be used in the further presentment.

A square symmetric matrix $\boldsymbol{r}_o$ of an order n,

$$\boldsymbol{r}_o = \left\|\left[r_o^{ij} \right]\right\| = \left\|\left[\frac{\partial^2 \mathsf{E}}{\partial q_i \partial q_j} (\mathbf{0}) \right]\right\|, \tag{1.17}$$

is called an *initial stiffness matrix of the system*, and, obviously, $\boldsymbol{r}_\lambda(0) = \boldsymbol{r}_o$. The self-balanced undeformed state is postulated to be stable, hence the initial stiffness matrix $\boldsymbol{r}_o$ is positive definite.

A vector $\boldsymbol{Q}$ that has n coordinates,

$$\boldsymbol{Q} = \left\|\left[Q^i \right]\right\| = \left\|\left[\frac{\partial \overline{\Pi}_s}{\partial q_i} (\mathbf{0}) \right]\right\|,$$

will be called a *unit load vector*.

Suppose the generalized coordinates q_i that participate in the expansion (1.7) are very small values in the state of equilibrium. Suppose they are so small that we can confine ourselves to using only lowest-power terms in the expansion (1.7) when searching for an equilibrium. This means the expression of energy E will contain only quadratic terms, and the expression of the external force potential, Π_s, will contain only linear terms. Then (1.7) is replaced by

$$\mathsf{E} = \frac{1}{2} \boldsymbol{q}^{\mathsf{T}} \boldsymbol{r}_o \boldsymbol{q}, \qquad \Pi_s = \lambda \boldsymbol{Q}^{\mathsf{T}} \boldsymbol{q}. \tag{1.18}$$

The equations of equilibrium become linear equations, and vector $\boldsymbol{q}_{\rm o}$ becomes equal to

$$\boldsymbol{q}_{\rm o} = \lambda \boldsymbol{r}_{\rm o}^{-1}\boldsymbol{Q}\,. \tag{1.19}$$

The respective mechanical system is *linear*. This is obvious because, as (1.19) shows, displacements $\boldsymbol{q}_{\rm o}$ are linearly dependent on the load.

The definition of the tangential stiffness matrix $\boldsymbol{r}_\lambda$ given by (1.11) shows that the problem of stability makes no sense within the scope of the linear analysis where $\boldsymbol{r}_\lambda = \boldsymbol{r}_{\rm o}$ and thus in any linear problem generally because the initial stiffness matrix of the system, $\boldsymbol{r}_{\rm o}$, is positive definite.

Now let us return to formulas (1.7) and keep in the expansion of E all terms up to third order inclusive, and in the expansion of $\Pi_{\rm s}$ all terms up to second order inclusive. In other words, we take into account one more term in each expansion comparing to the linear (first-order) analysis. After introducing the designations

$$a^{ijk} = \frac{\partial^3 \mathsf{E}}{\partial q_i \partial q_j \partial q_k}(\mathbf{0})\,, \qquad b^{ij} = \frac{\partial^2 \overline{\Pi}_s}{\partial q_i \partial q_j}(\mathbf{0})\,, \tag{1.20}$$

the expressions of E and $\Pi_{\rm s}$ become as simple as

$$\mathsf{E} = \frac{1}{2}\boldsymbol{q}^{\mathsf T}\boldsymbol{r}_{\rm o}\boldsymbol{q} + \frac{1}{6}\sum_{i=1}^{n}\sum_{j=1}^{n}\sum_{k=1}^{n} a^{ijk} q_i q_j q_k\,, \quad \Pi_s = \lambda \boldsymbol{Q}^{\mathsf T}\boldsymbol{q} + \frac{\lambda}{2}\boldsymbol{q}^{\mathsf T}\boldsymbol{b}\boldsymbol{q}\,, \tag{1.21}$$

where the square symmetric matrix $\boldsymbol{b}$ of the order n is made up of coefficients b^{ij}:

$$\boldsymbol{b} = \left\|\left[b^{ij} \right]\right\|.$$

Note that coefficients a^{ijk} of the "cubic" form $a^{ijk}q_i q_j q_k$ are symmetric with respect to any couple of their indexes, that is,

$$a^{ijk} = a^{jik} = a^{ikj} = a^{kji}\,. \tag{1.22}$$

This is true because the differentiation of a function with respect to its three arguments does not depend on the order of taking its derivatives.

By considering the formulas of single and double differentiation of E as given by (1.21), carefully tracking the differentiation process, and taking into account the symmetry of coefficients a^{ijk}, we find out that the derivative formulas can be generally represented as

$$\frac{\partial \mathsf{E}}{\partial q_i} = r_{\text{o}}^{ij} q_j + \frac{1}{2} a^{ijk} q_j q_k, \qquad \frac{\partial^2 \mathsf{E}}{\partial q_i \partial q_j} = r_{\text{o}}^{ij} + a^{ijk} q_k \tag{1.23}$$

where r_{o}^{ij} is the respective component of matrix $\boldsymbol{r}_{\text{o}}$. To shorten the notation, we use a tensor summation over repeated indexes.

We define a *linearized formulation* of the problem as one based on third order of approximation for E and on second order for Π_s, which produce the representations (1.21).

Based on representations (1.21), we can see that if coefficients a^{ijk} are not zero (a general case) then the equations of equilibrium like (1.3) are nonlinear. To see this, consider these equations of equilibrium:

$$\frac{\partial \mathsf{L}}{\partial q_i} = 0 \Rightarrow r_{\text{o}}^{ij} q_j + \frac{1}{2} a^{ijk} q_j q_k - \lambda Q^i - \lambda b^{ij} q_j = 0 \quad (i = 1, \ldots, n),$$

or, in another form,

$$(r_{\text{o}}^{ij} + \frac{1}{2} a^{ijk} q_k - \lambda b^{ij}) q_j = \lambda Q^i \quad (i = 1, \ldots, n). \tag{1.24}$$

Thus we have a system of square equations with respect to the components of vector $\boldsymbol{q}$. In a particular case, coefficients a^{ijk} may be equal to zero, then the set of equations becomes linear, and this can be written in the matrix form as

$$(\boldsymbol{r}_{\text{o}} - \lambda \boldsymbol{b}) \boldsymbol{q} = \lambda \boldsymbol{Q}. \tag{1.25}$$

If we manage to somehow find vector $\boldsymbol{q}_{\text{o}}$ which is a solution of (1.24), then this calculation procedure is called an *analysis in the deformed state*. The following two terms are responsible for it in the system of equations (1.24):

$$\frac{1}{2} a^{ijk} q_j q_k - \lambda b^{ij} q_j.$$

When we neglect these terms, we have a linear formulation of the problem and a solution from (1.19).

The deformed-state analysis is a lowest level on which we can take into account the geometrical nonlinearity: the equations of equilibrium are composed not in the original position of the system but in its deformed configuration. Of course, even when we have found a solution $\boldsymbol{q}_{\text{o}}$ of the system of equations (1.24), this does not mean we can avoid checking the stability of the state of equilibrium thus obtained.

However, an even simpler procedure exists for the stability check. The simplest problem formulation is as follows. The equations of equilibrium are composed for the linear problem, i.e. we adopt the approximation (1.18) to search for a state of equilibrium, so this approach produces (1.19). The stability of the equilibrium state is to be found out on the basis of the linearized approximations (1.21). According to (1.11), we have

$$\boldsymbol{r}_\lambda = \left\| \left[r_{\mathrm{o}}^{ij} - \lambda \left(b^{ij} - \frac{1}{2} a^{ijk} \overline{q}_{\mathrm{o}k} \right) \right] \right\| \tag{1.26}$$

where $\overline{q}_{\mathrm{o}k}$ are the components of the $\overline{\boldsymbol{q}}_{\mathrm{o}} = \boldsymbol{r}_{\mathrm{o}}^{-1}\boldsymbol{Q}$ vector, i.e. the same vector $\boldsymbol{q}_{\mathrm{o}}$ calculated at $\lambda = 1$.

Matrix $\boldsymbol{r}_G$

$$\boldsymbol{r}_G = \left\| \left[b^{ij} - \tfrac{1}{2} a^{ijk} \overline{q}_{\mathrm{o}k} \right] \right\| \tag{1.27}$$

is called a *geometric stiffness matrix.*

Thus, the problem of stability in its linearized formulation is an eigenvalue problem for the couple of matrices $\boldsymbol{r}_{\mathrm{o}}$ and $\boldsymbol{r}_G$ because (1.26) and (1.27) let us write

$$\boldsymbol{r}_\lambda = \boldsymbol{r}_{\mathrm{o}} - \lambda \boldsymbol{r}_G \,. \tag{1.28}$$

The geometric stiffness matrix $\boldsymbol{r}_G$ plays a most important role in the theory of stability of linearized mechanical systems. This matrix is an analogue of the mass matrix $\boldsymbol{m}$ which is used in the theory of natural mechanical oscillations. However, the eigenvalue problem in the stability analysis is more complex than that in the spectral analysis of natural oscillations. This complexity is due to matrix $\boldsymbol{r}_G$ which, unlike its counterpart (matrix $\boldsymbol{m}$) does not have to be positive definite.

11.1.3 Example 1

We'd better give some examples right now in order to illustrate our formal consideration of equations with visible mechanical images. We start by considering a mechanical system with two degrees of freedom shown in Fig. 11.1.

All sizes in this figure are given for an initial (undeformed) state of the system which is supposed to stay in equilibrium when there is no external load. This is the state we will count the generalized coordinates from.

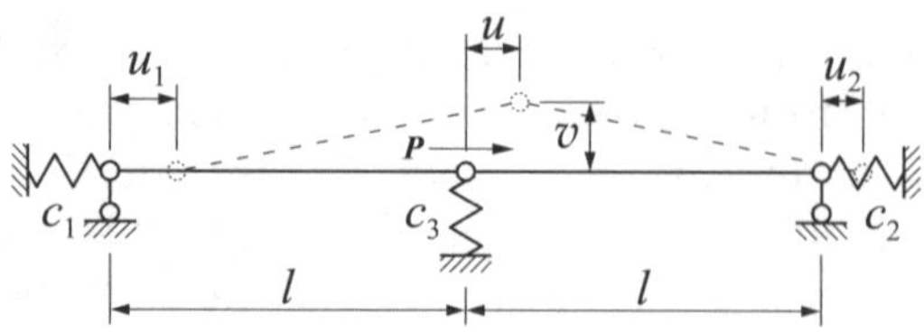

Fig. 11.1. Stability of a mechanical system that has two degrees of freedom

We assume the bars shown in Fig. 11.1 to be perfectly rigid, so all the strains are confined to three springs only; the stiffness characteristics of those are denoted by c_1, c_2, c_3, respectively.

The generalized coordinates of the system will be the horizontal, u, and vertical, v, displacements of the load application point, P, so that

$$\boldsymbol{q} = \|[u,v]\|^{\mathsf{T}}.$$

These two parameters indeed determine the position of the mechanical system in space unambiguously because the lengths of the bars are fixed. Denoting by u_1 and u_2 the respective value of the horizontal displacements of the end points of the first and second springs (Fig. 11.1 shows an imaginable deformed state of the system in a dash line), we derive this from the condition that the bars are non-deformable:

$$(l + u - u_1)^2 + v^2 = l^2\,, \qquad (l - u + u_2)^2 + v^2 = l^2\,,$$

hence

$$u_1 = l + u - l\sqrt{1-(v/l)^2}\,, \qquad u_1 = -l + u + l\sqrt{1-(v/l)^2}\,. \tag{1.29}$$

The potential energy of strain, E, of this system is a sum of energies accumulated in each of the three springs; it can be written as

$$\mathsf{E} = \frac{c_1 u_1^2}{2} + \frac{c_2 u_2^2}{2} + \frac{c_3 v^2}{2}\,. \tag{1.30}$$

Under the given conditions of loading, the external force potential is

$$\Pi_s = Pu = \lambda u. \tag{1.31}$$

Now we can write out equations of equlibrium of the type (1.9). According to the rule of indirect differentiation,

$$c_1 u_1 \frac{\partial u_1}{\partial u} + c_2 u_2 \frac{\partial u_2}{\partial u} - \lambda = 0\,, \quad c_1 u_1 \frac{\partial u_1}{\partial v} + c_2 u_2 \frac{\partial u_2}{\partial v} + c_3 v = 0\,. \tag{1.32}$$

The values u_1 and u_2 should be treated in these equations as functions of generalized coordinates u and v determined by the equation (1.29). Further, the differentiation of equations (1.29) gives

$$\frac{\partial u_1}{\partial u}=1, \quad \frac{\partial u_2}{\partial u}=1, \quad \frac{\partial u_1}{\partial v}=\frac{v/l}{\sqrt{1-(v/l)^2}}, \quad \frac{\partial u_2}{\partial v}=-\frac{v/l}{\sqrt{1-(v/l)^2}}. \tag{1.33}$$

Also, further we will need second partial derivatives of the same functions; we'd better write them out right now:

$$\frac{\partial^2 u_1}{\partial v^2}=\frac{1}{l\left[1-(v/l)^2\right]^{3/2}}, \quad \frac{\partial^2 u_2}{\partial v^2}=-\frac{1}{l\left[1-(v/l)^2\right]^{3/2}}. \tag{1.34}$$

Putting (1.33) in (1.32) produces equations of equilibrium in terms of displacements,

$$c_1 u_1 + c_2 u_2 = \lambda, \quad c_1 u_1 v - c_2 u_2 v + c_3 l v \sqrt{1-(v/l)^2} = 0. \tag{1.35}$$

One of solutions of the equilibrium equation system (1.35) is obvious. This solution is

$$u_0 = \frac{\lambda}{c_1 + c_2}, \quad v_0 = 0. \tag{1.36}$$

This one will be called the *main solution*. Obviously, the main solution exists at any value of the load parameter λ and at any (positive) values of the stiffness characteristics of the springs.

To complete the picture, we are going to find the other solutions of the system of equations (1.35). By assuming $v \neq 0$, we derive the following from (1.35):

$$u_1 = \frac{\lambda - c_3 l \sqrt{1-(v/l)^2}}{2c_1}, \quad u_2 = \frac{\lambda + c_3 l \sqrt{1-(v/l)^2}}{2c_2}. \tag{1.37}$$

To shorten the notation, we denote

$$x = \sqrt{1-(v/l)^2}, \tag{1.38}$$

and derive the following from (1.37) after putting in (1.29):

$$l + u - lx = \frac{\lambda - c_3 l x}{2c_1}, \quad -l + u + lx = \frac{\lambda + c_3 l x}{2c_2}.$$

We have

$$u = \frac{\lambda - c_3 lx}{4c_1} + \frac{\lambda + c_3 lx}{4c_2}, \qquad l(x-1) = -\frac{\lambda - c_3 lx}{4c_1} + \frac{\lambda + c_3 lx}{4c_2}.$$

Consequently,

$$x = \frac{4c_1c_2 + \lambda\frac{c_1 - c_2}{l}}{4c_1c_2 + c_3(c_1 + c_2)}, \quad u = \frac{c_3 l(c_1 - c_2) + \lambda\left[c_1 + c_2 + \frac{c_3(c_1^2 + c_2^2)}{2c_1c_2}\right]}{4c_1c_2 + c_3(c_1 + c_2)}, \tag{1.39}$$

and, according to (1.38),

$$v = l\sqrt{1 - x^2}.$$

Thus, two states of equilibrium can occur in the system: one (the main one) is defined by (1.36) and the other (the additional one) is defined by (1.39). Further we will be interested with the quality of only the main state of equilibrium. The question of stability of the nonlinear solution (1.39) is beyond our scope.

In order to find out whether the main state of equilibrium is stable or not, we should turn directly to the Lagrange–Dirichlet theorem. In particular, the Hessian matrix for the current example with two DOFs is

$$\mathbf{H} = \left\| \begin{bmatrix} \frac{\partial^2 \mathsf{L}}{\partial u^2} & \frac{\partial^2 \mathsf{L}}{\partial u \partial v} \\ \frac{\partial^2 \mathsf{L}}{\partial u \partial v} & \frac{\partial^2 \mathsf{L}}{\partial v^2} \end{bmatrix} \right\| =$$

$$= \left\| \begin{bmatrix} c_1 + c_2 & c_1\frac{\partial u_1}{\partial v} + c_2\frac{\partial u_2}{\partial v} \\ c_1\frac{\partial u_1}{\partial v} + c_2\frac{\partial u_2}{\partial v} & c_1\left[\left(\frac{\partial u_1}{\partial v}\right)^2 + u_1\frac{\partial^2 u_1}{\partial v^2}\right] + c_2\left[\left(\frac{\partial u_2}{\partial v}\right)^2 + u_2\frac{\partial^2 u_2}{\partial v^2}\right] + c_3 \end{bmatrix} \right\|.$$

When considering the stability of the main equilibrium state, we should use the solution (1.36) which gives the following if we take into account (1.33) and (1.34):

$$\left.\frac{\partial u_1}{\partial v}\right|_{v=0} = 0, \quad \left.\frac{\partial u_2}{\partial v}\right|_{v=0} = 0, \quad \left.\frac{\partial^2 u_1}{\partial v^2}\right|_{v=0} = \frac{1}{l}, \quad \left.\frac{\partial^2 u_2}{\partial v^2}\right|_{v=0} = -\frac{1}{l}.$$

Thus, the Hessian matrix is as follows in the main state of equilibrium:

$$\mathbf{H} = \begin{bmatrix} c_1 + c_2 & 0 \\ 0 & \frac{c_1 u_1}{l} - \frac{c_2 u_2}{l} + c_3 \end{bmatrix},$$

and now it is clear why the condition of positive definiteness of the Hessian matrix, **H**, hence the condition of stability of the main state, is equivalent to the requirement that the second diagonal element of the matrix be positive (because the first diagonal element is sure to be)

$$\frac{c_1 u_1}{l} - \frac{c_2 u_2}{l} + c_3 > 0 .$$

But the state of equilibrium from (1.29) and (1.36) has

$$u_1 = u_o = \frac{\lambda}{c_1 + c_2}, \qquad u_2 = u_o = \frac{\lambda}{c_1 + c_2}$$

which makes the stability condition

$$\lambda \frac{c_1 - c_2}{c_1 + c_2} + c_3 l > 0 . \tag{1.40}$$

This inequality shows that the main state of equilibrium is stable at any (positive) value of the load parameter λ if

$$c_1 \geq c_2 > 0 .$$

However, if the stiffness values of the first and second springs relate to each other differently,

$$0 < c_1 < c_2 ,$$

then the condition of equilibrium stability (1.40) holds for $\lambda < \lambda_{cr}$ where

$$\lambda_{cr} = c_3 l \frac{c_1 + c_2}{c_2 - c_1} . \tag{1.41}$$

The force value $\lambda = \lambda_{cr}$ that separates stable and unstable areas of the equilibrium is called a *critical load value*. Fig. 11.2 presents a graph of the dependency defined by (1.41). The dependence presented in the figure also takes into account the possibility for the system to lose its stability under negative loads, i.e. when the force P is opposite to the direction shown in Fig. 11.1.

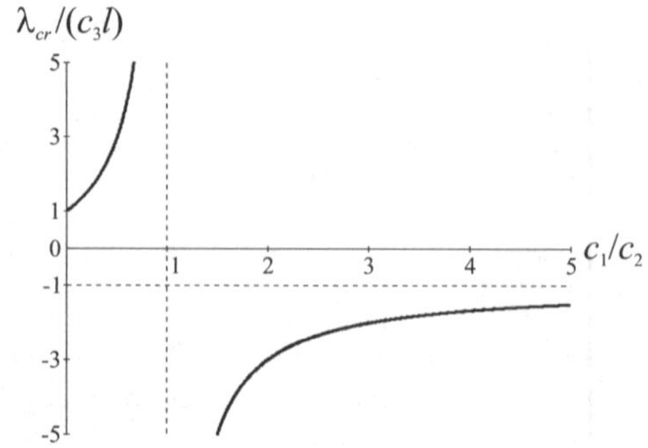

Fig. 11.2. The critical force vs. the ratio of the spring stiffness values

Now let us turn to Euler's concept for finding the critical load in the main state of equilibrium for our example. Suppose that, in addition to vector $\boldsymbol{q}_o = \|[u_o, v_o]\|^{\mathrm{T}}$ where u_o and v_o are defined for the main state by (1.27), there is an adjacent state of equilibrium defined by a perturbed vector of generalized displacements, $\boldsymbol{q}_o + \delta\boldsymbol{q} = \|[u_o + \delta u, v_o + \delta v]\|^{\mathrm{T}}$, where δu and δv are infinitesimal increments (variations) of the components of displacement vector $\boldsymbol{q}$. This means the equilibrium equations (1.33) hold together with the equations where the perturbed coordinates participate:

$$c_1(u_1 + \delta u_1) + c_2(u_2 + \delta u_2) = \lambda ,$$

$$c_1(u_1 + \delta u_1)(v + \delta v) - c_2(u_2 + \delta u_2)(v + \delta v) + c_3 l(v + \delta v)\sqrt{1 - \frac{(v + \delta v)^2}{l^2}} = 0 .$$

Substracting (1.35) from the above equations and seeing that $v = v_o = 0$ and $u_1 = u_2 = u_o$ in the state we are dealing with, we have

$$c_1\delta u_1 + c_2\delta u_2 = 0 , \quad c_1 u_o \delta v - c_2 u_o \delta v + c_3 l x \delta v = 0 . \tag{1.42}$$

The products of variations of the generalized coordinates are omitted from (1.42) being second-order values. According to (1.38) and (1.29),

$$\delta x = \frac{dx}{dv}\delta v = -\frac{v_o}{l^2\sqrt{1-(v_o/l)^2}}\delta v = 0 ,$$

$$\delta u_1 = \delta u - l\delta x = \delta u , \qquad \delta u_2 = \delta u - l\delta x = \delta u .$$

Also, $x = 1$ for our state of equilibrium. Therefore (1.42) transforms into

$$(c_1 + c_2)\delta u = 0 , \quad (c_1 - c_2)u_o\delta v + c_3 l\delta v = 0 . \tag{1.43}$$

The system of linear homogeneous equations (1.43) with respect to variations δu and δv has a nonzero solution only if the following condition holds:

$$(c_1 - c_2)u_o + c_3 l = 0,$$

which gives the Euler critical load λ_e after putting in the expression of u_o from (1.36),

$$\lambda_e = c_3 l \frac{c_1 + c_2}{c_2 - c_1}. \tag{1.44}$$

As we can see by comparing it with (1.41), the Euler critical force λ_e is equal to the critical load λ_{cr} that defines the loss of stability according to the Lagrange–Dirichlet theorem. This result is quite expectable.

If we keep only lower terms in the expansions of E and Π_s into the Taylor series in the vicinity of zero, we arrive at formulas that correspond to the linear analysis. The equations of equilibrium $\boldsymbol{r}_o\boldsymbol{q} = \lambda\boldsymbol{Q}$ which follow from this representation of the strain energy and the force potential correspond to the geometrically linear (first-order) analysis. The vector of displacements $\boldsymbol{q}_o = \lambda\boldsymbol{r}_o^{-1}\boldsymbol{Q}$, which conforms to the geometrically linear analysis, characterizes the deformed state of equilibrium of the linear system.

Let us see how the above-introduced notions manifest themselves in the example of a system shown in Fig. 11.1. We have

$$\boldsymbol{q} = \begin{bmatrix} u \\ v \end{bmatrix}, \quad \boldsymbol{r}_o = \begin{bmatrix} \partial^2\mathsf{E}/\partial u^2 & \partial^2\mathsf{E}/\partial u\partial v \\ \partial^2\mathsf{E}/\partial u\partial v & \partial^2\mathsf{E}/\partial v^2 \end{bmatrix}_{u=u_o, v=0} = \begin{bmatrix} c_1 + c_2 & 0 \\ 0 & c_3 \end{bmatrix},$$

$$\Pi_s = \lambda u, \qquad \boldsymbol{Q} = \begin{bmatrix} 1 \\ 0 \end{bmatrix}, \quad \boldsymbol{q}_o = \begin{bmatrix} u_o \\ v_o \end{bmatrix} = \lambda \begin{bmatrix} 1/(c_1 + c_2) \\ 0 \end{bmatrix}. \tag{1.45}$$

Comparing with (1.36) shows that the solution of the linear problem described by displacements $u_o = \lambda/(c_1 + c_2)$ and $v_o = 0$ coincides with the main deformed state of equilibrium.

As we have said before, the investigation of quality of the main equilibrium state should be based on expanding the E and Π_s functions in the vicinity of vector $\boldsymbol{q}_o$, not zero, where $\boldsymbol{q} = \boldsymbol{q}_o + \delta\boldsymbol{q}$. Recall that (1.11) defines the elements of matrix $\boldsymbol{r}_\lambda$,

$$\boldsymbol{r}_\lambda = \left\| \left[\frac{\partial^2 \mathsf{E}}{\partial q_i \partial q_j}(\boldsymbol{q}_o) - \lambda \frac{\partial^2 \overline{\Pi}_s}{\partial q_i \partial q_j}(\boldsymbol{q}_o) \right] \right\|,$$

as functionally dependent on parameter λ. We emphasize again that the second derivatives are taken at the n-dimensional "point" $\boldsymbol{q} = \boldsymbol{q}_o$ the components of which are also proportional to parameter λ as in (1.45).

In our problem with two degrees of freedom, in its main equilibrium state we have, as we already found out,

$$\boldsymbol{r}_\lambda = \left\| \begin{matrix} \partial^2 \mathsf{E}/\partial u^2 & \partial^2 \mathsf{E}/\partial u \partial v \\ \partial^2 \mathsf{E}/\partial u \partial v & \partial^2 \mathsf{E}/\partial v^2 \end{matrix} \right\|_{u=u_o, v=0} = \left\| \begin{matrix} c_1 + c_2 & 0 \\ 0 & \dfrac{c_1 - c_2}{l} u_o + c_3 \end{matrix} \right\| =$$

$$= \left\| \begin{matrix} c_1 + c_2 & 0 \\ 0 & \lambda \dfrac{c_1 - c_2}{l(c_1 + c_2)} + c_3 \end{matrix} \right\| = \left\| \begin{matrix} c_1 + c_2 & 0 \\ 0 & c_3 \end{matrix} \right\| - \lambda \left\| \begin{matrix} 0 & 0 \\ 0 & \dfrac{c_2 - c_1}{l(c_1 + c_2)} \end{matrix} \right\| = \boldsymbol{r}_o - \lambda \boldsymbol{r}_G .$$

Matrix $\boldsymbol{r}_G$,

$$\boldsymbol{r}_G = \left\| \begin{matrix} 0 & 0 \\ 0 & \dfrac{c_2 - c_1}{l(c_1 + c_2)} \end{matrix} \right\|,$$

is a *matrix of geometric stiffness* of the system.

The condition that the tangential stiffness matrix is no longer positive definite,

$$\boldsymbol{r}_\lambda = \boldsymbol{r}_o - \lambda \boldsymbol{r}_G, \tag{1.46}$$

gives the critical value of the load parameter according to Lagrange–Dirichlet,

$$\lambda_{cr} = c_3 \frac{c_1 + c_2}{c_2 - c_1},$$

which was established earlier by (1.41).

Now let us turn to a linearized formulation of stability for our example. By expanding functions (1.29) into a Taylor series in the vicinity of zero and keeping terms up to second order inclusive, we obtain

$$u_1 = u + \frac{v^2}{2l}, \qquad u_2 = u - \frac{v^2}{2l}.$$

It enables us to represent the expression of E from (1.30), where all terms up to third order are kept, as follows:

$$\mathsf{E} = \frac{c_1\left(u^2 + \dfrac{uv^2}{l}\right)}{2} + \frac{c_2\left(u^2 - \dfrac{uv^2}{l}\right)}{2} + \frac{c_3 v^2}{2}.$$

It follows from here and from (1.31) for the linearized problem that

$$\boldsymbol{r}_\lambda = \left\|\begin{bmatrix} \partial^2\mathsf{E}/\partial u^2 & \partial^2\mathsf{E}/\partial u\partial v \\ \partial^2\mathsf{E}/\partial u\partial v & \partial^2\mathsf{E}/\partial v^2 \end{bmatrix}\right\|_{u=u_o, v=v_o} = \left\|\begin{matrix} c_1 + c_2 & \dfrac{c_1 - c_2}{l} v_o \\ \dfrac{c_1 - c_2}{l} v_o & \dfrac{c_1 - c_2}{l} u_o + c_3 \end{matrix}\right\|.$$

We know that $u_o = \lambda/(c_1 + c_2)$ and $v_o = 0$ for the linear equilibrium state, therefore in the linearized problem we have the same formula (1.46) with the same matrices $\boldsymbol{r}_o$ and $\boldsymbol{r}_G$.

11.1.4 Example 2 – paradoxes in the stability analysis

The force potential in the preceding example was a linear function of the generalized coordinates of the system, and this fact made our reasoning simpler to a certain extent. However, the general expression of the force potential can be a more complicated function of the generalized coordinates. Now we are going to give an example where a natural choice of the generalized coordinates will produce a nonlinear expression of the force potential.

Let us consider a mechanical system with two degrees of freedom as shown in Fig. 11.3. The problem assumes the *AB* bar to be inflexible, but it can change its length due to a longitudinal deformation of a spring inserted in the bar and having the stiffness of c_2. All sizes in the figure are given for an initial (undeformed) state of the system, which is also assumed to be a stable equilibrium when there is no external load. This is the state we will count our generalized coordinates from.

The following generalized coordinates of the system will be convenient:

$$q_1 = \theta, \quad q_2 = \Delta, \tag{1.47}$$

where θ is an angle of rotation (a slope) of the rigid lever AB around the A point (which is positive when the lever rotates clockwise from its initial position), Δ is a shortening of the rotating lever AB. Let u and v be the respective horizontal and vertical displacements of point B (the positive directions of the displacements are shown in Fig. 11.3).

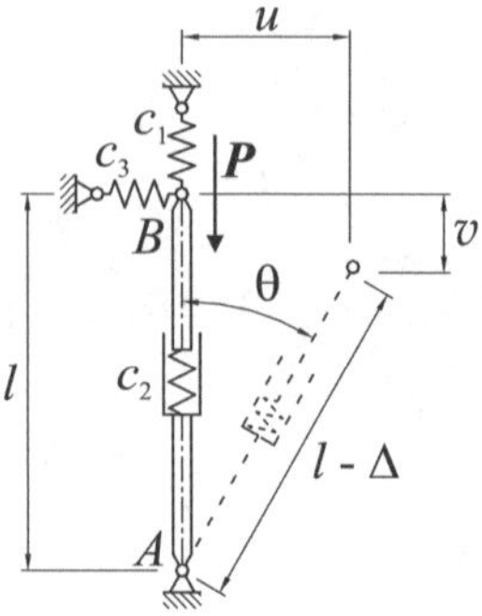

Fig. 11.3. An example of a mechanical system with two degrees of freedom

It is easy to see from geometrical considerations that in the general geometrically nonlinear case the kinematical parameters, u and v, are expressed via the generalized coordinates of the system, θ and Δ, as follows:

$$u=(l-\Delta)\sin\theta, \quad v=l-(l-\Delta)\cos\theta. \tag{1.48}$$

The basic state of equilibrium, the stability of which we want to find out, is described by kinematic parameters θ_o and Δ_o where

$$\theta_o=0, \quad \Delta_o=\frac{\lambda}{c_1+c_2}. \tag{1.49}$$

Obviously, the deformed state of equilibrium according to (1.49) conforms to the linear formulation of the problem.

First of all, we should analyze the stability of the equilibrium in the general geometrically nonlinear formulation. Let us begin with expressions of E and Π_s. We have

$$\mathsf{E}=c_1\frac{v^2}{2}+c_2\frac{\Delta^2}{2}+c_3\frac{u^2}{2}=c_1\frac{\left[l-(l-\Delta)\cos\theta\right]^2}{2}+c_2\frac{\Delta^2}{2}+c_3\frac{(l-\Delta)^2\sin^2\theta}{2}$$

and

$$\Pi_s=\lambda v=\lambda\left[l-(l-\Delta)\cos\theta\right].$$

The respective second derivatives of the full potential energy $\mathsf{L} = \mathsf{E} - \Pi_s$ with respect to the generalized coordinates at the points of our desirable state of equilibrium are

$$\frac{\partial^2 \mathsf{L}}{\partial \theta^2}(\theta_o, \Delta_o) = (l - \Delta_o)\left[c_1\Delta_o + c_3(l - \Delta_o) - \lambda\right] = \left(l - \frac{\lambda}{c_1 + c_2}\right)\left(c_3 l - \lambda\frac{c_2 + c_3}{c_1 + c_2}\right),$$

$$\frac{\partial^2 \mathsf{L}}{\partial \theta \partial \Delta}(\theta_o, \Delta_o) = 0, \quad \frac{\partial^2 \mathsf{L}}{\partial \Delta^2}(\theta_o, \Delta_o) = c_1 + c_2 .$$

Now we can write out a full expression for the tangential stiffness matrix in an arbitrary point of our state of equilibrium:

$$\boldsymbol{r}_\lambda = \left\| \begin{bmatrix} \dfrac{\partial^2 \mathsf{L}}{\partial \theta^2} & \dfrac{\partial^2 \mathsf{L}}{\partial \theta \partial \Delta} \\ \dfrac{\partial^2 \mathsf{L}}{\partial \theta \partial \Delta} & \dfrac{\partial^2 \mathsf{L}}{\partial \Delta^2} \end{bmatrix} \right\|_{\theta=\theta_o, \Delta=\Delta_o} = \left\| \begin{bmatrix} \left(l - \dfrac{\lambda}{c_1 + c_2}\right)\left(c_3 l - \lambda\dfrac{c_2 + c_3}{c_1 + c_2}\right) & 0 \\ 0 & c_1 + c_2 \end{bmatrix} \right\| .$$

Making the determinant of the tangential stiffness matrix equal to zero helps us find critical values of the load parameter. This condition produces two critical values:

$$\frac{\lambda'_{cr}}{(c_1 + c_2)l} = 1, \quad \frac{\lambda''_{cr}}{(c_1 + c_2)l} = \frac{c_3}{c_2 + c_3} .$$

Obviously, $\lambda''_{cr} < \lambda'_{cr}$, therefore the final critical value of the load parameter for this problem is the smaller one:

$$\lambda_{cr} = c_3 l \frac{c_1 + c_2}{c_2 + c_3} . \tag{1.50}$$

Let us take a closer look at this formula. In the limit case when the stiffness of the lower spring grows infinitely, $c_2 \to \infty$, this formula produces a well-known expression,

$$\lambda_{cr} = c_3 l , \tag{1.51}$$

which defines a critical longitudinal force for a perfectly rigid bar of the length l with one end of it fixed by a hinge and the other not being able to move laterally because of a spring of the stiffness c_3 .

However, in the opposite limit case when $c_2 \to 0$, formula (1.50) gives

$$\lambda_{cr} \to c_1 l , \tag{1.52}$$

and *this* looks suspicious. Intuition says the system should not lose its stability at all when $c_2 \to 0$, so there is a contradiction with (1.52). But this is just a first impression which can be deceptive.

The first thing to note is that the limit case $c_2 \to 0$ combines two opposite tendencies. One of them is that the longitudinal compressive force N,

$$N = \frac{\lambda c_2}{c_1 + c_2},$$

which appears in the *AB* bar, falls together with stiffness c_2. Obviously, this tendency increases the critical value of the load parameter. But there is another tendency, too. As stiffness c_2 decreases, the actual length of the *AB* bar decreases, too, and at the moment of loss of stability it becomes equal to l_*,

$$l_* = l - \frac{\lambda_{cr}}{c_1 + c_2}.$$

The shortening of the *AB* lever causes the critical value of the load parameter to lower. But the critical value of the longitudinal force in the *AB* bar is

$$N_{cr} = c_3 l_*,$$

which produces the same formula (1.50) after making the substitutions, and this fact brings back our belief in it.

So what is the matter? To clear up the things, we would like to return to the expression of the tangential stiffness matrix $\boldsymbol{r}_\lambda$ and note that the condition of its positive definiteness is equivalent to

$$(1 - \lambda'_{cr})(1 - \lambda''_{cr}) > 0. \qquad (1.53)$$

According to this condition, we define an area of stability and an area of instability of the system in the plane of two dimensionless parameters α and β by assuming

$$\alpha = \frac{c_2}{c_3}, \quad \beta = \frac{\lambda}{(c_1 + c_2)l}.$$

Fig. 11.4 shows the area of instability, where the inequality (1.53) does not hold, as a darkened region. The rest of the half-plane $\alpha \geq 0$ conforms to the stable equilibrium of the system. The meaning of the dash lines in this figure will be discussed later.

Let us imagine the process of increasing the load for a very small value of the spring's stiffness c_2, or, more exactly, for a small dimensionless parameter α. At the beginning the equilibrium remains stable, but as the load continues to grow we have to cross the lower boundary of the darkened area and thus make the equilibrium unstable.

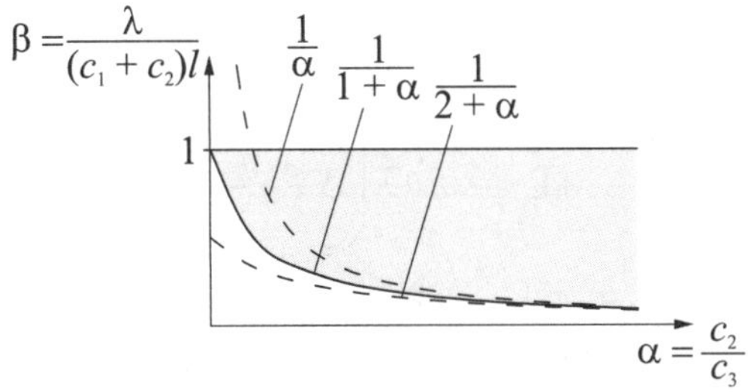

Fig. 11.4. Areas of stable and unstable equlibrium states of the system

By continuing to increase the load, we soon cross the second (horizontal) boundary of the instability area, and the system returns to a stable balanced state! The lower the α parameter is, the shorter the segment of the unstable equilibrium gets. In the limit case $c_2 = 0$ this segment becomes just an isolated point ($\alpha = 0, \beta = 1$). So what kind of equilibrium can we see in this secondary position?

Let us switch from formal mathematical transformations to the mechanical treatment of the area. It is easy to notice that the upper part of the stable equilibrium area conforms to such enormous deformations in the system that move the *B* point *below* the *A* point; the external force *P* travels down there, too. Thus, the *AB* bar is tensioned rather than compressed in that area. It takes one glance of a mechanician to conclude that this state of equilibrium is totally stable. Thus, the tendency (1.52) by no means contradicts our first intuitive apprehension of the stability of equilibrium at $c_2 = 0$.

As we can see, even this simple example helps confirm our introductory thesis of this chapter about the insidious peculiarity of the equilibrium statiblity problems.

Note that the previous derivation was based on the general nonlinear geometric relationships. Our next step is to construct a linearized formulation of the stability problem.

First of all, let us represent functions (1.48) as a Taylor series where only the second-order and lower terms are kept:

$$u = (l - \Delta)\theta, \quad v = \Delta + l\frac{\theta^2}{2}.$$

The potential strain energy E looks as follows, up to third-order terms:

$$\mathsf{E} = c_1 \frac{v^2}{2} + c_2 \frac{\Delta^2}{2} + c_3 \frac{u^2}{2} = c_1 \frac{\Delta^2 + l\Delta\theta^2}{2} + c_2 \frac{\Delta^2}{2} + c_3 \frac{l^2\theta^2 - 2l\Delta\theta^2}{2}.$$

At the same time, the force potential is represented by an expansion with the accuracy up to second-order terms, which gives

$$\Pi_s = \lambda v = \lambda \left(\Delta + \frac{l\theta^2}{2} \right).$$

Thus, in the linearized formulation of the problem we must use the following expression of the full potential energy:

$$\mathsf{L} = c_1 \frac{\Delta^2 + l\Delta\theta^2}{2} + c_2 \frac{\Delta^2}{2} + c_3 \frac{l^2\theta^2 - 2l\Delta\theta^2}{2} - \lambda \left(\Delta + \frac{l\theta^2}{2} \right).$$

Now we have

$$\frac{\partial^2 \mathsf{L}}{\partial \theta^2}(\theta_o, \Delta_o) = (c_1 - 2c_3) l \Delta_o + c_3 l^2 - \lambda l = l \left(c_3 l - \lambda \frac{c_2 + 2c_3}{c_1 + c_2} \right),$$

$$\frac{\partial^2 \mathsf{L}}{\partial \theta \partial \Delta}(\theta_o, \Delta_o) = 0, \quad \frac{\partial^2 \mathsf{L}}{\partial \Delta^2}(\theta_o, \Delta_o) = c_1 + c_2.$$

The condition of positive definiteness of $\boldsymbol{r}_\lambda$ produces an expression of the critical load in the linearized analysis:

$$\lambda_{cr} = c_3 l \frac{c_1 + c_2}{c_2 + 2c_3}. \tag{1.54}$$

But this expression is totally different from the "correct" formula (1.50)!

First of all, let us consider the limit cases again. As $c_2 \to \infty$, formula (1.54) of the critical load leads to a correct result, i.e. to the same formula (1.51). If we make c_2 tend to zero, we will have a halved result comparing to (1.52):

$$\lambda_{cr} \to c_1 l / 2. \tag{1.55}$$

What then, another paradox?

Rest assured there is no paradox at all. Indeed, why should we believe the critical loads calculated by the linearized method must always be equal to those of the nonlinear approach? There is no reason to believe so. But then we have the question to what extent the linearized models are applicable in the stability analysis.

Recall that the linearized formulation of the stability problem implies the check of stability of an equilibrium of a *linear* system, not just *any* equilibrium. Engineers know well that the error due to the linearization of the analysis is acceptable only when the relative strains in a structure's elements are much less than one, say of the order of magnitude 10^{-2} or even less. Therefore the linearization of the problem, both for the purpose of finding a state of equilibrium and for analyzing its stability, may be justified only for small relative strains[5]. Otherwise the error of the solution could get unacceptably large.

The smallness of the strains means in our problem that the relative shortening of bar *AB* is small. In other words, if we assume the linearization to be possible, therewith we imply that the following estimate holds:

$$\beta = \frac{\lambda}{(c_1 + c_2)l} << 1 .$$

In terms of dimensionless designations, the formula (1.54) can be rewritten as

$$\beta_{cr} = 1/(2+\alpha) ,$$

and its applicability implies, roughly speaking, that

$$\alpha = c_2 / c_3 > 100 .$$

Under these conditions, formula (1.50) rewritten in the same dimensionless designations,

$$\beta_{cr} = 1/(1+\alpha) ,$$

differs little from the approximate formula (1.54). The solution of the paradox is that passing to the limit as $c_2 \to \infty$ is justified while $c_2 \to 0$ is not allowed. In the latter case the condition of smallness of strains β, which is equivalent to a big α, is violated inevitably.

---

[5] It is exactly "*may be justified*" not simply "*is justified*". The smallness of the strains is a necessary condition for the linearization to be possible but not sufficient. As an example, we can consider a very shallow von Mises' truss the equilibrium state of which, determined by the linear theory, can be too rough an approximation of the true state defined by nonlinear equations.

A remark on a non-invariant critical load with respect to the choice of generalized coordinates

There is an important question that comes up with regard to the linearized analysis of stability. This question is somehow ignored in all books on stability that we know of. The question is why there is no invariance in the formulas of the critical load with respect to the choice of generalized coordinates. If we deal with the full nonlinear formulation, then the invariance obviously takes place at least when we investigate the stability of the same state of equilibrium which is an exact solution of the respective nonlinear equations. But we are interested with the linearized analysis which, as we already know, can produce approximate formulas of the critical load. Therefore it is reasonable to expect that the invariance can be lost in the linearized analysis.

This can be easily demonstrated by the example of the same system shown in Fig. 11.3. This time the generalized coordinates will be kinematical parameters u and v, and (1.47) will be replaced by

$$q_1 = u, \quad q_2 = v. \tag{1.56}$$

From the Pythagorean theorem,

$$(l-\Delta)^2 = u^2 + (l-v)^2 ,$$

we obtain this in a formal mathematical way:

$$\Delta = l - \sqrt{u^2 + (l-v)^2} = l\left(1 - \sqrt{1 - 2\frac{v}{l} + \frac{v^2}{l^2} + \frac{u^2}{l^2}}\right). \tag{1.57}$$

The minus sign only is kept intentionally before the square root in the above formula because it is this sign that corresponds to small strains when $\Delta \ll l$. Also, in the state of equilibrium of the linear system we should assume

$$u_o = 0, \quad v_o = \lambda/(c_1 + c_2) .$$

Expanding the function (1.57) into a Taylor series gives, up to quadratic terms:

$$\Delta = v - u^2/(2l) .$$

Now we can rewrite the expression of the system's full potential energy up to the third-order terms:

$$\mathsf{L}=c_1\frac{v^2}{2}+c_2\frac{\Delta^2}{2}+c_3\frac{u^2}{2}-\lambda v=c_1\frac{v^2}{2}+c_2\frac{v^2-\dfrac{vu^2}{l}}{2}+c_3\frac{u^2}{2}-\lambda v\,.$$

Calculation of second derivatives of the full potential strain energy in the state of equilibrium gives

$$\frac{\partial^2\mathsf{L}}{\partial u^2}(u_o,v_o)=c_3-c_2\frac{v_o}{l}\,,\qquad \frac{\partial^2\mathsf{L}}{\partial u\partial v}(u_o,v_o)=0\,,\qquad \frac{\partial^2\mathsf{L}}{\partial v^2}(u_o,v_o)=c_1+c_2\,.$$

Hence another expression of the critical load:

$$\lambda_{cr}=c_3 l\frac{c_1+c_2}{c_2}\,. \tag{1.58}$$

Rewriting this formula in a dimensionless form gives

$$\beta_{cr}=1/\alpha\,. \tag{1.59}$$

We already know that all three expressions for the dimensionless value of the critical load parameter actually give the same result, in spite of differences in their written forms, and differ only in the relative strain components which can be ignored comparing to one. Or,

$$1/\alpha\approx 1/(1+\alpha)\approx 1/(2+\alpha)$$

at sufficiently large values of α.

Nonetheless, the very formulas for the critical values of the load in the linearized problem formulation depend on the choice of the generalized coordinates. This is what we mean by saying there is no invariance in the formulas.

In Fig. 11.4, dashed lines depict two more curves that conform to the critical loads from two versions of the linearized problem formulations. As we can see, one of the curves is below the accurate solution of the nonlinear problem while the other is above. Hence a conclusion: the sign of the error introduced by the linearization is undefined.

* * *

The analysis of the above two examples might seem too cumbersome. This is indeed so, but we have an excuse: this complicated analysis is a sign of our great wish to demonstrate the technique of direct application of the Lagrange–Dirichlet theorem to the analysis of stability of equilibrium of a mechanical system or structure. The traditional way of calculating critical forces, which is based on a so-called static method, for systems with a finite number of DOFs has been described in sufficient detail in many books and textbooks such as [1], [18], [2]. Therefore we can omit demonstrations of its techniques in application to the above examples. Another incentive for us to discuss the technique of direct application of the Lagrange-Dirichlet theorem so thoroughly is the fact that this analysis is especially convincing in situations where so-called paradoxes take place. The situations like that could hardly be resolved within the scope of the purely static critical load analysis.

11.2 Variational description of critical loads

We have found that the problem of equilibrium stability in the linearized formulation is equivalent to a spectral problem for a couple of matrices one of which (the geometric stiffness matrix $\boldsymbol{r}_G$) is not necessarily positive definite.

The general characteristic equation,

$$|\boldsymbol{r}_0 - \lambda \boldsymbol{r}_G| = 0, \tag{2.1}$$

will have p positive roots and q negative ones so that $p + q \leq n$. The negative eigenvalues determine, according to their physical sense, the critical values of the load parameter when all the external forces alter their direction simultaneously.

Obviously, the most practical numbers are two lowest absolute critical values of the load parameter (of opposite signs). These two define an admissible range of the load values that ensures the stable equilibrium of a system[6]. However, it does not mean higher eigenvalues of the characteristic equation (2.1) have no practical sense, as it can be often read in various sources. Considerations and examples concerning this subject are given in our book [16].

Let a number λ_{cr} be one of the roots of the characteristic equation (2.1). It means the homogeneous system of equations (1.14) allows a nontrivial solution with respect to vector $\delta\boldsymbol{q}$ at $\lambda = \lambda_{cr}$. In its physical sense, the components of the respective eigenvector $\delta\boldsymbol{q}$ characterize such relationships between infinitesimal deviations of the generalized

[6] It can be enough in many cases to consider only positive eigenvalues of the characteristic equation (2.1) – when the sign of the load cannot alter. For example, the self-weight of a structure always has the same fixed direction according to the gravity. However, there can be other types of loads such as loads on a bridge caused by a braking rolling stock; those can act in either direction.

coordinates from the state of equilibrium which permit sign alteration of the second variation of the system's full potential energy.

The eigenvector $\delta\boldsymbol{q}$ is said to define a *mode of buckling (loss of stability) of the system*. Further we are going to drop the designation of the buckling mode as $\delta\boldsymbol{q}$ and use a more convenient designation of $\boldsymbol{z}$; this provides a symmetry with the natural oscillation analysis. Thus, in the new designations we have

$$(\boldsymbol{r}_{\mathrm{o}} - \lambda_{cr}\boldsymbol{r}_G)\boldsymbol{z} = \boldsymbol{0}. \tag{2.2}$$

To distinguish between positive and negative eigenvalues of the couple of matrices $\boldsymbol{r}_{\mathrm{o}}$ and $\boldsymbol{r}_G$, we order them separately by defining

$$\lambda_q^- \le \ldots \le \lambda_2^- \le \lambda_1^- < 0 < \lambda_1^+ \le \lambda_2^+ \le \ldots \le \lambda_p^+. \tag{2.3}$$

Clearly, at a given loading intensity λ the equilibrium is stable if $\lambda_1^- < \lambda < \lambda_1^+$. Therefore the open interval $(\lambda_1^-, \lambda_1^+)$ is called *an area of buckling (loss of stability)*.

The theory of equilibrium stability wields the term of a system's *degree of instability*. A state of equilibrium is said to have the degree of instability equal to s at a given λ if one of the two conditions holds:

$$\lambda_s^+ < \lambda < \lambda_{s+1}^+ \quad \text{or} \quad \lambda_{s+1}^- < \lambda < \lambda_s^-. \tag{2.4}$$

We will say that vector $\boldsymbol{z}_i^+$ is *i-th positive mode of buckling* of the system if

$$(\boldsymbol{r}_{\mathrm{o}} - \lambda_i^+\boldsymbol{r}_G)\boldsymbol{z}_i^+ = \boldsymbol{0} \quad (1 \le i \le p).$$

Similarly, we define vector $\boldsymbol{z}_i^-$ as *i-th negative mode of buckling* if

$$(\boldsymbol{r}_{\mathrm{o}} - \lambda_i^-\boldsymbol{r}_G)\boldsymbol{z}_i^- = \boldsymbol{0} \quad (1 \le i \le q).$$

We emphasize that these definitions relate the poisitivity (or negativity) of a buckling mode not to the signs of the components of the mode but to the sign of the respective eigenvalue for this mode.

11.2.1 A Rayleigh ratio and a recursive variational calculation of critical loads

Here we introduce a formal Rayleigh ratio for the equilbrium stability analysis by carrying this notion over from the frequency spectral analysis and using it to search for a spectrum of critical values of the load parameter. Similarly to (**10**.2.28), we have

$$\mathrm{r}(\boldsymbol{z}) = \frac{\boldsymbol{z}^{\mathrm{T}} \boldsymbol{r}_{\mathrm{o}} \boldsymbol{z}}{\boldsymbol{z}^{\mathrm{T}} \boldsymbol{r}_{G} \boldsymbol{z}} . \tag{2.5}$$

The critical difference between the Rayleigh ratio (2.5) and its frequency-analysis counterpart (**10**.2.28) is that formula (2.5) does not guarantee the positivity of its denominator.

We cannot, unfortunately, mimic our reasoning from the frequency analysis to find conditions of stationarity of this new Rayleigh ratio. The matter is that matrix $\boldsymbol{r}_G$ is not positive definite and therefore cannot generate a metric, so we cannot use the normalization condition $\boldsymbol{z}^{\mathrm{T}} \boldsymbol{r}_G \boldsymbol{z} = 1$ either. However, we have another opportunity at hand, and we will not be slack to use it.

Suppose the Rayleigh ratio takes a stationary value on a certain vector $\boldsymbol{z}$. This means

$$\frac{\partial \mathrm{r}}{\partial z_i} = 0 \quad (i = 1, \ldots, n) .$$

Representing formula (2.5) as

$$\mathrm{r}(\boldsymbol{z}) \boldsymbol{z}^{\mathrm{T}} \boldsymbol{r}_G \boldsymbol{z} = \boldsymbol{z}^{\mathrm{T}} \boldsymbol{r}_{\mathrm{o}} \boldsymbol{z}$$

and differentiating both parts of it with respect to z_i yields

$$\frac{\partial \mathrm{r}}{\partial z_i} \boldsymbol{z}^{\mathrm{T}} \boldsymbol{r}_G \boldsymbol{z} + 2\mathrm{r}\, r_G^{ij} z_j = 2 r_{\mathrm{o}}^{ij} z_j .$$

But $\partial \mathrm{r} / \partial z_i = 0$ at the point of stationarity, so at the same point we have

$$(\boldsymbol{r}_{\mathrm{o}} - \mathrm{r} \boldsymbol{r}_G) \boldsymbol{z} = \boldsymbol{0} . \tag{2.6}$$

Now it is clear that the stationary values of the Rayleigh ratio r are precisely equal to the eigenvalues of the couple of matrices $\boldsymbol{r}_{\mathrm{o}}$ and $\boldsymbol{r}_G$, and these eigenvalues are taken on the respective eigenvectors.

As we have already ordered all stationary values of r according to (2.3), therewith we have established the following two propositions, which concern lower critical values of the load parameter:

> *The lowest positive value of functional* $\mathrm{r}(\boldsymbol{z})$ *is taken on first positive eigenvector* $\boldsymbol{z}_1^+$, *and this value is equal to the least positive eigenvalue* λ_1^+ *of matrix* $\boldsymbol{r}_{\mathrm{o}}$ *with respect to matrix* $\boldsymbol{r}_G$, *i.e.*

$$\min \mathsf{r}(z) = \mathsf{r}(z_1^+) = \lambda_1^+ ,$$

where the minimum is searched for among all vectors z *that satisfy the condition* $z^\mathsf{T} \boldsymbol{r}_G z > 0$.

Similarly,

The lowest (by absolute value) negative value of functional $\mathsf{r}(z)$ *is taken on first negative eigenvector* z_1^-, *and this value is equal to the least (by absolute value) negative eigenvalue* λ_1^- *of matrix* $\boldsymbol{r}_\text{o}$ *with respect to matrix* $\boldsymbol{r}_G$, *i.e.*

$$\min |\mathsf{r}(z)| = -\mathsf{r}(z_1^-) = -\lambda_1^-$$

where the minimum is searched for among all vectors z *that satisfy the condition* $z^\mathsf{T} \boldsymbol{r}_G z < 0$.

Let λ_α and λ_β be two different eigenvalues (no matter whether they are of the same or opposite signs), and let z_α and z_β be their respective eigenvectors. This means

$$\boldsymbol{r}_\text{o} z_\alpha = \lambda_\alpha \boldsymbol{r}_G z_\alpha \quad \text{and} \quad \boldsymbol{r}_\text{o} z_\beta = \lambda_\beta \boldsymbol{r}_G z_\beta$$

We multiply the first of the equalities by z_β^T and the second by z_α^T on the left and carry λ over to the left-hand parts of the equalities:

$$\lambda_\alpha^{-1} z_\beta^\mathsf{T} \boldsymbol{r}_\text{o} z_\alpha = z_\beta^\mathsf{T} \boldsymbol{r}_G z_\alpha \quad \text{и} \quad \lambda_\beta^{-1} z_\alpha^\mathsf{T} \boldsymbol{r}_\text{o} z_\beta = z_\alpha^\mathsf{T} \boldsymbol{r}_G z_\beta .$$

The right-hand parts of these equalities are equal seeing that matrix $\boldsymbol{r}_G$ is symmetric. As matrix $\boldsymbol{r}_\text{o}$ is symmetric, too, hence

$$(\lambda_\alpha^{-1} - \lambda_\beta^{-1}) z_\beta^\mathsf{T} \boldsymbol{r}_\text{o} z_\alpha = 0 .$$

The eigenvalues λ_α and λ_β are different (and both nonzero), so the respective eigenvectors z_α and z_β are orthogonal in the metric generated by matrix $\boldsymbol{r}_\text{o}$, i.e. in the original energy metric,

$$(z_\beta, \boldsymbol{r}_\text{o} z_\alpha) = 0 . \tag{2.7}$$

Further we will say eigenvectors z_α and z_β are E-orthogonal meaning the equality (2.7).

The same eigenvectors also obey a similar equality:

$$(z_\beta, r_G z_\alpha) = 0\,. \tag{2.8}$$

However, the equality (2.8) cannot be interpreted in terms of orthogonality of vectors z_α and z_β because matrix r_G does not generate any metric.

As all $(p + q)$ eigenvectors are linearly independent, we can suggest that the relationships (2.7) and (2.8) hold, generally, for any couple of different eigenvectors z_α and z_β.

Returning to the proposition about the stationarity points of the Rayleigh ratio established above and considering the order of the eigenvalues defined by (2.3), we arrive at a recursive variational definition of the critical values of the load parameter:

The least positive value of functional $\mathsf{r}(z)$, *sought for in the set of vectors* z E-*othogonal to vectors* $z_1^+,\dots,z_{i-1}^+$, *is taken on i-th positive mode of buckling,* z_i^+, *and this least value is equal to i-th positive critical value of the load parameter,* λ_i^+, *that is,*

$$\min \mathsf{r}(z) = \mathsf{r}(z_i^+) = \lambda_i^+$$

under the conditions: $z^{\mathrm{T}} r_o z_1^+ = 0,\dots, z^{\mathrm{T}} r_o z_{i-1}^+ = 0$ *and* $z^{\mathrm{T}} r_G z > 0$.

Similarly,

The least (by absolute value) negative value of functional $\mathsf{r}(z)$, *sought for in the set of vectors* z E-*othogonal to vectors* $z_1^-,\dots,z_{i-1}^-$, *is taken on i-th negative mode of buckling,* z_i^-, *and this least (by absolute value) negative value is equal to i-th negative critical value of the load parameter,* λ_i^-, *that is,*

$$\min |\mathsf{r}(z)| = -\mathsf{r}(z_i^-) = -\lambda_i^-$$

under the conditions: $z^{\mathrm{T}} r_o z_1^- = 0,\dots, z^{\mathrm{T}} r_o z_{i-1}^- = 0$ *and* $z^{\mathrm{T}} r_G z < 0$.

The above recursive variational definitions of the critical loads are an extension of the Rayleigh–Weber variational principle onto the stability analysis of elastic systems.

11.2.2 A remark on the effect of constraints on the stability of a linearized elastic system

The effect of constraints on the stability of an elastic system is not so easily found out as it may seem from the first glance. The matter is that the stability analysis requires a more careful classification of the constraints than that needed for analyzing the effect of the constraints on the frequency spectrum of an elastic structure.

A point of view quite frequent among engineers – that the constraints always have a stabilizing effect – can lead to bad mistakes. To see this, take a look at an example shown in Fig. 11.1 and imagine that this system is subjected to an external rigid constraint that prevents the right end of the bar from moving. The mathematical interpretation of this constraint is

$$u_2 = 0\,. \tag{2.8}$$

The same constraint can be also treated as an infinite stiffness of the right spring, i.e. we can think that $c_2 \to \infty$ makes the system constrained. Passing to the limit in (1.41) gives a critical value of the load parameter $\tilde{\lambda}_{cr}$ for the constrained system:

$$\tilde{\lambda}_{cr} = c_3 l < \lambda_{cr} = c_3 l \frac{c_1 + c_2}{c_2 - c_1}\,.$$

In other words, the constraint of (2.8) has a destabilizing effect on the system. The same conclusion follows from the graph in Fig. 11.2 – the critical force falls together with the ratio c_1/c_2.

Why does it happen so and can we nonetheless indicate conditions under which constraints have a guaranteed stabilizing effect on a mechanical system?

Before we answer this question, it is useful to use the same example to analyze the effect of other constraints on the critical value of the load parameter. Now suppose that the system is subjected, unlike (2.8), to the constraint of the following kind:

$$u_1 = 0\,.$$

The mechanical meaning of this constraint is a prohibition of the longitudinal displacements at the left end of the left bar. Analysis of the same graph in Fig. 11.2 shows that even at $c_1 > c_2$, let alone $c_1 \to \infty$, the stability of the system is always stable under positive loads. This means the constraint $u_1 = 0$ is a stabilizing one for this particular state of equilibrium.

Finally, it is easy to see that the $v = 0$ constraint is also a stabilizing one because the system with it cannot lose its stability at all under any kinds of external actions.

We would like to note, returning to the destabilizing effect of the constraint (2.8), that that constraint both confines the potential mode of buckling and changes the very state of equilibrium. This means imposing the constraint (2.8) leads to a problem of stability of a totally different state of equilibrium[7].

We can distinguish constraints based on the time when they are imposed (conditionally). On one hand, there are constraints imposed on the system before the moment it is loaded. We will refer to such constraints as *prior constraints*. It is essential that a prior constraint is involved in the system's behavior in the course of its being loaded and, generally, affects the system's deformed state of equilibrium. However, there can be also the situation when a constraint is imposed on an already loaded and deformed system; this constraint will be unstressed. Such constraints will be referred to as *posterior constraints*. A posterior constraints does not change the state of equilibrium, but it confines a potential mode of buckling.

Of course, there can be prior constraints which do not change the state of equilibrium in any way. Following a nomenclature suggested by A.R. Rzhanitsin [19], this kind of constraint will be referred to as *an immobile constraint*. The constraint $v = 0$ in the above example is just this kind of constraint, while both $u_1 = 0$ and $u_2 = 0$ cannot be categorized as such.

Suppose the system is subjected to either a posterior constraint or a prior but immobile one. This means the system is in the same state of equilibrium described by vector $\boldsymbol{q}_o$ but is confined in its displacement variations $\delta\boldsymbol{q}$. According to our notation, such constraints are imposed on the components of vector $\boldsymbol{z}$ only. Further on, by the term 'constraint' we will mean this kind of constraints.

The general equation of such a constraint (supposed to be linear) is

$$a_1 z_1 + a_2 z_2 + \ldots + a_n z_n = 0 \tag{2.9}$$

with at least one of its coefficients a_i not equal to zero.

We reformulate the constraint equation (2.9) by introducing the constraint vector $\boldsymbol{b}$ like this:

[7] This is another evidence of the fact that the popular and seemingly harmless abridgement "*stability of a structure*" instead of the correct term "*stability of an equilibrium of a structure*" is not actually harmless at all.

$$\boldsymbol{b} = \boldsymbol{r}_0^{-1}\boldsymbol{a} \quad \text{where} \quad \boldsymbol{a} = \|[a_1,\ldots,a_n]\|^{\mathsf{T}}.$$

In these designations, the constraint equation (2.9) can be treated as a condition of E-orthogonality between the constraint vector $\boldsymbol{b}$ and vector $\boldsymbol{z}$:

$$\boldsymbol{b}^{\mathsf{T}}\boldsymbol{r}_0\boldsymbol{z} = (\boldsymbol{b}, \boldsymbol{z})_{\mathsf{E}} = 0. \tag{2.10}$$

We know from the recursive variational definition of the critical loads that $\lambda_1^+ = \min \mathsf{r}(\boldsymbol{z})$ for all $\boldsymbol{z} : \boldsymbol{z}^{\mathsf{T}}\boldsymbol{r}_G\boldsymbol{z} > 0$ and $-\lambda_1^- = \min|\mathsf{r}(\boldsymbol{z})|$ for all $\boldsymbol{z} : \boldsymbol{z}^{\mathsf{T}}\boldsymbol{r}_G\boldsymbol{z} < 0$. The same critical loads $\tilde{\lambda}_1^+$ and $\tilde{\lambda}_1^-$ for the system with the constraint (2.10) are the same minima but calculated under an additional condition, (2.10). But no additional limitation can lower the minimum. Hence

$$\tilde{\lambda}_1^+ \geq \lambda_1^+ \quad \text{and} \quad \tilde{\lambda}_1^- \leq \lambda_1^-. \tag{2.11}$$

Thus, imposing an immobile constraint can either extend the area of stability of the system or leave it unchanged. It means that the constraints of this kind are ones that stabilize the equilibrium of the structure. At the same time, prior constraints which are not immobile can be either stabilizing or destabilizing for the equilibrium.

In the same way as we did in the frequency analysis, we can give an independent variational definition of the critical loads and derive an analogue of the Routh theorem concerning the effect of constraints on the spectrum of critical forces. We are not going to do it now; instead, we suggest that the reader himself modify the respective formulations. It is not hard to do by using all stages of reasoning from Chapter **10** and take account of changes introduced into the eigenvalue spectrum by the presence of negative critical forces. We present only the final conclusion about the effect of the constraints:

By imposing posterior or immobile constraints on a mechanical system, the positive part of the spectrum of critical values of the load parameter can be shifted only to the right on the numerical axis, and the negative part of the spectrum can be shifted only to the left.

As for the notion of the Rayleigh ellipsoid used in the frequency spectrum analysis, it cannot be transferred to the stability analysis. The reason is obvious – the geometric stiffness matrix is not guaranteed to be positive definite.

11.2.3 Papkovich theorem of convexity of the stability area

The case when a structure is subjected to one fixed external action is rather an exception than a rule. Nearly always an engineer deals with a set of independent loads which can act in most various combinations. In this regard we encounter a problem how to choose a worst combination of the external actions both in the sense of strength and in the sense of a stability margin in states of equilibrium which correspond to various combinations of external loads.

We begin with a simple example. Imagine that an elastic system shown in Fig. 11.5 can be subjected to two independent forces P_1 and P_2, and the forces can vary within wide limits in their absolute value and can act in both directions. Three bars of the system are assumed to be perfectly rigid.

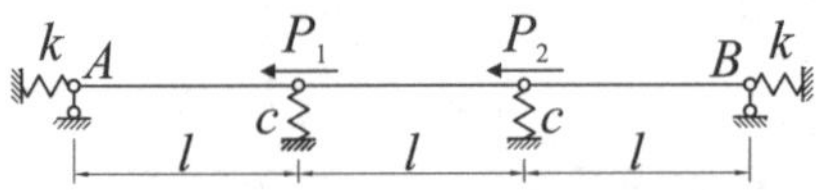

Fig. 11.5. The stability of equilibrium of an elastic system under two independent loads

If we fixate the ratio of the forces, P_1/P_2, then we can find a state of equilibrium and determine boundaries of the stability area for it following the standard procedure. Let us try to solve the problem generally by the Euler (static) method. We compose equations of equilibrium for the deformed state which is supposed to be feasible at the moment when the system loses its stability. Fig. 11.6 shows a hypothetic mode of buckling.

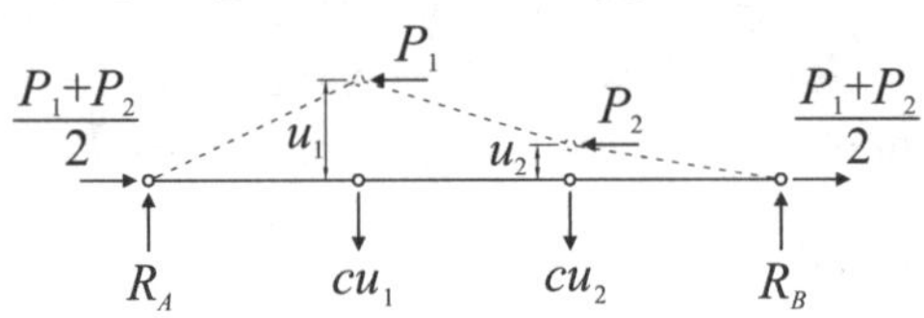

Fig. 11.6. Reactions in the constraints at the moment of loss of stability

In the linear formulation of the problem – one for which we are investigating the stability of equlibrium – the reactions in the horizontal springs of stiffness k will be both equal to $(P_1 + P_2)/2$ as shown in Fig. 11.6. Denoting the displacements of the hinged nodes above the vertical springs in the course of buckling by u_1 and u_2, we find that the respective reactions will be cu_1 and cu_2 where c is a stiffness of the vertical springs installed above the hinged nodes.

Reactions R_A and R_B of the vertically fixed supports (see Fig. 11.6) are easily determined from equations of equilibrium in terms of moments with respect to points A and B. We have

$$R_A = \frac{1}{3}\left(2c + \frac{P_1}{l}\right)u_1 + \frac{1}{3}\left(c + \frac{P_2}{l}\right)u_2, \quad R_B = \frac{1}{3}\left(c - \frac{P_1}{l}\right)u_1 + \frac{1}{3}\left(2c - \frac{P_2}{l}\right)u_2.$$

Now we use the fact that the moments are zero in the hinged nodes. This gives

$$R_A l - \frac{P_1 + P_2}{2}u_1 = 0, \quad R_B l + \frac{P_1 + P_2}{2}u_2 = 0.$$

Substituting in the above expressions of reactions R_A and R_B yields two simultaneous linear homogeneous equations for u_1 and u_2:

$$(4cl - P_1 - 3P_2)u_1 + (2cl + 2P_2)u_2 = 0, \ (2cl - 2P_1)u_1 + (4cl + 3P_1 + P_2)u_2 = 0.$$

Equating the determinant of the system to zero gives the following equality for the critical state:

$$4c^2l^2 + 4cl(P_1 - P_2) - (P_1 + P_2)^2 = 0. \quad (2.12)$$

It is convenient to introduce dimensionless force variables p_1 and p_2 by designating

$$p_1 = \frac{P_1}{cl}, \qquad p_2 = \frac{P_2}{cl}.$$

In terms of these, the equation (2.12) becomes

$$4 + 4(p_1 - p_2) - (p_1 + p_2)^2 = 0. \quad (2.13)$$

This equation defines a certain curve in the plane of parameters p_1 and p_2 as shown in Fig. 11.7. Clearly, a two-dimensional area Ω bounded by this curve is an area of stability of the system's equilibrium. The stability is unstable beyond this area. Fig. 11.7 shows the area of stability, Ω, the boundary of which satisfies the equation (2.13).

As we can see, point ($p_1 = 0$, $p_2 = 0$) belongs to the area of stability Ω – expectably, because the point conforms to the unloaded system. But the main thing is that the Ω area is a convex area.

By the way, note that the critical loads do not depend on the stiffness of the horizontal springs k[8].

---

[8] The only important thing here is that both horizontal springs (on the left and right ends of the system) have the same stiffness. If the stifness values were

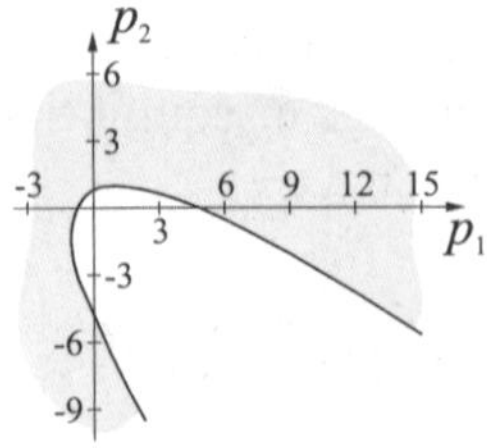

Fig. 11.7. An area of equilibrium stability

The property of convexity of the Ω area is not extrinsic to this problem. It turns out that the following very important theorem holds true; its author is P.F. Papkovich [14]:

> *When an elastic mechanical system is subjected to a combined loading, the area of its equilibrium stability, Ω, built in the space of loads within the scope of the linearized stability analysis, is a convex area that contains the coordinate origin.*

Papkovich was the first both to detect the fact of the convexity and to give a rigorous proof of that. The proof by Papkovich is, however, pretty complex, therefore other researchers tried subsequently to find other ways of validation. In particular, B.M. Broude [8] and A.R. Rzhanitsin [18] gave theirs.

---

related differently, the critical forces would be different, too. It should be noted also that the whole reasoning holds true only as long as the stiffnesses k are the same and <u>finite</u>. If the horizontal springs were replaced with rigid constraints, passing to the limit at $k \to \infty$ would be incorrect for the following reason. When the stiffness values are infinite, the relation between them is undefined, and so is the equlibrium state of the system, therefore the critical load is undefined as well. A similar example is given in a book by N.A. Alfutov [1] where those springs are replaced with rigid constraints, but the circumstance we indicate here was not noticed by the author of that book. By the way, Fig. 1.22 in the book by Alfutov [1] shows the stability area as closed while we have it open in Fig. 11.7. As can be easily seen, the conclusion that the Ω area is closed is a mistake for inextensible bars. To see this, imagine that forces P_1 and P_2 are equal in their value and opposite in their directions, $P_1 = -P_2$, so that they both extend the middle bar. This means the tension stresses remain completely within the limits of the middle bar, and the longitudinal force in the two extreme bars is zero, so the system cannot lose its stability with this distribution of the forces. The reasoning of Alfutov becomes correct if we assume the bars in the system to have an <u>equal</u>, perhaps very large but <u>finite</u> stiffness for tension/compression.

Before we start proving the Papkovich theorem, we would like to introduce a couple of definitions and establish one important auxiliary proposition.

Let a mechanical system be subjected to independent mechanical force actions in the number of m. Without limiting the generality, we can assume that an arbitrary combination of the loads is described by m-dimensional vector $\boldsymbol{P}$ where $\boldsymbol{P} = \|[P_1,\ldots,P_m]\|^{\mathsf{T}}$. We represent vector $\boldsymbol{P}$ as an expansion over unit vectors,

$$\boldsymbol{P} = \sum_{\alpha=1}^{m} P_\alpha \boldsymbol{e}_\alpha \,, \qquad \text{где } \boldsymbol{e}_\alpha = |\,[0,\ldots,0,1,0,\ldots,0]\,|^{\mathsf{T}} .$$

In other words, a unit vector $\boldsymbol{e}_\alpha$ has all its coordinates equal to zero except for α-th component which is equal to one.

Let us introduce the designation λ for the length of vector $\boldsymbol{P}$ in the load space,

$$\lambda = \sqrt{P_1^2 + \ldots + P_m^2} \; .$$

Now the load vector, $\boldsymbol{P}$, is represented as

$$\boldsymbol{P} = \lambda \sum_{\alpha=1}^{m} \overline{P}_\alpha \boldsymbol{e}_\alpha = \lambda \overline{\boldsymbol{P}} \,, \qquad \overline{\boldsymbol{P}} = \sum_{\alpha=1}^{m} \overline{P}_\alpha \boldsymbol{e}_\alpha \,, \qquad \overline{P}_\alpha = P_\alpha / \lambda \,,$$

where vector $\overline{\boldsymbol{P}}$ is by construction an arbitrary unit-length vector in the load space.

In the m-dimensional load space with axes $\boldsymbol{e}_1, \boldsymbol{e}_2, \ldots, \boldsymbol{e}_m$, we can find a critical value of the factor $\lambda = \lambda_{cr}$ for any fixed load defined by the components of a *unit load vector*, $\overline{\boldsymbol{P}} = \left\| \left[\overline{P}_1,\ldots,\overline{P}_m \right] \right\|^{\mathsf{T}}$, by which to multiply all components of the unit load vector to make the stability unstable in the linearized formulation of the problem. The geometrical interpretation of parameter λ_{cr} in the load space is a *critical length of the load vector.*

In this way the critical values of the load parameter become functions of the unit load vector, $\overline{\boldsymbol{P}}$,

$$\lambda_1^+ = \lambda_1^+(\overline{\boldsymbol{P}}) . \tag{2.14}$$

If suffices to find only the dependence between the first positive critical values and the load combination vector $\overline{\boldsymbol{P}}$. The negative critical values will not be lost in this way; they will be found automatically by changing the $\overline{\boldsymbol{P}}$ vector to its opposite $-\overline{\boldsymbol{P}}$.

The functional relationship (2.14) defines an $(m-1)$-dimensional surface Γ in the load space (the surface is not necessarily closed). This surface is described by the set of vectors $\lambda_1^+(\overline{\boldsymbol{P}})\overline{\boldsymbol{P}}$, and our task is to prove the proposition that this surface, Γ, bounds a convex m-dimensional area Ω that contains the origin of the coordinates. One possible geometrical interpretation for relationship (2.14) is a mapping of the surface of a unit-radius sphere from the load space onto the boundary Γ of the equilibrium stability area so that an arbirtrary unit-length vector $\overline{\boldsymbol{P}}$ is extended by the extension factor $\lambda_1^+(\overline{\boldsymbol{P}})$ and mapped into a vector the end of which belongs to Γ.

What has been said above implies that the functional relationship (2.14) is to be found from the following characteristic equation parametrized by $\overline{\boldsymbol{P}}$:

$$\det\left[\boldsymbol{r}_o - \lambda \boldsymbol{r}_G(\overline{\boldsymbol{P}})\right] = 0, \tag{2.15}$$

and, clearly, the initial stiffness matrix $\boldsymbol{r}_o$ does not depend on the load vector $\overline{\boldsymbol{P}}$.

The following proposition is true, which we will call a *lemma of linear dependence of the geometric stiffness matrix,* $\boldsymbol{r}_G$*, on vector* $\overline{\boldsymbol{P}}$. Viz.: the components of the matrix of geometric stiffness, $\boldsymbol{r}_G$, are linear homogeneous functions of the components of vector $\overline{\boldsymbol{P}}$.

To say it with math,

$$r_G^{ij} = \overline{P}_1\rho_1^{ij} + \overline{P}_2\rho_2^{ij} + \ldots + \overline{P}_m\rho_m^{ij}, \tag{2.16}$$

where ρ_α^{ij} are certain constants ($\alpha = 1,\ldots, m$). Obviously, matrix $\left\|\left[\rho_\alpha^{ij}\right]\right\|$ is a matrix of geometric stiffness that corresponds to the load upon the system in the form of unit vector $\boldsymbol{e}_\alpha$.

To prove the proposition, we use the general formula (1.27) for the components of the geometrical stiffness matrix,

$$r_G^{ij} = b^{ij} - \frac{1}{2}a^{ijk}\overline{q}_{ok}.$$

Apparently, the force potential Π_s for the combined load we deal with can be represented as a sum of the potentials of particular forces, i.e. as

$$\Pi_s = \lambda(\overline{P}_1 g_1 + \ldots + \overline{P}_m g_m),$$

where $g_\alpha = g_\alpha(\boldsymbol{q})$ are some functions of the generalized coordinates, $q_1,\ldots, q_n$. This means that, according to (1.20),

$$b^{ij} = \overline{P}_1 \frac{\partial^2 g_1}{\partial q_i \partial q_j} + \ldots + \overline{P}_m \frac{\partial^2 g_m}{\partial q_i \partial q_j},$$

and the values of b^{ij} are linear homogeneous functions of the components of the unit load vector, $\overline{\boldsymbol{P}}$.

Coefficients a^{ijk} depend only on energy E and do not depend on the load, according to their very definition in (1.20). Therefore we need only to prove that $\overline{q}_{\text{o}k}$ are linearly dependent on the components of the vector $\overline{\boldsymbol{P}}$. Earlier in section (**11**.1.2) we found out that $\overline{\boldsymbol{q}}_{\text{o}} = \boldsymbol{r}_{\text{o}}^{-1}\boldsymbol{Q}$, so it suffices to establish the linear dependence of the unit load vector, $\boldsymbol{Q}$, on vector $\overline{\boldsymbol{P}}$. Direct calculation of it gives

$$\boldsymbol{Q} = \left\| \left[Q^i \right] \right\| = \left\| \left[\frac{\partial \Pi_s}{\partial q_i}(\mathbf{0}) \right] \right\| = \left\| \begin{bmatrix} \sum\limits_{\alpha=1}^{m} \overline{P}_\alpha \dfrac{\partial g_\alpha}{\partial q_1} \\ \vdots \\ \sum\limits_{\alpha=1}^{m} \overline{P}_\alpha \dfrac{\partial g_\alpha}{\partial q_n} \end{bmatrix} \right\| = \sum_{\alpha=1}^{m} \left\{ \overline{P}_\alpha \left\| \begin{bmatrix} \dfrac{\partial g_\alpha}{\partial q_1} \\ \vdots \\ \dfrac{\partial g_\alpha}{\partial q_n} \end{bmatrix} \right\| \right\}.$$

Now it can be seen that the representation (2.16) takes place, which proves the proposition.

Returning directly to the Papkovich theorem, we would like to note that there is a number of different approaches to it. P.F. Papkovich himself based his reasoning on the following characteristic feature of an arbitrary convex area Ω. If any ray that comes from an arbitrary point $A \in \Omega$ crosses boundary Γ of area Ω once at the most, then area Ω is convex. The whole complexity of the proof done by Papkovich is concentrated in proving this property of the area of stability of equilibrium states in the load space.

The proofs by B.M. Broude and A.R. Rzhanitsin make use of a different characteristic feature of convexity of area Ω. The notion of a *base hyperplane* is introduced for the Ω area defined in the m-dimensional linear space $\mathbb{R}_m$.

Generally, a *hyperplane* in $\mathbb{R}_m$ is any affine set of dimensionality $m-1$. In particular, any straight line is a hyperplane in the space of dimensionality 2, and any two-dimensional plane is a hyperplane in the three-dimensional space, even if it does not cross the coordinate origin. Any hyperplane divides the space $\mathbb{R}_m$ into two m-dimensional half-spaces which lie on the opposite sides of the dividing hyperplane.

A hyperplane is called a *base one* for area Ω if at least one point of boundary Γ of area Ω belongs to the hyperplane while area Ω is wholly contained by one of the half-spaces created by this hyperplane.

Finally, area Ω is *convex* if every point of boundary Γ can be a point of a base hyperplane of the area. The last proposition is quite intuitive geometrically and sounds convincing for a mechanician. Readers who want a rigorous mathematical proof of this proposition are invited to get familiar with the branch of modern mathematics called "*Convex analysis*" [17], [10]. It is this characteristic feature of convexity of area Ω that B.M. Broude and A.R. Rzhanitsin use in their proofs, although they have slightly different approaches.

Following the logic by Rzhanitsin, the first step is to see that when a few constraints are imposed on the system (the constraints are such that confine the buckling mode but do not affect the state of equilibrium – i.e. posterior or immobile constraints according to our nomenclature), the stability area can only expand, never diminish. This follows immediately from results of Section **11**.2.2.

Further, let Γ be the boundary of the equilibrium state stability area in the load space. We take a point A that belongs to this boundary. Then the length of vector OA will be $\lambda_1^+(\bar{\boldsymbol{P}}_A)$, where $\bar{\boldsymbol{P}}_A$ is a unit vector codirectional with OA. Let vector z_1^+ be a mode of buckling that corresponds to eigenvalue λ_1^+, i.e.

$$(\boldsymbol{r}_o - \lambda_1^+ \boldsymbol{r}_G) z_1^+ = \boldsymbol{0}. \tag{2.17}$$

For this load pattern defined by vector $\bar{\boldsymbol{P}}_A$, we consider a system with one degree of freedom derived from the given system by imposing $(n-1)$ posterior constraints which prevent the buckling in any mode other than z_1^+. Clearly, this system with one degree of freedom will lose its stability of equilibrium at the same value of the critical load, λ_1^+. Let r_{*o} and r_{*G} be the initial stiffness matrix and the geometric stiffness matrix (first-order) for the system thus obtained with one degree of freedom. It means together with (2.17) we have

$$r_{*o} - \lambda_1^+ r_{*G} = 0. \tag{2.18}$$

According to the proposition proved earlier, number r_{*G} is a linear combination of the components of the unit load vector, and this permits to rewrite (2.18) as

$$r_{*o} - \lambda_1^+ \sum_{\alpha=1}^{m} \overline{P}_\alpha \rho_{*\alpha} = 0 \,. \tag{2.19}$$

Equation (2.19) defines a hyperplane in the load space, and all points of the space which conform to stable equilibrium states of the unidimensional system are on one side of the plane (when the left-hand part of (2.19) is positive) while all points of unstable states lie on the other side (when the left-hand part of (2.19) is negative). In other words, the equilibrium stability area for the unidimensional system is an m-dimensional half-space that lies on one side of a hyperplane crossing the point of interest, A. The half-space is defined unambgiously by the condition that it should contain the coordinate origin.

As the stability area Ω for the original n-dimensional system is just a part of the selected half-space in the m-dimensional load space (removing the constraints can only narrow the stability area), the whole Γ boundary is guaranteed to belong to the stability half-space.

This reasoning is good for every point A on boundary Γ. It means the hyperplanes are *base hyperplanes* with respect to area Ω. This ends the proof of the Papkovich theorem.

* * *

The great importance of the Papkovich theorem for applications is in the possibility to use its propositions for analyzing the stability of equilibrium states of a structure not for every thinkable combination of loads but for just a few particular load patterns. Next, the area of stability can be evaluated with a guaranteed sign of the error; in this process the true boundary of the area is approximated by a set of hyperplanes. See more details in [16].

11.2.4 The geometric stiffness matrix revisited

We have already mentioned that the geometric stiffness matrix of a system is one of key notions in the theory of stability of equilibria of mechanical systems. Therefore we deem it reasonable to discuss this notion a bit more carefully and to illustrate its application by an example of a simple mechanical system.

To do it, we return to our example 1 (see Fig. 11.1) and denote by N_1 the longitudinal force in bar 1 and by N_2 the longitudinal force in bar 2. These forces are supposed to be calculated in the equilibrium state of our interest. Apparently,

$$N_1 = c_1 u_o = P\frac{c_1}{c_1 + c_2}, \qquad N_2 = c_2 u_o = P\frac{c_2}{c_1 + c_2}.$$

Force N_1 is positive when bar 1 is extended, and force N_2 is positive when bar 2 is compressed.

We already know that matrices $\boldsymbol{r}_o$ and $\boldsymbol{r}_G$ in this problem look as follows:

$$\boldsymbol{r}_o = \begin{Vmatrix} c_1 + c_2 & 0 \\ 0 & c_3 \end{Vmatrix}, \qquad \boldsymbol{r}_G = \begin{Vmatrix} 0 & 0 \\ 0 & \dfrac{c_2 - c_1}{l(c_1 + c_2)} \end{Vmatrix},$$

where the first row and the first column of each relate to the longitudinal displacement, u, of the central node, and the second row/column relate to the lateral displacement, v, of the same node.

The structure and the contents of matrix $\boldsymbol{r}_o$ are quite clear. Its components correspond to an ordinary linear problem formulation. If we consider the equilibrium of each element of the system in a deformed state that conforms to its buckling mode, we will find (Fig. 11.8) that the lateral force $Q_1 = N_1 v/l$ develops in the left bar and the lateral force $Q_2 = N_2 v/l$ appears in the right bar.

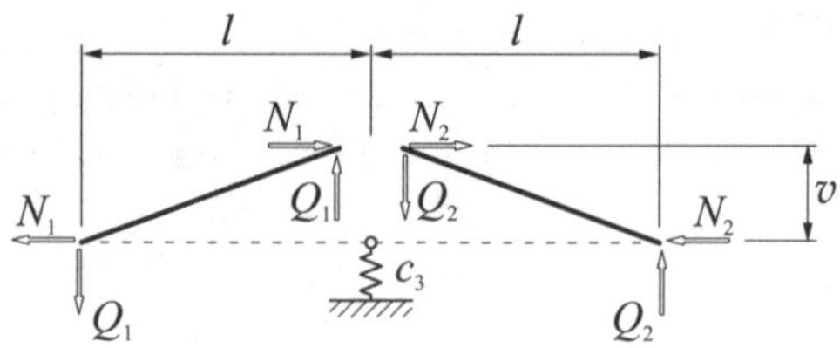

$$Q_1 = N_1 \frac{v}{l}, \quad Q_2 = N_2 \frac{v}{l}$$

Fig. 11.8. Obtaining the components of matrix $\boldsymbol{r}_G$ by the static method

Therefore, when the central node is deviated by a unit displacement, $v = 1$, the total reactive force upon this node, r_{22}, will consist of two terms:

$$r_{22} = c_3 + \frac{N_1}{l} - \frac{N_2}{l} = c_3 - P\frac{c_2 - c_1}{(c_1 + c_2)l}.$$

The second term in the right part of the formula is exactly the component r_{22} that relates to the geometric stiffness matrix.

This simple example demonstrates a traditional technique for obtaining the components of the geometric stiffness matrix – without using the expressions of the strain energy and the work of external forces, i.e. in a way based on the equations of equilibrium only. The equations of equilibrium must be composed for the deformed state of the system and must take into account those external forces that appear in the elements of the system in its state of equlibrium the stability of which is under consideration.

11.2.5 The stability of equilibrium under a non-force-type action

Up to this point, we dealt with the stability of equilibrium of a mechanical system under the action of forces only. An attentive reader may notice that the stability problems can be posed for kinematical actions, too. For example, the ends of a single bar can be forcedly pushed together by a given distance; the bar appears to be in equilibrium which, however, is not necessarily stable.

The problem is that the Lagrange–Dirichlet theorem's application is not straightforward in the case of the kinematical actions; some additional explanations are needed, at the least.

Our reasoning can be like this. We extend the set of degrees of freedom (1.1) to the number of m where $m > n$. The additional $s = m - n$ DOFs will be those for which nonzero displacements are specified as external actions. In our case there is no external force potential, Π_s, so

$$\mathsf{L} = \mathsf{E}(q_1,\dots,q_n, q_{n+1},\dots,q_m), \tag{2.20}$$

but the last s variables are not varied; instead, they are considered to be known values proportional to the load parameter, λ. To put it another way, we assume

$$q_{n+1} = \lambda \bar{q}_{n+1}, \ \dots, \ q_{n+s} = \lambda \bar{q}_{n+s}. \tag{2.21}$$

The score above denotes fixed values of the displacements that correspond to the unit load parameter, $\lambda = 1$.

We divide the $\boldsymbol{q}$ vector into two parts:

$$\boldsymbol{q} = \|[\boldsymbol{q}_1, \boldsymbol{q}_2]\|^{\mathsf{T}}, \quad \boldsymbol{q}_1 = \|[q_1,\dots, q_n]\|^{\mathsf{T}}, \quad \boldsymbol{q}_2 = \|[q_{n+1},\dots,q_{n+s}]\|^{\mathsf{T}}.$$

First vector $\boldsymbol{q}_1$ is made up of the varied displacements while the components of second vector $\boldsymbol{q}_2$ meet the requirements

$$\boldsymbol{q}_2 = \lambda \bar{\boldsymbol{q}}_2. \tag{2.22}$$

The equilibrium state of the system is found from the condition that the L function of m variables from (2.20) should be stationary with the additional conditions (2.21). The linear formulation of the problem (one that we are investigating the stability under the conditions of) gives

$$\mathsf{L} = \frac{1}{2}\left\|[\boldsymbol{q}_1, \boldsymbol{q}_2]\right\|^{\mathsf{T}} \left\|\begin{bmatrix} \boldsymbol{r}_{o11} & \boldsymbol{r}_{o12} \\ \boldsymbol{r}_{o21} & \boldsymbol{r}_{o22} \end{bmatrix}\right\| \left\|\begin{bmatrix} \boldsymbol{q}_1 \\ \boldsymbol{q}_2 \end{bmatrix}\right\| \tag{2.23}$$

where the components of the matrix blocks, $\boldsymbol{r}_{o11}$, $\boldsymbol{r}_{o12} = \boldsymbol{r}_{o21}^{\mathsf{T}}$, $\boldsymbol{r}_{o22}$, are defined by the common formula (1.17). The conditions of stationarity of L from (2.23) under the additional conditions (2.22) produce the following system of linear algebraic equations:

$$\left\|\begin{bmatrix} \boldsymbol{r}_{o11} & \boldsymbol{r}_{o12} & \mathbf{0} \\ \boldsymbol{r}_{o21} & \boldsymbol{r}_{o22} & \boldsymbol{I} \\ \mathbf{0} & \boldsymbol{I} & \mathbf{0} \end{bmatrix}\right\| \cdot \left\|\begin{bmatrix} \boldsymbol{q}_1 \\ \boldsymbol{q}_2 \\ \boldsymbol{\Lambda} \end{bmatrix}\right\| = \lambda \left\|\begin{bmatrix} \mathbf{0} \\ \mathbf{0} \\ \overline{\boldsymbol{q}}_2 \end{bmatrix}\right\| \tag{2.24}$$

with a vector of Lagangian multipliers, $\boldsymbol{\Lambda}$, of order m.

Solution of this system of equations gives the following formula for the components of vector $\boldsymbol{q}_{o1}$ that defines an equilibrium of the linear system under the kinematical actions specified by (2.22):

$$\boldsymbol{q}_{o1} = -\lambda \boldsymbol{r}_{o11}^{-1} \boldsymbol{r}_{o12} \overline{\boldsymbol{q}}_2 \,. \tag{2.25}$$

After introducing the designation

$$\overline{\boldsymbol{q}}_{o1} = -\boldsymbol{r}_{o11}^{-1} \boldsymbol{r}_{o12} \overline{\boldsymbol{q}}_2$$

we can rewrite (2.25) as

$$\boldsymbol{q}_{o1} = \lambda \overline{\boldsymbol{q}}_{o1} \,. \tag{2.26}$$

In order to find out the boundaries of the stability area for this equilibrium state, we can use the condition that Bolotin's stability functional should be positive definite (1.16). As the variable parameters are only the components of vector $\boldsymbol{q}_1$ in the number of n, we have the following condition of stability for our particular equilibrium:

$$\boldsymbol{z}^{\mathsf{T}} \boldsymbol{r}_{\lambda 11} \boldsymbol{z} > 0$$

for any vector $\boldsymbol{z}$ of dimensionality n. Matrix $\boldsymbol{r}_{\lambda 11}$ is a tangential stiffness matrix of the system, with dimensions $n \times n$, which conforms to the variations $\delta q_1 = z_1, \ldots, \delta q_n = z_n$ of the generalized displacements. Also,

$$r_{\lambda 11}^{ij} = \frac{\partial^2 \mathsf{L}}{\partial q_i \partial q_j}(\boldsymbol{q}_{o1}, \lambda \overline{\boldsymbol{q}}_2), \qquad i, j = 1, \dots, n \,. \tag{2.27}$$

The linearized formulation of this stability problem gives

$$\boldsymbol{r}_{\lambda 11} = \boldsymbol{r}_{o11} - \lambda \boldsymbol{r}_{G11} \,. \tag{2.28}$$

It is easy to notice from the general formula (1.27) of the components of the geometric stiffness matrix that coefficients b^{ij} are identical to zero in our case because there is no force potential Π_s. Therefore, according to (1.20), (1.21), and (1.27),

$$r_{G11}^{ij} = \frac{1}{2} \sum_{k=1}^{n} \frac{\partial^3 \mathsf{E}}{\partial q_i \partial q_j \partial q_k}(\mathbf{0}) \overline{q}_{o1k} \,, \tag{2.29}$$

where $\overline{q}_{o1k}$ is k-th component of vector $\overline{\boldsymbol{q}}_{o1}$ $(k = 1, \dots, n)$.

Now we are going to demonstrate the technique of solving the problem of stability under a kinematical action by the example shown in Fig. 11.3. However, now we will take as an external action a forced vertical displacement of the upper node of the spring the stiffnes of which is c_1, rather than force P. The length of this displacement will be denoted by λ. Using the designations from (1.47), we have

$$\boldsymbol{q}_1 = \|[\theta, \Delta]\|^{\mathsf{T}}, \qquad \boldsymbol{q}_2 = \lambda \left\|[1]\right\|, \qquad n = 2, \ s = 1.$$

Apparently, the equilibrium state of the system is described by the following displacements in the linear formulation:

$$\theta_o = 0, \qquad \Delta_o = \frac{c_1}{c_1 + c_2} \lambda \,. \tag{2.30}$$

Next, the expression of the strain energy will be

$$\mathsf{L} = \mathsf{E} = c_1 \frac{(v - \lambda)^2}{2} + c_2 \frac{\Delta^2}{2} + c_3 \frac{u^2}{2} \,,$$

which gives the following after putting in the expressions of u and v from (1.48):

$$\mathsf{L} = c_1 \frac{\left[l - (l - \Delta)\cos\theta - \lambda\right]^2}{2} + c_2 \frac{\Delta^2}{2} + c_3 \frac{(l - \Delta)^2 \sin^2\theta}{2} \,.$$

We determine the components of the tangential stiffness matrix immediately, without using the linearized formulation of the problem. These are

$$\frac{\partial^2 \mathsf{L}}{\partial \theta^2}(\theta_o, \Delta_o) = c_1(\Delta_o - \lambda)(l - \Delta_o) + c_3(l - \Delta_o)^2 =$$

$$= (l - \Delta_o)\left[c_1(\Delta_o - \lambda) + c_3(l - \Delta_o)\right],$$

$$\frac{\partial^2 \mathsf{L}}{\partial \theta \partial \Delta}(\theta_o, \Delta_o) = 0, \quad \frac{\partial^2 \mathsf{L}}{\partial \Delta^2}(\theta_o, \Delta_o) = c_1 + c_2 .$$

According to (2.27),

$$\boldsymbol{r}_{\lambda 11} = \left\| \begin{bmatrix} (l - \Delta_o)\left[c_1(\Delta_o - \lambda) + c_3(l - \Delta_o)\right] & 0 \\ 0 & c_1 + c_2 \end{bmatrix} \right\| .$$

After replacing Δ_o with its value from (2.30), we use the positive definiteness of matrix $\boldsymbol{r}_{\lambda 11}$ to arrive formally at two critical values of the external action parameter,

$$\frac{\lambda'_{cr}}{l} = \frac{c_1 + c_2}{c_1}, \qquad \frac{\lambda''_{cr}}{l} = \frac{c_3(c_1 + c_2)}{c_1(c_3 + c_2)} .$$

We introduce dimensionless parameters by assuming

$$\alpha = \frac{c_2}{c_3}, \qquad \beta = \frac{\lambda}{l}\frac{c_1}{c_1 + c_2} . \tag{2.31}$$

It is easy to see that the boundary of the stable equilibrium area for the given kinematic action in the plane of the above dimensionless parameters is defined by the conditions

$$\beta'_{cr} = 1, \qquad \beta''_{cr} = 1/(1+\alpha) .$$

Now it is clear that the boundaries between the areas of stable and unstable states of equilibrium are described by the same graphs in Fig. 11.4 but parameter β in that figure should be defined by the second of the formulas (2.31).

In order to conclude this section, we would like to note that more possible non-force-type actions exist that may cause the loss of stability. Those include an initial strain caused by heat/frost, shrinkage or creep of concrete in ferroconcrete/composite structures. Further details are omitted.

11.3 Geometrically nonlinear problems in elasticity

In the linear elasticity, the components of the strain tensor are expressed via the components of the displacement vector by linear relationships (**4**.2.1–*b*). But this degree of accuracy of the geometric relationships is not enough for the equilibrium stability analysis, therefore we have to take into account additional quadratic terms of the displacements. In this regard we would like to briefly discuss a geometrically nonlinear formulation of problems in elasticity.

11.3.1 Geometric equations

As is known, the three-dimensional geometric equations of the theory of elasticity which express a relation between the strains and the displacements are as follows in standard designations, provided the elongations and shears are small comparing to one [12][9]

$$\varepsilon_{xx} = \frac{\partial u}{\partial x} + \frac{1}{2}\left[\left(\frac{\partial u}{\partial x}\right)^2 + \left(\frac{\partial v}{\partial x}\right)^2 + \left(\frac{\partial w}{\partial x}\right)^2\right], \quad \ldots$$

$$\varepsilon_{xy} = \frac{1}{2}\left(\frac{\partial u}{\partial y} + \frac{\partial v}{\partial x} + \frac{\partial u}{\partial x}\frac{\partial u}{\partial y} + \frac{\partial v}{\partial x}\frac{\partial v}{\partial y} + \frac{\partial w}{\partial x}\frac{\partial w}{\partial y}\right), \quad \ldots \tag{3.1}$$

(3.1) presents two of six geometric relationships; the other four are derived from (3.1) by cyclic permutation of the letters as $x \to y \to z \to x$ and $u \to v \to w \to u$.

We can introduce an index notation by assuming, as usual,

$$x_1 = x,\ x_2 = y,\ x_3 = z, \quad \text{and} \quad u_1 = u,\ u_2 = v,\ u_3 = w,$$

to rewrite the nonlinear geometrical equations (3.1) in a tensor form more convenient for understanding and transforming:

---

[9] Note that, unlike (3.1), Novozhilov's expressions of shear strains ε_{xy}, ε_{yz}, ε_{zx} [12] do not include the multiplier of ½. This is not a mistake; in [12] the same designations are used for the shears, which we denoted by γ_{xy}, γ_{yz}, γ_{zx} in Chapter **4**, rather than for the shear components of the strain tensor. Also note that we confine ourselves to considering such simplifications of the geometric equations that follow from the assumption of smallness of *all* components of the strain tensor comparing to one. We omit the consideration of large strains which can be practically important for rubber-like materials.

$$\varepsilon_{ij} = \frac{1}{2}(u_{i,j} + u_{j,i}) + \frac{1}{2}\delta^{km} u_{k,i} u_{m,j} \,. \tag{3.2}$$

Kronecker's delta δ^{km} is introduced here in order to keep our convention of summing over repeated indexes placed on different levels. As we can see, the relationships (3.2) are supplemented with additional terms quadratically dependent on the displacements, unlike the linear elasticity.

In order to distinguish between the components of the stress and strain tensors in the linear and geometrically nonlinear elastic analysis, we will use the following notation further in this chapter:

- τ^{ij}, e_{ij} are the components of the stress and strain tensors in the linear elastic analysis;
- σ^{ij}, ε_{ij} are the components of the stress and strain tensors in the geometrically nonlinear elastic analysis.

It means we have

$$e_{ij} = \frac{1}{2}(u_{i,j} + u_{j,i}) \,. \tag{3.3}$$

The two strain tensors, ε_{ij} and e_{ij}, are related as

$$\varepsilon_{ij} = e_{ij} + f_{ij} \,, \tag{3.4}$$

where f_{ij} is a part of tensor ε_{ij} which is quadratic with respect to the displacements,

$$f_{ij} = \frac{1}{2}\delta^{km} u_{k,i} u_{m,j} \,. \tag{3.5}$$

It is useful to introduce a so-called *antisymmetric rotation (slope) tensor* with the components

$$\omega_{ij} = \frac{1}{2}(u_{i,j} - u_{j,i}) \,, \tag{3.6}$$

so that

$$u_{i,j} = e_{ij} + \omega_{ij} \,. \tag{3.7}$$

Putting the expansion (3.7) in (3.4) and (3.5) gives

$$\varepsilon_{ij} = e_{ij} + \frac{1}{2}\delta^{km}\left(e_{ki} + \omega_{ki}\right)\left(e_{mj} + \omega_{mj}\right). \tag{3.8}$$

As is known [12], the nonlinear geometrical relationships (3.1) or, equivalently, (3.2) hold true for the general case but only when the elongations and the shears are small. It means all components of the $\|[\varepsilon_{ij}]\|$ tensor are much less than one by absolute value. In other words, for all combinations of indexes the following estimate must hold:

$$|\varepsilon_{ij}| << 1 . \tag{3.9}$$

Further simplifications of the geometrical relationships are based on an additional assumption that the ω_{ij} parameters are small, too. The latter can be then identified with average slopes of an infinitesimal element of volume that surrounds a point of the elastic medium with respect to appropriate axes. Denoting the components of the slopes as ω_x, ω_y, ω_z, we can express them via the components of the rotation tensor (see [12]) as

$$\omega_x = \omega_{32}, \quad \omega_y = \omega_{13}, \quad \omega_z = \omega_{21}.$$

So we assume in addition to (3.9) that the estimate holds:

$$|\omega_{ij}| << 1 . \tag{3.10}$$

Smallness of the slopes in comparison to one does not yet guarantee the equivalence of the order of magnitude between the slopes and the strains. Therefore the nearest simplifying assumption is a set of the following estimates:

$$|\varepsilon_{ij}| << 1 , \quad |\omega_{ij}| << 1 , \quad |e_{ij}| << |\omega_{ij}| . \tag{3.11}$$

It is essential that the slopes, being small comparing to one, can exceed by their order of magnitude the kinematical parameters e_{ij}. The last estimate in (3.11) postulates, actually, that the e_{ij} values have the same order of smallness as the slope squares, that is,

$$| e_{ij} | \sim \omega_{ij}^2 . \tag{3.12}$$

But then, as we can see from (3.8) and the estimates (3.11), the following formulas are quite acceptable by accuracy:

$$\varepsilon_{ij} = e_{ij} + \frac{1}{2}\delta^{km}\omega_{ki}\omega_{mj} . \tag{3.13}$$

The representation of the strains in (3.13) keeps only the terms the order of smallness of which corresponds to first power of e_{ij}.

Expressions of variations of the strain tensor components

For our further presentment we will need to track how the strain tensor components change vs. the vector of displacements **u** that causes the strains. In other words, we suppose that the vector of displacements, **u**, gets infinitesimal increments of its components, i.e. it changes from **u** to **u** + δ**u**. Apparently, this changes the strain tensor **ε**, too, making **ε** + δ**ε** out of it[10].

As any arbitrary component ε_{ij} of the strain tensor depends on the components of the displacement vector with the order not higher than quadratic, we can use a Taylor series expansion and obtain a general formula for increments $\delta\varepsilon_{ij}$:

$$\delta\varepsilon_{ij} = \frac{d\varepsilon_{ij}}{du_k}\delta u_k + \frac{1}{2}\frac{d^2\varepsilon_{ij}}{du_k du_m}\delta u_k \delta u_m .$$

The second term above contains the products of variations, $\delta u_k \delta u_m$, which are values of second order of smallness comparing to linear terms δu_k, therefore they can be omitted. Considering the representation of ε_{ij} as a sum in (3.4), we can write

$$\delta\varepsilon_{ij} = \delta e_{ij} + \delta f_{ij} , \tag{3.14}$$

where, according to (3. 5),

$$\delta f_{ij} = \frac{1}{2}\delta^{km}(u_{k,i}\delta u_{m,j} + u_{m,j}\delta u_{k,i}) . \tag{3.15}$$

Further we will need second variations of f_{ij}, too. Clearly,

$$\delta^2 f_{ij} = \delta^{km}\delta u_{k,i}\delta u_{m,j} . \tag{3.16}$$

For the simplified geometrical relationships (3.13), which hold true also for small slopes $|\varepsilon_{ij}| \sim \omega_{ij}^2$, we have the following instead of (3.15) and (3.16):

$$\delta f_{ij} = \frac{1}{2}\delta^{km}(\omega_{ki}\delta\omega_{mj} + \omega_{mj}\delta\omega_{ki}) , \qquad \delta^2 f_{ij} = \delta^{km}\delta\omega_{ki}\delta\omega_{mj} . \tag{3.17}$$

11.3.2 Equations of equilibrium and static boundary conditions

The geometrically nonlinear elastic analysis implies that the equations of equilibrium must be composed for the deformed state of the elastic body. These equations can be derived following a typical procedure when an

---

[10] We hope the reader's eyes are able to distinguish between the resembling designations of Kronecker's delta δ^{ij} and the symbol of variation δ with no indexes.

infinitesimal volume is extracted from an elastic body and the equilibrium equations in projections onto the respective axes are composed for it in its deformed state [12]. However, we can use a different approach and derive the same equations following a variational procedure (see, for example, [3]).

We remind that our consideration is confined to an elastic material where a linear physical relation between the stresses and the strains holds:

$$\sigma^{ij} = C^{ijkm}\varepsilon_{km}\,. \tag{3.18}$$

As we are talking about geometrically nonlinear formulations, we should understand the components of both the stress tensor and the strain tensor in the above notation as referring to the Lagrangian basis in the deformed state of the system. The Lagrangian basis will be specific to every point of the body, so it can be only a local Lagrangian basis.

This is the general expression of the full potential energy of the elastic body, L:

$$\mathsf{L} = \frac{1}{2}\int\limits_{\Omega} C^{ijkm}\varepsilon_{ij}\varepsilon_{km}\,d\Omega - \int\limits_{\Omega}\overline{X}^{i}u_{i}\,d\Omega - \oint\limits_{\Gamma_p}\overline{p}^{i}u_{i}\,d\Gamma\,. \tag{3.19}$$

To simplify the notation, we suppose here and futher that there are no mixed boundary conditions and the whole boundary Γ of area Ω occupied by the elastic body consists of two parts as in (**1**.2.10). To interpret this formula correctly, note that the components of the external volumetric, $\overline{X}^{i}$, and surface, $\overline{p}^{i}$, forces together with the vector of displacements, u_i, are assumed to refer to the Eulerian basis which is supposed to coincide with the Lagrangian one when there is no deformation. This confusion of the bases for different terms in the same functional should not confound the reader because each of the integrals in (3.19) is a scalar invariant with respect to the basis while the tensor representation of each scalar can refer to its own basis.

Now let us derive the equations of equilibrium as conditions of stationarity of the Lagrange functional, L, on the set of physically and kinematically admissible fields. By varying L and taking the elasticity relationships (3.18) into account, we produce this:

$$\delta\mathsf{L} = \int\limits_{\Omega} C^{ijkm}\varepsilon_{km}\delta\varepsilon_{ij}\,d\Omega - \int\limits_{\Omega}\overline{X}^{i}\delta u_{i}\,d\Omega - \oint\limits_{\Gamma_p}\overline{p}^{i}\delta u_{i}\,d\Gamma =$$

$$= \int\limits_{\Omega}\sigma^{ij}\delta\varepsilon_{ij}\,d\Omega - \int\limits_{\Omega}\overline{X}^{i}\delta u_{i}\,d\Omega - \oint\limits_{\Gamma_p}\overline{p}^{i}\delta u_{i}\,d\Gamma\,. \tag{3.20}$$

The condition of $\delta \mathsf{L} = 0$ is a principle of virtual displacements in its mechanical interpretation. The expressions of strains ε_{ij}, as well as those of their variations $\delta\varepsilon_{ij}$, will depend on how the geometrical nonlinearity is taken into account.

We divide the first of the integrals in the right-hand part of (3.20) into two components using the representation (3.14):

$$\int_{\Omega} \sigma^{ij} \delta\varepsilon_{ij} d\Omega = \int_{\Omega} \sigma^{ij} \delta e_{ij} d\Omega + \int_{\Omega} \sigma^{ij} \delta f_{ij} d\Omega . \tag{3.21}$$

Next, the Gauss–Ostrogradsky formula gives

$$\int_{\Omega} \sigma^{ij} \delta e_{ij} d\Omega = -\int_{\Omega} \sigma^{ij}_{,j} \delta u_i d\Omega + \oint_{\Gamma_p} \sigma^{ij} n_j \delta u_i d\Gamma . \tag{3.22}$$

Also,

$$\int_{\Omega} \sigma^{ij} \delta f_{ij} d\Omega = \frac{1}{2} \int_{\Omega} \sigma^{ij} \delta^{km} (u_{k,i} \delta u_{m,j} + u_{m,j} \delta u_{k,i}) d\Omega =$$

$$= -\delta^{km} \int_{\Omega} \left(\sigma^{ij} u_{k,i} \right)_{,j} \delta u_m d\Omega + \delta^{km} \oint_{\Gamma_p} \sigma^{ij} u_{k,i} n_j \delta u_m d\Gamma =$$

$$= -\delta^{ki} \int_{\Omega} \left(\sigma^{mj} u_{k,m} \right)_{,j} \delta u_i d\Omega + \delta^{ki} \oint_{\Gamma_p} \sigma^{mj} u_{k,m} n_j \delta u_i d\Gamma . \tag{3.23}$$

The transformations of (3.23) use a standard technique of replacing the umbral indexes, which leaves the result intact. Also, these transformations make use of the symmetry of the stress tensor σ^{ij} which follows from (3.18) and the symmetry of the tensor of elastic constants C^{ijkm} – see relationships (**1**.2.3). Note also (3.22) and (3.23) use the fact that $\delta \boldsymbol{u} = \mathbf{0}$ on Γ_u because of kinematical admissibility of the displacement fields.

Substituting (3.22) and (3.23) in (3.20) and equating $\delta \mathsf{L}$ to zero yields the Euler equation and natural boundary conditions,

$$-\sigma^{ij}_{,j} - \delta^{ki} \left(\sigma^{mj} u_{k,m} \right)_{,j} - \bar{X}^i = 0 \in \Omega,$$

$$(\sigma^{ij} + \delta^{ki} \sigma^{mj} u_{k,m}) n_j = \bar{p}^i \in \Gamma_p , \tag{3.24}$$

treated from the physical standpoint as equations of equilibrium in the volume of the body, Ω, and as static boundary conditions on boundary Γ_p, respectively.

Now let us turn to the geometrically nonlinear equations of equilibrium for small slopes the squares of which have the same order of smallness as the elongations and the shears. According to (3.17), this version of the geometrically nonlinear theory gives the following instead of (3.23):

$$\int_{\Omega} \sigma^{ij} \delta f_{ij} d\Omega = \frac{1}{2} \int_{\Omega} \sigma^{ij} \delta^{km} (\omega_{ki} \delta\omega_{mj} + \omega_{mj} \delta\omega_{ki}) d\Omega = \delta^{km} \int_{\Omega} \sigma^{ij} \omega_{ki} \delta\omega_{mj} d\Omega =$$

$$= \frac{1}{2} \delta^{km} \int_{\Omega} \sigma^{ij} \omega_{ki} (\delta u_{m,j} - \delta u_{j,m}) d\Omega =$$

$$= -\frac{1}{2} \int_{\Omega} \left[\left(\delta^{ki} \sigma^{mj} - \delta^{km} \sigma^{ij} \right) \omega_{kj} \right]_{,l} \delta u_i d\Omega + \frac{1}{2} \oint_{\Gamma_p} \left(\delta^{ki} \sigma^{mj} - \delta^{km} \sigma^{ij} \right) \omega_{kj} n_m \delta u_i d\Gamma .$$

The final form of the equations of equilibrium and the static boundary conditions for this geometrically nonlinear theory is as follows.

$$-\sigma^{ij}_{,j} - \frac{1}{2} \left[\left(\delta^{ki} \sigma^{mj} - \delta^{km} \sigma^{ij} \right) \omega_{kj} \right]_{,m} - \bar{X}^i = 0 \in \Omega,$$

$$\sigma^{ij} n_j + \frac{1}{2} \left(\delta^{ki} \sigma^{mj} - \delta^{km} \sigma^{ij} \right) \omega_{kj} n_m = \bar{p}^i \in \Gamma_p . \tag{3.25}$$

Now let us concentrate all the governing equations of the geometrically nonlinear theory of elasticity that we have obtained in a convenient Table 11.1.

Table 11.1

Geometrically nonlinear equations of elasticity		Geometrically linear equations of elasticity	
$\mid \varepsilon_{ij} \mid << 1$	$\mid \varepsilon_{ij} \mid << 1,\ \mid e_{ij} \mid \sim \omega_{ij}^2$	$\mid \varepsilon_{ij} \mid << 1,\ \mid e_{ij} \mid \sim \mid \omega_{ij} \mid$	Estimate of order of smallness
$-\sigma^{ij}_{,j} - \delta^{ki}\left(\sigma^{mj} u_{k,m}\right)_{,j} - \overline{X}^i = 0$	$-\sigma^{ij}_{,j} - \frac{1}{2}\left[\left(\delta^{ki}\sigma^{mj} - \delta^{km}\sigma^{ij}\right)\omega_{kj}\right]_{,m} - \overline{X}^i = 0$	$-\tau^{ij}_{,j} = \overline{X}^i$	Equations of equilibrium
$\varepsilon_{ij} = \frac{1}{2}(u_{i,j} + u_{j,i}) + \frac{1}{2}\delta^{km} u_{k,i} u_{m,j}$	$\varepsilon_{ij} = e_{ij} + \frac{1}{2}\delta^{km}\omega_{ki}\omega_{mj}$	$e_{ij} = \frac{1}{2}(u_{i,j} + u_{j,i})$	Geometrical equations
$\sigma^{ij} = C^{ijkm}\varepsilon_{km}$	$\sigma^{ij} = C^{ijkm}\varepsilon_{km}$	$\tau^{ij} = C^{ijkm} e_{km}$	Physical equations
$(\sigma^{ij} + \delta^{ki}\sigma^{mj} u_{k,m}) n_j = \overline{p}^i$	$\sigma^{ij} n_j + \frac{1}{2}\left(\delta^{ki}\sigma^{mj} - \delta^{km}\sigma^{ij}\right)\omega_{kj} n_m = \overline{p}^i$	$\tau^{ij} n_j = \overline{p}^i$	Static boundary conditions
$u_i = \overline{u}_i$	$u_i = \overline{u}_i$	$u_i = \overline{u}_i$	Kinematic boundary conditions

11.4 Stability of equilibrium of an elastic body

Suppose we know the state of equilibrium for our mechanical system. Let the displacements of the system in this state be described by vector **u**. By formally extending the Lagrange–Dirichlet theorem onto systems with an infinite number of DOFs, we use the minimum of functional L of the system's full potential energy in a particular state of equilibrium as a criterion of stability of this state.

Let us assume all the external loads to grow proportionally to one parameter λ. Understanding $\bar{\boldsymbol{X}}$ and $\bar{\boldsymbol{p}}$ as external force actions that correspond to the unit value of the load parameter, $\lambda = 1$, we set down this general expression of the Lagrange functional:

$$\mathsf{L} = \frac{1}{2}\int_\Omega \boldsymbol{C}\boldsymbol{\varepsilon} \cdot \boldsymbol{\varepsilon} d\Omega - \lambda\int_\Omega \bar{\boldsymbol{X}}^{\mathsf{T}}\mathbf{u} d\Omega - \lambda\oint_{\Gamma_p} \bar{\boldsymbol{p}}^{\mathsf{T}}\mathbf{u} d\Gamma =$$

$$= \frac{1}{2}\int_\Omega \boldsymbol{C}(\boldsymbol{e}+\boldsymbol{f}) \cdot (\boldsymbol{e}+\boldsymbol{f}) d\Omega - \lambda\int_\Omega \bar{\boldsymbol{X}}^{\mathsf{T}}\mathbf{u} d\Omega - \lambda\oint_{\Gamma_p} \bar{\boldsymbol{p}}^{\mathsf{T}}\mathbf{u} d\Gamma . \tag{4.1}$$

The displacements **u** correspond to the state of equilibrium of the system, and this means the first variation of functional L is zero:

$$\delta\mathsf{L} = \int_\Omega \boldsymbol{C}\boldsymbol{\varepsilon}(\mathbf{u}) \cdot \delta\boldsymbol{\varepsilon}(\mathbf{u}) d\Omega - \lambda\int_\Omega \bar{\boldsymbol{X}}^{\mathsf{T}}\delta\mathbf{u} d\Omega - \lambda\oint_{\Gamma_p} \bar{\boldsymbol{p}}^{\mathsf{T}}\delta\mathbf{u} d\Gamma = 0 \tag{4.2}$$

for any homogeneously kinematically admissible variations of the displacements $\delta\mathbf{u}$. The vector argument in parentheses after strains $\boldsymbol{\varepsilon}$ indicates displacements which create those strains.

A minimum takes place in points of stationarity of L if the second variation of the functional is positive; thus, the criterion of equilibrium stability is $\delta^2\mathsf{L} > 0$. The second variation $\delta^2\mathsf{L}$ is equal to $\delta^2\mathsf{E}$, or

$$\delta^2\mathsf{L} = \delta^2\mathsf{E} = \delta(\delta\mathsf{E}) = \delta\left(\int_\Omega \boldsymbol{C}(\boldsymbol{e}+\boldsymbol{f}) \cdot (\delta\boldsymbol{e}+\delta\boldsymbol{f}) d\Omega\right) =$$

$$= \int_\Omega \boldsymbol{C}(\delta\boldsymbol{e}+\delta\boldsymbol{f}) \cdot (\delta\boldsymbol{e}+\delta\boldsymbol{f}) d\Omega + \int_\Omega \boldsymbol{C}(\boldsymbol{e}+\boldsymbol{f}) \cdot \delta^2 \boldsymbol{f} d\Omega , \tag{4.3}$$

because vector $\boldsymbol{e}$ depends linearly on the displacements and thus its second variation with respect to the displacement field vanishes.

But $\boldsymbol{C}(\boldsymbol{e}+\boldsymbol{f})=\boldsymbol{\sigma}$ where $\boldsymbol{\sigma}$ is a stress vector in the equilibrium state of interest. Apparently, this vector depends on the load intensity λ (nonlinearly in a general case), so

$$\boldsymbol{\sigma}=\boldsymbol{\sigma}(\lambda)\,. \tag{4.4}$$

The variation of the displacements, $\delta\mathbf{u}$, will be denoted by $\mathbf{v}$ for a while, for the sake of convenience. Second variation $\delta^2\mathsf{L}$, treated as a functional of vector $\mathbf{v}$, is nothing but a quadratic functional of stability, S; its being positive definite guarantees the stability of equilibrium that corresponds to the displacement vector $\mathbf{u}$.

Thus, (4.3) produces a general expression of the equilibrium stability functional:

$$\mathsf{S}(\mathbf{v})=\frac{1}{2}\int_\Omega \boldsymbol{C}\delta\boldsymbol{\varepsilon}\cdot\delta\boldsymbol{\varepsilon}\,d\Omega+\frac{1}{2}\int_\Omega\boldsymbol{\sigma}\cdot\delta^2\boldsymbol{f}\,d\Omega\,. \tag{4.5}$$

The multiplier ½, inessential for the stationarity points of the above functional, has been introduced for convenience of subsequent physical interpretation. Note also that our nomenclature of designations prescribes to understand $\delta\varepsilon_{ij}$ and $\delta^2 f_{ij}$ as

$$\delta\varepsilon_{ij}=\frac{1}{2}(v_{i,j}+v_{j,i})+\frac{1}{2}\delta^{km}(u_{k,i}v_{m,j}+u_{m,j}v_{k,i})\,,\quad \delta^2 f_{ij}=\delta^{km}v_{k,i}v_{m,j}\,. \tag{4.6}$$

As a result of substituting (4.6) in (4.5) and using the symmetry of the elasticity coefficient tensor $\boldsymbol{C}$, we have this final form of the stability functional:

$$\mathsf{B}_1(\mathbf{v})=$$

$$=\frac{1}{2}\int_\Omega C^{ijkm}\left(v_{i,j}v_{k,m}+2\delta^{pr}u_{r,i}v_{p,j}v_{k,m}+\delta^{pr}\delta^{qs}u_{r,i}u_{s,k}v_{p,j}v_{q,m}\right)d\Omega+$$

$$+\frac{\delta^{km}}{2}\int_\Omega\sigma^{ij}v_{k,i}v_{m,j}\,d\Omega\,. \tag{4.7}$$

This is a general form established for the first time by V.V. Bolotin [5], [7], hence the designation of $\mathsf{B}_1(\mathbf{v})$.

In the case when the estimates (3.11) hold, we should consider (3.17) and replace (4.6) with

$$\delta\varepsilon_{ij}=\frac{1}{2}(v_{i,j}+v_{j,i})+\frac{1}{2}\delta^{km}[\varpi_{k\,i}\omega_{mj}+\varpi_{m\,i}\omega_{kj}],\qquad \delta^2 f_{ij}=\delta^{km}\omega_{ki}\omega_{mj}\,, \tag{4.8}$$

where ϖ_{ij} denote the components of the rotation tensor for the original state of equilibrium, and ω_{ij} denote the components of the rotation tensor defined by the displacement vector **v**. In other words,

$$\varpi_{ij} = \frac{1}{2}\left(u_{i,j} - u_{j,i}\right), \qquad \omega_{ij} = \frac{1}{2}\left(v_{i,j} - v_{j,i}\right). \tag{4.9}$$

Next, inserting (4.8) in (4.5) produces the following form of the stability functional:

$$\mathsf{N}_1(\mathbf{v}) =$$

$$= \frac{1}{2}\int_\Omega C^{ijkm}\left(v_{i,j}v_{k,m} + 2v_{i,j}\delta^{pr}\varpi_{rm}\omega_{pk} + \delta^{pr}\delta^{qs}\varpi_{ri}\varpi_{sk}\omega_{pj}\omega_{qm}\right)d\Omega +$$

$$+\frac{\delta^{km}}{2}\int_\Omega \sigma^{ij}\omega_{ki}\omega_{mj}\,d\Omega\,. \tag{4.10}$$

Functional $\mathsf{N}_1(\mathbf{v})$ corresponds to the formulation of the stability analysis problem considered by Novozhilov, so the designation of $\mathsf{N}_1(\mathbf{v})$ is a tribute to the author of [12].

11.4.1 A linearized formulation of the equilibrium stability problem for an elastic body

Now let us consider a formulation of the linearized problem of equilibrium stability for general elastic bodies by generalizing the procedure described in Section **11**.1.2 for systems with a finite number of degrees of freedom.

First of all, we assume that the initial state of equilibrium to analyze the stability of has been obtained using the linear analysis. This assumption is quite admissible in most applications because [12]:

> "*slopes ϖ_x, ϖ_y, ϖ_z, which conform to the initial mode of equilibrium of a body, can be either equal to zero or of the same order as the elongations and shears*".

According to our earlier notation, the tensor $\boldsymbol{\tau}$ is a stress tensor for the initial state of equilibrium and the tensor $\boldsymbol{e}$ is its respective strain tensor. The latter means that

$$\boldsymbol{\tau} = \boldsymbol{Ce} \qquad \text{and} \qquad \boldsymbol{\tau} = \lambda\overline{\boldsymbol{\tau}}\,. \tag{4.11}$$

The obvious symbol $\overline{\boldsymbol{\tau}}$ is used to denote a stress tensor determined by the linear analysis for the unit value of the load parameter, $\lambda = 1$.

Exactly as we did with systems that had a finite number of DOFs, we keep all terms containing the displacements raised to third or lower power in the expression of the strain energy, in order to construct a linearized formulation of the equilibrium stability problem. This procedure gives

$$\mathsf{L}=\frac{1}{2}\int_{\Omega}\boldsymbol{Ce}\cdot(\boldsymbol{e}+2\boldsymbol{f})d\Omega-\lambda\int_{\Omega}\overline{\boldsymbol{X}}^{\mathsf{T}}\mathbf{u}d\Omega-\lambda\oint_{\Gamma_p}\overline{\boldsymbol{p}}^{\mathsf{T}}\mathbf{u}d\Gamma\,, \tag{4.12}$$

and the expression of second variation $\delta^2\mathsf{L}$ becomes

$$\delta^2\mathsf{L}=\int_{\Omega}\boldsymbol{C}\delta\boldsymbol{e}\cdot(\delta\boldsymbol{e}+2\delta\boldsymbol{f})d\Omega+\int_{\Omega}\boldsymbol{Ce}\cdot\delta^2\boldsymbol{f}d\Omega\,. \tag{4.13}$$

With the above designations, the stability functional for the linearized problem formulation is

$$\mathsf{B}_2(\mathbf{v})=\frac{1}{2}\int_{\Omega}C^{ijkm}\left(\upsilon_{i,j}\upsilon_{k,m}+2\delta^{pr}u_{r,i}\upsilon_{p,j}\upsilon_{k,m}\right)d\Omega+\frac{\lambda}{2}\int_{\Omega}\overline{\tau}^{ij}\delta^{km}\upsilon_{k,i}\upsilon_{m,j}d\Omega\,.$$

We have assumed that the original stress-and-strain distribution in the system obeys the linear elasticity relationships. It means that both stresses τ^{ij} and displacements u_i are proportional to the load parameter λ. That is, we can define

$$u_i=\lambda\overline{u}_i$$

where $\overline{u}_i$ are components of the displacement vector of the system's initial state, which correspond to the unit value of the load parameter, $\lambda = 1$. Now it turns out that the stability functional for the linearized analysis is

$$\mathsf{B}_2(\mathbf{v})=\frac{1}{2}\int_{\Omega}C^{ijkm}\upsilon_{i,j}\upsilon_{k,m}d\Omega+$$

$$+\frac{\lambda}{2}\int_{\Omega}\left(\overline{\tau}^{ij}\delta^{km}\upsilon_{k,i}\upsilon_{m,j}+2C^{ijkm}\delta^{pr}\overline{u}_{r,i}\upsilon_{p,j}\upsilon_{k,m}\right)d\Omega\,. \tag{4.14}$$

A further simplification that ignores the role played by the initial displacements **u** produces the following expression for the stability functional,

$$\mathsf{B}_3(\mathbf{v})=\frac{1}{2}\int_{\Omega}C^{ijkm}\upsilon_{i,j}\upsilon_{k,m}d\Omega+\frac{\lambda}{2}\int_{\Omega}\overline{\tau}^{ij}\delta^{km}\upsilon_{k,i}\upsilon_{m,j}d\Omega\,. \tag{4.15}$$

To change from functional B_2 to functional B_3, we just omit the term $2C^{ijkm}\delta^{pr}\bar{u}_{r,i}v_{p,j}v_{k,m}$ in the integrand of B_2, and this approach is sometimes interpreted mechanically as the following condition:

> the linearized formulation of the equilibrium stability problem assumes that "*an elastic body in its initial state is treated as being stressed but undeformed*" [1].

This condition permits to formally assume the components of the initial displacement vector $\bar{u}_i$ in the formula (4.14) to be equal to zero, which gives (4.15).

Functional B_3 is known well in the theory of elastic equilibrium stability, although experts do not agree on the authorship of the functional. For example, N.A. Alfutov [1] attributes this functional to Bryan while V.V. Bolotin [7] deems it fair to recognize the authorship of Trefftz. As for functional B_2, it was derived for the first time by the same V.V. Bolotin, although his method of derivation in [7] is different from one we present here.

If we use the simplified functional N_1 from (4.10) in the linearized analysis following Novozhilov's approach, we will replace (4.14) with functional N_2 and (4.15) with functional N_3 :

$$N_2(\mathbf{v}) = \frac{1}{2}\int_\Omega C^{ijkm} v_{i,j} v_{k,m} d\Omega +$$

$$+\frac{\lambda}{2}\int_\Omega \left(\bar{\tau}^{ij}\delta^{km}\omega_{ki}\omega_{mj} + 2C^{ijkm} v_{i,j}\delta^{pr}\bar{\varpi}_{pm}\omega_{rk}\right) d\Omega,$$

$$N_3(\mathbf{v}) = \frac{1}{2}\int_\Omega C^{ijkm} v_{i,j} v_{k,m} d\Omega + \frac{\lambda}{2}\int_\Omega \bar{\tau}^{ij}\delta^{km}\omega_{ki}\omega_{mj} d\Omega \tag{4.16}$$

where $\bar{\varpi}_{pm}$ are the components of the rotation tensor for the original state of equilibrium calculated for the unit value of the load parameter, $\lambda = 1$, that is,

$$\bar{\varpi}_{pm} = \frac{1}{2}\left(\bar{u}_{p,m} - \bar{u}_{m,p}\right).$$

Functionals N_2 and N_3 conform to a linearized formulation of the equilibrium stability problem for thin-walled bodies which was considered and solidly validated by V.V. Novozhilov [12].

To observe historical justice, we think we should refer to the equilibrium stability functionals N_2 and N_3 as *Novozhilov functionals.*

A remark on a mechanical interpretation of particular terms in the stability functional

As can be easily seen, the integral

$$\mathsf{E}(\mathbf{v}) = \frac{1}{2}\int_\Omega C^{ijkm} v_{i,j} v_{k,m} d\Omega = \frac{1}{2}\int_\Omega C^{ijkm}\left(\frac{v_{i,j}+v_{j,i}}{2}\right)\left(\frac{v_{k,m}+v_{m,k}}{2}\right) d\Omega$$

is formally coincident with the expression of the strain energy accumulated by an elastic body on displacements v_i. We emphasize (because it is really important) that energy $\mathsf{E}(\mathbf{v})$ is calculated in the same way as in the geometrically linear analysis. We can write it as follows in the general operator form:

$$\mathsf{E}(\mathbf{v}) = \frac{1}{2}(CA\mathbf{v}, A\mathbf{v}). \tag{4.17}$$

At the same time, the integrals of the kind

$$\mathsf{A_B}(\mathbf{v}) = -\frac{\lambda}{2}\int_\Omega \overline{\tau}^{ij}\delta^{km} v_{k,i} v_{m,j} d\Omega, \qquad \mathsf{A_N}(\mathbf{v}) = -\frac{\lambda}{2}\int_\Omega \overline{\tau}^{ij}\delta^{km}\omega_{ki}\omega_{mj} d\Omega \tag{4.18}$$

are named in scientific, engineering, and educational literature, according to the tradition (unluckily, a totally incorrect one), as a *negative work of external forces on displacements* v_i. The obvious inadmissibility of this mechanical interpretation of functionals $\mathsf{A}(\mathbf{v})$ was noticed by V.V. Bolotin back in 1965 [5]; he stated justly that the functionals

$$\mathsf{B_3}(\mathbf{v}) = \mathsf{E}(\mathbf{v}) - \mathsf{A_B}(\mathbf{v}), \qquad \mathsf{N_3}(\mathbf{v}) = \mathsf{E}(\mathbf{v}) - \mathsf{A_N}(\mathbf{v}) \tag{4.19}$$

were in essense the second variation of the full potential energy in the vicinity of the state of equilibrium.

Nonetheless, functionals $\mathsf{A}(\mathbf{v})$ can and should receive a certain mechanical interpretation. We transform the integrals in (4.18) by using the Gauss–Ostrogradsky formula:

$$\mathsf{A_B}(\mathbf{v}) = -\frac{\lambda}{2}\int_\Omega \overline{\tau}^{ij}\delta^{km} v_{k,i} v_{m,j} d\Omega =$$

$$= \frac{\lambda}{2}\int_\Omega \delta^{km}\left(\overline{\tau}^{ij} v_{k,i}\right)_{,j} v_m d\Omega - \frac{\lambda}{2}\oint_\Gamma \overline{\tau}^{ij}\delta^{km} v_{k,i} n_j v_m d\Gamma \tag{4.20}$$

and

$$\mathsf{A_N}(\mathbf{v}) = -\frac{\lambda}{2}\int_\Omega \overline{\tau}^{ij}\delta^{km}\omega_{ki}\omega_{mj}\,d\Omega = \frac{\lambda}{2}\int_\Omega \frac{\left(\delta^{km}\overline{\tau}^{ij}\omega_{ki} - \delta^{kj}\overline{\tau}^{im}\omega_{ki}\right)_{,j}}{2} v_m\,d\Omega -$$

$$-\frac{\lambda}{2}\oint_\Gamma \frac{\delta^{km}\overline{\tau}^{ij}\omega_{ki} - \delta^{kj}\overline{\tau}^{im}\omega_{ki}}{2} n_j v_m\,d\Gamma\,. \tag{4.21}$$

Clearly, the terms

$$X_B^m = \lambda\overline{X}_B^m = \lambda\delta^{km}\left(\overline{\tau}^{ij}v_{k,i}\right)_{,j},\ X_N^m = \lambda\overline{X}_N^m = \frac{1}{2}\left(\delta^{km}\overline{\tau}^{ij}\omega_{ki} - \delta^{kj}\overline{\tau}^{im}\omega_{ki}\right)_{,j} \tag{4.22}$$

can be treated as a conditional volumetric load with the components X_B^m or X_N^m that depends on the components of vector $\mathbf{v}$. Similarly, the terms

$$p_B^m = \lambda\overline{p}_B^m = -\lambda\overline{\tau}^{ij}\delta^{km}v_{k,i}n_j,\ \ p_N^m = \lambda\overline{p}_N^m = -\lambda\frac{1}{2}\left(\delta^{km}\overline{\tau}^{ij} - \delta^{kj}\overline{\tau}^{im}\right)\omega_{ki}n_j \tag{4.23}$$

can be interpreted as a conditional surface (edge) load.

A conditional load of this kind is often introduced to derive differential equations of structural equilibrium stability. Sometimes this load is called "*effective*" [21], sometimes "*equivalent*", sometimes "*fictitious*" [6]. No matter how it is called, this conditional load is introduced as a result of a certain colourable reasoning based on half-intuitive static and geometric considerations rather than on a rigorous mathematical theory. It is the formulas (4.22) and (4.23) that give this notion a clear mathematical sense.

But then integrals $\mathsf{A}(\mathbf{v})$, being represented as

$$\mathsf{A_B}(\mathbf{v}) = \frac{\lambda}{2}\int_\Omega \overline{X}_B^m v_m\,d\Omega + \frac{\lambda}{2}\oint_\Gamma \overline{p}_B^m v_m\,d\Gamma\,,$$

$$\mathsf{A_N}(\mathbf{v}) = \frac{\lambda}{2}\int_\Omega \overline{X}_N^m v_m\,d\Omega + \frac{\lambda}{2}\oint_\Gamma \overline{p}_N^m v_m\,d\Gamma\,,$$

can be treated from the mechanical standpoint simply as a *work of the equivalent load on the displacements determined by vector* $\mathbf{v}$. The factor of ½ participates in the expression of this work because the equivalent load depends on the displacements $\mathbf{v}$ linearly according to (4.22) and (4.23).

Criteria of a critical state in a system

Obviously, the equilibrium of an elastic system is stable when the load parameter is equal to zero, because at $\lambda = 0$ any of the stability functionals

$\mathsf{S}(\mathbf{v})$ is a positive functional for certain, i.e. $\mathsf{S}(\mathbf{v}) > 0$ for any (nonzero) vector $\mathbf{v}$. Moreover, the equilibrium is stable also in a certain vicinity of the point $\lambda = 0$. However, when the load grows further, it is not unlikely that at some $\lambda = \lambda_{cr}$ the stability functional $\mathsf{S}(\mathbf{v})$ can lose its positivity; this is the way to define the boundary of a stability area of the system.

$\mathsf{S}(\mathbf{v})$ is a homogeneous quadratic functional of $\mathbf{v}$, hence it can be represented in the most general way as

$$\mathsf{S}(\mathbf{v}) = \tfrac{1}{2}\,(\boldsymbol{S}\mathbf{v},\mathbf{v}) \tag{4.24}$$

where $\boldsymbol{S}$ is a linear operator defined in the set of kinematically admissible fields of displacements, vectors $\mathbf{v}$. We will call it a *stability operator.*

Note that the stability operator depends parametrically on the load intensity λ. Clearly, the requirement of obligatory positive definiteness of functional $\mathsf{S}(\mathbf{v})$ is equivalent to the requirement that the stability operator $\boldsymbol{S}$ should be positive definite. A moment when the stability operator $\boldsymbol{S}$ loses in positive definiteness (at $\lambda = \lambda_{cr}$) corresponds to the critical state. This means the first variation of the stability functional is zero in the critical state because the condition of existence of a nonzero solution of the homogeneous equation,

$$\boldsymbol{S}\mathbf{v} = \mathbf{0}, \tag{4.25}$$

is equivalent to the condition

$$\delta\mathsf{S}(\mathbf{v}) = 0. \tag{4.26}$$

Vector $\mathbf{v}$ found from (4.25) or (4.26) defines a *mode of buckling* of the system, and the $\mathbf{v}$ vector itself is called an *eigenvector that conforms to eigenvalue* $\lambda = \lambda_{cr}$.

As the $\mathsf{A}(\mathbf{v})$ integral depends on the load parameter λ linearly, we can write

$$\mathsf{A}(\mathbf{v}) = \lambda\,\bar{\mathsf{A}}(\mathbf{v}),$$

where $\bar{\mathsf{A}}(\mathbf{v})$ is, obviously, the work of the equivalent unit load on the mode of buckling $\mathbf{v}$.

So, let us know a critical value of the load value, λ_{cr}, and its respective mode of buckling, $\mathbf{v}$. Thne by definition this vector makes $\mathsf{S}(\mathbf{v}) = 0$:

$$\mathsf{E}(\mathbf{v}) - \lambda_{cr}\,\bar{\mathsf{A}}(\mathbf{v}) = 0. \tag{4.27}$$

In other words, in the critical state of the system its potential strain energy accumulated during the deformation in the buckling mode is equal to the work of the effective load on the displacements defined by eigenvector $\mathbf{v}$.

While the condition (4.26) is a variational-type criterion, the equality (4.27), strictly speaking, is not. The equality (4.27) can be shown to be an immediate corollary of the variational criterion (4.26) in the formal mathematical sense. It suffices to recall that the stability functional $\mathsf{S}(\mathbf{v})$ is a homogeneous quadratic functional.

We should note once again that the equality (4.27) follows from (4.26) while the inverse is, generally, not true. In other words, (4.27) does hold in a critical state of the system, but the mere fact that the equality $\mathsf{E}(\mathbf{v}) - \lambda\bar{\mathsf{A}}(\mathbf{v}) = 0$ takes place for a vector $\mathbf{v}$ and for a value λ does not imply necessarily that this particular λ and this particular vector $\mathbf{v}$ are a critical value of the load parameter and a buckling mode, respectively. If the inverse *were* true, then taking an arbitrary vector $\mathbf{v}$ such that $\bar{\mathsf{A}}(\mathbf{v}) \neq 0$ and calculating λ as $\lambda = \mathsf{E}(\mathbf{v})/\bar{\mathsf{A}}(\mathbf{v})$ would make us think that $\lambda = \lambda_{cr}$ and that vector $\mathbf{v}$ defines a buckling mode of the system. This is, obviously, not the case, therefore the condition (4.26) can be called a critical-state criterion while (4.27) is just an equality that always holds in a critical state. We draw the reader's attention to this simple fact only because literature abounds in partially correct propositions where the equality (4.27) is treated as a criterion[11].

Taking first variation of the Bryan–Trefftz stability functional $\mathsf{B}_3(\mathbf{v})$ from (4.15) with respect to vector $\mathbf{v}$ (in the field of homogeneously kinematically admissible displacements) and equating it to zero produces, after applying the Gauss–Ostrogradsky formula, the following Euler equations and natural boundary conditions:

$$\left(C^{ijkm} v_{k,m}\right)_{,j} + \lambda\left(\overline{\tau}^{mj}\delta^{ki} v_{k,m}\right)_{,j} = 0 \in \Omega,$$

$$\left(C^{ijkm} + \lambda\overline{\tau}^{mj}\delta^{ki}\right) v_{k,m} n_j = 0 \in \Gamma_p. \tag{4.28}$$

The formulation of the spectral problem in the differential form of (4.28) is known as a Sturm–Liouville problem in mathematics. So we can say that the linearized formulation of the elastic stability analysis is mathematically equivalent to the Sturm–Liouville problem. It is also clear that the

[11] S.P. Timoshenko in his works often makes the equality (4.27) sound like a criterion. We do not want to say that Timoshenko makes mistakes in his results by applying the equality (4.27) to particular equilibrium stability problems. However, the correct final result is not the only important thing; it is equally important to make the considerations which produce that result mathematically rigorous and theoretically convincing.

differential equations and static-type boundary conditions in (4.28), together with the homogeneous kinematical boundary conditions

$$v_k = 0 \in \Gamma_u,$$

are nothing but the expanded operator equation (4.25) in application to the Bryan–Trefftz functional.

Using the definition of the equivalent load from (4.22) and (4.23), we can rewrite (4.28) as

$$-\left(C^{ijkm} v_{k,m}\right)_{,j} - \lambda \bar{X}_B^i = 0 \in \Omega, \qquad C^{ijkm} v_{k,m} n_j - \lambda \bar{p}_B^i = 0 \in \Gamma_p. \tag{4.29}$$

Now it is easy to see that one possible mechanical sense of the equations (4.29) is that they are equations of equilibrium and static boundary conditions for an elastic body subjected to the equivalent load. The equations must follow the geometrically linear problem formulation.

The same technique for another functional, Novozhilov's functional $\mathsf{N}_3(\mathbf{v})$, produces the following Sturm–Liouville boundary-value problem:

$$\left(C^{ijkm} v_{k,m}\right)_{,j} + \lambda \frac{1}{2}\left(\delta^{ki} \bar{\tau}^{mj} \omega_{km} - \delta^{kj} \bar{\tau}^{im} \omega_{km}\right)_{,j} = 0 \in \Omega,$$

$$C^{ijkm} v_{k,m} n_j + \lambda \frac{1}{2}\left(\delta^{ki} \bar{\tau}^{mj} - \delta^{kj} \bar{\tau}^{im}\right) \omega_{km} n_j = 0 \in \Gamma_p. \tag{4.30}$$

Finally, we can introduce the Rayleigh ratio for the linearized elastic equilibrium stability analysis as

$$\mathsf{r} = \frac{\mathsf{E}(\mathbf{v})}{\bar{\mathsf{A}}(\mathbf{v})}. \tag{4.31}$$

Also, it is easy to show that the criterion for the critical state to occur can be written as a condition of stationarity of the Rayleigh ratio as a functional of vector $\mathbf{v}$ on the kinematically admissible fields of displacements,

$$\delta \mathsf{r} = 0. \tag{4.32}$$

No formal proof is presented here; the reader can do it by herself.

The criterion (4.32) helps find the modes of buckling, $\mathbf{v}$, for the system. The critical value of the load parameter that conforms to this particular mdoe of buckling is calculated as a Rayleigh ratio on the mode $\mathbf{v}$,

$$\lambda_{cr} = \mathsf{r}(\mathbf{v}). \tag{4.33}$$

11.4.2 The Ritz method

One of most effective ways to reduce a problem of equilibrium stability for an elastic body to the same problem for a finite-dimensional system is the Ritz method.

Exactly as we did in Section **9**.1, we introduce a system of linearly independent displacement vectors $V_1, V_2,\ldots, V_n$. In addition to their linear independence, we demand that they belong to a homogeneously kinematically admissible field of displacements. As for the requirement of smoothness, it suffices that there exist finite integrals $\mathsf{E}(V_i)$ and $\mathsf{A}(V_i)$, $i = 1,\ldots, n$. As before, vector functions $V_1, V_2,\ldots, V_n$ that satisfy the said conditions will be called *base functions*.

We expand the desirable mode of buckling of the system, **v**, over the base vector functions by postulating an approximate equality:

$$\mathbf{v} = \sum_{i=1}^{n} z_i V_i \,. \tag{4.34}$$

Any of the criteria (4.26) or (4.32) can be used to determine the desirable coefficients z_i.

We choose to apply the criterion (4.26), so we insert the approximations (4.34) in there. This makes the equilibrium stability functional, $\mathsf{S}(\mathbf{v}) = \mathsf{E}(\mathbf{v}) - \mathsf{A}(\mathbf{v})$, a quadratic form of vector z:

$$\mathsf{S}(z) = \frac{1}{2} z^{\mathsf{T}} (\boldsymbol{r}_{\mathrm{o}} - \lambda \boldsymbol{r}_G) z \,, \tag{4.35}$$

where $\boldsymbol{r}_{\mathrm{o}}$ and $\boldsymbol{r}_G$ are the respective matrices of initial stiffness and of geometric stiffness for the system with a finite (and equal to n) number of DOFs, and vector z is composed of the sought-for coefficients z_i

$$z = \|[z_1,\ldots,z_n]\|^{\mathsf{T}} \,. \tag{4.36}$$

The components of the initial stiffness matrix, $r_{\mathrm{o}ij}$, are defined as

$$r_{\mathrm{o}ij} = (\boldsymbol{C}\boldsymbol{A}V_i, \boldsymbol{A}V_j) \qquad (i, j = 1,\ldots, n) \,. \tag{4.37}$$

Here, the matrix differential operator $\boldsymbol{A}$ is a known geometry operator that maps the displacement vector into the strain vector in the linear elasticity. The components of the geometric stiffness matrix r_{Gij} are calculated in different ways depending on a particular form of the stability functional which is used.

Each of the base vector functions $V_1, V_2,\ldots, V_n$ consists of three components, so

$$V_i = \begin{Vmatrix} V_{i1} \\ V_{i2} \\ V_{i3} \end{Vmatrix} \quad (i = 1,\dots, n)\,, \qquad v_k = \sum_{i=1}^{n} z_i V_{ik} \quad (k = 1, 2, 3)\,.$$

It is easy to see that with the Bryan–Treffz functional the formula for the r_{Gij} components is

$$r_{Gij} = \int_{\Omega} \overline{\tau}^{pq} \delta^{km} V_{ik,p} V_{jm,q} d\Omega \quad (i, j = 1,\dots, n)\,, \quad (p, q, k, m = 1, 2, 3). \tag{4.38}$$

To write a general formula of the components of the geometric stiffness matrix based on the Novozhilov functional, it is convenient to represent the expression of functional $\mathsf{A_N}(\mathbf{v})$ from (4.18) in matrix form. We denote

$$\overline{\boldsymbol{\tau}} = \begin{Vmatrix} \overline{\tau}^{11} \\ \overline{\tau}^{22} \\ \overline{\tau}^{33} \\ \overline{\tau}^{12} \\ \overline{\tau}^{23} \\ \overline{\tau}^{31} \end{Vmatrix} = \begin{Vmatrix} \overline{\tau}^{xx} \\ \overline{\tau}^{yy} \\ \overline{\tau}^{zz} \\ \overline{\tau}^{xy} \\ \overline{\tau}^{yz} \\ \overline{\tau}^{zx} \end{Vmatrix},$$

$$\boldsymbol{\varphi} = \begin{Vmatrix} \delta^{km}\omega_{k1}\omega_{m1} \\ \delta^{km}\omega_{k2}\omega_{m2} \\ \delta^{km}\omega_{k3}\omega_{m3} \\ 2\delta^{km}\omega_{k1}\omega_{m2} \\ 2\delta^{km}\omega_{k2}\omega_{m3} \\ 2\delta^{km}\omega_{k3}\omega_{m1} \end{Vmatrix} = \begin{Vmatrix} \omega_{21}^2 + \omega_{31}^2 \\ \omega_{32}^2 + \omega_{12}^2 \\ \omega_{13}^2 + \omega_{23}^2 \\ 2\omega_{31}\omega_{32} \\ 2\omega_{12}\omega_{13} \\ 2\omega_{23}\omega_{21} \end{Vmatrix} = \begin{Vmatrix} \omega_z^2 + \omega_y^2 \\ \omega_x^2 + \omega_z^2 \\ \omega_y^2 + \omega_x^2 \\ -2\omega_y\omega_x \\ -2\omega_z\omega_y \\ -2\omega_x\omega_z \end{Vmatrix}, \tag{4.39}$$

and now can rewrite the $\mathsf{A_N}(\mathbf{v})$ functional as

$$\mathsf{A_N}(\mathbf{v}) = -\frac{\lambda}{2} \int_{\Omega} \overline{\boldsymbol{\tau}}^{\mathsf{T}} \boldsymbol{\varphi}\, d\Omega\,. \tag{4.40}$$

Obviously, the components of vector $\boldsymbol{\varphi}$ are quadratically dependent on the components of vector $\mathbf{v}$. It means putting the approximations (4.34) in $\boldsymbol{\varphi}(\mathbf{v})$ makes the $\mathsf{A_N}$ integral from (4.40) a homogeneous quadratic form of coefficients z_i. The coefficients of this quadratic form are exactly the components of the geometric stiffness matrix, r_{Gij}. Hence the general formula, considering that the components of the initial stress vector $\overline{\tau}$ in (4.39) do not depend on z_i:

$$r_{Gij} = \int_{\Omega} \overline{\tau}^{\mathsf{T}} \frac{\partial^2 \varphi}{\partial z_i \partial z_j} d\Omega \quad (i, j = 1,\ldots, n)\,. \tag{4.41}$$

Now let us suppose that matrices $\boldsymbol{r}_0$ and $\boldsymbol{r}_G$ are already constructed. Then the criterion (4.26) becomes

$$\frac{\partial \mathsf{S}(\boldsymbol{z})}{\partial z_i} = 0 \;\; (i = 1,\ldots, n)\,.$$

The latter condition is met if

$$(\boldsymbol{r}_0 - \lambda \boldsymbol{r}_G)\boldsymbol{z} = \boldsymbol{0}\,, \tag{4.42}$$

which is the same as the already known condition (2.2) for a critical state to come up in a system with a finite number of DOFs.

All the reasoning could be done for the variational criterion in the form of (4.32). The reader can make sure by herself that the result will be the same finite-dimensional spectral problem (4.42) for the couple of matrices $\boldsymbol{r}_0$ and $\boldsymbol{r}_G$.

By the way, if we insert the Ritz approximations in (4.27), this equality becomes

$$\boldsymbol{z}^{\mathsf{T}}(\boldsymbol{r}_0 - \lambda \boldsymbol{r}_G)\boldsymbol{z} = 0\,.$$

The latter equality does no more than gives a hint that the $\boldsymbol{z}$ vector is orthogonal to the $(\boldsymbol{r}_0 - \lambda \boldsymbol{r}_G)\boldsymbol{z}$ vector in the system's critical state. The equality (4.42) is a stronger result which does not follow directly from the above.

11.5 Stability of equilibrium in particular classes of problems

Based on various forms of the stability functional, $\mathsf{S}(\mathbf{v})$, we will try to derive formulations of the stability problem in application to some particular conditions and configurations.

11.5.1 Stability of equilibrium of a bar in the engineering theory of bars

We start our consideration by a simple case of equilibrium stability for a rectilinear-axis bar (the planar case). All reasoning and calculations will be very detailed in this simple problem, in order to be able in the future to omit similar ones where necessary.

We make the longitudinal axis of the bar, X, coincident with the line of center of gravity of the bar's cross-sections; it is assumed to be straight. For the sake of simplicity, we suppose that the bar is subjected to axial loads only, so only the longitudinal force $N = \lambda\bar{N}$ appears in the initial stressed state of the bar, where $\bar{N} = \bar{N}(x)$ is a longitudinal force that corresponds to the unit load parameter. To confine ourselves to the planar analysis, we will suppose the bar is fixed in some way against the buckling in the (X,Z)-plane. Axes Y and Z are assumed to be principal central axes of inertia of the bar's cross-sections, and their directions are not variable along the bar.

We represent the vector of displacements, $\mathbf{v}$, of an arbitrary point with the coordinates (x,y,z), which defines the buckling mode, by its components as

$$\mathbf{v} = \|[U,V]\|^{\mathrm{T}},$$

where U,V are the components of vector $\mathbf{v}$ with respect to the set of axes (X,Y).

We introduce the notation

$$u(x) = U(x,0), \qquad v(x) = V(x,0).$$

In other words, u, v are the components of the displacement vector of the points on the axis of the bar.

Taking into account the kinematic hypotheses of the engineering theory of tension/compression/bending of bars and using geometric considerations, we can write

$$U = u - v'y\,, \qquad V = v\,. \tag{5.1}$$

Here and further, the stroke means taking the derivative with respect to coordinate x. Thus it turns out the engineering theory defines the bar's kinematics by two functions of coordinate x, namely

$$u = u(x), \quad v = v(x).$$

Based on the results of Section **4**.4 and Section **4**.5, we can immediate write out an expression of functional $\mathsf{E}(\mathbf{v})$ according to (4.17). It is

$$\mathsf{E}(\mathbf{v}) = \frac{1}{2}\int_0^l \left[EAu'^2 + EIv''^2 \right] dx \tag{5.2}$$

where l is the bar's length, A is its cross-section's area, and I is its cross-section's moment of inertia.

To derive a variational formulation for the stability of equilibrium of the bar, we have yet to handle the $\mathsf{A}(\mathbf{v})$ functional using the general formulas (4.18).

First of all, note that the expression of functional $\mathsf{A}(\mathbf{v})$ contains components of the initial unit stress tensor, $\overline{\tau}^{ij}$. According to (4.11), these components are based on Hooke's law and the initial strain tensor, $\boldsymbol{e}$. In our case the engineering theory says the only nonzero component of the initial strain tensor, $\boldsymbol{e}$, is the e_{xx} component that describes the normal stresses in the cross-sections of the bar – we are dealing with the stability of equilibrium of a bar subjected to axial longitudinal loads. Obviously, the respective nonzero component, $\overline{\tau}^{11} = \overline{\tau}^{xx}$, of the initial unit stress tensor is a well-known expression,

$$\overline{\tau}^{xx} = \frac{\overline{N}}{A}. \tag{5.3}$$

We have

$$\omega_{11} = 0, \quad \omega_{12} = -\omega_{21} = \frac{1}{2}\left(\frac{\partial U}{\partial y} - \frac{\partial V}{\partial x} \right) = -v'. \tag{5.4}$$

Next, we put the above in (4.18) and integrate over the area of cross-section of the bar, A. The result is

$$\mathsf{A}_\mathsf{N}(\mathbf{v}) = -\frac{\lambda}{2}\int_0^l \int_A \overline{\tau}^{xx} \omega_{21}^2 \, dA\, dx = -\frac{\lambda}{2}\int_0^l \overline{N} v'^2 \, dx,$$

so, according to (4.19), the Novozhilov stability functional in our problem becomes

$$\mathsf{N}_3(\mathbf{v}) = \mathsf{E}(\mathbf{v}) - \mathsf{A}_\mathsf{N}(\mathbf{v}) = \frac{1}{2}\int_0^l \left[EAu'^2 + EIv''^2 + \lambda \overline{N} v'^2 \right] dx. \tag{5.5}$$

As we can see, this functional falls apart into two independent parts. The first part is a functional of the longitudinal displacement, u, of the bar's axis, and it is always nonzero for any function $u = u(x)$. The second part depends on the lateral displacements, $v = v(x)$, of the bar's axis. It is this second part that can change the sign of the $\mathsf{N}_3(\mathbf{v})$ functional at a certain (critical) value of the load parameter, $\lambda = \lambda_{cr}$. It is also obvious that the minimum critical value, $\lambda = \lambda_{cr}$, corresponds to $u' = 0$, i.e. $u = Const$. So actually it suffices to keep only such terms in the stability functional that depend on the v function. Consequently, we can take

$$\mathsf{N}_3(v) = \frac{1}{2}\int_0^l \left[EIv''^2 + \lambda \bar{N} v'^2 \right] dx \,. \tag{5.6}$$

The functional (5.6) is known well in the theory of stability of bars. Taking the variation of the functional with respect to homogeneously kinematically admissible fields of displacements and equaling its first variation to zero leads to the following fourth-order differential equation:

$$\left(EIv''\right)'' - \lambda\left(\bar{N}v'\right)' = 0 \,, \tag{5.7}$$

a nonzero solution of which under the respective homogeneous boundary conditions defines a buckling mode for the bar.

By the way, the equality (4.27) obliged to hold in the critical state of the system implies that for $\bar{N} \geq 0$ (pure tension along the bar) the buckling may occur only if λ is negative. And vice versa, if the longitudinal force is compressive, $\bar{N} \leq 0$, then the buckling occurs at a positive load parameter λ solely. The physical meaning is obvious.

If we replace the Novozhilov functional, N_3, with the Bryan–Trefftz one, B_3, then the general formula (4.18) will give

$$\mathsf{A_B}(\mathbf{v}) = -\frac{\lambda}{2}\int_0^l \int_A \bar{\tau}^{xx}\left(U_{,x}^2 + V_{,x}^2\right) dA dx = -\frac{\lambda}{2}\int_0^l \frac{\bar{N}}{A} \iint_A \left[\left(u' - v''y\right)^2 + v'^2 \right] dA dx \,,$$

and the integration of the latter over the cross-section's area A yields

$$\mathsf{A_B}(\mathbf{v}) = \frac{1}{2}\int_0^l \lambda \bar{N}\left(u'^2 + r^2 v''^2 + v'^2\right) dx$$

where $r^2 = I/A$ is the squared radius of inertia of the bar's cross-section. The result is the following Bryan–Trefftz stability functional:

$$\mathsf{B}_3(v) = \frac{1}{2}\int_0^l \left[EIv''^2 + \lambda \bar{N}\left(r^2 v''^2 + v'^2\right) \right] dx \,. \tag{5.8}$$

As we can see, the Bryan–Trefftz stability functional differs from the Novozhilov functional by an additional term, $r^2 v''^2$. It is easy to see that removing this additional term $r^2 v''^2$ from the integrand of (5.8) is equivalent to neglecting the value of r^2/s^2 comparing to one, where s

denotes a "*characteristic scale of variability*" of v' – see [6]. This scale can be understood as a half-wavelength that characterizes the variance in v' along the bar's axis. This assumption is apparently consistent with the basic definition of a bar – a three-dimensional body extended in one direction for which the ratio of the characteristic size of its cross-section to the characteristic scale of variability of the stress-and-strain distribution along the extended size is greatly less than one. If this estimate does not take place, then we should question the very applicability of the theory of bars to the case. Thus, ignoring the said term in the integrand of (5.8) is fully consistent with the basic assumptions of the bar theory. Then the Bryan–Trefftz functional becomes just identical to the Novozhilov functional.

11.5.2 Stability of equilibrium of a Timoshenko bar

Now let us turn to the stability of equilibrium of the same bar where we will try to allow for shear strains (using the Timoshenko theory of bars). As we will see a bit later, this seemingly simple problem has peculiar pitfalls of a theoretical nature which were not yet discussed in literature as far as we know.

Instead of (5.2), we have the following expression of the bar's strain energy in its mode of buckling $\mathbf{v}$:

$$\mathsf{E}(\mathbf{v}) = \frac{1}{2}\int_0^l \left[EAu'^2 + EI\theta'^2 + GF(v' - \theta)^2 \right] dx\,, \tag{5.9}$$

where we remind that F is a so-called shear area of the bar's cross-section.

Also, the kinematics of the bar is now described by the following relationships instead of (5.1):

$$U = u - \theta\, y\,, \qquad V = v\,, \tag{5.10}$$

where $\theta = \theta(x)$ is an independently calculated slope of the bar's cross-section – see Section **4**.6. Therefore, instead of (5.4) and on the basis of (5.10), we have

$$\omega_{11} = 0\,, \quad \omega_{12} = -\omega_{21} = \frac{1}{2}\left(\frac{\partial U}{\partial y} - \frac{\partial V}{\partial x} \right) = -\frac{v' + \theta}{2}\,. \tag{5.11}$$

Substituting and integrating over the cross-section's area gives

$$\mathsf{A_N}(\mathbf{v}) = -\frac{\lambda}{2}\int_0^l \int_A \overline{\tau}^{xx} \omega_{21}^2 \, dA dx = -\frac{\lambda}{2}\int_0^l \overline{N}\frac{(v'+\theta)^2}{4}\, dx .$$

The same reasoning as in the preceding section produces the following Novozhilov functional for the stability of the Timoshenko bar:

$$\mathsf{N_3}(v,\theta) = \frac{1}{2}\int_0^l \left[EI\theta'^2 + GF(v'-\theta)^2 + \lambda\overline{N}\frac{(v'+\theta)^2}{4} \right] dx . \tag{5.12}$$

The Euler equations for it are

$$-(EI\theta')' - GF(v'-\theta) + \frac{\lambda}{4}\overline{N}(v'+\theta) = 0 ,$$

$$\left[GF(v'-\theta)\right]' + \frac{\lambda}{4}\left[\overline{N}(v'+\theta)\right]' = 0 .$$

Let us consider a bar of a constant section loaded by a constant compressive force, $\overline{N} = -1$. We rewrite each of the equations of this system expressing v via θ. So we have

$$v' = -\frac{4EI}{4GF+\lambda}\theta'' + \frac{4GF-\lambda}{4GF+\lambda}\theta , \qquad v'' = \frac{4GF+\lambda}{4GF-\lambda}\theta' . \tag{5.13}$$

Now, differentiating the first of them and subtracting the second yields a third-order differential equation with respect to function θ,

$$\theta''' + \frac{4GF\lambda}{EI(4GF-\lambda)}\theta' = 0 . \tag{5.14}$$

By denoting

$$a^2 = \frac{EI}{GF}, \quad \mu = \frac{\lambda}{GF}, \quad k^2 = \frac{4\mu}{a^2(4-\mu)} \tag{5.15}$$

we make (5.14) look like

$$\theta''' + k^2\theta' = 0$$

with its general solution being

$$\theta = C_1 \sin kx + C_2 \cos kx + C_3 . \tag{5.16}$$

Placing θ from (5.16) in the first of the equations of (5.13) and integrating once gives

$$v = \frac{4a^2k^2 + 4 - \mu}{(4+\mu)k}\left(-C_1 \cos kx + C_2 \sin kx\right) + C_3 x + C_4 . \tag{5.17}$$

As an example, we take a bar of a length l with the boundary conditions of hinged (simple) support on its ends, i.e.

$$\theta'(0) = 0, \quad \theta'(l) = 0, \quad v(0) = 0, \quad v(l) = 0 .$$

The given boundary conditions help find the four constants C_i needed to know the solution. These are

$$C_1 = 0, \quad C_2 \sin kl = 0, \quad C_3 = 0, \quad C_4 = 0,$$

so a nonzero solution for v and θ is possible only when $kl = n\pi$ at any integer n. Hence, using (5.15),

$$\frac{n^2\pi^2}{l^2} = \frac{4\mu_{cr}}{a^2(4-\mu_{cr})} \quad \text{and} \quad \lambda_{cr} = \frac{EI}{1 + \dfrac{a^2n^2\pi^2}{4l^2}} \frac{n^2\pi^2}{l^2} .$$

Obviously, the minimum critical value of the load parameter, $\lambda = \lambda_{cr}$, is achieved at $n = 1$, so

$$\lambda_{cr} = \frac{\pi^2 EI}{l^2} \frac{1}{1 + \dfrac{a^2\pi^2}{4l^2}} . \tag{5.18}$$

In the limit case when $GF \to \infty$, the a^2 parameter tends to zero and the critical load becomes equal to the well-known value

$$\lambda_{cr} = \frac{\pi^2 EI}{l^2} .$$

As we can see, the formula (5.12) and its subsequent corollaries up to (5.18) are based on a consistent mathematical analysis that makes use of the Novozhilov functional. However, the literature on engineering traditionally uses a slightly different stability functional for the Timoshenko bar (see [1], for example):

$$\mathsf{S}(v,\theta) = \frac{1}{2}\int_0^l \left[EI\theta'^2 + GF(v' - \theta)^2 + \lambda \bar{N} v'^2 \right] dx , \tag{5.19}$$

which is equivalent to using this expression of the slopes:

$$\omega_{12} = -\omega_{21} = -v'$$

instead of earlier (5.11).

The equations of equilibrium for the neutral state in terms of displacements, which follow from the above stability functional, are:

$$-(EI\theta')' - GF(v' - \theta) = 0,$$

$$\left[GF(v' - \theta)\right]' + \lambda\left[\bar{N}v'\right]' = 0.$$

Further calculations yield a critical value of the load parameter obtained by Engesser back in 1891 (see also [1]):

$$\lambda_{cr} = \frac{\pi^2 EI}{l^2} \frac{1}{1 + \dfrac{a^2\pi^2}{l^2}}. \tag{5.20}$$

As we can see, the formulas (5.18) and (5.20) are significantly different in their form. The stability functional from (5.12) has been derived by rigorous mathematics, which is not the case for (5.19) and its corollary (5.20). But, actually, the dimensionless parameter $a^2\pi^2/l^2$ is a value much less than one, at least for common bars made of an isotropic material. Therefore the addition to one in the denominators of (5.18) and (5.20) can be simply ignored in view of practical applications.

Of course, it should be taken into account that the Timoshenko theory is sometimes applied to compound bars/beams where the $a^2\pi^2/l^2$ parameter cannot be deemed small in comparison with one. Then the difference in the results calculated by these two formulas can be significant. However, as A.R. Rzhanitsyn showed [18], the very extension of the Timoshenko theory onto the case of a compound bar is already a rough approximation that can distort the results noticeably. A neater theory of compound bars not based on the Timoshenko theory has been created by the same Rzhanitsin [18], and the interested reader is invited to get familiar with that by herself.

Another, second variation of applicability of the Timoshenko theory of bars exists. We mean bars made of materials with drastically anisotropic properties for which the estimate $G \ll E$ takes place. For example, the same book [1] presents estimates of the $a^2\pi^2/l^2$ parameter for a three-layered bar where the material of the innermost layer (filling) has much weaker stiffness characteristics in comparison with those of the outer layers. It is shown there also that the dimensionless parameter $a^2\pi^2/l^2$ can be not so small in that case. Then the application of (5.20) can give a noticeably different result from that calculated by (5.18).

The approach to the Timoshenko bar stability analysis where the equilibrium stability functional is written as (5.19), is somehow very popular[12], therefore we feel it necessary to do a more careful analysis of the situation.

Let us try to evaluate an unknown error immanent in the preceding analysis due to the use of the simplified stability functional by Novozhilov. To do it, we take a more accurate stability functional, one by Bryan–Trefftz, $\mathsf{B}_3(v,\theta)$. So we have

$$\mathsf{A}_\mathsf{B}(\mathbf{v}) = -\frac{\lambda}{2}\int_0^l \int_A \bar{\tau}^{xx}\left(U_{,x}^2 + V_{,x}^2\right) dA dx\,. \tag{5.21}$$

According to (5.9),

$$U_{,x} = u' - \theta' y\,, \quad V_{,x} = v'\,.$$

Inserting the above in the integrand of $\mathsf{A}_\mathsf{B}(\mathbf{v})$ and then integrating over the area of the bar's cross-section gives

$$\mathsf{A}_\mathsf{B}(\mathbf{v}) = -\frac{\lambda}{2}\int_0^l \bar{N}\left(u'^2 + \theta'^2 r^2 + v'^2\right) dx\,, \quad r^2 = \frac{I}{A}\,.$$

The final form of the Bryan–Treffz stability functional is

$$\mathsf{B}_3(v,\theta) = \frac{1}{2}\int_0^l \left[EI\theta'^2 + GF(v'-\theta)^2 + \lambda\bar{N}(\theta'^2 r^2 + v'^2)\right] dx\,. \tag{5.22}$$

The Euler equations of this functional are

$$(EI\theta')' - GF(v'-\theta) - \lambda(\bar{N}r^2\theta')' = 0\,,$$

$$\left[GF(v'-\theta)\right]' + \lambda(\bar{N}v')' = 0\,.$$

For the constant-section bar at $\bar{N} = -1$, we can write these equations as

$$v' = -\frac{EI - \lambda r^2}{GF}\theta'' + \theta\,, \qquad v'' = \frac{GF}{GF-\lambda}\theta'\,. \tag{5.23}$$

Hence an equation that contains only function θ:

$$\frac{EI - \lambda r^2}{GF}\theta''' + \frac{\lambda}{GF-\lambda}\theta' = 0\,.$$

---

[12] After Engesser, the formula (5.20) was given by many authors: A.S Volmir [22], S.P. Timoshenko [20], A.R. Rzhanitsin [18], N.A. Alfutov [1] etc.

In the designations of (5.15),

$$\theta''' + \frac{\mu}{(1-\mu)(a^2 - \mu r^2)}\theta' = 0 .$$

The general solution of this equation is of the form (5.16) but

$$k^2 = \frac{\mu}{(1-\mu)(a^2 - \mu r^2)} . \tag{5.24}$$

Putting (5.16) in the first equation of the system (5.23) and integrating the result gives

$$v = \frac{(a^2 - \mu)k^2 + 1}{k}\left(-C_1 \cos kx + C_2 \sin kx\right) + C_3 x + C_4 .$$

The boundary conditions help find the integration constants:

$$C_1 = 0 , \;\; C_2 \sin kl = 0 , \;\; C_3 = 0 , \;\; C_4 = 0 ,$$

wherefrom $k_{cr} = \pi/l$. To determine μ_{cr}, we use (5.24) and arrive at the quadratic equation

$$\mu^2 - \left(1 + \frac{\xi}{t} + \frac{1}{t}\right)\mu + \frac{\xi}{t} = 0 , \tag{5.25}$$

where we introduce additional dimensionless parameters

$$\xi = \frac{\pi^2 a^2}{l^2} , \quad t = \frac{\pi^2 r^2}{l^2} . \tag{5.26}$$

The least root of the quadratic equation (5.25) determines μ_{cr},

$$\mu_{cr} = \frac{1}{2}\left(1 + \frac{\xi}{t} + \frac{1}{t}\right) - \sqrt{\frac{1}{4}\left(1 + \frac{\xi}{t} + \frac{1}{t}\right)^2 - \frac{\xi}{t}} .$$

So we have

$$\lambda_{cr} = \frac{\pi^2 EI}{l^2} \frac{\left(1 + \frac{\xi}{t} + \frac{1}{t}\right) - \sqrt{\left(1 + \frac{\xi}{t} + \frac{1}{t}\right)^2 - \frac{4\xi}{t}}}{2\xi} . \tag{5.27}$$

Let us, for example, take a fairly typical value of $t = \pi^2 r^2/l^2 = 0.02$. Then the relation beteen the dimensionless factor at $\pi^2 EI/l^2$ in the expression of λ_{cr} and the dimensionless parameter $\xi = \pi^2 a^2/l^2$ will look as shown in

Fig. 11.9 by curve 3. In the same figure, curve 1 conforms to the solution of (5.18) and curve 2 to the solution of (5.20).

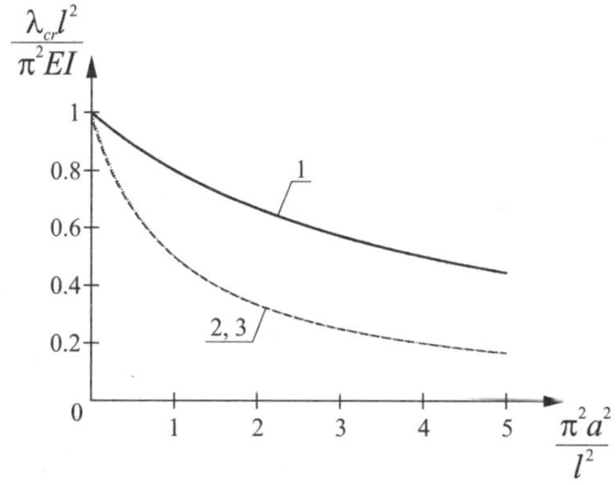

Fig. 11.9. The critical load vs. the dimensionless parameter $\xi = \pi^2 a^2/l^2$ for the Timoshenko bar

As we can see, curve 2 and curve 3 are nearly coincident. This is not accidental; we can suggest an explanation. Parameter t is small comparing to one in any case because this is a purely geometrical parameter and we are dealing with a bar. Thus, we can omit this parameter when it is added to one and turn the formula (5.27) to

$$\lambda_{cr} = \frac{\pi^2 EI}{l^2} \frac{(1+\xi) - \sqrt{(1+\xi)^2 - 4\xi t}}{2\xi t} \simeq \frac{\pi^2 EI}{l^2} \frac{(1+\xi)\left[1 - 1 + \dfrac{2\xi t}{(1+\xi)^2}\right]}{2\xi t} =$$

$$= \frac{\pi^2 EI}{l^2} \frac{1}{1+\xi},$$

which is an exact replica of the Engesser formula (5.20).

In this regard, the stability functional (5.19) from which the Engesser formula follows, can be justified as follows. This functional is derived from the Bryan–Treffz functional (5.22) by removing the term $\theta'^2 r^2$. This simplification is legitimate because the following estimate holds:

$$EI\theta'^2 + \lambda \bar{N} \theta'^2 r^2 = EI\theta'^2 \left(1 + \frac{\lambda \bar{N}}{EA}\right) \simeq EI\theta'^2, \tag{5.28}$$

where the addition to one is the relative longitudinal strain in the bar's cross-sections so it does not exceed the strain in the bar's critical state. The latter value is greatly less than one by definition.

Thus, the Novozhilov functional is applicable to the Timoshenko bars made of an isotropic material; however, when $G \ll E$, it can produce significant errors. At the same time, the traditional functional (5.19) derived from the Bryan–Trefftz functional by omitting small values works for any case.

11.5.3 Stability of equilibrium of a Kirchhoff–Love plate

The state of plane stress in a sufficiently thin plate the bending of which is consistent with the Kirchhoff–Love theory has the following expression of the strain energy in the buckling mode **v** according to (**5**.1.21):

$$\mathsf{E} = \frac{1}{2}\int_{\Omega} D\left[w_{,xx}^2 + 2\nu w_{,xx} w_{,yy} + w_{,yy}^2 + 2(1-\nu) w_{,xy}^2 \right] d\Omega . \tag{5.29}$$

We represent the vector function of displacements **v**, of an arbitrary point of the plate with the coordinates (x,y,z), that defines the buckling mode by its components as

$$\mathbf{v} = \|[U,V,W]\|^{\mathsf{T}} ,$$

where U,V,W are components of vector **v** with respect to the system of axes (X,Y). Next, we denote

$$u(x,y) = U(x,y,0), \qquad v(x,y) = V(x, y,0), \quad w(x,y) = W(x, y,0).$$

So, u,v,w are components of the displacement vector function of the points from the plate's median surface.

Using these designations and formulas (**5**.1.2)–(**5**.1.4), we can write

$$U = u - z w_{,x} , \qquad V = v - z w_{,y} , \qquad W = w. \tag{5.30}$$

According to the engineering theory of plates, the only nonzero components of the initial strain tensor, $\boldsymbol{e}$, are e_{xx}, e_{yy}, e_{xy}. The respective nonzero components $\overline{\tau}^{11} = \overline{\tau}^{xx}$, $\overline{\tau}^{22} = \overline{\tau}^{yy}$, $\overline{\tau}^{12} = \overline{\tau}^{xy}$ of the initial unit stress tensor in the plane stress are

$$\overline{\tau}^{xx} = \frac{\overline{N}^{xx}}{h}, \quad \overline{\tau}^{yy} = \frac{\overline{N}^{yy}}{h}, \quad \overline{\tau}^{xy} = \frac{\overline{N}^{xy}}{h}. \tag{5.31}$$

Further,

$$\omega_{11}=0\,,\quad \omega_{22}=0\,,\quad \omega_{33}=0\,,$$

$$\omega_{12}=-\omega_{21}=\frac{1}{2}\left(\frac{\partial U}{\partial y}-\frac{\partial V}{\partial x}\right)=\frac{1}{2}(u_{,y}-v_{,x})\,,$$

$$\omega_{13}=-\omega_{31}=\frac{1}{2}\left(\frac{\partial U}{\partial z}-\frac{\partial W}{\partial x}\right)=-w_{,x}\,,$$

$$\omega_{23}=-\omega_{32}=\frac{1}{2}\left(\frac{\partial V}{\partial z}-\frac{\partial W}{\partial y}\right)=-w_{,y}\,. \tag{5.32}$$

We insert this in (4.18) and integrate over the thickness of the plate, h. The result is

$$\mathsf{A}_{\mathsf{N}}(\mathbf{v})=-\frac{\lambda}{2}\int\limits_{\Omega}\int_{-h/2}^{h/2}\left[\overline{\tau}^{xx}\left(\omega_{21}^2+\omega_{31}^2\right)+2\overline{\tau}^{xy}\omega_{31}\omega_{32}+\overline{\tau}^{yy}\left(\omega_{12}^2+\omega_{32}^2\right)\right]dzd\Omega=$$

$$-\frac{\lambda}{2}\int\limits_{\Omega}\left[\overline{N}^{xx}\left(\frac{(u_{,y}-v_{,x})^2}{4}+w_{,x}^2\right)+2\overline{N}^{xy}w_{,x}w_{,y}+\overline{N}^{yy}\left(\frac{(u_{,y}-v_{,x})^2}{4}+w_{,y}^2\right)\right]d\Omega\,.$$

Removing the terms that depend on tangential displacements u and v from the expression of $\mathsf{A}_{\mathsf{N}}(\mathbf{v})$ gives the final equilibrium stability functional for the Kirchhoff–Love plates:

$$\mathsf{N}_3(w)=\frac{1}{2}\int\limits_{\Omega}D\left[w_{,xx}^2+2\nu w_{,xx}w_{,yy}+w_{,yy}^2+2(1-\nu)w_{,xy}^2\right]d\Omega+$$

$$+\frac{\lambda}{2}\int\limits_{\Omega}\left(\overline{N}^{xx}w_{,x}^2+2\overline{N}^{xy}w_{,x}w_{,y}+\overline{N}^{yy}w_{,y}^2\right)d\Omega\,. \tag{5.33}$$

This functional is also known well and was used many times to analyze the stability of equilibrium of various plates [20].

11.5.4 Stability of equilibrium of a Reissner plate

When working with the medium-thickness plates, we should replace (5.29) with (**5**.3.46). Inserting functions θ_x and θ_y instead of functions φ_x и φ_y as in (**5**.3.15) gives the following expression of the plate's strain energy in its buckling mode **v**[13]:

$$\mathsf{E}(w,\theta_x,\theta_y) = \frac{5}{24}\int_\Omega \frac{Eh}{1+\nu}\left[(w_{,x}+\theta_y)^2+(w_{,y}-\theta_x)^2\right]d\Omega +$$

$$+\frac{1}{2}\int_\Omega D\left[\theta_{y,x}^2 - 2\nu\theta_{y,x}\theta_{x,y} + \theta_{x,y}^2 + \frac{1-\nu}{2}(-\theta_{y,y}+\theta_{x,x})^2\right]d\Omega . \tag{5.34}$$

where θ_x, θ_y are slopes of the normal to the plate's median surface which are independent of lateral deflections w, D is a cylindrical stiffness of the plate.

For the Reissner's plate, the kinematical relationships (5.30) should be replaced with – see (**5**.3.6) –

$$U = u + z\theta_y , \quad V = v - z\theta_x , \qquad W = w , \tag{5.35}$$

therefore

$$\omega_{12} = -\omega_{21} = \frac{1}{2}\left(\frac{\partial U}{\partial y} - \frac{\partial V}{\partial x}\right) = \frac{u_{,y} - v_{,x} + (\theta_{x,x}+\theta_{y,y})z}{2}, \tag{5.36}$$

$$\omega_{13} = -\omega_{31} = \frac{1}{2}\left(\frac{\partial U}{\partial z} - \frac{\partial W}{\partial x}\right) = \frac{1}{2}(\theta_y - w_{,x}),$$

$$\omega_{23} = -\omega_{32} = \frac{1}{2}\left(\frac{\partial V}{\partial z} - \frac{\partial W}{\partial y}\right) = \frac{1}{2}(-\theta_x - w_{,y}) .$$

Thus, the proper substitution and integration over the thickness of the plate gives

$$\mathsf{A_N}(\mathbf{v}) = -\frac{\lambda}{2}\int_\Omega\int_{-h/2}^{h/2}\left[\bar\tau^{xx}\left(\omega_{21}^2+\omega_{31}^2\right)+2\bar\tau^{xy}\omega_{31}\omega_{32}+\bar\tau^{yy}\left(\omega_{12}^2+\omega_{32}^2\right)\right]dz d\Omega =$$

$$= -\frac{\lambda}{2}\int_\Omega \bar N^{xx}\left[\frac{(u_{,y}-v_{,x})^2}{4} + \frac{(\theta_{x,x}+\theta_{y,y})^2 h^2}{48} + \frac{(\theta_y - w_{,x})^2}{4}\right]d\Omega -$$

[13] Chapter 5 uses the symbol of λ to denote a different quantity. The formula (**5**.3.29) defines it as $\lambda = 5(1\text{-}\nu)/h^2$. Therefore, in order to avoid the conflict of notation, we do not use this symbol in (5.34) up here.

$$-\frac{\lambda}{2}\int_{\Omega}\bar{N}^{yy}\left[\frac{(u_{,y}-v_{,x})^2}{4}+\frac{(\theta_{x,x}+\theta_{y,y})^2h^2}{48}+\frac{(\theta_x+w_{,y})^2}{4}\right]d\Omega+$$

$$+\frac{\lambda}{2}\int_{\Omega}\bar{N}^{xy}\frac{(\theta_y-w_{,x})(\theta_x+w_{,y})}{2}d\Omega.$$

Let us omit the quantities dependent on tangential displacements u and v because they are insignificant for the buckling. Also, we use the reasoning smilar to that concerning the Timoshenko bar to derive the estimates

$$\frac{(\theta_{x,x}^2+\theta_{y,y}^2)h^2}{12}<<\frac{(\theta_y-w_{,x})^2}{4},\quad \frac{(\theta_{x,x}^2+\theta_{y,y}^2)h^2}{12}<<\frac{(\theta_x+w_{,y})^2}{4}.$$

So, this is the final Novozhilov stability functional for the medium-thickness plates:

$$\mathsf{N}_3(w,\theta_x,\theta_y)=\frac{5}{24}\int_{\Omega}\frac{Eh}{1+\nu}\left[(w_{,x}+\theta_y)^2+(w_{,y}-\theta_x)^2\right]d\Omega+$$

$$+\frac{1}{2}\int_{\Omega}D\left[\theta_{y,x}^2-2\nu\theta_{y,x}\theta_{x,y}+\theta_{x,y}^2+\frac{1-\nu}{2}(-\theta_{y,y}+\theta_{x,x})^2\right]d\Omega+$$

$$+\frac{\lambda}{8}\int_{\Omega}\left[\bar{N}^{xx}(\theta_y-w_{,x})^2+\bar{N}^{yy}(\theta_x+w_{,y})^2-2\bar{N}^{xy}(\theta_y-w_{,x})(\theta_x+w_{,y})\right]d\Omega \quad (5.37)$$

If we base our consideration on the Bryan–Trefftz functional, we have

$$\mathsf{A}_B(\mathbf{v})=-\frac{\lambda}{2}\int_{\Omega}\int_{-h/2}^{h/2}\left[\bar{\tau}^{xx}\left(U_{,x}^2+V_{,x}^2+W_{,x}^2\right)+\bar{\tau}^{yy}\left(U_{,y}^2+V_{,y}^2+W_{,y}^2\right)\right]dzd\Omega-$$

$$-\frac{\lambda}{2}\int_{\Omega}\int_{-h/2}^{h/2}2\bar{\tau}^{xy}\left(U_{,x}U_{,y}+V_{,x}V_{,y}+W_{,x}W_{,y}\right)dzd\Omega.$$

Inserting the expressions of (5.35), integrating over z from $-h/2$ to $h/2$, and removing terms that depend on tangential displacements u and v gives

$$\mathsf{A}_B(w,\theta_x,\theta_y)=-\frac{\lambda}{2}\int_{\Omega}\bar{N}^{xx}\left[\frac{(\theta_{y,x}^2+\theta_{x,x}^2)h^2}{12}+w_{,x}^2\right]d\Omega-$$

$$-\frac{\lambda}{2}\int_{\Omega}\bar{N}^{yy}\left[\frac{(\theta_{y,y}^2+\theta_{x,y}^2)h^2}{12}+w_{,y}^2\right]d\Omega-$$

$$\frac{\lambda}{2}\int_{\Omega} 2\bar{N}^{xy}\left[\frac{(\theta_{y,x}\theta_{y,y}+\theta_{x,x}\theta_{x,y})h^2}{12}+w_{,x}w_{,y}\right]d\Omega\,,$$

and this permits to build the Bryan–Trefftz functional by assuming

$$\mathsf{B}_3(w,\theta_x,\theta_y) = \mathsf{E}(w,\theta_x,\theta_y) - \mathsf{A}_\mathsf{B}(w,\theta_x,\theta_y)$$

and taking the expression of E from (5.30). Again we have the estimates

$$\frac{(\theta_{y,x}^2+\theta_{x,x}^2)h^2}{12} << w_{,x}^2\,, \qquad \frac{(\theta_{y,y}^2+\theta_{x,y}^2)h^2}{12} << w_{,y}^2\,,$$

$$\frac{(\theta_{y,x}\theta_{y,y}+\theta_{x,x}\theta_{x,y})h^2}{12} << w_{,x}w_{,y}\,,$$

therefore the final form of the stability functional for the Reissner plate will be

$$\mathsf{S}(w,\theta_x,\theta_y) = \frac{5}{24}\int_{\Omega}\frac{Eh}{1+\nu}\left[(w_{,x}+\theta_y)^2+(w_{,y}-\theta_x)^2\right]d\Omega +$$

$$+\frac{1}{2}\int_{\Omega} D\left[\theta_{y,x}^2 - 2\nu\theta_{y,x}\theta_{x,y}+\theta_{x,y}^2+\frac{1-\nu}{2}(-\theta_{y,y}+\theta_{x,x})^2\right]d\Omega +$$

$$+\frac{\lambda}{2}\int_{\Omega}\left(\bar{N}^{xx}w_{,x}^2+\bar{N}^{yy}w_{,y}^2+2\bar{N}^{xy}w_{,x}w_{,y}\right)d\Omega\,. \tag{5.38}$$

The stability functional of the type(5.38) for the Reissner plate is a counterpart of (5.19) for the Timoshenko bar and works for any case including an anisotropic material with a weak shear rigidity. It goes without saying that the first two integrals in (5.38) which describe the potential strain energy $\mathsf{E}(w,\theta_x,\theta_y)$ should be written in a slightly different form for anisotropic plates (to account for anisotropy). Details are omitted.

11.5.5 Stability of equilibrium of a thin-walled bar

Equations of equilibrium stability derived for open-profile thin-walled bars by V.Z. Vlasov [21] provoked a vivid discussion in their time. The main inducement for that discussion was a poor decisivity of a method of derivation of the equilibrium stability equations that Vlasov chose to use.

First, in his derivation Vlasov did not start from the formal conditions for an equilibrium to be stable, which produce homogeneous equations; instead, he based his consideration on so-called strain equations and then eliminated the right-hand parts of those to determine the critical forces. Some of the arguments advanced by the participants of the discussion and the respective references are given in the often-cited book by Panovko and Gubanova [13].

Second, Vlasov borrowed a trick from Timoshenko based on the notion of the effective load to construct the equilibrium equations for the bar's deformed state, but he did not give this notion a clear mathematical treatment. Therefore some people tried to revise Vlasov's equations, mainly their parametric terms. A mathematically flawless, holeproof method for constructing the equations of stability for thin-walled bars, which did not leave any opportunities for subsequent argument about the decisivity of the derivation and was based exactly on the variational stability criterion, was suggested by V.V. Bolotin [5] back in 1965.

We are not going to consider the stability of equilibrium of a general thin-walled bar; we confine ourselves to a simplest case of stability of a centrally compressed thin-walled bar[14]. Referring to the formulas of Chapter **6**, here we present basic relationships of the theory of open-profile thin-walled bars which we need in this section.

So, following the notation of Chapter **6**, we define the components of the displacement vector **v** of an arbitrary point of the bar's profile with respect to the global system of axes, X,Y,Z, as follows

$$\mathbf{v} = \|[U, V, W]\|^{\mathrm{T}} .$$

These displacements, being functions of two coordinates – longitudinal coordinate x and arc coordinate s, are known to be defined by four functions of only the x coordinate:

$$\xi = \xi\,(x), \quad \eta = \eta\,(x), \quad \zeta = \zeta\,(x), \quad \theta = \theta(x),$$

where:

- ξ is a longitudinal displacement of the center of gravity of the bar's cross-section, which is imaginably rigidly connected to the zero point of the profile, O ;
- η, ζ are lateral displacements (along the principal central axes of inertia of the bar's cross-section, Y and Z respectively) of principal pole P of the profile which coincides with the bending center and is imaginably rigidly connected with the cross-section's center of gravity;

[14] The general analysis of equilibrium stability of thin-walled bars is by itself of great interest in both theory and applications. First of all, we should indicate original works by Vlasov [21] as a fountainhead. The same problems are discussed in a popular form in many other sources such as Volmir [22].

- θ is a slope of the bar's cross-section with respect to the longitudinal axis *X*.

The following relations take place:

$$U = \xi - \eta' y - \zeta' z - \theta'\omega\,,$$

$$V = \eta - (z - z_P)\theta\,, \qquad W = \zeta + (y - y_P)\theta\,, \tag{5.39}$$

where:

- ω is a sectorial coordinate of the current point of the profile with the arc coordinate *s*;
- *y*, *z* are global coordinate of the same point;
- y_P, z_P are global coordinates of the pole, P.

Here and further a stroke means taking the derivative with respect to longitudinal coordinate *x*.

The expression of the potential strain energy of the open-profile thin-walled bar is as follows – see (**6**.2.71):

$$\mathsf{E} = \frac{1}{2}\int_0^L \left(EA\xi'^2 + GI_x\theta'^2 + EI_z\eta''^2 + EI_y\zeta''^2 + EI_\omega\theta''^2 \right) dx\,. \tag{5.40}$$

The length of the thin-walled bar is denoted, as in Chapter **6**, by *L*, and the *l* symbol will be reserved for the length of the profile. The initial normal strains, $\overline{\tau}^{xx}$, in the cross-section of the centrally compressed bar are

$$\overline{\tau}^{xx} = \frac{\overline{N}}{A}\,.$$

Meaning to use the Bryan–Treffz stability functional, we define the A_B functional according to (4.18)

$$\mathsf{A}_\mathsf{B}(\mathbf{v}) = -\frac{\lambda}{2}\int_0^L \int_l \overline{\tau}^{xx} \left(U_{,x}^2 + V_{,x}^2 + W_{,x}^2 \right) h\,ds\,dx\,.$$

We make use of the hypothesis of no moment in the bar's shell longitudinally, so the normal stresses $\overline{\tau}^{xx}$ are distributed evenly over the thickness of the profile, *h*. Obviously, we should substitute the expressions of the integrand terms from (5.39). Let's calculate the internal integral by taking each of the three terms separately:

$$\int_l \overline{\tau}^{xx} U_{,x}^2 h\,ds = \int_l \frac{\overline{N}}{A} (\xi' - \eta'' y - \zeta'' z - \theta''\omega)^2\, h\,ds =$$

$$= \bar{N}\left(\xi'^2 + \eta''^2 r_z^2 + \zeta''^2 r_y^2 + \theta''^2 r_\omega^2\right),$$

$$\int_l \bar{\tau}^{xx} V_{,x}^2 h ds = \int_l \frac{\bar{N}}{A}(\eta' + z_P\theta' - z\theta')^2 \, h ds = \bar{N}\left[(\eta' + z_P\theta')^2 + \theta'^2 r_y^2\right],$$

$$\int_l \bar{\tau}^{xx} W_{,x}^2 h ds = \int_l \frac{\bar{N}}{A}(\zeta' - y_P\theta' + y\theta')^2 \, h ds = \bar{N}\left[(\zeta' - y_P\theta')^2 + \theta'^2 r_z^2\right],$$

where

$$r_y^2 = \frac{I_y}{A}, \qquad r_z^2 = \frac{I_z}{A}, \qquad r_\omega^2 = \frac{I_\omega}{A}.$$

Therefore

$$\mathsf{A}_{\mathsf{B}}(\mathbf{v}) =$$

$$-\frac{\lambda}{2}\int_0^L \bar{N}\left[\xi'^2 + \eta''^2 r_z^2 + \zeta''^2 r_y^2 + \theta''^2 r_\omega^2 + (\eta' + z_P\theta')^2 + (\zeta' - y_P\theta')^2 + \theta'^2\rho^2\right]dx$$

where ρ is a polar radius of inertia of the cross-section, $\rho^2 = r_y^2 + r_z^2$.

Considering obvious estimates

$$\eta''^2 r_z^2 << \eta'^2, \qquad \zeta''^2 r_y^2 << \zeta'^2$$

and omitting insignificant terms depending on ξ' gives the stability functional

$$\mathsf{S} = \frac{1}{2}\int_0^L \left(GI_x\theta'^2 + EI_z\eta''^2 + EI_y\zeta''^2 + EI_\omega\theta''^2\right)dx +$$

$$+\frac{\lambda}{2}\int_0^L \bar{N}\left[(\eta' + z_P\theta')^2 + (\zeta' - y_P\theta')^2 + \theta''^2 r_\omega^2 + \theta'^2\rho^2\right]dx.$$

This functional can be simplified, too, if we notice that

$$EI_\omega + \lambda\bar{N}r_\omega^2 = EI_\omega\left(1 - \frac{\lambda\bar{N}}{EA}\right) \simeq EI_\omega.$$

So finally we have

$$\mathsf{S}=\frac{1}{2}\int_0^L\left(GI_x\theta'^2+EI_z\eta''^2+EI_y\zeta''^2+EI_\omega\theta''^2\right)dx+$$

$$+\frac{\lambda}{2}\int_0^L\bar{N}\left[(\eta'+z_P\theta')^2+(\zeta'-y_P\theta')^2+\theta'^2\rho^2\right]dx. \quad (5.41)$$

The conditions of stationarity for that functional produce the following simultaneous equations:

$$\left(EI_z\eta''\right)''-\lambda\left[\bar{N}(\eta'+z_P\theta')\right]'=0,\ \left(EI_y\zeta''\right)''-\lambda\left[\bar{N}(\zeta'-y_P\theta')\right]'=0,$$

$$\left(EI_\omega\theta''\right)''-\left(GI_x\theta'\right)'-\lambda\left[\bar{N}(\eta'+z_P\theta'\right]'z_P+\lambda\left[\bar{N}(\zeta'-y_P\theta'\right]'y_P-$$

$$-\lambda\left(\bar{N}\theta'\rho^2\right)'=0. \quad (5.42)$$

These equations look like the following for a bar of a constant cross-section subjected to a lengthwise-constant compressive force (we assume $\bar{N}=-1$):

$$EI_z\eta^{IV}+\lambda(\eta''+z_P\theta'')=0, \qquad EI_y\zeta^{IV}+\lambda(\zeta''-y_P\theta'')=0,$$

$$EI_\omega\theta^{IV}-GI_x\theta''+\lambda\left(z_P\eta''-y_P\zeta''+a^2\theta''\right)=0, \quad (5.43)$$

where we denote

$$a^2=y_P^2+z_P^2+\rho^2. \quad (5.44)$$

Integration of the system of equations (5.43) is not discussed here. Just not that this system is identical to a slightly more general one derived by Vlasov in another way and in other designations – see the equations (1.10) in page 249 of [21]. To assure yourself of it, assume the initial bending moments in Vlasov's equations equal to zero.

11.6 Mixed functionals in the stability analysis

Variational formulations for the problems of equilibrium stability can be derived using mixed functionals just as well as any other ones.

Let $\boldsymbol{\sigma}$ be a stress tensor that corresponds to displacements $\mathbf{v}$, i.e. let

$$\boldsymbol{\sigma}=\boldsymbol{CA}\mathbf{v}. \quad (6.1)$$

We introduce a mixed functional built similarly to the Reissner one. The quasi-static re-formulation of the problem will be convenient here, exactly as it was in the frequency spectrum problem for deriving the Reissner functional where the inertia forces were treated as external loads. The role of the inertia forces is played in the equilibrium stability problems by an equivalent load the concept of which was defined in Section **11**.4.1. Therefore we can use immediately the formulas (**10**.4.2) and (**10**.4.3) replacing the work of the inertia forces with the work of the equivalent load. But the work of the equivalent load is nothing but the functional $\mathsf{A}(\mathbf{v}) = \lambda\,\bar{\mathsf{A}}(\mathbf{v})$. As a result, we have the following expressions for two forms of the Reissner-type functionals in application to the stability of equilibrium:

$$\mathsf{R}_1(\boldsymbol{\sigma}, \mathbf{v}) = \tfrac{1}{2}(\boldsymbol{C}^{-1}\boldsymbol{\sigma}, \boldsymbol{\sigma}) - (\boldsymbol{A}\mathbf{u}, \boldsymbol{\sigma}) - \lambda\,\bar{\mathsf{A}}(\mathbf{v}), \tag{6.2}$$

$$\mathsf{R}_2(\boldsymbol{\sigma}, \mathbf{v}) = \tfrac{1}{2}(\boldsymbol{C}^{-1}\boldsymbol{\sigma}, \boldsymbol{\sigma}) - (\boldsymbol{A}^{\mathsf{T}}\boldsymbol{\sigma}, \mathbf{u}) - \lambda\,\bar{\mathsf{A}}(\mathbf{v})\,. \tag{6.3}$$

For the sake of simplicity, these functionals are assumed to act on homogeneously kinematically admissible displacement fields and homogeneously statically semi-admissible stress fields. This permits to exclude non-integral terms from the expressions of the mixed functionals. Also, the formulas (6.2) and (6.3) in comparison to (**10**.4.2) and (**10**.4.3) lack terms that depend on the elastic medium. We omit them, too, by assuming $\boldsymbol{K} = \boldsymbol{O}$.

We can also use the Ritz method for the mixed functionals, just as we did in the frequency spectrum analysis. The method of two functionals is equally applicable to the stability analysis, too. The technique of this procedure is borrowed from Chapter **10**. However, theorems and estimates presented there do not work for the stability analysis. Clearly, the reason why those estimates cannot be extended onto the equilibrium stability analysis is again the lack of guaranteed positive definiteness of the geometric stiffness matrix for a finite-dimensional system made discrete by the Ritz method.

Let us consider a classic problem of stability of equilibrium of a compressed bar, to take an example. We already know that

$$\mathsf{A}(\mathbf{v}) = \frac{\lambda}{2}\int_0^l \bar{N} v'^2\,dx\,,$$

so the first form of the Reissner functional becomes

$$\mathsf{R}_1(M,v)=\frac{1}{2}\int_0^l \frac{M^2}{EI}dx+\int_0^l v''Mdx-\frac{\lambda}{2}\int_0^l \bar{N}v'^2 dx$$

where M are bending moments that develop in the bar's cross-sections at the moment its stability is lost.

By moving one derivative from the deflections onto the moments in the second of the integrals using integration by parts, we obtain the third form of the Reissner functional,

$$\mathsf{R}_3(M,v)=\frac{1}{2}\int_0^l \frac{M^2}{EI}dx-\int_0^l v'M'dx-\frac{\lambda}{2}\int_0^l \bar{N}v'^2 dx\,. \tag{6.4}$$

This form is convenient in that all needed functions participate in the integrands with their derivatives no higher than first. Note that the non-integral term in (6.4) vanishes again due to our limitation of the domain of the mixed functionals.

The conditions of stationarity for this functional produce the following Euler equations:

$$\frac{M}{EI}+v''=0\,,\qquad M''+\lambda\left(\bar{N}v'\right)'=0\,,$$

or, after the M moments are excluded,

$$\left(EIv''\right)''-\lambda\left(\bar{N}v'\right)'=0$$

This coincides with the equation (5.7) derived earlier.

Ann example of the application of the functional (6.4) to a problem of stability of equilibrium of a compressed bar (in the finite element analysis style) can be found in our paper [9]. The same paper demonstrates that the mixed forms of the finite element method are able, just as in the frequency spectral analysis, to bring about spurious critical loads in the semse discussed in detail in Section **10**.4.3.

11.7 Final comments to Chapter 11

At the beginning of Section **11**.3 we said that the geometrically linear (first-order) formulation was not sufficient for the equilibrium stability analysis. We used the principle of virtual displacements and the Lagrange variational principle based on it without any additional comments both in that place and in application to nonlinear problems. But this principle was formulated in Chapter **1**, strictly speaking, as applicable to a purely linear elastic analysis. However, the extended treatment of the principle of virtual displacements is also well-known, totally legitimate, and described in books on structural mechanics; see [12], for example. The only thing to do in order to extend this fundamental variational principle onto the nonlinear analysis is to re-formulate it in the following slightly modified form:

For a deformed mechanical system to stay in equilibrium in a certain state, it is necessary and sufficient that the total virtual work of all external and internal forces of this state on any homogeneously kinematically admissible infinitesimal displacements be equal to zero.

The displacement variations we worked with in this chapter actually belong to the said homogeneous kinematically admissible infinitesimal displacements.

There is another issue worth metnioning. As we told at the beginning of Section **11**.4, the application of the Lagrange–Dirichlet theorem to systems with an infinite number of degrees of freedom is a bit too formalistic. The matter is when we try to generalize the notion of the equilibrium stability onto systems with an infinite number of DOFs, we start having peculiar theoretical dfficulties. This fact is mentioned in the well-known book by H. Ziegler [23] where he refers to rescarches done by Shield, Green, and Koiter. However, that same Ziegler justifies the generalization by stating that "*presently we don't really have much of a choice*".

Note also an important fact: our formulations of the equilibrium stability problems are confined to the case of dead load. Were it otherwise, the external load components with respect to the Euler coordinate system would be dependent on the displacements. But then we would have to vary the load, too, when varying the basic functionals. This formulation of the stability problems is unquestionably feasible and often encountered in practice, even if the load is conservative. A simplest example is a hydrostatic pressure always directed along the normal to the boundary surface of an elastic body in the latter's deformed state. Of course, the load of this origin is conservative, but it is not a dead load. What generalizations need to be made in order to apply the equilibrium stability theory to the case of conservative but not dead external forces – this quesiton is beyond our scope of discussion.

References

1. Alfutov NA (1978) Foundations of stability analysis of elastic systems (in Russian). Mashinostroyeniye, Moscow

2. Besukhov NI, Luzhin OV (1963) Structural stability and dynamics in examples and problems (in Russian). Stroyizdat, Moscow
3. Bolotin VV (1963) Nonconservative problems of the theory of elastic stability. Pergamon Press, New York
4. Bolotin VV (1965) On the notion of stability in strucutral mechanics (in Russian). In collected papers: *Problems of stabiltiy in structural mechanics*. Stroyizdat, Moscow, pp 6–27
5. Bolotin VV (1965) On reducing three-dimensional elastic stability problems to one and two dimensions (in Russian). In collected papers: *Problems of stabiltiy in structural mechanics*. Stroyizdat, Moscow, pp 186–196
6. Bolotin VV (1966) On estimation of errors in equations of applied theory of elasticity (in Russian). In collected papers: *Structural mechanics*. Stroyizdat, Moscow, pp 297–302
7. Bolotin VV (1973) On variational principles of the theory of elastic stability (in Russian). In the book: *Problems in mechanics of deformable solids*. Sudostroyeniye, Leningrad, pp 83–88
8. Broude BM (1949) The stability of plates in steel structural elements (in Russian). Ministry of Machine Building, Moscow
9. Ispolov YG, Slivker VI (1984) On an effect that appears in applications of the mixed form of the finite element method (in Russian). Structural mechanics and analysis, 1: 43–48
10. Leichtweiss K (1980) Konvexe Mengen. Springer Verlag, Berlin
11. Loitsiansky LG, Lurie AI (1955) A course of theoretical mechanics (in Russian). Vol. 2. Gostekhteorizdat, Moscow
12. Novozhilov VV (1953) Foundations of nonlinear theory of elasticity. Graylock Press, Rochester, N.Y.
13. Panovko YG, Gubanova II (1987) Stability and oscillations of elastic systems (in Russian). Nauka, Moscow
14. Papkovich PF (1963) Works on structural mechanics of ships. In 4 volumes. Vol.4 Sudostroyeniye, Leningrad
15. Perelmuter AV, Slivker VI (2002) On an Error of a Mysteriuos Nature that happens in Software when Analyzing Mechanical Systems for Buckling. Proceedings of 15th Nordic Seminar on Computational Mechanics. Aalborg, Denmark, pp 229–232
16. Perelmuter AV, Slivker VI (2003) Numerical Structural Analysis. Springer-Verlag, Berlin Heidelberg New York
17. Rockafellar RT (1970) Convex analysis. Princeton University Press
18. Rzhanitsin AR (1955) Stability of equlibrium of elastic systems (in Russian). Gostekhteorizdat, Moscow
19. Rzhanitsin AR (1964) On some properties of an equilibrium stability area (in russian). Structural mechanics and analysis, 1: 1–8
20. Timoshenko SP, Gere JM (1961) Theory of elastic stability. McGraw-Hill, New York
21. Vlasov VZ Selected works. Vol.2. Thin-walled elastic bars (in Russian). USSR Acad. Sci., Moscow

22. Volmir AS (1963) The stability of elastic systems (in Russian). Physmathgiz, Moscow
23. Ziegler H (1968) Principles of structural stability. Blaisdell Publishing Company, Waltham, MA

CONCLUSION

The understanding of mathematics is not acquired in a painlessly entertaining way – nor you can acquire, for example, the sense of music by only reading journals (however bright and exciting) if you do not learn to listen *attentively and intently. Similarly, you cannot advance in mathematics without actively touching the very* essense *of the living mathematical science.*

Moiseyev NN (1979) Mathematics makes an experiment (in Russian). Nauka, Moscow

Is it only mathematics that the words of the epigraph are applicable to? No, they seem to apply to a lot of other sciences and fields of knowledge equally well. However, in the author's opinion, the science of mechanics is not the last to claim the right for the replacement *mathematics* → *mechanics* in the above, very relevant phrase.

When writing, the author was in continuous imaginable conversation with the potential reader of this book and so did his best to not complicate the presentment where possible – on one hand. On the other hand, no bedtime entertainment was intended either.

As the book was being written, its layout, contents, and structure of the knowledge presented were undergoing numerous changes. Many original intentions of the author taken by him as guidelines from the very beginning were afterwards altered or dropped, sometimes ruefully. Among the reasons are the limited volume of the book, pressure of time, re-evaluation of a comparative educational value of particular theoretical aspects. It happens all the time that the intention is separated from the implementation by a long distance which does not leave the author's design intact. The result is that I did what I did, no more, no less.[1]

This is what I would like to notice from the original plan that ultimately was omitted from the book:

---

[1] An immortal tag comes back unwittingly: *We strived for the best, but did our usual.*

- basics of theory of shells;
- a theory of compound bars by A.R. Rzhanitsin;
- thin-walled bars with a curvilinear axis;
- dynamic problems not limited to the spectral analysis;
- consideration of the stability of elastic equilibrium from a more general standpoint;
- some aspects of the finite element method as a variation of the Ritz method where special coordinate functions are employed, in application to the problematics of structural mechanics.

The author wishes he could present, in a form similar to this book, a discussion on at least some of the problems of mechanics of solids and structures which are formulated above. Anyway, this work is yet for the future.

* * *

The preface to the last book by A.I. Lurie says[2]: "*I realize the shortcomings of this work quite clearly. No author is free from this feeling. I just did what I could*". Let me join the words said by the outstanding mechanician.

December 2004,
Vladimir Slivker

[2] Lurie AI (1980) Nonlinear Theory of Elasticity (English translation). New York: North-Holland

APPENDIX

A. THE LEGENDRE AND FRIEDRICHS TRANSFORMS

A.1 The Legendre transform in the finite-dimensional case

Following I.I. Goldenblatt [4], [5], we consider a scalar function $\mathsf{E} = \mathsf{E}(\boldsymbol{q})$ of a vector argument $\boldsymbol{q} = \|[q_1,\ldots,q_n]\|^{\mathrm{T}}$ the total differential of which can be represented as

$$d\mathsf{E} = P_1 dq_1 + \ldots + P_n dq_n \tag{A.1}$$

where

$$P_i = \frac{\partial \mathsf{E}}{\partial q_i} = P_i(\boldsymbol{q}) \qquad (i = 1,\ldots, n) \tag{A.2}$$

are some functions of the same vector argument, $\boldsymbol{q}$.

We will call variables $q_1,\ldots, q_n$ *principal* with respect to function E, and the variables $P_1,\ldots,P_n$ *conjugate*. The function E itself is called a *generating function* with respect to the principal variables.

We assume the system of equations (A.2) to be solvable in terms of the principal variables.

We can select another principal set of n variables from the comprehensive set of $2n$ variables $q_1,\ldots, q_n$ and $P_1,\ldots, P_n$ with the only limitation: a combination of the variables is admissible if and only if for each index j from 1 to n the combination includes only one of the two variables: either q_j or P_j.

Any arbitrary combination of the variables which is acceptable as a principal one can be represented as $P_1,\ldots, P_k, q_{k+1},\ldots, q_n,$ up to the renumbering of the variables. The set of such variables makes up a new vector, $\boldsymbol{V}$, of the principal variables,

$$V = \|[v_1,\ldots, v_k, v_{k+1},\ldots, v_n]\|^{\mathsf{T}} = \|[P_1,\ldots,P_k, q_{k+1},\ldots,q_n]\|^{\mathsf{T}}. \qquad \text{(A.3)}$$

By solving the first k equations of the system (A.2) with respect to $q_1,\ldots, q_k$, we can obtain expressions for $q_1,\ldots, q_k$ as functions of new variables $P_1,\ldots,P_k, q_{k+1},\ldots, q_n$:

$$q_j = f_j(V) \qquad (j = 1,\ldots, k)\,. \qquad \text{(A.4)}$$

The relation between vectors $\boldsymbol{q}$ and V can be represented as

$$\boldsymbol{q} = \|[f_1(V),\ldots, f_k(V), v_{k+1},\ldots, v_n]\|^{\mathsf{T}} = \boldsymbol{F}(V). \qquad \text{(A.5)}$$

The relationships (A.5) permit to treat E as a function of the new principal variables:

$$\mathsf{E} = \mathsf{E}(\boldsymbol{F}(V)).$$

However, the $\mathsf{E} = \mathsf{E}(\boldsymbol{F}(V))$ function is not a generating function for the new combination of the principal variables. To obtain a new generating function $\mathsf{E}_k = \mathsf{E}_k(V)$ for the group of variables (A.3), we can make use of a so-called *Legendre transform*,

$$\mathsf{E}_k(V) = P_1q_1 + \ldots + P_kq_k \; - \mathsf{E}(\boldsymbol{q}) =$$
$$= P_1 f_1(V) + \ldots + P_k f_k(V) - \mathsf{E}(\boldsymbol{F}(V)). \qquad \text{(A.6)}$$

And indeed, the full differential $d\mathsf{E}_k$ is

$$d\mathsf{E}_k = P_1dq_1 + \ldots + P_kdq_k + q_1dP_1 + \ldots + q_kdP_k - d\mathsf{E} =$$
$$= q_1dP_1 + \ldots + q_kdP_k - P_{k+1}dq_{k+1} - \ldots - P_ndq_n$$

hence,

$$\frac{\partial \mathsf{E}_k}{\partial P_i} = q_i, \qquad \frac{\partial \mathsf{E}_k}{\partial q_j} = -P_j \quad (i = 1,\ldots, k; \quad j = k+1,\ldots, n)\,. \qquad \text{(A.7)}$$

Thus, E_k is really a generating function with respect to the new principal variables V. It is also clear that the vector of conjugate variables with respect to the new ones is the vector

$$\boldsymbol{W} = \|[q_1,\ldots, q_k, -P_{k+1},\ldots, -P_n]\|^{\mathsf{T}} = \|[w_1,\ldots, w_k, w_{k+1},\ldots, w_n]\|^{\mathsf{T}}.$$

In particular, at $k = n$ we have

$$\mathsf{E}_n = P_1q_1 + \ldots + P_nq_n - \mathsf{E}(\boldsymbol{q}) = \mathsf{E}_n(\boldsymbol{P})\,, \qquad \text{(A.8)}$$

and

$$\frac{\partial \mathsf{E}_n}{\partial P_i} = q_i \quad (i = 1, \ldots, n) . \tag{A.9}$$

Thus, the Legendre transform at $k = n$ produces such a generating function E_n for which the vector of principal variables and that of conjugate variables switch places. This particular case will be called a *complete* Legendre transform, to distibguish it from a *partial* Legendre transform for $k < n$.

If we treat the $\boldsymbol{q}$ vector mechanically as a vector of generalized displacements and the generating function E as a strain energy, then the vector of conjugate variables $\boldsymbol{P}$ will have the meaning of a vector of generalized forces that conform to the respective generalized displacements $q_1,\ldots,q_n$. Then the relationships (A.2) constitute nothing but the Lagrange theorem, and (A.9) are the Castigliano theorem in application to a so-called *complementary potential energy* E_n.

Let us discuss the field most interesting for us here, the linear structural analysis where the generating function E is a strain energy of a mechanical system and a homogeneous (assumed to be positive definite) quadratic form of generalized displacements $\boldsymbol{q}$,

$$\mathsf{E} = \tfrac{1}{2}\, \boldsymbol{q}^{\mathsf{T}} \boldsymbol{R}\, \boldsymbol{q} , \tag{A.10}$$

and

$$\boldsymbol{P} = \boldsymbol{R}\, \boldsymbol{q} . \tag{A.11}$$

The symmetric positive definite matrix $\boldsymbol{R}$ has the meaning of a stiffness matrix of the structure.

The full Legendre transform from (A.8), i.e. at $k = n$, produces this in the linear analysis:

$$\mathsf{E}_n = \boldsymbol{P}^{\mathsf{T}} \boldsymbol{q} - \tfrac{1}{2}\, \boldsymbol{q}^{\mathsf{T}} \boldsymbol{R}\, \boldsymbol{q} = \tfrac{1}{2}\, \boldsymbol{P}^{\mathsf{T}} \boldsymbol{R}^{-1} \boldsymbol{P} = \mathsf{E} , \tag{A.12}$$

so the strain energy of the system, E, and the additional potential energy E_n do not differ when their argument is the solution of the linear problem.

Let us consider the Legendre transform (A.6) and find out how the matrix of the quadratic form changes if we replace the generating function E with the generating function E_k. We represent vector $\boldsymbol{q}$ of the principal variables and vector $\boldsymbol{P}$ of the conjugate variables as two subvectors of the respective dimensions k and $(n–k)$:

$$\boldsymbol{q} = \|[\, \boldsymbol{q}_1, \boldsymbol{q}_2]\|^{\mathsf{T}}, \qquad \boldsymbol{q}_1 = \|[q_1,\ldots, q_k]\|^{\mathsf{T}}, \qquad \boldsymbol{q}_2 = \|[q_{k+1},\ldots, q_n]\|^{\mathsf{T}},$$

$$\boldsymbol{P} = \|[\boldsymbol{P}_1, \boldsymbol{P}_2]\|^{\mathrm{T}}, \qquad \boldsymbol{P}_1 = \|[P_1,\ldots, P_k]\|^{\mathrm{T}}, \qquad \boldsymbol{P}_2 = \|[P_{k+1},\ldots, P_n]\|^{\mathrm{T}},$$

and matrix $\boldsymbol{R}$ of the quadratic form E in the block form

$$\boldsymbol{R} = \left\|\begin{bmatrix} \boldsymbol{R}_{11} & \boldsymbol{R}_{12} \\ \boldsymbol{R}_{21} & \boldsymbol{R}_{22} \end{bmatrix}\right\|.$$

With these designations and the symmetry $\boldsymbol{R}_{21} = \boldsymbol{R}_{12}^{\mathrm{T}}$, the formulas (A.10) and (A.11) become

$$\mathsf{E} = \tfrac{1}{2}\, \boldsymbol{q}^{\mathrm{T}} \boldsymbol{R}\, \boldsymbol{q} = \tfrac{1}{2} \boldsymbol{q}_1^{\mathrm{T}} \boldsymbol{R}_{11} \boldsymbol{q}_1 + \boldsymbol{q}_1^{\mathrm{T}} \boldsymbol{R}_{12} \boldsymbol{q}_2 + \tfrac{1}{2} \boldsymbol{q}_2^{\mathrm{T}} \boldsymbol{R}_{22} \boldsymbol{q}_2 \ ,$$

$$\boldsymbol{P}_1 = \boldsymbol{R}_{11} \boldsymbol{q}_1 + \boldsymbol{R}_{12} \boldsymbol{q}_2 , \quad \boldsymbol{P}_2 = \boldsymbol{R}_{21} \boldsymbol{q}_1 + \boldsymbol{R}_{22} \boldsymbol{q}_2 . \tag{A.13}$$

Noting that $\boldsymbol{q}_1 = \boldsymbol{R}_{11}^{-1} \boldsymbol{P}_1 - \boldsymbol{R}_{11}^{-1} \boldsymbol{R}_{12} \boldsymbol{q}_2$, we express vector $\boldsymbol{q}$ via the vector of the new principal variables,

$$\boldsymbol{V} = \|[P_1,\ldots,P_k, q_{k+1},\ldots,q_n]\|^{\mathrm{T}} = \|[\boldsymbol{P}_1, \boldsymbol{q}_2]\|^{\mathrm{T}}.$$

We have

$$\boldsymbol{q} = \left\|\begin{bmatrix} \boldsymbol{q}_1 \\ \boldsymbol{q}_2 \end{bmatrix}\right\| = \left\|\begin{bmatrix} \boldsymbol{R}_{11}^{-1} & -\boldsymbol{R}_{11}^{-1} \boldsymbol{R}_{12} \\ \boldsymbol{O} & \boldsymbol{I} \end{bmatrix}\right\| \left\|\begin{bmatrix} \boldsymbol{P}_1 \\ \boldsymbol{q}_2 \end{bmatrix}\right\| = \boldsymbol{H}\boldsymbol{V},$$

$$\text{where} \quad \boldsymbol{H} = \left\|\begin{bmatrix} \boldsymbol{R}_{11}^{-1} & -\boldsymbol{R}_{11}^{-1} \boldsymbol{R}_{12} \\ \boldsymbol{O} & \boldsymbol{I} \end{bmatrix}\right\|. \tag{A.14}$$

With the new principal variables, the strain energy of the system, E, can be represented as a quadratic form:

$$\mathsf{E} = \tfrac{1}{2}\, \boldsymbol{q}^{\mathrm{T}} \boldsymbol{R}\, \boldsymbol{q} = \tfrac{1}{2}\, \boldsymbol{V}^{\mathrm{T}} \boldsymbol{H}^{\mathrm{T}} \boldsymbol{R}\, \boldsymbol{H} \boldsymbol{V} . \tag{A.15}$$

Now it turns out that the matrix of quadratic form $\mathsf{E}(\boldsymbol{V})$ has a block diagonal structure

$$\boldsymbol{H}^{\mathrm{T}} \boldsymbol{R} \boldsymbol{H} = \left\|\begin{bmatrix} \boldsymbol{R}_{11}^{-1} & \boldsymbol{O} \\ \boldsymbol{O} & \boldsymbol{R}_{22} - \boldsymbol{R}_{21} \boldsymbol{R}_{11}^{-1} \boldsymbol{R}_{12} \end{bmatrix}\right\|. \tag{A.16}$$

This is not accidental; there is a neat mechanical interpretation of this fact as a manifestation of the field orthogonality theorem – see Section **1**.4.3.

And indeed, the block diagonal structure of matrix $\boldsymbol{H}^{\mathrm{T}} \boldsymbol{R} \boldsymbol{H}$ permits to represent the strain energy of the system, E, as a sum of two energies,

$$\mathsf{E} = \mathsf{E}_P + \mathsf{E}_q$$

$$\text{where} \quad \mathsf{E}_P = \tfrac{1}{2}\boldsymbol{P}_1^{\mathsf{T}}\boldsymbol{R}_{11}^{-1}\boldsymbol{P}_1\,, \quad \mathsf{E}_q = \tfrac{1}{2}\boldsymbol{q}_2^{\mathsf{T}}(\boldsymbol{R}_{22} - \boldsymbol{R}_{21}\boldsymbol{R}_{11}^{-1}\boldsymbol{R}_{12})\boldsymbol{q}_2^{\mathsf{T}}\,. \tag{A.17}$$

Clearly, E_P is the strain energy of the system subjected to purely force actions in the form of given external generalized forces $P_1,\ldots, P_k$. Consequently, this state is homogeneously kinematically admissible. At the same time, E_q is the strain energy of the same system subjected to purely kinematic actions in the form of given generalized displacements $q_{k+1},\ldots, q_n$, and thus this state is homogeneously statically admissible.

Let us take a look at the structure of quadratic form E_k created by the partial Legendre transform. Based on (A.6),

$$\mathsf{E}_k(\boldsymbol{V}) = P_1q_1 + \ldots + P_kq_k - \mathsf{E}(\boldsymbol{q}) = \boldsymbol{P}_1^{\mathsf{T}}\boldsymbol{q}_1 - \tfrac{1}{2}\,\boldsymbol{q}^{\mathsf{T}}\boldsymbol{R}\,\boldsymbol{q} = \boldsymbol{P}_1^{\mathsf{T}}\boldsymbol{q}_1 - \tfrac{1}{2}\,\boldsymbol{V}^{\mathsf{T}}\boldsymbol{H}^{\mathsf{T}}\boldsymbol{R}\boldsymbol{H}\boldsymbol{V}\,.$$

It is further clear that

$$\boldsymbol{P}_1 = \|[\boldsymbol{I}\;\boldsymbol{O}]\|\boldsymbol{V}, \qquad \boldsymbol{q}_1 = \|[\boldsymbol{I}\;\boldsymbol{O}]\|\,\boldsymbol{H}\boldsymbol{V}.$$

The $\boldsymbol{P}_1^{\mathsf{T}}\boldsymbol{q}_1$ expression is a scalar, so it does not change by transposition and permits the symmetrization

$$\boldsymbol{P}_1^{\mathsf{T}}\boldsymbol{q}_1 = \tfrac{1}{2}\,\boldsymbol{P}_1^{\mathsf{T}}\boldsymbol{q}_1 + \tfrac{1}{2}\,\boldsymbol{q}_1^{\mathsf{T}}\boldsymbol{P}_1 = \tfrac{1}{2}\boldsymbol{V}^{\mathsf{T}}\left\|\begin{bmatrix}\boldsymbol{I}\\ \boldsymbol{O}\end{bmatrix}\right\|\|[\boldsymbol{I}\;\boldsymbol{O}]\|\,\boldsymbol{H}\boldsymbol{V} + \tfrac{1}{2}\boldsymbol{V}^{\mathsf{T}}\boldsymbol{H}^{\mathsf{T}}\left\|\begin{bmatrix}\boldsymbol{I}\\ \boldsymbol{O}\end{bmatrix}\right\|\|[\boldsymbol{I}\;\boldsymbol{O}]\|\boldsymbol{V} =$$

$$= \tfrac{1}{2}\boldsymbol{V}^{\mathsf{T}}\left\|\begin{bmatrix}2\boldsymbol{H}_{11} & \boldsymbol{H}_{12}\\ \boldsymbol{H}_{21} & \boldsymbol{O}\end{bmatrix}\right\|\boldsymbol{V} = \tfrac{1}{2}\boldsymbol{V}^{\mathsf{T}}\left\|\begin{bmatrix}2\boldsymbol{R}_{11}^{-1} & -\boldsymbol{R}_{11}^{-1}\boldsymbol{R}_{12}\\ -\boldsymbol{R}_{21}\boldsymbol{R}_{11}^{-1} & \boldsymbol{O}\end{bmatrix}\right\|\boldsymbol{V}\,.$$

Therefore

$$\mathsf{E}_k(\boldsymbol{V}) =$$

$$= \frac{1}{2}\boldsymbol{V}^{\mathsf{T}}\left\{\left\|\begin{bmatrix}2\boldsymbol{R}_{11}^{-1} & -\boldsymbol{R}_{11}^{-1}\boldsymbol{R}_{12}\\ -\boldsymbol{R}_{21}\boldsymbol{R}_{11}^{-1} & \boldsymbol{O}\end{bmatrix}\right\| - \left\|\begin{bmatrix}\boldsymbol{R}_{11}^{-1} & \boldsymbol{O}\\ \boldsymbol{O} & \boldsymbol{R}_{22} - \boldsymbol{R}_{21}\boldsymbol{R}_{11}^{-1}\boldsymbol{R}_{12}\end{bmatrix}\right\|\right\}\boldsymbol{V} =$$

$$= \tfrac{1}{2}\boldsymbol{V}^{\mathsf{T}}\boldsymbol{G}\boldsymbol{V}\,, \tag{A.18}$$

and matrix $\boldsymbol{G}$ of quadratic form $\mathsf{E}_k(\boldsymbol{V})$ is[1]

[1] It is interesting to notice that there is a close affinity between the partial Legendre transform and the so-called *Gauss-Jordan elimination* in the theory of matrices [11]. As is immediately clear from (A.19), matrix $\boldsymbol{G}$ is derived from the original matrix $\boldsymbol{R}$ by k successive Gauss-Jordan elimination steps with the governing elements on the main diagonal. The technique of the Gauss-Jordan

$$G = \left\| \begin{bmatrix} G_{11} & G_{12} \\ G_{21} & G_{22} \end{bmatrix} \right\| = \left\| \begin{bmatrix} R_{11}^{-1} & -R_{11}^{-1}R_{12} \\ -R_{21}R_{11}^{-1} & -R_{22} + R_{21}R_{11}^{-1}R_{12} \end{bmatrix} \right\| . \quad \text{(A.19)}$$

The expanded quadratic form $\mathsf{E}_k(V)$ looks like

$$\mathsf{E}_k(V) = \tfrac{1}{2}\, P_1^{\mathsf{T}} R_{11}^{-1} P_1 - \tfrac{1}{2}\, P_1^{\mathsf{T}} R_{11}^{-1} R_{12} q_2 -$$

$$- \tfrac{1}{2}\, q_2^{\mathsf{T}} R_{21} R_{11}^{-1} P_1 + \tfrac{1}{2}\, q_2^{\mathsf{T}} (-R_{22} + R_{21} R_{11}^{-1} R_{12}) q_2 . \quad \text{(A.20)}$$

Recalling (A.7), we can derive matrix expressions for the subvectors of vector $W = \|[q_1, P_2]\|^{\mathsf{T}}$ conjugate to vector $V = \|[P_1, q_2]\|^{\mathsf{T}}$,

$$\frac{\partial \mathsf{E}_k}{\partial P_1} = q_1 = R_{11}^{-1} P_1 - R_{11}^{-1} R_{12} q_2 ,$$

$$-\frac{\partial \mathsf{E}_k}{\partial q_2} = P_2 = R_{21} R_{11}^{-1} P_1 + (R_{22} - R_{21} R_{11}^{-1} R_{12}) q_2 . \quad \text{(A.21)}$$

In the source [4] mentioned above, I.I. Goldenblatt treats the set of all possible generating functions E_k as a set of various mixed potentials of structural mechanics for systems with a finite number of degrees of freeedom. Clearly, the full set of all possible variations of the Legendre transform for a system with n degrees of freedom permits to construct various potentials of structural mechanics for this system in the total number of 2^n.

A mixed method of analysis which is known in mechanics of bar structures as the Gvozdiov method can be described in its variational aspect as an implementation of the stationarity conditions for one of mixed potentials of structural mechanics.

To see this, suppose we have chosen a principal system for the mixed method, and this system (structure) is subjected to unknown sought-for actions: generalized forces $X_1,\ldots, X_k$ and generalized displacements $Z_{k+1},\ldots, Z_n$. In this system all generalized displacements q_i in the directions of forces X_i ($i = 1,\ldots, k$) must be zero. In the same principal system, all

---

elimination is a generally useful tool widely employed in structural analysis. For example, a generalized Gauss-Jordan elimination helps easily develop a unified procedure for constructing the stiffness matrix of an arbitrary finite element with partly removed constraints if the stiffness matrix is known for the same element with all constraints present [8]. The application of the Gauss-Jordan elimination to systems with unilateral constraints is peculiar and unexpected [6]. A detailed description of the Gauss-Jordan elimination and its use in structural mechanics can be found in [9].

forces P_j ($j = k + 1,..., n$) treated as reactions in the constraints of the principal system must be zero. So, according to (A.7) and with a possible replacement of the designations, we arrive at the following equations of the mixed method:

$$\frac{\partial \mathsf{E}_k}{\partial X_i} = 0 \quad (i = 1,\ldots, k)\,, \qquad \frac{\partial \mathsf{E}_k}{\partial Z_j} = 0 \quad (j = k+1,\ldots, n)\,.$$

These equations are a corollary to the variational principle of stationarity of the mixed potential, E_k, in structural mechanics:

$$\mathsf{E}_k = \mathsf{E}_k(X_1,\ldots,X_k, Z_{k+1},\ldots,Z_n)\,.$$

The historical reason why the Legendre transform came up was not the needs of mechanics of solids and structures – it was invented to transform equations of analytical mechanics. The Legendre transform is used in analytical mechanics to handle equations of motion of particle systems – to validate a transition from the velocities as principal variables to the momenta as conjugate variables, and vice versa. This method is one of ways to establish a relation between the Lagrangian and Gamiltonian formalism in analytical mechanics. This can be learned in more detail from a brilliant-style book by C. Lanczos [7].

A.2 The Legendre transform in the general case

Perhaps the most interesting thing about the Legendre transform is how it works for a general variational formulation of elastic problems.

Following [10], we will demonstrate the use of the Legendre transform to switch from the Lagrange functional to the Reissner one. So, we begin with the Lagrange functional but write it in a slightly changed form – as one depending on the strains and displacements,

$$\mathsf{L}(\boldsymbol{\varepsilon},\mathbf{u}) = ½(\boldsymbol{\varepsilon}^{\mathsf{T}}, \boldsymbol{C}\boldsymbol{\varepsilon}) - (\overline{\boldsymbol{X}}, \mathbf{u}) - (\boldsymbol{E}_p\overline{\boldsymbol{p}}, \boldsymbol{E}_p\boldsymbol{H}_u\mathbf{u})_\Gamma\,. \tag{A.22}$$

The minimum of the Lagrangian functional, L, should be sought for under the additional conditions

$$\boldsymbol{\varepsilon} = \boldsymbol{A}\mathbf{u} \in \Omega\,, \qquad \boldsymbol{E}_u\boldsymbol{H}_u\mathbf{u} = \overline{\boldsymbol{u}} \in \Gamma\,. \tag{A.23}$$

Using the method of undetermined Lagrange multipliers, we switch from the minimization of L from (A.22) under the conditions (A.23) to a free variational problem for functional J,

$$\mathsf{J}(\boldsymbol{\varepsilon},\mathbf{u},\boldsymbol{\Lambda},\boldsymbol{\lambda}) =$$

$$= \int\limits_{\Omega}\left[M(\boldsymbol{\varepsilon},\mathbf{u}) + \boldsymbol{\Lambda}^{\mathsf{T}}(A\mathbf{u}-\boldsymbol{\varepsilon})\right]d\Omega - \int\limits_{\Gamma}(\boldsymbol{E}_p\boldsymbol{H}_u\mathbf{u})^{\mathsf{T}}\,\overline{\boldsymbol{p}}\,d\Gamma - \int\limits_{\Gamma}\boldsymbol{\lambda}^{\mathsf{T}}\boldsymbol{E}_u(\boldsymbol{H}_u\mathbf{u}-\overline{\boldsymbol{u}})\,d\Gamma,$$

with the vector Lagrangian multipliers $\boldsymbol{\Lambda}$ and $\boldsymbol{\lambda}$. Here we designate

$$M(\boldsymbol{\varepsilon},\mathbf{u}) = \tfrac{1}{2}\boldsymbol{\varepsilon}^{\mathsf{T}}\boldsymbol{C}\boldsymbol{\varepsilon} - \mathbf{u}^{\mathsf{T}}\overline{\boldsymbol{X}}. \qquad \text{(A.24)}$$

After taking the first variation of $\mathsf{J}(\boldsymbol{\varepsilon},\mathbf{u},\boldsymbol{\Lambda},\boldsymbol{\lambda})$ and equaling it to zero, we will have this, in particular:

$$\boldsymbol{\Lambda} = \frac{\partial M}{\partial \boldsymbol{\varepsilon}} = \boldsymbol{C}\boldsymbol{\varepsilon} \in \Omega, \qquad \boldsymbol{E}_p(\boldsymbol{\lambda} - \boldsymbol{H}_\sigma\boldsymbol{\Lambda}) = \mathbf{0} \in \Gamma. \qquad \text{(A.25)}$$

Now we can use (A.25) to exclude the Lagrangian multipliers from the $\mathsf{J}(\boldsymbol{\varepsilon},\mathbf{u},\boldsymbol{\Lambda},\boldsymbol{\lambda})$ functional. The result is a variational problem where we need to find stationarity conditions for functional $\mathsf{F}(\boldsymbol{\varepsilon},\mathbf{u}) = \mathsf{J}(\boldsymbol{\varepsilon},\mathbf{u},\boldsymbol{\Lambda}(\boldsymbol{\varepsilon},\mathbf{u}),\boldsymbol{\lambda}(\boldsymbol{\varepsilon},\mathbf{u}))$, i.e.

$$\mathsf{F}(\boldsymbol{\varepsilon},\mathbf{u}) = \int\limits_{\Omega}\left[M(\boldsymbol{\varepsilon},\mathbf{u}) + (A\mathbf{u}-\boldsymbol{\varepsilon})^{\mathsf{T}}\frac{\partial M}{\partial \boldsymbol{\varepsilon}}\right]d\Omega - \int\limits_{\Gamma}(\boldsymbol{E}_p\boldsymbol{H}_u\mathbf{u})^{\mathsf{T}}\,\overline{\boldsymbol{p}}\,d\Gamma -$$

$$- \int\limits_{\Gamma}\boldsymbol{E}_u(\boldsymbol{H}_u\mathbf{u}-\overline{\boldsymbol{u}})^{\mathsf{T}}\boldsymbol{H}_\sigma\frac{\partial M}{\partial \boldsymbol{\varepsilon}}\,d\Gamma.$$

Along with function $M(\boldsymbol{\varepsilon},\mathbf{u})$, we would like to introduce a new function, $N(\boldsymbol{\sigma},\mathbf{u})$,

$$N(\boldsymbol{\sigma},\mathbf{u}) = \boldsymbol{\sigma}^{\mathsf{T}}\boldsymbol{\varepsilon} - M(\boldsymbol{\varepsilon},\mathbf{u}), \qquad \text{(A.26)}$$

where

$$\boldsymbol{\sigma} = \boldsymbol{C}\boldsymbol{\varepsilon}.$$

But then, according to (A.26) and (A.24),

$$N(\boldsymbol{\sigma},\mathbf{u}) = \boldsymbol{\sigma}^{\mathsf{T}}\boldsymbol{C}^{-1}\boldsymbol{\sigma} - M(\boldsymbol{\varepsilon},\mathbf{u}) = \tfrac{1}{2}\,\boldsymbol{\sigma}^{\mathsf{T}}\boldsymbol{C}^{-1}\boldsymbol{\sigma} + \mathbf{u}^{\mathsf{T}}\overline{\boldsymbol{X}}. \qquad \text{(A.27)}$$

The formulas (A.24) and (A.27) are the Legendre transform's proper for two functions $M(\boldsymbol{\varepsilon},\mathbf{u})$ and $N(\boldsymbol{\sigma},\mathbf{u})$, and for variables $\boldsymbol{\varepsilon}$ and $\boldsymbol{\sigma}$. This transform can be conveniently written as two mutually convertible relationships:

$\boldsymbol{\sigma} = \dfrac{\partial M}{\partial \boldsymbol{\varepsilon}},$ $N(\boldsymbol{\sigma},\mathbf{u}) = \boldsymbol{\sigma}^{\mathsf{T}}\boldsymbol{\varepsilon} - M(\boldsymbol{\varepsilon},\mathbf{u}),$ $\boldsymbol{\varepsilon} = \boldsymbol{C}^{-1}\boldsymbol{\sigma},$	$\boldsymbol{\varepsilon} = \dfrac{\partial N}{\partial \boldsymbol{\sigma}},$ $M(\boldsymbol{\varepsilon},\mathbf{u}) = \boldsymbol{\sigma}^{\mathsf{T}}\boldsymbol{\varepsilon} - N(\boldsymbol{\sigma},\mathbf{u}),$ $\boldsymbol{\sigma} = \boldsymbol{C}\boldsymbol{\varepsilon}.$

(A.28)

The Legendre transform (A.28) permits now to reformulate the variational problem for functional $\mathsf{F}(\boldsymbol{\varepsilon},\mathbf{u})$. The integrand expression of this functional is transformed using (A.28) as follows:

$$M(\boldsymbol{\varepsilon},\mathbf{u}) + \boldsymbol{\sigma}^{\mathsf{T}}(A\mathbf{u} - \boldsymbol{\varepsilon}) = \boldsymbol{\sigma}^{\mathsf{T}}A\mathbf{u} - N(\boldsymbol{\sigma},\mathbf{u}),$$

and appropriate substitutions produce a functional $\mathsf{G}(\boldsymbol{\sigma},\mathbf{u})$,

$$\mathsf{G}(\boldsymbol{\sigma},\mathbf{u}) =$$

$$= \int_{\Omega}\left[\boldsymbol{\sigma}^{\mathsf{T}}A\mathbf{u} - N(\boldsymbol{\sigma},\mathbf{u})\right]d\Omega - \int_{\Gamma}(\boldsymbol{E}_p\boldsymbol{H}_u\mathbf{u})^{\mathsf{T}}\overline{\boldsymbol{p}}d\Gamma - \int_{\Gamma}\boldsymbol{E}_u(\boldsymbol{H}_u\mathbf{u} - \overline{\boldsymbol{u}})^{\mathsf{T}}\boldsymbol{H}_\sigma\boldsymbol{\sigma}d\Gamma =$$

$$= \int_{\Omega}\left[\boldsymbol{\sigma}^{\mathsf{T}}A\mathbf{u} - \frac{1}{2}\boldsymbol{\sigma}^{\mathsf{T}}C^{-1}\boldsymbol{\sigma} - \mathbf{u}^{\mathsf{T}}\overline{X}\right]d\Omega - \int_{\Gamma}(\boldsymbol{E}_p\boldsymbol{H}_u\mathbf{u})^{\mathsf{T}}\overline{\boldsymbol{p}}d\Gamma -$$

$$- \int_{\Gamma}\boldsymbol{E}_u(\boldsymbol{H}_u\mathbf{u} - \overline{\boldsymbol{u}})^{\mathsf{T}}\boldsymbol{H}_\sigma\boldsymbol{\sigma}d\Gamma.$$

As we can see now, functional $\mathsf{G}(\boldsymbol{\sigma},\mathbf{u})$ is nothing but the first form of the Reissner functional (up to the sign) – see (**3**.1.4). Thus, the Legendre transform of this kind turns the Lagrangian functional into the first form of the Reissner functional.

A.3 The Friedrichs transform

The theory of transforms of variational problems makes use of a transform developed by Friedrichs [3]. The Friedrichs transform became widely known, apparently, after a classic of mathematical physics [2] had been published. This transform can be divided into two phases. The first phase is to supplement a functional, the stationarity of which is under consideration, with all or some additional relationships that the varied functions have to satisfy using the Lagrange multiplier technique. In this way the variational problem with additional conditions is transformed into an unconditional problem that depends on a larger number of the sought-for functions. Conditions of stationarity of this new functional produce Euler equations and natural boundary conditions that establish a relationship between the old variables and the Lagrange multipliers. The second phase is to return to a conditional variational problem but in a different form where all (or some) old variables are excluded from the functional using the previous relationships. As a result, the functional keeps only the Lagrange multipliers, but now they also have to meet some additional requirements.

The Friedrichs transform in the finite-dimensional case

We start by showing a sequence of operations to use the Friedrichs transform in a finite-dimensional variational problem. In order to do this, we consider a mechanical system with n degrees of freedom the full potential energy of which, L, is

$$\mathsf{L} = \tfrac{1}{2}\, \boldsymbol{q}^{\mathsf{T}} \boldsymbol{R}\, \boldsymbol{q} - \boldsymbol{q}^{\mathsf{T}} \boldsymbol{P} . \tag{A.29}$$

The vector of displacements $\boldsymbol{q}$ and the vector of external forces $\boldsymbol{P}$ have the dimensionality of n. Suppose the given system is transformed into a new mechanical system by imposing k additional constraints, the matrix equation of which is

$$\boldsymbol{B}\boldsymbol{q} = \boldsymbol{0} . \tag{A.30}$$

Matrix $\boldsymbol{B}$ of the size $(k{\times}n)$ and the rank k is called a constraint matrix.

To determine the state of equilibrium of this new system, we pose a problem of a conditional extremum of functional L under the additional conditions (A.30). After introducing a vector of Lagrangian multipliers, $\boldsymbol{\Lambda}$, of the dimensionality k, we have a free variational problem for functional $\mathsf{J}(\boldsymbol{q}, \boldsymbol{\Lambda})$:

$$\mathsf{J}(\boldsymbol{q}, \boldsymbol{\Lambda}) = \tfrac{1}{2}\, \boldsymbol{q}^{\mathsf{T}} \boldsymbol{R}\, \boldsymbol{q} - \boldsymbol{q}^{\mathsf{T}} \boldsymbol{P} + \boldsymbol{\Lambda}^{\mathsf{T}} \boldsymbol{B}\boldsymbol{q} . \tag{A.31}$$

The conditions of stationarity of $\mathsf{J}(\boldsymbol{q}, \boldsymbol{\Lambda})$ produce the equations

$$\boldsymbol{R}\, \boldsymbol{q} + \boldsymbol{B}^{\mathsf{T}} \boldsymbol{\Lambda} = \boldsymbol{P}, \qquad \boldsymbol{B}\boldsymbol{q} = \boldsymbol{0} , \tag{A.32}$$

wherefrom

$$\boldsymbol{q} = -\, \boldsymbol{R}^{-1} \boldsymbol{B}^{\mathsf{T}} \boldsymbol{\Lambda} + \boldsymbol{R}^{-1} \boldsymbol{P} . \tag{A.33}$$

The obvious mechanical intepretation of the Lagrangian multipliers defines them as generalized forces $\boldsymbol{S}$ (reactions) in the new constraints, i.e.

$$\boldsymbol{\Lambda} = \boldsymbol{S} .$$

Replacing vector $\boldsymbol{q}$ in functional $\mathsf{J}(\boldsymbol{q}, \boldsymbol{\Lambda})$ with its expression from (A.33) and $\boldsymbol{\Lambda}$ with the vector of reactions in the constraints, $\boldsymbol{S}$, we get a functional that depends solely on the force vector $\boldsymbol{S}$, which looks as follows (up to an additive constant):

$$\mathsf{J}(\boldsymbol{q}(\boldsymbol{S}), \boldsymbol{S}) = -\, \tfrac{1}{2}\, \boldsymbol{S}^{\mathsf{T}} (\boldsymbol{B}\boldsymbol{R}^{-1} \boldsymbol{B}^{\mathsf{T}}) \boldsymbol{S} + \boldsymbol{P}^{\mathsf{T}} \boldsymbol{R}^{-1} \boldsymbol{B}^{\mathsf{T}} \boldsymbol{S} . \tag{A.34}$$

This functional can be treated (up to a sign) as a Castigliano functional that depends on the sought-for stresses/forces, $\boldsymbol{S}$, in the constraints, that is,

$$\mathsf{K}(\boldsymbol{S}) = \tfrac{1}{2}\, \boldsymbol{S}^{\mathsf{T}} (\boldsymbol{B}\boldsymbol{R}^{-1} \boldsymbol{B}^{\mathsf{T}}) \boldsymbol{S} - \boldsymbol{P}^{\mathsf{T}} \boldsymbol{R}^{-1} \boldsymbol{B}^{\mathsf{T}} \boldsymbol{S} . \tag{A.35}$$

Functional $\mathsf{K}(\boldsymbol{S})$ is defined on the set of vectors $\boldsymbol{S}$ that satisfy the additional conditions

$$\boldsymbol{BR}^{-1}\boldsymbol{B}^{\mathsf{T}}\boldsymbol{S} - \boldsymbol{BR}^{-1}\boldsymbol{P} = \boldsymbol{0}; \tag{A.36}$$

these conditions can be interpreted from the physical standpoint as equations of equilibrium for the reactions in the new constraints. The conditions (A.36) follow from the requirements (A.30) if vector $\boldsymbol{q}$ is replaced with its expression from (A.33).

The Friedrichs transform in the general case

Finally, we demonstrate the technique of the Friedrichs transform for an infinite-dimensional elastic problem using an elementary example and the Castigliano functional. For the simplicity of notation, we confine our consideration to only kinematical and homogeneous boundary conditions on the whole boundary Γ of area Ω. Then the standard Castiglianian formulation of the variational problem (no elastic foundation is present, $\boldsymbol{K} = \boldsymbol{O}$) requires that we search for a minimum of the following functional, see (**2**.3.1):

$$\mathsf{K}(\boldsymbol{\sigma}) = \tfrac{1}{2}\,(\boldsymbol{C}^{-1}\boldsymbol{\sigma}, \boldsymbol{\sigma}) \tag{A.37}$$

under additional conditions which the sought-for stresses must satisfy:

$$\bar{\boldsymbol{X}} = \boldsymbol{A}^{\mathsf{T}}\boldsymbol{\sigma} \in \Omega\,. \tag{A.38}$$

By introducing the vector of functional Lagrangian multipliers, $\boldsymbol{\Lambda}$, we arrive at an unconditional variational problem that requires the search for a stationary point of functional J that depends both on stresses $\boldsymbol{\sigma}$ and on the Lagrangian multipliers,

$$\mathsf{J}(\boldsymbol{\sigma}, \boldsymbol{\Lambda}) = \tfrac{1}{2}\,(\boldsymbol{C}^{-1}\boldsymbol{\sigma}, \boldsymbol{\sigma}) + (\boldsymbol{\Lambda},\ \bar{\boldsymbol{X}} - \boldsymbol{A}^{\mathsf{T}}\boldsymbol{\sigma})\ . \tag{A.39}$$

The $\boldsymbol{A}^{\mathsf{T}}$ operator obeys the basic integral identity (**1**.2.17), i.e.

$$(\boldsymbol{\Lambda}, \boldsymbol{A}^{\mathsf{T}}\boldsymbol{\sigma}) = (\boldsymbol{A}\boldsymbol{\Lambda}, \boldsymbol{\sigma}) - (\boldsymbol{H}_{\sigma}\boldsymbol{\sigma}, \boldsymbol{H}_{u}\boldsymbol{\Lambda})_{\Gamma}\ .$$

This equality permits to turn functional J into

$$\mathsf{J}(\boldsymbol{\sigma}, \boldsymbol{\Lambda}) = \tfrac{1}{2}\,(\boldsymbol{C}^{-1}\boldsymbol{\sigma}, \boldsymbol{\sigma}) - (\boldsymbol{A}\boldsymbol{\Lambda}, \boldsymbol{\sigma}) + (\boldsymbol{\Lambda}, \bar{\boldsymbol{X}}) + (\boldsymbol{H}_{\sigma}\boldsymbol{\sigma}, \boldsymbol{H}_{u}\boldsymbol{\Lambda})_{\Gamma}\ . \tag{A.40}$$

Clearly, the Euler equations for functional $\mathsf{J}(\boldsymbol{\sigma},\boldsymbol{\Lambda})$ are

$$\boldsymbol{C}^{-1}\boldsymbol{\sigma} = \boldsymbol{A}\boldsymbol{\Lambda} \in \Omega\,, \quad \boldsymbol{A}^{\mathsf{T}}\boldsymbol{\sigma} = \bar{\boldsymbol{X}} \in \Omega\,, \tag{A.41}$$

and its natural boundary conditions are

$$\boldsymbol{H}_{u}\boldsymbol{\Lambda} = \boldsymbol{0} \in \Gamma\,. \tag{A.42}$$

(A.41) and (A.42) make it clear that the Lagrangian multipliers, $\mathbf{\Lambda}$, can be identified with displacements **u**.

To begin the second phase of the Friedrichs transform, we replace the Lagrangian multipliers $\mathbf{\Lambda}$ in the $\mathsf{J}(\boldsymbol{\sigma},\mathbf{\Lambda})$ functional with displacements **u**, and stresses $\boldsymbol{\sigma}$ with their expressions from (A.41). This produces a functional, $\mathsf{J}(\boldsymbol{C}\boldsymbol{A}\mathbf{u},\mathbf{u})$, that depends on the Lagrangian multipliers, **u**, solely:

$$\mathsf{J}(\boldsymbol{C}\boldsymbol{A}\mathbf{u},\mathbf{u}) = -\,{\textstyle\frac{1}{2}}\,(\boldsymbol{C}\boldsymbol{A}\mathbf{u},\boldsymbol{A}\mathbf{u}) + (\mathbf{u},\bar{\boldsymbol{X}}).$$

The stationarity for this functional should be sought for among displacements $\mathbf{u} = \mathbf{\Lambda}$ which satisfy the homogeneousl kinematical boundary conditions (A.42). Now it is clear that functional $\mathsf{J}(\mathbf{u})$ coincides (up to a sign) with the conventional Lagrange functional L. Thus, the Friedrichs transform permits to use a formal technique to switch from the Castiglianian variational formulation to the Lagrangian one.

To give a slightly more complicated but illustrative example of using the Friedrichs transform, we recommend a paper by V.V. Bolotin [1] where this transform was used to derive a stability functional for an elastic body in a rigorous mathematical manner; this functional did not depend, formally, on the components of the initial stressed state.

* * *

To conclude the appendix, we would like to note that both the Legendre transform and the Friedrichs transform belong to the class of *involutory* transforms. It means that when either transformation is applied to the already transformed problem for the second time, we get back to the original formulation of the problem.

References

1. Bolotin VV (1973) On variational principles of the theory of elastic stability (in RussianI. In the book: Problems in mechanics of deformable solids. Sudostroyeniye, Leningrad, pp 83–88
2. Courant R, Gilbert D (1953) Methods of Mathematical Physics. Vol. I. Interscience, New York
3. Friedrichs K (1929) Ein Verfahren der Variationsrechnung. Nachr. der Gesellschaft der Wissensschaften, Göttingen
4. Goldenblatt II (1957) Extreme and variational principles in structural theory (in Russian). In coll. papers: Structural mechanics in USSR in 1917 through 1957, ed. by I.M. Rabinovich. Stroyizdat, Moscow pp 264–278
5. Goldenblatt II (1969) Nonlinear problems in elasticity (in Russian). Physmathgiz, Moscow

6. Gordeyev VN (1967) An algorithm for analysis of systems with unilateral redundant constraints (in Russian). In collected proceedings of IV All-Union conference on application of mathematical machines to strucutral mechanics. Inst. of Cybernetics, Acad. Sci. of Ukraine, Kiev
7. Lanczos C (1966) The variational principles of mechanics. University of Toronto Press, Toronto
8. Parikov VI, Slivker VI (1983) The stiffness matrix for a finite element when the element is nonrigidly fixed to nodes (in Russian). Bulletin of B.Vedeneyev's All-Union Research Institute for Water Engineering, coll. papers, Vol. 164. Static and dynamic analysis of constructions of water, thermal, and nuclear power plants. Energoatomizdat, Leningrad pp 20–28
9. Perelmuter AV, Slivker VI (2003) Numerical Structural Analysis. Springer-Verlag, Berlin Heidelberg New York
10. Rozin LA (1978) Variational formulations of problems for elastic systems (in Russian). Leningrad State Univ.
11. Zukhovitsky SI, Avdeyeva LI (1967) Linear and convex programming (in Russian). Physmathgiz, Moscow

B. TANGENTIAL STRESSES IN THE BENDING OF BARS

B.1 Tangential stresses in the bending of straight bars

It may seem there is no place for any peculiar and previously unnoticed effects on such a trodden spot of the science of strength as the law of distribution of the tangential stresses in a flexural beam. We agree there is not indeed. Therefore it is even more surprising that most courses on strength of material give the derivation of the well-known Zhuravsky formula, which looks like this in stadard designations:

$$\tau = \frac{QS}{Ib}, \tag{B.1}$$

without even mentioning the limitations which have to be taken into account to make sure the formula works[2]. By the way, it is this formula that constitutes a basis for logical reasoning which is used to allow for shear components in the displacements of a flexural beam.

We start by formulating those limitations explicitly; they are obviously independent and complementary with respect to three earlier propositions

[2] There are, however, some exceptions such as [2] and [1]. Here (in Section B.1) we follow mainly the presentment of [4].

that constitute together a so-called planar-sections hypothesis (see Section **4**.3).

So, let X be the longitudinal axis of a bar that contains the centers of gravity of the bar's cross-sections; let axes Y and Z be principal central axes of inertia of the cross-sections. We postulate that the Y and Z axes do not rotate between two adjacent cross-sections or, as it is sometimes said, the bar does not have a natural (immanent) twist.

To be particular, we consider the bending in the (X,Y) plane and make the following assumptions:

- the tangential stresses along the Y axis are distributed evenly over the width of the cross-section. In other words, we assume function $\tau^{xy} = \tau^{xy}(x,y)$ to be independent of coordinate z. The τ^{xz} component is continuous in the elementary theory of bending;
- the set of external actions upon the bar does not contain any distributed moment load m in the section of interest;
- the cross-section of the bar does not change along it in the vicinity of the section of interest.

For the beginning, let us take only the first of this assumption – that the tangential stresses, τ^{xy}, are evenly distributed over the width of the bar. The question how to find the "horizontal" tangential stresses, τ^{xz}, will be discussed later.

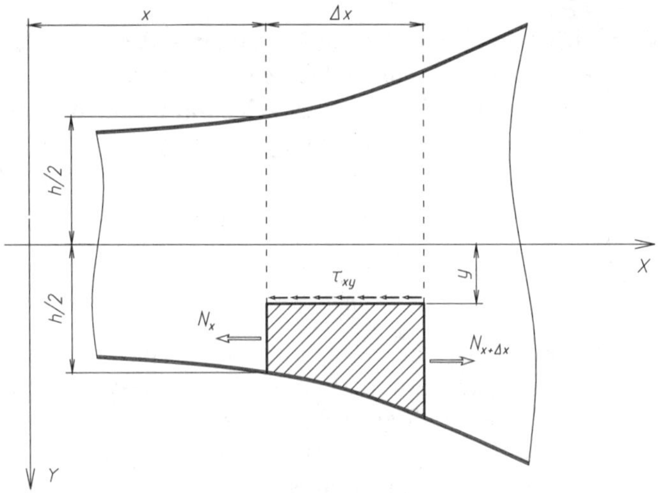

Fig. B.1.

The planar-sections hypothesis implies that each cross-section of the bar behaves like a perfectly rigid, non-deformable solid. This means the shear strains, γ, are identical to zero and thus the respective tangential stresses, τ, cannot be determined from the physical relation between strains γ and stresses τ – Hooke's law for a linear material. However, it is still possible to get the tangential stresses from the equilibrium equations. To do it, we extract an elementary piece of the bar between two cross-sections that have the respective coordinates x and $x + \Delta x$, which is cut off from the rest of the bar by a horizontal plane at the distance y from the (X,Z)-plane. Fig. B.1 shows the extracted element of the bar hatched.

The full force, N_x, in the left cross-section of the extracted element is given by a simplest consideration:

$$N_x = \int_{A_0} \sigma^x dA = \int_{A_0} \frac{My}{I_z} dA = \frac{M}{I_z} \int_{A_0} y dA = \frac{MS_{zo}}{I_z} .$$

Here A_0 and S_{zo} are the respective area and static moment of the left cross-section of the cut-off piece of the bar with respect to the neutral axis Z of the whole cross-section. The full force $N_{x+\Delta x}$ in the right cross-section (coordinate $x + \Delta x$) is equal to

$$N_{x+\Delta x} = N_x + \frac{dN_x}{dx} \Delta x.$$

The resulting force created by normal stresses σ^x and applied to the extracted element of the bar will be

$$\frac{dN_x}{dx} \Delta x .$$

Assuming the length of the extracted piece, Δx, to be a value of first order of smallness and dropping small values of higher orders, we can determine the resultant of the tangential stresses τ_{xy} that act on the horizontal section of the extracted piece as $\tau_{xy} b \Delta x$ where $b = b(x,y)$ is the width of the bar's crosss-section – generally, a function of two coordinates x and y.

Supposing the exterior surface of the bar has no load component along X and making up an equation of equilibrium of the extracted element in the projection onto this axis, we come up with the formula

$$\tau^{xy} = \frac{1}{b}\frac{dN_x}{dx} = \frac{1}{b}\frac{d}{dx}\left(\frac{MS_{zo}}{I_z}\right) . \qquad \text{(B.2)}$$

Now, recalling (**4**.5.11) that says $M' = Q + m$ we get

$$\tau^{xy} = \frac{1}{b}\left[\frac{(Q+m)S_{zo}}{I_z} + M\frac{d}{dx}\left(\frac{S_{zo}}{I_z}\right)\right].$$

Let's discuss the terms in this formula in more detail. First of all, note that we did not yet consider the way the moment load m is created; more exactly, we implied that such a load was created by force couples applied directly to the longitudinal axis. However, there is another, more natural method to create the moment load which, as we will see, will have the external distributed moment m completely excluded from the formula.

To simplify the reasoning, we suppose that the Z axis is an axis of symmetry of the bar's cross-section. Then the moment load m can be represented as a result of longitudinal load q_x where q_x is an intensity per unit of volume of the bar's material. We postulate an even distribution of the load over the width of the cross-section. As for its distribution over the height, we will assume function $q_x = q_x(x,y)$ to be an antisymmetric function of coordinate y, that is,

$$q_x(x,y) = -\,q_x(x,-y)\,. \tag{B.3}$$

It is clear that

$$m = -\int_{-h/2}^{h/2} q_x(x,y)b(y)y\,dy \tag{B.4}$$

where $h = h(x)$ is the height of the cross-section.

The q_x load must be taken into account also in the equations of equilibrium of the extracted piece of the bar, so instead of (B.2) we have

$$\tau^{xy} = \frac{1}{b}\frac{dN_x}{dx} + \frac{1}{b}\int_y^{h/2} q_x(x,y)b(y)dy = \frac{1}{b}\frac{d}{dx}\left(\frac{MS_{zo}}{I_z}\right) + \frac{1}{b}\int_y^{h/2} q_x(x,y)dy\,,$$

wherefrom proper substitutions and transformations produce the final result:

$$\tau^{xy} = \frac{1}{b}\left[\frac{QS_{zo}}{I_z} + M\frac{d}{dx}\left(\frac{S_{zo}}{I_z}\right) + \frac{mS_{zo}}{I_z} + \int_y^{h/2} q_x(x,y)b(y)dy\right]. \tag{B.5}$$

If now we take $q_x = 0$ and assume the bar to have a constant cross-section, therewith we will simplify the general formula (B.5) into the already familiar Zhuravsky formula (B.1).

We can further show that the two last terms in brackets in (B.5) should be omitted under the assumptions of the Bernoulli–Euler theory of bending. We represent the longitudinal load $q_x = q_x(x,y)$ as an expansion into a power series over the y coordinate; the function is antisymmetric with respect to y, so the series will contain only terms of odd powers. We have

$$q_x(x,y) = \alpha_1(x)y + \alpha_3(x)y^3 + \ldots \tag{B.6}$$

The ellipsis denotes the other terms of the series, ones of higher powers of y. Actually, the Bernoulli–Euler theory of bending requires that only the first term (linear w.r. to y) should be kept and all the others omitted.

According to the planar-sections hypothesis, each cross-section has only one degree of freedom in bending with respect to displacements u along X, which is a rigid rotation of the whole section by an angle θ about Z. In other words,

$$u = -\theta y\,.$$

This means the first term of the series (B.6) covers all generalized external forces allowed by the Euler–Bernoulli model and comprising the degrees of freedom of the sections which conform to the longitudinal displacements. So, within this theory of bending, we should assume

$$q_x(x,y) = \alpha_1(x)y\,. \tag{B.7}$$

The sought-for functional coefficient $\alpha_1(x)$ can be calculated by multiplying the equality (B.7) by $b(y)y$ and then integrating over the height of the section, which together with (B.4) gives

$$\alpha_1(x) = -\frac{m}{I_z}\,. \tag{B.8}$$

The result is

$$\int_y^{h/2} q_x(x,y)b(y)dy = -\frac{m}{I_z}\int_y^{h/2} b(y)y dy = -\frac{mS_{oz}}{I_z}\,,$$

and the two last terms in (B.5) cancel each other. Therefore

$$\tau^{xy} = \frac{1}{b}\left[\frac{QS_{zo}}{I_z} + M\frac{d}{dx}\left(\frac{S_{zo}}{I_z}\right)\right]. \tag{B.9}$$

It is easy to show that

$$\int_A \frac{S_{zo}}{b} dA = I_z \tag{B.10}$$

where A is an area of the bar's cross-section.

To see this, use the integration by parts:[3]

$$\int_A \frac{S_{zo}}{b} dA = \int_{-h/2}^{h/2} S_{zo} dy = S_{zo} y\Big|_{-h/2}^{h/2} - \int_{-h/2}^{h/2} y \frac{dS_{zo}}{dy} dy .$$

The static moment of the whole section with respect to axis Z is zero, that is, $S_{zo}(x,h/2) = 0$ and $S_{zo}(x,-h/2) = 0$, so the non-integral term in the right part of the formula vanishes. Next,

$$\frac{dS_{zo}}{dy} = \frac{d}{dy} \int_y^{h/2} b(x,y) y dy = -b(x,y) y , \qquad \text{(B.11)}$$

which gives (B.10) after proper substitutions.

The formula (B.9) is of course neater concerning the *distribution* of the tangential stresses over the section than the simpler Zhuravsky formula. However, if we want to know just an *averaged* shear in the section, then the second integral term in the advanced formula (B.9) can be shown to make a zero contribution to the total integral $\int_A \tau^{xy} dA$.

We have

$$\int_A \frac{1}{b(x,y)} \frac{d}{dx}\left(\frac{S_{zo}}{I_z}\right) dF = \frac{d}{dx}\left(\frac{1}{I_z} \int_{-h/2}^{h/2} S_{zo} dy\right) = \frac{d}{dx}\left(\frac{1}{I_z} I_z\right) = 0 .$$

Thus, the integral contribution of the second term in (B.9) to the section-averaged shear is zero. This is a key reason why the simple formula by Zhuravsky can be the basis for calculating the averaged shear in the planar-sections-based theory of bars. It is this consideration that permits us to use the Zhuravsky formula to derive equations of bending for a Timoshenko bar. However, popular books on strength of materials do not give any comments or explanations about their use of the Zhuravsky formula for constructing the theory of bending of Timoshenko bars. This fact can be treated only as a logical fault that can perplex an inquiring student.

It is of interest to discuss briefly the question how the horizontal tangential stress, τ^{xz}, is distributed. It turns out the answer even to this question can be found in the elementary bending theory. Following a relatively new work by V.D. Kharlab [3], we determine those stresses from the differential equation of equilibrium:

---

[3] Note that the y varies between $-h/2$ and $h/2$ because we assume the bar's section to be symmetric with respect to axis Z.

$$\frac{\partial\tau^{xz}}{\partial z}=-\frac{\partial\sigma^{x}}{\partial x}-\frac{\partial\tau^{xy}}{\partial y}=$$

$$=-\frac{\partial}{\partial x}\left(\frac{My}{I_z}\right)-\frac{\partial}{\partial y}\left[\frac{QS_{zo}}{bI_z}+\frac{M}{b}\frac{d}{dx}\left(\frac{S_{zo}}{I_z}\right)\right]. \tag{B.12}$$

We transform the equation in this way:

$$\frac{\partial\tau^{xz}}{\partial z}=-\frac{Q}{I_z}y-My\frac{\partial}{\partial x}\left(\frac{1}{I_z}\right)-\frac{Q}{I_z}\frac{\partial}{\partial y}\left(\frac{S_{zo}}{b}\right)-$$

$$-M\frac{\partial}{\partial y}\left[\frac{1}{b}\frac{\partial}{\partial x}\left(\frac{S_{zo}}{I_z}\right)\right]. \tag{B.13}$$

Using (B.11) and denoting $b'=\partial b/\partial y$, we have

$$\frac{\partial}{\partial y}\left(\frac{S_{zo}}{b}\right)=-y-\frac{S_{zo}b'}{b^2},\quad \frac{\partial}{\partial y}\left[\frac{1}{b}\frac{\partial}{\partial x}\left(\frac{S_{zo}}{I_z}\right)\right]=-\frac{b'}{b^2}\frac{\partial}{\partial x}\left(\frac{S_{zo}}{I_z}\right)-\frac{y}{b}\frac{\partial}{\partial x}\left(\frac{b}{I_z}\right),$$

which, when inserted in (B.13) and transformed, gives

$$\frac{\partial\tau^{xz}}{\partial z}=\tau^{xy}\frac{b'}{b}+My\left[\frac{1}{b}\frac{\partial}{\partial x}\left(\frac{b}{I_z}\right)-\frac{\partial}{\partial x}\left(\frac{1}{I_z}\right)\right].$$

Seeing that

$$\frac{\partial}{\partial x}\left(\frac{1}{I_z}\right)=\frac{\partial}{\partial x}\left(\frac{b}{bI_z}\right)=\frac{1}{b}\frac{\partial}{\partial x}\left(\frac{b}{I_z}\right)-\frac{1}{bI_z}\frac{\partial b}{\partial x},$$

we have the following expression of the derivative $\partial\tau^{xz}/\partial z$:

$$\frac{\partial\tau^{xz}}{\partial z}=\frac{\tau^{xy}}{b}\frac{\partial b}{\partial y}+\frac{My}{bI_z}\frac{\partial b}{\partial x}. \tag{B.14}$$

Noting that our assumption makes all the components in the right-hand part of (B.14) independent of z and assuming $\tau^{xz}(x,y,0)=0$ gives a simple final formula for the horizontal tangential stresses,

$$\tau^{xz}=(\tau^{xy}\frac{\partial b}{\partial y}+\sigma^{x}\frac{\partial b}{\partial x})\frac{z}{b} \tag{B.15}$$

The formula (B.15) differs from the respective one in [3] by the last term that depends on stresses σ^x. In other words, Kharlab's formula

$$\tau^{xz} = \tau^{xy} \frac{\partial b}{\partial y} \frac{z}{b} \tag{B.16}$$

is a particular case of (B.15) and follows from (B.15) only when the cross-section's width does not change over the length of the bar. The reason is that V.D. Kharlab based his reasoning on the simplified Zhuravsky formula of tangential stress τ^{xy} rather than on the refined formula (B.9).

We take as an example a round bar of a variable radius $r = r(x)$. Obviously, $b^2(x,y)/4 + y^2 = r^2(x)$, so

$$\frac{\partial b}{\partial y} = -\frac{4y}{b}, \qquad \frac{\partial b}{\partial x} = 4r'\frac{r}{b},$$

where we denote $r' = \partial r/\partial x$.

Next, it is easy to find that

$$I = \frac{\pi r^4}{4}, \qquad S_o = \frac{b^3}{12},$$

consequently,

$$\frac{S_o}{I} = \frac{b^3}{3\pi r^4}, \qquad \frac{\partial}{\partial x}\left(\frac{S_o}{I}\right) = \frac{b^2}{\pi r^4}\left(\frac{\partial b}{\partial x} - \frac{4br'}{r}\right) = \frac{b^2 r'}{I}\left(\frac{r}{b} - \frac{b}{3r}\right).$$

Inserting this in (B.9) gives

$$\tau^{xy} = \frac{b^2}{12I}\left[Q + M\frac{12r'}{b}\left(\frac{r}{b} - \frac{b}{3r}\right)\right].$$

Now it is easy to find tangential stresses τ^{xz}, too, by using (B.15). The result is

$$\tau^{xz} = \frac{yz}{3I}\left(-Q + M\frac{4r'}{r}\right),$$

which differs from the similar formula in [3] by the additional term that depends on the bending moment.

B.2 Tangential stresses in the bending of curvilinear bars

Again, as with the rectilinear-axis bar, we extract an elementary piece from a curvilinear bar, located between the cross-section with the arc coordinate s on the left and the cross-section with the arc coordinate $s+\Delta s$ on the right (Fig. B.2). This elementary piece of the bar, which hatched in Fig. B.2, is also cut off from the rest of the bar by a curvilinear surface $z = Const$ where z is the normal coordinate as defined in Section **4**.7.

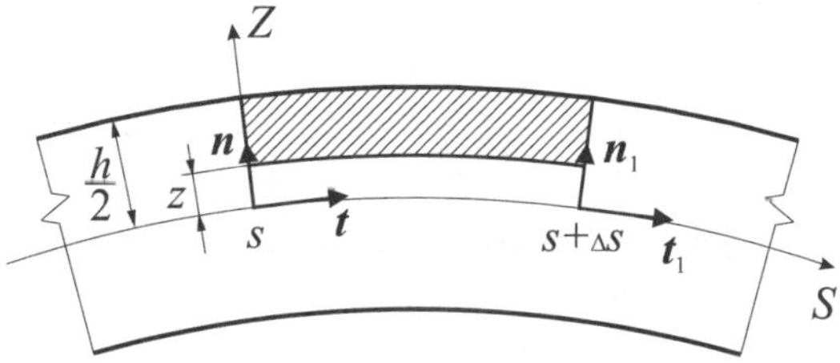

Fig. B.2.

The left section of the extracted element is subjected to normal stresses σ which together create a longitudinal force N_0 and a bending moment M_0 with respect to the Y axis. The right section is respectively subjected to the normal stresses

$$\sigma + \frac{d\sigma}{ds}\Delta s .$$

Normal stresses σ can be represented as follows, depending on what theory of curvilinear bars is used:

Table B.1

small-curvature bars	*medium-curvature bars*	*big-curvature bars*
$\sigma = \frac{N}{A} + \frac{Mz}{I}$	$\sigma = \frac{N + kM}{A} + \frac{Mz}{I} - \frac{kMz^2}{I}$	$\sigma = \frac{N + kM}{A} + \frac{M}{I_\rho}\frac{z}{1+kz}$

Further we will find useful the following notation of the geometrical characteristics of the left-hand cross-section:

$$A_0 = \int_z^{h/2} b dz , \quad S_0 = \int_{A_0} z dA , \quad I_0 = \int_{A_0} z^2 dA ,$$

$$S_{\rho 0} = \int_{A_0} \frac{z}{1+kz} dA , \quad I_{\rho 0} = \int_{A_0} \frac{z^2}{1+kz} dA , \quad I_{20} = \int_{A_0} z^3 dA , \qquad \text{(B.17)}$$

where A_o is the area of the section.

The following formulas hold:

$$\int_A \frac{A_o}{b} dA = A\frac{h}{2}, \qquad \int_A \frac{S_o}{b} dA = I, \qquad \int_A \frac{zS_o}{b} dA = 0, \qquad \int_A \frac{I_o}{b} dA = I\frac{h}{2}. \qquad \text{(B.18)}$$

The proof is based on the integration by parts and can be easily reproduced using two points:

1) the formulas of differentiation of the geometrical characteristics:

$$\frac{dA_o}{dz} = -b(z), \qquad \frac{dS_o}{dz} = -zb(z), \qquad \frac{dI_o}{dz} = -z^2 b(z);$$

2) the symmetry of the bar's cross-section with respect to axis Y, hence

$$\int_{-h/2}^{h/2} z^n b dz = 0 \quad \text{for any odd } n.$$

The following estimates for the comparative orders of magnitude take place:

$$\frac{S_o}{A_o} < h, \qquad \frac{I_o}{S_o} < h, \qquad \frac{I_{2o}}{S_o} < h^2. \qquad \text{(B.19)}$$

It is convenient to use the equation of equilibrium for the extracted element of the curvilinear bar *in moments* rather than in projections onto the horizontal axis as we did for a straight-axis bar. More exactly, we set it down that the moments of all forces acting on the element with respect to the curvature center of the bar's axis should be equal to zero[4]. So, denoting the s coordinate derivative by a stroke and recalling that $k = 1/\rho$, we have

$$-\int_z^{h/2} \sigma b(\rho + z)dz + \int_z^{h/2} [\sigma b + (\sigma b)'\Delta s](\rho + \rho'\Delta s + z)dz - \tau^{sz} b(1+kz)(\rho+z) = 0.$$

After canceling out, we keep only terms of at most first order of smallness with respect to Δs and have

$$\int_z^{h/2} (\sigma b)'(\rho + z)dz + \int_z^{h/2} \sigma b\rho' dz - \tau^{sz} b\rho(1+kz)^2 = 0.$$

[4] Generally, when the curvature varies along the bar, the locations of the curvature centers for the left and right sections are different. It can be shown, however, that ignoring this fact is equivalent to the removal of values of second and higher orders of smallness with respect to Δs from the equation of moments.

The integrand in the first term of the above formula can be conveniently represented as

$$\frac{d(\sigma b)}{ds}(\rho+z)=\frac{d[\sigma b(\rho+z)]}{ds}-\sigma b\frac{d\rho}{ds}.$$

This makes it possible to place the operation of taking the derivative with respect to s outside the integral. As the curvature of the bar's axis, k, does not depend on z, we have this final general expression of the tangential stress in the cross-section of a variable-profile curvilinear bar:

$$\tau^{sz}=\frac{k}{b(1+kz)^2}\frac{d}{ds}\left[\frac{1}{k}\int_z^{h/2}\sigma b(1+kz)dz\right]. \tag{B.20}$$

After substitutions and integrations where we use formulas from Table B.1, we obtain the working formulas for tangential stress τ^{sz} depending on what theory is used to describe a particular curvilinear bar.

Small-curvature bars ($kh<<1$)

$$\tau^{sz}=\frac{k}{b}\frac{d}{ds}\left(N\frac{A_o+kS_o}{kA}+M\frac{S_o+kI_o}{kI}\right).$$

In full accordance with the order of accuracy used in the theory of small-curvature bars and with the estimates (B.19), this formula becomes simpler as

$$\tau^{sz}=\frac{k}{b}\frac{d}{ds}\left(N\frac{A_o}{kA}+M\frac{S_o}{kI}\right). \tag{B.21}$$

Now we use the Kirchhoff equations of equilibrium (**4**.7.30) for the curvilinear bars:

$$N'=-kQ-q_t,\qquad kN=Q'+q_n,\qquad M'=Q-m. \tag{B.22}$$

Taking the derivative in (B.21) and using the expressions of N' and M' from (B.22) gives

$$\tau^{sz}=\frac{1}{b}\left[Q\left(\frac{S_o}{I}-\frac{kA_o}{A}\right)+Nk\left(\frac{A_o}{A}\right)'+Mk\left(\frac{S_o}{kI}\right)'-q_t\frac{A_o}{A}-m\frac{S_o}{I}\right]. \tag{B.23}$$

If now we calculate the integral characteristic of the tangential stresses, τ^{sz}, over the whole section of the bar, it should be apparently equal to the shear force, Q. However, even the first term in (B.23) where (B.18) are substituted already gives as much as

$$Q\int_A \frac{1}{b}\left(\frac{S_o}{I}-\frac{kA_o}{A}\right)dA = Q(1-\frac{kh}{2}) \approx Q\,,$$

so all the other terms total to a zero contribution to the general shear force. Therefore we can confine ourselves to the same old Zhuravsky formula and determine the section-average shear γ as

$$\tau^{sz} = \frac{QS_o}{bI}\,.$$

Obviously, the values of the tangential stress in particular 'fibers' of the cross-section can be calculated more precisely by the refined formula (B.21).

Medium-curvature bars ($k^2h^2 \ll 1$)

Using (B.20) as a basis and considering the second column of Table B.1 gives

$$\tau^{sz} = \frac{k}{b(1+kz)^2}\frac{d}{ds}\left[N\frac{A_o+kS_o}{kA}+M\left(\frac{A_o+kS_o}{A}+\frac{S_o-k^2I_{2o}}{kI}\right)\right]. \qquad \text{(B.24)}$$

Within the scope of accuracy adopted for the medium-curvature bars, we assume

$$\frac{1}{(1+kz)^2} \approx 1-2kz\,,$$

which permits, together with the estimates from (B.19), to simplify (B.24) into the following:

$$\tau^{sz} = \frac{k(1-2kz)}{b}\frac{d}{ds}\left[N\frac{A_o+kS_o}{kA}+M\left(\frac{A_o+kS_o}{A}+\frac{S_o}{kI}\right)\right]. \qquad \text{(B.25)}$$

Taking the derivative with respect to s in (B.25) and using the Kirchhoff equations (B.22) helps derive the final formula of the tangential stresses, τ^{sz}, in application to the medium-curvature bars:

$$\tau^{sz} = \frac{1-2kz}{b} \times$$

$$\times \left[Q\frac{S_o}{I} + Nk\left(\frac{A_o + kS_o}{kA}\right)' + Mk\left(\frac{A_o + kS_o}{A} + \frac{S_o}{kI}\right)' - q_t \frac{A_o + kS_o}{A} - mk\left(\frac{A_o + kS_o}{A} + \frac{S_o}{kI}\right)\right].$$

Integration of τ^{sz} over the whole cross-section's area, A, makes the first term of the above expression equal to

$$Q\int_A \frac{1-2kz}{b}\frac{S_o}{I}dA = Q,$$

where we take (B.18) into account. Thus, for the medium-curvature bars the Zhuravsky formula becomes

$$\tau^{sz} = \frac{QS_o}{bI}(1-2kz). \tag{B.26}$$

Big-curvature bars

Using (B.20) as a basis and considering the third column of Table B.1 gives

$$\tau^{sz} = \frac{k}{b(1+kz)^2}\frac{d}{ds}\left[N\frac{A_o + kS_o}{kA} + M\left(\frac{A_o + kS_o}{A} + \frac{S_o}{kI_\rho}\right)\right]. \tag{B.27}$$

By taking the derivative and considering the Kirchhoff equations (B.22), we transform the last formula into

$$\tau^{sz} = \frac{1}{b(1+kz)^2} \times$$

$$\times \left[Q\frac{S_o}{I_\rho} + Nk\left(\frac{A_o + kS_o}{kA}\right)' + Mk\left(\frac{A_o + kS_o}{A} + \frac{S_o}{kI_\rho}\right)' - q_t \frac{A_o + kS_o}{A} - mk\left(\frac{A_o + kS_o}{A} + \frac{S_o}{kI_\rho}\right)\right].$$

Integration of τ^{sz} over the whole cross-section's area A gives the following for the first term of the above expression:

$$Q\int_A \frac{1}{b(1+kz)^2}\frac{S_o}{I_\rho}dA = \frac{Q}{I_\rho}\int_{-h/2}^{h/2}\frac{S_o}{(1+kz)^2}dz.$$

Integration by parts makes the following out of the last integral:

$$\int_{-h/2}^{h/2} \frac{S_o}{(1+kz)^2} dz = -\frac{S_o}{k(1+kz)}\bigg|_{-h/2}^{h/2} - \frac{1}{k}\int_{-h/2}^{h/2} \frac{zb}{1+kz} dz .$$

The non-integral term is zero because the static moment of the cut-off part of the section is zero at the upper and lower integration limits. The remaining integral is a static moment S_ρ of the reduced section – see (**4**.7.38). The result is

$$Q\int_A \frac{1}{b(1+kz)^2}\frac{S_o}{I_\rho} dA = -\frac{QS_\rho}{I_\rho k} = Q .$$

Thus, for the big-curvature bars the Zhuravsky formula turns into

$$\tau^{sz} = \frac{QS_o}{bI_\rho(1+kz)^2} . \qquad \text{(B.28)}$$

References

1. Birger IA, Mavliutov RR (1986) Strength of materials (in Russian). Physmathgiz, Moscow
2. Gastev VA (1959) A brief course of strength of materials (in Russian). Physmathgiz, Moscow
3. Kharlab VD (1991) On tangential stresses in the elementary theory of planar bending (in Russian). In: Researches in mechanics of strucutral constructions and materials. Proceedings of Leningrad Institute of Civil Engineering, pp 92–95
4. Slivker VI (2002) On tangential stresses in the bending of bars (in Russian). Researches in mechanics of strucutral constructions and materials. Topical proceedings of universities. St.-Petersburg State Acad. of Civil Engng and Architec., pp 90–96

C. INTEGRALS OF PRODUCTS OF FUNCTIONS

We often have to deal with integrals of products of two functions in structural mechanics, such as

$$J = \int_0^l f_1(x) f_2(x) dx .$$

In particular, integrals of this kind are used when we apply the Mohr formula to bar structures, when we calculate the geometrical chaarcteristics of the cross-sections of thin-walled bars, and in other situations.

	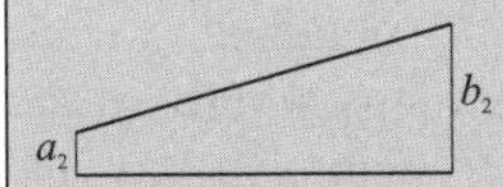	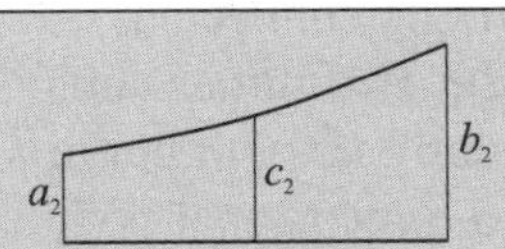
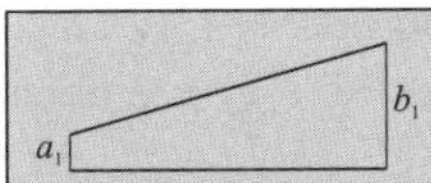	$\frac{l}{6}[2a_1a_2+2b_1b_2+$ $+a_1b_2+a_2b_1]$	$\frac{l}{6}[a_1a_2+b_1b_2+2c_2(a_1+b_1)]$
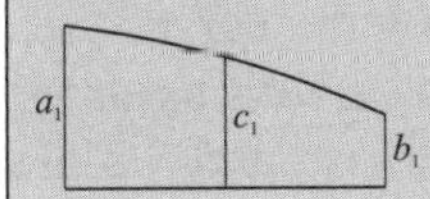	$\frac{l}{6}[a_1a_2+b_1b_2+$ $+2c_1(a_2+b_2)]$	$\frac{l}{6}[2a_1a_2+2b_1b_2+c_1(a_2+b_2)+$ $+c_2(a_1+b_1)-\frac{a_1b_2+a_2b_1}{2}+8c_1c_2]$

If functions f_1 and f_2 are arbitrary, we have to resort to numerical integration. However, in most typical situations the integrands are polynomials, and the integrals can be calculated analytically. In the case when the integrand multipliers are polynomials of second order at the most, the formulas become especially simple, so we think we should present them here as a table for the convenience of referencing. The figures and the formulas denote by c_1 and c_2 the values of the square parabolas in the middle point of the interval $[0,l]$.

D. CIRCULATION OF TANGENTIAL STRESSES

Section **6**.1.2 dedicated to the Saint-Venant problem of the pure torsion of a prismatic bar presents a well-known Bredt theorem of circulation of the tangential stress. We intend to formulate this theorem here is a slightly more general form[5].

D.1 The generalized Bredt theorem

Let us consider a stressed state of a three-dimensional elastic body with its position in space defined in a right-hand Cartesian coordinate system (X,Y,Z) which we will call a global coordinate system. We choose an arbitrary point M of the body with a coordinate x and cut the body apart by a plane that passes through the M point and is orthogonal to the X axis. We

[5] The contents of this appendix follows the article [5].

assume that a planar two-dimensional area created by this kind of intersection can have holes (orifices) in a general case, i.e. it does not have to be singly connected.

In the two-dimensional area thus obtained, we choose a closed and sufficiently smooth curve Γ that belongs wholly to the body. The area of the two-dimensional body contained within curve Γ will be denoted by A, and the total area comprised by the closed curve Γ will be denoted by $\Omega/2$. It is clear that $A \leq \Omega/2$, and the equality is achieved when the Γ curve does not have orifices inside – as shown in Fig. D.1.

We can introduce a curvilinear orthogonal system of coordinates in the vicinity of curve Γ on the (Y,Z)-plane and associate it with the curve – see Appendix F. The direction of the arc coordinate s on this curve is assumed positive when the closed contour is traced counterclockwise and we are looking from the positive direction of the X-axis.

The stressed state of the body in an arbitrary point that belongs to area A, on the plane the normal to which is coincident with the X axis, is defined by a normal stress σ^{xx} and a vector of tangential stress $\boldsymbol{\tau}$ in the (Y,Z)-plane.

We denote by τ^{xs} the projection of the vector of tangential stress $\boldsymbol{\tau}$ onto the direction of the tangent to Γ; the positive direction of the tangent is defined by a unit vector $\boldsymbol{t}$ that looks towards the increasing arc coordinate s. To put it another way, $\tau^{xs} = \boldsymbol{\tau} \cdot \boldsymbol{t}$.

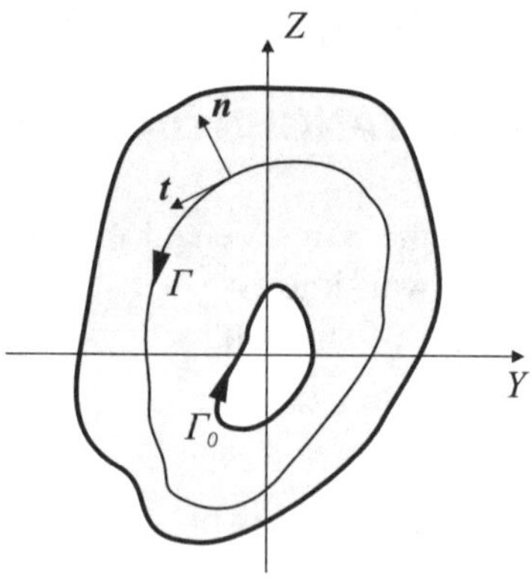

Fig. D.1. Intersection between the body and the (Y,Z) plane

The integral J such that

$$J = \oint_{\Gamma} \tau^{xs} ds \tag{D.1}$$

is called *a circulation of the tangential stress* along the Γ curve.

Let U,V,W be the components of the displacement of an arbitrary point of the body with respect to the global system of axes, X,Y,Z.

For points of the body that belong to the planar area A in some vicinity of the Γ curve, the displacement components can be specified also in a local coordinate system of axes with unit vectors $(\boldsymbol{i}_x, \boldsymbol{n}, \boldsymbol{t})$ where $\boldsymbol{n}$ is a unit vector of the external normal to the boundary Γ (Fig. D.1). The displacement components in the local axes will be denoted by u,v,w. Thus, denoting the unit vectors of axes X,Y,Z as $\boldsymbol{i}_x$, $\boldsymbol{i}_y$, $\boldsymbol{i}_z$, we have

$$u\boldsymbol{i}_x + v\boldsymbol{t} + w\boldsymbol{n} = \mathrm{U}\boldsymbol{i}_x + \mathrm{V}\boldsymbol{i}_y + \mathrm{W}\boldsymbol{i}_z \,. \tag{D.2}$$

The connection between the components of the displacement vector in the global and local coordinate systems is established by the formulas

$$\begin{aligned} &u = \mathrm{U}, \qquad v = \mathrm{V}t_y + \mathrm{W}t_z\,, \qquad w = \mathrm{V}t_z - \mathrm{W}t_y\,, \\ &\mathrm{U} = u, \qquad \mathrm{V} = vt_y + wt_z, \qquad \mathrm{W} = vt_z - wt_y\,, \end{aligned} \tag{D.3}$$

where t_y and t_z are cosines of the unit vector $\boldsymbol{t}$ with respect to axes Y and Z.

Hooke's law for the tangential stress τ^{xs} gives

$$\tau^{xs} = G\left(\frac{\partial u}{\partial s} + \frac{\partial v}{\partial x}\right).$$

The requirement of unambiguous longitudinal displacements u implies the requirement of

$$\oint_\Gamma \frac{\partial u}{\partial s} ds = 0\,,$$

which permits to write our circulation also in the form

$$J = G\oint_\Gamma \frac{\partial v}{\partial x} ds = G\frac{\partial}{\partial x}\oint_\Gamma v ds = G\frac{\partial}{\partial x}\oint_\Gamma \left(\mathrm{V}t_y + \mathrm{W}t_z\right) ds\,. \tag{D.4}$$

Suppose first that the process of deformation of the closed curve Γ obeys the so-called *rigid contour condition.* We say the curve Γ obeys the rigid contour condition when the projection of the curve in the deformed state onto the (Y,Z)-plane is identical to the original (undeformed) curve Γ, possibly moved and rotated as a rigid solid with respect to its undeformed position.

Let $\mathbf{v} = \mathbf{v}(s)$ be a vector of displacements of the Γ contour's points in the (Y,Z)-plane. The mathematical form of the rigid contour condition is

$$\mathbf{v} = \mathbf{v}_\mathrm{P} + \boldsymbol{\theta}\times\boldsymbol{\rho}\,, \tag{D.5}$$

where:

- $\mathbf{v}_P$ is a vector of translational displacements of the Γ contour in the (Y,Z)-plane, which can be treated as a vector of displacements of an arbitrarily chosen point P (a pole) of the (Y,Z)-plane rigidly attached to the curve Γ;
- $\boldsymbol{\theta} = \theta \boldsymbol{i}_x$ is a vector of slope of the Γ curve with respect to the pole P, where θ is an angle of this slope (rotation);
- $\boldsymbol{\rho} = \boldsymbol{\rho}(s)$ is a vector that goes from the pole P to the current point of the curve Γ.

Obviously, the tangential displacement υ introduced earlier is the projection of the $\mathbf{v}$ vector onto the direction of the $\boldsymbol{t}$ unit vector. To express it differently,

$$\upsilon = \mathbf{v}\cdot\boldsymbol{t} = \boldsymbol{t}\cdot\mathbf{v}_P + \boldsymbol{t}\cdot(\boldsymbol{\theta}\times\boldsymbol{\rho}) = \boldsymbol{t}\cdot\mathbf{v}_P + \theta\boldsymbol{\rho}\cdot(\boldsymbol{t}\times\boldsymbol{i}_x) = \boldsymbol{t}\cdot\mathbf{v}_P + \theta\boldsymbol{\rho}\cdot\boldsymbol{n}\,. \tag{D.6}$$

Here we use well-known properties of the mixed product of three vectors, $\boldsymbol{a}\cdot(\boldsymbol{b}\times\boldsymbol{c}) = \boldsymbol{c}\cdot(\boldsymbol{a}\times\boldsymbol{b})$ – see [2], for example.

Now we have

$$\frac{\partial \upsilon}{\partial x} = \boldsymbol{t}\cdot\frac{d\mathbf{v}_P}{dx} + \theta'\boldsymbol{\rho}\cdot\boldsymbol{n} \tag{D.7}$$

where θ' is the derivative with respect to x of the slope θ.

It can be easily seen that

$$\oint_\Gamma \boldsymbol{t}ds = \mathbf{0}\,, \qquad \oint_\Gamma (\boldsymbol{n}\cdot\boldsymbol{\rho})ds = \Omega\,,$$

where Ω is the doubled area bounded by Γ.

Inserting the above in (D.4) gives a well-known Bredt formula for the circulation of the tangential stress [1], which looks as follows in the modern formulation:

$$J = \theta' G\Omega \tag{D.8}$$

and is sometimes called a Bredt theorem. The tradition of courses on elasticity refers the Bredt formula to the torsion of bars solely.

However, as we have just seen, the Bredt formula requires only the rigid contour condition to hold. The particular stress distribution is of no importance as long as the condition is met. For example, the rigid contour condition holds for an orthotropic solid the elasticity moduli of which in one of the orthotropy planes are so large that can be deemed infinite. Sometimes situations when the rigidity contour condition is met are caused by a kinematical hypothesis used to build a particular applied theory. For example, any variation of the engineering theory of thin-walled bars is based on the assumption that the profile is nondeformable in its plane – which *is* the rigid contour condition. By the way, the Bredt formula shows that when there is no twist ($\theta' = 0$) the circulation of the tangential stresses around any closed contour is identical to zero.

Now let us see whether we can generalize the Bredt theorem onto the case when the rigid contour condition is no longer guaranteed.

Let ω_x be the X-axis component of the slope vector of an arbitrary infinitesimal solid element in the vicinity of the current point of the body. The theory of elasticity says [4] that

$$\omega_x = \frac{1}{2}\left(\frac{\partial W}{\partial y} - \frac{\partial V}{\partial z}\right). \tag{D.9}$$

We denote by Θ the average value of slope ω_x in the A area, that is,

$$\Theta = \frac{1}{A}\iint_A \omega_x dA. \tag{D.10}$$

We denote by the symbol Γ_0 the interior boundary, i.e. the boundary of an orifice (or the united boundaries of all orifices) contained in the area which the Γ contour comprises. The positive direction along the interior boundary traces Γ_0 clockwise if we look from the positive direction of the X-axis (Fig. D.1).

By replacing ω_x in (D.10) with its expression from (D.9) and using the Gauss–Ostrogradsky formula, we obtain the following, where our sign conventions for the contour tracing are used:

$$\Theta = \frac{1}{2A}\iint_A \left(\frac{\partial W}{\partial y} - \frac{\partial V}{\partial z}\right) dA = \frac{1}{2A}\left[\oint_\Gamma \left(Wn_y - Vn_z\right) ds + \oint_{\Gamma_0} \left(Wn_y - Vn_z\right) ds\right],$$

n_y and n_z being the direction cosines of unit vector $\boldsymbol{n}$ with respect to axes Y and Z.

Geometrical considerations make it clear that

$$n_y = t_z, \qquad n_z = -t_y,$$

therefore, with designations of (D.3),

$$\Theta = \frac{1}{2A}\left[\oint_\Gamma \left(Wt_z + Vt_y\right) ds + \oint_{\Gamma_0} \left(Wt_z + Vt_y\right) ds\right] =$$

$$= \frac{1}{2A}\left[\oint_{\Gamma} v ds + \oint_{\Gamma_0} v ds\right]. \tag{D.11}$$

According to (D.4), we can write

$$\oint_{\Gamma} \frac{\partial v}{\partial x} ds = \frac{1}{G} J, \qquad \oint_{\Gamma_0} \frac{\partial v}{\partial x} ds = -\frac{1}{G} J_0, \tag{D.12}$$

where J_0 is a circulation of the tangential stress around the contour Γ_0, and the minus sign stands there because the calculation of the circulation on Γ_0 is based on the direction of contour tracing opposite to that shown in Fig. D.1.

Thus, the following formula holds true for the general case:

$$J - J_0 = G(2A)\,\Theta', \tag{D.13}$$

which generalizes the Bredt formula.

In a particular case when the area is singly connected and there is no boundary Γ_0, we have $2A = \Omega$, and the formula of circulation J becomes a usual Bredt formula,

$$J = G\Omega\,\Theta', \tag{D.14}$$

where we should emphasize Θ is an average angle of slope in the area bounded by contour Γ.

Finally, when the rigid contour condition holds for any closed curve Γ fully contained in the cross-section x = Const, the average slope Θ will be identical to slope θ of the cross-section in the (Y,Z)-plane as a rigid solid. Therefore (D.13) will become

$$J - J_0 = G(2A)\theta', \tag{D.15}$$

which corresponds to the earlier Bredt formula (D.8) because that gives the following when applied to each of the closed contours Γ and Γ_0 separately:

$$J = \theta' G\Omega, \qquad J_0 = \theta' G(\Omega - 2A).$$

To conclude the section, we would like to note that an inhomogeneous material of the body requires the Bredt theorem of circulation of the tangential stress to be generalized to a *theorem of circulation of shear strains* [3]. In this latter case the analytical form of the theorem is as follows instead of (D.8):

$$\theta'\Omega = \oint_{\Gamma} \frac{\tau^{xs}}{G} ds .$$

References

1. Bredt R (1896) Kritische Bemerkungen zur Drehungselastizität. Zeitschrift des Vereines deutscher Ingenieure, Bd 40, N29: 813–817
2. Kochin NE (1965) Vector calculus and basics of tensor calculus (in Russian). Nauka, Moscow
3. Lomakin VA (1976) Theory of elasticity of inhomogeneous solids. University Press, Moscow
4. Novozhilov VV (1958) Theory of elasticity (in Russian). Sudpromgiz, Leningrad
5. Slivker VI (2005) On circulation of tangential stresses (in Russian). Bulletin of civil engineers, Saint-Petersburg, N1(2), pp 49-52

E. CONSERVATIVE EXTERNAL FORCES

The engineer encounters the notion of force everywhere and every moment without often thinking what are their sources and how the forces behave during the deformation of a mechanical system. Of course, major mistakes are mostly avoided, and the day is saved by intuition based on analogies and experience.

The history of structural mechanics as a science, however, contains cautionary examples which show how the intuition and analogies can let one down. Let us recall an already classic problem of stability of a cantilever bar under a follower force. The primary conclusion that the bar was stable under any value of the force was a mistake, but it was confuted only after the follower force had been noticed not to belong to the class of conservative loads, so the static Euler method was not really legitimate in application to this problem [1].

The fundamental concept of the force is made familiar to the students of engineering early, when they study a course of theoretical mechanics. When this concept is introduced in the course, it presents also a classification of various forces that exist. In particular, here in structural mechanics we are interested with an important feature used to classify forces into conservative and nonconservative.

Usually the conservative forces are such that can be found from an energy conservation condition, i.e. they have a potential. The very term 'conservative' implies they have an energy-conservative character in the mechanical sense because they are associated with the absence of external sources or consumers of energy.

Further, the same course of theoretical mechanics shows that the work of a force having a potential which is spent to move the force's application point from one spatial position to another does not depend on the path between these positions; it depends solely on the start and end positions of the material point to which the force is applied.

Many courses of theoretical mechanics mistake this latter property of the force that has a potential for the definition of a conservative force. The fact that the two definitions are not totally equivalent is sometimes dissembled. To clear things up, imagine a force different from zero the work of which, nevertheless, is always zero and thus does not depend on the path. Obviously, this force does not have a potential. There are such forces in the nature; the simplest example is a Coriolis force. This circumstance is brightly explained by H. Ziegler [2].

The class of the conservative forces can be further divided into so-called *dead forces* and others (which do not have any special entitlement). A conservative force applied to a mechanical system is thought to be dead if in the course of deformation of the system the material point of its application remains the same, and so do both the direction and the absolute value of the force. In other words, a conservative force is a dead force if two following conditions are met:

- the Lagrangian (material) coordinates of the force application point do not change in the course of deformation;
- the projections of the force onto the axes of the Euler (spatial) coordinate system remain unchanged in the course of deformation.

The simplest and most popular example of a dead force is the gravity force, i.e. the force caused by universe attraction.

The dead forces are easiest to analyze. The reason is that the work of a dead force has the simplest form possible: a scalar product of the force vector by the vector of full displacement of the material point of its application.

E.1 Some cases of the behavior of external forces

Let us analyze the behavior of a force P applied to a planar deformable solid. Let this force be applied to a node of the system in its undeformed state. The node is understood as a material body of an infinitesimal size which, unlike a material point, has both two translational degrees of freedom and one rotational DOF.

Now imagine how the node occupies a new position in the deformed state of the system by having acquired displacements u and v, and how it has rotated by an angle θ as shown in Fig. E.1.

We will deem the behavior of force P to depend on the displacements of the node. It is described by the following four functions:

$$\xi = \xi(u, v, \theta), \quad \eta = \eta(u, v, \theta), \quad P_x = P_x(u, v, \theta), \quad P_y = P_y(u, v, \theta), \qquad \text{(E.1)}$$

where the ξ and η parameters define a locus of points to which the force can be applied in the deformed state, and P_x and P_y are projections of the force in the deformed state onto the respective coordinate axes. Fig. E.1 shows the new geometric position of force P and the material node to which the force was applied initially (by dash lines).

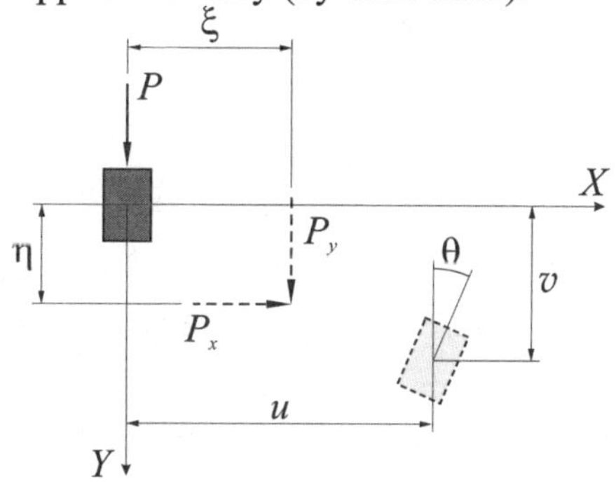

Fig. E.1. The behavior of an external force that depends on displacements

According to the above definition, the force P can be classified as a dead load if and only if

$$\xi = u, \qquad \eta = v, \qquad P_x = 0, \qquad P_y = P. \qquad \text{(E.2)}$$

However, right now we are more interested with general conditions that the functions in (E.1) must satisfy in order for the P force to be legitimately conservative. Clearly, in the case of (E.2) the same conditions must be met because the dead load is a particular case of the conservative load.

Let us calculate the increment of the work, dA, done by force P on the increments of the node's displacements. We carry the force over to the node and apply the horizontal force P_x, the vertical force P_y, and the moment M to it. In this carry-over we have

$$M = P_x(v - \eta) - P_y(u - \xi), \qquad \text{(E.3)}$$

so the work can be also written as

$$dA = P_x\, du + P_y\, dv + M\, d\theta. \tag{E.4}$$

The P force is conservative by definition if the increment of the work, dA, is a total differential. Mathematical analysis says that in order for the differential expression (E.4) to be a total differential, it is necessary and sufficient that functions P_x, P_y and M of three variables u, v, θ satisfy the conditions

$$\frac{\partial P_x}{\partial v} = \frac{\partial P_y}{\partial u}, \qquad \frac{\partial P_y}{\partial \theta} = \frac{\partial M}{\partial v}, \qquad \frac{\partial M}{\partial u} = \frac{\partial P_x}{\partial \theta}.$$

Inserting the expression of moment M from (E.3) gives the following three analytical conditions for the P force to be conservative:

$$\frac{\partial P_x}{\partial v} = \frac{\partial P_y}{\partial u},$$

$$\frac{\partial P_y}{\partial \theta} = \frac{\partial P_x}{\partial v}(v-\eta) + P_x\left(1-\frac{\partial \eta}{\partial v}\right) - \frac{\partial P_y}{\partial v}(u-\xi) + P_y\frac{\partial \xi}{\partial v},$$

$$\frac{\partial P_x}{\partial \theta} = \frac{\partial P_x}{\partial u}(v-\eta) - P_x\frac{\partial \eta}{\partial u} - \frac{\partial P_y}{\partial u}(u-\xi) - P_y\left(1-\frac{\partial \xi}{\partial u}\right). \tag{E.5}$$

Let us take as an example some load transfer patterns on a cantilever bar shown in Fig. E.2.

Let us do a formal analysis of these force transfer patterns based on the analytical criterion (E.5).

Pattern *a* corresponds to the dead force P which is obviously conservative because the functions (E.2) satisfy the conditions.

Pattern *b* conforms to the behavior of a load directed toward a fixed point in space. This load can be implemented with a heavy weight, a rope and ideal pulleys to transfer the force to the upper end of the post. It is mechanically evident that the force is conservative. The formal mathematical expressions for the functions of the type (E.1) are

$$\xi = u, \quad \eta = v, \quad P_x = -P\sin\varphi, \; P_y = P\cos\varphi, \tag{E.6}$$

where

$$\mathrm{tg}\varphi = \frac{u}{h-v}. \tag{E.7}$$

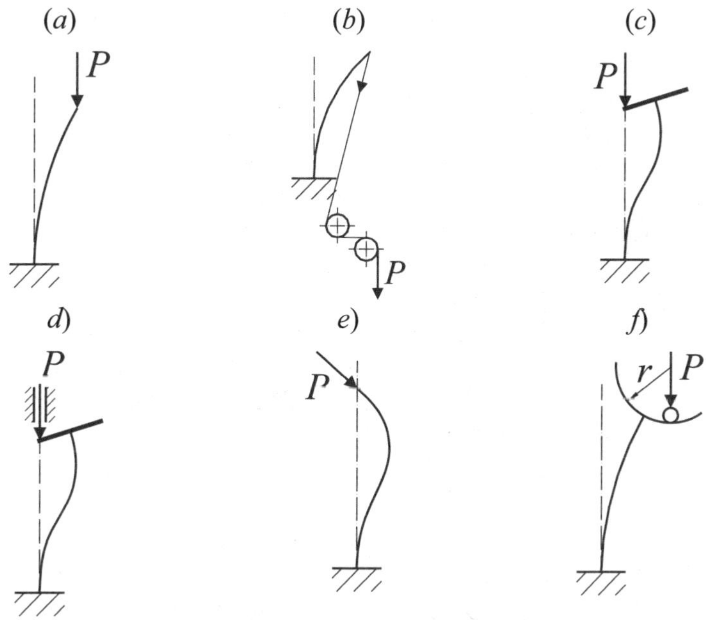

Fig. E.2. Various load patterns

We can see immediately that the second and third conditions of (E.5) are met. Next, taking the derivatives of (E.7) with respect to u and v gives

$$\frac{1}{\cos^2 \varphi}\frac{\partial \varphi}{\partial u} = \frac{1}{h-v}, \qquad \frac{1}{\cos^2 \varphi}\frac{\partial \varphi}{\partial v} = \frac{u}{(h-v)^2},$$

so, consequently,

$$\frac{\partial P_x}{\partial v} = -P\cos\varphi\frac{\partial \varphi}{\partial v} = -P\frac{u\cos^3\varphi}{(h-v)^2} = -P\frac{\sin\varphi\cos^2\varphi}{h-v},$$

$$\frac{\partial P_y}{\partial u} = -P\sin\varphi\frac{\partial \varphi}{\partial u} = -P\frac{\sin\varphi\cos^2\varphi}{h-v}.$$

The first of the conditions in (E.5) is met, too.

In the load pattern c (this force action pattern is known as a Reut problem), the force is always applied along the initial vertical axis. The physical implementation of such a force can be imagined as the pressure of a gas or liquid jet, the spatial deviation of which due to the bar's deformation can be ignored. The behavior of the force is defined by the functions

$$\xi = 0, \qquad \eta = v - u\,\mathrm{tg}\theta, \qquad P_x = 0\,, \qquad P_y = P\,. \tag{E.8}$$

It is easy to check that the first and second requirements from (E.5) are met. However, the third relationship is equivalent to $0 = 1$ for this case, which is not true. The conclusion is that the force is not conservative.

Pattern *d* is different from the preceding one in that the force is transferred to the post via a rigid disk that slides in a holder and pushes a rigid frame on the upper end of the post with its edge. The edge of the disk is assumed to slide freely along the top frame, which is acceptable only if there is no friction between the frame and the disk. It means the actual force applied to the bar via the frame is always orthogonal to the axis of the frame; the slope of the latter is identical to that of the top section of the post. In this case, apparently,

$$\xi = 0, \qquad \eta = v - u\,\mathrm{tg}\theta, \qquad P_x = -P\mathrm{tg}\theta\,, \qquad P_y = P\,. \tag{E.9}$$

A direct substitution helps make sure all three requirements from (E.5) are met. the force is conservative.

Pattern *e* is known as a follower force (Beck's problem). The behavior of the force is described by the functions

$$\xi = u, \qquad \eta = v, \qquad P_x = -P\sin\theta\,, \qquad P_y = P\cos\theta\,. \tag{E.10}$$

The check shows that the first and third conditions of (E.5) for the force to be conservative are met; however, the second one is violated. The force is not conservative.

According to pattern *f*, the force is applied always to the lowest point of the cup of the radius *r*, which is rigidly attached to the end of the post. The behavior of the *P* force is described by the equations

$$\xi = u + r\sin\theta, \qquad \eta = v + r(1-\cos\theta), \qquad P_x = 0\,, \qquad P_y = P\,. \tag{E.11}$$

The conditions (E.5) are observed, so the force is conservative.

Of course, with the force behavior patterns discussed above, we could check the conservative status of the external forces by simpler means on the basis of mechanical considerations. At the same time, the approach presented above is both general and analytically convincing.

E.2 A remark on a hydrostatic load

Sometimes the analysis can have pecularities under a load which is directed always along the normal to the surface of a body and remains so directed to the deformed surface, too. This kind of load is usually called a *hydrostatic load*. In cases when such a load is really the pressure of a still fluid, its conservative nature is self-evident. However, there are worse known cases when this behavior of a load makes it nonconservative and thus it can no longer be implemented by the pressure of the fluid. This fact was indicated by V.V. Bolotin [1].

To see this, let us consider a planar curvilinear bar loaded by a unifrom pressure q along the normal to the bar's axis in the latter's deformed state. The arc length element of this bar is ds before deformation and $(1 + \varepsilon_0)ds$ after that, where ε_0 is a relative longitudinal strain at the level of the bar's axis. But then the vector of distributed load can be written as follows, up to linear terms, after the deformation of the bar:

$$\boldsymbol{q} = \left\| \begin{bmatrix} q_n \\ q_t \end{bmatrix} \right\| = \left\| \begin{bmatrix} q \\ q\sin\theta \end{bmatrix} \right\| = \left\| \begin{bmatrix} q \\ q\theta \end{bmatrix} \right\| \qquad \text{(E.12)}$$

where θ is the slope of the bar's cross-section.

The formulas from Table **4**.1 show that the following general relations hold independently of the order of smallness of the bar's axis curvature:

$$\varepsilon_0 = v' + kw \quad , \quad \theta = -w' + kv \, . \qquad \text{(E.13)}$$

Here, we use the notation of Chapter **4** according to which v is a tangential displacement of the bar's axis, w is a normal displacement of the bar's axis, k is an initial curvature of the bar's axis.

Now we can write the expression of the increment of the work of external forces on admissible variations of displacements δw and δv, and the result is as follows, up to terms linear with respect to the displacements:

$$\delta A = q \int_{s_1}^{s_2} \left[(1+\varepsilon_0)\delta w + \theta\delta v \right] ds =$$

$$= q \int_{s_1}^{s_2} \left[(1 + v' + kw)\delta w + (-w' + kv)\delta v \right] ds \, . \qquad \text{(E.14)}$$

Here s_1 and s_2 denote arc coordinates of the boundaries of the loaded part of the bar.

Using integration by parts

$$\int_{s_1}^{s_2} v'\delta w ds = \left[v\delta w \right]\Big|_{s_1}^{s_2} - \int_{s_1}^{s_2} v\delta w' ds \, ,$$

we represent (E.14) as

$$\delta \mathsf{A} = q\int_{s_1}^{s_2}\left[\delta\left(w+\frac{kw^2}{2}\right)-\delta(vw')+\delta\left(\frac{kv^2}{2}\right)\right]ds+\left[v\delta w\right]_{s_1}^{s_2}. \qquad \text{(E.15)}$$

Now it is clear that, when the non-integral term is zero, the increment of the work, δA, is the variation of the potential

$$q\int_{s_1}^{s_2}\left(w+\frac{kw^2}{2}-vw'+\frac{kv^2}{2}\right)ds,$$

and thus the load is conservative.

As we can see, the key role in the classification of the hydrostatic load as a conservative force is played by the boundary conditions at $s = s_1$ and $s = s_2$. If either the normal displacement w or the tangential displacement v becomes zero on each end of the loaded part of the curvilinear bar, then the load is conservative because the non-integral term indicated above vanishes. But if there is a free edge on at least one end of the bar's loaded part, then the hydrostatic load is generally nonconservative. Obviously, such a load cannot be implemented as the pressure of a still fluid.

References

1. Bolotin VV (1963) Nonconservative Problems of the Theory of Elastic Stability. Pergamon Press, New York
2. Ziegler H (1968) Principles of structural stability. Blaisdell Publishing Company, Waltham, MA

F. CURVILINEAR COORDINATES

In the course of structural analysis activities, one often has to derive various formulas in curvilinear coordinates. The general theory of curvilinear coordinates and their applications to practical problems of mechanics of solids and structures are described in great detail in many well-known books[6]; however, we deem it convenient for the reader to make a short presentation of linear elasticity relationships related to the curvilinear coordinates in the context of this book.

Let (X_1,X_2,X_3) be a right-hand Cartesian coordinate system with its origin at a point O. This coordinate system will be further called a *global coordinate system*. The position of an arbitrary point $\mathrm{M} = \mathrm{M}(x_1,x_2,x_3)$ is

[6] See [1], [3], [4].

defined in the three-dimensional space by three scalar parameters, x_1,x_2,x_3. These scalars are components of a three-dimensional vector **M** that goes from point O to point M that we are interested with. To express it mathematically,

$$\mathbf{M} = x_1\boldsymbol{i}_1 + x_2\boldsymbol{i}_2 + x_3\boldsymbol{i}_3\ , \tag{F.1}$$

where $\boldsymbol{i}_1, \boldsymbol{i}_2, \boldsymbol{i}_3$ are vectors of the unit length codirectional with the respective coordinate axes. These unit vectors define the same basis as the X_1,X_2,X_3 axes.

In many application problems of mechanics of solids and structures it may be convenient not to use the Cartesian coordinates to describe the positions of points of the medium; instead, a more general way of handling the coordinates can be used. This approach is based on the notion of a curvilinear coordinate system.

We introduce three independent parameters $\alpha_1,\alpha_2,\alpha_3$ so that there can be a one-to-one mapping (bijection) between the three Cartesian coordinates x_1,x_2,x_3 and the three parameters $\alpha_1,\alpha_2,\alpha_3$; the bijection is defined by the functional relationships:

$$x_1 = x_1(\alpha_1,\alpha_2,\alpha_3), \qquad x_2 = x_2(\alpha_1,\alpha_2,\alpha_3), \qquad x_3 = x_3(\alpha_1,\alpha_2,\alpha_3), \tag{F.2}$$

or, inverted,

$$\alpha_1 = \alpha_1(x_1,x_2,x_3), \qquad \alpha_2 = \alpha_2(x_1,x_2,x_3), \qquad \alpha_3 = \alpha_3(x_1,x_2,x_3). \tag{F.3}$$

As we know from mathematical analysis, in order for the relationships (F.2) to be solvable with respect to $\alpha_1,\alpha_2,\alpha_3$, i.e. in order to be able to derive (F.3) from (F.2), it is necessary and sufficient that the determinant, J, of the Jacobi matrix, **J**, for the transformation (F.2), which is also called a *Jacobian* determinant,

$$\mathbf{J} = \left\| \begin{matrix} \dfrac{\partial x_1}{\partial \alpha_1} & \dfrac{\partial x_1}{\partial \alpha_2} & \dfrac{\partial x_1}{\partial \alpha_3} \\ \dfrac{\partial x_2}{\partial \alpha_1} & \dfrac{\partial x_2}{\partial \alpha_2} & \dfrac{\partial x_2}{\partial \alpha_3} \\ \dfrac{\partial x_3}{\partial \alpha_1} & \dfrac{\partial x_3}{\partial \alpha_2} & \dfrac{\partial x_3}{\partial \alpha_3} \end{matrix} \right\|, \qquad \mathrm{J} = \det \mathbf{J} \equiv \frac{\partial(x_1,x_2,x_3)}{\partial(\alpha_1,\alpha_2,\alpha_3)} \neq 0$$

be different from zero. Together with the Jacobian determinant of the transform (F.2), the determinant J^{-1} of the inverse transform (F.3) is also nonzero and equal to

$$\mathbf{J}^{-1} = \left\| \begin{bmatrix} \dfrac{\partial\alpha_1}{\partial x_1} & \dfrac{\partial\alpha_1}{\partial x_2} & \dfrac{\partial\alpha_1}{\partial x_3} \\ \dfrac{\partial\alpha_2}{\partial x_1} & \dfrac{\partial\alpha_2}{\partial x_2} & \dfrac{\partial\alpha_2}{\partial x_3} \\ \dfrac{\partial\alpha_3}{\partial x_1} & \dfrac{\partial\alpha_3}{\partial x_2} & \dfrac{\partial\alpha_3}{\partial x_3} \end{bmatrix} \right\|, \qquad J^{-1} = \det \mathbf{J}^{-1} \equiv \frac{\partial(\alpha_1,\alpha_2,\alpha_3)}{\partial(x_1,x_2,x_3)} \neq 0.$$

It can happen so that the transforms of this kind are not bijectional in the whole three-dimensional space $\mathbb{R}^3$ but only in some limited area $\Omega \subset \mathbb{R}^3$. Then the transforms (F.2) and (F.3) will be valid only in area Ω rather than the whole space because only there the bijection between the coordinates x_1,x_2,x_3 and the three parameters $\alpha_1,\alpha_2,\alpha_3$ is guaranteed. In addition to this limitation of the domain, we will assume also the functions (F.2) and (F.3) to be smooth enough. More exactly, we will demand that the functions be at least twice continuously differentiable in area Ω.

If we assume $\alpha_1 = Const$ in (F.2), those formulas will define a surface in the three-dimensional space; such a surface is usually called a *coordinate surface*. In the same way we define two other families of the coordinates surfaces, $\alpha_2 = Const$ and $\alpha_3 = Const$. These surfaces make up *families* of surfaces because each coordinate surface of one family conforms to a value of the constant. An intersection of two coordinate surfaces (such as $\alpha_2 = Const$ and $\alpha_3 = Const$) is a curve in space which we usually call a *coordinate curve*, α_1. The coordinate curves α_2 and α_3 are defined similarly.

As the functions (F.2) and (F.3) are one-valued, each point M of the area Ω belongs to precisely one representative of each coordinate surface family. This is a geometrical image for the point M's position to be definable unambiguously by the three introduced parameters, $\alpha_1,\alpha_2,\alpha_3$. The **M** vector can be treated now as a vector function of $\alpha_1,\alpha_2,\alpha_3$,

$$\mathbf{M} = \mathbf{M}(\alpha_1,\alpha_2,\alpha_3)\,. \tag{F.4}$$

A set of parameters $\alpha_1,\alpha_2,\alpha_3$ that satisfies all the requirements listed above is called a *curvilinear coordinate system*.

Let us introduce unit vectors $\boldsymbol{e}_1$, $\boldsymbol{e}_2$, $\boldsymbol{e}_3$ which are tangential to the coordinate curves at the point M and directed toward the increasing values of variables $\alpha_1,\alpha_2,\alpha_3$ (Fig. F.1). The triple of unit vectors $\boldsymbol{e}_1$, $\boldsymbol{e}_2$, $\boldsymbol{e}_3$ together is called a *local basis*, or a *coordinate basis*, or a *moving basis*. The latter means the directions of the local basis vectors changes together with the coordinates of the point M where this basis is defined.

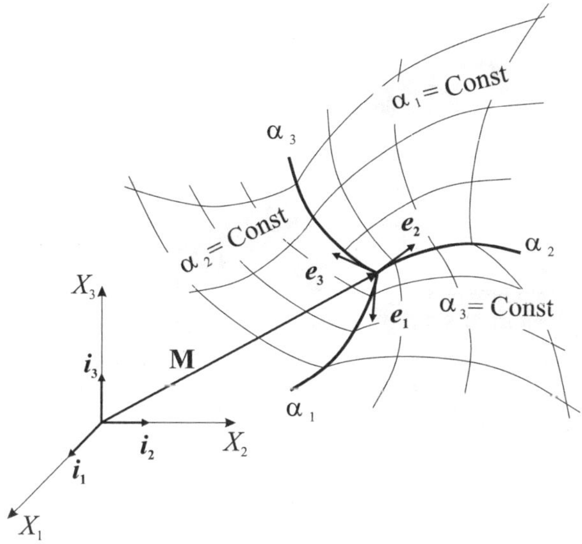

Fig. F.1. Curvilinear coordinates and a local basis

Let us take a partial derivative of the vector function $\mathbf{M}(\alpha_1,\alpha_2,\alpha_3)$ with respect to α_1 at fixed values of α_2 and α_3. When the two parameters are fixed and only α_1 varies, the **M** vector will slide along the coordinate curve α_1, so the partial derivative $\partial\mathbf{M}/\partial\alpha_1$ will be tangential to the coordinate curve α_1 toward bigger α_1[7]. It means

$$\frac{\partial\mathbf{M}}{\partial\alpha_1} = H_1\boldsymbol{e}_1\,, \tag{F.5}$$

where H_1 is the length of the vector $\partial\mathbf{M}/\partial\alpha_1$. From (F.1) we have

$$\frac{\partial\mathbf{M}}{\partial\alpha_1} = \frac{\partial x_1}{\partial\alpha_1}\boldsymbol{i}_1 + \frac{\partial x_2}{\partial\alpha_1}\boldsymbol{i}_2 + \frac{\partial x_3}{\partial\alpha_1}\boldsymbol{i}_3\,, \tag{F.6}$$

and making scalar products of both parts of the equality (F.5) gives

$$H_1^2 = \left(\frac{\partial x_1}{\partial\alpha_1}\right)^2 + \left(\frac{\partial x_2}{\partial\alpha_1}\right)^2 + \left(\frac{\partial x_3}{\partial\alpha_1}\right)^2.$$

Similar formulas can be obtained for the partial derivatives of **M** with respect to α_2 and α_3. So we have

---

[7] In the nomenclature of vector analysis, the coordinate curve α_1 is a *hodograph* of vector **M** at the fixed α_2 and α_3.

$$\boldsymbol{e}_i = \frac{\partial \mathbf{M}}{H_i \partial \alpha_i}, \qquad H_i^2 = \left(\frac{\partial x_1}{\partial \alpha_i}\right)^2 + \left(\frac{\partial x_2}{\partial \alpha_i}\right)^2 + \left(\frac{\partial x_3}{\partial \alpha_i}\right)^2, \quad (i = 1,2,3). \qquad \text{(F.7)}$$

The theory of curvilinear coordinates calls the above parameters H_1, H_2, H_3 the *Lame parameters*; they play a most important part in the theory.

Recalling the structure of the Jacobi matrix for the direct transform (F.2), we note that the square of i-th Lame parameter is equal to the sum of squares of the elements of i-th column of **J**.

Further, if coordinate α_1 gets an increment $d\alpha_1$ and the rest curvilinear coordinates of point M do not change, then the end of the **M** vector will move along the coordinate curve α_1 by a distance ds_1 such that

$$ds_1 = \left|\frac{\partial \mathbf{M}}{\partial \alpha_1}\right| d\alpha_1 = H_1 d\alpha_1 .$$

Hence a simple geometrical sense of the Lame parameters. Each Lame parameter H_i is a ratio between the arc length increment ds_i on the coordinate curve α_i and the respective increment of the curvilinear coordinate $d\alpha_i$.

F.1 Orthogonal curvilinear coordinates

When orthogonal curvilinear coordinates are used, the working formulas are especially simple, therefore we confine our consideration to this kind of coordinates only.

A system of coordinates is called *orthogonal* if its coordinate curves cross one another at right angles. This means the unit vectors $\boldsymbol{e}_1$, $\boldsymbol{e}_2$, $\boldsymbol{e}_3$ are mutually orthogonal. In the vector form, the orthogonality means that the scalar product of any two different unit vectors of the basis is zero,

$$\boldsymbol{e}_i \cdot \boldsymbol{e}_j = \delta_{ij},$$

where δ_{ij} is Kronecker's delta. From (F.5) and (F.6) we find that the condition of orthogonality between $\boldsymbol{e}_i$ and $\boldsymbol{e}_j$ (at $i \neq j$) in coordinates is

$$\frac{\partial x_1}{\partial \alpha_i}\frac{\partial x_1}{\partial \alpha_j} + \frac{\partial x_2}{\partial \alpha_i}\frac{\partial x_2}{\partial \alpha_j} + \frac{\partial x_3}{\partial \alpha_i}\frac{\partial x_3}{\partial \alpha_j} = 0. \qquad \text{(F.8)}$$

In other words, the sum of pairwise products of the elements in two different columns of the Jacobi matrix for an orthogonal curvilinear coordinate system is zero.

To be definite, we will assume further the triple of vectors $\boldsymbol{e}_1$, $\boldsymbol{e}_2$, $\boldsymbol{e}_3$ to be right-hand, i.e.

$$\boldsymbol{e}_3 = \boldsymbol{e}_1 \times \boldsymbol{e}_2 .$$

This requirement is not vexatious for functions in (F.2) and (F.3); if the basis is left-hand, we need just to reorder the curvilinear coordinates or alter the sign of one of parameters $\alpha_1,\alpha_2,\alpha_3$.

We have indicated that, although the length of the local basis vectors is always equal to one, their directions can change when the M point changes its location in space. Consequently, the local basis vectors are vector functions of the curvilinear coordinates,

$$\boldsymbol{e}_1 = \boldsymbol{e}_1(\alpha_1,\alpha_2,\alpha_3), \qquad \boldsymbol{e}_2 = \boldsymbol{e}_2(\alpha_1,\alpha_2,\alpha_3), \qquad \boldsymbol{e}_3 = \boldsymbol{e}_3(\alpha_1,\alpha_2,\alpha_3). \tag{F.9}$$

If we want to know how "fast" the local basis changes with the curvilinear coordinates of point M, we need to be able to take derivatives of its vector functions.

The derivative of a vector function with respect to a scalar argument is again a vector, so it can be expanded oven the local basis. We take, for example, vector $\boldsymbol{e}_1$ and expand the derivatives of it over the axes of the basis by assuming

$$\frac{\partial \boldsymbol{e}_1}{\partial \alpha_1} = a_{11}\boldsymbol{e}_1 + a_{12}\boldsymbol{e}_2 + a_{13}\boldsymbol{e}_3 ,$$

$$\frac{\partial \boldsymbol{e}_1}{\partial \alpha_2} = a_{21}\boldsymbol{e}_1 + a_{22}\boldsymbol{e}_2 + a_{23}\boldsymbol{e}_3 ,$$

$$\frac{\partial \boldsymbol{e}_1}{\partial \alpha_3} = a_{31}\boldsymbol{e}_1 + a_{32}\boldsymbol{e}_2 + a_{33}\boldsymbol{e}_3 . \tag{F.10}$$

Now we have a problem how to determine 9 coefficients a_{ij} in the above expansions. More exactly, we would like to express these coefficients via the Lame parameters introduced earlier. First, we differentiate both parts of the equality $\boldsymbol{e}_1 \cdot \boldsymbol{e}_1 = 1$ with respect to α_i, to get

$$\frac{\partial \boldsymbol{e}_1}{\partial \alpha_i} \cdot \boldsymbol{e}_1 = 0 \qquad (i = 1,2,3) . \tag{F.11}$$

Thus, the derivative of the basis vector with respect to a curvilinear coordinate is orthogonal to the vector itself. By making the scalar product of each equality in (F.10) with vector $\boldsymbol{e}_1$ and considering the orthogonality of the vectors as in (F.11), we find immediately that

$$a_{11} = a_{21} = a_{31} = 0\,. \tag{F.12}$$

If we differentiate the equality $\boldsymbol{e}_1 \cdot \boldsymbol{e}_2 = 0$ with respect to α_1 (it holds because the basis vectors are orthogonal), we will have

$$\frac{\partial \boldsymbol{e}_1}{\partial \alpha_1} \cdot \boldsymbol{e}_2 + \frac{\partial \boldsymbol{e}_2}{\partial \alpha_1} \cdot \boldsymbol{e}_1 = 0\,,$$

or, using (F.7) and (F.10),

$$a_{12} + \frac{\partial}{\partial \alpha_1}\left(\frac{\partial \mathbf{M}}{H_2 \partial \alpha_2}\right) \cdot \boldsymbol{e}_1 = 0\,. \tag{F.13}$$

Now let us do a chain of transformations

$$\frac{\partial}{\partial \alpha_1}\left(\frac{\partial \mathbf{M}}{H_2 \partial \alpha_2}\right) = \frac{\partial}{\partial \alpha_1}\left(\frac{1}{H_2}\right)\frac{\partial \mathbf{M}}{\partial \alpha_2} + \frac{1}{H_2}\frac{\partial^2 \mathbf{M}}{\partial \alpha_1 \partial \alpha_2} =$$

$$= -\frac{1}{H_2}\frac{\partial H_2}{\partial \alpha_1}\boldsymbol{e}_2 + \frac{1}{H_2}\frac{\partial (H_1 \boldsymbol{e}_1)}{\partial \alpha_2} = -\frac{1}{H_2}\frac{\partial H_2}{\partial \alpha_1}\boldsymbol{e}_2 + \frac{H_1}{H_2}\frac{\partial \boldsymbol{e}_1}{\partial \alpha_2} + \frac{1}{H_2}\frac{\partial H_1}{\partial \alpha_2}\boldsymbol{e}_1\,.$$

We make a scalar product of this equality with $\boldsymbol{e}_1$ and take into account that the $\partial \boldsymbol{e}_1/\partial \alpha_2$ vector is orthogonal to $\boldsymbol{e}_1$ according to (F.11). The result is

$$\frac{\partial}{\partial \alpha_1}\left(\frac{\partial \mathbf{M}}{H_2 \partial \alpha_2}\right) \cdot \boldsymbol{e}_1 = \frac{1}{H_2}\frac{\partial H_1}{\partial \alpha_2}\,.$$

Putting this expression in (F.13) produces an expression of coefficient a_{12} via the Lame parameter,

$$a_{12} = -\frac{1}{H_2}\frac{\partial H_1}{\partial \alpha_2}\,.$$

In the same way we find that

$$a_{13} = -\frac{1}{H_3}\frac{\partial H_1}{\partial \alpha_3}, \quad a_{22} = \frac{1}{H_1}\frac{\partial H_2}{\partial \alpha_1}, \quad a_{23} = 0, \quad a_{32} = 0, \quad a_{33} = \frac{1}{H_1}\frac{\partial H_3}{\partial \alpha_1}\,.$$

The final form of the relationships (F.10) is

$$\begin{bmatrix} \dfrac{\partial \boldsymbol{e}_1}{\partial \alpha_1} \\ \dfrac{\partial \boldsymbol{e}_1}{\partial \alpha_2} \\ \dfrac{\partial \boldsymbol{e}_1}{\partial \alpha_3} \end{bmatrix} = \begin{bmatrix} 0 & -\dfrac{1}{H_2}\dfrac{\partial H_1}{\partial \alpha_2} & -\dfrac{1}{H_3}\dfrac{\partial H_1}{\partial \alpha_3} \\ 0 & \dfrac{1}{H_1}\dfrac{\partial H_2}{\partial \alpha_1} & 0 \\ 0 & 0 & \dfrac{1}{H_1}\dfrac{\partial H_3}{\partial \alpha_1} \end{bmatrix} \begin{bmatrix} \boldsymbol{e}_1 \\ \boldsymbol{e}_2 \\ \boldsymbol{e}_3 \end{bmatrix}. \qquad \text{(F.14-1)}$$

The formulas for differentiating the other two vectors of the local basis are derived in the same way. But these formulas can be written out immediately by shuffling thc indexes in (F.14-1). Here are the formulas for $\boldsymbol{e}_2$

$$\begin{bmatrix} \dfrac{\partial \boldsymbol{e}_2}{\partial \alpha_1} \\ \dfrac{\partial \boldsymbol{e}_2}{\partial \alpha_2} \\ \dfrac{\partial \boldsymbol{e}_2}{\partial \alpha_3} \end{bmatrix} = \begin{bmatrix} \dfrac{1}{H_2}\dfrac{\partial H_1}{\partial \alpha_2} & 0 & 0 \\ -\dfrac{1}{H_1}\dfrac{\partial H_2}{\partial \alpha_1} & 0 & -\dfrac{1}{H_3}\dfrac{\partial H_2}{\partial \alpha_3} \\ 0 & 0 & \dfrac{1}{H_2}\dfrac{\partial H_3}{\partial \alpha_2} \end{bmatrix} \begin{bmatrix} \boldsymbol{e}_1 \\ \boldsymbol{e}_2 \\ \boldsymbol{e}_3 \end{bmatrix}, \qquad \text{(F.14-2)}$$

and for $\boldsymbol{e}_3$

$$\begin{bmatrix} \dfrac{\partial \boldsymbol{e}_3}{\partial \alpha_1} \\ \dfrac{\partial \boldsymbol{e}_3}{\partial \alpha_2} \\ \dfrac{\partial \boldsymbol{e}_3}{\partial \alpha_3} \end{bmatrix} = \begin{bmatrix} \dfrac{1}{H_3}\dfrac{\partial H_1}{\partial \alpha_3} & 0 & 0 \\ 0 & \dfrac{1}{H_3}\dfrac{\partial H_2}{\partial \alpha_3} & 0 \\ -\dfrac{1}{H_1}\dfrac{\partial H_3}{\partial \alpha_1} & -\dfrac{1}{H_2}\dfrac{\partial H_3}{\partial \alpha_2} & 0 \end{bmatrix} \begin{bmatrix} \boldsymbol{e}_1 \\ \boldsymbol{e}_2 \\ \boldsymbol{e}_3 \end{bmatrix}. \qquad \text{(F.14-3)}$$

The formulas (F.14) establish rules for differentiating any arbitrary vector function $\mathbf{V} = \mathbf{V}(\alpha_1,\alpha_2,\alpha_3)$ because any vector $\mathbf{V}$ can be represented by its expansion over the axes of the local basis,

$$\mathbf{V} = \mathrm{V}_1\boldsymbol{e}_1 + \mathrm{V}_2\boldsymbol{e}_2 + \mathrm{V}_3\boldsymbol{e}_3 \,.$$

F.2 Differentiation with respect to curvilinear coordinates

Let an area Ω, in which the coordinate transformations (F.2) and (F.3) are defined, have a scalar function specified on it – say, function $w = w(x_1, x_2, x_3)$. We are interested here with the possibility to represent the derivatives $\partial w/x_i$ via the derivatives of the same function in an orthogonal curvilinear coordinate system, i.e. via $\partial w/\partial\alpha_1$, $\partial w/\partial\alpha_2$, $\partial w/\partial\alpha_3$ and the cosines of mutual orientation angles for the local and global basis.

We introduce a matrix $\mathbf{\Lambda}$ composed of the direction cosines of the local basis axes with respect to the global basis axes, i.e.

$$\mathbf{\Lambda} = \left\| \begin{matrix} \lambda_{11} & \lambda_{12} & \lambda_{13} \\ \lambda_{21} & \lambda_{22} & \lambda_{23} \\ \lambda_{31} & \lambda_{32} & \lambda_{33} \end{matrix} \right\|, \qquad \lambda_{ij} = \boldsymbol{i}_i \cdot \boldsymbol{e}_j \,. \tag{F.15}$$

Knowing matrix $\mathbf{\Lambda}$, we can write out an expansion of the global basis vectors over those of the local basis,

$$\boldsymbol{i}_1 = \lambda_{11}\boldsymbol{e}_1 + \lambda_{12}\boldsymbol{e}_2 + \lambda_{13}\boldsymbol{e}_3 \,,$$

$$\boldsymbol{i}_2 = \lambda_{21}\boldsymbol{e}_1 + \lambda_{22}\boldsymbol{e}_2 + \lambda_{23}\boldsymbol{e}_3 \,,$$

$$\boldsymbol{i}_3 = \lambda_{31}\boldsymbol{e}_1 + \lambda_{32}\boldsymbol{e}_2 + \lambda_{33}\boldsymbol{e}_3 \,.$$

As the local basis varies along the curve (it is a moving basis), the components of matrix $\mathbf{\Lambda}$ will be different for different points of area Ω.

The indirect differentiation rules give

$$\frac{\partial w}{\partial x_i} = \frac{\partial w}{\partial \alpha_1}\frac{\partial \alpha_1}{\partial x_i} + \frac{\partial w}{\partial \alpha_2}\frac{\partial \alpha_2}{\partial x_i} + \frac{\partial w}{\partial \alpha_3}\frac{\partial \alpha_3}{\partial x_i} \,. \tag{F.16}$$

Therefore all we need is to establish a relation between derivatives $\partial\alpha_i/\partial x_j$ and the direction cosines, λ_{ij}, introduced above.

Turning to formula (F.6) and making a scalar product of it with vector $\boldsymbol{i}_i$ gives

$$\frac{\partial \mathbf{M}}{\partial \alpha_1} \cdot \boldsymbol{i}_i = \frac{\partial x_i}{\partial \alpha_1} \,.$$

According to (F.7), $\partial\mathbf{M}/\partial\alpha_1 = H_1\boldsymbol{e}_1$, so we derive from here and (F.15) that

$$\frac{\partial x_i}{\partial \alpha_1} = H_1\lambda_{i1} \,.$$

This one is apparently true for both the first column of the Jacobi matrix and the the other two. All we need is to replace index 1 by 2 or 3. Thus, we have three relationships:

$$\frac{\partial x_i}{\partial \alpha_1} = H_1\lambda_{i1}\,, \qquad \frac{\partial x_i}{\partial \alpha_2} = H_2\lambda_{i2}\,, \qquad \frac{\partial x_i}{\partial \alpha_3} = H_3\lambda_{i3} \quad (i = 1,2,3)\,. \qquad \text{(F.17)}$$

If we take the **M** vector's components in the global basis – see (F.1), we have

$$\frac{\partial \mathbf{M}}{\partial x_1} = \boldsymbol{i}_1\,.$$

However, the derivative $\partial\mathbf{M}/\partial x_1$ can be expanded also like this:

$$\boldsymbol{i}_1 = \frac{\partial \mathbf{M}}{\partial x_1} = \frac{\partial \mathbf{M}}{\partial \alpha_1}\frac{\partial \alpha_1}{\partial x_1} + \frac{\partial \mathbf{M}}{\partial \alpha_2}\frac{\partial \alpha_2}{\partial x_1} + \frac{\partial \mathbf{M}}{\partial \alpha_3}\frac{\partial \alpha_3}{\partial x_1} =$$

$$= H_1\boldsymbol{e}_1\frac{\partial \alpha_1}{\partial x_1} + H_2\boldsymbol{e}_2\frac{\partial \alpha_2}{\partial x_1} + H_3\boldsymbol{e}_3\frac{\partial \alpha_3}{\partial x_1}\,.$$

By making the scalar products of both parts with $\boldsymbol{e}_j$, we get an expression for the derivative $\partial\alpha_j/\partial x_1$. The final formulas are

$$\frac{\partial \alpha_j}{\partial x_1} = \frac{\lambda_{1j}}{H_j}\,, \qquad \frac{\partial \alpha_j}{\partial x_2} = \frac{\lambda_{2j}}{H_j}\,, \qquad \frac{\partial \alpha_j}{\partial x_3} = \frac{\lambda_{3j}}{H_j} \quad (j = 1,2,3),\ (j\,!)\,. \qquad \text{(F.18)}$$

Returning to (F.16), we rewrite it like this:

$$\frac{\partial w}{\partial x_i} = \frac{\lambda_{i1}}{H_1}\frac{\partial w}{\partial \alpha_1} + \frac{\lambda_{i2}}{H_2}\frac{\partial w}{\partial \alpha_2} + \frac{\lambda_{i3}}{H_3}\frac{\partial w}{\partial \alpha_3}\,. \qquad \text{(F.19)}$$

The formula (F.19) represents the derivatives of the scalar function w with respect to the global coordinates via the derivatives of the same function with respect to the local coordinates, provided the Lame parameters and the matrix of direction cosines, $\mathbf{\Lambda}$, are known.

Together with (F.19), it is useful to have an inverse formula that makes a transition in the differentiation procedure from the global coordinates to the local ones. We have

$$\frac{\partial w}{\partial \alpha_i} = \frac{\partial w}{\partial x_1}\frac{\partial x_1}{\partial \alpha_i} + \frac{\partial w}{\partial x_2}\frac{\partial x_2}{\partial \alpha_i} + \frac{\partial w}{\partial x_3}\frac{\partial x_3}{\partial \alpha_i} =$$

$$= \lambda_{1i} H_i \frac{\partial w}{\partial x_1} + \lambda_{2i} H_i \frac{\partial w}{\partial x_2} + \lambda_{3i} H_i \frac{\partial w}{\partial x_3} \quad (i\,!)\,. \qquad \text{(F.20)}$$

The formulas (F.19) and (F.20) can be transformed to a conventional tensor form. To do it, we introduce the designations of p^{ij} and q^{ij} for the elements of the Jacobi matrix of the direct transformation (F.2) $\mathbf{J} = \|[p^{ij}]\|$ and the inverse one (F.3) $\mathbf{J}^{-1} = \|[q^{ij}]\|$. By definition,

$$p^{ij} = \frac{\partial x_i}{\partial \alpha_j} = \lambda_{ij} H_j \quad (j\,!)\,, \qquad q^{ij} = \frac{\partial \alpha_i}{\partial x_j} = \frac{\lambda_{ji}}{H_i} \quad (i\,!)\,. \qquad \text{(F.21)}$$

With the tensors introduced above, the formulas (F.19) and (F.20) become convenient for tensor index transformations:

$$\frac{\partial w}{\partial x_i} = q^{ji} \frac{\partial w}{\partial \alpha_j}, \qquad \frac{\partial w}{\partial \alpha_i} = p^{ji} \frac{\partial w}{\partial x_j}. \qquad \text{(F.22)}$$

Some problems require that we be able to transform second derivatives of the same function in addition to first ones. From (F.22) we have

$$\frac{\partial^2 w}{\partial x_j \partial x_i} = q^{lj} \frac{\partial \left(q^{ki} \dfrac{\partial w}{\partial \alpha_k} \right)}{\partial \alpha_l} = q^{lj} \left(q^{ki} \frac{\partial^2 w}{\partial \alpha_l \partial \alpha_k} + \frac{\partial q^{ki}}{\partial \alpha_l} \frac{\partial w}{\partial \alpha_k} \right) \qquad \text{(F.23)}$$

and

$$\frac{\partial^2 w}{\partial \alpha_j \partial \alpha_i} = \frac{\partial \left(p^{ki} \dfrac{\partial w}{\partial x_k} \right)}{\partial \alpha_j} = \frac{\partial p^{ki}}{\partial \alpha_j} \frac{\partial w}{\partial x_k} + p^{ki} p^{lj} \frac{\partial^2 w}{\partial x_l \partial x_k}. \qquad \text{(F.24)}$$

In order to be able to apply these formulas to a particular problem, we would need first to obtain the expressions of the components of tensors $\{p^{ij}\}$ and $\{q^{ij}\}$ together with their derivatives in the curvilinear cordinates $\alpha_1,\alpha_2,\alpha_3$.

F.3 Formulas for strain components in a curvilinear orthogonal coordinate system

We consider a deformation of a three-dimensional body the spatial position of which is defined in an orthogonal curvilinear coordinate system, $\alpha_1,\alpha_2,\alpha_3$. Let us find out the location of a selected point M after the deformation. Let the point get a displacement **u** when the body is subjected to the deformation. We consider the displacement vector, **u**, of the point M in the local basis:

$$\mathbf{u} = u_1 \boldsymbol{e}_1 + u_2 \boldsymbol{e}_2 + u_3 \boldsymbol{e}_3 \,. \tag{F.25}$$

After the deformation of the body, the spatial location of point M is defined by the vector $\mathbf{M} + \mathbf{u}$. Let the M point lie on a coordinate curve of α_1. We take a point N on this curve that has the coordinates of $(\alpha_1+d\alpha_1,\alpha_2,\alpha_3)$ and is immediately adjacent to M. The distance between the M and N points along this curve before the deformation is $ds_1 = H_1 d\alpha_1$. After the deformation, the spatial positions of these two points will be defined by vectors $\mathbf{M}'$ and $\mathbf{N}'$ where

$$\mathbf{M}' = \mathbf{M} + \mathbf{u}\,, \qquad \mathbf{N}' = \mathbf{N} + \mathbf{u} + d\mathbf{u}\,.$$

The displacement vector for point N gets the increment of $d\mathbf{u}$ due to the increment of α_1 only, therefore

$$d\mathbf{u} = \frac{\partial \mathbf{u}}{\partial \alpha_1} d\alpha_1 \,.$$

Considering the rules of differentiation of the basis vectors in (F.14), we get this by making proper substitutions:

$$d\mathbf{u} = \left(\frac{\partial u_1}{\partial \alpha_1} + \frac{1}{H_2}\frac{\partial H_1}{\partial \alpha_2} u_2 + \frac{1}{H_3}\frac{\partial H_1}{\partial \alpha_3} u_3 \right) \boldsymbol{e}_1 d\alpha_1 +$$

$$+ \left(\frac{\partial u_2}{\partial \alpha_1} - \frac{1}{H_2}\frac{\partial H_1}{\partial \alpha_2} u_1 \right) \boldsymbol{e}_2 d\alpha_1 + \left(\frac{\partial u_3}{\partial \alpha_1} - \frac{1}{H_3}\frac{\partial H_1}{\partial \alpha_3} u_1 \right) \boldsymbol{e}_3 d\alpha_1 \,. \tag{F.26}$$

As can be seen from (F.26), the distance between the points of interest in the direction of coordinate α_1 has increased due to the deformation of the body by

$$\left(\frac{\partial u_1}{\partial \alpha_1} + \frac{1}{H_2}\frac{\partial H_1}{\partial \alpha_2} u_2 + \frac{1}{H_3}\frac{\partial H_1}{\partial \alpha_3} u_3 \right) d\alpha_1 \,.$$

Composing a ratio of this elongation to the original length $ds_1 = H_1 d\alpha_1$ helps determine (as far as the linear strain theory is concerned) the relative elongation ε_{11} in the direction of coordinate α_1

$$\varepsilon_{11} = \frac{1}{H_1}\frac{\partial u_1}{\partial \alpha_1} + \frac{1}{H_1 H_2}\frac{\partial H_1}{\partial \alpha_2}u_2 + \frac{1}{H_1 H_3}\frac{\partial H_1}{\partial \alpha_3}u_3 . \tag{F.27}$$

Starting off by the same formula (F.26), we will find the projection of angle $\gamma_{12}^{(1)}$ between the directions of coordinate curves α_1 in the deformed and undeformed states onto the plane of vectors $\boldsymbol{e}_1$ and $\boldsymbol{e}_2$. Assuming the angles to be small and thus making the tangent of the sought-for angle identical to the angle itself, we have

$$\gamma_{12}^{(1)} = \left[\left(\frac{\partial u_2}{\partial \alpha_1} - \frac{1}{H_2}\frac{\partial H_1}{\partial \alpha_2}u_1\right)d\alpha_1\right] / (H_1 d\alpha_1) = \frac{1}{H_1}\frac{\partial u_2}{\partial \alpha_1} - \frac{1}{H_1 H_2}\frac{\partial H_1}{\partial \alpha_2}u_1 .$$

Similarly, we put down an expression for the projection of angle $\gamma_{21}^{(2)}$ between the directions of the coordinate curves α_2 onto the same plane $(\boldsymbol{e}_1,\boldsymbol{e}_2)$ in both the deformed and undeformed states,

$$\gamma_{21}^{(2)} = \frac{1}{H_2}\frac{\partial u_1}{\partial \alpha_2} - \frac{1}{H_1 H_2}\frac{\partial H_2}{\partial \alpha_1}u_2 .$$

Summing the $\gamma_{12}^{(1)}$ and $\gamma_{21}^{(2)}$ angles gives the full shear angle, γ_{12}, between the coordinate curves α_1 and α_2; after some transformations we have

$$\gamma_{12} = \frac{H_2}{H_1}\frac{\partial}{\partial \alpha_1}\left(\frac{u_2}{H_2}\right) + \frac{H_1}{H_2}\frac{\partial}{\partial \alpha_2}\left(\frac{u_1}{H_1}\right). \tag{F.28}$$

Having the formulas (F.27) and (F.28) for the elongations in the direction of α_1 and the shear angles between the coordinate curves α_1 and α_2, it is easy to derive the rest of the formulas for the elongation and shear strains by permutation of the indexes. Here are all these formulas for the convenience of referencing:

$$\varepsilon_{11} = \frac{1}{H_1}\frac{\partial u_1}{\partial \alpha_1} + \frac{1}{H_1 H_2}\frac{\partial H_1}{\partial \alpha_2}u_2 + \frac{1}{H_1 H_3}\frac{\partial H_1}{\partial \alpha_3}u_3 ,$$

$$\varepsilon_{22} = \frac{1}{H_2}\frac{\partial u_2}{\partial \alpha_2} + \frac{1}{H_2 H_3}\frac{\partial H_2}{\partial \alpha_3}u_3 + \frac{1}{H_2 H_1}\frac{\partial H_2}{\partial \alpha_1}u_1 ,$$

$$\varepsilon_{33} = \frac{1}{H_3}\frac{\partial u_3}{\partial \alpha_3} + \frac{1}{H_1 H_3}\frac{\partial H_3}{\partial \alpha_1}u_1 + \frac{1}{H_3 H_2}\frac{\partial H_3}{\partial \alpha_2}u_2 ,$$

$$\gamma_{12} = \gamma_{21} = \frac{H_2}{H_1}\frac{\partial}{\partial \alpha_1}\left(\frac{u_2}{H_2}\right) + \frac{H_1}{H_2}\frac{\partial}{\partial \alpha_2}\left(\frac{u_1}{H_1}\right),$$

$$\gamma_{13} = \gamma_{31} = \frac{H_3}{H_1}\frac{\partial}{\partial \alpha_1}\left(\frac{u_3}{H_3}\right) + \frac{H_1}{H_3}\frac{\partial}{\partial \alpha_3}\left(\frac{u_1}{H_1}\right),$$

$$\gamma_{23} = \gamma_{32} = \frac{H_2}{H_3}\frac{\partial}{\partial \alpha_3}\left(\frac{u_2}{H_2}\right) + \frac{H_3}{H_2}\frac{\partial}{\partial \alpha_2}\left(\frac{u_3}{H_3}\right). \qquad \text{(F.29)}$$

F.4 Curvilinear coordinates on a plane, associated with a planar curve

One particular case of an orthogonal curvilinear coordinate system defined on a plane is of special interest. Let the (X,Y)-plane have a sufficiently smooth curve Γ defined on it [8]. The equation of this curve can be specified in a variety of ways – for example, parametrically:

$$x = X(s) , \qquad y = Y(s) . \qquad \text{(F.30)}$$

Limiting the interval of the parameter variation to, say,

$$s \in [0,L],$$

defines our working segment of curve Γ[9]. Of course, parameter s can have a lot of different meanings. We, however, prefer it to be the geometric length of curve Γ counted from the origin (zero point) to the current point,

---

[8] Here, we drop the indexing of the global Cartesian coordinates and switch to the traditional designations of the X,Y axes with unit vectors $\boldsymbol{i}_x$ and $\boldsymbol{i}_y$. Also, we will complement the two-dimensional coordinates (X,Y) imaginably to get the three-dimensional right-hand triple (X,Y,Z).

[9] It is not unlikely that Γ is a closed curve. In that case L is understood as a full length of this closed curve.

M(s), which belongs to the curve. Our origin point on the curve will be a point of the plane M_0 that has global coordinates $X(0)$, $Y(0)$. We assume that when parameter s grows from zero to L, this is a positive direction of movement along curve Γ. The s parameter can be treated now as a single coordinate that defines the position of the current point M on curve Γ unambiguously. Therefore parameter s will be called an *arc coordinate.*

In each point M(s), curve Γ has such a characteristic value as curvature k which is one inverse to the radius of curvature, $\rho = 1/k$, of curve Γ in the same current point M(s).

Let a unit-length vector $\boldsymbol{t}$ be directed tangentially to curve Γ. The positive direction for this vector will be the same as the positive one for moving along Γ, that is, such that makes parameter s increase. Now we define vector $\boldsymbol{n}$ so that it satisfy a set of requirements:

$$\boldsymbol{n} \cdot \boldsymbol{n} = 1, \qquad \boldsymbol{n} \cdot \boldsymbol{t} = 0, \qquad \boldsymbol{n} \times \boldsymbol{t} = \boldsymbol{i}_x \times \boldsymbol{i}_y\,. \tag{F.31}$$

Obviously, the unit vector $\boldsymbol{n}$ codirectional with the normal to Γ and defined by the above requirements is unambiguous. Now we can define a sign convention for curvature k. We will deem it positive when a vector that goes from the curvature center to the current point M on curve Γ has the same direction as the local unit vector $\boldsymbol{n}$ at the point M.

An ordered couple of vectors ($\boldsymbol{n}, \boldsymbol{t}$) will be called a *local basis associated with curve* Γ. The mutual orientation of the local basis and the global one is established by the relations

$$\boldsymbol{n} = n_x\boldsymbol{i}_x + n_y\boldsymbol{i}_y\,, \qquad \boldsymbol{t} = t_x\boldsymbol{i}_x + t_y\boldsymbol{i}_y\,, \tag{F.32}$$

or

$$\boldsymbol{i}_x = n_x\boldsymbol{n} + t_x\boldsymbol{t}\,, \qquad \boldsymbol{i}_y = n_y\boldsymbol{n} + t_y\boldsymbol{t}\,. \tag{F.33}$$

Here n_x, n_y are direction cosines of normal $\boldsymbol{n}$, and t_x, t_y are direction cosines of unit vector $\boldsymbol{t}$.

Notice immediately from Fig. F.2 that the components of vectors $\boldsymbol{n}$ и $\boldsymbol{t}$ in the (X,Y) coordinate system are

$$n_x = \cos\varphi, \quad n_y = \sin\varphi, \qquad t_x = -\sin\varphi, \quad t_y = \cos\varphi,$$

so

$$t_x = -n_y\,, \qquad t_y = n_x\,. \tag{F.34}$$

Here φ is an angle between the X-axis and normal $\boldsymbol{n}$, which is positive when the X-axis rotates about the Z-axis *clockwise* to coincide with axis $\boldsymbol{n}$, if we look along the positive direction of Z.

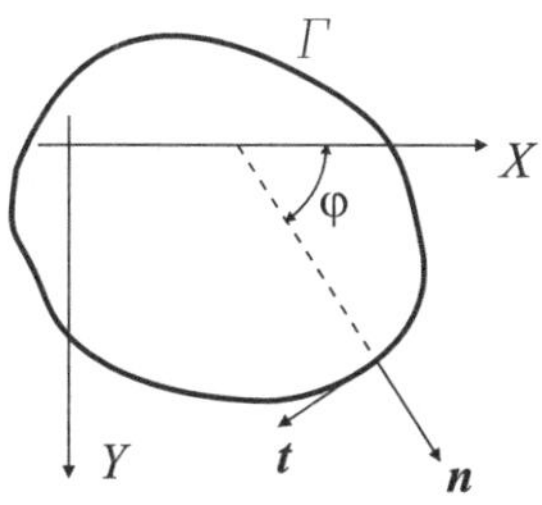

Fig. F.2. A local basis on curve Γ

We introduce a second-rank tensor, **ω**, referred to as a *rotation tensor*, with the components of ω^{ij} defined as follows

$$\omega^{ij} = \begin{cases} 0, & if\ i = j \\ -1, & if\ i = 1\ and\ j = 2, \\ 1, & if\ i = 2\ and\ j = 1 \end{cases} \tag{F.35}$$

or, in the matrix form,

$$\boldsymbol{\omega} = \left\| \begin{bmatrix} 0 & -1 \\ 1 & 0 \end{bmatrix} \right\|. \tag{F.36}$$

The rotation matrix, **ω**, helps represent the connection between unit vectors ***n*** and ***t*** as defined by (F.34) in the form

$$\left\| \begin{bmatrix} t_x \\ t_y \end{bmatrix} \right\| = \left\| \begin{bmatrix} 0 & -1 \\ 1 & 0 \end{bmatrix} \right\| \left\| \begin{bmatrix} n_x \\ n_y \end{bmatrix} \right\|, \quad \text{or} \quad \boldsymbol{t} = \boldsymbol{\omega n}, \tag{F.37}$$

and the inverse matrix of **ω** coincides with the transposed matrix, $\boldsymbol{\omega}^{\mathsf{T}}$, which is in its turn the same as –**ω**, i.e.

$$\boldsymbol{\omega}^{-1} = \boldsymbol{\omega}^{\mathsf{T}} = -\boldsymbol{\omega}, \quad \text{and therefore } \boldsymbol{n} = -\boldsymbol{\omega t}. \tag{F.38}$$

Manipulations with the tensors can be more convenient using the index form of the same relationships,

$$t_i = \omega^{ij} n_j = -\omega^{ji} n_j, \quad n_i = -\omega^{ij} t_j = \omega^{ji} t_j; \tag{F.39}$$

this is used where appropriate in the main text of the book. Geometrically, the ***t*** vector is derived from the ***n*** vector by a simple rotation of the latter by angle $\pi/2$, hence the name of the rotation matrix (or tensor).

Let us draw the normal, ***n***, to curve Γ at the point M(s) $\in \Gamma$ and consider a point N on this normal that has the coordinates of (n,s) in the local basis. Parameter n the meaning of which is the distance between the points M and N will be called a *normal coordinate* of point N. The normal coordinate, n, is positive when the MN vector is codirectional with vector $\boldsymbol{n}$ (Fig. F.3).

Let the N point have global coordinates (x,y).

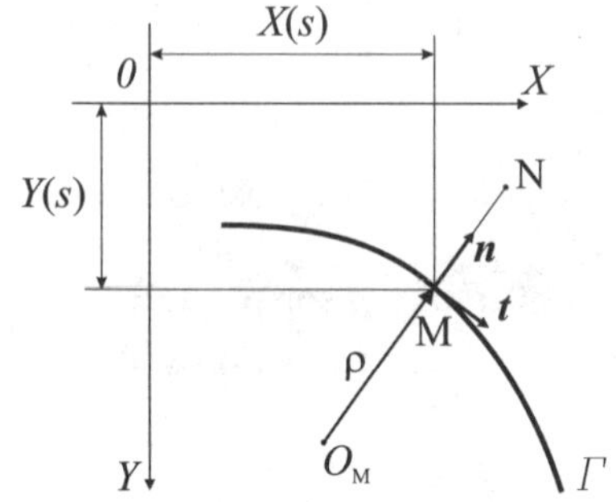

Fig. F.3. Orthogonal curvilinear coordinates associated with curve Γ

The following relation exists between these coordinates (Fig. F.3)

$$x = X(s) + nn_x\,, \qquad y = Y(s) + nn_y\,.$$

The normal, n, and arc, s, coordinates can be used as orthogonal curvilinear coordinates, α_1 and α_2, in a certain area Ω in the vicinity of curve Γ where the Jacobian determinant J of the transformation

$$x = X(\alpha_2) + \alpha_1 n_x(\alpha_2)\,, \qquad y = Y(\alpha_2) + \alpha_1 n_y(\alpha_2) \tag{F.40}$$

is different from zero:

$$\mathbf{J} = \left\| \begin{bmatrix} \dfrac{\partial x}{\partial \alpha_1} & \dfrac{\partial x}{\partial \alpha_2} \\ \dfrac{\partial y}{\partial \alpha_1} & \dfrac{\partial y}{\partial \alpha_2} \end{bmatrix} \right\| = \left\| \begin{bmatrix} n_x & X' + nn'_x \\ n_y & Y' + nn'_y \end{bmatrix} \right\|, \quad \mathrm{J} = \det \mathbf{J}. \tag{F.41}$$

Here the stroke denotes differentiation with respect to arc coordinate s. For some time, let's not find out formal requirements for the Jacobian determinant J in (F.41) to be different from zero. We just note that for a sufficiently smooth curve Γ there is always a vicinity of it where the correspondence between the Cartesian coordinates of its points and the above curvilinear coordinates α_1,α_2 is one-to-one. It is for this vicinity of Γ that we deem the transformations (F.41) defined.

The relevant name for the set of parameters α_1, α_2 will be *orthogonal curvilinear coordinates associated with curve* Γ.

It is easy to notice that in these coordinates the Lame parameters are equal to $H_1 = 1$ and $H_2 = 1+kn$. To see this, consider how $d\alpha_1 = dn = ds_1$ when we move along the normal, so $H_1 = 1$. The arc length element ds_2 along the coordinate curve α_2 (at $\alpha_1 = n = Const$) is equal to

$$ds_2 = \frac{\rho + n}{\rho} d\alpha_2 = (1+kn)d\alpha_2$$

where $\rho = 1/k$ is a radius of curvature of curve Γ at the point with the local coordinates $(0,s)$[10]. This length of the arc element is equal to ds only at $\alpha_1 = n = 0$, i.e. directly on Γ. Thus,

$$H_1 = 1, \qquad H_2 = 1+kn. \tag{F.42}$$

It is easy to notice that in our designations we have

$$\lambda_{11} = \lambda_{xn} = n_x\,, \quad \lambda_{21} = \lambda_{yn} = n_y\,, \qquad \lambda_{12} = \lambda_{xs} = t_x\,, \quad \lambda_{22} = \lambda_{ys} = t_y\,,$$

so, from (F.21) and (F.42) it follows that

$$\mathbf{J} = \|[p^{ij}]\| = \left\|\begin{bmatrix} n_x & t_x(1+kn) \\ n_y & t_y(1+kn) \end{bmatrix}\right\|, \mathbf{J}^{-1} = \|[q^{ij}]\| = \left\|\begin{bmatrix} n_x & n_y \\ \dfrac{t_x}{(1+kn)} & \dfrac{t_y}{(1+kn)} \end{bmatrix}\right\|. \tag{F.43}$$

By the way, these formulas immediately produce a simple expression of the Jacobian determinant J:

$$\mathrm{J} = (n_x\, t_y - n_y\, t_x)(1+kn)\,.$$

Due to (F.34), $n_x t_y - n_y t_x = n_x n_x + n_y n_y = 1$. Hence the Jacobian determinan, J, is nonzero for values of the normal coordinate n at which $1 + kn \neq 0$.

Now let us consider the formulas (F.22) for mutual transitions between differentiation with respect to the curvilinear coordinates and to the original Cartesian ones. The first of the formulas is as follows for our case:

$$\frac{\partial w}{\partial x} = q^{11}\frac{\partial w}{\partial \alpha_1} + q^{21}\frac{\partial w}{\partial \alpha_2} = n_x\frac{\partial w}{\partial n} + \frac{t_x}{1+kn}\frac{\partial w}{\partial s},$$

$$\frac{\partial w}{\partial y} = q^{12}\frac{\partial w}{\partial \alpha_1} + q^{22}\frac{\partial w}{\partial \alpha_2} = n_y\frac{\partial w}{\partial n} + \frac{t_y}{1+kn}\frac{\partial w}{\partial s}, \tag{F.44}$$

[10] The curvature $k = 1/\rho$ is positive when the vector $O_M M$ is codirectional with vector $\boldsymbol{n}$ (Fig. F.2).

To derive formulas for first derivatives at the points of curve Γ (this case if of the biggest interest), we should assume $n = 0$ in (F.44) and in this way get

$$\frac{\partial w}{\partial x} = n_x \frac{\partial w}{\partial n} + t_x \frac{\partial w}{\partial s}, \qquad \frac{\partial w}{\partial y} = n_y \frac{\partial w}{\partial n} + t_y \frac{\partial w}{\partial s}. \tag{F.45}$$

Turning to the second of formulas in (F.22), we have

$$\frac{\partial w}{\partial n} = p^{11} \frac{\partial w}{\partial x_1} + p^{21} \frac{\partial w}{\partial x_2} = n_x \frac{\partial w}{\partial x} + n_y \frac{\partial w}{\partial y},$$

$$\frac{\partial w}{\partial s} = p^{12} \frac{\partial w}{\partial x_1} + p^{22} \frac{\partial w}{\partial x_2} = \left(t_x \frac{\partial w}{\partial x} + t_y \frac{\partial w}{\partial y} \right)(1 + kn), \tag{F.46}$$

so directly on curve Γ we have

$$\frac{\partial w}{\partial n} = n_x \frac{\partial w}{\partial x} + n_y \frac{\partial w}{\partial y}, \qquad \frac{\partial w}{\partial s} = t_x \frac{\partial w}{\partial x} + t_y \frac{\partial w}{\partial y}. \tag{F.47}$$

If we introduce another rotation matrix which we denote by $\boldsymbol{\omega}_\varphi$ and which rotates the vector by φ rather than $\pi/2$ (see Fig. F.2),

$$\boldsymbol{\omega}_\varphi = \left\| \begin{bmatrix} n_x & t_x \\ n_y & t_y \end{bmatrix} \right\| = \left\| \begin{bmatrix} n_x & -n_y \\ n_y & n_x \end{bmatrix} \right\|, \qquad \boldsymbol{\omega}_\varphi^{-1} = \boldsymbol{\omega}_\varphi^{\mathsf{T}}, \tag{F.48}$$

then the linear relationship between the differentiation operators on curve Γ in the global and local coordinate systems from (F.45) and (F.47) can be written in the following symbolic matrix form:

$$\left\| \begin{bmatrix} \partial/\partial n \\ \partial/\partial s \end{bmatrix} \right\| = \boldsymbol{\omega}_\varphi^{\mathsf{T}} \left\| \begin{bmatrix} \partial/\partial x \\ \partial/\partial y \end{bmatrix} \right\|, \qquad \left\| \begin{bmatrix} \partial/\partial x \\ \partial/\partial y \end{bmatrix} \right\| = \boldsymbol{\omega}_\varphi \left\| \begin{bmatrix} \partial/\partial n \\ \partial/\partial s \end{bmatrix} \right\|. \tag{F.49}$$

Further, formulas (F.14-1) and (F.14-2) for differentiating the vectors of the local basis, $\boldsymbol{n}$ and $\boldsymbol{t}$, become as follows after the proper substitutions:

$$\left\| \begin{bmatrix} \dfrac{\partial \boldsymbol{n}}{\partial n} \\[2mm] \dfrac{\partial \boldsymbol{n}}{\partial s} \end{bmatrix} \right\| = \left\| \begin{bmatrix} 0 & 0 \\ 0 & k \end{bmatrix} \right\| \left\| \begin{bmatrix} \boldsymbol{n} \\ \boldsymbol{t} \end{bmatrix} \right\|, \qquad \left\| \begin{bmatrix} \dfrac{\partial \boldsymbol{t}}{\partial n} \\[2mm] \dfrac{\partial \boldsymbol{t}}{\partial s} \end{bmatrix} \right\| = \left\| \begin{bmatrix} 0 & 0 \\ -k & 0 \end{bmatrix} \right\| \left\| \begin{bmatrix} \boldsymbol{n} \\ \boldsymbol{t} \end{bmatrix} \right\|. \tag{F.50}$$

This is evident because when we move in the normal direction of n the unit vectors of the local basis, $\boldsymbol{n}$ and $\boldsymbol{t}$, acquire only translational displacements (no rotation), so

$$\frac{\partial n_x}{\partial n} = \frac{\partial n_y}{\partial n} = \frac{\partial t_x}{\partial n} = \frac{\partial t_y}{\partial n} = 0 . \tag{F.51}$$

When we move along curve Γ, these unit vectors rotate and the second rows in the matrix relationships (F.50) are well-known Frenet formulas [2] in application to the planar curves,

$$\frac{\partial \boldsymbol{n}}{\partial s} = k\boldsymbol{t}, \qquad \frac{\partial \boldsymbol{t}}{\partial s} = -k\boldsymbol{n} .$$

This is how it looks in components:

$$\frac{\partial n_x}{\partial s} = kt_x, \quad \frac{\partial n_y}{\partial s} = kt_y, \qquad \frac{\partial t_x}{\partial s} = -kn_x, \quad \frac{\partial t_y}{\partial s} = -kn_y . \tag{F.52}$$

From (F.43) we derive expressions of parameters p^{ij} and q^{ij}, which are

$$\frac{\partial \boldsymbol{p}}{\partial \alpha_1} = \frac{\partial \boldsymbol{p}}{\partial n} = \left[\begin{bmatrix} 0 & kt_x \\ 0 & kt_y \end{bmatrix}\right], \qquad \frac{\partial \boldsymbol{p}}{\partial \alpha_2} = \frac{\partial \boldsymbol{p}}{\partial s} = \left[\begin{bmatrix} kt_x & -kn_x(1+kn) + nt_x \dfrac{\partial k}{\partial s} \\ kt_y & -kn_y(1+kn) + nt_y \dfrac{\partial k}{\partial s} \end{bmatrix}\right],$$

and

$$\frac{\partial \boldsymbol{q}}{\partial \alpha_1} = \frac{\partial \boldsymbol{q}}{\partial n} = \left[\begin{bmatrix} 0 & 0 \\ -\dfrac{kt_x}{(1+kn)^2} & -\dfrac{kt_y}{(1+kn)^2} \end{bmatrix}\right],$$

$$\frac{\partial \boldsymbol{q}}{\partial \alpha_2} = \frac{\partial \boldsymbol{q}}{\partial s} = \left[\begin{bmatrix} kt_x & kt_y \\ \left(-\dfrac{kn_x}{1+kn} - \dfrac{nt_x}{(1+kn)^2}\dfrac{\partial k}{\partial s}\right) & \left(-\dfrac{kn_y}{1+kn} - \dfrac{nt_y}{(1+kn)^2}\dfrac{\partial k}{\partial s}\right) \end{bmatrix}\right].$$

Now we are able to derive explicit formulas of transition from differentiation with respect to x and y to double differentiation with respect to coordinates n, s, and vice versa. We can either replace q^{ij} and p^{ij} in (F.23) and (F.24) with their new expressions or derive them directly. The second method is preferred here. We start by (F.44) and have

$$\frac{\partial^2 w}{\partial x^2} = \frac{\partial}{\partial x}\left(\frac{\partial w}{\partial x}\right) = \left(n_x \frac{\partial}{\partial n} + \frac{t_x}{1+kn}\frac{\partial}{\partial s}\right)\left(n_x \frac{\partial w}{\partial n} + \frac{t_x}{1+kn}\frac{\partial w}{\partial s}\right) =$$

$$= n_x^2 \frac{\partial^2 w}{\partial n^2} + \frac{2n_x t_x}{1+kn}\frac{\partial^2 w}{\partial n \partial s} + \frac{t_x^2}{(1+kn)^2}\frac{\partial^2 w}{\partial s^2} +$$

$$+ n_x \frac{\partial}{\partial n}\left(\frac{t_x}{1+kn}\right)\frac{\partial w}{\partial s} + \frac{t_x}{1+kn}\frac{\partial n_x}{\partial s}\frac{\partial w}{\partial n} + \frac{t_x}{1+kn}\frac{\partial}{\partial s}\left(\frac{t_x}{1+kn}\right)\frac{\partial w}{\partial s}.$$

Next,

$$\frac{\partial}{\partial n}\left(\frac{t_x}{1+kn}\right) = -\frac{kt_x}{(1+kn)^2}, \quad \frac{\partial}{\partial s}\left(\frac{t_x}{1+kn}\right) = -\frac{kn_x}{1+kn} - \frac{t_x}{(1+kn)^2} n \frac{dk}{ds},$$

which turns into the following after substitutions and transformations:

$$\frac{\partial^2 w}{\partial x^2} = n_x^2 \frac{\partial^2 w}{\partial n^2} + \frac{2n_x t_x}{1+kn}\frac{\partial^2 w}{\partial n \partial s} + \frac{t_x^2}{(1+kn)^2}\frac{\partial^2 w}{\partial s^2} +$$

$$+ \frac{kt_x^2}{1+kn}\frac{\partial w}{\partial n} - \frac{t_x}{(1+kn)^2}\left(2kn_x + \frac{nt_x}{1+kn}\frac{dk}{ds}\right)\frac{\partial w}{\partial s}. \tag{F.53}$$

The formula of $\partial^2 w/\partial y^2$ is derived from (F.53) by simply replacing subscript $_x$ with $_y$. The only thing to get is the formula for second mixed derivative. We have

$$\frac{\partial^2 w}{\partial x \partial y} = \frac{\partial}{\partial x}\left(\frac{\partial w}{\partial y}\right) = \left(n_x \frac{\partial}{\partial n} + \frac{t_x}{1+kn}\frac{\partial}{\partial s}\right)\left(n_y \frac{\partial w}{\partial n} + \frac{t_y}{1+kn}\frac{\partial w}{\partial s}\right) =$$

$$= n_x n_y \frac{\partial^2 w}{\partial n^2} + \frac{n_x t_y + n_y t_x}{1+kn}\frac{\partial^2 w}{\partial n \partial s} + \frac{t_x t_y}{(1+kn)^2}\frac{\partial^2 w}{\partial s^2} +$$

$$+ n_x \frac{\partial}{\partial n}\left(\frac{t_y}{1+kn}\right)\frac{\partial w}{\partial s} + \frac{t_x}{1+kn}\frac{\partial n_y}{\partial s}\frac{\partial w}{\partial n} + \frac{t_x}{1+kn}\frac{\partial}{\partial s}\left(\frac{t_y}{1+kn}\right)\frac{\partial w}{\partial s}.$$

By using the substitutes

$$\frac{\partial}{\partial n}\left(\frac{t_y}{1+kn}\right) = -\frac{kt_y}{(1+kn)^2}, \quad \frac{\partial}{\partial s}\left(\frac{t_y}{1+kn}\right) = -\frac{kn_y}{1+kn} - \frac{t_y}{(1+kn)^2} n \frac{dk}{ds},$$

we transform the previous formula for the mixed derivative into

$$\frac{\partial^2 w}{\partial x \partial y} = n_x n_y \frac{\partial^2 w}{\partial n^2} + \frac{n_x t_y + n_y t_x}{1+kn} \frac{\partial^2 w}{\partial n \partial s} + \frac{t_x t_y}{(1+kn)^2} \frac{\partial^2 w}{\partial s^2} +$$

$$+ \frac{k t_x t_y}{1+kn} \frac{\partial w}{\partial n} - \frac{1}{(1+kn)^2} \left(k(n_x t_y + n_y t_x) + \frac{n t_x t_y}{1+kn} \frac{dk}{ds} \right) \frac{\partial w}{\partial s} . \qquad \text{(F.54)}$$

And again, if we want to know the values of second derivatives right on the Γ curve which is usually the case, then assuming $n = 0$ in (F.53) and (F.54) gives the following on Γ:

$$\frac{\partial^2 w}{\partial x^2} = n_x^2 \frac{\partial^2 w}{\partial n^2} + 2 n_x t_x \frac{\partial^2 w}{\partial n \partial s} + t_x^2 \frac{\partial^2 w}{\partial s^2} + k t_x^2 \frac{\partial w}{\partial n} - 2 k n_x t_x \frac{\partial w}{\partial s} ,$$

$$\frac{\partial^2 w}{\partial y^2} = n_y^2 \frac{\partial^2 w}{\partial n^2} + 2 n_y t_y \frac{\partial^2 w}{\partial n \partial s} + t_y^2 \frac{\partial^2 w}{\partial s^2} + k t_y^2 \frac{\partial w}{\partial n} - 2 k n_y t_y \frac{\partial w}{\partial s} ,$$

$$\frac{\partial^2 w}{\partial x \partial y} = n_x n_y \frac{\partial^2 w}{\partial n^2} + (n_x t_y + n_y t_x) \frac{\partial^2 w}{\partial n \partial s} + t_x t_y \frac{\partial^2 w}{\partial s^2} +$$

$$+ k t_x t_y \frac{\partial w}{\partial n} - k(n_x t_y + n_y t_x) \frac{\partial w}{\partial s} . \qquad \text{(F.55)}$$

To complete the story, below we present also inverse formulas – those for finding second derivatives $w_{,nn}$, $w_{,ss}$, $w_{,ns}$ via the derivatives of the same function with respect to the global coordinates. We start by the general formula (F.46) to get

$$\frac{\partial^2 w}{\partial n^2} = \frac{\partial}{\partial n} \left(\frac{\partial w}{\partial x} n_x + \frac{\partial w}{\partial y} n_y \right) = \frac{\partial^2 w}{\partial n \partial x} n_x + \frac{\partial^2 w}{\partial n \partial y} n_y =$$

$$= \left(\frac{\partial^2 w}{\partial x^2} n_x + \frac{\partial^2 w}{\partial y \partial x} n_y \right) n_x + \left(\frac{\partial^2 w}{\partial x \partial y} n_x + \frac{\partial^2 w}{\partial y^2} n_y \right) n_y =$$

$$\frac{\partial^2 w}{\partial x^2} n_x^2 + 2 \frac{\partial^2 w}{\partial x \partial y} n_x n_y + \frac{\partial^2 w}{\partial y^2} n_y^2$$

and

$$\frac{\partial^2 w}{\partial s^2} = \frac{\partial}{\partial s} \left[\left(\frac{\partial w}{\partial x} t_x + \frac{\partial w}{\partial y} t_y \right) (1+kn) \right] =$$

$$=\left(\frac{\partial^2 w}{\partial s\partial x}t_x+\frac{\partial^2 w}{\partial s\partial y}t_y-k\frac{\partial w}{\partial x}n_x-k\frac{\partial w}{\partial y}n_y\right)(1+kn)+\left(\frac{\partial w}{\partial x}t_x+\frac{\partial w}{\partial y}t_y\right)\frac{dk}{ds}n=$$

$$=\left(\frac{\partial^2 w}{\partial x^2}t_x^2+2\frac{\partial^2 w}{\partial x\partial y}t_xt_y+\frac{\partial^2 w}{\partial y^2}t_y^2\right)(1+kn)^2-\left(\frac{\partial w}{\partial x}n_x+\frac{\partial w}{\partial y}n_y\right)k(1+kn)+$$

$$+\left(\frac{\partial w}{\partial x}t_x+\frac{\partial w}{\partial y}t_y\right)\frac{dk}{ds}n,$$

and also

$$\frac{\partial^2 w}{\partial n\partial s}=\frac{\partial}{\partial n}\left[\left(\frac{\partial w}{\partial x}t_x+\frac{\partial w}{\partial y}t_y\right)(1+kn)\right]=$$

$$=\left(\frac{\partial^2 w}{\partial n\partial x}t_x+\frac{\partial^2 w}{\partial n\partial y}t_y\right)(1+kn)+\left(\frac{\partial w}{\partial x}t_x+\frac{\partial w}{\partial y}t_y\right)k=$$

$$=\left(\frac{\partial^2 w}{\partial x^2}n_xt_x+\frac{\partial^2 w}{\partial x\partial y}(n_yt_x+n_xt_y)+\frac{\partial^2 w}{\partial y^2}n_yt_y\right)(1+kn)+\left(\frac{\partial w}{\partial x}t_x+\frac{\partial w}{\partial y}t_y\right)k.$$

We assume $n = 0$, and directly on the Γ curve we have

$$\frac{\partial^2 w}{\partial n^2}=\frac{\partial^2 w}{\partial x^2}n_x^2+2\frac{\partial^2 w}{\partial x\partial y}n_xn_y+\frac{\partial^2 w}{\partial y^2}n_y^2,$$

$$\frac{\partial^2 w}{\partial s^2}=\left(\frac{\partial^2 w}{\partial x^2}t_x^2+2\frac{\partial^2 w}{\partial x\partial y}t_xt_y+\frac{\partial^2 w}{\partial y^2}t_y^2\right)-\left(\frac{\partial w}{\partial x}n_x+\frac{\partial w}{\partial y}n_y\right)k,$$

$$\frac{\partial^2 w}{\partial n\partial s}=\frac{\partial^2 w}{\partial x^2}n_xt_x+\frac{\partial^2 w}{\partial x\partial y}(n_yt_x+n_xt_y)+\frac{\partial^2 w}{\partial y^2}n_yt_y+\left(\frac{\partial w}{\partial x}t_x+\frac{\partial w}{\partial y}t_y\right)k. \quad \text{(F.56)}$$

We would like to present also a tensor index form of the formula (F.45), (F.47), (F.55), and (F.56) which can be useful in transformations. Assuming $X_1 = X$, $X_2 = Y$, we can rewrite all the above as (F.57), (F.58) and (F.59),

$$w_{,i}=w_{,n}n_i+w_{,s}t_i, \qquad w_{,n}=w_{,i}n^i, \qquad w_{,s}=w_{,i}t^i, \quad \text{(F.57)}$$

$$w_{,ij}=w_{,nn}n_in_j+w_{,ns}(n_it_j+n_jt_i)+w_{,ss}t_it_j+kw_{,n}t_it_j-kw_{,s}(n_it_j+n_jt_i), \text{(F.58)}$$

$$w_{,nn} = w_{,ij} n^i n^j , \quad w_{,ss} = w_{,ij} t^i t^j - k w_{,i} n^i , \quad w_{,ns} = w_{,ij} n^i t^j + k w_{,i} t^i . \tag{F.59}$$

The above form is convenient to use and remember.

There is a useful formula for the Laplace operator in coordinates (*n*,*s*). From (F.55) we have

$$\nabla^2 w = \frac{\partial^2 w}{\partial x^2} + \frac{\partial^2 w}{\partial y^2} = \frac{\partial^2 w}{\partial n^2} + \frac{\partial^2 w}{\partial s^2} + k \frac{\partial w}{\partial n} .$$

F.4.1 Formulas for strains in the (n,s) coordinates

We denote the components of the displacement vector, **u**, in the local basis as w and v by assuming

$$\mathbf{u} = w\boldsymbol{n} + v\boldsymbol{t} . \tag{F.60}$$

This notation for the displacements seems unusual because the first component is denoted by w and the second by v. The justification for it is our wish to stay consistent with the traditional notation of the displacements in the curvilinear bars where the v symbol is used commonly to denote tangential displacements and w to denote the displacements normal to the bar's axis.

Taking the general formulas (F.29) for the strain components in orthogonal curvilinear coordinates and doing all the needed substitutions and transformations will give

$$\varepsilon_{nn} = \frac{\partial w}{\partial n} ,$$

$$\varepsilon_{ss} = \frac{1}{1+kn} \frac{\partial v}{\partial s} + \frac{k}{1+kn} w ,$$

$$\gamma_{ns} = \gamma_{sn} = \frac{\partial v}{\partial n} - \frac{k}{1+kn} v + \frac{1}{1+kn} \frac{\partial w}{\partial s} . \tag{F.61}$$

References

1. Green AE, Zerna W (1954) Theoretical elasticity. Clarendon press, Oxford
2. Kochin NE (1965) Vector calculus and basics of tensor calculus (in Russian). Nauka, Moscow
3. Novozhilov VV (1958) Theory of elasticity (in Russian). Sudpromgiz, Leningrad

4. Vlasov VZ (1962) Selected works. Volume 1 (in Russian). USSR Acad. Sci. Press, Moscow

G. SECTORIAL CHARACTERISTICS OF CROSS-SECTIONS OF THIN-WALLED BARS

G.1 Sectorial characteristics of thin-walled open profiles

Back in Section **6**.2, we introduced the notion of a sectorial coordinate ω (or, as it is sometimes referred to, a sectorial area) for a thin-walled open profile. With the designations of Section **6**.2, the sectorial coordinate of the points of a bar's profile is defined mathematically as the integral

$$\omega(s) = \int_0^s \rho ds \tag{G.1}$$

with the variable upper limit; hence it is a function of the arc coordinate, s.

First of all, we would like to return to the sign convention established in Section **6**.2 for the increment of the sectorial coordinate, $d\omega$, along the profile. It is convenient, however, to reformulate this sign convention in an equivalent way:

> *The increment of the sectorial coordinate, $d\omega$, is deemed positive if the* $\boldsymbol{R}_{\mathrm{PM}}$ *vector* (see Fig. 6.4) *rotates around the pole P counterclockwise as we move along the profile and look at the vector from the positive direction of the longitudinal axis, X.*

It can be checked easily that the above sign convention is indeed equivalent to that from Section **6**.2.

Further, as we indicated in Section **6**.2, the position of the principal pole and the zero sectorial point are defined by the conditions (**6**.2.17),

$$\int_l h\omega ds = 0\,, \qquad \int_l h\omega y ds = 0\,, \qquad \int_l h\omega z ds = 0\,, \tag{G.2}$$

where we should remind l is a full length of the profile and y and z are current coordinates of the profile's point in the system of axes (Y,Z). These axes are thought to be principal central axes of inertia of the thin-walled bar's section.

G.1.1 Determining the location of the principal pole

For the beginning, let us see how the sectorial coordinate varies when the location of the pole changes.

We take two arbitrary poles, A and B, with the respective coordinates (y_A, z_A) and (y_B, z_B) as shown in Fig. G.1.

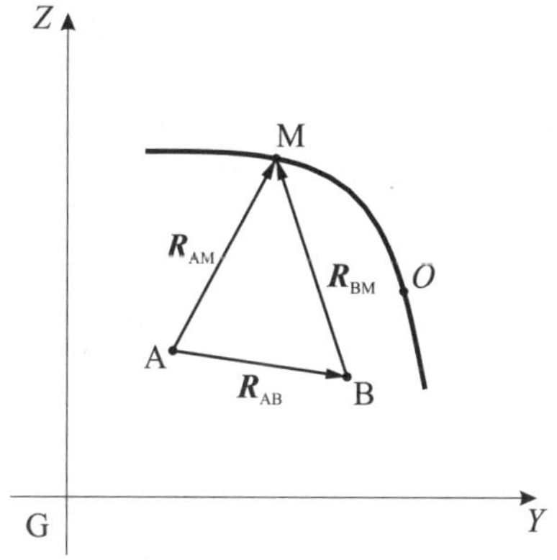

Fig. G.1. Variation of the sectorial coordinate when the pole is changing its location

The origin of the arc coordinate s will be an arbitrary point O of the profile, with its coordinates (y_0,z_0). Each of the two poles defines for each point M of the profile with the arc coordinate s its particular sectorial coordinate, ω_A or ω_B, so that

$$\omega_A(s) = \int_0^s \boldsymbol{R}_{AM} \cdot \boldsymbol{n} ds, \qquad \omega_B(s) = \int_0^s \boldsymbol{R}_{BM} \cdot \boldsymbol{n} ds, \tag{G.3}$$

where $\boldsymbol{R}_{AM}$ or $\boldsymbol{R}_{BM}$ is a vector that goes from the respective pole A or pole B to the current point of the profile, M. Hence

$$\omega_A(s) - \omega_B(s) = \int_0^s \boldsymbol{R}_{AB} \cdot \boldsymbol{n} ds = \boldsymbol{R}_{AB} \cdot \int_0^s \boldsymbol{n} ds, \tag{G.4}$$

because $\boldsymbol{R}_{AB} + \boldsymbol{R}_{BM} = \boldsymbol{R}_{AM}$, and vector $\boldsymbol{R}_{AB} = (y_B - y_A)\boldsymbol{i}_y + (z_B - z_A)\boldsymbol{i}_z$ does not depend on the arc coordinate, s. Further,

$$\boldsymbol{n} = n_y \boldsymbol{i}_y + n_z \boldsymbol{i}_z = t_z \boldsymbol{i}_y - t_y \boldsymbol{i}_z,$$

so the integral of vector $\boldsymbol{n}$ is

$$\int_0^s \boldsymbol{n} ds = \boldsymbol{i}_y \int_0^s t_z ds - \boldsymbol{i}_z \int_0^s t_y ds = (z - z_0)\boldsymbol{i}_y - (y - y_0)\boldsymbol{i}_z, \tag{G.5}$$

and therefore

$$\omega_A(s)-\omega_B(s)=(y_B-y_A)(z-z_0)-(z_B-z_A)(y-y_0)\,. \tag{G.6}$$

First of all, we would like to show that the conditions (G.2) define the only possible location of the principal pole. It means that both points A and B cannot be principal poles at the same time.

To prove this, we multiply (G.6) by hy and integrate over the arc coordinate along the whole contour l. Considering (G.2) and supposing that both points A and B are principal poles together, we have zero in the left-hand part. Therefore, recalling that the Y and Z axes are principal central axes of inertia of the bar's cross-section, we have

$$(z_B-z_A)I_z=0,$$

hence $z_B = z_A$ because the moment of inertia I_z is not zero. The similar reasoning proves $y_B = y_A$.

Now let point B be the principal pole P, i.e. $y_B = y_P$ and $z_B = z_P$ where (y_P, z_P) are coordinates of the principal pole P. At the same time, let point A remain a certain arbirtrarily chosen point in the (Y,Z)-plane. The equality (G.6) still holds, so again we can multiply it first by hz and then by hy and integrate over the arc coordinate along the whole profile to get

$$\int_l \omega_A(s)hz\,ds=(y_P-y_A)I_y\,,\qquad \int_l \omega_A(s)hy\,ds=-(z_P-z_A)I_z\,,$$

therefore

$$y_P=y_A+\frac{1}{I_y}\int_l \omega_A(s)hz\,ds\,,\qquad z_P=z_A-\frac{1}{I_z}\int_l \omega_A(s)hy\,ds\,. \tag{G.7}$$

The formulas (G.7) define the coordinates of the principal pole if we know:

- the location of some arbitrarily selected point of the profile, O, which works as an origin for the arc coordinate s;
- the sectorial area $\omega_A(s)$ calculated with respect to an arbitrary pole A with its coordinates (y_A, z_A) and the given origin O;
- the moments of inertia of the cross-section, I_y and I_z.

As we can see, the coordinates of the principal pole are determined by the second and third conditions in (G.2) and do not depend on the first condition.

G.1.2 Determining the location of the profile's zero point

Let us show now that the first condition in (G.2), which has remained unused yet, is enough to determine the location of the zero point O on the profile.

In order to do it, we want to see how the sectorial coordinate of the profile's points varies vs. the change of the origin of the arc coordinate, i.e. when the zero point is moved. We choose two arbitrary points O_1 and O_2 on the profile and denote the 'arc coordinate distance' between those by s_{12} as shown in Fig. G.2.

Let the arc coordinate of a certain (current) point M of the profile with its origin at O_1 be s_1, and let the coordinate of the same point with its origin at O_2 be s_2. Clearly, $s_{12} = s_2 - s_1$. To be definite, we will assume point O_2 to have a larger arc coordinate on the profile whatever the origin is, as shown in Fig. G.2.

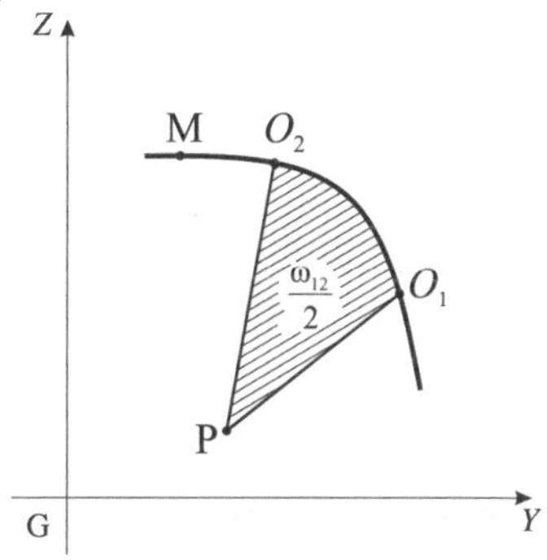

Fig. G.2. Changing the origin of the arc coordinate

Note that the location of the principal pole P is already known to us. We denote by ω_1(M) and ω_2(M) the sectorial coordinates calculated with the arc coordinate's origin at the respective points O_1 and O_2. The definition of the sectorial coordinate gives

$$\omega_1(\mathrm{M}) = \int_{O_1}^{\mathrm{M}} \rho ds\,, \qquad \omega_2(\mathrm{M}) = \int_{O_2}^{\mathrm{M}} \rho ds\,. \tag{G.8}$$

Now it turns out that changing the location of the arc coordinate's origin and keeping the location of the pole P makes the sectorial area just shift by a certain constant value because

$$\omega_1(\mathrm{M}) - \omega_2(\mathrm{M}) = \int_{O_1}^{O_2} \rho ds = \omega_{12}\,. \tag{G.9}$$

Clearly, constant ω_{12} is a doubled area of the sector hatched in Fig. G.2. We want to select this constant in such way that the O_2 point be a real zero point of the profile. This means we use the first condition in (G.2),

$$\int_l h\omega_2(\mathrm{M})ds = 0\,,$$

and derive from (G.9) that

$$\int_l h\omega_1(\mathrm{M})ds = \omega_{12}\int_l hds = \omega_{12}A$$

where A is the cross-section's area. Thus, for the point O_2 to be the zero point of the profile, it suffices to assign the ω_{12} constant so as to satisfy the equality

$$\omega_{12} = \frac{1}{A}\int_l h\omega_1(\mathrm{M})ds\,. \tag{G.10}$$

Having determined the ω_{12} constant from (G.10), we easily calculate the principal sectorial coordinate, $\omega(\mathrm{M})$, by shifting function $\omega_1(\mathrm{M})$ by this constant. So we have

$$\omega(\mathrm{M}) = \omega_1(\mathrm{M}) - \omega_{12} = \omega_1(\mathrm{M}) - \frac{1}{A}\int_l h\omega_1(\mathrm{M})ds\,. \tag{G.11}$$

The principal zero point O of the profile can be any point where the sectorial coordinate $\omega(\mathrm{M})$ calculated by (G.11) is zero. It is easy to see that, unlike the principal pole, the zero point cannot be unambiguous under this condition. There can be multiple, or even an infinite number of, points of this kind. But this is not critical for us, so the zero point can be any one the sectorial coordinate of which is zero.

If we know the coordinates of the principal pole, (y_P, z_P), and those of the zero point, (y_0, z_0), and if we also know the sectorial coordinate $\omega_A(\mathrm{M})$ of an arbitrary point of the profile built for the pole A and the zero point O, then the principal sectorial coordinate of the same point, $\omega(\mathrm{M})$, can be determined from (G.6) that gives

$$\omega(s) = \omega_A(s) - (y_P - y_A)(z - z_0) + (z_P - z_A)(y - y_0)\,. \tag{G.12}$$

We would like to formulate some general properties of the cross-section concerning the locations of the principal pole and the principal zero point:

- if the cross-section of the bar is symmetric with respect to an axis, then the principal pole belongs to this axis, just as the center of gravity does. The zero point belongs to it either.
- if there are two axes of symmetry, then both the principal pole and the zero point are at the intersection of the axes.

The above propositions follow immediately from the conditions (G.2).

G.1.3 The principal pole and the zero point as parameters of minimization of the sectorial moment of inertia of a bar's profile

The students of subjects related to strength of materials are supposed to know well the extreme properties of the principal central axes of inertia which make the moments of inertia of a cross-section take minimum and maximum values. The sectorial moment of inertia of a thin-walled bar's cross-section seems to possess similar extremality properties, too, although the literature sources known to the author do not describe any for some strange reason.

Let I_ω be a sectorial moment of inertia of the cross-section calculated for the principal pole P and the zero point O. We denote by $I_{\omega\mathrm{A}}$ the sectorial moment of inertia of the same cross-section with the same zero point O but calculated for a pole located in some other point, A,

$$I_\omega = \int_l \omega^2 hds\,, \qquad I_{\omega\mathrm{A}} = \int_l \omega_\mathrm{A}^2 hds\,.$$

Using the equality (G.12), we obtain the following relationships between these quantities:

$$I_{\omega\mathrm{A}} = I_\omega + 2\int_l \omega[(y_\mathrm{P} - y_\mathrm{A})(z - z_0) - (z_\mathrm{P} - z_\mathrm{A})(y - y_0)]hds +$$

$$+ \int_l [(y_\mathrm{P} - y_\mathrm{A})(z - z_0) - (z_\mathrm{P} - z_\mathrm{A})(y - y_0)]^2 hds\,. \qquad \text{(G.13)}$$

It is easy to see that

$$\int_l \omega[(y_\mathrm{P} - y_\mathrm{A})(z - z_0) - (z_\mathrm{P} - z_\mathrm{A})(y - y_0)]hds = 0\,.$$

This equality follows immediately from the conditions (G.2).

We denote the last integral in the formula (G.13) by J. Direct calculation gives

$$J = \int_l [(y_P - y_A)(z - z_0) - (z_P - z_A)(y - y_0)]^2 \, hds =$$

$$= (y_P - y_A)^2 (I_y + z_0^2 A) - 2(y_P - y_A)(z_P - z_A) y_0 z_0 A +$$

$$+ (z_P - z_A)^2 (I_z + y_0^2 A),$$

where A is the area of the bar's cross-section, I_y and I_z are moments of inertia of the section with respect to its principal central axes of inertia. The latter equation follows right from the condition that the (Y,Z)-axes should be principal central axes of inertia.

Let b^2 be the square of the length of vector $\boldsymbol{R}_{PA}$, i.e.

$$b^2 = (y_P - y_A)^2 + (z_P - z_A)^2.$$

We denote by α the angle between vector $\boldsymbol{R}_{PA}$ and the Y axis. Using these designations, we write the previous formula as

$$J = b^2 [(I_y + z_0^2 A)\cos^2\alpha - y_0 z_0 A \sin 2\alpha + (I_z + y_0^2 A)\sin^2\alpha], \qquad \text{(G.14)}$$

and

$$I_{\omega A} = I_\omega + J. \qquad \text{(G.15)}$$

The expression in the parentheses in the formula (G.14) is a moment of inertia of the section taken with respect to an axis parallel to vector $\boldsymbol{R}_{PA}$ and passing through the zero point, O, of the profile. So, we understand immediately that the value of J is not negative and is zero if and only if $b^2 = 0$. Therefore

$$I_\omega < I_{\omega A}. \qquad \text{(G.16)}$$

Now, let $I_{\omega 1}$ be a sectorial moment of inertia calculated for the principal pole, P, and some point of the profile, O_1, from which we count the arc coordinate s. According to (G.11), we can write

$$I_{\omega 1} = \int_l \left[\omega + \frac{1}{A} \int_l \omega_1 hds \right]^2 hds = I_\omega + \frac{1}{A} \left(\int_l \omega_1 hds \right)^2.$$

Now it becomes clear that if we assign a point O_1 different from the zero point O to be the origin of the arc coordinate, the sectorial moment of inertia of the section will increase. In other words, the following estimate takes place:

$$I_\omega < I_{\omega 1}. \tag{G.17}$$

Based on the estimates (G.16) and (G.17), we can formulate the following extremality property of the sectorial moment of inertia:

> *Among all points of the plane of a thin-walled bar's cross-section, the principal pole P and the zero point O of the profile are peculiar in that the sectorial moment of inertia, I_ω, based on these points takes the least value possible.*

G.1.4 A remark on a foil profile

In Section **6**.2.8 we noted that a foil (foliate) profile (Fig. 6.10) belongs to a class of non-warped profiles for which the sectorial coordinate is identical to zero, $\omega = 0$.

As an illustration of using the extremality property of the sectorial moment of inertia, we would like to show that the bending center P of the foil profile is located at the cross-section of all its leafs (edges). If it is indeed so, the equality $\omega = 0$ will become obvious.

According to the definition of the sectorial moment of inertia, we have $I_\omega \geq 0$. If the pole P is located at the intersection of the leafs, then $I_\omega = 0$, and this value is a minimum. Considering the extremality property of I_ω and the uniqueness of the principal pole's position, we conclude that the point P where all the leafs intersect is the center of bending.

G.1.5 An example

Our example will be a thin-walled profile in the shape of a circular arc of a radius R and with the opening angle of 2α (Fig. G.3). Let the thickness of the profile's wall, h, be constant along the whole profile. The center of gravity, G, of this cross-section belongs to the axis of symmetry and lies at the distance of c from the center of the circle:

$$c = R\frac{\sin\alpha}{\alpha}. \tag{G.18}$$

The area and the moments of inertia of the section with respect to its principal central axes are, respectively,

$$A = 2R\alpha h,$$

$$I_y = 2hR^3 \int_0^\alpha \cos^2 \varphi d\varphi - Ac^2 = hR^3 \frac{2\alpha^2 + \alpha \sin 2\alpha - 4\sin^2 \alpha}{2\alpha},$$

$$I_z = 2hR^3 \int_0^\alpha \sin^2 \varphi d\varphi = hR^3 \frac{2\alpha - \sin 2\alpha}{2}. \tag{G.19}$$

The most convenient point A is the center of circle. Obviously,

$$\omega_A(M) = \varphi R^2. \tag{G.20}$$

The plus sign is taken because the radius vector $\boldsymbol{R}_{AM}$ rotates counterclockwise when we move in the positive direction of s.

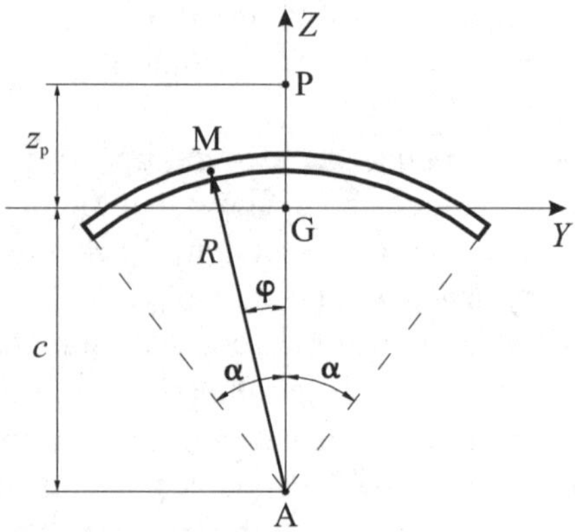

Fig. G.3. A thin-walled section of a round shape

Using formulas (G.7) and seeing that $z_A = -c$, we have

$$z_P = z_A - \frac{1}{I_z} \int_l \omega_A(s) hy ds = R\left[2\frac{\sin\alpha - \alpha\cos\alpha}{\alpha - \sin\alpha\cos\alpha} - \frac{\sin\alpha}{\alpha}\right]. \tag{G.21}$$

Also, the symmetry of the bar's section with respect to axis Y makes the y_P coordinate of the principal pole equal to zero.

Now we can obtain the principal sectorial coordinate of each point M of the profile. To do it, we can use the formula (G.12) assuming $y_0 = 0$, $z_0 = R - c$. The result is

$$\omega(\varphi) = \varphi R^2 + (z_P + c)y = R^2\left[\varphi - 2\frac{\sin\alpha - \alpha\cos\alpha}{\alpha - \sin\alpha\cos\alpha}\sin\varphi\right]. \tag{G.22}$$

The sectorial moment of inertia of the cross-section, I_ω, is

$$I_\omega = 2R^5 h \int_0^\alpha \left[\varphi - 2\frac{\sin\alpha - \alpha\cos\alpha}{\alpha - \sin\alpha\cos\alpha} \sin\varphi \right]^2 d\varphi . \tag{G.23}$$

To complete the story, we are going to derive formulas for the geometric characteristics of the cut-off part of the section; these are sometimes needed for determining the tangential stresses by formula (**6**.2.61). So, from (**6**.2.30) we derive the following

$$A_\mathrm{o}(s) = Rh \int_{-\alpha}^{\varphi} d\varphi = Rh(\varphi + \alpha),$$

$$S_{\mathrm{o}y}(\varphi) = Rh \int_{-\alpha}^{\varphi} z d\varphi = Rh \int_{-\alpha}^{\varphi} \left(R\cos\varphi - c\right) d\varphi = R^2 h \left[\sin\varphi - \frac{\sin\alpha}{\alpha}\varphi \right],$$

$$S_{\mathrm{o}z}(\varphi) = Rh \int_{-\alpha}^{\varphi} y d\varphi = -R^2 h \int_{-\alpha}^{\varphi} \sin\varphi d\varphi = R^2 h \left(\cos\varphi - \cos\alpha\right),$$

$$S_{\mathrm{o}\omega}(\varphi) = Rh \int_{-\alpha}^{\varphi} \omega d\varphi =$$

$$= R^3 h \left[\frac{\varphi^2 - \alpha^2}{2} + 2\frac{\sin\alpha - \alpha\cos\alpha}{\alpha - \sin\alpha\cos\alpha} (\cos\varphi - \cos\alpha) \right]. \tag{G.24}$$

Having determined the geometrical characteristics of the cut-off part of the section from (G.24), we are already able to calculate the components of the matrix of shape factors for the section. Omitting elementary but toilsome transformations, we present final tabulated results (see Tables G.1 and G.2) for some values of angle α.

Table G.1

2α	I_ω/R^5h		μ_{zz}	μ_{zy}	$\mu_{z\omega}$	μ_{yy}	$\mu_{y\omega}$	$\mu_{\omega\omega}$
$\pi/4$	1.854×10^{-6}		28.452	0	0	1.239	4.160	144.480
$\pi/2$	2.473×10^{-4}		7.634	0	0	1.365	1.694	31.540
π	3.738×10^{-2}		2.537	0	0	2	0	4.928
$3\pi/2$	0.833		1.822	0	0	3.589	-1.113	1.556
2π	8.104		2	0	0	6	-2	1.099

Table G.2

2α	I_y/R^3h	I_z/R^3h	ν_{zz}	ν_{zy}	$\nu_{z\omega}$	ν_{yy}	$\nu_{y\omega}$	$\nu_{\omega\omega}$
$\pi/4$	4.060×10^{-4}	3.915×10^{-2}	0.035	0	0	0.893	-0.026	0.00766
$\pi/2$	1.216×10^{-2}	0.285	0.131	0	0	0.785	-0.042	0.034
π	0.298	1.571	0.394	0	0	0.5	0	0.203
$3\pi/2$	1.432	2.856	0.549	0	0	0.358	0.256	0.826
2π	3.142	3.142	0.5	0	0	0.424	0.771	2.313

These results can be useful as validation data for testing software that calculates geometric properties of thin-walled profiles[11].

G.2 Cross-sections of a combined profile

We present here, for the convenience of referencing, basic formulas for calculation of the sectorial characteristics of the combined-profile thin-walled bars. Based on (**8**.1.4), (**8**.1.14), (**8**.1.15), (**8**.1.16), (**8**.1.17), (**8**.1.18), we have

$$\alpha(s) = \frac{I_d}{\Omega} p(s) - \omega(s), \qquad p(s) = \int_0^s \frac{\kappa ds}{gh}, \qquad I_d = \frac{\Omega^2}{\oint \frac{ds}{gh}}, \tag{G.25}$$

$$\omega_0 = \frac{1}{A} \int_l \alpha(s) ehds, \tag{G.26}$$

$$\varpi = \omega_0 - \alpha, \tag{G.27}$$

$$\int_l \varpi ehds = 0, \tag{G.28}$$

$$\int_l \varpi ehyds = 0, \qquad \int_l \varpi ehzds = 0. \tag{G.29}$$

G.2.1 Determining the position of the principal pole

First of all, we would like to establish formulas for recalculating the generalized sectorial coordinate ϖ when the location of the pole is changed.

Suppose we have two points, A and B, specified in the (Y,Z)-plane which can be treated as two different poles. Indexing the functions of the arc coordinate with the symbols of the respective poles, we can write the following on the basis of (G.25) and (G.6):

$$\alpha_A - \alpha_B = -(\omega_A - \omega_B) = -(y_B - y_A)(z - z_0) + (z_B - z_A)(y - y_0). \tag{G.30}$$

---

[11] These calculations were done by the student D.V. Dereviankin at the author's request.

From (G.26) we determine constants ω_{0A} and ω_{0B}, or, more exactly, the difference between the constants:

$$\omega_{0A} - \omega_{0B} = z_0(y_B - y_A) - y_0(z_B - z_A). \tag{G.31}$$

Substracting (G.30) from (G.31) and using (G.27) gives the difference of the generalized sectorial coordinates calculated for the same current point M but for two different poles,

$$\varpi_A - \varpi_B = z(y_B - y_A) - y(z_B - z_A). \tag{G.32}$$

Now let us superpose the B point and the principal pole P by assuming $y_B = y_P$, $z_B = z_P$. Multiplying (G.32) first by *ehz* and then by *ehy*, integrating over the whole profile, and taking the orthogonality conditions (G.29) into account gives the coordinates of the principal pole:

$$y_P = y_A + \frac{1}{I_y}\oint \varpi_A(s)ehzds\,, \qquad z_P = z_A - \frac{1}{I_z}\oint \varpi_A(s)ehyds\,. \tag{G.33}$$

As we can see, these formulas are identical to (G.7) which we derived earlier for open profiles. After finding the coordinates of the principal pole, the diagram of the generalized sectorial coordinate ϖ can be determined easily from (G.32) where point B should be replaced by pole P; this gives

$$\varpi = \varpi_A - z(y_P - y_A) + y(z_P - z_A). \tag{G.34}$$

G.2.2 The principal pole as a parameter for minimization of the sectorial inertia moment of a profile

Exactly as in the case of an open profile, a general combined profile has the following property of extremality of the sectorial moment of inertia:

Among all points of the plane of the combined profile of a thin-walled bar, the principal pole, P, is peculiar in that the sectorial moment of inertia, I_ϖ, based on it takes the least value possible.

Let I_ϖ be a sectorial moment of inertia of the section calculated for the principal pole P. We denote by $I_{\omega A}$ a sectorial moment of inertia of the same section calculated for a pole which is placed in another point A. To put it another way,

$$I_\varpi = \int_l \varpi^2 ehds\,, \qquad I_{\varpi A} = \int_l \varpi_A^2 ehds\,.$$

Using the equality (G.34) and taking the requirements of (G.29) into account, we obtain the following relationship between these values:

$$I_{\varpi A} = \int_l [\varpi + z(y_P - y_A) - y(z_P - z_A)]^2 \, ehds =$$

$$I_{\varpi} + (y_P - y_A)^2 I_y + (z_P - z_A)^2 I_z .$$

Now it is clear that

$$I_{\varpi A} \geq I_{\varpi} ,$$

and the equality takes place only when the A point coincides with the principal pole, P.

Doing the same analysis as one for the foil profile in Section **G**.1.4, we derive a corollary – a simple proof of the fact that the bending center, P, coincides with the center of the inscribed circle for a class of closed profiles made of a homogeneous materials such as one presented in Fig. 7.3. To do the proof, it suffices to see that the generalized sectorial coordinate, ϖ_P, built on the center of the inscribed circle as a pole is identical to zero.

We will not dwell any longer on the technique of calculation of the rest of physical and mechanical characteristics of the thin-walled sections. The calculations are fairly laborious and tiresome, so for practice we recommend to use available software where all the needed data are calculated automatically. This is the way to go with complicated multiple-contour sections. The engineer, nonetheless, should have a clear understanding of principles that the algorithms for calculating the section's properties are based on, in order to apply them consciously and to perform an evaluation check of the results obtained with computer software – at least qualitatively, at a glance. This is the reason why the current appendix is included in this book.

AUTHOR INDEX

SUBJECT INDEX[1]

[1] The index gives only pages where definitions for the respective terms are introduced.

Foundations of Engineering Mechanics

Series Editors: Vladimir I. Babitsky, Loughborough University, UK
Jens Wittenburg, Karlsruhe University, Germany

Further volumes of this series can be found on our homepage: springer.com

(Continued from page ii)

Alfutov, N.A.
Stability of Elastic Structures, 2000
ISBN 3-540-65700-2

Astashev, V.K., Babitsky, V.I., Kolovsky, M.Z., Birkett, N.
Dynamics and Control of Machines, 2000
ISBN 3-540-63722-2

Kolovsky, M.Z., Evgrafov, A.N., Semenov Y.A, Slousch, A.V.
Advanced Theory of Mechanisms and Machines, 2000
ISBN 3-540-67168-4

Printing: Krips bv, Meppel
Binding: Stürtz, Würzburg